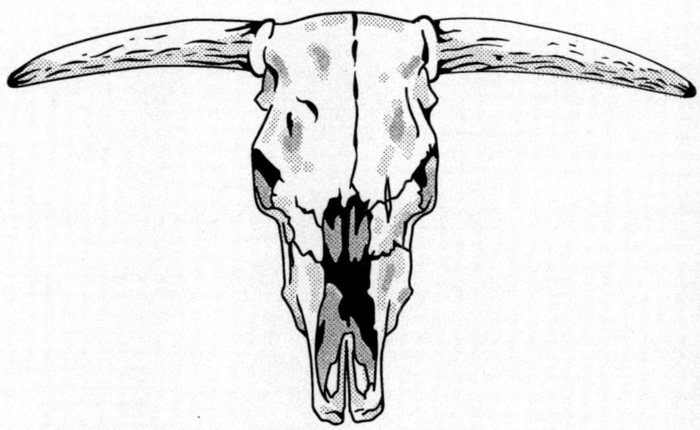

Landscape History of Hadramawt

The Jo Anne Stolaroff Cotsen Prize Imprint honors outstanding studies in archaeology to commemorate a special person whose appreciation for scholarship was recognized by all whose lives she touched.

UCLA COTSEN INSTITUTE OF ARCHAEOLOGY PRESS
Monumenta Archaeologica

Volume 42

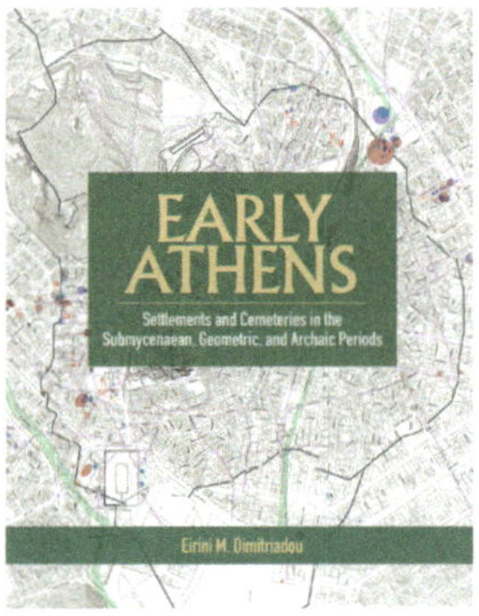

Early Athens: Settlements and Cemeteries in the Submycenaean, Geometric, and Archaic Periods

By Eirini M. Dimitriadou

Volume 41

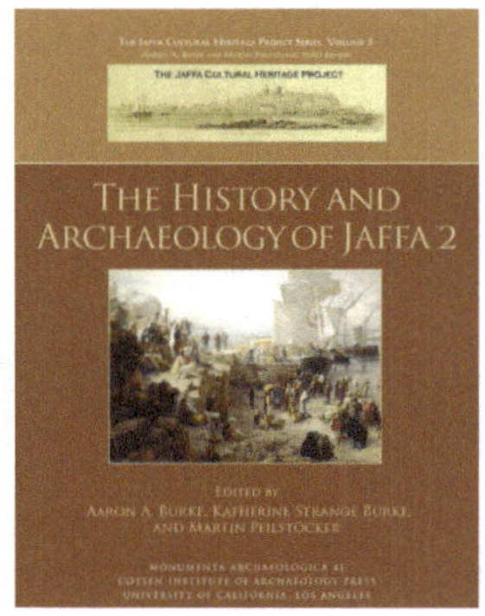

The History and Archaeology of Jaffa 2

Edited by Aaron A. Burke, Katherine Strange Burke, and Martin Peilstöcker

Volume 40

Tangatatau Rockshelter: The Evolution of an Eastern Polynesian Socio-Ecosystem

Edited by Patrick Vinton Kirch

Volume 39

The Desert Fayum Reinvestigated: The Early to Mid-Holocene Landscape Archaeology of the Fayum North Shore, Egypt

Edited by Simon J. Holdaway and Willeke Wendrich

Volume 38

The Archaeology of Grotta Scaloria: Ritual in Neolithic Southeast Italy

Edited by Ernestine S. Elster, Eugenia Isetti, John Robb, and Antonella Traverso

Volume 37

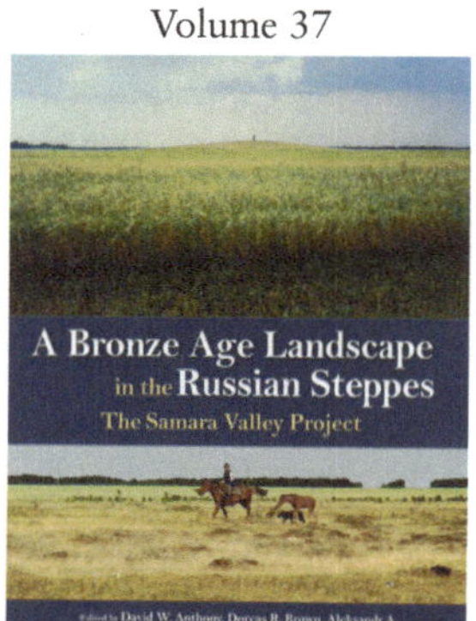

A Bronze Age Landscape in the Russian Steppes: The Samara Valley Project

Edited by David W. Anthony, Dorcas R. Brown, Aleksandr A. Khokhlov, Pavel F. Kuznetsov, and Oleg D. Mochalov

Volume 36

Rural Archaeology in Early Urban Northern Mesopotamia: Excavations at Tell al-Raqa'i

Edited by Glenn M. Schwartz

Volume 35

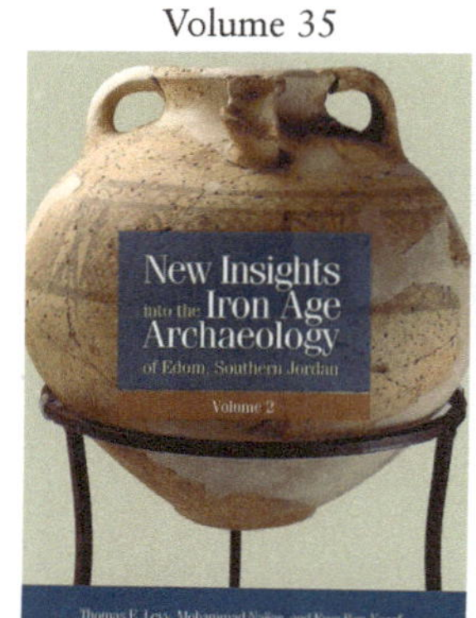

New Insights into the Iron Age Archaeology of Edom, Southern Jordan

By Thomas E. Levy, Mohammad Najjar, and Erez Ben-Yosef

Landscape History of Hadramawt

The Roots of Agriculture in Southern Arabia (RASA) Project 1998–2008

Edited by
Joy McCorriston and Michael J. Harrower

UCLA COTSEN INSTITUTE OF ARCHAEOLOGY PRESS

The Cotsen Institute of Archaeology Press is the publishing unit of the Cotsen Institute of Archaeology at UCLA, a premier research organization dedicated to the creation, dissemination, and conservation of archaeological knowledge and heritage. It is home to both the Interdepartmental Archaeology Graduate Program and the UCLA/Getty Master's Program in the Conservation of Archaeological and Ethnographic Materials. The Cotsen Institute provides a forum for innovative faculty research, graduate education, and public programs at UCLA in an effort to positively impact the academic, local and global communities. Established in 1973, the Cotsen Institute is at the forefront of archaeological research, education, conservation and publication, and is an active contributor to interdisciplinary research at UCLA.

The Cotsen Institute of Archaeology Press specializes in producing high-quality academic volumes in nine different series, including *Monumenta Archaeologica, Monographs, World Heritage and Monuments, Cotsen Advanced Seminars,* and *Ideas, Debates, and Perspectives.* Through a generous endowment by Lloyd E. Cotsen, longtime Institute volunteer and benefactor, the Press makes the fruits of archaeological research accessible to scholars, professionals, students, and the general public. Our archaeological publications receive critical acclaim in both academic communities and the public at large.

Edited by Peg Goldstein
Designed by Sally Boylan

Library of Congress Cataloging-in-Publication Data

Names: McCorriston, Joy, 1961- editor. | Harrower, Michael J., editor.
Title: Landscape history of Hadramawt : the roots of agriculture in southern Arabia (RASA project 1998-2008) / [edited by] Joy McCorriston, Michael J. Harrower.
Description: Los Angeles : The Cotsen Institute of Archaeology Press, 2020.
Includes bibliographical references and index. | Summary: "The rugged highlands of southern Yemen are one of the less archaeologically explored regions of the Near East. This final report of survey and excavations by the Roots of Agriculture in Southern Arabia (RASA) Project addresses the development of food production and human landscapes, topics of enduring interest as scholarly conceptualizations of the Anthropocene take shape. Along with data from Manayzah, site of the earliest dated remains of clearly domesticated animals in Arabia, the volume also documents some of the earliest water management technologies in Arabia, thereby anchoring regional dates for the beginnings of pastoralism and of potential farming. The authors argue that the initial Holocene inhabitants of Wādī Sanā were Arabian hunters who adopted limited pastoral stock in small social groups, then expanded their social collectives through sacrifice and feasts in a sustained pastoral landscape. This volume will be of interest to a wide audience of archaeologists including not only those working in Arabia, but more broadly those interested in the ancient Near East, Africa, South Asia, and in Holocene landscape histories generally"-- Provided by publisher.
Identifiers: LCCN 2019051939 | ISBN 9781950446124 (cloth)
Subjects: LCSH: Landscape archaeology--Yemen (Republic)--Ḥaḍramawt (Province) | Excavations (Archaeology)--Yemen (Republic)--Ḥaḍramawt (Province) | Agriculture--Yemen (Republic)--Ḥaḍramawt (Province) | Pastoral systems, Prehistoric--Yemen (Republic)--Ḥaḍramawt (Province) | Ḥaḍramawt (Yemen : Province)--Antiquities.
Classification: LCC DS247.Y43 L36 2020 | DDC 939.4/9--dc23
LC record available at https://lccn.loc.gov/2019051939

To the people of Yemen,
with gratitude

Contents

List of Figures

List of Tables

Arabic Foreword

مُقدّمة

أعمال بعثة (RASA) من عام 1998 2008م

الأهمّية العملية والعلمية

شهدت حضرموت بعد استقلال جنوب اليمن عام 1967م أعمال حفريات علمية أثرية موسمية طويلة، ابتدأت من البعثة الفرنسية عام 1970م، ثم تبعتها أعمال البعثة اليمنية السوفييتية المُجمّعيّة المشتركة للآثار والدراسات التاريخية الثقافية والأثنوغرافية عام 1983م. وقد شملت أبحاث البعثتين مجالات العصور الحجرية القديمة والحديثة، بالإضافة إلى أعمال تنقيباتهما الأساسية عن مستوطنات حضارة مملكة حضرموت الكلاسيكية القديمة. وقد توقّفت كلتاهما عام 2007م، نتيجة للظروف الأمنية التي تمّر بها اليمن منذ ذلك التاريخ وحتى الآن.

وفي عام 1998م بدأت أعمال بعثة (RASA) الأثرية المخصصة كلياً للمسح والتنقيب والبحث عن بدايات المناشط الزراعية في جنوب الجزيرة العربية السابقة لقيام الحضارة العربية الجنوبية القديمة المعروفة، وقد ترأس أعمال البعثة كل من السيدة البروفيسور عالمة الآثار والأنثروبولوجيا "جوي ميكيريستون Joy McCorriston والبروفيسور "إيريك أوشيس Eric Oches المختص بالجيولوجيا القديمة والمظهر الطوبوغرافي والبيئة القديمة.

ولقد كانت بعثة (RASA) منذ بداياتها بعثة عالمية بحق، فقد اعتمدت على مشاركة علماء من الولايات المتحدة الأمريكية، وألمانيا، وانجلترا، وفرنسا، وكندا، واليمن. واستفاد منها الكادر اليمني من هيئة الآثار بصنعاء وفرعها في حضرموت، وطلّاب الدراسات العليا من أمريكا وكندا وفرنسا، استفادةً قصوى من التدريب على عمليات الحفر والمسح الأثري المُمنهج باستخدام التقنية الجغرافية الحديثة في مسح جميع المواقع والمعالم، وتوزيع المنطقة إلى مربعات تُضمُّ إلى بعضها بعضا وتسجيل كل معلم أثري ضمن هذا المربع في بطاقة خاصة به بعد التوصيف العام للمربع المراد مسحه في البطاقة العامة المخصصة لذلك وأخذ قياساته ووصفه وتصويره... إلخ، واستخدام جهاز ال (GPS) عالي الدّقة لتحديد كافة المواقع والمعالم الأثرية، وتحديد أماكنها عن طريق صور الأقمار الصناعية.

وقد تمّ مسح أجزاء كبيرة من وديان "شُمْلِيّة"، و"شُمَيْلِيّة"، و"خُزْمَة"، والجزء العلوي والأوسط من وادي "سناء"، وفروع جانبية كثيرة تصبّ فيه ومساحات كبيرة من الجول الجنوبي الشرقي من حضرموت، وفروع من وادي عدم... وغيرها.

وشملت أعمال المسح والتوثيق معالم ومواقع أثرية بلغت مئات عديدة، تمّ تصنيفها إلى أنواع، ولكل نوع استمارة خاصة به تحوي أدقّ التفاصيل عنه، من أصغر "موقد" أو "مضبى" لا يتعدّى قطره 70 80 سم، مروراً بالمقابر العرومية التي يبلغ قطرها 5 أمتار وامتداد ذيولها الحجرية إلى عشرات الأمتار في بعض الحالات، حتى محطّات العصور "النيوليثية" وورشها وبقايا منشآت الري البسيطة، والمستوطنات السكنية الدائرية وشبه الدائرية، والمنشآت الطقوسية البيضوية للسكّان "الرُحل" و"أشباه الرُحل" منذ الألف السابع قبل الميلاد حتى تخوم العصر الحضاري في الألفية الثانية قبل الميلاد وما بعدها.

وتمّ التنقيب كُلياً أو جزئياً، أو لأخذ عينات من الفحم وذلك لأجل فحص "براديو كربون 14" بهدف تحديد تواريخ عشرات المعالم

الصغيرة المُتنوّعة، وبعض المستوطنات الكبيرة والمُنشآت الطقوسية، والمقابر... وغيرها.

وقد ترافدت عِدة علوم لإيضاح صورة البيئة القديمة التي عاشت فيها تلك الأقوام في العصور "النيوليثية الحديثة" ، و"البرونزية"، ومظهرها الطوبغرافي القديم، في ترابط مع علم الآثار: كعلم "البيئة القديمة والآثار Paleoecology and Archaeology، وعلم العظام الحيواني الأثري "Zoorchaeology Archaeozoology وإعادة تصوّر مجاري الوديان، ومستوى ارتفاع التربة الطينية، وتوضيح المنظر البيئي العام للوديان التي جرى البحث فيها عن آثار العصر "البرونزي"، وما شهدته من فترات مَطِيرة وفترات جفاف... الخ. كما تمّ التعرّف على الغطاء النباتي، والبيئة، ونوعية الغذاء، والحيوانات المُدجّنة... إلخ.

وقد تمّت الاستعانة بالأبحاث والمقاربات الأثنو أثرية لغوية، ودراسة وملاحظة مناشط وطرق حياة السكّان الحاليين المُترحّلين موسمياً إلى مناطق محدودة، مع الأخذ بعين الاعتبار التطوّر الذي حدث عبر التاريخ لهذا النمط من المناشط المرتبطة بالحياة شبه الرعوية المعتمدة على تربية الماشية، والزراعة المحدودة من مياه الأمطار المُوجّهة عبر منشآت ريٍّ غير مُعقّدة "الشُروج"، والاستقرار في مناطق صغيرة ثابتة تُشكّل نقطة انطلاق لممارسة بعض المناشط الرعوية الزراعية البسيطة، والنقل التجاري... إلخ.

كل هذه المؤشرات والدلائل الأثرية الكثيرة التي قام بها فريق مُتكامل في بعثة (RASA) تُؤكد أن منطقة الجول وبدايات الوديان الفرعية الجانبية كانت مجالاً واسعاً لنشاط الرعاة شبه الرُحل في عصور ما قبل التاريخ وعلى تخوم عصر قيام مملكة حضرموت القديمة.

وبهذا نستطيع القول ختاما بأنّ كُل المؤشرات تدّل على المنبع المحلّي للحضارة العربية الجنوبية في حضرموت، وامتداد مؤثراتها من الشرق والشمال الشرقي في المناطق الممتدة على طول منطقة عمان الحالية والخليج العربي و"اليمامة"، والتي هي بحاجة إلى مزيد من البحث الأثري على طول خطوط هذه الامتدادات، وكذا بحث امتداد العصر البرونزي في مناطق اليمن الشمالية الغربية (المختلفة عن منطقة حضرموت)، التي لا يُستبعد أن تكون على طول الضفة الغربية للشرق الإفريقي المقابل.

إن أعمال بعثة (RASA) أزاحت جزءاً من الضباب الكثيف حول أصول الحضارة الزراعية المعروفة في حضرموت، التي هي بحاجة إلى أعمال أثرية أكثر تعمّقاً وامتداداً في المستقبل المنظور، سواءً في حضرموت نفسها، أو في المنطقة الممتدة من ظفار حتى الخليج العربي والجزء الأوسط من شرق الجزيرة العربية القريب من إمارات الخليج العربي، مع عدم إغفال المؤثرات اللاحقة من منطقة الجنوب العربي للجزيرة العربية، التي قامت فيها ممالك حضرية قديمة متزامنة أيضاً مع مملكة حضرموت، ومشتركة معها في الوعاء الحضاري العام.

د. عبدالعزيز جعفر بن عقيل

المدير العام السابق للهيئة العامة للآثار والمتاحف

فرع حضرموت المكلا

وعضو البعثة الأثرية (RASA) (1996 2008م)

المكلا، أكتوبر 2016م.

Foreword

Since the independence of the People's Democratic Republic of Yemen (South Yemen) in 1967, Hadramawt has witnessed many prolonged seasonal archaeological and scientific excavations, from the French archaeological mission in 1970 to the Yemeni–Soviet Joint Mission for Archaeology and Historical-Cultural and Ethnographic Studies in 1983. The research of these two missions covered the Paleolithic and the Neolithic eras and exploration of the settlements of the ancient Kingdom of Hadramawt (circa 800 BCE–300 CE). But due to the security situation in Yemen, the works of the two missions came to an end in 2008.

Headed by archaeologist Joy McCorriston and geologist Eric Oches, the RASA Project began in 1998, excavating and surveying the early beginnings and roots of agricultural activities in Southern Arabia prior to the well-known South Arabian civilization.

The mission of RASA was really an international one, as it included participants from the United States, Germany, Britain, France, Canada, and Yemen and benefited the Yemeni cadre working for the General Organization for Antiques and Museums and its branch in Hadramawt. RASA also benefited many graduate students from the United States, France, and Canada, who obtained knowledge of and training on archaeological survey and excavation using advanced geospatial technologies, dividing the survey region into blocks—to be gathered together afterward—and registering the data of each archaeological feature discovered within each block on a dedicated form after writing down a general description of the survey block on another dedicated general form. These forms included measurements, description, and images of the specified survey block; and the documentation was linked with RASA's use of high-accuracy GPS to pinpoint the location of the archaeological features and sites and to map these through satellite imagery.

Large parts of Shumlya and Shumailya wadis (valleys), the upper and middle parts of Wādī Sanā and many of its sub-wadis, the southeast of Hadramawt, and a number of Wādī ʿIdim`s sub-branches have been surveyed.

The work of surveying and documentation covered hundreds of archaeological features and sites, which have been divided into categories. Each category was described on a special form, with detailed descriptions and specifications of the category, from the smallest ovens (*madhābiḥ*) of a diameter of less than 80 cm to high circular tombs of 5 m in diameter and stone pile tails extending to dozens of meters in some cases. Other categories include the remains of irrigation structures and facilities and the stone-tool workshops of Neolithic age, the circular and semicircular residential settlements, and the ritual oval structures belonging to the nomadic and seminomadic populations dating from 7000 years BP to the early beginnings of the civilization in the second millennium BCE. Excavation—in whole or in part—has been carried out to take samples for radiocarbon dating of dozens of varied small archaeological features, some big installations and ritual structures, and tombs.

Many scientific specializations, such as paleoecology, archaeology, and archaeozoology, contributed to clarifying this ancient environment in which people lived in the Neolithic and the Bronze Age. Such research has studied topography, reconstructed the courses of wadis, modeled levels of the silt terraces, described the general environmental landscape of the wadis, excavated for monuments belonging to the so-called Bronze Age, and reconstructed the rain and drought seasons of the wadis. Furthermore, the vegetation, environmental resources, wild foods, and domestic animals have been identified.

Inaddition to using earlier ethnolinguistic-archaeological studies and approaches, the RASA team has studied and observed the lifestyles and activities of the current inhabitants, who used to travel seasonally to a few locations, taking into consideration the developments these activities witnessed throughout the history of semipastoral life centered around livestock. Some of these developments included limited-scale agriculture dependent on rainfall controlled by simple irrigation structures (*shrūj*). Other developments involved settling down in small areas as an initiation for practicing simple pastoral-agricultural activities. Likewise, commercial transport evolved over time.

All the archaeological indicators the RASA team has identified affirm that the Southern Jol and its wadis witnessed extensive activities practiced by seminomadic herders throughout prehistoric ages to the emergence of the ancient Kingdom of Hadramawt.

In conclusion, all indicators point to the source of the South Arabian civilization evolving in Hadramawt and extending to the east and northeast to the regions stretching along present-day Oman, the Arabian Gulf, and Al-Yamāmah (in tenth- to thirteenth-century CE Saudi Arabia). In all areas, more excavations are required, as is the case with the influences of the Bronze Age in Yemen, thought to reach to the northeast parts of Yemen (different from Hadramawt) and, perhaps, East Africa.

The works of the RASA Project have moved away much of the fog engulfing the sources of known agricultural civilization in Hadramawt. Now is needed deeper and larger works of archaeological excavation in the near future, either in Hadramawt or along the area extending from Dhofar to the Arabian Gulf and the middle parts of the eastern Arabian Peninsula. Nor should one neglect subsequent influences coming from the southwest of the Arabian Peninsula, where ancient urban kingdoms contemporary with the Kingdom of Hadramawt had common general features in terms of the concept of civilizations and concurrently emerged.

—ʿAbdalʿazīz Jaʿafar Bin ʻAqīl, member of RASA (1996–2008) and former director general of the General Organization for Antiques and Museums, Hadramawt branch
Mukalla, October 2016

Acknowledgments

Over 12 years of fieldwork and 10 years of subsequent analysis and publication, numerous individuals and institutions have provided critical support for this project. While the responsibility and conclusions are ours, the authors and editors, we owe immense thanks to each other and to all the contributors to fieldwork and this volume. As a team, we have all labored for years to bring this work to final publication, and each team member's ideas and energy have been an immense source of inspiration and professional enrichment to the others.

Joy McCorriston especially wishes to thank the General Organization of Antiquities and Museums, Yemen; CanOxy-Yemen; Canadian Nexen Petroleum Yemen (CNPY); the U.S. National Science Foundation; the American Institute for Yemeni Studies (AIYS); the University of Minnesota; and The Ohio State University for logistical and financial support. This volume is based upon work financially supported by the U.S. National Science Foundation under grants SBR-9711270, SBR-9903555, BCS-0211497, BCS-0332278, and BCS-0624268 (to McCorriston; McCorriston and Oches; McCorriston and Harrower; and McCorriston, Goel, and Brzezinska); by a multi-country fellowship from the Council of American Overseas Research Centers (CAORC) (McCorriston); by two NMERTA fellowships from AIYS (McCorriston); by a McKnight Land-Grant Professorship and a Grant in Aid of Research from the University of Minnesota (McCorriston); by a Seed Grant from The Ohio State University (McCorriston); by a Wenner Gren Foundation International Collaborative Research Grant (Bin ʿAqīl and McCorriston); by a grant from the Foundation for Exploration and Research on Cultural Origins (FERCO) (McCorriston); and by a Community Development Grant from CNPY in Calgary (McCorriston). Any opinions, findings, and conclusions or recommendations expressed in this material are those of the authors and do not necessarily reflect the views of the National Science Foundation.

Harrower is particularly grateful to The Ohio State University, University of Michigan, University of Toronto, University of California–Los Angeles, and Johns Hopkins University for their support during the nearly 20 years in which his involvement in RASA research, analysis, writing, and editing were under way. Harrower's research with RASA was supported by a U.S. National Science Foundation Dissertation Improvement Grant with McCorriston (BCS-0332278), Social Sciences and Humanities Research Council of Canada (SSHRC) doctoral and postdoctoral fellowships, and postdoctoral fellowship funding from the Cotsen Institute of Archaeology at UCLA. These institutions and in particular Henry Wright, Joyce Marcus, Ted Banning, Michael Chazan, Charles Stanish, Elizabeth Carter, Willeke Wendrich, Hans Barnard, Alan Shapiro, and Glenn Schwartz deserve special thanks for their advice, guidance, and support.

The RASA team gratefully acknowledges the efforts of GOAM officials, especially Dr. Yūsuf ʿAbdulla, Dr. ʿAbdalʿazīz Bin ʿAqīl, Dr. Ahmed Shugaʿa Al Deen, ʿAbdal-Baset Nʿoman, and Ahmed Shemsan, in facilitating logistics and permission to conduct research. (Names are spelled in English orthography as preferred by the individuals named.)

The team thanks Neil Bennett, Kevin Tracy, Mohamad Lardy, Phil McGregor, Bob Simpson, Muhammed Bin Naibhan, Eric Bolton, Saleem Dakeek, Ed Erlendson, Ernie Flynn, Ray Harrison, Muhammed Omar Al-Jabrī, Walīd Jizrawī, Dan O'Byrne, Hassan Samawy, Ken Spraggs, Dave Smith, Kevin Marlow, Muhamad Mushgary, Ahmad Ahgary, and many others for their extraordinary hospitality and support. We recognize the efforts and ethical standards maintained by CanOxy/CPNY and their employees in stewardship of Yemen's archaeological resources and for supporting our research.

The directors of AIYS—Noha Sadek, Marta Colburn, and Chris Edens—and their staff offered unstinting support, and AIYS provided us a welcome base in Sana'a.

Thanks also to other researchers in Yemen who shared reprints, comments, and information, especially Burkhard Vogt, Tony Wilkinson, Alexander Sedov, Juris Zarins, Ietha Al-ʿmari, Mac Gibson, Selma Radi, Frank Braemer, Jérémie Schiettecatte, Henry Thompson, Sergey Frantsouzoff, Axelle Rouguelle, and Ed Keall. Joy McCorriston and Eric Oches thank Subir Banerjee, Margaret Davis, Emi Ito, and our colleagues at the University of Minnesota, who offered us so much encouragement in the early years of this research.

Many undergraduates have helped with this project over the years, and we especially acknowledge here Rachel Bursheim, Elizabeth Townsend, Nate Donaldson, Jamie McCarthy, Nisha Patel, Petra Creamer, Jennifer Everhart, Matthew Millard, Katherine Kelley, and Timothy Everhart. The field team was immensely enriched by the repeated and generous participation of Catherine Heyne, whose ideas and experience in survey, excavation, and documentation greatly enhanced the research and whose influence appears throughout the volume.

Joy McCorriston gratefully thanks Frank Braemer, Martine Regert, Anne-Marie Gomez, Patricia Anderson, Liliane Meignon, Fréderique Bertoncello, Gourguen Davtian, and colleagues and students at Cultures et Environnement Préhistoire, Antiquité, Moyen Âge (CEPAM-UMR7264 CNRS), who welcomed and hosted her for a delectable sabbatical, during which much of this volume was written. Perhaps it might have finished sooner with less Provençale rosé, but the *cafés noisettes* in Nice certainly helped balance the writing!

We thank the professionals who helped us produce this volume. Artist Clara Hickman has patiently and steadily corrected most illustrations and produced some of them from field drawings, for which we are deeply indebted, and we also thank Emma Lagan for her assistance. Archaeological photographer Jules Angel digitized and manipulated many of the photographs to clarify their archaeological content. While her name appears on few captions, her pervasive efforts have lent continuous quality and consistency to the archive and this volume. Ann Woodward and Lael Ensor-Bennett of the Johns Hopkins University Visual Resources Collection deserve thanks for efforts in editing many color photographs. We also had crucial help from Alessia Prioletta, who reviewed and revised our field readings of South Arabian graffiti and agreed to coauthor one chapter. Jason Weimar contributed a table of graffiti characters. With a generous gift of her time, Paige Paulsen edited references. Joe Roe contributed statistical support for the analysis of cattle skulls. With discipline, judicious insight, and clarity, Kevin Johnston of Scholaris Editing transformed a rather rough manuscript into corrected, formatted form. Not least, we are extremely grateful to the Cotsen Institute of Archaeology Press, Aaron Burke, Randi Danforth, Peg Goldstein, Sally Boylan, reviewers, and support staff, for their attentive commitment to making this report the best it can be.

Louise Martin would like especially to thank Joy McCorriston, who not only invited her to study the Kheshiya cattle skulls and organized the research trip to Mukalla to do so, but also provided all manner of logistical and academic support in Yemen and thereafter. Lisa Usman gave the project the benefit of her wisdom and experience by undertaking the conservation of the cattle skulls in Yemen; we cannot thank her enough for joining the team. Lisa and Louise were financially supported during the research trip to Mukalla by University College London's Dean's Travel Fund. In Yemen, the staff of the Mukalla Museum were welcoming and resourceful, and we thank them for their patience in hosting us. ʿAbdalʿazīz Bin ʿAqīl provided endless amounts of support in Mukalla, for which we are forever grateful. In London, Caroline Grigson shared on many occasions her vast knowledge and expertise on cattle, and especially cattle skulls, which was incredibly helpful but also fun. Caroline also allowed Louise to use her own unpublished osteometrics of Danish cattle (*Bos primigenius*) collections (used in figure 11.22). Louis Chaix very kindly shared his Kerma cattle metrical database (used in figures 11.23, 11.24, and 11.25), for which Louise is very grateful. Stuart Laidlaw (UCL Institute of Archaeology) converted amateur photographs into crisp publication images, and we thank him for his time and skill. Lastly, Louise thanks Joseph (Jojo) Martin for his early participation and patience in the project, even if he does not remember it.

Finally, with nostalgia and deep appreciation for the times and tasks we shared, the things we learned from each other, and our collective accomplishments, we acknowledge the people of Yemen for their tremendous hospitality and all those who participated in RASA fieldwork in Yemen, 1996–2008:

Field Participants 1996

Joy McCorriston
Ingrid Hehmeyer
ʿAbdalʿazīz Bin ʿAqīl
Ietha Al-ʿmari
Samīr Al-Qadasī,
Nasser Habtour
Sheila McNally
Sulaiman and other bedouin hosts from the al-Bāhāj clan in Mudhaynab

Field Participants 1998

Joy McCorriston
Eric Oches
Zachariah Johnson
Russanne Low
Louise Martin
Pieter Vlah
ʿAbdalʿazīz Bin ʿAqīl
ʿAbdal-Baset Nʿoman
ʿAbdallah Nasser Ṣarām
Aḥmad Nagī
Ṣaleh Sālim Ḥuṣn Al-Ḥakīmī,
ʿAli Ḥasan
Nasser Al-ʿAlīy and six bedouin workmen from the Al-ʿAlīy clan in Wādī Sanā

Field Participants 2000

Joy McCorriston
Eric Oches
ʿAbdalʿazīz Bin ʿAqīl
Michael Harrower
Dawn Walter
Catherine Heyne
ʿAbdal-Baset Nʿoman
Khālid Bā-Dhufāry
Stephen DeVogel
ʿAbdallah Nasser Ṣarām
ʿAbdal Khalak
Muḥammad Al-Hijāzī
Aḥmad Nagī
ʿAli Ḥasan
Nasser Al-ʿAlīy
ʿUbayd Al-ʿAlīy
Saʿid Al-ʿAlīy

Field Participants 2004

Joy McCorriston
Eric Oches
ʿAbdalʿazīz Bin ʿAqīl
Michael Harrower
Joshua Anderson
Catherine Heyne
Tara Steimer-Herbet
Rémy Crassard
Julien Espagne
ʿAbdal-Baset Nʿoman
ʿAbdal Mnain
Khālid Bā-Dhufāry
ʿAbdallah Nasser Ṣarām
Saʿad Al-ʿAjiyly
Nasser Al-ʿAlīy
ʿUbayd Al-ʿAlīy
Saʿid Al-ʿAlīy
Rotating crews of three soldiers

Field Participants 2005

Joy McCorriston
Eric Oches
Michael Harrower
Khālid Bā-Dhufāry
Catherine Heyne
Ietha Al-ʿmari
Tara Steimer-Herbet
Rémy Crassard
Julien Espagne
Mohammed Sinnah
Nisha Patel
Margaret Wilson
Ghufrān Aḥmad
Kirk Sander
Scott Anderson
Ramzy Ladeh

ʿAbdallah Nasser Ṣarām
ʿAbdalKarīm Al-Burkānī
ʿAli Alwān
Muḥammad Alwān
Nasser Al-ʿAlīy
ʿUbayd Al-ʿAlīy
ʿAbdallah Al-ʿAlīy
Rotating crews of Al-ʿAlīy workmen
Rotating crews of three soldiers

Field Participants 2006

Joy McCorriston
ʿAbdalʿazīz Bin ʿAqīl
Louise Martin
Lisa Usman
Khālid Bā-Atwa

Field Participants 2008

Joy McCorriston
ʿAbdalʿazīz Bin ʿAqīl
Michael Harrower
Khālid Bā-Dhufāry
Tara Steimer-Herbet
Catherine Heyne
ʿAbdalKarīm Al-Burkānī
Matthew Senn
Jennifer Everhart
Kimberly Williams
ʿAbdallah Nasser Ṣarām
ʿAli Alwān
Muḥammad Alwān
Timothy Archer
Prem Goel

Archaeologists' camp in Wādī Sanā, 2005.

Editors and Contributors

Editors

Joy McCorriston

With a research profile focused on the prehistoric Near East, Joy McCorriston teaches archaeology, prehistory, and human environment as a professor in the Department of Anthropology at The Ohio State University. Trained as an undergraduate at University of London's Institute of Archaeology, McCorriston completed her PhD in anthropology at Yale University in 1992. After a postdoctoral position at the Smithsonian Institution, she joined the University of Minnesota as assistant professor. During her early research career she focused on the transition to agriculture and its regional development, with publications in *American Anthropologist*, *Antiquity*, and *Journal of Field Archaeology*. Her field research led her to the Khabur River salvage excavations of northern Mesopotamia (Syria), where she sampled many small tells for plant remains, integrating her results with faunal and chronological studies to recognize a new Bronze Age system of regional specialization in agricultural production and exchange, work she published in *Current Anthropology* and *Journal of Archaeological Science*. In 1996 McCorriston began research in southern Yemen with a regional survey for habitation sites that might yield plant remains and clues to agricultural transitions and early sedentism. McCorriston initiated and codirected the RASA (Roots of Agriculture in Southern Arabia) Project over seven seasons from 1998 to 2008. Funded by multiple grants from the U.S. National Science Foundation, American Institute for Yemeni Studies, Wenner Gren Foundation, University of Minnesota, and The Ohio State University, the team was also handsomely supported by Nexen, a local oil company. Since 2009, McCorriston's team has been working on the monuments and settlements of nomadic pastoralists in Dhofar, Oman, funded in part by a Dynamics of Coupled Natural and Human System grant from the U.S. National Science Foundation. The RASA work has inspired a new generation, as former students and collaborators now lead their own projects and funding in Oman, Ethiopia, Ghana, Indonesia, Switzerland, Lebanon, Kuwait, Jordan, and Saudi Arabia.

Michael J. Harrower

Michael Harrower is an associate professor of archaeology in the Department of Near Eastern Studies at Johns Hopkins University. His current research concentrates on long-term histories, from the beginnings of agriculture through the rise of ancient states across Southern Arabia and the Horn of Africa. His interests in archaeological theory center on the interface of scientific and humanistic perspectives in archaeology as expressed in studies of environmental and social landscapes. Thematically, his recent research concentrates on spatial, political, and ideological dynamics of water. Methodologically, he is a specialist in geographic information systems (GIS) mapping, satellite imagery, advanced GPS, and ancient irrigation. In 1999 Harrower joined the RASA Project as a graduate student. He eventually served as a RASA codirector; his contributions involved helping formulate survey and test excavation strategies and field directing RASA investigations. His contributions to RASA have been published in major journals, including *Current Anthropology*, *Journal of Archaeological Science*, and *Journal of Arid Environments*. Using a combination of archaeological survey and satellite imagery analysis, Harrower's recent NASA-funded fieldwork in Oman and Ethiopia concentrates on long-term water histories of early civilizations. In Ethiopia, investigations have focused on Pre-Aksumite through Late Aksumite (1000 BCE to 700 CE) settlement patterns and excavations of the newly discovered ancient town of Beta Samati. In Oman, research has involved wide-ranging archaeological survey, 3D modeling of a megalithic Bronze Age (Umm an-Nar Period) tower monument, and most recently investigations and mapping of trade in copper and chlorite during the Bronze Age, Iron Age, and Islamic era.

Contributors

Joshua Anderson
Independent Consultant and Geologist
Memphis, TN

Cecilia Anderung
Department of Evolutionary Biology
Uppsala University

ʿAbdalʿazīz Bin ʿAqīl
Regional Director
General Organization for Antiquities and Museums, Hadramawt

Kenneth Cole
School of Earth Sciences and Environmental Sustainability
Northern Arizona University

Rémy Crassard
Chercheur
CNRS–CEFAS
French Center for Archaeology and Social Sciences

Thomas S. Dye
President and CEO
T. S. Dye & Colleagues, Archaeologists, Inc.

Jennifer L. Everhart
Department of Anthropology
Washington University, St. Louis

Anders Götherström
Professor
Institutionen för arkeologi och antikens kultur
Stockholm University

Michael J. Harrower
Associate Professor
Department of Near Eastern Studies
Johns Hopkins University

Louise Martin
Reader
UCL Institute of Archaeology

Joy McCorriston
Professor
Department of Anthropology
The Ohio State University

Eric A. Oches
Professor and Chair
Department of Natural and Applied Science
Bentley University

Alessia Prioletta
CNRS-UMR 8167 Orient et Méditerranée

Joseph Roe
Researcher
University of Copenhagen

Tara Steimer-Herbet
Lecturer
Laboratoire d'archéologie préhistorique et anthropologie
Département F.-A. Forel des sciences de l'environnement et de l'eau
University of Geneva

Lisa Usman
Head of Conservation
Plowden and Smith

Dawn Walter (Gagliano)
Principal Investigator-Archaeologist
ASC Group

Kimberly Williams
Department of Anthropology
Temple University

Glossary Conventions

A discerning reader will notice some seeming inconsistencies in our use of Arabic words and names. For example, why use wadi in some places and wādī in others? We have made every effort to be systematic. This book follows the Library of Congress (LoC) convention for Roman/Latin transliteration of Arabic orthography. Where a location has been previously published as an archaeological site name not following LoC transliteration (for example, Shabwa) or not following any standard transliteration (for example, Bint al Methul), that published name has been retained. So for such sites (such as Wadi Zerqa), we use wadi. Where we refer to place-names and correct transliteration, we use wādī.

Certain place names—Mukalla, Sana'a, Dhofar, Yemen—have common usage in English transliterations that do not conform to LoC conventions. We use them here judiciously for reading ease.

Personal names pose a problem for strict consistency in orthography. In accordance with LoC convention, we do not capitalize an initial aleph (ʾ); we make exceptions for personal names. In references, we use the name as published. Where there are two or more published versions (for example, 'Abdalaziz Bin 'Aqil and ʿAbdalʿazīz Bin ʿAqīl), we have kept the version as published in the references but used transliteration that most closely conforms to LoC when the person's name appears in text. Where names have not been previously published but given to us in Latin characters by their namesake or an Arabic speaker, we use the name as given to us and signed by that person (for example, Ietha Al-ʿmari, ʿAbdalBaset Nʿoman), even with nonstandard capitalization and mixed orthography. Where we use a name as we heard it, we have to our best ability used correct LoC orthography. Our choice reflects respect for the bearer and his or her choice of orthography—after all, Sarah, Sara, and Sarai all may choose how they spell their names.

Other conventions we have observed:

- In tables and data, we follow the format xxx-xxx (for example, 045-001), but in the text, the zeros are omitted (for example, SU45-1, C7-1, W5-1)
- Metric units have been abbreviated without periods: 18 cm, 3 m, 12 km, 28 ha.
- We use *rockart* rather than *rock art*.

Unless otherwise specified, map coordinates in the book are provided in Universal Transverse Mercator (UTM) Zone 39 North WGS84 (World Geodetic System 1984), with elevation values in mean sea level (MSL) using the EGM96 (Earth Gravity Model 1996) geoid. In some cases, this involved converting elevation values originally recorded as height above ellipsoid (HAE). For example, near the Khuzmum, for excavations at 151-1 and 37-3, elevation values originally recorded in notebooks in HAE were converted to MSL (EGM96) by adding 19.14 m.

Part I

Research Objectives, Geological and Environmental Context

Chapter 1
Introduction to Research

Joy McCorriston

News travels fast and far over the desert; word of our arrival had preceded us. Late in the afternoon, the shadows reach halfway up the bare hills of middle Wādī Sanā, and the curving slopes give off a deceptive allure. In a university office, one forgets the midday glare off a billion grains of quartz, trapped in hard silt terraces laid down in the bygone days of flooding. And one forgets the blinding polish of stream-rolled limestone cobbles, white as old bones in the season of desiccation. Forgotten, too, are the moments of thirst and danger—a broken axle far from passible roads, the rumble of a flash flood in the darkness, or a double gunshot warning. But our guides had not forgotten us, so as our cranky Toyota Land Cruiser reached the valley floor at al-Faqqāsh one unremarkable January afternoon, there stood ʿUbayd Al-ʿAlīy. ʿUbayd is one of the Ḥumūm bedouin who sparsely populate Wādī Sanā, and as we later discovered, he was living in a rockshelter some 10 km to the south. He did not just happen to be at the foot of the pass that afternoon and pointed to his finest apparel, a battered wristwatch, to show that we were late. ʿUbayd wanted to be hired again by the archaeologists, and he intended also to secure work for his brother Nasser. Standing by the track, he wore his only shirt, frayed at neck and sleeves, over a dusty *futa* wrapped at his waist, with plastic Chinese flip-flops from the local Yemeni military supply shack where local bedouin could find cigarettes, sugar, tea, cheap sandals, and, it would seem, news (figure 1.1).

Or perhaps ʿUbayd Al-ʿAlīy had not heard we were coming from the Yemeni army network of radios and walkie-talkies. Our preparations in Mukalla, seven hours away over the mountains, had included hiring another Ḥumūmi tribesman as driver and guide. Perhaps the departures of ʿAbdallah and Khālid from Shiḥr to join us had echoed up the mountain passes. Had our Land Cruisers and occupants been noted as we passed by the hamlets of Dafīsh and al-ʿUlayb? By the next day, when we brought in a 3-ton truck with camping equipment and supplies, Nasser had appeared from 20 km up-wadi, and shortly thereafter our old guard, ʿAli Al-ʿAlīy, hurried in from his winter camp in Wādī Hadramawt, 80 km to the northeast. He had heard that "the Doctura" had arrived.

In light of what we had learned over more than 10 years of research in Yemen, this news relay is unsurprising. Yemen's landscape is a social construction, tied together by the relationships and recollections of its inhabitants. It has probably always been so in these arid mountains, where farming is a precarious risk and herding not much safer (figure 1.2). Even as people live in widely dispersed communities and many families move with their herds, every household needs social backup. In this context, news is a precious resource and can be exchanged, hoarded, and brokered against trust, reciprocity, obligation, support, and conformity to social norms. These are an intangible calculus of human decision-making and as critical as material resources in the behavioral ecology that shapes long-term landscape histories.

Figure 1.1. ʿUbayd Al-ʿAlīy with a borrowed camel in Wādī Sanā. The military checkpoint at al-Faqqāsh appears in the background, at right. *Photograph by Joy McCorriston.*

Research Problem

It was through existing social networks that news and stock of the earliest domesticated plants and animals spread into Arabia thousands of years ago, but until our RASA (Roots of Agriculture in Southern Arabia) Project, there had been limited archaeological research on this process. For most of the century after botanists first identified the ancient Near East as the originating point of transformative domestications, archaeologists sought to explain the domestication process and associated cultural and environmental changes at the end of the Pleistocene and beginning of the Holocene. Because sedentism has had an important conceptual place in explanations of transitions to farming (Harris 1989; Keeley 1995; Rosenberg 1998; Smith 2001), the earliest sedentary communities have seemed archaeological hallmarks of human populations most likely to adopt plant food production. In the Near East, archaeologists have often focused upon Natufian hamlets and the earliest Neolithic villages of the Levant and Taurus mountains, finding domesticated plants and farming tools (Asouti and Fuller 2013; Bar-Yosef and Belfer-Cohen 1989, 1992; Belfer-Cohen and Bar-Yosef 2000; Byrd 2005; McCorriston and Hole 1991; Simmons 2007; cf. Willcox et al. 2008). An underlying assumption has been that a basic Irano-Levantine agricultural package of domesticated plants, animals, and technologies emerged at 10,000 to 8000 cal BP and subsequently spread to new habitats across Eurasia (Harris 1996; Hole 1984). Evidence for the earliest animal domestication now suggests that it occurred independent of plant domestication, several times, and that sedentism played no discernable role (Martin and Edwards 2013; Zeder and Hesse 2000). As the evidence for domestications in different regions and cultural groups has grown, the concept of a Near Eastern package of plant and animal domesticates spreading with farmers and new villages across Eurasia (Ammerman and Cavalli-Sforza 1984; Bellwood 2005; Harris 1996) has been revised and refined (e.g., Bocquet-Appel et al. 2012; Colledge et al. 2004, 2013; Zeder 2009). Relatively new technologies of ancient DNA analysis now suggest multiple domestication sites for some plants and animals (Bradley et al. 1996; Hanotte et al. 2002; Larson et al. 2005), while the spread of farming engaged both migrations and local adoptions (Czekaj-Zastawny et al. 2013; cf. Bellwood 2005; Zvelebil 2006).

Figure 1.2. Middle Wādī Sanā research area looking east up Wādī as-Shumlyah. *Photograph by Joy McCorriston.*

Neolithic Geography

Does the spread of food production into Eurasia, however well documented, truly offer a universal model for this process? When the RASA Project began in 1996, we asked: What are the roots of agriculture in Southern Arabia, and how does the archaeological record from this region inform a broader understanding of the spread of farming out of the heartland of crop domestications? The archaeological record shows that the Early Holocene saw great expansion of domesticates into new environments. This spread has been extensively modeled (e.g., Ammerman and Cavalli-Sforza 1984; Bellwood 2005; Rindos 1984), highlighting the importance of factors such as population pressure (Rosenberg 1998), resource diversity (Gregg 1988; Keeley 1995; Rindos 1984; Zeder 2012) and reliability (Ingold 1980; Winterhalder and Kennett 2006), and social significance (Hastorf 1998; Hayden 1992; Marshall 1990). If one shifts archaeological focus from the much-studied Fertile Crescent to center a map on the Indian Ocean/Arabian Sea, then the mountain ranges of Arabia's western and southern coastlines suggest a new landscape for the spread of domesticates from multiple centers (figure 1.3). Southern Arabia is separated from the Fertile Crescent by a formidable desert and from two continents by water. These are on the one hand important geographical and ecological barriers to the spread of domesticates. On the other hand, people in Southern Arabia had the choice of multiple sets of domesticates adapted to different geographical regions. These factors suggest that the spread of domesticates, farming, and technologies into Arabia may have occurred under circumstances and processes much different than elsewhere in Eurasia.

The Neolithic package of domesticates consolidated in the northern and western Fertile Crescent lies within a northern Mediterranean climate and ecological system to which domesticates were adapted—their transfer to Arabia meant passage through new climate systems. Wheat, barley, flax, and pulses like chickpeas and lentils were the first farmed plants, with cattle, pigs, sheep, and goats as Near Eastern animal domesticates (Clutton-Brock 1999; Zohary and Hopf 1988). All of these were easily transferred to new Mediterranean lands in Europe, beyond which were no alternate domesticated species or food-producing peoples in adjacent continental habitats. The pioneering farming of continental Europe's new farmers is

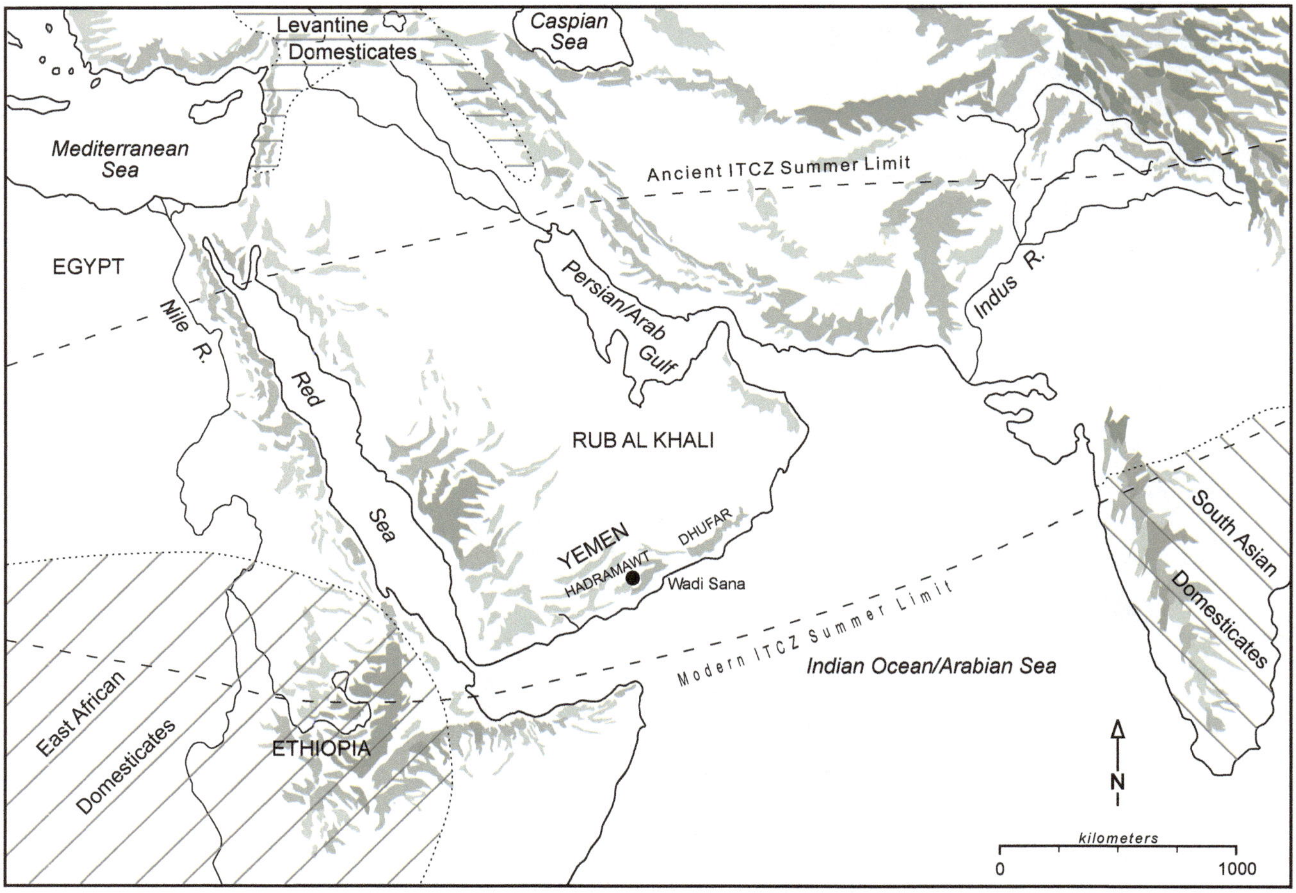

Figure 1.3. Map of the Arabian Peninsula and neighboring regions with potential domesticates. Dashed lines indicate the InterTropical Convergence Zone (ITCZ) where wind belts of the northern and southern hemispheres meet, with significant implications for rainfall. It shifts over the course of a year, but has also shifted historically as a function of planetary climate drivers. *Illustration by Jarrod Burks.*

well documented (Barker 1985; Colledge et al. 2004; Harris 1996). In Southern Arabia, circumstances differed. Early Holocene foragers were potential recipients of Pre-Pottery Neolithic B (PPNB) food-producing strategies introduced from the northeast (Dreschler 2007) or via South Asia. But Arabia's Neolithic peoples faced inherent ecological challenges in adopting many of these domesticates, especially in the case of plants adapted to winter rainfall, dry summers, and subtropical daylight regimes (McCorriston 2006) (figure 1.4).

Domesticates introduced to Arabia encountered not only new climates but also a new social landscape. Evidence from PPNB steppe dwellers in the southern Levant strongly suggests that many PPNB groups relied primarily on hunting, with little evidence of herding until the Late PPNB (cf. Henry 1994; Henry and Beaver 2014; Horowitz et al. 2000; Martin 1999; Martin and Edwards 2013; Martin et al. 2010; Tchernov and Bar Yosef 1982). Thus they were unlikely to carry and exchange farming strategies southward, even as new evidence emerges for contact between different cultural groups (Crassard et al. 2013). It seems increasingly evident that Arabia's post-Pleistocene landscape was repopulated by foraging-hunting native people from Arabian refugia (Al-Abri et al. 2012; Černý et al. 2011; Crassard and Dreschler 2013; Rose et al. 2013) rather than by advancing PPNB herders (Dreschler 2007), whose technologies appear no farther south than the middle Gulf region (Charpentier and Crassard 2013).

In Southern Arabia, Early Holocene foragers would have potentially had access to African and/or South Asian cattle and millets available at different times. New evidence of a sustained transfer of African obsidian into Arabia offers a clear trail of the networks along which less tangible information and domesticates might have passed (Khalidi et al. 2013). African millets and cattle were not available as an integrated package for Early Holocene Southern Arabian hunter-foragers, for African records show that plants and animals were domesticated at different times over 8,000 years (Amblard 1996; Boardman 1999; Bower 1991; Close and Wendorf 1992; Gautier 1984; Peters 1986; Rowley-Conwy et al. 1999; Stemler 1990; Wasylikowa 1993; Wasylikowa and Dahlberg 1999). African domesticates include sorghum, millets, and perhaps donkeys (Marshall and Hildebrand 2002; Marshall and Weissbrod 2011) and cattle (Bradley et al. 1996; Gifford-Gonzalez 2005). While emerging in different locales at different times along the fringes of the Sahara (e.g., D'Andrea et al. 2001; Rowley-Conwy et al. 1999) and in the East African highlands, these domesticated resources and the knowledge of how to cultivate them offered an important potential for exchange and a reserve for migrating people. Domesticated cattle appeared early in the Saharan Neolithic, around 8,000 years ago (Gifford-Gonzalez 2005). Sorghum was reportedly introduced to Arabia 5,000 years ago (Cleuziou and Costantini 1980; Cleuziou and Tosi 1997; Costantini 1984, 1990; Soderstrom 1969) and to South Asia thereafter (Boivin and Fuller 2009; Costantini 1981; Weber 1998). But the evidence for domesticated sorghum in Arabia before the medieval era is controversial (Potts 1994; Willcox 1994).

To Arabia's east lies the Indian subcontinent, with local domestications and food producers in the southwestern Neolithic. From 6,000 years ago, native cattle herders farmed indigenous gram (small legumes) and millets, and eventually acquired cotton adapted to the southern monsoon climate. An increasing range of archaeological studies demonstrates early traffic and the movement of domesticates across the Arabian Sea (e.g., Boivin and Fuller 2009; Khalidi et al. 2010; Matthews 2002; Weber 1998), but there remain questions about the history of their transfer and the roles of intervening lands and peoples. Arabian culture history still holds important keys to understanding the transfer of information and technologies across sparsely populated regional networks at the epicenter of a multi-pole region where different packages of domesticates were potentially available.

Food Production and Climate Change

In this arena, the RASA archaeological and paleoecological research project sought evidence of the processes by which domesticates were first introduced and the social and environmental landscapes into which they were adopted. Changing Middle Holocene (circa 6000–5000 cal BP) climatic patterns in Southern Arabia forced people in the early stages of developing agricultural subsistence economies to adapt their cultivation technologies, water management practices, and herding strategies to an increasingly arid landscape (Lézine et al. 2010). Food production emerged in a dynamic of climate change, with environmental and human responses. Terrestrial proxy records of Late Pleistocene and Holocene climate change in the Southern Arabian Peninsula include lacustrine deposits and higher lake levels (Lézine et al. 2007; McClure 1976, 1984; Roberts and Wright 1993), stable isotope measurements in speleothems and groundwater carbonates (Clark and Fontes 1990; Fleitmann et al. 2003, 2007, 2011; Van Rampelbergh et al. 2013), and paleosols (Brinkmann 1996; Fedele 1990; Parker and Goudie 2008; Wilkinson 1997). These point toward a period of increased moisture during the Early to Middle Holocene, from about 10,000 to 6000 cal BP (Fleitmann et al. 2007, 2011). Bordering regions of eastern Africa and western India provide additional data in support of a period of increased wetness during that interval (e.g., Abell and Hoelzmann 2000; Gasse 2002; Lézine 2009; Rawat et al. 2015; Tierney and deMenocal 2013).

Climate models combined with paleoclimate proxy data indicate that precipitation over the Arabian Peninsula is largely controlled by the strength of the southwest Indian monsoon (Marzin et al. 2013; Prell and Kutzbach 1987; Roberts and Wright 1993; Zhao et al. 2005), controlled by precession-forced northern hemisphere insolation variations and late- and post-glacial boundary conditions (sea surface temperatures, ice volume, meltwater fluxes, albedo, CO_2) (summarized by Braconnot et al. 2007; Gupta et al. 2003; Kutzbach et al. 2008; Marzin and Braconnot 2009; Marzin et al. 2013; Neff et al. 2001). Periods of abrupt change in moisture balance in continental paleoclimate records from the Arabian Peninsula and adjacent regions coincide with shifts in the Inter-Tropical Convergence Zone (ITCZ) and monsoon intensity documented in continuous sediment cores from the Indian Ocean and the Arabian Sea (Bassinot et al. 2011; Tierney and deMenocal 2013). General circulation models (GCMs) of climate and regional paleoenvironmental proxy records of the Middle Holocene indicate a major climatic shift, with the southward displacement of the ITCZ and retreat of the southwest Indian monsoon, accompanied by a decline in precipitation across the southern Arabian Peninsula 6,000 to 5,000 years ago (Fleitmann et al. 2003, 2007, 2011; Lézine et al. 2007; Marzin et al. 2013; McClure 1976, 1984; Prell and Kutzbach 1987; Roberts and Wright 1993).

Other GCM studies indicate that on a broad, regional scale, initial monsoonal enhancement was relatively abrupt, while decline in monsoonal precipitation was gradual (Kutzbach et al. 1996; Naqvi and Fairbanks 1996). This would have had an important effect on the region's vegetation and on local-scale human adaptations

Figure 1.4. Wheat farmed by irrigation under date palms in Wādī Ḥaḍramawt. Irrigation is the only way winter wheats grow in the Southern Arabian climate with its summer rainfall. *Photograph by Joy McCorriston.*

that had evolved in the more humid phase. The paleoenvironmental record investigated in the Wādī Sanā–Wādī ʿIdim region of southern Yemen offered local-scale reconstruction of environmental adjustments in the context of human adaptive response during this important Middle Holocene period of changes in monsoon-driven precipitation in the region (Berger et al. 2012; Harrower et al. 2012). This paleoenvironmental record, combined with the archaeological evidence in the region, provided an ideal opportunity to test the research hypothesis that Middle Holocene climatic and environmental changes prompted Southern Arabian foragers to adopt new domesticates and commit themselves to food production.

Neolithization in the Changing Sahara: A Close Parallel?

Which domesticates might they choose? The cultural and paleoenvironmental record of resources, domesticates, choices, and technologies in the highlands of southern Yemen offers evidence to test and refine a hypothetical Middle Holocene climate-driven transition to food production. This hypothesis stems from one developed in Egypt and the Sudan. Archaeological studies in the eastern Sahara suggest that the spread of Levantine domesticates southward into subtropical arid ecosystems was an adoption by local foragers/cattle herdsmen evading expanding Sahelian deserts (Close and Wendorf 1992; Gifford-Gonzales and Hanotte 2013; Hassan et al. 2001; Madella et al. 2014; Marshall and Hildebrand 2002; Phillipps et al. 2012; Schild and Wendorf 2001). Desertification in Egypt likely occurred as a result of the same mid-Holocene shift in the monsoon that marks the end of the African humid period (deMenocal 2015). The general mechanism and timing of this climate change is no longer disputed (Berkea et al. 2012), but its regional and local effects may have differed according to geographic and population parameters. Therefore, one might expect that the models advanced to explain adoption in the Nile Valley of a Levantine package of domesticates—sheep, goats, pigs, winter cereals such as wheat and barley, legumes, and flax—require modification in an Arabian theater. The Arabian case adds to an understanding of how foragers responded to desertification, an issue of broad anthropological significance in the reconfiguration of agriculture in the Old World and New (Kusimba 2003; Minnis 1992). Also, Arabia's geography, with equidistant potential influences from the Levant, East Africa, and South Asia, emphasizes how little we understand about why one package (Levantine) seemingly spread in adjacent regions like Egypt and South Asia long before their peoples developed indigenous domesticates.

The hypothesis that Middle Holocene monsoon recession and subsequent aridification prompted the adoption of domesticates has been difficult to test in Egypt, where the nature of heavy alluvial overburden along the Nile Valley has obscured most early sites, and none show a transition to farming (Phillipps et al. 2012; Wetterstrom 1993). Saharan foragers might have incorporated one or more elements of the Levantine package rather than adopting plants, animals, and technologies together (Holdaway et al. 2010; Phillipps et al. 2012; Vermeersch et al. 1996; Wenke et al. 1988). It remains unclear how they reconciled such elements with keeping African domesticated cattle (Close and Wendorf 1992) or whether cattle were indeed a local domestication (Gautier 1984; Gifford-Gonzales 2005; Gifford-Gonzales and Hanotte 2013). Unlike the Nile Valley and Saharan oases, the archaeological record in highland Southern Arabia has been largely preserved without the obstruction of heavy flood deposits or dense populations in subsequent millennia. Thus, by testing in Arabia the hypothesis that a mid-Holocene climate change limited existing subsistence options while opening opportunities to incorporate Levantine domesticates, archaeologists may contribute more sophisticated models for the spread of farming after initial domestication.

Specific Research Questions

The RASA team sought to address the following questions:

1. What were the first domesticates adopted into the South Arabian highlands and when did these introductions occur?
2. What were the local and regional paleoenvironments of the time?
3. Were these domesticates introduced by migrant peoples arriving into new ecological niches or were these domesticates predominantly transferred to extant Arabian hunter-foragers?
4. To what extent was the selection and adoption of domesticates constrained by existing cultural structures?
5. How did adoptions of new domesticates physically and socially affect human landscapes?

Social Networks and Food Production

An Arabian Context

The Arabian Desert of the Early Holocene was no lonely wasteland but supported sparse populations who well understood the energetic payoff of cooperative hunting. A classic Levantine-centric bias in research has viewed the steppe margins of the Fertile Crescent as a marginal zone lacking the attractively reliable resources of Mediterranean woodlands (Belfer-Cohen and Bar-Yosef 2000; Henry 1989; McCorriston and Hole 1991), where crop plants were domesticated, plausibly in the social context of a household sharing the fruits of food production and foraging in a relatively crowded land (e.g., Asouti and Fuller 2013; Byrd 2005; Peterson 2002; Simmons 2007; Stiner et al. 2000). So-called desert margins supported sometimes vast herds of animals (Betts et al. 2013; Legge and Rowley-Conwy 1987; Martin 1999; Tchernov and Bar-Yosef 1982), whose attractiveness to hunters may have well outranked that of nuts and seeds. Recent research in the desert peripheries of the Levant suggests populations stable over long periods (Maher et al. 2012a, 2012b) who engaged in cooperative hunting (Arav 2015; Crassard et al. 2015; Helms and Betts 1987) and needed to maintain the social apparatus integral to it.

Arabia was populated, albeit sparsely, from the beginning of the Holocene (Cleuziou and Tosi 1997; Charpentier 2008; Lézine and Cleuziou 2012; Wilkinson 2009) and perhaps throughout the hyper-arid Terminal Pleistocene (Al Abri et al. 2012; Černý et al. 2011; Rose 2010; Rose et al. 2013). The sparse and patchy resources of an arid landscape imposed constraints and localized opportunities on foragers, who sustained mobility in the interior and along the coasts, sometimes passing seasonally between them (e.g., Berger et al. 2013; Dreschler 2010; Lancaster and Lancaster 1999). Throughout much of the Early Holocene, lakes persisted in desert areas that today are devoid of permanent standing moisture (Lézine et al. 2010; Parker and Preston 2008), providing a focal point for hunters and prey of the arid interior (Inizan et al. 1997; McClure 1988). Whatever the socioeconomic strategies practiced by Arabian foraging populations, domesticates were not introduced into a vacuum (Inizan et al. 1997; Maher et al. 2012a, 2012b), and the emerging evidence of contacts between Northern Arabian populations and Levantine ones suggests that there were existing cultural and social templates within Arabia from the end of the Pleistocene (e.g. Charpentier and Crassard 2013; Crassard 2008; Crassard et al. 2013).

The first centuries of the Arabian Early Holocene have left few material traces of these foraging-hunting groups, with little evidence of their social lives (Charpentier 2008). Small and intermittently occupied encampments (e.g., Cremaschi and Negrino 2005; Inizan et al. 1997; Rose et al. 2013; Uerpmann 1992) suggest mobile, small hunting parties, perhaps one or several small families. Undoubtedly a wide array of their organic technologies—using feathers, hide, thongs, fibers, dyes, wood carvings, skin, symbols, and tokens—has not survived. Stone tools do occur from the first quarter of the Holocene (Hilbert 2013), and these universal testimonials of human passage suggest an indigenous Arabian cultural template (Charpentier and Crassard 2013; Crassard 2008), with a growing emphasis on individual skill and prowess at the beginning of the Holocene (McCorriston 2013). If collective hunting, maybe with nets and dogs, appeared later, it might reflect or have influenced broader social bonds and wider social collectives. The uptake of domesticates, with their delayed returns, future discounting, and inherent risks (Alvard and Kuznar 2001; Rindos 1984; Russell 1988) must have depended not only on whether and where they could be adapted to Arabian ecosystems but also whether and how ownership or territorial imperatives of domestic stock-keeping could be accommodated in an existing Arabian social environment.

To generate the data needed to address these questions of the spread of domesticates out of the Levant (or elsewhere), the variability of adaptation processes in different regions, and the parameters of social constraints, the RASA team pursued specific research objectives, whose outcomes appear in following chapters.

(1) RASA sought to reconstruct human economic behaviors in the Southern Arabian uplands during the wetter Early–Middle Holocene. During that time, hunting, burning cover, herding, various mobilities, and water management strategies were practiced, as RASA research has shown for the Wādī Sanā (Martin et al. 2009; McCorriston et al. 2005, 2012) and as is now apparent in a wider Arabian sphere (Magee 2014). By collecting samples of animal, plant, and lithic remains stratified in and around structures and shelters in Wādī Sanā, RASA sought to determine whether animal bones were from domestic herds and whether any plant remains might indicate cultivation in conjunction with water management technology. Animal and plant remains also address when caprines were introduced and whether local resources (for example, dates, *Ziziphus* fruits, wild grasses, and wild camels) were cultivated or managed. Lithic tool forms can be suggestive of function, and lithic materials indicate exchanges and contacts outside people's closest communities.

(2) RASA sought to provide regional context for high-resolution archaeological and palaeoecological records from Wādī Sanā by comparing sequences from different highland areas. To supplement RASA focus on a single highland drainage, Wādī Sanā, the project also investigated selected sites in Wādī ʿIdim. Records from Wādī ʿIdim and Wādī Sanā combined can build insight into environmental variability and examine how representative is the record RASA has found in the main drainage of Wādī Sanā. Important sites in Wādī ʿIdim, such as the sanctuary with anthropomorphic statuettes and stelae (chapter 15), have no known correlates in Wādī Sanā. With appropriate chronologies (dated hearths, tombs, monuments, and occupation sites), researchers can test the hypothesis that people abandoned areas without active springs as precipitation declined. If Middle Holocene aridification prompted populations to concentrate near water sources, then we expect to document less activity in middle Wādī Sanā, where there are no springs, offset by Middle to Late Holocene activities where springs continued to flow, such as Ghayl Bin Yumain (upper Wādī Sanā) and Ghayl ʿUmar (Wādī ʿIdim).

(3) RASA sought to develop a chronology for the RASA sedimentary paleorecord of environmental changes associated with the Middle Holocene climatological shift in precipitation. A southward shift in the ITCZ and associated monsoon decline over the Arabian Peninsula played a significant role in changing precipitation patterns reflected in Middle Holocene paleorecords from Arabia (Lézine et al. 2010; Parker and Preston 2008). Continuous sediment records in Wādī Sanā and fossil spring deposits in Wādī ʿIdim provide sedimentary and geochemical proxies of local environmental histories and processes. These data can be used to reconstruct environments of human occupation and activity in the region and as local spatially distributed proxies of the shift to hyper-arid conditions characteristic of Southern Arabia during the last 5,000 years.

Archaeology in Southern Yemen

Most of Southern Arabia is a hyper-arid zone. As such, it forms a southern extension of the great Syrian–Arabian desert that has long connected the mobile peoples of the Levant, Zagros, and Arabia in times of greater moisture and divided them through aridity. Southern Arabia's prevailing climate, driven by the Southwest Asian monsoon, and its fauna and flora have greater affinities with Saharan and East African environments than with the Mediterranean and continental climates at the Syrian–Arabian desert's northern rim. Culturally too, Southern Arabia boasts a different diversity, with unique culture histories stretching back into the Pleistocene era.

The first documentation of archaeological remains of southern Yemen appeared in travelogues and histories of the medieval era and are seldom accessed by Western scholars. Al-Hamdānī ([945] 1989) wrote of the tribes and places of pre-Islamic Yemen. In the 1200s Ibn Al-Mujawīr ([1204/5–1291/2] 1986) described the monuments he saw as he passed through the eastern provinces on a circuitous haj. Each commented on the ruins of ages past, and throughout these Islamic geographies and histories run the filaments of folklore tangled into Qur'anic accounts of Jāhiliyah, the age before revelation to the Prophet. Eastern Ḥaḍramawt was the Qur'anic land of ʿĀd, whose wicked people were swept away by a burning wind, leaving only the demonic spirits in Bi'r ("Well of") Barhūt, and whose glittering city, Iram of the Pillars, sank beneath the sands. Long into the twentieth century, these areas were inaccessible to foreigners. Much of Arabia remained a mysterious destination, whose vast Empty Quarter lured the last explorers with a penchant for alternating heat and cold, thirst, hardship, and adventure.

Explorers and archaeologists alike have also shared an expectation that ancient settlements must have existed where, in our era, mobile pastoralists claim domain. Such is the myth of lost civilizations and the trope of Western dominance over a degraded East, depicted in stunning detail by Orientalist painters like David Roberts and Gustav Bauerfeind. Scholars of the late 1800s traveled in disguise to glimpse ruins and inscriptions in Arabia's northern and western desert fringes (e.g., Burton 1964 [1878]; Doughty 1979 [1888]; Halévy 1873, 1877), and through their accounts, Petra, Najrān, and the South Arabian kingdoms came to Western view, set against the diminutive figures of bedouin wanderers in a timeless desert. Farther south and east, Mabel and Theodore Bent (1900) painfully followed the centuries-old directions of Macrīzī (Noskowÿj 1866) to reach the interior towns of Hadramawt, but they were unable to reach fabled Bi'r Barhūt, where it was said demons issued from the earth. By the 1930s, explorers vied to cross Arabia's great interior deserts. They had heard the bedouin speak of stricken civilizations buried in the sands there (Philby 1939; Thesiger 1959; Thomas 1938).

In Southern Arabia, travelers struggled to reach Shabwa, ancient capital of the Hadramawt Kingdom, in a race that mirrored colonialist political rivalries and local resistances to them (Helfritz 1935; Philby 1939; Stark 1953). Harry St. John Philby, in the employ of King ʿAbdalʿazīz Ibn Saʿūd and much to the chagrin of his German and British rivals, succeeded in describing Shabwa's ruins where others had met hostility and obstacles. Freya Stark relied on her modest presence as a lone traveler and on British pacification of the Aden Protectorate to afford her safe passage and shelter. Van der Meulen (1932, 1947) pressed safe passage with heavy bags of silver Austrian thalers. He was accompanied by the scholarly Hermann von Wissmann, who collected equally heavy inscriptions in Old South Arabian languages. A pioneering, single-winter archaeological expedition at the first-century BCE Moon Temple at Hureidha (Caton-Thompson 1944) was preceded by a mountainous journey through terrain geologically and archaeologically unexplored (Caton-Thompson and Gardiner 1939). As director of antiquities in the Aden Protectorate, Brian Doe (1971, 1983) cataloged and classified the sites and most prominent monuments he found through the 1960s. Throughout all these experiences, archaeological discovery depended on local informants and local knowledge to identify site locations and traditional intersite routes.

In the 1950s, an expedition from the Smithsonian Institution documented many prehistoric sites within the Wādī Ḥaḍramawt itself. None of these discoveries suggested architecture or settlement, and all remains were presumably left by mobile people (Van Beek et al. 1964). In the final decades of the twentieth century, a Soviet–Yemeni expedition at the former Hadramitic center of Raybun (Sedov and Griaznavich 1996) and a French team at Shabwa (Breton 1991) began to lay out the cultural-historical framework of the first-millennium BCE Hadramitc Kingdom and to venture into the surrounding regions of the urban centers (Amirkhanov 1997; Inizan and Ortlieb 1987, Pirenne 1990; Sedov and Griaznavich1996). They found irrigation works, tombs, rockart, inscriptions, and the lithic scatters and hearths that marked former campsites of mobile people. By the mid-1990s, North and South Yemen had unified, fought a brief civil war in which union was preserved, and opened the south to oil and archaeological exploration. An American team began studying pre-Islamic Juja in Wādī Ḥaḍramawt itself (Hansen et al. 2004), and the pioneering Canadian Occidental Yemen (CANOXY) oil company commissioned an archaeological survey in advance of pipeline construction (Vogt and Sedov 1994). But little else was known in a vast and archaeologically unsurveyed province.

After Yemen's unification in 1990, new expeditions appeared. Archaeological survey revealed settlements in the lower reaches of major tributaries to the Wādī Ḥaḍramawt (Sedov 1996a, 1996b), and an American expedition to al-Ghaydah documented medieval Islamic settlement there (Newton 2009). French research in the eastern Wādī Ḥaḍramawt probed a sequence of occupation at Iron Age Makaynūn (Mouton et al. 2011) and a network of contemporary dependent communities (Schiettecatte 2010). A coastal survey and excavations at Shiḥr (Hardy-Guilbert 2005) and Sharma (Rougeulle 2001) have focused on latter histories of the early Islamic and medieval ports and coastline, but the dissected plateaus of the Southern Jol, between the coast and the Wādī Ḥaḍramawt interior, and its Northern Jol counterpart (Braemer et al. 2003) have seen little archaeological activity.

Wādī Sanā and the Southern Jol

Wādī Sanā is one of the least accessible of the north-ward-draining dry rivers of an uplifted plateau, (or *jol*). The remote location and sparse population of Wādī Sanā have contributed to the preservation of a rich archaeological record. Today's inhabitants are mobile and pasture mostly goats, with a few sheep and camels, in the seasonal flush and thorny scrub of an arid ecosystem.

Apart from a small settlement at Ghayl Bin Yumain, where springs and wells in a shallow water table have supported modest date palm orchards on sediment terraces and fans, there is no settlement until one reaches the village of Sanā itself, on an 80 km route to the north (figure 1.5). Sanā consists of a handful of homes and a mosque near the confluence of Wādī Sanā and Wādī Ḥaḍramawt, whose eastern springs feed a tiny perennial stream into Wādī Masilah. Since there are few springs in the narrow drainage, there are few perennial sources of water. In a few spots, a rock basin (*qīr*) or an enhanced vernal pool (*krif*) supplies sufficient water to sustain flocks, but these sources are variable from year to year and may be drained in a few weeks, forcing the bedouin and their flocks to move on (figure 1.6).

Inhabitants of the northern reach of the Wādī Sanā today belong to the al-Manāhīl bedouin clans of the north, while in the southern drainage, bedouin are Ḥumūm tribesmen, with closer social ties along the coast and southern escarpment. According to ʿAbdalʿazīz Bin ʿAqīl (2004 personal communication), local memory is of Wādī Sanā as a recent eastern border between Ḥumūm and Jābirī tribespeople to the west, so that when tensions arose between the tribes, Wādī Sanā was seldom visited.

Modern bedouin have returned to Wādī Sanā, using rockshelters carpeted with their predecessors' mats of packed goat dung, which they set alight and char to cleanse

Figure 1.5. Mosque at Ghayl Bin Yumain. *Photograph by Joy McCorriston.*

for reoccupation. They build no houses, use no tents, and construct only low semicircular dry-stone windbreaks and small pens for goats. The most favored winter rock-shelters face east, so they have early warmth and are in shade during the afternoon. Goats huddle close by at night, but by day they are herded across the stony lower slopes, picking at the last brittle browse and clambering into sparse scrub in the wadi for tender leaves. They are diminutive and hardy, with kids no bigger than tiny spaniels. Herders seal the youngest kids in crude stone bins capped with a few flat rocks and branches to keep them from wandering with their mothers, who search far for food. What the goats leave behind, a few camels strip, deftly lipping around thorns of acacia and *'ilb* (*Ziziphus spina-christi*). Herders sometimes pollard higher trees for fodder, and excursions from Ghayl Bin Yumain strip the countryside of woody branches to make charcoal. Whatever remains is food for termites, which devour the deadwood of living trees.

Life takes place in the wadi bottoms and along its lower slopes. North of Ghayl Bin Yumain's dying date groves, today there is no steady agriculture in the narrow canyon of Wādī Sanā and its tributaries. Outside the cleft of Wādī Sanā's main drainage, the slopes climb to a desolate plateau, where there has been no soil for thousands of years; the stony surface neither holds water nor supports vegetation. Human traffic has crossed these plateaus for centuries, leaving behind only the heat-shattered clasts of rock-filled hearths or the patinated debris from countless knapping episodes. A major east–west route climbs out of middle Wādī Sanā at today's military checkpoint, al–Faqqāsh, and the same route leads "all the way to Oman" as it heads east. In the first research seasons, the team watched heavy trucks lumber toward Oman, carrying massive loads of cumin. In subsequent seasons the trucks flowed westward, loaded with used vehicles to be sold in Yemen's souks. And by the end of the RASA research, the trucks groaned with drilling equipment and heavy machinery headed east again to build oil fields in the eastern Masila sector.

The Impact of Oil Exploration

The first exploration for petroleum in post-unification southern Yemen used tracks that follow ancient routes. Many ancient tracks have now been modified beyond recognition, with some graded to make passage for major traffic. At the other extreme, some routes have been blocked to control vehicle access to oil fields and infrastructure. Some of the earliest CGG (Compagnie Générale de Géophysique) prospecting in the early 1990s in Wādī Sanā brought heavy equipment across the *jol* and graded the plateau path into the middle Wādī Sanā. There are no prominent traces of ancient human passage remaining along its route, but the well-worn foot track, leading out of the wadi bottom to the high plateau, survives today beside the SU110 rockshelter (N 1738930, E 328155), where the feet of people and camels have smoothed sections of bedrock to a high polish. The road grader chose a slightly different route for vehicle traffic farther north, below what is now the military checkpoint, al-Faqqāsh. The eastward part of this same trail suffered destructive bulldozer swipes that have effaced forever parts of the 2,000-year-old trilith monuments that line the route up Wādī as-Shumlyah and toward Oman. Sometimes it appears the operator just took a detour for fun, or boredom, or the power to destroy, for the road itself bypasses some monuments reduced to rubble by a deliberate destruction serving no infrastructural purpose.

In middle Wādī Sanā, there has been limited physical impact in the wadi bottom and the fragile archaeological remains that cover its lower slopes and terraces. Early prospecting by CGG left narrow drill holes, a few spent charges, and perhaps some charges unspent with their rusting wires still sprawling across the lower terraces. An abandoned well hole, capped and protruding from a raw platform of bulldozed fill, marks the confluence of Wādī Sanā and its northern tributary, Wādī Ḥimayrī. The early failures to find drillable oil left Wādī Sanā mostly to the bedouin and their flocks, who continue to camp and browse along the barren slopes.

The first company to strike extractable oil in the Southern Jol was CANOXY-Yemen (Canadian Occidental, Yemen), later to reincorporate as Canadian Nexen Petroleum Yemen, Ltd. (NEXEN, or CPNY). With successful drilling in its Masila Block concession, CANOXY built an extraction infrastructure around its upland CPF (central processing facility), situated on the high plateau far from the small villages of the Raidāt (Ghayl Bin Yumain, Harou, Dafīsh, al-'Ulayb, Risib). These communities exist where broad drainages funnel enough surface water for floodwater farming in the summer months, and small populations have settled around cisterns and springs. The presence of CANOXY's CPF has had significant effects on local populations, even as the company has sought to mitigate friction and provide benefits to its neighbors. For more than a decade, CANOXY- and NEXEN-contracted doctors have run a local health clinic. Deliveries of school supplies and furniture, a cell phone tower for a regional signal, and maintenance of a graded road serving villages along the way have brought some welcome amenities. But there has been tension too, over local access, water usage,

Figure 1.6. An al-ʿAlīy bedouin woman tending goats in middle Wādī Sanā; view toward the south. *Photograph by Joy McCorriston.*

alleged drainage into Harou's fragile farmland, and the inequalities of the means and wealth of foreigners, whom locals see extracting resources from southern Yemen's impoverished highlands.

One of the biggest impacts of CANOXY's infrastructural investment has been the laying and maintenance of southern Yemen's first oil pipeline, which carries crude from the plateau to a coastal terminal. The pipe itself lies mostly on (graded) ground surface and has had limited impact on antiquities, but the footprint of pipeline access roads and subsequent reroutings over an ancient mountain pass have destroyed some ancient monuments and brought modern traffic and unwitting damage to others. Well aware of this potential damage, CANOXY completed a cultural resource impact survey in 1994 before construction of the pipeline. The survey was carried out over six months by Burkhard Vogt and Alexander Sedov on behalf of the German Archaeological Institute in Sana'a. Although still unpublished, the survey report includes a map of several hundred archaeological sites along the pipeline route and along major routes through the Southern Jol near CANOXY's CPF. Many of the sites destroyed by the pipeline construction were excavated and thoroughly described in the mitigation survey, and NEXEN has continued its policy of assisting local archaeologists for survey and rescue documentation. An unpublished text describes a "Hadramawt Megalithic Complex" (Vogt and Sedov 1994), and the locations and questions highlighted by Vogt proved very influential in the selection of Wādī Sanā for further archaeological research.

By the time the RASA team arrived in 1998 to develop more survey and excavation, CANOXY had hired a subcontractor to prospect for groundwater and drill wells to supply the local population. In Wādī Sanā, a new, pump-driven water well was operating, attracting more bedouin than had been there in decades and allowing them to stay longer. Several small building projects appeared in middle Wādī Sanā: some bulldozing to break slabs from limestone strata, a one-room building quickly abandoned, and a few tentative efforts at floodwater farming. By 2004 the pump was broken, the parts were plundered, and the bedouin had moved on. Ever since, water comes weekly in a huge tanker that crawls along the cobble bed track up Wādī Sanā from Ghayl Bin Yumain. This water also comes from NEXEN wells, maintained further afield and distributed—to plastic barrels—to each bedouin-occupied rockshelter by a fee-collecting entrepreneur.

Finally, the activities of other exploration and drilling have carved their traces on Wādī Sanā. Oil Search Inc. widened the track up Wādī as-Shumlyah, doing further damage to the triliths. After 5,000 years intact, the dolmen beside this track has lost its roof to looters and vandals in the last 10. Uprights that have stood for 7,000 years have been knocked flat, or targeted and shattered by gunfire. There is new graffiti beside the old. Foreigners and bedouin pick over the lithic scatters, plucking up the diagnostic points and tools for collectors, leaving mute debitage for archaeologists to decipher. Where carved stelae once stood over tumuli, they have been ripped out and carried off, whether to adorn a garden or sell to a collector is anyone's guess. One finds them sometimes in caches by the roadside, out of context and awaiting a night pickup.

Amid this destruction of cultural heritage there is a more positive facet to oil exploration, for the roads and support networks allow archaeologists to access areas that have been all but impossible to visit. In the 1930s, Hermann Von Wissman passed through Ghayl Bin Yumain and removed half an inscription from Qārah Ḥabshiyah, a high fortress controlling the southern access to Wādī Sanā's narrow canyon. He could stay only hours to probe the ruins, but 2004–2005 excavations in the nearby site of Manayzah lasted weeks over two seasons of excavation, thanks to the infrastructure and supplies provided by CANOXY and NEXEN. The generosity of these successive companies, compatible with the Responsible Care program adopted by NEXEN in an attempt to conserve natural and cultural resources, supported archaeological fieldwork. The RASA Project enjoyed CANOXY and NEXEN support in the form of hospitality, housing, logistics, supplies, and in-kind donations of repairs, equipment, emergency services, and fuel, and with outright financial contributions, including a Community Relations Grant from Calgary headquarters and the Sana'a head office. This grant pays, in part, for this and other publications of RASA results, after supporting field and laboratory analysis for several years.

Not only has NEXEN supported RASA research, but it has engaged local archaeologists to assess and document areas to be impacted by expanding oil production. While the footprint of petroleum extraction creates a permanent and massive overlay on the physical landscape of the Southern Jol, this footprint is nevertheless only the latest of human forces, physical and social, that have shaped the highland landscape and left an indelible testimony to human presence in Arabia. In only a few years, Yemen's oil will reportedly run out, but the remains of oil production belong to its cultural heritage as surely as the Middle Paleolithic cores dropped on its high plateaus many thousands of years ago.

Archaeology and the Law in Yemen

The modernization of the infrastructure of Yemen, which naturally included various development projects—roads and bridges, dams, ports, desalination plants, oil and gas pipelines, and various constructions—presented through the 1990s and early twenty-first century a unique opportunity for the discovery of archaeological sites. Although the military conflicts in Yemen since 2015 have inflicted immense damage on cultural heritage, the prior peacetime framework also had problems. Infrastructure opened access and uncovered sites, but these projects also posed significant dangers for the preservation of archaeological remains.

Through his direction of a cultural impact assessment in the vanguard of a gas pipeline, archaeologist Rémy Crassard developed detailed knowledge of Yemen's protection laws, also provided in the field by ʿAbdalʿazīz Bin ʿAqīl. Both are large contributors to RASA research and this volume. Yemeni Law 21 of 1994, on antiquities, established the necessary provisions for the implementation of archaeological work, but the law has some shortcomings, particularly in its protection and mitigating studies of sites. Any vestige older than 200 years is considered archaeological whether above- or belowground. A national agency, GOAM (General Organization for Antiquities and Museums), is responsible for antiquities and for movable and fixed archaeological remains. GOAM must ensure their protection, preservation, and inventory and disseminate knowledge to the general public. These archaeological remains are state property (Law 21, articles 1–5). Archaeological remains are not property of the discoverer or the owner of the land. Archaeological excavations and surveys are also regulated; they must be performed by professional archaeologists (articles 6–9). Article 12 of this law states very clearly that when development works are carried out, the state must protect buildings and ancient archaeological remains. The law also notes that the developer must take account of these remains in projected construction. The next article obliges every developer to inform GOAM of new construction projects, with GOAM authorized to stop any development that could have a negative impact on archaeological remains. Articles 23–28 stipulate that only GOAM is empowered to conduct excavations in Yemen. Foreign teams and expeditions or other Yemeni research institutes may also be allowed to work, always through agreements with GOAM. Finally, trafficking in antiquities is severely punishable (articles 29–35 for traffic; articles 36–42 for the applicable sanctions), with up to five years' imprisonment and heavy fines. Safeguarding destroyed sites or sites under threat can also be achieved with rescue excavations, still covered by this antiquities law.

Public Awareness of Archaeological Heritage

Yemen has far to go in its protections and public awareness of archaeological heritage, despite the legal framework and committed efforts of a dedicated few. Formal protections for antiquities as national heritage and written into national law are seldom understood and applied in rural areas; nor are the antiquities themselves readily recognized. Where they are recognized as having value, they often offer a source of private income rather than a common heritage asset. Yemen as a whole is inordinately rich in heritage traditions, architectural and archaeological treasures that can provide the infrastructure for a sustainable tourism economy. But several significant factors impede widespread conservation of Yemen's archaeological heritage as a resource held in public trust.

First of all, Yemen is currently and has repeatedly been the site of conflicts that have destroyed national treasures. The tragic gutting of the Zinjibar Museum following takeover of the town is a latter example of the devastation of Yemen's treasures through many recent conflicts. The Saudi-led coalition of Gulf Forces airstrikes have massively damaged archaeological treasures at Baraqish, Ma'rib, Dhamār, and the Old City in Sana'a. In Hadramawt, a civil conflict between north and south in 1994 resulted in damage to the Mukalla Museum in the Sultan's Palace. The history of its subsequent neglect and patchy repairs is emblematic of the general situation in which the national budget has provided less support than needed for a conservation infrastructure of Yemen's monuments and antiquities. In the civil chaos of 1994, Mukalla Museum exhibits were pillaged, the building damaged, and archives ravaged, a sad fate that has recurred in subsequent conflicts. Although there have been periodic partial repairs and furnishings, the overall structure remains in desperate need of major repair and a long-term maintenance plan.

Antiquities in the ground might, one could suppose, be safer, but the international antiquities market provides financial incentive for looting. The education system has failed to instill a sense of common heritage in rural areas. Although archaeology can play a key role in the formulation of a national identity, its potential in this arena is underexploited in Yemen, today highly fragmented into tribal and ethnic-religious factions. The icons of high Arabian civilization appear on Yemeni currency, but the actual sites have been damaged by looters and by builders quarrying ruins for stone.

The political structure of Yemen itself constitutes a pervasive challenge in conserving archaeological antiquities. Tension between centrifugal tribes and centralizing states characterizes many contemporary and historical Middle Eastern societies (Khoury and Kostiner 1991), of which Yemen is but one example. Antiquities fall within the property rights contested in the theater of political integration of tribe and state in Yemen. For example, there are two major types of legal systems, *'urf* and Sharia, commonly recognized throughout Yemen. *'Urf* expresses widely embraced tribal custom, with use-rights and tenure determined through tribal and social affiliation. Therefore tribal groups in areas weakly controlled by the central government tend to view antiquities within their regions as tribal property, available to members of the tribal community, and the use of archaeological remains for tribal members' buildings or sales may provoke no particular sanction. Mandates from the central government, whether for in situ conservation, halting looting, research excavations, or tourism access benefiting exogenous guides provokes resistance, as such governance contravenes *'urf*. This tension is especially apparent where local tribesmen—seldom the most educated and experienced guides—perceive other Yemenis benefiting from the tourism sector in tribal territory.

Sharia plays a contravening role. In general, according to Sharia principles, unimproved resources like grazing land are available to all comers, but where someone makes an investment to improve a resource, such as digging a well, installing a pump, and building a cistern, then he may enjoy proprietorship of that resource. Working for the World Bank, ethnographer ʿAbdalʿazīz Bin ʿAqīl articulated the entangled principles of ʻurf and Sharia as they apply to new farming initiatives benefiting social groups outside of tribal affiliation (Bin ʻAqīl, 2000 personal communication). These entangled legal precepts apply also to the conservation of archaeological sites, seen by tribesmen as their exclusive domain and by the national government as a national trust.

Overlay on this conflict an educational system with an epistemological emphasis very different from Western education and pedagogical approaches consistent with Islamic epistemology. In this framework, Yemen's education system leads to a significant perception of archaeological remains as unimportant. This difference in perspective is evident in the experiences and especially the audience questions Joy McCorriston has had in public presentations in Yemen, leading to an important understanding of the challenges in archaeological heritage conservation.

Figure 1.7. Joy McCorriston giving at talk at the Mukalla Cultural Forum in 2000. *Photograph by ʿAbdalʿazīz Bin ʿAqīl.*

"But what does Qur'ān say about your research?"

Over the years, I have been asked to give a number of public talks about RASA research and findings, and I've found myself addressing a shed filled with overall-clad petroleum workers, a *qāt*-enhanced gathering of cross-legged intellectuals, or a solemn array of university students (figure 1.7). In Yemen, the first question is always the same: "What does Qur'ān say…"

I used to deflect this question, demurring that as a non-Muslim, I am unentitled to comment about Qur'ān. But I realized there was a structural parallel linking this question to the differences in how I develop a presentation and how my closest Yemeni colleague speaks on the same topic. When ʿAbdalʿazīz Bin ʿAqīl gives a public presentation, a good two-thirds is much like the "Archaeology in Southern Yemen" section above, with a scant comment on what our research has actually added. Nor does he get the same question on Qur'ān. My presentations usually begin with a scientific question and its broader intellectual context. We have radically different approaches to explaining the past, and they stem from very different epistemological approaches.

Mine is scientific, and Yemenis expect instead an approach based on precedence or jurisprudence. As revealed knowledge, Qur'ān holds the highest esteem of Muslims and maintains capstone status. Revelation trumps all worldly knowledge, so ultimate authority stems from Qur'ān. As the early Muslim *ʿummah* discovered, there is much worldly knowledge, like geography and natural science, that is not exhaustively discussed in Qur'ān. Jurisprudence, with its weight on precedence and its evaluation of the attribution and proximity of knowledge (to the sayings, or hadīth, of the Prophet), is an alternate and lesser epistemology that nonetheless offers a route to knowledge. In this framework, my colleague's lengthy review of the archaeological endeavors of our predecessors in Hadramawt is an appropriate way to validate our own research findings.

"But what does Qur'ān say about your research?" That question targets my exclusive reliance on lesser epistemology, science, without using even the scaffolding of jurisprudence. When I stopped demurring and really considered what Qur'ān says about archaeology, this persistent question revealed important challenges in archaeological heritage conservation. Qur'ān mentions the ruins of former civilizations as the remains of Jāhiliyah, the Age of

Ignorance (of revealed Qur'ān). Qur'ān points to the remains of wicked people destroyed by Allāh for refusing the warnings of prophets like Hūd and Sālah. Ancient cities were wasted, and they sank beneath the sands. As the apogee of Islamic knowledge, Qur'ān's specific mention of archaeological remains in this context offers little incentive for their conservation or integration into a national consensual identity linked to the past. Against this backdrop, the best intentions of foreign governments and development agencies to conserve pre-Islamic monuments, display antiquities, train guides, and develop tourism infrastructure beat weak waves on a strong shore. It is little wonder that militant Islamists seek to obliterate some of the more prominent ruins and have no reverence for museums or for the network of small-scale monuments and occupational traces that shaped the human landscape of the Southern Jol. But as Yemen emerges from the cataclysmic grip of civil conflict, Yemenis attuned to precedence will recognize that they build a peaceful future not on a tabula rasa wiped clean by war but in a social and physical landscape inimically the outcome of history.

Looking Ahead

In this volume, the multinational, multidisciplinary RASA team has pooled its talents and resources to document our research in the Southern Jol from 1996 to 2008. This book is divided into six parts:

Part 1: Research Objectives; Geological and Environmental Context

In addition to the present chapter, which introduces the book, chapter 2 serves as a background to the geology, climate, flora, and fauna. Chapter 3 reviews research on the paleohydrology, geomorphology, and paleoecology.

Part 2: Archaeological Survey: Methods and Basic Results

Chapter 4 describes early, generalized survey approaches. Chapter 5 reviews topic-specific survey approaches concentrating on lithics, water management, and small-scale monuments. Chapter 6 presents an overview assessment and quantitative analysis of survey results.

Part 3: Pleistocene to Early Holocene: Hunter-Foragers and the Introduction of Domesticates

Chapter 7 presents the surface evidence for Pleistocene occupation long before the introduction of domesticates. The earliest Holocene occupations appear in chapter 8, which documents excavations at Manayzah (occupied 7550–5415 BCE according to Bayesian analysis), the earliest known pastoralist site in Hadramawt. In chapter 9 we describe excavations and collections at other rockshelter and open-air sites roughly contemporary with Manayzah.

Part 4: Middle Holocene: A Pastoralist Landscape

Chapter 10 presents results of excavations of an impressive ring of cattle skulls and associated platform monument at Kheshiya. Chapter 11 presents the zooarchaeological analyses of those finds. Chapter 12 describes excavations of other Neolithic platform monuments (4620–4175 BCE), an important Arabian phenomenon first recognized through the work of the RASA team in Wādī Sanā. Chapter 13 documents water management structures that first appeared before 3800 BC and their hydrological contexts.

Part 5: Middle to Late Holocene: The Social Life of Pastoralists

Chapter 14 reviews survey and excavations of small-scale monuments that developed from the earlier RASA work in Hadramawt. Ritual practices in the early Arabian Bronze Age engaged an anthropomorphic iconography, described in the context of Wādī 'Idim's Rawk shrine (chapter 15). While not the specific target of our investigations, Wādī 'Idim's house sites at Munayder (last occupied in the middle second millennium BCE) (chapter 16) offer an intriguing parallel to human settlement around springs in Wādī Sanā and a hint at habitations in a human landscape largely defined by monuments. Chapter 17 reports the graffiti and rockart recorded by the RASA Project.

Part 6: Synthesis and Conclusions

In chapter 18 we present a regional chronological model and Bayesian analysis of the radiocarbon and optically stimulated luminescence ages as the basis for new chronological insight into Hadramawt's prehistory. Chapter 19 uses this chronology to summarize our findings as a human landscape history in Wādī Sanā and the Southern Jol.

References

Abell, Paul I, and Philipp Hoelzmann

2000 Holocene Palaeoclimates in Northwestern Sudan: Stable Isotope Studies on Molluscs. *Global and Planetary Change* 26(1):1–12.

Al-Abri, Abdulrahim, Eliška Podgorná, Jeffrey I. Rose, Luísa Pereira, Connie J. Mulligan, Nuno M. Silva, Riad Bayoumi, Pedro Soares, and Viktor Černý

2012 Pleistocene-Holocene Boundary in Southern Arabia from the Perspective of Human mtDNA Variation. *American Journal of Physical Anthropology* 149(2):291–98.

Al-Hamdani, Hasan I. A.
[945] 1989 *Ṣifat Jazīrat al-'Arab / ta'līf al-Ḥasan ibn Aḥmad ibn Ya'qūb al-Hamdānī.* Ministry of Culture and Sciences, Dār al-Shu'ūn al-Thaqāfīyah al-'Āmmah "Āfāq 'Arabīyah," Baghdad.

Alvard, Michael S., and Lawrence Kuznar
2001 Deferred Harvests: The Transition from Hunting to Animal Husbandry. *American Anthropologist* 103(2):295–311.

Amblard, Sylvie
1996 Agricultural Evidence and the Interpretation of the Dhars Tichitt and Oualata, South-Eastern Mauritania. In *Aspects of African Archaeology: Papers from the 10th Congress of the Pan-African Association for Prehistory and Related Studies*, edited by Gilbert Pwiti and Robert Soper, pp. 421–27. University of Zimbabwe Publications, Harare.

Amirkhanov, Hizri A.
1997 *The Neolithic and Post-Neolithic of the Hadramaut and Mahra.* Scientific World, Moscow.

Ammerman, Albert J., and Luigi Luca Cavalli-Sforza
1984 *The Neolithic Transition and the Genetics of Populations in Europe.* Princeton University Press, Princeton, NJ.

Arav, Reuma
2015 High-Resolution Documentation, 3-D modeling and Analysis of "Desert Kites." *Journal of Archaeological Science* 57:302–14.

Asouti, Eleni, and Dorian Q. Fuller
2013 A Contextual Approach to the Emergence of Agriculture in Southwest Asia: Reconstructing Early Neolithic Plant-Food Production. *Current Anthropology* 54(3):299–345.

Barker, Graeme
1985 *Prehistoric Farming in Europe*. Cambridge University Press, Cambridge.

Bar-Yosef, Ofer, and Anna Belfer-Cohen
1989 The Origins of Sedentism and Farming Communities in the Levant. *Journal of World Prehistory* 3(4):447–98.
1992 From Foraging to Farming in the Mediterranean Levant. In *Transitions to Agriculture in Prehistory*, edited by Anne B. Gebauer and Douglas T. Price, pp. 21–48. Monographs in World Prehistory 4. Prehistory Press, Madison, WI.

Bassinot, F. C., C. Marzin, P. Braconnot, O. Marti, E. Mathien-Blard, F. Lombard, and L. Bopp
2011 Holocene Evolution of Summer Winds and Marine Productivity in the Tropical Indian Ocean in Response to Insolation Forcing: Data-Model Comparison. *Climate of the Past* 7(3):815–29.

Belfer-Cohen, Anna, and Ofer Bar-Yosef
2000 Early Sedentism in the Near East: A Bumpy Ride to Village Life. In *Life in Neolithic Farming Communities*, edited by Ian Kuijt, pp. 19–38. Kluwer Academic, London.

Bellwood, Peter S.
2005 *First Farmers: The Origins of Agricultural Societies.* Blackwell, Malden, MA.

Bent, Theodore, and Mabel Bent
1900 *Southern Arabia.* Smith, Elder and Company, London.

Berger, Jean-François, Jean-Paul Bravard, Louise Purdue, Anne Benoist, Michel Mouton, and Frank Braemer
2012 Rivers of the Hadramawt Watershed (Yemen) during the Holocene: Clues of Late Functioning. *Quaternary International* 266(0):142–61.

Berger, J. F., V. Charpentier, R. Crassard, C. Martin, G. Davtian, and J. A. López-Sáez
2013 The Dynamics of Mangrove Ecosystems, Changes in Sea Level and the Strategies of Neolithic Settlements along the Coast of Oman (6000–3000 cal BC). *Journal of Archaeological Science* 40(7):3087–3104.

Berkea, Melissa A., Thomas C. Johnson, Josef P. Werneb, Stefan Schoutenc, and Jaap S. Sinninghe Damstéc
2012 A Mid-Holocene Thermal Maximum at the End of the African Humid Period. *Earth and Planetary Science Letters* 351:95–104.

Betts, A. V. G., Dawn N. Cropper, Louise A. Martin, and Carole McCartney
2013 *The Later Prehistory of the Badia.* Oxbow Books, Oxford.

Boardman, Sheila
1999 The Agricultural Foundation of the Aksumite Empire, Ethiopia. In *The Exploitation of Plant Resources in Ancient Africa*, edited by Marijke van der Veen, pp. 137–47. Kluwer Academic Plenum, New York.

Bocquet-Appel, Jean-Pierre, Stephan Naji, Marc Vander Linden, and Janusz Kozlowski
2012 Understanding the Rates of Expansion of the Farming System in Europe. *Journal of Archaeological Science* 39(2):531–46.

Boivin, Nicole, and Dorian Q. Fuller
2009 Shell Middens, Ships, and Seeds: Exploring Coastal Subsistence, Maritime Trade, and the Dispersal of Domesticates in and around the Ancient Arabian Peninsula. *Journal of World Prehistory* 22(2):113–80.

Bower, John
1991 The Pastoral Neolithic of East Africa. *Journal of World Prehistory* 5:49–82.

Braconnot, P., B. Otto-Bliesner, S. Harrison, S. Joussaume, J. Y. Peterchmitt, A. Abe-Ouchi, M. Crucifix, E. Driesschaert, T. Fichefet, C. D. Hewitt, M. Kageyama, A. Kitoh, A. Laine, M. F. Loutre, O. Marti, U. Merkel, G. Ramstein, P. Valdes, S. L. Weber, Y. Yu, and Y. Zhao

2007 Results of PMIP2 Coupled Simulations of the Mid-Holocene and Last Glacial Maximum. Part 2, Feedbacks with Emphasis on the Location of the ITCZ and Mid- and High-Latitudes Heat Budget. *Climate of the Past* 3(2):279–96.

Bradley, Daniel G., David E. MacHugh, Patrick Cunningham, and Ronan T. Loftus

1996 Mitochondrial Diversity and the Origins of African and European Cattle. *Proceedings of the National Academy* 93(10):5131–36.

Braemer, Frank, Serge Cleuziou, and Tara Steimer

2003 Dolmen-Like Structures: Some Unusual Funerary Monuments in Yemen. *Proceedings of the Seminar for Arabian Studies* 33:169–82.

Breton, Jean-François

1991 *Fouilles de Shabwa II: rapports préliminaires*. Syria 68. Paul Geuthner, Paris.

Brinkmann, Robert

1996 Pedological Characteristics of Anthrosols in the al-Jadidah Basin of Wadi al-Jubah, and Native Sediments in Wadi al-Ajwirah, Yemen Arab Republic. In *The Wadi Jubah Archaeological Project.* Vol. 5, *Environmental Research in Support of Archaeological Investigations in the Yemen Arab Republic, 1982–1987*, edited by Maurice J. Grolier, Robert Brinkmann, and James A. Blakely, pp. 45–211. American Foundation for the Study of Man, Washington, DC.

Burton, Richard F.

1964 [1878] *Personal Narrative of a Pilgrimage to al-Madinah and Meccah, by Sir Richard F. Burton*. Dover, New York.

Byrd, Brian F.

2005 Reassessing the Emergence of Village Life in the Near East. *Journal of Archaeological Research* 13(3):231–90.

Caton-Thompson, Gertrude

1944 *The Tombs and Moon Temple of Hureidha (Hadhramaut)*. Oxford University Press, Oxford.

Caton-Thompson, Gertrude, and Elinor W. Gardner

1939 Climate, Irrigation, and Early Man in the Hadhramaut. *Geographical Journal* 93:18–38.

Černý, Viktor, Connie J. Mulligan, Verónica Fernandes, Nuno M. Silva, Farida Alshamali, Amy Non, Nourdin Harich, Lotfi Cherni, Amel El Gaaied Ben Ammar, Ali Al-Meeri, and Luísa Pereira

2011 Internal Diversification of Mitochondrial Haplogroup R0a Reveals Post-Last Glacial Maximum Demographic Expansions in South Arabia. *Molecular Biology and Evolution* 28(1):71–78.

Charpentier, Vincent

2008 Hunter-Gatherers of the "Empty Quarter of the Early Holocene" to the Last Neolithic Societies: Chronology of the Late Prehistory of South-eastern Arabia (8000–3100 BC). *Proceedings of the Seminar for Arabian Studies* 38:93–115.

Charpentier, Vincent, and Rémy Crassard

2013 Back to Fasad . . . and the PPNB Controversy. Questioning a Levantine Origin for Arabian Early Holocene Projectile Points Technology. *Arabian Archaeology and Epigraphy* 24(1):28–36.

Clark, Ian D., and Jean-Charles Fontes

1990 Paleoclimatic Reconstruction in Northern Oman Based on Carbonates from Hyperalkaline Groundwaters. *Quaternary Research* 33(3):320–36.

Cleuziou, Serge, and Lorenzo Costantini

1980 Premiers éléments sur l'agriculture protohistorique de l'Arabie orientale. *Paléorient* 6:245–51.

Cleuziou, Serge, and Maurizio Tosi

1997 Hommes, climats et environnements de la Péninsule arabique à l'Holocène. *Paléorient* 23(2):121–35.

Close, Angela, and Fred Wendorf

1992 The Beginnings of Food Production in the Eastern Sahara. In *Transitions to Agriculture in Prehistory*, edited by Anne Birgitte Gebauer and T. Douglas Price, pp. 63–72. Prehistory Press, Madison, WI.

Clutton-Brock, Juliet

1999 *A Natural History of Domesticated Mammals*. Cambridge University Press, Cambridge.

Colledge, Susan, James Conolly, Keith Dobney, Katie Manning, and Stephen Shennan (editors)

2013 *The Origin and Spread of Domestic Animals in Southwest Asia and Europe*. Left Coast Press, Walnut Creek, CA.

Colledge, Sue, James Conolly, and Stephen Shennan

2004 Archaeobotanical Evidence for the Spread of Farming in the Eastern Mediterranean. *Current Anthropology* 45:S35.

Costantini, Lorenzo

1981 Palaeoethnobotany at Pirak, A Contribution to the 2nd Millennium BC Agriculture of the Sibi-Kaachi Plain, Pakistan. In *South Asian Archaeology 1979*, edited by Herbert Härtel, pp. 272–79. Reimer Verlag, Berlin.

1984 Plant Impressions in Bronze Age Pottery from Yemen Arab Republic. *East and West* 34:107–15.

1990 Ecology and Farming of the Protohistoric Communities in the Central Yemeni Highlands. In *The Bronze Age Culture of Hawlan at-Tiyal and al Hada (Republic of Yemen)*, edited by Alessandro de Maigret, pp. 187–204. Istituto del' Medio ed Estremo Oriente, Rome.

Crassard, Rémy

2008 La préhistoire du Yémen: diffusions et diversités locales, à travers l'étude d'industries lithiques du Hadramawt. British Archaeological Reports 1842. Archaeopress, Oxford.

Crassard, Rémy, Olivier Barge, Charles-Edmond Bichot, Jacques Élie Brochier, Jwana Chahoud, Marie-Laure Chambrade, Christine Chataigner, Kamel Madi, and Emmanuelle Régagnon

2015 Addressing the Desert Kites Phenomenon and Its Global Range through a Multi-proxy Approach. *Journal of Archaeological Method and Theory* 22(4):1093–1121.

Crassard, Rémy, and Philipp Drechsler

2013 Towards New Paradigms: Multiple Pathways for the Arabian Neolithic. *Arabian Archaeology and Epigraphy* 24(1):3–8.

Crassard, Rémy, Michael D. Petraglia, Adrian G. Parker, Ash Parton, Richard G. Roberts, Zenobia Jacobs, Abdullah Alsharekh, Abdulaziz Al-Omari, Paul Breeze, Nick A. Drake, Huw S. Groucutt, Richard Jennings, Emmanuelle Régagnon, and Ceri Shipton

2013 Beyond the Levant: First Evidence of a Pre-Pottery Neolithic Incursion into the Nefud Desert, Saudi Arabia. *PLoS ONE* 8(7), https://doi.org/10.1371/journal.pone.0068061, accessed January 12, 2015.

Cremaschi, Mauro, and Fabio Negrino

2005 Evidence for an Abrupt Climatic Change at 8700 ^{14}C yr BP in Rockshelters and Caves of Gebel Qara (Dhofar-Oman): Palaeoenvironmental Implications. *Geoarchaeology* 20:559–79.

Czekaj-Zastawny, Agnieszka, Jacek Kabacinski, Thomas Terberger, and Jolanta Ilkiewicz

2013 Relations of Mesolithic Hunter-Gatherers of Pomerania (Poland) with Neolithic Cultures of Central Europe. *Journal of Field Archaeology* 38(3):195–209.

D'Andrea, A. C., J. Casey, and M. Klee

2001 Archaeobotanical Evidence for Pearl Millet (*Pennisetum glaucum*) in Sub-Saharan West Africa. *Antiquity* 75:341–48.

deMenocal, Peter B.

2015 Palaeoclimate: End of the African Humid Period. *Nature Geoscience* 8(86–87).

Doe, Brian

1971 *South Arabian Antiquities*. Thames and Hudson, London.

1983 *Monuments of South Arabia*. Oleander Press, London.

Doughty, Charles Montagu, and Edward Garnett

1979 [1888] *Travels in Arabia Deserta*. 3rd ed. Dover, New York.

Dreschler, Philipp

2007 The Neolithic Dispersal into Arabia. *Proceedings of the Seminar for Arabian Studies* 37:93–109.

2010 Life at the End of the Holocene Moist Phase in South-East Arabia—the Late Neolithic Site of Jebel Thanais 1 (JTH1). *Arabian Archaeology and Epigraphy* 21(2):81–95.

Fedele, Francesco G.

1990 Man, Land, and Climate: Emerging Interactions from the Holocene of the Yemen Highlands. In *Man's Role in the Shaping of the Eastern Mediterranean Landscape*, edited by Sytze Bottema, Gerte Enjes-Nieborg, and Willem Van Zeist, pp. 31–42. A. A. Balkema, Rotterdam.

Fleitmann, Dominik, Stephen J. Burns, Augusto Mangini, Manfred Mudelsee, Jan Kramers, Igor Villa, Ulrich Neff, Abdulkarim Al-Subbary, Annett Buettner, Dorothea Hippler, and Albert Matter

2007 Holocene ITCZ and Indian Monsoon Dynamics Recorded in Stalagmites from Oman and Yemen (Socotra). *Quaternary Science Reviews* 26(1–2):170–88.

Fleitmann, Dominik, Stephen J. Burns, Manfred Mudelsee, Ulrich Neff, Jan Kramers, Augusto Mangini, and Albert Matter

2003 Holocene Forcing of the Indian Monsoon Recorded in a Stalagmite from Southern Oman. *Science* 300(5626):1737–39.

Fleitmann, Dominik, Stephen J. Burns, Marek Pekala, Augusto Mangini, Abdulkarim Al-Subbary, Mohammad Al-Aowah, Jan Kramers, and Albert Matter

2011 Holocene and Pleistocene Pluvial Periods in Yemen, Southern Arabia. *Quaternary Science Reviews* 30(7):783–87.

Gasse, Françoise

2002 Diatom-Inferred Salinity and Carbonate Oxygen Isotopes in Holocene Waterbodies of the Western Sahara and Sahel (Africa). *Quaternary Science Reviews* 21(7):737–67.

Gautier, Achille

1984 The Fauna of the Neolithic Site of Kadero (Central Sudan). In *The Origins and Early Development of Food Producing Cultures in N.E. Africa*, edited by Lech Krzyzaniak and Michel Kobusiewicz, pp. 317–19. Polish Academy of Sciences, Poznàn.

Gifford-Gonzalez, Diane P.
2005 Pastoralism and Its Consequences. In *African Archaeology: A Critical Introduction*, edited by Ann B. Stahl, pp. 187–224. Blackwell, Oxford.

Gifford-Gonzalez, Diane, and Olivier Hanotte
2013 Domesticating Animals in Africa. In *The Oxford Handbook of African Archaeology*, pp. 491–506. Oxford University Press, Oxford.

Gregg, Susan A.
1988 *Foragers and Farmers: Population Interaction and Agricultural Expansion in Prehistoric Europe*. University of Chicago Press, Chicago.

Gupta, Anil K., David M. Anderson, and Jonathan T. Overpeck
2003 Abrupt Changes in the Asian Southwest Monsoon during the Holocene and Their Links to the North Atlantic Ocean. *Nature* 221(6921):354–57.

Halévy, Joseph
1873 Voyage au Nedjrân. *Bulletin de la Société de Géographie de Paris*, 6th series:600–04.
1877 De Sana au Nedjrân. *Bulletin de la Société de Géographie de Paris*, 6th series:466–79.

Hanotte, Olivier
2002 African Pastoralism: Genetic Imprints of Origins and Migrations. *Science* 296(5566):336–39.

Hansen, Donald P., Edward L. Ochsenschlager, and Selma Al-Radi
2004 Excavations at Jujah, Shibam, Wadi Hadhramawt. *Arabian Archaeology and Epigraphy* 15(1):43–67.

Hardy-Guilbert, Claire
2005 The Harbour of al-Shihr, Hadramawt, Yemen: Sources and Archaeological Data on Trade. *Proceedings of the Seminar for Arabian Studies* 35:71–85.

Harris, David R.
1989 An Evolutionary Continuum of People-Plant Interaction. In *Foraging and Farming: The Evolution of Plant Exploitation*, edited by David R. Harris and Gordon C. Hillman, pp. 11–26. Unwin Hyman, London.
1996 *The Origins and Spread of Agriculture and Pastoralism in Eurasia*. Smithsonian Institution Press, Washington, DC.

Harrower, Michael J., E. A. Oches, and J. McCorriston
2012 Hydro-Geospatial Analysis of Ancient Pastoral/Agro-Pastoral Landscapes along Wādī Sanā (Yemen). *Journal of Arid Environments* 86:131–38.

Hassan, F. A., B. Barich, M. Mahmoud, and M. A. Hemdan
2001 Holocene Playa Deposits of Farafra Oasis, Egypt, and Their Palaeoclimatic and Geoarchaeological Significance. *Geoarchaeology* 16(1):29–46.

Hastorf, Christine A.
1998 The Cultural Life of Early Domestic Plant Use. *Antiquity* 72(278):773–82.

Hayden, Brian
1992 Contrasting Expectation in Theories of Domestication. In *Transitions to Agriculture in Prehistory*, edited by Anne B. Gebauer and T. D. Price, pp. 11–18. Prehistory Press, Madison, WI.

Helfritz, Hans
1935 The First Crossing of Southwestern Arabia. *Geographical Review* 25(3):395–407.

Helms, Svend, and Alison Betts
1987 Desert "Kites" of the Badiyat esh-Sham and North Arabia. *Paléorient* 13(1):41–67.

Henry, Donald O.
1989 From Foraging to Agriculture: The Levant at the End of the Ice Age. University of Pennsylvania Press, Philadelphia.
1994 Prehistoric Cultural Ecology in Southern Jordan. *Science* 265(5170):336–41.

Henry, Donald O., and Joseph E. Beaver (editors)
2014 *The Sands of Time: The Desert Neolithic Settlement at Ayn Abū Nukhayla*. Ex Oriente, Berlin.

Hilbert, Yamandu H.
2013 Khamseen rockshelter and the Late Palaeolithic-Neolithic Transition in Dhofar. *Arabian Archaeology and Epigraphy* 24(1):51–58.

Holdaway, Simon, Willeke Wendrich, and Rebecca Phillipps
2010 Identifying Low-Level Food Producers: Detecting Mobility from Lithics. *Antiquity* 84(323):185–94.

Hole, Frank
1984 A Reassessment of the Neolithic Revolution. *Paléorient* 10(2):49–60.

Horowitz, Liora K., Eitan Tchernov, Pierre Ducos, Cornelius Becker, Angela von den Dreisch, Louise Martin, and Andrew Garrard
2000 Animal Domestication in the Southern Levant. *Paléorient* 25(2):63–80.

Ibn al-Mujawar, Yusuf ibn Yuqub
[1204/5–1291/2] 1986 *Ṣifāt Bilād al-Yamin wa Makka wa baʿad al-Hijāz al-muṣamman Tārīkh al-Muṣtabṣir*. Oskar Lufgren, Beirut.

Ingold, Tim
1980 *Hunters, Pastoralists, and Ranchers: Reindeer Economies and Their Transformations*. Cambridge University Press, Cambridge.

Inizan, Marie-Louise, Anne-Marie Lézine, Bruno Marcolongo, Jean-François Saliège, Christian Robert, and Frédéric Werth
1997 Paléolacs et peuplements holocènes du Yémen: le Ramlat As-Sabat'ayn. *Paléorient* 23(2):137–49.

Inizan, Marie-Louise, and L. Ortlieb
1987 Préhistoire dans la région de Shabwa au Yémen du sud (R.D.P Yémen). *Paléorient* 13:5–22.

Keeley, Lawrence
1995 Protoagricultural Practices among Hunter-Gatherers: A Cross-Cultural Survey. In *Last Hunters—First Farmers*, edited by Douglas T. Price and Anne B. Gebauer, pp. 243–72. School of American Research Press, Santa Fe, NM.

Khalidi, Lamya, Marie-Louise Inizan, Bernard Gratuze, and Rémy Crassard
2013 Considering the Arabian Neolithic through a Reconstitution of Interregional Obsidian Distribution Patterns in the Region. *Arabian Archaeology and Epigraphy* 24(1):59–67.

Khalidi, Lamya, Clive Oppenheimer, Bernard Gratuze, Sophie Boucetta, Ali Sanabani, and Ahmed al-Mosabi
2010 Obsidian Sources in Highland Yemen and Their Relevance to Archaeological Research in the Red Sea Region. *Journal of Archaeological Science* 37(9):2332–45.

Khoury, Philip S., and Joseph Kostiner
1991 *Tribes and State Formation in the Middle East*. University of California Press, Berkeley.

Kusimba, Sibel Barut
2003 *African Foragers: Environment, Technology, Interactions*. AltaMira Press, Walnut Creek, CA.

Kutzbach, J. E., G. Bonan, J. Foley, and S. P. Harrison
1996 Vegetation and Soil Feedbacks on the Response of the African Monsoon to Orbital Forcing in the Early to Middle Holocene. *Nature* 384:623–26.

Kutzbach, J. E., Xiaodong Liu, Zhengyu Liu, and Guangshan Chen
2008 Simulation of the Evolutionary Response of Global Summer Monsoons to Orbital Forcing Over the Past 280,000 Years. *Climate Dynamics* 30:567–79.

Lancaster, William, and Fidelity Lancaster
1999 Identities and Economies: Mountain and Coastal Ras al Khaimah. *Proceedings of the Seminar for Arabian Studies* 29:89–94.

Larson, Greger, Keith Dobney, Umberto Albarella, Meiying Fang, Elizabeth Matisoo-Smith, Judith Robins, Stewart Lowden, Heather Finlayson, Tina Brand, Eske Willerslev, Peter Rowley-Conwy, Leif Andersson, and Alan Cooper
2005 Worldwide Phylogeography of Wild Boar Reveals Multiple Centers of Pig Domestication. *Science* 307(5715):1618–21.

Legge, Anthony J., and Peter Rowley-Conwy
1987 Gazelle Killing in Stone Age Syria. *Scientific American* 257(a):88–95.

Lézine, Anne-Marie
2009 Timing of Vegetation Changes at the End of the Holocene Humid Period in Desert Areas at the Northern Edge of the Atlantic and Indian Monsoon Systems. *Comptes rendus—Geoscience* 341:750–59.

Lézine, Anne-Marie, and Serge Cleuziou
2012 Climate Change and Human Occupation in the Southern Arabian Lowlands during the Late Quaternary. In *Aux marges de l'archéologie: hommage à Serge Cleuziou*, edited by Serge Cleuziou, Jessica Giraud, and Guillaume Gernez, pp. 329–42. Travaux de la Maison Réné Ginouves 16. Editions De Boccard, Paris.

Lézine, Anne-Marie, Christian Robert, Serge Cleuziou, Marie-Louise Inizan, Frank Braemer, Jean-François Saliège, Florence Sylvestre, Jean-Jacques Tiercelin, Rémy Crassard, Sophie Méry, Vincent Charpentier, and Tara Steimer-Herbet
2010 Climate Change and Human Occupation in the Southern Arabian Lowlands during the Last Deglaciation and the Holocene. *Global and Planetary Change* 72(4):412–28.

Lézine, Anne-Marie, Jean-Jacques Tiercelin, Christian Robert, Jean-François Saliège, Serge Cleuziou, Marie-Louise Inizan, and Frank Braemer
2007 Centennial to Millennial-Scale Variability of the Indian Monsoon during the Early Holocene from a Sediment, Pollen, and Isotope Record from the Desert of Yemen. *Palaeogeography, Palaeoclimatology, Palaeoecology* 243(3–4):235–49.

Madella, Marco, Juan José García-Granero, Welmoed A. Out, Philippa Ryan, and Donatella Usai
2014 Microbotanical Evidence of Domestic Cereals in Africa 7000 Years Ago. *PLoS ONE* 9(10):e110177, https://doi.org/10.1371/journal.pone.0110177, accessed January 22, 2015.

Magee, Peter
2014 *The Archaeology of Prehistoric Arabia: Adaptation and Social Formation from the Neolithic to the Iron Age*. Cambridge University Press, Cambridge..

Maher, Lisa S., Danielle Macdonald, Matthew D. Jones, Louise Martin, and Jay T. Stock
2012a Twenty Thousand Year-Old Huts at a Hunter-Gatherer Settlement in Eastern Jordan. *PLoS ONE* 7(2):e31447, https://doi.org/10.1371/journal.pone.0031447, accessed January 7, 2015.

Maher, Lisa S., Tobias Richter, and Jay T. Stock
2012b The Pre-Natufian Epipaleolithic: Long-Term Behavioral Trends in the Levant. *Evolutionary Anthropology* 221(2):69–81.

Marshall, Fiona B.
1990 Origins of Specialized Pastoral Production in East Africa. *American Anthropologist* 92:873–94.

Marshall, Fiona, and Elisabeth Hildebrand
2002 Cattle before Crops: The Beginnings of Food Production in Africa. *Journal of World Prehistory* 16(2):99–143.

Marshall, Fiona, and Lior Weissbrod
2011 Domestication Processes and Morphological Change. *Current Anthropology* S52:S397–S413.

Martin, Louise
1999 Mammal Remains from the Eastern Jordanian Neolithic, and the Nature of Caprine Herding in the Steppe. *Paléorient* 25(2):87–104.

Martin, Louise A., and Yvonne Edwards
2013 Diverse Strategies: Evaluating the Appearance and Spread of Domestic Caprines in the Southern Levant. In *The Origin and Spread of Domestic Animals in Southwest Asia and Europe*, edited by Susan Colledge, James Conolly, Keith Dobney, Katie Manning, and Stephen Shennan, pp. 49–82. Left Coast Press, Walnut Creek, CA.

Martin, Louise, Yvonne Edwards, and Andrew N. Garrard
2010 Hunting Practices at an Eastern Jordanian Epipalaeolithic Aggregation Site: The Case of Kharaneh IV. *Levant* 42(2):107–35.

Martin, Louise, Joy McCorriston, and Rémy Crassard
2009 Early Arabian Pastoralism at Manayzah in Wādī Ṣanā, Ḥaḍramawt. *Proceedings of the Seminar for Arabian Studies* 39:271–82.

Marzin, Charline and Pascale Braconnot
2009 Variations of Indian and African Monsoons Induced by Insolation Changes at 6 and 9.5 kyr BP. *Climate Dynamics* 33(2–3):215–31.

Marzin, Charline, Pascale Braconnot, and Masa Kageyama
2013 Relative Impacts of Insolation Changes, Meltwater Fluxes, and Ice Sheets on African and Asian Monsoons during the Holocene. *Climate Dynamics* 41(9–10):2267–86.

Matthews, Roger
2002 Zebu: Harbingers of Doom in Bronze Age Western Asia? *Antiquity* 76(292).

McClure, Harold A.
1976 Radiocarbon Chronology of Late Quaternary Lakes in the Arabian Desert. *Nature* 263:755–56.
1984 Late Quaternary Palaeoenvironments of the Rub' Al Khali. PhD dissertation, University College London.
1988 Late Quaternary Paleogeography and Landscape Evolution of the Rub 'Al Khali. In *Araby the Blest*, edited by Daniel T. Potts, pp. 9–13. University of Copenhagen Carston Niebuhr Institute Publications 7. Tusculanum Press, Copenhagen.

McCorriston, Joy
2006 Breaking the Rainfall Barrier and the Tropical Spread of Near Eastern Agriculture into Southern Arabia. In *Behavioral Ecology and the Transition to Agriculture*, edited by Douglas J. Kennett and Bruce Winterhalder., pp. 217–36. University of California Press, Berkeley.
2013 The Neolithic in Arabia: A View from the South. *Arabian Archaeology and Epigraphy* 24(1):68–72.

McCorriston, Joy, Michael J. Harrower, Louise A. Martin, and Eric A. Oches
2012 Cattle Cults of the Arabian Neolithic and Early Territorial Societies. *American Anthropologist* 114:45–63.

McCorriston, Joy, Michael Harrower, Eric Oches, Abdalaziz Bin 'Aqil, Catherine Heyne, AbdalBaset Noman, Rémy Crassard, Khalid BaDhofary, and Joshua Anderson
2005 Foraging Economies and Population in the Middle Holocene Highlands of Southern Yemen. *Proceedings of the Seminar for Arabian Studies* 35:143–54.

McCorriston, Joy, and Frank Hole
1991 The Ecology of Seasonal Stress and the Origins of Agriculture in the Near East. *American Anthropologist* 93(1):46–69.

Minnis, Paul
1992 Earliest Plant Cultivation in the Desert Borderlands of North America. In *The Origins of Agriculture: An International Perspective*, edited by Nancy L. Benco, C. Wesley Cowan, and Patty Jo Watson, pp. 121–41. Smithsonian Institution Press, Washington, DC.

Mouton, Michel, Anne Benoist, and Jérémie Schiettecatte
2011 Makaynûn and Its Territory: The Formation of an Urban Centre during the South Arabian Period in the Hadramawt. *Arabian Archaeology and Epigraphy* 22(2):155–65.

Naqvi, Wajih A., and Richard G. Fairbanks
1996 A 27,000-Year Record of Red Sea Outflow: Implication for Timing of Post-Glacial Monsoon Intensification. *Geophysical Research Letters* 23(12):1501–04.

Neff, U., S. J. Burns, A. Mangini, M. Mudelsee, D. Fleitmann, and A. Matter
2001 Strong Coherence between Solar Variability and the Monsoon in Oman between 9 and 6 kyr Ago. *Nature* 411:290–93.

Newton, Lynne S.
2009 *A Landscape of Pilgrimage and Trade in Wadi Masila, Yemen: al-Qisha and Qabr Hud in the Islamic Period.* British Archaeological Reports, International Series 1899. Archaeopress, Oxford.

Noskowÿj, Paul B.
1866 Maqrizii "De Valle Hadhramaut" Arabice Editus et Illustratus. PhD dissertation, Frederick Willheim University, Berlin.

Parker, Adrian G., and Andrew S. Goudie
2008 Geomorphological and Palaeoenvironmental Investigations in the Southeastern Arabian Gulf Region and the Implication for the Archaeology of the Region. *Geomorphology* 101(3):458–70.

Parker, Adrian G., and Gareth W. Preston

2008 Early to Middle Holocene Climate Change and Vegetation Dynamics in the Northern Oman Peninsula: An Ecogeographical Synthesis of the Jebel al-Buhais Environment. In *The Natural Environment of Jebel al-Buhais: Past and Present*, edited by Hans-Peter Uerpmann, Margarethe Uerpmann, and Sabab Abboud Jasim, pp. 71–82. Kerns Verlag, Tübingen.

Peters, Joris

1986 A Revision of the Faunal Remains from Two Central Sudanese Sites: Khartoum Hospital and Esh Shaheinab. *Archaeozoologia Mélanges* 11:33.

Peterson, Jane

2002 *Sexual Revolutions: Gender and Labor at the Dawn of Agriculture*. AltaMira Press, Walnut Creek, CA.

Philby, Harry St. John B.

1939 *Sheba's Daughters, Being a Record of Travel in Southern Arabia*. Methuen, London.

Phillipps, Rebecca, Simon Holdaway, Willeke Wendrich, and René Cappers

2012 Mid-Holocene Occupation of Egypt and Global Climatic Change. *Quaternary International* 251:64–76.

Pirenne, Jacqueline

1990 *Les Témoins écrits de la région de Shabwa et l'histoire*. Paul Geuthner, Paris.

Potts, Daniel T.

1994 Contributions to the Agrarian History of Eastern Arabia. Vol. 2, The Cultivars. *Arabian Archaeology and Epigraphy* 5:236–75.

Prell, Warren L., and John E. Kutzbach

1987 Monsoon Variability over the Past 150,000 Years. *Journal of Geophysical Research: Atmospheres* 92(D7):8411–25.

Rawat, Suman, Anil K. Gupta, S. J.Sangode, Priyeshu Srivastava, and H. C. Nainwal

2015 Late Pleistocene-Holocene Vegetation and Indian Summer Monsoon Record from the Lahaul, Northwest Himalaya, India. *Quaternary Science Reviews* 114.

Rindos, David

1984 *The Origins of Agriculture: An Evolutionary Perspective*. Academic Press, Orlando.

Roberts, Neil, and Henry E. Wright Jr.

1993 Vegetational, Lake-Level, and Climatic History of the Near East and Southwest Asia. In *Global Climates since the Last Glacial Maximum*, edited by H. E. Wright Jr., John E. Kutzbach, Thompson Webb III, Warren A. Ruddiman, F. A. Street-Perrott, and P. J. Bartlein, pp. 194–220. University of Minnesota Press, Minneapolis.

Rose, Jeffrey I.

2010 New Light on Human Prehistory in the Arabo-Persian Gulf Oasis. *Current Anthropology* 51(6):849–83.

Rose, Jeffrey I., Viktor Černý, and Riad Bayoumi

2013 Tabula Rasa or Refugia? Using Genetic Data to Assess the Peopling of Arabia. *Arabian Archaeology and Epigraphy* 24(1):95–101.

Rosenberg, Michael

1998 Cheating at Musical Chairs: Territoriality and Sedentism in an Evolutionary Context. *Current Anthropology* 39(5):653–81.

Rougeulle, Axelle

2001 Notes on Pre- and Early Islamic Harbours of Hadramawt (Yemen). *Proceedings of the Seminar for Arabian Studies* 31:203–14.

Rowley-Conwy, Peter, William J. Deakin, and Charles H. Shaw

1999 Ancient DNA from Sorghum: The Evidence from Qasr Ibrim, Egyptian Nubia. In *The Exploitation of Plant Resources in Ancient Africa*, edited by Marijke van der Veen, pp. 55–61. Kluwer Academic Plenum, New York.

Russell, Kenneth W.

1988 *After Eden: The Behavioral Ecology of Early Food Production in the Near East and North Africa*. British Archaeological Reports, International Series, 391. British Archaeological Reports, Oxford.

Schiettecatte, Jérémie

2010 Urbanization and Settlement Pattern in Ḥaḍramawt (1st mill. BC). Bulletin of Archaeology, University of Kanazawa (28):11–28.

Schild, Romuald, and Fred Wendorf

2001 Geoarchaeology of the Holocene Climatic Optimum at Nabta Playa, Southwestern Desert, Egypt. *Geoarchaeology* 16(1):7–28.

Sedov, Alexander V.

1996a Al-Guraf in the Wādī 'Idim: Notes on an Archaeological Map of the Ḥaḍramawt, 2. *Arabian Archaeology and Epigraphy* 7:53–62.

1996b Monuments of the Wādī al-`Ayn: Notes on an Archaeological Map of the Ḥaḍramawt, 3. *Arabian Archaeology and Epigraphy* 7:253–78.

Sedov, Alexander V., and P. A. Griaznevich (editors)

1996 *Raybun Settlement (1983–1987 Excavations)*. Preliminary Reports of the Soviet–Yemeni Joint Complex Expedition 2. Russian Academy, Moscow.

Simmons, Alan H.

2007 *The Neolithic Revolution in the Near East: Transforming the Human Landscape*. University of Arizona Press, Tucson.

Smith, Bruce D.
2001 Low-Level Food Production. *Journal of Archaeological Research* 9(1):1–43.
Soderstrom, J. S.
1969 Appendix II: Impressions of Cereals and Other Plants in the Pottery of Hajar Bin Humeid. In *Hajar Bin Humeid: Investigations at a Pre-Islamic Site in Saudi Arabia*, edited by Gus W. Van Beek, pp. 399–408. Johns Hopkins University Press, Baltimore.
Stark, Freya
1953 *The Coast of Incense*. John Murray, London.
Stemler, Ann
1990 A Scanning Electron Microscope Analysis of Plant Impressions in Pottery from the Sites of Kadero, El Zakiab, Um Direiwa, and El Kadada. *Archaeologie du Nil Moyen* 4:87–105.
Stiner, Mary C., Natalie D. Munro, and Todd A. Surovell
2000 The Tortoise and the Hare: Small-Game Use, the Broad-Spectrum Revolution, and Paleolithic Demography. *Current Anthropology* 41(1):39–73.
Tchernov, Eitan, and Ofer Bar-Yosef
1982 Animal Exploitation in the Pre-Pottery Neolithic B Period at Wadi Tbeik, Southern Sinai. *Paléorient* 8(2):17–37.
Thesiger, Wilfred
1959 *Arabian Sands*. E. P. Dutton, New York.
Thomas, Bertram
1938 *Arabia Felix*. Readers' Union, London.
Tierney, Jessica E., and Peter B. deMenocal
2013 Abrupt Shifts in Horn of Africa Hydroclimate since the Last Glacial Maximum. *Science* 342(6160):843–46.
Uerpmann, Margarethe
1992 Structuring the Late Stone Age of Southeastern Arabia. *Arabian Archaeology and Epigraphy* 3(2):65–109.
Van Beek, Gus W., Glen H. Cole, and W. F. Jamme
1964 *An Archaeological Reconnaissance in Hadhramaut, South Arabia: A Preliminary Report*. Bulletin 4587. Smithsonian Institution, Washington, DC.
Van der Meulen, D.
1932 Hadhramaut: Some of Its Mysteries Unveiled. John Murray, London.
1947 *Aden to the Hadhramaut*. John Murray, London.
Van Rampelbergh, Maïté, Dominik Fleitmannb, Sophie Verheydena, Hai Chenge, Lawrence Edwardsf, Peter De Geestg, David De Vleeschouwera, Stephen J. Burnsh, Albert Matterb, Philippe Claeysa, and Eddy Keppens
2013 Mid- to Late Holocene Indian Ocean Monsoon Variability Recorded in Four Speleothems from Socotra Island, Yemen. *Quaternary Science Reviews* 65:129–42.
Vermeersch, Pierre, Philip Van Peer, Jan Moeyersons, and Willem Van Neer
1996 Neolithic Occupation of the Sodmein Area, Red Sea Mountains, Egypt. In *Aspects of African Archaeology: Papers from the 10th Congress of the PanAfrican Association for Prehistory and Related Studies 1996*, edited by Gilbert Pwiti and R. C. Soper, pp. 411–19. University of Zimbabwe, Harare.
Vogt, Burkhard, and Alexander V. Sedov
1994 *Surveys and Rescue Excavations in the Hadhramawt Governate, Republic of Yemen*. Manuscript on file. Deutches Archäologisches Institut, Sana'a.
Wasylikowa, Krystyna
1993 Plant Macrofossils from the Archaeological Sites of Uan Muhuggiag and Ti-n-Torha, Southwestern Libya. In *Environmental Change and Human Culture in the Nile Basin and Northern Africa until the Second Millennium BC, Studies in African Archaeology*, edited by Lech Krzyzaniak, Michel Kobusiewics, and J. Alexander, pp. 21–43. Poznàn Archaeological Museum, Poznàn, Poland.
Wasylikowa, Krystyna, and Jeff Dahlberg
1999 Sorghum in the Economy of the Early Neolithic Nomadic Tribes at Nabta Playa, Southern Egypt. In *The Exploitation of Plant Resources in Ancient Africa*, edited by Marijke van der Veen, pp. 11–31. Kluwer Academic Plenum, New York.
Weber, Steven A.
1998 Out of Africa: The Initial Impact of Millets in South Asia. *Current Anthropology* 39(2):267.
Wenke, Robert, Janet Long, and Paul Buck
1988 Epipaleolithic and Neolithic Subsistence and Settlement in the Fayyum Oasis of Egypt. *Journal of Field Archaeology* 15:29–51.
Wetterstrom, Wilma
1993 Foraging and Farming in Egypt: The Transition from Hunting and Gathering to Horticulture in the Nile Valley. In *The Archaeology of Africa: Food, Metals and Towns*, edited by Thurston Shaw, Paul Sinclair, Bassey Andah and Alex Okpoko, pp. 165–226. Routledge, London.
Wilkinson, Tony J.
1997 Holocene Environments of the High Plateau, Yemen: Recent Geoarchaeological Investigations. *Geoarchaeology* 12(8):833–64.
2009 Environment and Long-Term Population Trends in Southwest Arabia. In *The Evolution of Human Populations in Arabia*, edited by Michael D. Petraglia and Jeffrey I. Rose, pp. 51–68. Springer, Dordrecht.
Willcox, George
1994 Archaeobotanical Finds. In *Qala'at al Bahrain*. Vol. 1, *The Northern City Wall and the Islamic Fortress*, edited

by Flemming Hojlund and H. Hellmuth Andersen, pp. 459–62. Jutland Archaeological Society, Aarhus, Denmark.

Willcox, George, Sandra Fornite, and Linda Herveiux

2008 Early Holocene Cultivation before Domestication in Northern Syria. *Vegetation History and Archaeobotany* 17(3):313–25.

Winterhalder, Bruce, and Douglas J. Kennett

2006 Behavioral Ecology and the Transition from Hunting and Gathering to Agriculture. In *Behavioral Ecology and the Transition to Agriculture*, edited by Bruce Winterhalder and Douglas J. Kennett, pp. 20–63. University of California Press, Berkeley.

Zeder, Melinda A.

2009 The Neolithic Macro-(R)Evolution: Macroevolutionary Theory and the Study of Culture Change. *Journal of Archaeological Research* 17(1):1–63.

2012 The Domestication of Animals. *Journal of Anthropological Research* 68(2):161–90.

Zeder, Melinda A., and Brian Hesse

2000 The Initial Domestication of Goats (Capra hircus) in the Zagros Mountains 10,000 Years Ago. *Science* 287:2254–57.

Zhao, Y., P. Braconnot, O. Marti, S. P. Harrison, C. Hewitt, A. Kitoh, Z. Liu, U. Mikolajewicz, B. Otto-Bliesner, and S. L. Weber

2005 A Multi-Model Analysis of the Role of the Ocean on the African and Indian Monsoon During the Mid-Holocene. *Climate Dynamics* 25(7):777.

Zohary, Daniel, and Maria Hopf

1988 *Domestication of Plants in the Old World: The Origin and Spread of Cultivated Plants in West Asia, Europe, and the Nile Valley*. Clarendon Press, Oxford University Press, Oxford.

Zvelebil, Marek

2006 Mobility, Contact, and Exchange in the Baltic Sea Basin 6000–2000 BC. *Journal of Anthropological Archaeology* 25(2):178–92.

Chapter 2

Geological and Environmental Background

Joy McCorriston

Geology

Southern Arabia includes the southern margin of the Rub' al Khali Desert, including the sandy Ramlah as-Ṣab'atayn Desert basin that drains eastward into the Wādī Ḥaḍramawt–Wādī Masilah (figure 2.1). In the east, the Hadramawt–Masila drainage is fed by catchments spanning the arid mountainous governorates of South Yemen—Shabwa and Hadramawt. These governorates, together with Yemen's Mahra Governorate and Dhofar (Sultanate of Oman), form the Southern Arabian highlands, which include the mountains west of central Oman's lowland desert.

Southern Yemen lies adjacent to the tectonically active Africa–Somalia–Arabia triple junction, which extends southwest through the East African Rift System, northward through the Red Sea Rift, and east through the Gulf of Aden spreading center. Initially separating from the African plate in the Late Oligocene epoch, about 25 million years ago, resulting in the opening of the Red Sea and the Gulf of Aden, the southwestern Arabian Peninsula continues to experience uplift, volcanism, and seismic activity, which have shaped the present-day landscape.

In Southern Arabia, faulting and erosion have produced a dissected and often inaccessible landscape. Overlying Precambrian basement rocks, a transgressive sequence of sediments is represented by Jurassic, Cretaceous, and Paleogene sandstones, limestones, shales, and evaporites deposited in Mesozoic rift basins that formed during the breakup of Gondwana. Eocene uplift, tilting, and compression formed a series of roughly west–east–trending faults and broad basins and arches, resulting in the high plateaus known as the Northern and Southern Jol, separated by the eastward trending Wādī Ḥaḍramawt–Wādī Masilah drainage. Bedrock exposed in deep canyons incised into the upland plateaus includes thick layers of Upper Paleocene–Lower Eocene Umm er-Radhuma Formation limestone, overlain by Middle Eocene Jeza Formation shale and limestone, capped by isolated remnants of Upper Eocene Rus Formation gypsum, shale, and limestone. Travelers in Wādī Ḥaḍramawt readily recognize the great white cliffs of the Umm er-Radhuma Formation framing the drainages (figure 2.2).

A series of long east–west normal faults produced the impressively steep limestone escarpment parallel to the Southern Arabian coastline. In most places, this escarpment forms a steep barrier to the plateau known as the Southern Jol (Beydoun 1966). The plateau tilts northward and eastward 150 by 500 km, with the drainage divides approaching its southern escarpment. Consequently, long wadis (for example, Wādī Sanā and Wādī 'Idim) have formed by incision and headward erosion and flow northward as tributaries to the west–east Wādī Ḥaḍramawt.

Weathering and erosion have differentially impacted the rock formations of the plateau. Where the elevations are highest due to faulting and uplift in the Southern Jol, overlying shales, marls, thin limestones, and gypsums of the Eocene Jeza and Rus Formations have been exposed to

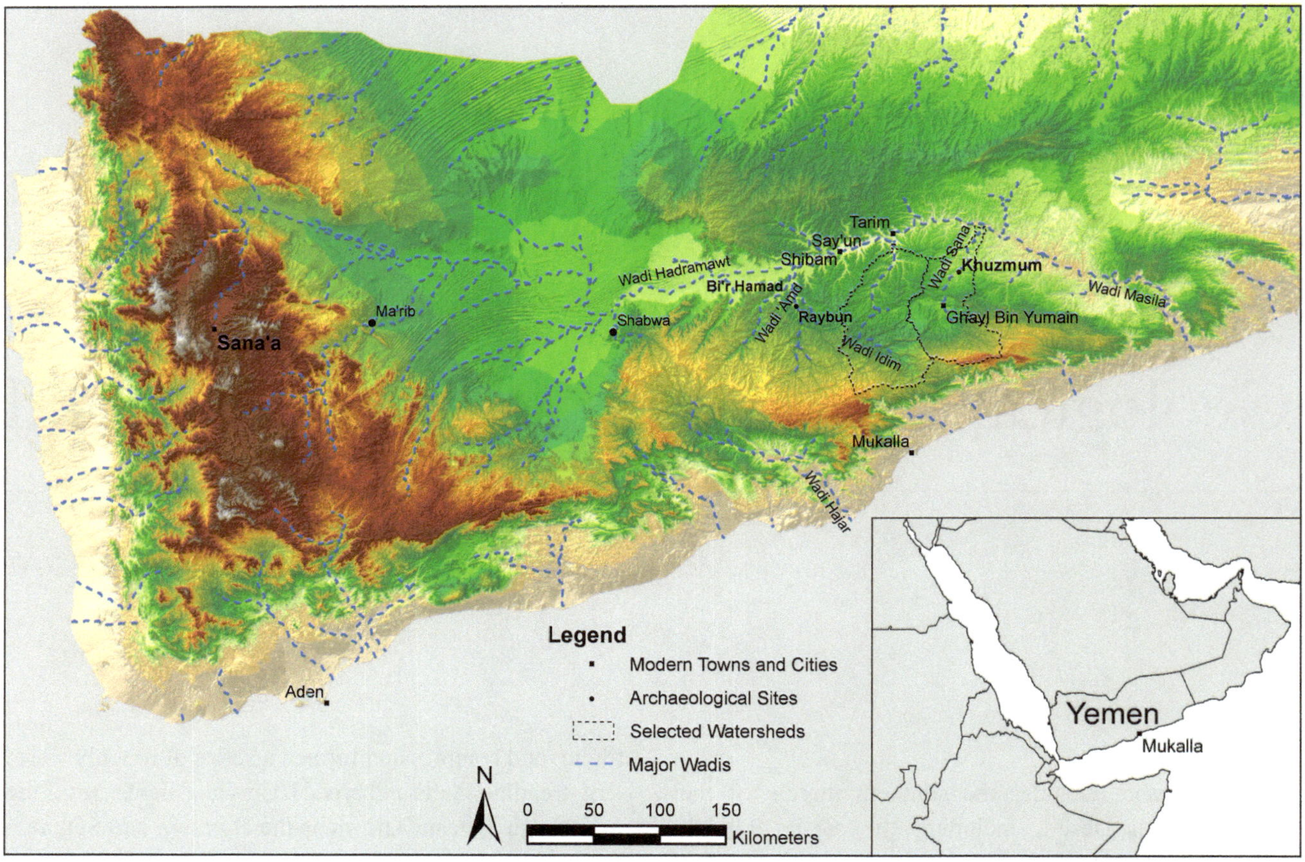

Figure 2.1. Map of Southern Arabia showing the location of Wādī Sanā and Wādī ʿIdim. Topography and hydrology based on Shuttle Radar Topography Mission (SRTM), version 4. *Illustration by Michael Harrower.*

the greater coastal precipitation and have been reduced or eroded altogether (figure 2.3). Because of the regional tilt, Jeza and Rus strata are more widespread in the region of the Hadramawt Arch. Aeolian and alluvial action have deposited thick sediments derived from upland Jeza and Rus exposures into wadis dissecting underlying Umm er-Radhuma limestone. Thick Umm er-Radhuma Formation limestones have resisted further erosion, resulting in flattened terrain with deeply entrenched wadi channels. Wādī Ḥaḍramawt and its eastward extension, Wādī Masilah, flow through the gentle depression of the Say'un-Masila basin. To the north, in the region of the Hadramawt Arch, the lower-lying Jeza Formation exposures are less weathered, first forming broad, shallow basins (Raidah), with Rus Formation buttes in a mesa-marked landscape. Concentration of drainage waters from many tributaries has deeply incised wadi channels flowing into Wādī Ḥaḍramawt, cutting through the Paleocene limestone and into the underlying shales and sandstones of the Cretaceous Mukalla Formation (Beydoun 1966; Caton-Thompson and Gardner 1939). This northern extreme of the Southern Jol retains deeply incised Neogene drainage channels in-filled and terraced during the Quaternary. These are the northward-flowing Wadi Do'an, Wādī Bin ʿAlī, Wādī ʿIdim, and Wādī Sanā.

The limestone contains bands of chert, long valued by humans for making tools and evident as an important prehistoric resource. Some chert appears as in situ outcrops and seams, while other sources of chert are the weathered nodules surviving erosion as smoothed cobbles in wadi beds. Weathering of the highly friable shales of the Jeza and Rus cap has generated much of the sand and silt deposited as wadi fill. While much of this fill is likely derived from the top of the plateau, silicaceous sands from the northern deserts, transported by winter winds, also contributed to wadi fill.

Today an important source of water, and therefore an attraction for settlement, are springs that seep laterally through karstified zones within the Umm er-Radhuma limestones (figure 2.4). Such seeps emerge at the lower cliffs of deeply incised northern wadis (and in faults in the southern escarpment), and surface water may flow and pond over northern wadi bottoms where redeposited clays from Jeza Formation shales create an impermeable base (cf. Caton-Thompson and Gardner 1939).

Climatic Conditions

The Southern Jol catches fog and seasonal (spring and summer) precipitation near the coastal escarpment (Blanchet et al. 1997), with rare inland rain. A Hadramawt proverb says that "a bedouin woman will give birth twice before she bathes," in acknowledgement of the rarity of standing water. Rain typically occurs on the escarpment without any clouds visible to those in the interior. Unimpeded by permeable soil, runoff flows inland (northward) in wadis that drain the uplifted and tilted plateau into the Wādī Ḥaḍramawt. The rush of flash flooding, or *sayl*, along the dry boulders of a wadi bed sounds like a dozen great transport trucks careering over the rocks (figure 2.5). It can send bystanders scrambling up the slopes and can remove anything in its path.

Today extreme aridity with less than 100 mm of rainfall characterizes much of the Southern Jol, whose southern, faulted escarpment nonetheless captures diurnal fog along with spring and summer precipitation. As the land heats up, convection causes inland flow of cooler air bringing moisture from the oceans, which infrequently falls as rain at higher elevations on the escarpment. In Dhofar (western Oman), moisture is a heavy mist in summer months, when summer monsoon circulation causes upwelling and circulation of cold waters into the western Arabian Sea, strong onshore winds, and precipitation as the moisture-laden air is carried over high elevations. The effect is less pronounced in Hadramawt. Along the South Arabian coastal highlands, diurnal mists are present in winter months, when local thermal contrast can force orographic precipitation of moisture-rich coastal air at sunset and sunrise.

Past climates were without question more clement. Much of the erosion forming deep wadis must have taken place in the more mesic Neogene period, and the paths of the great northward-flowing wadis conform to those ancient constraints. As further examined in chapter 3, Holocene climate changes are of greater significance to recent human history. The early and mid-Holocene saw enhanced Southwest Asian monsoonal airflow (Marzin et al. 2013; Prell and Kutzbach 1987; Zhao et al. 2005). Regional land-based records show heightened Early Holocene moisture with a gradual decline in precipitation across Arabia at 6000–5000 cal BP (Fleitmann et al. 2003, 2007, 2011; Lézine 2009; Lézine et al. 2007, 2010).

Flora and Ecosystems

Climatic Impacts on Vegetation

Southern Arabia's flora can mostly be characterized as dry-tropical arborescent vegetation of paleotropical origin (for example, *Acacia–Commiphora* woodlands) (Kürschner 1998; Zohary 1973). Pockets of mesic African relicts (including *Boswellia sacra*, *Carissa edulis, Dodonaea viscosa, Sarcostemma virimnale, Maerua crassifolia, and Moringa peregrina*) appear in the Southern and Southwestern Arabian montane vegetation through historic isolation after the rifting of the Red Sea (Kürschner 1998). Enhanced monsoonal precipitation, with greater surface

Figure 2.2. Wādī Ḥaḍramawt Umm er-Radhuma Paleocene limestone forms sheer cliffs, a distinctive feature of the central Hadramawt Valley and the mouths of its tributaries. *Photograph by Joy McCorriston.*

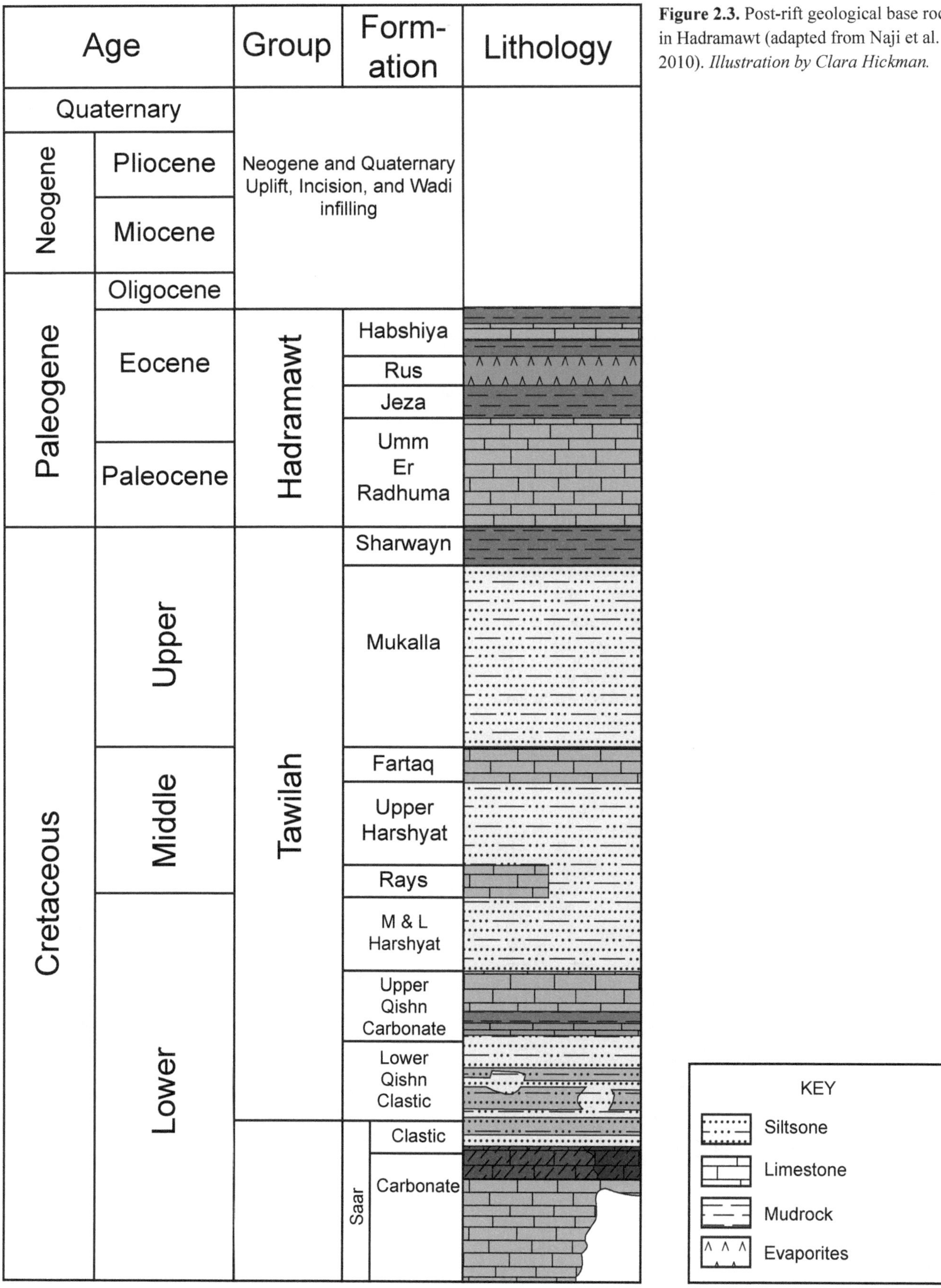

Figure 2.3. Post-rift geological base rocks in Hadramawt (adapted from Naji et al. 2010). *Illustration by Clara Hickman.*

Figure 2.4. Springs at Ghayl ʿUmar, Wādī ʿIdim, February 1998. *Photograph by Joy McCorriston.*

flow for manipulation and cultivation and with greater recharging of aquifers in the Early Holocene, would have had an important effect on water availability for an established mixed shrubland and grassland formation. As precipitation declined in the mid-Holocene, it may possibly also have become more concentrated in two short annual peaks (Ellis and Galvin 1994; Harrower et al. 2012).

The effects of climate change may have varied across different vegetation types and regions. Pollen records are broadly regional in scope when extracted from large lake basins such as the Rub' al Khali playas, and such paleorecords suggest little vegetation difference from today across the Arabian Desert throughout moister climates (Lézine et al. 2007, 2010; Whitney 1983). Far from the inland playas, vegetation and cover may have changed more substantially. Palynological records closely linked with significant geomorphological changes along the easternmost stretches of Southern Arabia's coasts show major changes in Holocene coastal vegetation. One of the most significant developments was the reduction and disappearance of coastal mangrove stands and brackish lagoons after 5000 cal BP, probably significantly linked to sea level rise and changes in coastal geomorphology (Berger et al. 2013). Based on geomorphological and sedimentological sequences, it is clear that enhanced Early Holocene moisture in the western highlands of Yemen sustained perennial lakes (Davies 2006; Weiss and Brunner 2010) in foggy basins with permanent plant cover. Stable land surfaces with abundant plant cover resulted in soil formation, today fossilized as paleosols embedded in the erosional deposition from subsequent arid Middle to Late Holocene conditions (Fedele 1990; Wilkinson 1997). Rising populations, grazing pressures from herded domesticates, and the introduction of farming at 5200 cal BP brought substantial anthropogenic pressures to the native vegetation of the western highlands (Wilkinson 2003), ultimately transforming it to a landscape of elaborately terraced slopes, craggy rock faces, gallery forests in steep drainages, and sedimented plains intensively settled, tilled, and grazed. Comparable climatic and anthropogenic forces would have affected vegetation in the Southern Arabian highlands, which never captured the quantity of rain shed over western Yemen.

Figure 2.5. Rare early spring *sayl* in Wādī Sanā, March 2005. *Photograph by Joy McCorriston.*

Although there are few vegetation paleorecords of any type for Southern Arabia's mountain zone, one may infer that vegetative changes accompanying Holocene climate shifts were significant at local scales, most especially changes in spatial distributions of plant communities, as occurred in the eastern coastal zones. In particular, the drying of springs as rain-fed recharge tapered off after 6000 cal BP surely created greater distances between verdant patches in an arid landscape, challenging humans and animals to adjust. Likewise, a shift in the balance of precipitation and evaporation, evident in the disappearance of interior lakes, would also have affected vernal pools forming in highland basins. Communities of moisture-loving plants at the shrinking margins would have been fewer, of shorter duration, and of greater spacing and interannual uncertainty. Chapter 3 reports in-depth analysis of RASA records of paleoclimate and vegetation in the Wādī Sanā.

Modern Plant Communities

Today vegetation in the Southern Jol contains many species (table 2.1) that could have been more prolific in the past and reflects both climatic and anthropogenic factors that shape its character and extent (Deil and al Grifi 1998). Along the escarpment and high plateau, where precipitation and above all mist supply regular moisture, vegetation is most dense, with thickets in wadis, open shrubland on slopes, and heavy lichen growth on stony surfaces. Scattered trees include *Acacia tortilis*, *A. ehrenbergiana*, *A. mellifera*, *Commiphora habessinica*, and *Maerua crassifolia*, while shrubs and succulents include *Dodonaea viscosa*, *Euphorbia* spp., *Adenium obesum*, *Jatropha glandulosa*, *Cissus quadrifolia*, and *Aloe* spp. Beyond the orographic precipitation zone of the southern escarpment, open scrubland thins to bare, rocky slopes, with scattered myrrh trees (*Commiphora* spp.) clinging to rocky ground. Other vegetation shelters in the wadi bottoms, which are moisture-enhanced by the rare flash flood and occasional springs. Springs occur at the interfaces where porous rock overlies hard substrates in cliff walls. In remote wadis, only lightly frequented by herders, the flora gives a good impression of the range of plant life that may thrive even with erratic interannual rainfall and spate runoff. Trees and shrubs include multiple *Acacia* species, the ban tree *Moringa peregrina*, native frankincense *Boswellia sacra*, *Maerua crassifolia*, *Anogeissus bentii*, *Cadaba heterotricha*, *Grewia erythraea*, *Lycium shawii*, *Indigofera* spp., and *Delonix regia*. Where deep roots can access underground water, especially along the wadi bottoms, *Ficus salicifolia* and *Tamarix nilotica* can thrive. Grasses and annuals such as *Cleome brachycarpa*, *Cassia* italica, and *Tephrosia* spp., offer intermittent forage.

Table 2.1. Botanical checklist for Wādī Sanā, 1998–2008.

Family	Genus	Species (if known)	Author
Acanthaceae	*Barleria*	sp.	
	Bentia	*fruticulosa*	Rolfe
	Blepharis	*ciliaris*	(L.) B.L. Burtt.
	Ecbolium	*viride*	(Forssk.) Alston
	Ruellia	*praetermissa*	Lindau
Aizoaceae	*Glinus*	*lotoides*	L.
Amaranthaceae	*Aerva*	*javanica*	(Burnm.f.) Schultes
	Amaranthus	*graecizans*	L.
	Digera	*muricata*	(L.) Mart.
Apocynaceae	*Rhazya*	*stricta*	Decne.
Asclepiadaceae	*Calotropis*	*procera*	(Ait.) Ait. f.
	Caralluma	sp.	
	Leptadenia	*pyrotechnica*	(Forssk.) Decne
Balanitaceae	*Balanites*	*aegyptiaca*	(L.) Del.
Boraginaceae	*Arnebia*	*hispidissima*	(Lehm.) D.C.
	Heliotropium	*longiflorum*	(ADC) Steud. & Hochst. ex Bunge
	Heliotropium	*ramosissimum*	(lehm) Sieb. ex D.C.
Burseraceae	*Commiphora*	cf. *myrrha*	(Nees.) Engl.
	Boswellia	*sacra*	Flueck
	Commiphora	*gileadensis*	(L.) C.Chr.
Capparaceae	*Cadaba*	*heterotricha*	Stocks
	Capparis	*spinosa*	L.
	Cleome	*ambylocarpa*	Barr. & Murb.
	Cleome	*droserifolia*	Del
	Cleome	*scaposa*	D.C.
	Dipterygium	*glaucum*	Decne
	Maerua	*crassifolia*	Forssk.
Caryophyllaceae	*Cometes*	*abysinnica*	R.Br.
	Herniaria	*hirsuta*	L.
Chenopodiaceae	*Chenopodium*	*murale*	L.
	Cornulaca	*monocantha*	Del.
	Salsola	*bottae*	(Jaub et Spach.) Boiss.
Combretaceae	*Anogeissus*	*bentii*	Baker
Compositae	*Atractylis*	*kentrophylloides*	(Baker) F.G. Davies
	Centaurea	*pseudosiniaca*	Czrep.
	Chrysanthellum	*americanum*	(L.) Vatke.
	Eclipta	*prostrata*	(L.) L.

Table 2.1. Botanical checklist for Wādī Sanā, 1998–2008. *(continued)*

Family	Genus	Species (if known)	Author
	Flaveria	*trinervia*	(Spreng.) Mohr.
	Hochstettera	*schimperi*	D.C.
	Launea	*capitata*	(Spreng.) Dandy
	Pulicaria	*crispa*	(Forssk.) Oliv.
Convolvulaceae	*Convolvulus*	*arvensis*	L.
	Convolvulus	*fatmensis*	Kunze
	Convolvulus	*glomeratus*	Choisy
	Convolvulus	*prostratus*	Forssk.
Cruciferae	cf. *Eremobium*	*aegyptiacum*	(Spreng.) Boiss.
	Farsetia	*linearis*	Decne ex Boiss.
	Farsetia	*longisiliqua*	Decne
	Schouwia	*purpurea*	(Forssk.) Schweinf.
Cucurbitaceae	*Citrullus*	*colocynthus*	(L.) Schrad.
	Cucumis	*figarei*	Naud.
Cyperaceae	*Cyperus*	cf. *esculentus*	L.
	Cyperus	*rotundus*	L.
Euphorbiaceae	*Chrozophora*	*oblongifolia*	(Del.) A. Juss. ex Spreng.
	Euphorbia	*granulata*	Forssk.
	Jatropha	*spinosa*	Vahl.
Geraniaceae	*Erodium*	*malacoides*	(L.) Willd.
Gramineae	*Aristida*	*congesta*	Roem. & Schultes
	Aristida	*mutabilis*	Trin. & Rupr.
	Aristida	*triticoides*	Henr.
	Enneapogon	*scoparius*	Stapf.
	Cynodon	*dactylon*	(L.) Pers.
	Hordeum	*sativum*	L.
	Sorghum	*bicolor*	(L.) Moench
	Sorghum	*x drummondii*	(Nees ex Steud.) Millsp. et Chase
	Stipagrostis	*ciliata*	(Desf.) DeWinter
	Tragus	*berteronianus*	Schult.
	Triraphis	*pumilio*	R. Br.
	Triticum	*aestivum*	L.
Labiatae	*Ocimum*	*forskolei*	Benth.
Leguminosae	*Acacia*	*ehrenbergiana*	Hayne
	Acacia	*hamulosa*	Benth.
	Acacia	*mellifera*	(Vahl.) Benth.
	Acacia	*oerfota*	(Forssk.) Schweinf.
	Cassia	*holosericea*	Fres.
	Cassia	*italica*	(Mill.) Lam. ex Steud.
	Crotalaria	*aegyptiaca*	Benth.
	Crotalaria	cf. *senegalensis*	(Pers.) D.C.
	Crotalaria	*emarginella*	Vatke.
	Crotalaria	*oocarpa*	Bak.
	Crotalaria	sp.	
	Delonix	*elata*	(L.) Gamble
	Indigofera	*articulata*	Gouan.

Family	Genus	Species (if known)	Author
	Indigofera	*oblongifolia*	Forssk.
	Indigofera	sp.	
	Indigofera	*spinosa*	Forssk.
	Rhynchosia	*memnonia*	(Del.) D.C.
	Tephrosia	cf. *pumila*	(Lam.) Pers.
	Tephrosia	*apollinea*	(Del.) D.C.
Malvaceae	*Abutilon*	*fruticosum*	Guill. & Perr.
	Abutilon	*pannosum*	(Forst.f.) Schletcht.
	Pavonia	*triloba*	Guill. & Perr.
	Senra	*incana*	Cav.
	Sida	*ovata*	Forssk.
Meninspermiaceae	*Cocculus*	*pendulus*	(J.R. & D. Forst) Diels
Moraceae	*Ficus*	*salicifolia*	Vahl.
Nyctaginaceae	*Commicarpus*	cf. *helenae*	(J.A.Schultes) Meikle
Oxalidaceae	*Oxalis*	*corniculata*	L.
Polygalaceae	*Polygala*	*erioptera*	D.C.
	Polygala	*senensis*	Kloszsch.
Resedaceae	*Ochradenus*	*baccatus*	Del.
	Reseda	*sphenocleoides*	Deflers
Rhamnaceae	*Ziziphus*	*leucodermis*	(Baker) O. Schwartz
	Ziziphus	*spina-christi*	(L.) Willd.
Rubiaceae	*Kohautia*	sp.	
	Pavetta	cf. *longiflora*	Vahl.
Rutaceae	*Haplophyllum*	*tuberculatum*	(Forssk.) A.Juss.
Salvadoraceae	*Salvadora*	*persica*	(L.) Garcin.
Scrophulariaceae	*Anticharis*	*glandulosa*	Asch.
	Anticharis	*linearis*	(Benth.) Hochst. ex Aschers
	Aptosimum	nr. *pumilium*	Benth.
	Lindenbergia	*indica*	(L.) Kuntze
	Schweinfurthia	*papilionacea*	(L.) Boiss.
Solanaceae	*Datura*	*innoxia*	Mill.
	Lycium	*shawii*	Roem. & Schultes
	Solanum	*coagulans*	Forssk.
	Solanum	*nigrum*	L.
	Solanum	sp.	
Sterculiaceae	*Hermannia*	*paniculata*	Franch.
Tamaricaceae	*Tamarix*	cf. *aphylla*	(L.) Karst.
Tiliaceae	*Corchorus*	*depressus*	(L.) Stocks.
	Grewia	*erythraea*	Schweinf.
Urticaceae	*Forsskalea*	*tenacissima*	L.
Violaceae	*Viola*	*cinerea*	Boiss.
Vitaceae	*Cyphostemma*	*ternatum*	(Forssk.) Descoings
Zygophyllaceae	*Fagonia*	*ovalifolia*	Hadidi
	Tribulus	*bimucronatus*	Viv.
	Tribulus	cf. *omanensis*	H. Hosni ex Hadidi
	Zygophyllum	cf. *hamiense*	Schweinf.
	Zygophyllum	*simplex*	L.

Anthropogenic Factors in Modern Vegetation

Regional climate models predict that in the Southern Jol, xerophytic woods/scrub and warm grass/shrub vegetation would have characterized the Early Holocene (Kutzbach et al. 1996). Yet local environments rather than panregional effects or global change most closely affect and respond to human decision-making in historical ecology (Orlove 1980). In most areas, wadis have been severely impacted by grazing, leaving only the thorny, deciduous, or unpalatable species, such as multiple types of *Acacia*, *Zygophyllum simplex*, *Ficus salicifolia*, *Rhazya stricta*, and *Pergularia tomentosa* (figure 2.6).

Humans manipulating past water flow may have increased vegetation for grazers and browsers and sustained crops raised in scattered, small, optimal locations such as the sediments at the mouths of small tributaries and shallow basins. Runoff water management and fertile sediments make it possible to farm discrete pockets of the *jol*, and the broad basins of Raidah al-Maʿārrah and Kawr Saybān support small permanent villages (for example, al-ʻUlayb, Dafīsh, Ghayl Bin Yumain, Risib, and Bayn al-Jibāl). Where there are springs and permanent surface water, there is abundant vegetation, without exception today heavily impacted by humans. Reeds (*Phragmites australis*) and *Tamarix nilotica* are the most obvious, with cultivated date palms (*Phoenix dactylifera*), toothbrush plant (*Salvadorea persica*), *dom* (*Ziziphus spina-christi*), *Calotropis procera*, and *Balanites aegyptiaca* all enjoying anthropogenic enhancement (figure 2.7). *Blumea* sp. and *Salvadorea persica* grow along irrigation canals, while field weeds include *Anagallis arvensis*, *Cyperus rotundus*, *Tribulus terrestris*, *Glinus lotoides*, *Cynodon dactylon*, *Portulaca quadrifida*, and *Aizoon hispanicum*.

Fauna and Ecosystems

Young men with guns, of whom there are many on the Southern Jol, will shoot wild animals—for food, for safety, for sport, and sometimes in collective hunts that echo with history and cultural tradition. Of the large wild mammals, few remain (table 2.2). Gazelles (*Gazella gazelle*) and ibex (*Capra nubiana*) make tasty prey. Baboons (*Papio hamadryas*) and hyenas (*Hyaena hyaena*) are unwelcome in human proximity, yet a few baboons still raid the gardens of date palm oases. Arabian leopards (*Panthera pardus nimr*) and wolves (*Canis lupus arabs*) are nearly gone. It is doubtful whether tahr (*Arabitragus jayakari*) or onagers

Figure 2.6. Goats browsing thorny *Lycium shawii* in Wādī Sanā. *Photograph by Joy McCorriston.*

Figure 2.7. Date palm orchards at Ghayl ʿUmar, Wādī ʿIdim. *Photograph by Joy McCorriston.*

(*Equus hermionus*), both still wild in neighboring Oman, ever extended their range into the Southern Jol. In the wadis, fennec foxes (*Vulpes zerda*) still hunt snakes, jerboas (Dipodidae), mice, and large lizards. Shepherds continue to maintain remote leopard traps, and the occasional hyena lurks near human outposts. Tasty hyraxes (*Procavia capensis*), big rodent-size shy cousins to the elephant, are reputedly gone; Nasser Al-ʿAlīy (personal communication 2000) says, "The eagles got them all." At dusk one can sometimes still startle a few gazelles.

The great seasonal winter bird migrations have tapered, in part due to intensive trapping in their northern summer ranges and in part because their winter habitats in Arabia have experienced increasing human pressure, such as the draining and damming of the coastal brackish inlets and orchard hunting with firearms. Flamingoes (*Phoenicopterus minor*) no longer cluster in Mukalla's *khor*. Local partridge (*Alectoris melanocephala, Ammoperdix heyi*) and sand grouse (*Pterocles* spp.) make good eating, and smaller bird targets are also good for practice, while the colorful bee-eaters (*Merops orientalis*) are despised by beekeepers. Certainly the wild fauna was not always so sparse in the Southern Jol. Like the interior deserts of the Early Holocene, the Jol once supported greater plant and animal biomass, with a probable exception in the human populations, which were likely always lower than today.

There is a rich ethnographic and historical record of traditional hunting in Arabia. Open game drives and traps used in the interior deserts and northern steppes were not precisely replicated in the mountainous contours of the Southern Jol, so there are no local versions of the northern desert kites (so-called from their shape seen from the air) used to funnel game to waiting hunters (Betts and Burke 2015; Crassard et al. 2015). The introduction of motorized vehicles to interior Arabia in the early twentieth century had severe consequences for herd animals like oryx (*Oryx leucoryx*) and gazelles, overhunted to near extinction from rapidly moving vehicles (Philby 1939). But outracing prey

Table 2.2. Faunal checklist for the Southern Jol.

Family	Genus	Species	Sighted	Common name
Mammals				
Antilopinae	*Gazella*	*dorcas*	Southern Jol, upper plateau 1998	Dorcas gazelle
Bovidae (Caprinae)	*Capra*	*aegagrus hircus*	widely, wadis	goat
Bovidae (Caprinae)	*Capra*	*nubiana*	Southern Jol, upper plateau 1998, fresh kill	ibex
Bovidae (Caprinae)	*Ovis*	*aries*	widely, wadis	sheep
Camelidae	*Camelus*	*dromedarius*	widely	camel
Canidae	*Vulpes*	*rueppellii*	Wādī Sanā 2004, 2005	Rüppell's fox
Erinaceidae	*Paraechinus*	*aethiopicus*	skins?	Ethiopian hedgehog
Felidae	*Felis*	*silvestris lybica*	Wādī Sanā, nocturnal, not clear sighting	wildcat
Felidae	*Panthera*	*pardus nimr*	Wādī ʿIdim, reputed, trapped 2008	leopard
Hyaenidae	*Hyaena*	*hyaena*	Southern Jol, upper plateau 1998	hyena
Hystricidae	*Hystrix*	*indica?*	Wādī Sanā , 2000	porcupine
Muridae	*Acomys*	*dimidiatus?*	Wādī Sanā	mouse
Muridae	*Gerbilus*	*cheesmani?*	Wādī Sanā	gerbil
Papionini	*Papio*	*hamadryas*	Wādī ʿIdim, 1998	baboon
Procaviidae	*Procavia*	*capensis*	Wādī Sanā, reputed	hyrax
Birds				
Ardeidae	*Egretta*	*gularis*	Wādī ʿIdim, 1998	western reef heron
Accipitridae	*Neophron*	*percnopterus*	Southern Jol, upper plateau, 1998	Egyptian vulture
Phasianidae	*Alectoris*	*melanocephala*	Wādī ʿIdim, 1998	Arabian partridge
Phasianidae	*Ammoperdix*	*heyi*	Wādī Sanā, 2000, 2005; not clearly sighted	sand partridge
Pteroclididae	*Pterocles*	*lichtensteinii*	Wādī ʿIdim, 1998	Lichtenstein's sandgrouse
Columbidae	*Columba*	*livia*	Southern Jol, upper plateau 2004	(feral) pigeon/rock dove
Columbidae	*Streptopelia*	*semitorquata*	Wādī ʿIdim, 1998; Southern Jol upper plateau 2005	red-eyed dove
Apodidae	*Apus*	*affinis*	Wādī Sanā , 1998	little swift
Meropidae	*Merops*	*orientalis*	Wādī Sanā, 2004, 2005; Wādī Hadramawt1996, 1998	little green bee-eater
Upupidae	*Upupa*	*epops*	Wādī Sanā, 2005	hoopoe
Alaudidae	*Eremopterix*	*nigriceps*	Wādī Sanā, 2000	black-crowned finch lark
Motacillidae	*Anthus*	*similis*	Wādī Sanā, 1998	long-billed pipit
Motacillidae	*Mortacilla*	*alba*	Southern Jol, upper plateau 2000, 2004, 2005	white wagtail
Pycnonotidae	*Pycnonotus*	*xanthopygos*	Wādī Sanā, 2004, 2005	yellow-vented bulbul
Muscicapidae	*Oenanthe*	*monacha*	Wādī ʿIdim, 1998	hooded wheatear
Scotocercidae	*Scotocerca*	*inquieta*	Wādī Sanā, 2005	scrub warbler
Sylviidae	*Sylvia*	*leucomelaena*	Southern Jol, upper plateau, 2004, 2005	Arabian warbler
Nectariniidae	*Nectarinia*	*habessinica*	Wādī ʿIdim, 2000	shining sunbird
Nectariniidae	*Nectarinia*	*osea*	Wādī Sanā, 2000	Palestine sunbird
Nectariniidae	*Anthreptes*	*metallicus*	Wādī ʿIdim, 1998	Nile Valley sunbird
Laniidae	*Lanius*	*excubitor*	Wādī Sanā, 2000	great grey shrike
Passeridae	*Passer*	*euchlorus*	Wādī ʿIdim, 1998; Southern Jol, upper plateau 2004	Arabian golden sparrow
Ploceidae	*Ploceus*	*galbula*	Wādī Hadramawt, 1996	Rüppell's weaver
Reptiles				
Agamidae	*Uromastyx*	not identified	Wādī Sanā	dhub
Gekkonidae	not identified	not identified	Wādī Sanā	geckos
Viperidae	multiple	not identified	Wādī Sanā	snake

in cars was impossible in the mountains, where the first motorcars were transported piece by piece on camelback to the interior Wādī Ḥaḍramawt, where a few roads could accommodate them. The falcon-assisted hunt for bustards (*Chlamydotis undulata*) and hares (*Lepus capensis*) from camel- or horseback is an eastern Gulf tradition foreign to the highlands of Southern Arabia. In the Southern Jol, where the old flint-matchlock muskets made little impact on wildlife populations (figure 2.8), the introduction of repeating guns, like the ubiquitous Kalashnikov rifle, has taken a drastic toll.

Native hunting tradition in Southern Arabia revolves around the ibex, whose curved horns confer immense prestige to the owners of mudbrick homes they adorn. Anthropologists and Arabists have reported the ibex hunt as an ancient tradition described in inscriptions of the Hadramawt Kingdom (500 BCE–150 CE). Political leaders held the authority to preside over ritual hunts and to perform sacrifice of the captured ibex (Loudine 1990:98; Ryckmans 1976). In antiquity the ritual hunt ensured the fecundity of pasture animals or the prosperity of an undertaking or construction (Ryckmans 1976). An ibex hunt ending in sacrifice could mark the succession of a king and would have been an effective ritual to showcase his role as a political unifier integrating (highland) tribespeople into a federated kingdom centered in the lowlands. A collective hunt for ibex persisted into modern times in Hadramawt (Ryckmans 1976; Sergeant 1976) and seems closely associated with pan-Arabian, pre-Islamic *istiqāʿ* (intercession for rain) at high places where ibex live (Bin ʿAqīl 2004; Daum 2015; Rodinov 1997). Men gather to drive an ibex into a net, and the slain animal is triumphantly returned to the lowland community among much festivity and traditional dances. There are few ibex left to hunt these days, and the associated prestige means that solitary huntsmen or sporting groups avidly seek the last survivors. Visitors to Mukalla's Al-Aḥgāf Hotel could see one of the last ibex, its cranium and horns on display in the lobby, shot by the owner in 1998.

Figure 2.8. Ṣalaḥ Al-ʿAlīy with his matchlock rifle, inherited from his father and once used against British bombers "pacifying" tribal conflicts. Using lead from the market (probably fishing sinkers) Ṣalaḥ melted and formed his own bullets. The tinder and flint for ignition come from the wadi. The gunpowder is made from special plant ash (supplying potassium nitrate), charcoal, and sulfur-rich earth from the springs at Ghayl Bin Yumain. *Photograph by Joy McCorriston.*

There are very few collective hunts driving game, the ibex hunt being the notable example. More common is the use of hunting blinds, which can be spotted at the high vantage of cliff edges overlooking narrow passes. Some hunting blinds have pillaged building stone from the nearest ready source—an archaeological Bronze Age tomb. Locals recall how these blinds have been used to ambush not only fleet and shy gazelles but also human enemies, whether raiders, the raided, or, in recent years, soldiers in civil conflict. Men recall clearly when they last waited on a moonless night with contraband guns or household firearms for the Toyota Hilux pickups bearing enemy soldiers, their machine gun advantage sapped in unfamiliar terrain. Even a foreigner will eventually hear the stories of bedouin fathers and uncles waiting for raiders coming up the valley or planning to ambush caravans foolish enough to pass without a *murāfiq* (tribal guide). And then there is poaching—something inviting costly retribution. None will admit to it, but animals do go missing and the attendant penalties and conflicts suggest that poaching others' livestock occurs.

The wildlife has been gradually replaced by a superabundance of domesticated herd animals—principally goats, a few sheep, camels, and, in villages, the occasional cow or donkey. Zebu (*Bos taurus indicus*) can still be found in the towns, but most are imported from Somali herds to supply beef in urban markets. In the days before roads, trucks hauling feed, refrigeration, and the current market economy, transport in the mountains was handled by camel caravan. Camels and cattle were also richly valued for dairy products, which alongside rice (from Java), dried fish, and dates, formed the dietary base. Herd camels and cattle needed supplements to the available forage, and a lively trade was in sun-dried sardines. These were cured on Hadramawt's greasy beaches in the winter and packed into the interior on camel caravans for cattle and camel feed. Once, the Ḥumūm bedouin kept large numbers of transport camels constantly waiting outside the defensive walls of Hadramawt's coastal cities. Blue from head to toe in indigo dye rubbed off traditional cotton wraps, their long tribal locks flowing from thin leather headbands, these bedouin and their complaining camels were the picturesque and unruly lifelines to the interior. Bedouin were not allowed inside the urban perimeter, and bitter feelings still persist from the riot and massacre of Ḥumūm bedouin outside the gates of Shiḥr more than half a century ago.

Nowadays, camels and cattle are herd animals of the wealthy. They do not thrive together and represent different sectors of the pastoral economy. Cattle have largely disappeared from Hadramawt, although traditional dance and ritual hint at their former significance (Dostal 1983). In neighboring Dhofar, inhabitants of the narrow escarpment and plateau continue to keep cattle in numbers that severely stress the vegetation and soils (Janzen 1986). While goats may be herded with cattle, camels may not, for they require different feed, ranges, and watering (Russell 1988).

Camel herds have dropped since the days when they provided all transport, for the cost of acquiring and keeping them is no longer readily offset by their economic value. They remain important beasts of labor in settlements and towns, turning oil presses, raising water, and crushing lime. Among rural herders, they are markers of social prestige; they provide traditional transport for items like charcoal, wild indigo, and frankincense; and they are few. The combined economic changes with development of road transport and industry (fish canneries, petroleum extraction, and tourism) have at the same time obviated the need for camel caravans and eliminated the primary income source for the Ḥumūm and other bedouin. Now their major income comes from scarce wage labor and the sale of livestock, especially at the season of sacrifice, but profits must be discounted by the cost of getting animals to market, and most families will buy a goat only for sacrifice. Bedouin are poorer than they once were, and their ability to build and maintain herds faces tough constraints. Instead of camels, which reproduce slowly and require larger investments, most bedouin now count their stock of tiny white goats and hairy sheep, living cautiously on milk, baked sorghum dough, rice, preserved fish, dates, and sugared tea.

Goats are the most numerous herd animals today in the Southern Jol and are readily adapted to the stony paths and steep slopes, where a thin flush of annuals clings at the end of the summer monsoon season. When vernal pools, or *krif*, develop (summer–autumn) in the uplands near the coastal escarpment, goat herders can maintain herds among the foggy thickets nearby. Otherwise the daily watering of goats tethers them and their herders near springs, seeps, and rock pools, or *qīr*, in the wadi bottoms, and near the peripheries of date palm orchards and cultivation. Summer is a season when goat herders can scatter more widely to take advantage of monsoonal moisture, but in winter months they must concentrate near water, a pattern that has changed with the advent of machinery and development programs. Roads and petroleum exploration have made remote areas more accessible year-round. Machine-dug wells, pumps, and water deliveries have greatly extended the range of goat herders in the past decades, even as camel herds have dwindled and the larger wild fauna have been extinguished.

Conclusions

Today the Southern Jol retains the vestiges of its historic flora and fauna on a rocky substrate formed under oceans of the Cretaceous and Paleogene eras. Subsequently, Neogene rivers carved deep channels through the uplifted limestone, shaping basins and upland contours for the eons to come. For the past 10,000 years, plant diversity and taxa have probably changed little, and the greatest changes have most likely been of scale and distribution rather than of flora and association. Wild animal populations are an extension of African fauna and have been severely impacted by changes in climate and moisture distribution, and most particularly by human predation in recent decades. In the meantime, goat herding has expanded with the range of humans, causing overgrazing in already fragile arid ecosystems.

References

Berger, J. F., V. Charpentier, R. Crassard, C. Martin, G. Davtian, and J. A. López-Sáez

2013 The Dynamics of Mangrove Ecosystems, Changes in Sea Level and the Strategies of Neolithic Settlements along the Coast of Oman (6000–3000 cal BC). *Journal of Archaeological Science* 40(7):3087–104.

Betts, Alison, and David Burke

2015 Desert Kites in Jordan: A New Appraisal. *Arabian Archaeology and Epigraphy* 26(2):74–94.

Beydoun, Ziad R.

1966 Eastern Aden Protectorate and Part of Dhufar. In *Geology of the Arabian Peninsula*, edited by K. M. Al Naqib, pp. 1–49. Professional Paper 560H. U.S. Government Printing Office, Washington, DC.

Bin ʿAqīl, ʿAbdalraḥmān J.

2004 *Qaniṣun al-Waʿlun fī Ḥaḍramawt*. ʿAsmah al-Thaqafih al-ʿArabiyah, Sana'a.

Blanchet, Guy, Paul Sanlaville, and Myriam Traboulsi

1997 Le Moyen-Orient de 20 000 ans BP à 6 000 ans BP Essai de reconstitution paléoclimatique. *Paléorient* 23(2):187–96.

Caton-Thompson, Gertrude, and Elinor W. Gardner

1939 Climate, Irrigation, and Early Man in the Hadhramaut. *Geographical Journal* 93:18–38.

Crassard, Rémy, Olivier Barge, Charles-Edmond Bichot, Jacques Élie Brochier, Jwana Chahoud, Marie-Laure Chambrade, Christine Chataigner, Kamel Madi, Emmanuelle Régagnon, Hamida Seba, and Emmanuelle Vila

2015 Addressing the Desert Kites Phenomenon and Its Global Range through a Multi-Proxy Approach. *Journal of Archaeological Method and Theory* 22(4):1093–121.

Daum, W.

2015 Qabr Hūd Revisited: The Pre-Islamic Religion of Ḥaḍramawt, Yemen, and Mecca. In *Pre-Islamic South Arabia and Its Neighbors: New Developments of Research; Proceedings of the 17th Rencontres Sabéenes Held in Paris, 6–8 June 2013*, edited by Mounir Arbach and Jérémie Schiettecatte, pp. 49–72. BAR International Series 2740. Archaeopress, Oxford.

Davies, Caroline P.

2006 Holocene Paleoclimates of Southern Arabia from Lacustrine Deposits of the Dhamar Highlands, Yemen. *Quaternary Research* 66(3):454–64.

Deil, Ulrich, and Abdul-Nasser al Grifi

1998 Montane and Wadi Vegetation. In *Vegetation of the Arabian Peninsula*, edited by Shahina A. Ghazanfar and Martin J. Fisher, pp. 125–74. Kluwer Academic Publishers, Dordrecht.

Dostal, Walter

1983 Some Remarks on the Ritual Significance of the Bull in Pre-Islamic Times. In *Arabian and Islamic Studies: Articles Presented to R. B. Sergeant*, edited by Robin Leonard Bidwell, G. R. Smith, and R. B. Serjeant, pp. 196–213. Longman, London.

Ellis, Jim, and Kathleen A. Galvin

1994 Climate Patterns and Land-Use Practices in the Dry Zones of Africa. *Bioscience* 44(5).

Fedele, Francesco G.

1990 Man, Land, and Climate: Emerging Interactions from the Holocene of the Yemen Highlands. In *Man's Role in the Shaping of the Eastern Mediterranean Landscape*, edited by Sytze Bottema, Gerte Enjes-Nieborg, and Willem Van Zeist, pp. 31–42. A. A. Balkema, Rotterdam.

Fleitmann, Dominik, Stephen J. Burns, Augusto Mangini, Manfred Mudelsee, Jan Kramers, Igor Villa, Ulrich Neff, Abdulkarim Al-Subbary, Annett Buettner, Dorothea Hippler, and Albert Matter

2007 Holocene ITCZ and Indian Monsoon Dynamics Recorded in Stalagmites from Oman and Yemen (Socotra). *Quaternary Science Reviews* 26(1–2):170–88.

Fleitmann, Dominik, Stephen J. Burns, Manfred Mudelsee, Ulrich Neff, Jan Kramers, Augusto Mangini, and Albert Matter

2003 Holocene Forcing of the Indian Monsoon Recorded in a Stalagmite from Southern Oman. *Science* 300(5626):1737–39.

Fleitmann, Dominik, Stephen J. Burns, Marek Pekalab, Augusto Mangini, Abdulkarim Al-Subbary, Mohammad Al-Aowah, Jan Kramers, and Albert Matter

2011 Holocene and Pleistocene Pluvial Periods in Yemen, Southern Arabia. *Quaternary Science Reviews* 30(7–8):783–87.

Harrower, Michael J., Eric A. Oches, and Joy McCorriston

2012 Hydro-Geospatial Analysis of Ancient Pastoral/Agro-Pastoral Landscapes along Wādī Sanā (Yemen). *Journal of Arid Environments* 86:131–38.

Janzen, Jörg

1986 *Nomads in the Sultanate of Oman: Tradition and Development in Dhofar*. Westview Press, Boulder.

Kürschner, Harald

1998 Biogeography and Introduction to Vegetation. In *Vegetation of the Arabian Peninsula*, edited by Shahina A. Ghazanfar and Martin J. Fisher, pp. 63–98. Kluwer Academic, Dordrecht.

Kutzbach, J. E., G. Bonan, J. Foley, and S. P. Harrison

1996 Vegetation and Soil Feedbacks on the Response of the African Monsoon to Orbital Forcing in the Early to Middle Holocene. *Nature* 384:623–26.

Lézine, Anne-Marie

2009 Timing of Vegetation Changes at the End of the Holocene Humid Period in Desert Areas at the Northern Edge of the Atlantic and Indian Monsoon Systems. *Comptes Rendus de l'Académie des Sciences: Geoscience* 341:750–59.

Lézine, Anne-Marie, Christian Robert, Serge Cleuziou, Marie-Louise Inizan, Frank Braemer, Jean-François Saliège, Florence Sylvestre, Jean-Jacques Tiercelin, Rémy Crassard, Sophie Méry, Vincent Charpentier, and Tara Steimer-Herbet

2010 Climate Change and Human Occupation in the Southern Arabian Lowlands during the Last Deglaciation and the Holocene. *Global and Planetary Change* 72(4):412–28.

Lézine, Anne-Marie, Jean-Jacques Tiercelin, Christian Robert, Jean-François Saliège, Serge Cleuziou, Marie-Louise Inizan, and Frank Braemer

2007 Centennial to Millennial-Scale Variability of the Indian Monsoon during the Early Holocene from a Sediment, Pollen, and Isotope Record from the Desert of Yemen. *Palaeogeography, Palaeoclimatology, Palaeoecology* 243(3–4):235–49.

Loudine, A. G.

1990 Le banquet rituel dans l'état de Saba. *Proceedings of the Seminar for Arabian Studies* 20:95–100.

Marzin, Charline, Pascale Braconnot, and Masa Kageyama

2013 Relative Impacts of Insolation Changes, Meltwater Fluxes, and Ice Sheets on African and Asian Monsoons during the Holocene. *Climate Dynamics* 41(9–10):2267–86.

Naji, Hassan S., Mohammad H. Hakimi, Mohammed Khalil, and Farooq A. Sharief

2010 Stratigraphy, Deposition, and Structural Framework of the Cretaceous (Review) and 3D Geological Model of the Lower Cretaceous Reservoirs, Masila Oil Field, Yemen. *Arabian Journal of Geosciences* 3(3):221–248.

Orlove, Benjamin

1980 Ecological Anthropology. *Annual Review of Anthropology* 9:235–73.

Philby, Harry St. John B.

1939 *Sheba's Daughters*. Methuen, London.

Prell, Warren L., and John E. Kutzbach

1987 Monsoon Variability over the Past 150,000 years. *Journal of Geophysical Research: Atmospheres* 92(D7):8411–25.

Rodinov, Mikhail A.

1997 Mawlā Maṭar and Other *awlīya*': On Social Functions of Religious Places in Western Ḥaḍramawt. *Mare Erythræum* 1: 107–14. Staatiches Museum für Völkerkunde, Munich.

Russell, Kenneth W.

1988 *After Eden: The Behavioral Ecology of Early Food Production in the Near East and North Africa*. British Archaeological Reports, Oxford.

Ryckmans, Jacques

1976 La chasse rituelle dans l'Arabie du sud ancienne. In *Al-Bahit. Festscrift Joseph Henninger*, pp. 259–308. Studia Instituti Anthropos 28. Ed. Verlag des Anthropos Instituts, Bonn.

Sergeant, Robert B.

1976 *South Arabian Hunt*. Luzac, London.

Weiss, Christian, and Ueli Brunner

2010 Lacustrine to Palustrine Sediments from the Sirwah Oasis, Yemen: Sedimentological Implications for the Local Paleoecology and Archeology. *Neues Jahrbuch für Geologie und Paläontologie—Abhandlungen* 257(2):129–45.

Whitney, John W.

1983 *Erosional History and Surficial Geology of Western Saudi Arabia*. Ministry of Petroleum and Mineral Resources, Jiddah.

Wilkinson, Tony J.

1997 Holocene Environments of the High Plateau, Yemen: Recent Geoarchaeological Investigations. *Geoarchaeology* 12(8):833–64.

2003 The Organization of Settlement in Highland Yemen during the Bronze and Iron Ages. *Proceedings of the Seminar for Arabian Studies* 33:157–68.

Zhao, Y., P. Braconnot, O. Marti, S. P. Harrison, C. Hewitt, A. Kitoh, Z. Liu, U. Mikolajewicz, B. Otto-Bliesner, and S. L. Weber

2005 A Multi-Model Analysis of the Role of the Ocean on the African and Indian Monsoon during the Mid-Holocene. *Climate Dynamics* 25(7):777.

Zohary, Michael

1973 *Geobotanical Foundations of the Middle East*. Vols. 1 and 2. Gustav Fischer Verlag, Stuttgart.

Chapter 3

Paleohydrology, Geomorphology, and Paleoecology

Eric A. Oches, Joshua Anderson, Joy McCorriston, Kenneth Cole, and Michael J. Harrower

Long-term human activity and landscape history of the Hadramawt are deeply interconnected with paleohydrology, geomorphology, and paleoecology. As north-flowing tributaries to the Wādī Ḥaḍramawt–Wādī Masilah system, Wādī Sanā and Wādī ʿIdim preserve valuable archives of Late Pleistocene and Holocene environmental changes. As introduced in chapter 2, a more northerly position of the ITCZ during the Early Holocene brought enhanced monsoonal airflow and summer moisture across Arabia, accompanied by increases in both the magnitude and seasonal duration of stream flow, lake and wetland expansion, enhanced groundwater recharge, spring discharge and tufa formation, sedimentation, and other paleohydrologic changes (Berger et al. 2012; Fleitmann et al. 2007; Lézine et al. 2007, 2010; Matter et al. 2015).

To reconstruct the paleohydrologic and geomorphic response to Holocene climate change in the RASA study region, we carried out a survey of sedimentary sections exposed by wadi incision. We measured geomorphic transects across wadi sections, sampled materials for radiocarbon and luminescence dating, and mapped sediments and tufa deposits at representative localities in the upper and middle basin of Wādī Sanā, at the confluence of Wādī Sanā and Wādī as-Shumlyah, and in the middle reaches of Wādī ʿIdim between the villages of Rāwik and Sāh (figures 3.1 and 3.2) during the 1998, 2000, 2004, and 2005 field seasons.

Sediment History: Structural and Paleoclimatic Controls

Middle Holocene climate change forced significant environmental responses and influenced human activities throughout Southern Arabia. Climate models and proxy data indicate that climate along the southern Arabian Peninsula changed from a moist phase, spanning the Early to Middle Holocene, to an arid phase, which persisted for the last roughly 5,000 years. A weakening and southward shift of the Indian summer monsoon (ISM), forced by precession-driven northern hemisphere insolation variations, is suggested as the mechanism for the abrupt shift to more arid conditions (Kutzbach et al. 2008; Marzin and Braconnot, 2009; Marzin et al. 2013). Geoarchaeological evidence suggests that agriculture and pastoralism were more widespread and evolved alongside the development of irrigation technologies during a period when rainfall was shifting from more plentiful than today to a period of greater aridity, around 4,500 ^{14}C years ago (Crassard et al. 2006; Harrower 2006; Harrower et al. 2012; Lézine et al. 2010; McCorriston et al. 2002, 2012). Here we investigate the surficial record of the dynamic fluvial response to the Late Quaternary climate shift and reconstruct the geochronology and the geomorphic evolution of a portion of the roughly 125 km length of Wādī Sanā and a limited stretch of middle Wādī ʿIdim.

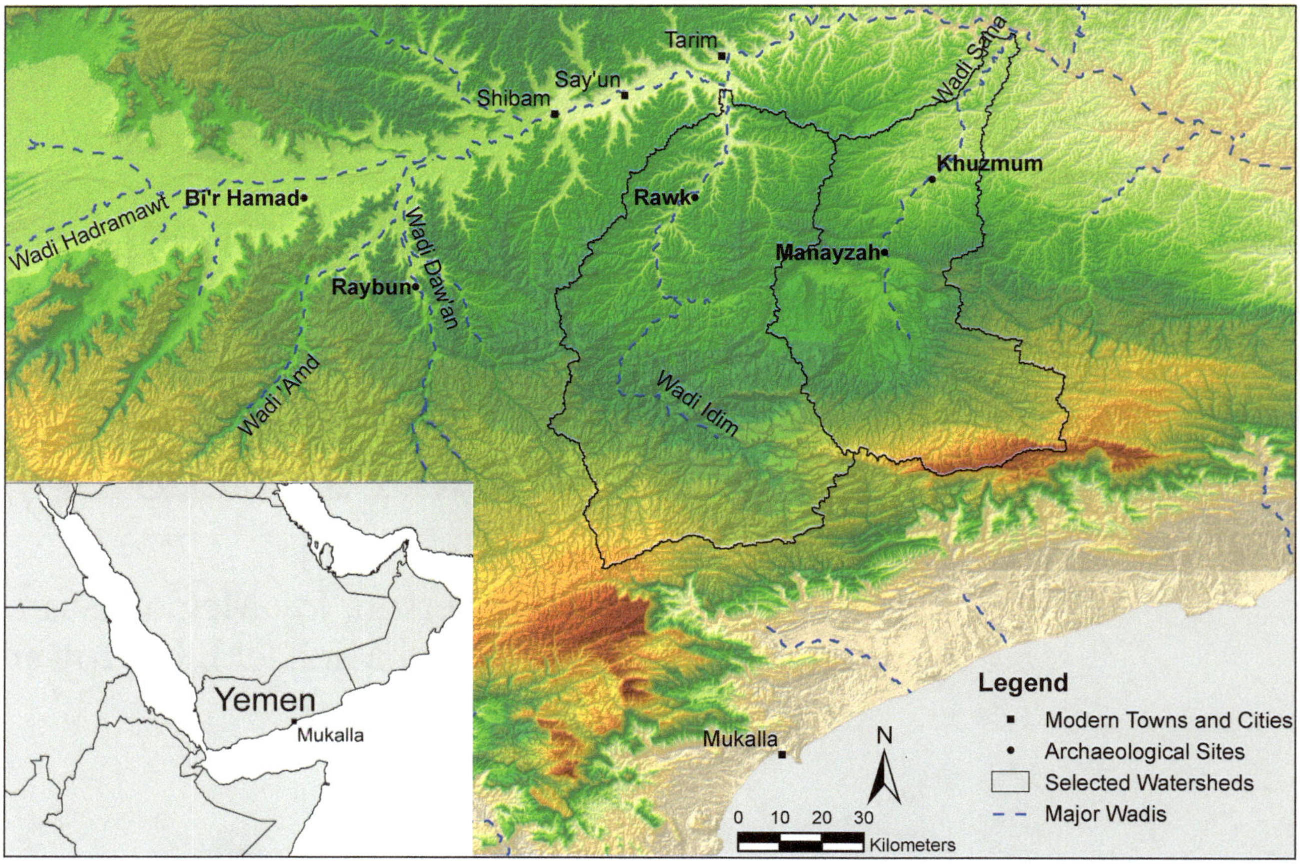

Figure 3.1. Map of Wādī ʿIdim and Wādī Sanā, Hadramawt, Yemen. Topography and hydrology based on Shuttle Radar Topography Mission (SRTM) version 4. *Illustration by Michael Harrower.*

Using high-accuracy (differential and kinematic) GPS survey, combined with analysis of the sedimentary record, we have developed a paleohydrologic reconstruction of Wādī Sanā to provide a context for understanding how fluvial landscapes, hydrologic regime, and human activity reacted to changing Middle Holocene climates. Radiocarbon and luminescence dating of remnant silt terraces suggests that fine-grained sediment began accumulating on an older (Late Pleistocene) coarse cobble surface between 12,000 and 7,000 years ago and continued aggrading until about 5,000 years ago. Paralleling the climate shift, Wādī Sanā began incising and eroding the thick sediment infilling about 5,000 years ago, a process that has continued to the present time. Field reconnaissance and analysis reveals structural and lithologic controls on the source and availability of these fluvial sediments for downstream deposition during the Late Pleistocene and Holocene. We propose that a change in hydrologic regime, driven by the ITCZ shift, is the cause of the Middle Holocene channel adjustment from an aggradational to an incising mode in Wādī Sanā. The changing conditions decreased groundwater recharge, causing a rapid drying of spring-fed wetlands and shallow lakes in Wādī Sanā and Wādī ʿIdim. Only small, isolated remnants of once expansive springs provide water for irrigated agriculture and settlements in a few discrete locations in upper Wādī Sanā and Wādī ʿIdim today.

Wādī Sanā and Wādī ʿIdim Hydrology and Geomorphology

As introduced in chapter 2, the Southern Jol, 150 km (north–south) by 500 km (east–west), dips to the north and east and drains into Wādī Ḥaḍramawt–Wādī Masilah and eventually into the Indian Ocean. Wādī ʿIdim and Wādī Sanā are two of the main drainages of the Southern Jol, with watersheds of 5,400 and 3,900 km^2, respectively (figure 3.1). Wādī ʿIdim flows northward to Tarīm, where it joins Wādī Ḥaḍramawt (which is renamed Wādī Masilah to the east). Wādī Sanā forms a teardrop-shaped watershed that empties into Wādī Masilah at the village of Sanā.

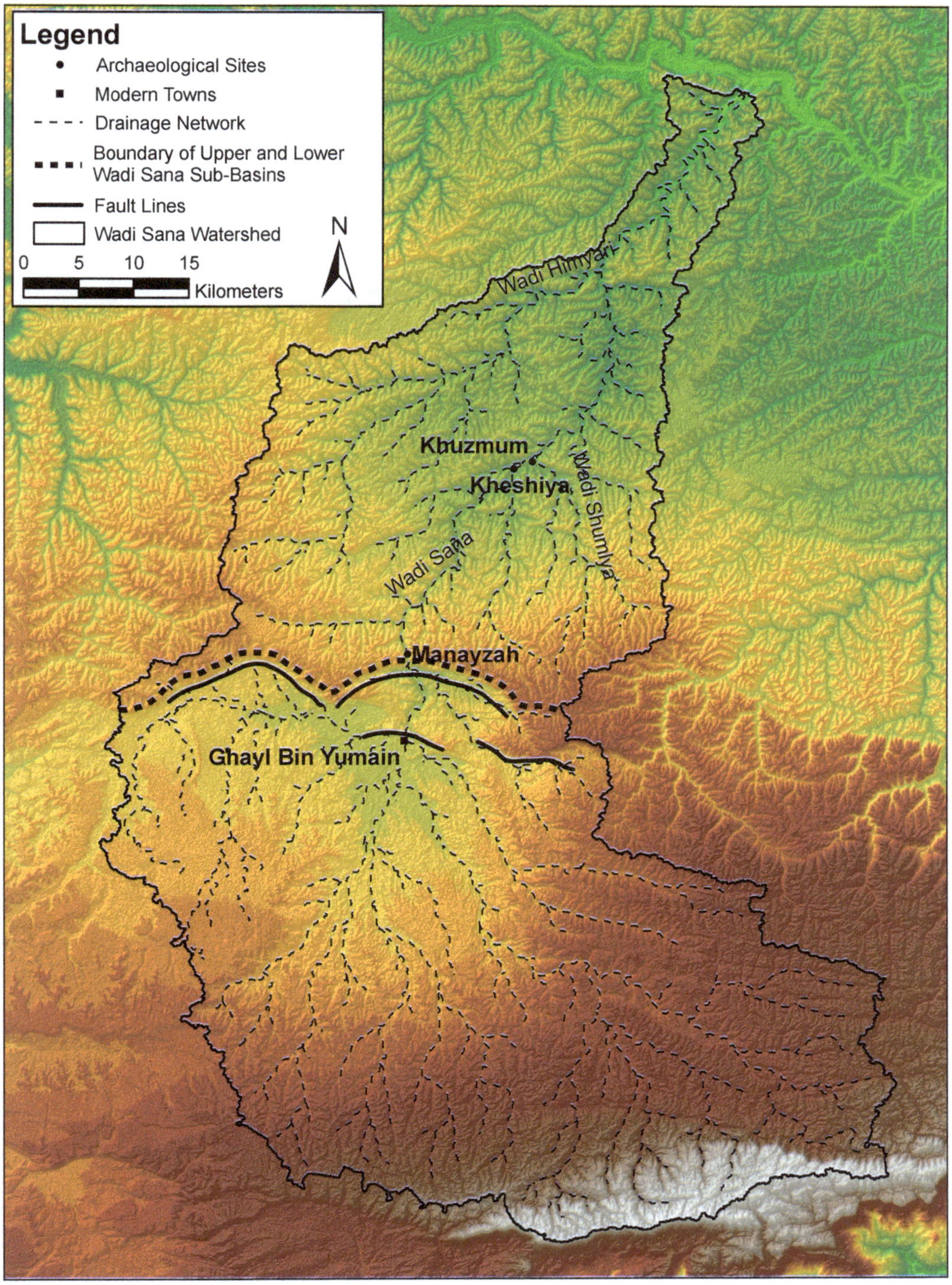

Figure 3.2. Map of Wādī Sanā drainage basin, showing detailed drainage network. Normal faulting, indicated by solid lines, separates the basin into upper and lower Wādī Sanā sub-basins. Topography and hydrology based on Advanced Spaceborne Thermal Emission and Reflection Radiometer (ASTER) Global Digital Elevation Model (GDEM) version 2. *Illustration by Michael Harrower.*

With respect to geomorphology, erosion impacted rock units differently, which in turn shaped resulting sedimentation. The Rus and Jeza formations, consisting of Eocene shales, marls, thin limestone and gypsum deposits, comprise the highest elevations in the southern regions of the plateau (figure 3.3). Erosion of the Rus and Jeza Formations produced sediment that was transported and deposited by aeolian and fluvial action into thick sequences in the wadi bottoms. Underlying the Jeza Fm is the more indurated Paleocene Umm er-Radhuma Formation limestone. That unit eroded more slowly to produce bare rock plateaus, karstic dissolution features, and entrenched wadis.

A series of east–west–trending normal faults, located north of the town of Ghayl bin Yumain, divides the wadi

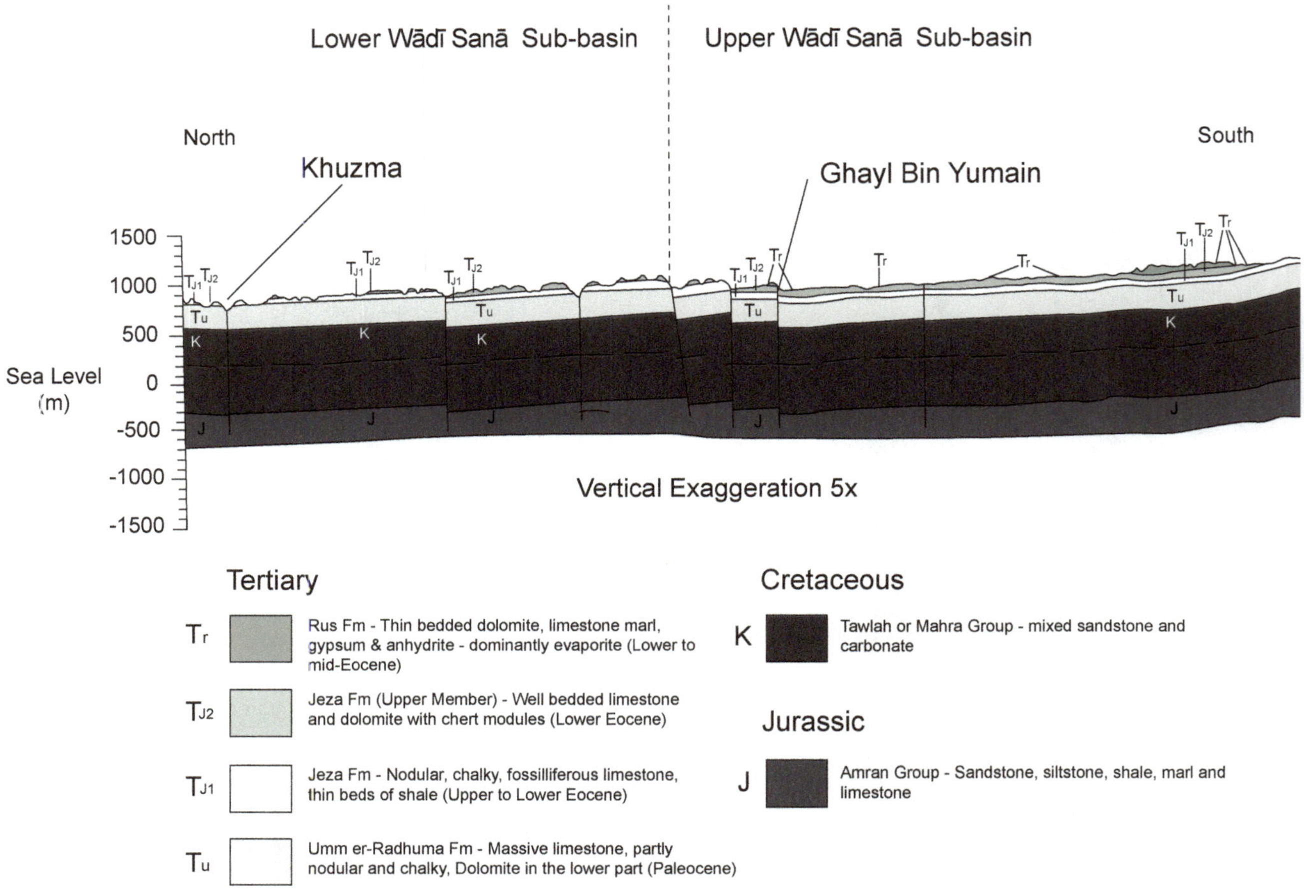

Figure 3.3. North–south geologic cross section paralleling Wādī Sanā, with lithologies indicated. Normal faulting separates the broad low-relief basin of upper Wādī Sanā from the deeply incised canyon at the head of lower Wādī Sanā. *Illustration by Joshua Anderson and Clara Hickman, based on Komex International 1997.*

drainage into an upper and lower basin (figures 2.2 and 2.3). The faulting produced two different geomorphic landscapes on the north (footwall) block and on the south (hanging wall) block. Wādī ʿIdim is a similar deeply incised drainage, flanked by up to 200 m high cliffs of Umm er-Radhuma limestone.

The upper basin of Wādī Sanā, south of the faults, is an uplifted sedimentary basin with components of the Rus and Jeza Formations still present in the upstream reaches of the drainage network. Large volumes of carbonate and siliciclastic sediment are located within this sedimentary basin (figure 3.4a–b). The lower basin, north of the fault system, is distinguished by uplifted terraces of Jeza and Umm er-Radhuma limestone, incised by fluvial downcutting, creating steep-walled canyons and a series of step-terraces (figures 3.5 and 3.6). Normal faulting created a structural restriction at the boundary between the upper and lower basins, resulting in significant differences in the landscapes, sediment production and storage, and fluvial geomorphology in each basin. This structural restriction, interacting with precipitation, also regulates sediment flux between the upper and lower basins of Wādī Sanā. Changing sedimentary and fluvial processes in the Wādī Sanā basin are ultimately driven by Late Pleistocene and Holocene climate variations and associated shifts in the strength and penetration of the Indian summer monsoon across the study area.

Geomorphology of Wadi Landforms

In 2004 we surveyed channel cross-section transects at 26 localities along the roughly 40 km extent of the middle reach of Wādī Sanā. Twenty of these cross sections were mapped using a differential-corrected Trimble Pro-XRS GPS. Given problems with GPS data accuracy, 10 were remeasured with a more sophisticated Trimble 5700 kinematic GPS system in 2008 (see Harrower et al. 2012). GPS-based survey provided the geographic data necessary to map relevant fluvial geomorphic surfaces

a

b

Figures 3.4a–b. Expansive silt deposits and source lithologies within the upper Wādī Sanā sub-basin. Both (a) eroded silt terraces in the Wādī Ḥarū and (b) eroding Rus Formation lithologies provide wadi silt source material in Wādī Jizah. *Photographs by (a) Joy McCorriston and (b) Michael Harrower.*

Figure 3.5. Deeply incised canyon at the upstream reach of the lower Wādī Sanā sub-basin. *Photograph by Eric Oches.*

and characterize channel hydraulic geometry at representative sections across Wādī Sanā. Geomorphic surfaces, reflecting stages of deposition, fluvial incision, and dissection of the wadi sediments, were identified and mapped, and the resulting classification scheme was used in geoarchaeological and paleoenvironmental analysis. Landforms and geomorphic surfaces are grouped into seven classes (Harrower et al. 2002), represented schematically in figure 3.7 and described below.

Plateau (P): Upland bedrock surfaces above bedrock slopes and cliffs, covered in primarily angular, carbonate, small-cobble-size clasts. The plateau landform is the gently southwest–northeast–dipping surface into which down-cutting created wadi systems throughout the region. The plateau is typically about 900 m elevation near the Wādī ʿIdim mouth and about 750 m elevation at the mouth of Wādī Sanā. Distant headwaters in both watersheds rise to about 1200–1400 m elevation, with some high points reaching up to 1900 m, where faulting separates the Southern Jol highlands from the coastal lowlands. The bedrock plateau is eroded into uplifted Umm er-Radhuma and Jeza Formations, with occasional isolated knobs of Rus Formation lithologies.

Bedrock Slope (BS): Greater than 15-degree slope or cliff (sometimes partially covered in talus and/or scree) that separates upland plateaus from all other classes. These slopes are most often exposures of the Umm er-Radhuma bedrock that form steep angles, extending from plateaus above down to the wadi silts and gravel terrace sediments on the wadi floor. Continuous erosional notches that we define as paleo-stage indicators (PSI) are occasionally found inscribed into BS cliff lines along Wādī Sanā. We hypothesize that PSIs are caused by Early to Middle Holocene flooding and maximum wadi surface elevation above the present channel floor. Caves and notches along the PSI contain upper wadi silt unit sediments representing the uppermost distribution of fine-grained sediments and maximum elevation of floodplain sedimentation (figure 3.8). Ages on sediments deposited within the PSI notches, described below, indicate that these maximum flood deposits correspond to the latest phase of sedimentation in Wādī Sanā.

Scree Slope (SS): Angular clasts of talus and scree often of a low (< 20-degree) gradient that often separate the plateau and bedrock slopes above from the terraces, wadi sediments, and wadi channels below.

Gravel Terrace (GT): Sub-rounded to rounded clasts, often capping wadi silts and adjacent to wadi channels. Gravel terraces form a laterally continuous bed ranging from pebbles to boulders, found at the base of plateaus, bedrock slopes, and scree slopes. The gravel terrace is the lowest stratigraphic unit in the wadi sediment sequence and extends below the depth of observable channel infilling.

Bedrock Terrace (BT): The youngest low-angle (< 5 degrees) or horizontal bedrock surface, often found between wadi sediments and/or upland landforms, and frequently covered in primarily small-cobble-size carbonate clasts. Bedrock terrace landforms are essentially a locally lower-elevation expression of the Umm er-Radhuma Formation limestone bedrock.

Wadi Channel (WC): The lowest and most fluvially active area, often demarcated by whitish-gray, rounded cobbles, boulders, and more vegetation. Wadi channel landforms consist of the present thalweg of active channel gravels (ACG), incising into and reworking underlying gravel terrace sediments. Sediment particles within the ACG range in size from pebbles (a few centimeters in diameter) to boulders (tens of centimeters in diameter) and reflect very high-energy transport during flash flood events.

Wadi Silts (WS): Pinkish-tan areas of very fine sand and silt above wadi channels, they often contain isolated lenses and a scattered cover of gravel. Wadi silts are interpreted as floodplain and slackwater deposits, comprising dominantly carbonate, with secondary siliclastic grains, ranging in size from silt to sand. Wadi silts stratigraphically overlie the GT unit and are occasionally preserved

Figure 3.6. Lower Wādī Sanā at the confluence with tributary Wādī as-Shumlyah, showing wide active channel gravels, expansive wadi silts, and a series of bedrock terraces characteristic of middle–lower Wādī Sanā. *Photograph by Eric Oches.*

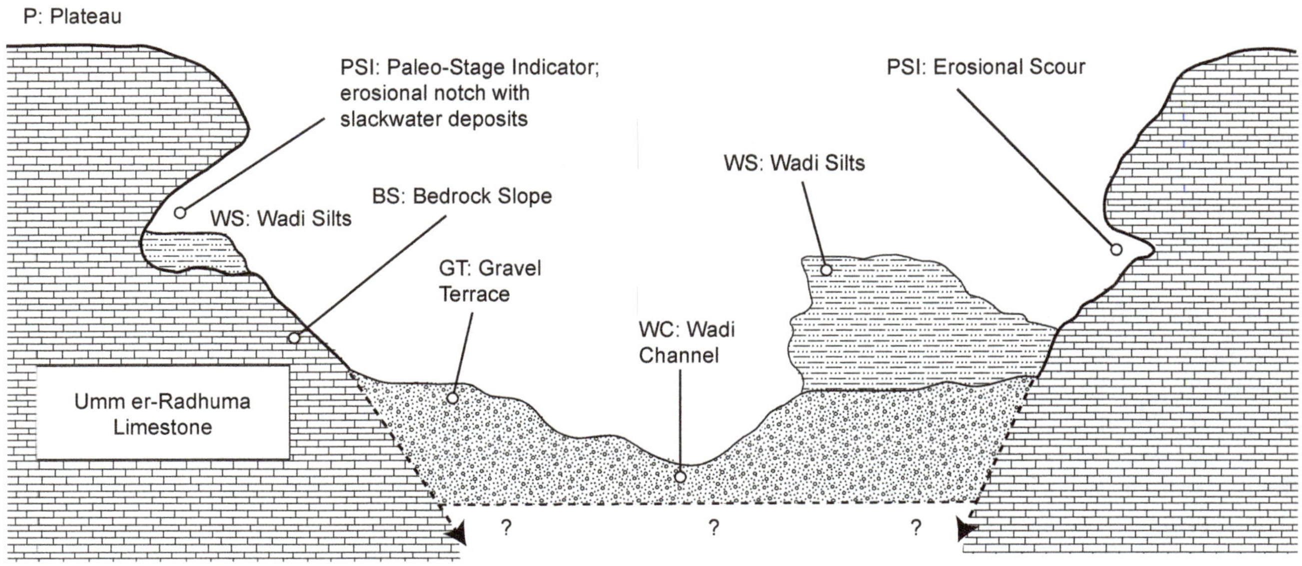

Figure 3.7. A generalized view of Wādī Sanā cross-section stratigraphy. *Illustration by Joshua Anderson and Clara Hickman.*

within erosional notches and dissolution cavities found throughout the wadi canyon walls. Wadi silts were separated into three major stratigraphic subdivisions to estimate the timing of deposition. The stratigraphically highest wadi silt sediments, found in small caves and PSI erosional notches along the canyon wall, are interpreted to reflect the upper limit of deposition and are classified as the upper wadi silt unit. Thick sequences of wadi silts found in wadi bottoms are divided into middle and lower wadi silt units, based on their stratigraphic positions.

Geochronology of Wādī Sanā Sedimentation

To reconstruct the timing and dynamics of fluvial aggradation and incision, and the ages of the geomorphic surfaces defined in Wādī Sanā, we measured, described, and sampled representative sediment profiles exposed through down-cutting of the present-day channel. We collected 24 samples for AMS radiocarbon dating and 10 sediment samples for optically stimulated luminescence (OSL) dating to provide age constraints on the longitudinal and vertical extent of wadi silts distributed throughout the reach. Radiocarbon samples consisted of 22 charcoal fragments collected either from archaeological hearths buried in wadi silts or as disseminated fragments from discrete levels within the wadi silts. We also collected two samples of terrestrial gastropod shells for radiocarbon dating (table 3.1). Ten OSL samples were collected from selected sediment profiles and provide ages when sediment profiles lack organic material suitable for radiocarbon dating (table 3.2). The University of Arizona AMS Radiocarbon Lab analyzed the majority of radiocarbon samples (indicated as AA lab numbers in table 3.1.) Woods Hole Oceanographic Institution and Beta Analytic analyzed other reported samples (shown as OS and Beta lab numbers in table 3.1.) Beta Analytic analyzed three of our reported samples (shown as Beta lab numbers in table 3.1.) Radiocarbon ages are calibrated using the OxCal version 4.2.4 radiocarbon calibration program (Bronk Ramsey 2009; Reimer et al. 2013; see https://c14.arch.ox.ac.uk/) and are reported as cal BP. OSL ages are calculated and reported to be directly comparable with calibrated radiocarbon ages. OSL analyses were performed at the Leibnitz Institute of Applied Geosciences, in Hannover, Germany, under the supervision of Dr. Manfred Frechen. Selected sediment profile locations along Wādī Sanā for which geochronological data are reported are shown in figure 3.9.

Radiocarbon samples were collected from interpreted slackwater deposits preserved within small caves and erosional notches along the bedrock slopes flanking the wadis and from wadi silt terraces that stratigraphically overlie the gravel terrace geomorphic surface. Seven OSL samples were collected from sand lenses within gravel terraces. Three OSL samples were collected from wadi silt sections to provide age constraints when no organic material was available for ^{14}C dating. The dominantly carbonate composition of wadi silt sediments, with only minor amounts of feldspar and quartz, increases the range of error of OSL dating results. At one section, a suite of samples consisting of charcoal and gastropod shell material was collected in combination with an OSL sediment sample from the same stratigraphic layer in an attempt to better understand the

Figure 3.8. Paleo-stage indicator represented by erosional notch in limestone wall, filled with overbank silts, representing highest level of wadi silt deposition during aggradational phase. *Photograph by Eric Oches.*

geochronological accuracy of the different sample materials and techniques.

Most Wādī Sanā geochronological samples were from discrete intervals in a single sediment exposure, which offered little chance to check internal consistency and reproducibility in age determination. However, we had one opportunity to compare radiocarbon age estimates on charcoal and shell plus an OSL age on fluvial sand from the same horizontal position within a single profile (samples 04WS-3a, b, c). At one locality (P2000-8), we collected multiple samples in stratigraphic superposition, allowing us to evaluate stratigraphic consistency and potentially assess sedimentation rates.

Comparing three dated samples from a narrow horizontal interval at Section 04-WS3 reveals that the charcoal (7428–7177 cal BP) and OSL (7040 ± 1100) samples are in general geochronologic agreement, while the gastropod shell (12,375–11,759 cal BP) yields a much older age estimate. Given the calcareous substrate, uptake of older carbon by the gastropod shell during growth is likely to be the source of this discrepancy. Alternatively, the shell might have been reworked from older sediments and does not represent the true sediment age. In both cases in which gastropod shells were radiocarbon dated, resulting age estimates were several thousand years older than expected, based on stratigraphic context and comparison with charcoal-based ages from similar sedimentary environments.

Profile P2000-8 is a stratigraphic exposure with about 6 m of wadi silts overlying about 1 m of gravel terrace sediments at the confluence of Wādī as-Shumlyah and Wādī Sanā, offering an opportunity to compare two radiocarbon ages and three OSL ages in stratigraphic superposition. Radiocarbon sample AA-38380 is charcoal from 0.25 m below the top of the wadi silt exposure and dates to 6413–6121 cal BP. Charcoal sample AA-38381 comes from wadi silt 1.7 m lower in the section and dates to 7287–6991 cal BP, suggesting an average sedimentation rate of about 1.5–3 cm per year. This high sedimentation rate possibly reflects a change in stream capacity, as Wādī as-Shumlyah converged with Wādī Sanā, resulting in the deposition of suspended sediment load and accumulation of wadi silts in this confluence region.

Table 3.1. Radiocarbon ages for paleohydrologic investigation in Wādī Sanā and Wādī ʿIdim (locations in UTM Zone 39 North WGS1984).

Map ID	Lab #	Year	Site	Off-site ID	Northing	Easting	Material
Wādī Sanā ^{14}C Samples Collected from Wadi Silts for Geologic Investigation							
7	OS16947	1998	Wādī Sanā profile	98-Hearth 13	1745744.6	336705.1	charcoal
41	OS16958	1998	Wādī Sanā -PSI sediment	Cave Site 1, 0.30–0.40	1742313.2	331838.4	charcoal
44	OS18691	1998	Wādī Sanā -PSI sediment	Cave Site 2	1742607.1	331745.9	charcoal
37	AA61078	2004	Wādī Sanā -PSI sediment	04-WS-4(1)	1735714.6	328875.5	charcoal
42	AA59756	2004	Wādī Sanā -PSI sediment	04WS-17(1)	1740346.4	328512.2	charcoal
43	AA59757	2004	Wādī Sanā -PSI sediment	04WS-17(4)	1740346.4	328512.2	charcoal
51	Beta-208494	2005	Wādī Sanā profile	05-WSX-16A2	1744568.6	334346.5	charcoal
52	AA59763	2004	Wādī Sanā profile	04WS-7(0.7)	1744605.3	337148.7	charcoal
53	AA59761	2004	Wādī Sanā profile	04-WS-6	1744471.4	334263.6	charcoal
54	AA38380	2000	Wādī Sanā profile	P2000-8A-0.25	1745663.3	336468.8	charcoal
63	OS16934	1998	Wādī Sanā profile	98-Hearth 10	1744738.1	336888.9	charcoal
65	AA61077	2004	Wādī Sanā profile	04-WS-18	1735240.2	333386.7	charcoal
67	Beta-208495	2005	Wādī Sanā profile	05-WSX-19W	1744629.1	334393.3	charcoal
69	AA59760	2004	Wādī Sanā profile	04WS-7(3.6)	1744605.3	337148.7	charcoal
70	OS16933	2/9/1998	Wādī Sanā profile	98-Hearth 2	1744597.4	337220.1	charcoal
71	OS16689	2/14/98	Wādī Sanā profile	98-WS2-+.45	1744579.3	337173.5	charcoal
74	AA59762	2004	Wādī Sanā profile	04-WS-8	1744553.9	334330.3	charcoal
78	OS16950	1998	Wādī Sanā profile	98-Hearth 14	1745744.6	336705.1	charcoal
79	OS16935	5/28/1998	Wādī Sanā profile	98-Hearth 16	1744561.7	336699.9	charcoal
83	AA38381	2000	Wādī Sanā profile	P2000-8A-1.95	1745663.3	336468.8	charcoal
85	AA59764	2004	Wādī Sanā profile	04WS-3(a)	1740963.9	329772.5	charcoal
96	AA59765	2004	Wādī Sanā profile	04WS-10(4b)	1749986.8	341256.4	charcoal
98	AA59768	2004	Wādī Sanā profile	04WS-3(b)	174096.9	329772.5	shell
97	Beta-208490	2005	Wādī Sanā profile	05-GBY02-1.0	1722792.6	324904.7	shell

^{14}C Year BP	Cal BP 2-Sigma	Comments
680 +/- 35	**684–559**	13, Khuzmah, charcoal in vial split for ^{14}C, date seems incompatible with stratigraphic position, resample desirable.
4610 +/- 45	**5568–5067**	Wādī Sanā Cave I, 0.30–0.40 below top of silt, LRC 1090.
4800 +/- 60	**5651–5327**	Wādī Sanā Cave II, layer 0.25 m below top of silt, LRC 1093.
4545 +/- 45	**5435–5046**	Wādī Sanā, uppermost wadi silt in PSI infilling.
4633 +/- 40	**5569–5296**	Wādī Sanā, PSI wadi silts, 20 cm below surface, youngest silt deposition.
4721 +/- 56	**5585–5321**	Wādī Sanā, PSI wadi silts, 90 cm below surface, youngest silt deposition.
5320 +/- 50	**6271–5948**	Charcoal in channel fill clay (paleochannel), 50 cm below surface, Wādī Sanā near cattle skulls.
5329 +/- 42	**6269–5992**	Wādī Shumlyah, uppermost burned horizon in 98-WS3 profile, 0.7 m below surface.
5402 +/- 42	**6294–6020**	Top edge of paleochannel filling.
5485 +/- 64	**6413–6121**	Khuzma as-Shumlya, close to top of profile, dates end of sedimentation/level to which modern surface has eroded.
5750 +/- 45	**6659–6443**	Shumlya Hearth 10, 22/II LRC 1063, no further charcoal available.
5765 +/- 45	**6665–6452**	Wādī Sanā tributary, middle of wadi silt deposition.
5800 +/- 50	**6731–6488**	Charcoal from hearth adjacent to paleochannel, Wādī Sanā , near cattle skulls, about 20 cm below surface.
5842 +/- 43	**6772–6508**	Wādī Shumlyah, lower exposed silt in 98-WS3 profile, 3.6 m below surface.
5870 +/- 45	**6792–6561**	Shumlyah Hearth 2, LRC 1062.
5880 +/- 55	**6851–6547**	98 WS 2; +45 cm; sampled 4/98.
5970 +/- 72	**6995–6645**	Basal infill of channel.
6070 +/- 40	**7151–6793**	Shumlyah Hearth 14, 98/14 in silt section northeast of Khuzma as-Shumlya.
6080 +/- 55	**7156–6795**	Shumlyah Hearth 16, LRC 1082.
6246 +/- 58	**7287–6991**	Khuzma as-Shumlya, terrace profile 2000-8; 1.95 m below top of silts (modern surface).
6387 +/- 61	**7428–7177**	Wādī Sanā , charcoal in hearth near base of wadi silts (onset of silt deposit).
9252 +/- 52	**10,565–10,268**	Wādī Sanā , single large charcoal fragment from lower third of wadi silts (1.4 m above base, 4 m below top).
10254 +/- 55	**12,375–11,759**	Wādī Sanā , shells from base of wadi silts (onset of silt deposit).
10220 +/- 40	**12,097–11,768**	Aquatic gastropod shell, 1.0 m below surface in tufa-capped marl, about 500 m north of Ghayl bin Yumain *(Melanoides tuberculata).*

Table 3.1. Radiocarbon ages for paleohydrologic investigation in Wādī Sanā and Wādī ʿIdim (locations in UTM Zone 39 North WGS1984). *(continued)*

Map ID	Lab #	Year	Site	Off-site ID	Northing	Easting	Material
Wādī ʿIdim ^{14}C Samples Collected to Date Tufa Formation and Sediment Aggradation							
29	Beta-208493	2005	Wādī ʿIdim tufa	05-WIY-03D	1735536.3	273981.8	charcoal
49	AA38384	2000	Wādī ʿIdim tufa	00-WI-10	1736704.5	274127.5	charcoal
50	AA38385	2000	Wādī ʿIdim tufa	00-WI-12	1735785	274020	charcoal
86	AA38383	2000	Wādī ʿIdim tufa	00-WI-9	1737632.3	274694.3	charcoal
87	AA38382	2000	Wādī ʿIdim tufa	00-WI-5	1739629.6	275496.7	charcoal
35	Beta-208492	2005	Wādī ʿIdim tufa	05-WIY-01-915	1739386.2	275483.6	charcoal
75	Beta-208491	2005	Wādī ʿIdim tufa	05-WIY-01-180	1739386.2	275483.6	charcoal

Table 3.2. Optically stimulated luminescence ages on fluvial sediments in Wādī Sanā (locations in UTM Zone 39 North WGS1984).

Map ID	Field Number	Northing	Easting	Material	Depth below Ground Surface	Elevation (meters)	Lab Number	U (ppm)	Th (ppm)
99	04WS-3c	1740963.9	329772.5	sandy silt	4 m	726	658	1.91 ± 0.02	4.79 ± 0.05
100	04WS-11	1745674.8	336174.5	sand lens	6.5 m	694	659	1.75 ± 0.02	2.34 ± 0.02
101	04WS-13	1750173.9	341411.3	sand lens	4.5 m	664	660	1.56 ± 0.02	1.17 ± 0.01
102	04WS-14	1753665.3	343185.3	sand lens	1.5 m	655	661	1.75 ± 0.02	1.43 ± 0.01
103	04WS-15	1762582.4	341802.4	sand lens	6.5 m	620	662	1.89 ± 0.02	2.29 ± 0.02
104	04WS-19	1742299.3	333711.9	sand lens	4 m	718	663	1.77 ± 0.02	2.57 ± 0.03
105	04WS-22-1	1745674.6	336183.2	sandy silt	8 m	710	664	1.91 ± 0.02	3.98 ± 0.04
106	04WS-22-2	1745674.6	336183.2	sand lens	9 m	710	665	2.06 ± 0.02	2.54 ± 0.03
107	04WS-22-3	1745674.6	336783.2	sand lens	9.5 m	710	666	2.02 ± 0.02	2.11 ± 0.02
108	04WS-23	1740908.8	329736.4	sand lens	4 m	707	667	1.83 ± 0.02	3.32 ± 0.03

^{14}C Year BP	Cal BP 2-Sigma	Comments
4400 +/- 70	**5285-4848**	Charcoal in sand lens, 2.5 m below surface of tufa exposure, southernmost extent of tufa mounds in Wādī ʿIdim.
5280 +/- 52	**6191–5928**	Wādī ʿIdim, charcoal embedded in tufas from southern extent of fossil spring deposits.
5288 +/- 52	**6202–5932**	Wādī ʿIdim, charcoal embedded in tufas from southern extent of fossil spring deposits.
6586 +/- 56	**7573–7424**	Wādī ʿIdim, charcoal embedded in tufas from northern extent of fossil spring deposits.
6859 +/- 57	**7825–7590**	Wādī ʿIdim, charcoal embedded in tufas from northern extent of fossil spring deposits.
4520 +/- 40	**5310–5046**	Charcoal in hearth in wadi sand/silt with interbedded marl and tufa, 9.15 m above wadi cobbles, northerly range of tufa, Wādī ʿIdim.
5970 +/- 50	**6930–6676**	Charcoal in cross-bedded sand and gravel, interbedded with marl and tufa, 1.8 m above wadi cobbles, northerly range of tufa, Wādī ʿIdim.

K (%)	H2O (%)	Cosmic (μGy/a)	Dose Rate (Gy/ka)	Paleodose (Gy)	IRSL Age (year BP)	Environmental Setting
0.71 ± 0.02	3.5 ± 1.5	144 ± 7	2.30 ± 0.18	16.2 ± 2.2	7040 ± 1,100	Sand in channel gravel within lower wadi silt, about 20 cm below 04-WS3 Hearth (^{14}C).
0.42 ± 0.01	2 ± 1	122 ± 6	1.64 ± 0.13	17.4 ± 2.0	10,600 ± 1,500	Sand lens in gravel terrace, 1 m below contact with overlying wadi silt.
0.21 ± 0.01	2 ± 1	138 ± 7	1.20 ± 0.10	28.0 ± 3.1	22,400 ± 3,300	Sand lens in gravel terrace, 1.5 m below contact with wadi silt; about 3 m below 04WS10-4 (^{14}C).
0.19 ± 0.01	2 ± 1	169 ± 9	1.33 ± 0.12	18.6 ± 3.2	14,000 ± 2,700	Sand lens in gravel terrace, 1.0 m below contact with overlying wadi silt.
0.38 ± 0.01	2 ± 1	120 ± 6	1.65 ± 0.14	19.2 ± 2.1	11,700 ± 1,600	Sand lens in gravel terrace, 0.5 m below contact with overlying wadi silt; mouth of Wādī Ḥimayrī.
0.48 ± 0.01	2 ± 1	144 ± 7	1.76 ± 0.14	11.5 ± 2.3	6500 ± 1,400	Sand lens in gravel terrace, 1.0 m below contact with overlying wadi silt; mouth of small tributary to Wādī Sanā .
0.73 ± 0.02	3.5 ± 1.5	111 ± 6	2.19 ± 0.16	15.6 ± 2.4	7140 ± 1,220	P2000-8 basal wadi silt.
0.52 ± 0.01	3.5 ± 1.5	106 ± 5	1.89 ± 0.15	17.4 ± 1.2	9230 ± 970	P2000-8 sand at transition between overlying wadi silt and underlying gravel terrace.
0.41 ± 0.01	2 ± 1	101 ± 5	1.69 ± 0.14	18.9 ± 1.8	11,200 ± 1,400	P2000-8 sand lens in upper gravel terrace underlying wadi silt.
0.53 ± 0.01	2 ± 1	144 ± 7	1.93 ± 0.15	17.7 ± 4.2	9160 ± 2,290	Sand lens, about 0.5 m below wadi silt/gravel terrace contact.

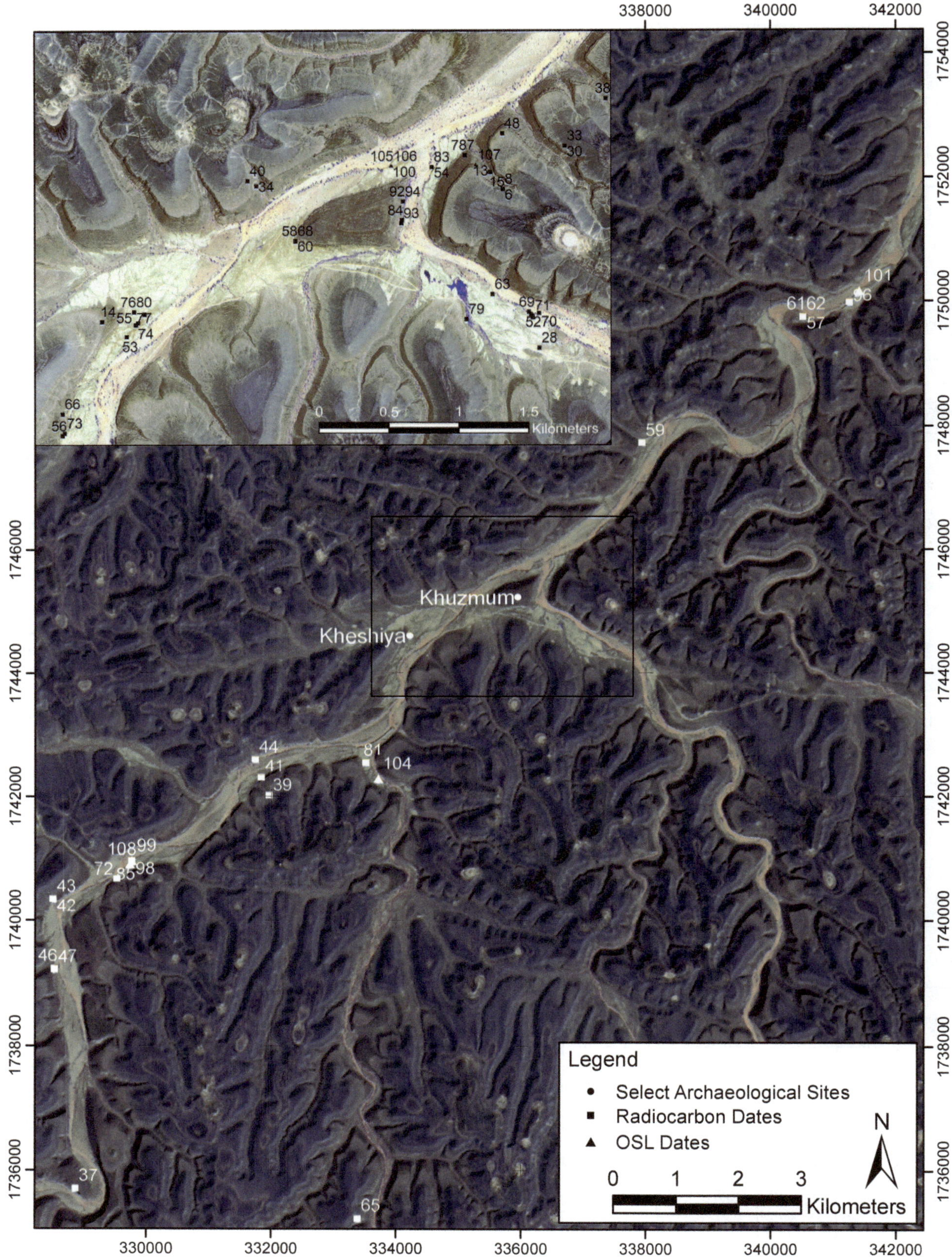

Figure 3.9. Map of radiocarbon and optimally stimulated luminescence (OSL) date sample locations along Wādī Sanā (UTM Zone 39 North WGS1984). Additional details provided in tables 3.1 and 3.2 and in chapter 18. *Illustration by Michael Harrower.*

At the base of Section P2000-8, luminescence samples were collected from three levels straddling the WS/GT contact: 04-WS22-3 is a sand lens in the uppermost gravel terrace and dates to 11,200 ± 1400 years BP; 04-WS22-2 is fluvial sand at the transition between overlying wadi silt and the underlying gravel terrace and yields an OSL age of 9230 ± 970 years BP; 04-WS22-1 is basal wadi silt, giving an OSL age of 7140 ± 1220 years BP. The stratigraphic consistency among these three sample ages suggests that gravel terrace deposition ended sometime after 11,000 years ago and that wadi silt accumulation commenced sometime around 9,000 years ago at this site.

Geochronological results from Wādī Sanā show the spatial distribution of age estimates along the studied reach. Figure 3.10 presents a frequency distribution of those ages, which are further discussed in the Bayesian model in chapter 18. Radiocarbon and OSL results are summarized below from the lowest (earliest) to the highest (latest) stratigraphic levels. The succession of stratigraphic units progresses from the gravel terrace up through the lower and middle wadi silt terrace, capped by paleo-stage indicator wadi silts stranded in erosional notches and dissolution cavities on the bedrock slope high above the present wadi floor.

Fluvial sediments exposed through incision reveal an abrupt contact between the gravel terrace (GT) and wadi silt (WS) stratigraphic units throughout the study area. Dating results from six OSL samples taken from sand lenses in the upper part of exposed GT sediment ranged from 22,400 ± 3300 (04-WS-13) to 9160 ± 2290 (04-WS-23) years BP. A younger GT age of 6550 ± 1400 (04-WS-19) years BP was measured on a sand lens in the upper terrace gravels in a small tributary to Wādī Sanā. It is likely that higher energy flow persisted later in small basin, steeper-gradient tributaries well after fine-grained deposition dominated in the

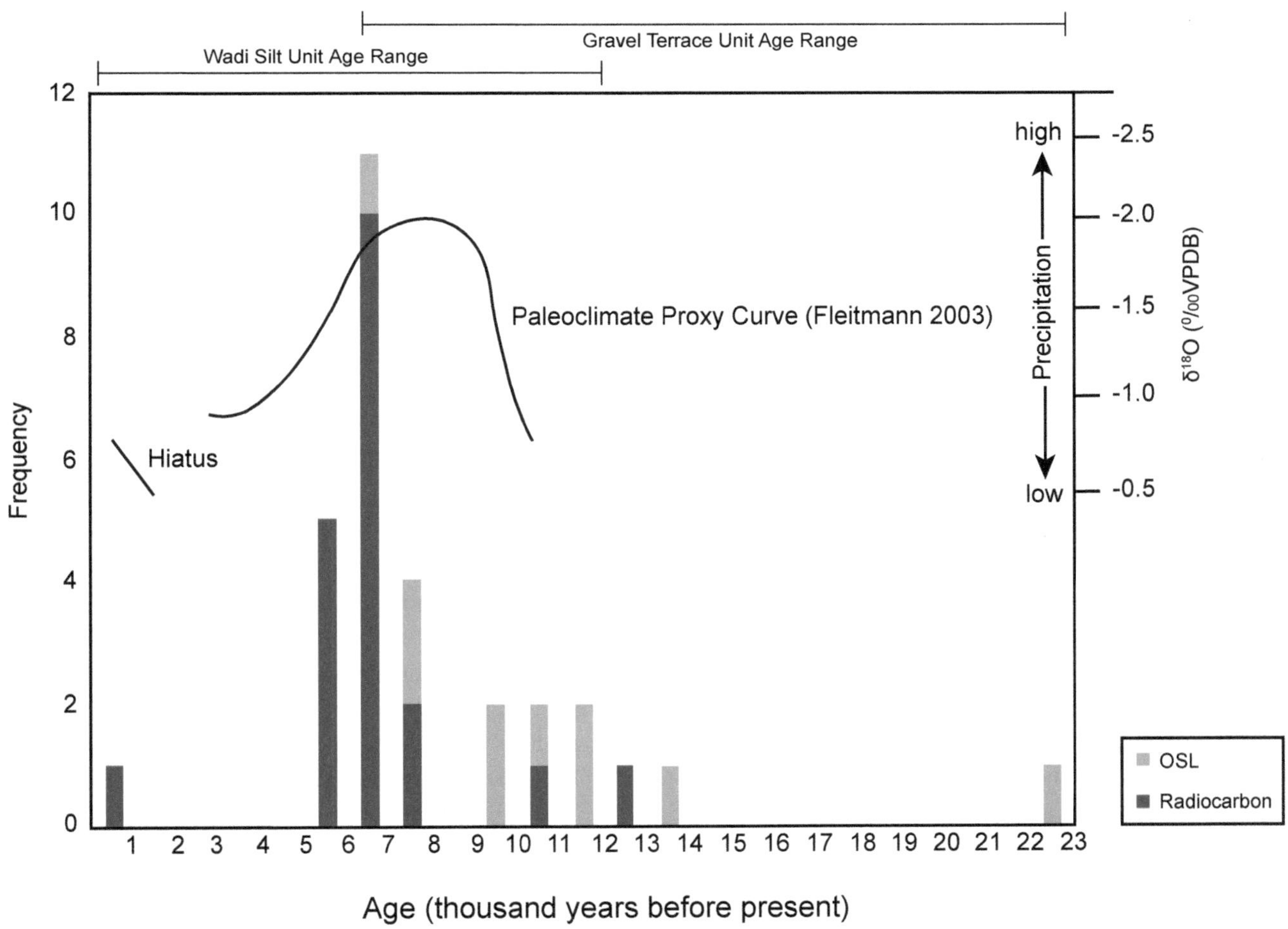

Figure 3.10. Graph showing frequency distribution of ^{14}C ages measured on charcoal and OSL ages measured on silts from wadi silts and gravel terrace deposits in Wādī Sanā. Superimposed on age distribution is a paleoprecipitation proxy curve from Fleitman et al. 2003. *Illustration by Joshua Anderson and Clara Hickman.*

Figure 3.11. Natural exposure in Wādī Sanā showing Holocene wadi silts overlying Late Pleistocene gravel terrace sediments. *Photograph by Eric Oches.*

main Wādī Sanā channel. OSL samples from basal wadi silts yield ages ranging from 9230 ± 970 (04-WS2-22-2) to 7040 ± 1100 (04-WS-3c) (table 3.2). Two charcoal samples from the lower WS yielded calibrated ^{14}C age estimates of 10,565–10,268 (04-WS-10-4b) and 7428–7177 (04-WS-3a) cal BP. Together, these OSL and ^{14}C ages indicate that high-energy reworking of braided channel gravels was the dominant process in Wādī Sanā from the latest Pleistocene until about 9,000 years ago, after which slackwater deposits accumulated on the gravel terrace surface. The contact between the GT and lower WS, therefore, represents the beginning of wadi silt deposition at about the Pleistocene–Holocene transition (figure 3.11).

Wādī Sanā sediment profiles interpreted in the field to represent the middle to upper WS layers were dated, and ages are correlative throughout the study reach of Wādī Sanā. Fourteen charcoal samples were collected from middle wadi silts, and ^{14}C age estimates ranged from 7287–6991 (P2000-8A-1.95) to 6271–5948 (05-WSX-16A2) cal BP. Charcoal from a hearth buried in middle wadi silts was dated as 684–559 cal BP (98-Hearth-13). That age is considered anomalous and could be related to sub-modern campfire intrusion into the surface sediments or modern contamination of a shallowly buried archaeological horizon.

Five charcoal samples were taken from upper wadi silts throughout the middle Wādī Sanā reach and range in age from 5585–5321 (04-WS17 [4]) to 5435–5046 (04-WS4) cal BP. These samples were in remnant sediments stranded in dissolution notches or small caves high on the limestone walls of Wādī Sanā and are considered paleo-stage indicators. (Refer to figures 3.7, 3.8, 3.12, and 3.13.) Voids in the limestone above the PSI elevation did not contain sediments, suggesting that these samples represent the upper elevation limit of Middle Holocene slackwater deposition, prior to the shift to incision and down-cutting by the wadi

Figure 3.12. Dissolution notches and small sediment-filled caves serve as paleo-stage indicators in Wādī Sanā. *Photograph by Eric Oches.*

channel. Wādī Sanā therefore represents a rather unusual environment in which the late-stage, high-water level is so clearly preserved in multiple locations, represented by erosional notches often infilled with alluvial sediments.

Considered in the context of stratigraphy and geomorphic setting, radiocarbon and luminescence ages allow for the assessment of the timing and mechanism of wadi infilling and incision from the Late Pleistocene through the Holocene. Four periods of sedimentation are indicated, beginning with Late Pleistocene through Early Holocene deposition of the gravel terrace cobbles. For deposition, ages range from about 22,000 years ago to as recent as about 9,000 years ago. The younger ages/latest stages of deposition are probably associated with higher-energy environments in the mouth of small tributaries to the main Wādī Sanā channel, which may have continued to rework the coarse cobbles, while the main channel had transitioned to sand and silt accumulation.

Limited age determinations on basal wadi silt units indicate that this unit was actively accumulating by about 10,000 years ago and continued through the Middle Holocene. Middle wadi silts begin above the zone of interbedded channel gravels, sands, and basal wadi silts. Deposition of the middle wadi silts unit began about 7,300 years ago and continued until about 5,900 years ago, which appears to be the maximum age for the wadi silt surface in the main wadi channel. Upper wadi silts, which are represented by the remnant silts preserved in the paleo-stage indicators—erosional and dissolution notches along the bedrock slope, high above the present wadi floor—date to as recent as 5435–5046 cal BP (04-WS4), the youngest ages determined for wadi silt deposition in Wādī Sanā. Shortly after that time, the system appears to have transitioned rather abruptly to an incision mode, and sediments have been eroded and transported out of the upper and middle reaches of Wādī Sanā for the last roughly 5,000 years.

Figure 3.13. A semi-continuous, sub-horizontal erosional notch filled with high-water wadi silts serves as a paleo-stage indicator in Wādī Sanā. *Photograph by Eric Oches.*

Springs, Wetlands, and Lacustrine Sediments

The middle reach of Wādī ʿIdim is currently marked by numerous villages and date palm oases supported by spring-fed stream flow (figure 3.14). A limited survey of the approximately 25 km length between the villages of Sāh and Rāwik revealed numerous fossil springs in the form of tufa mounds and interbedded fossiliferous sand and lacustrine sediments. In the 2000 and 2005 field seasons, we recorded GPS locations of tufa mounds and sampled materials for radiocarbon and luminescence dating to reconstruct the timing and extent of former spring discharge and wetland expansion. Throughout the surveyed extent of middle Wādī ʿIdim, we identified 20 discrete outcrops of calcareous tufa formations and fossiliferous marl sediments interbedded with cross-bedded sands interpreted as having formed in lacustrine or wetland environments (table 3.3). Radiocarbon ages of charcoal recovered from sand lenses within tufa mounds range from about 7825 to 4850 cal BP, suggesting that enhanced spring discharge and associated tufa formation in Wādī ʿIdim occurred during the same time period as maximum sedimentation of wadi silt deposits in Wādī Sanā.

The interval of active spring discharge and associated lacustrine environments suggests a period of enhanced groundwater recharge from regional precipitation, which declined abruptly around 5,000 years ago. In a recent reinterpretation of regional lacustrine records, Enzel et al. (2015) suggested that many previously reconstructed lake systems in Southern Arabia were actually shallow, closed-basin wetlands that would have required significantly less hydrologic inputs to sustain. Enzel et al. (2015) suggest that a small intensification of the southwest summer monsoon, rather than a strong shift in ITCZ position and ISM strength, can explain the shift in paleohydrologic conditions represented in Southern Arabian Holocene sedimentary records. We suggest, however, that the extent and duration of groundwater discharge represented by the numerous fossil spring deposits in Wādī ʿIdim and upper Wādī Sanā would have required considerable enhancement in precipitation and associated groundwater recharge to support the regional expansion of spring-fed wetlands and shallow lakes in the RASA study area.

Figure 3.14. Ghayl ʿUmar springs in Wādī ʿIdim, 2004. *Photograph by Michael Harrower.*

Table 3.3. Wādī ʿIdim tufa survey: GPS coordinates of tufa mounds and related sediments, 2000, 2005 field season (locations in UTM Zone 39 North WGS1984).

Field Number	Northing	Easting	Comments
Sāh	1723652	269555	Sāh village - southern (upstream) limit of modern palm groves
Ghayl ʿUmar	1734512	270749	Approx. northern (downstream) limit of modern palm groves
00-WI-01	1738828	275411	tufa mound directly below Munayder site
00-WI-02	1738828	275411	Same outcrop as 00-WI-01, opposite side (E-side; 01 is W-side)
00-WI-03	1739713	275468	Northern limit of tufas observed in 2000.
00-WI-04	1739630	275497	Sample of banded tufa collected
00-WI-05	1739630	275497	Samples of degraded shells & charcoal (same site as 00-WI-04)
00-WI-06	1739630	275497	More shells & charcoal (same site as 00-WI-05)
00-WI-x1	1739343	275530	No samples collected; no field number assigned
00-WI-07	1739304	275491	snail shells collected
00-WI-08	1738635	275338	snails and charcoal collected
00-WI-09	1737633	274694	tufa with sample of charcoal and snails
00-WI-H2O	1732696	270497	present-day spring on bedrock slope near Ghayl ʿUmar
00-WI-10	1736704	274128	shells, charcoal in marl, channel sands, gravel
00-WI-11	1736186	273965	Thick tufa mound, large spherical CaCO3 forms, against BS
00-WI-12	1735786	274020	shells, charcoal in marl, channel sands, gravel
05-WIY-01	1737542	275466	interbedded lacustrine/marsh sediments, wadi silts, marl, tufa
05-WIY-02	1739776	275506	tufa mound with sand lens, snails, charcoal; northern tufa limit
05-WIY-03	1735536	273982	extensive tufa+marl across tributary mouth; southern tufa limit
05-WIY-04	1735799	274037	tufa+marl+sediment at north end of tributary mouth tufas
05-WIY-05	1735853	274053	small tufa ledge capping wadi gravels across tributary mouth
05-WIY-06	1735830	274055	small laminated tufa dam across gulley channel
05-WIY-07	1736123	273970	extensive tufa complex, attached to bedrock -- spring source?

Channel History at Kheshiya

Although flowing water was concentrated mainly in active wadi channels and discharging from springs, we observed evidence of standing water on the wadi silt surface adjacent to the primary channel in various locations, which we interpret as oxbow lakes and wetlands. During the 2005 field season, we identified one such feature within the wadi silts adjacent to the cattle skull site at Kheshiya (SU151-1; see chapter 10), in central Wādī Sanā, in the form of a sinuous lens of organic-rich, laminated sandy silt. This traceable sedimentary feature is approximately 30 m wide and extends in two segments: the WSX section can be traced on the ground and in QuickBird satellite imagery for approximately 600 to 800 m; the WS2X section, about 1.5 km northeast, can be followed on land and in satellite imagery for about 200 m. Charcoal samples have been dated from the top and base of these dark, organic-rich sediment bands, yielding radiocarbon ages of 5400 ± 40 and 5970 ± 70 ^{14}C BP, respectively. We interpret both as abandoned channel fills from the mid-Holocene period of increased precipitation and suggest that the paleochannels would have existed as an available water source for human subsistence during an approximately 600-year time period. Abundant archaeological materials are identified on and within the expansive wadi silts that the channel flowed across, suggesting that people were living and processing food near a quiet channel, oxbow lake, or wetland close to the active wadi flow some short distance to the south.

Climatic versus Structural Controls on Wādī Sanā Sedimentation

Significant questions arise as to the origins of the fine-grained sediment that infilled Wādī Sanā and the relationships between climatic conditions, sediment sources, and transport mechanisms. Through field reconnaissance and map analysis, we ruled out the upland terraces and plateaus flanking the main Wādī Sanā channel as the source of fine-grained sediment. Adjacent uplands are covered in thin desert pavements overlying Jeza and Umm er Radhuma limestones, offering only a thin cover of fine-grained sand and silt that could have been transported by wind and water into the incised channels of Wādī Sanā and its tributaries. Field observations upstream of the study reach, however, showed significantly different geomorphology and bedrock lithologies south of the village of Ghayl bin Yumain. In that region, date palm farming and other agricultural activities are presently carried out across a broad, uplifted sediment-filled basin surrounded by highlands consisting of erodible limestone, shales, marls, and evaporites of the Rus and Jeza Formations. Normal faulting separates the broad low-relief basin of upper Wādī Sanā from the deeply incised canyon at the head of lower Wādī Sanā. The canyon represents a structural restriction that was occasionally blocked by a tufa-formed dam, creating an expansive, shallow, spring-fed lake and wetlands stretching south to Ghayl bin Yumain.

The Rus and Jeza Formations, summarized by Beydoun (1966) as sources of thick wadi infilling throughout the Southern Jol, are located upstream of the studied reach of Wādī Sanā, and outcrops of those units are separated from the lower drainage system by a series of normal faults. Climate change and geologic structures separating the upper and lower sub-basins are proposed here as controls on the geomorphology and paleohydrology of Wādī Sanā.

A conceptual model describing the hypothesized evolution of Wādī Sanā throughout the Early Holocene wet phase and Late Holocene dry phase is illustrated in figure 3.15. In this model the upper and lower sub-basins, which are separated by normal faulting, are hydrologically connected during the wetter Early Holocene phase, while the sub-basins operate somewhat independently during

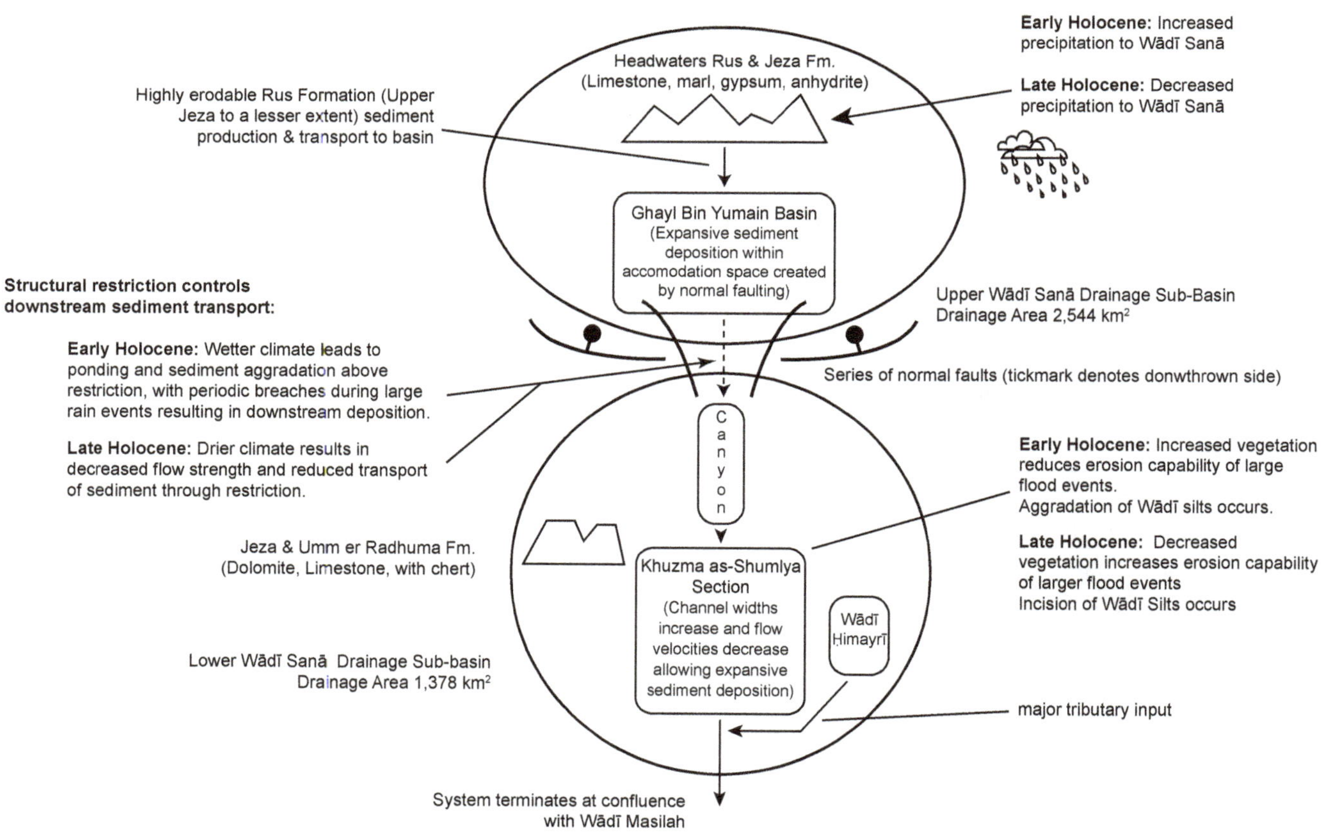

Figure 3.15. A conceptual model to explain sediment production, erosion, and transport during the Early and Late Holocene. *Illustration by Joshua Anderson and Clara Hickman.*

the drier climatic phase of the Middle and Late Holocene. As discussed previously, the Early to Middle Holocene throughout Southern Arabia is documented to have had greater seasonal distribution and total annual precipitation than occurs presently (e.g., Fleitmann et al. 2007). Increased stream flow throughout the structurally distinct upper drainage sub-basin would have increased sediment production and transport.

The structural restriction created by normal faulting at the boundary between the upper and lower sub-basins of Wādī Sanā caused internal drainage and aggradation of sediments into a fault-bounded region located at the terminus of the upper sub-basin. Increased precipitation, sufficient to enhance stream flow above a critical topographic threshold, could have caused a breach of the sediments damming the restriction between the upper and lower sub-basins, resulting in an episodic connection of upper and lower Wādī Sanā. Massive volumes of fine-grained sediments would have subsequently been transported downstream through the narrow canyon section and onward to areas where the present silt terraces are observed. The narrow canyon system at the normal faults would have initially received the large flood discharges with high sediment yield and relatively high flow velocities. With increased velocities throughout the canyon sections, little to no sediment deposition occurred until the cross-sectional areas increased farther downstream. Areas where the Wādī Sanā channel is wide and velocities decreased resulted in the deposition of the fine-grained sediment load. An unusual situation exists within the canyon of upper-middle Wādī Sanā, where the archaeological site Manayzah, dating to 9200–7600 cal BP, would have been expected to be washed away by the torrential floodwaters flowing past the site. It is possible that the indurated calcite crust capping the site, formed by springs flowing out of the adjacent canyon walls, may have protected the slightly elevated site at the channel margin from erosion and removal by floodwaters.

The Wādī Sanā–Wādī as-Shumlyah confluence at the Khuzma is a wide, flat segment where sediment deposition would have occurred whenever the sediment-laden source waters breached the upstream topographic barrier, flushed through the narrow canyon, and distributed the sediment load across the wide channel bottoms in the middle and lower reaches of Wādī Sanā. Numerous weak paleosols observed throughout exposed wadi silt profiles suggest that sediment aggradation occurred in repeated flooding events interrupted by periods of stabilization.

The geochronological record of alluvial sediment accumulation terminates at approximately 5000 cal BP. Paleoclimate proxy data illustrated in figure 3.10 document a gradual shift, beginning around 8000 years BP, toward current arid conditions in the Middle Holocene (Fleitmann et al. 2003, 2007). A shift in hydrologic regime toward drier conditions would eliminate the ability of the fluvial system to surpass the critical precipitation threshold necessary to erode and transport large volumes of fine-grained material downstream. Vegetation cover would have also decreased across the entire basin, leading to greater erosivity of the fluvial sediments previously deposited.

Middle to Late Holocene reduction in precipitation and stream flow would have lowered rates of sediment production and transport in the upper sub-basin. The decrease in stream roughness, due to a reduction of vegetation cover and drier soils would have led to incision during typical flow events and rapid erosion and incision during low-frequency, high-magnitude flood events. We hypothesize that those drier regime flow events caused incision and produced the dissected floodplain geomorphology presently observed throughout the lower Wādī Sanā sub-basin.

In summary, we suggest that the structural geologic features, coupled with different Early and Middle Holocene precipitation regimes, controlled sediment production, erosion, and downstream transport from the source materials in the upper Wādī Sanā sub-basin to the broad slackwater deposits filling the channels of the lower sub-basin. The upper drainage basin contains the appropriate source lithologies, and the normal faulting formed a physical barrier between the upper and lower basins, creating accommodation space for temporary sediment accumulation within the basin surrounding Ghayl bin Yumain. Increasing Early Holocene seasonal rainfall mobilized the stored sediment, flushing it downstream through the fault block barrier. The same sediment accumulated throughout the middle and lower reaches of Wādī Sanā as slackwater deposits along the active wadi channel.

Paleovegetation

While fossil pollen data preserved in lake catchments and bogs certainly remains one of the best understood and most widely used proxies of paleovegetation (e.g., Birks and Deacon 1973; Bottema 1991; Davis et al. 2005; Gasse 2002; Lézine 2009), the Arabian Peninsula has had few such catchments since the Early Holocene (Davies 2006; Inizan et al. 1997; Parker and Goudie 2008; cf. Enzel et al. 2015; Hoorn and Cremaschi 2004; Lézine et al. 2002). In Arabia, regional vegetation histories must also rely on less conventional proxy data records. With records that are spatially and temporally discontinuous, paleovegetation reconstruction may be sometimes localized, sometimes regional, and sometimes an inferential interpolation among records. A RASA-generated vegetation history of Wādī Sanā in the Southern Jol draws upon broader regional reconstructions and relies on locally available proxies with uneven chronological and spatial resolution.

The nearest terrestrial palynological series is from the Al-Hawa lake, one of a handful of desert oases across the Early Holocene Arabian Peninsula (Parker et al. 2006; Wright and Roberts 1993). Al-Hawa lies in the Ramlah as-Sabʿatayn basin of the Jawf–Hadramawt drainage (E 701665, N 1755093, UTM Zone 38 North) and was last filled episodically during the period 12,000 to 7500 cal BP (Lézine et al. 2007). Interpreted in the context of multiproxy lacustrine sedimentology, isotopic ratios, and aquatic fauna, fossil pollen show a continuous presence of semiarid, steppe-like vegetation in the lake's immediate vicinity. Lake accumulations from enhanced runoff supported reed swamps with enhanced mosses and ferns (Lézine et al. 2007:247). Regional pollen rain from arid tropical highlands includes *Podocarpus*, currently absent from the Arabian Peninsula, and *Juniperus*, *Myrica*, and *Erica* types almost certainly transported long distances from relict forests (Lézine et al. 2007:246). This forest, the semi-evergreen Afro-montaine woodland habitat of *Juniperus procera*, scarcely survives in Arabia today, but its range can be estimated by a proxy plant, the shrub *Justicia areysiana. J. areysiana* today grows in geographically isolated populations in highland southern Yemen and southern Oman. DNA analysis shows significant genetic distances between *J. areysiana* stands and suggests a climate-induced retreat to highland isolates some 0.8–1.8 million cal BP (Meister et al. 2006), long before the Early Holocene Al-Hawa lake formed.

If the evergreen Afro-montaine woodlands had mostly retreated to high ground and fragmented during the Pleistocene era, fragmentation of drought-deciduous monsoon forest of the lower elevations is more recent. *Anogeissus dhofarica* today extends across the slopes of the Dhofar escarpment and probably also once covered the plateau grasslands, now anthropogenically transformed for grazing cattle (Kürschner et al. 2004:586–87; Sale 1980). Highly fragmented relicts of the *Hybantho durae–Anogeissetum* association in the Fartak Mountains of Mahra and the Jebal Samhan in Dhofar attest to its once broader expanse (Kürschner et al. 2004:593). Genetic diversity within patches of *Anogeissus dhofarica* is relatively low, and the genetic distance between highly isolated patches suggests that the forest was recently contiguous and that modern fragmentation stems from Holocene changes in monsoon intensity (Oberprieler et al. 2009), including the same 6 kya decline in annual rainfall that dried the Al-Hawa lake.

This same well-documented and widespread weakening of the monsoon in the Southern Jol also must have affected relict woodland patches, whose remnants are still evident with the *Anogeissus benthii* trees in the wadis of Hadramawt and eastern Shabwa today (Oberprieler et al. 2009:41). Thus one can reconstruct from highly fragmented relicts a once more widespread drought-deciduous woodland of *Anogeissus benthii.* Weakened monsoons could produce a rapid vegetative transformation to grassland on thin soil (Hildebrandt and Eltahir 2008) or extend the semi-desert marked by *Boswellia sacra* (Hildebrandt et al. 2007; Oberprieler et al. 2009). While the immediate vicinity of the Al-Hawa lake was continuously a semiarid steppe, the mountain ranges of the Southern Jol had variable cover, the parameters of which were susceptible to slight changes in rainfall and to the impacts of human pastoralists.

Wādī Sanā Proxies

The local impacts of climatic changes and human landscapes were probably variable and need to be reconstructed through local proxies. While local effects may have been quite dramatic, they may nevertheless have been changes in density rather than in character of vegetation. Localized changes in Wādī Sanā vegetation may not have registered in a faraway lake catchment like Al-Hawa. The RASA Project investigated multiple local proxies from Wādī Sanā and Wādī ʿIdim. These proxies—charred plant remains such as seeds and charcoal, phytoliths, pollen, and desiccated plant remains—have been preserved by different taphonomic factors with variable spatial and chronological parameters.

Charred Plant Remains—Hearths, Burned Dung Mats

To recover charred plant remains, the RASA team relied on an Ankara-style barrel flotation tank provided with the support of the machine shop and camp administration at Canadian Nexen Petroleum Yemen's central processing facility. In 1998 they built the tank to our specifications and thereafter adjusted it. We faced logistical restrictions on water availability, on available labor for flotation, and on sediment transport to the flotation tank. With such restrictions, we decided to target deposits, relying on geomorphological insights into sedimentation processes and site formation to select samples (table 3.4). RASA excavators selected hearths and ashy layers as most likely to yield charred plant assemblages in primary cultural context. Excavations at sites described in later chapters—Manayzah, Khuzmum rockshelters, Kheshiya and Khuzmum platforms and trilith hearths—revealed many layers and features targeted for flotation. In survey and testing elsewhere, hearths and ashy layers also show clearly in natural terrace profiles. Many sediment flotation samples contained tiny ostracods and soil gastropods from the alternate flooding and paleosol episodes that characterized the Early to Middle Holocene sediment terrace formation. Ashy layers may result from flooding and erosion of hearths, but they also occur as primary contexts where vegetation stands or dung mats in rockshelters were burned. Samples from hearths sometimes contained glassy nodules naturally formed with the heating of sediment particles of sand, limestone, and wood ash.

The RASA team bagged and floated between 0.5 and 12 liters of sediment per sample. To multiple samples, a precounted measure of 50 poppy seeds was added to check recovery rates, which were unfortunately low (12–25 percent). An experienced crew member, Catherine Heyne, did most of the flotation and sorted heavy fraction materials in Yemen, recovering shell, chipped stone, bone, and heavy charcoal. Light fractions were dried in the shade, bagged, and exported to the United States for sorting under a binocular light microscope using 6x–10x magnification. Many were sorted by Heyne; others by student volunteers under the supervision of Joy McCorriston. For identifications, McCorriston relied on the botanical reference collections made by Heyne in Wādī Sanā and McCorriston's wider Near Eastern and subtropical reference collections and floras. McCorriston is responsible for most of the botanical identifications of reference collections.

In most flotation samples, fragile charcoals survived poorly and suffered further fragmentation during processing; shattering also happened with seeds like *Ziziphus* drupes. To supplement flotation, excavators also handpicked charcoal fragments. These were wrapped individually in separate foils, and excavators sought to recover up to 20 separate wood pieces for each sampled context. Handpicking provided a nonrandomized method to select some larger fragments and ensure wood identification. Smaller fragments from flotation were sorted first, with larger handpicked specimens serving to support and supplement identifications.

Because the charred materials come from a variety of different sites and ages in the Wādī Sanā system, results of analysis are reported within chapters that describe the archaeological contexts and other associations. In this review of paleovegetation, the results of system-wide analysis are more relevant.

An analysis of wood charcoals from hearths included up to 20 identifiable specimens (some samples had fewer) from a total of 20 hearths to study species availability, fuel choice, and land use management. The hearths were of two cultural types, clearly differentiated by their construction styles. Radiocarbon ages on several charcoal fragments helped establish a regional chronology of hearths. This chrono-stratigraphic work on select examples provides a model for dating hearth features relative to the wadi silt sedimentation phase of the Early Holocene throughout Wādī Sanā (chapter 18). Charcoal analysis found that fragments reliably could be assigned to one of six wood genera—*Acacia* sp., *Anogeissus* sp., *Cadaba* sp., *Ficus* sp., *Tamarix* sp., and *Ziziphus* sp. We found no statistically significant difference ($p = 0.76$) between the means of diversity indices of charcoal assemblages in hearths of the Early Holocene and hearths of the Middle Holocene (circa 8000–5000 cal BP) (Kimiaie and McCorriston 2014). Our results suggest that whatever anthropogenic- or climate-driven shifts in woody vegetation might have occurred, human fuel choice remained consistent through time. It is most likely that these wood charcoals represent *Acacia hamulosa* Benth., *Acacia odorata* Desv., *Anogeissus bentii* Baker., *Cadaba heterotricha* Stocks., *Ficus salicifolia* Vahl., *Tamarix nilotica*, and *Ziziphus leucodermis* (Baker) O. Schwartz, all of which grow in Wādī Sanā today (chapter 2; table 2.5).

Qualitative observations also point to the stability of vegetation distribution. A single plant of *Cyphostemma crinitum* (Planch.) Desc., a highly visible grape-like vine with fleshy green leaves, grows beside a rockshelter today as the only recorded incidence of this plant in all of Wādī Sanā. In all the charcoal samples identified, the sole incidence of (unmistakable) *Cyphostemma* sp. charcoal came from a burned dung mat in that rockshelter (1998 CS-2), dated to 5500 cal BP. Upstream and opposite the site of Manayzah, the only *Delonix elata* tree in Wādī Sanā today shades a stony terrace hugging the cliff face. And the only fragment of wood charcoal identifiable as either *Delonix* sp. or *Acacia* sp. came from Manayzah, across the stream channel. While these coincidences remain anecdotal, they are consistent with archaeological evidence for long-term continuity of woody taxa distributions and human uses of them in Wādī Sanā (Kimiaie and McCorriston 2014).

Burned dung mats in rockshelters can preserve charred seeds, as apparently was the case in at least two shelters (1998-CS 1 and 1998-CS 2) with terminal sedimentation deposits protected by rock overhang from subsequent erosion. Flotation of several ashy layers within these terminal sedimentation deposits yielded some of the richest assemblages of charred seeds and non-wood plant macroremains from Wādī Sanā (McCorriston et al. 2002:80). There were also morphologically distinct pellets of ova-caprine dung,

Table 3.4. RASA flotation samples 1998–2005. No samples were floated in 2008.

Sample #	Site or Area	Context	Seeds	Wood	Dung
2000-045-1-A-004 Bag x	Khuzmum rockshelter	hearth in NW corner of Quad A			
2000-045-1-A-006 Bag x	Khuzmum rockshelter	hearth		x	
2000-045-1-A-005 Bag x	Khuzmum rockshelter	ashy lens with many rootlets in excavation unit Quad A			
2000-045-1-A-008 Bag 44	Khuzmum rockshelter	hearth			
2000-045-1-A-009 Bag 43	Khuzmum rockshelter	Hearth A		x	
2000-045-1-A-009 Bag 42	Khuzmum rockshelter	Hearth B	x	x	
2000-045-1-A-009 Bag 45	Khuzmum rockshelter	Hearth C		x	
2000-045-1-A-013 Bag 1	Khuzmum rockshelter	layer underlying rockshelter occupation			
2000-044-3	Khuzmum rockshelter	hearth in natural terrace profile			
2000-044-4	Khuzmum rockshelter	hearth in natural terrace profile			
2000-044-6	Khuzmum rockshelter	hearth in natural terrace profile			
2000-044-7	Khuzmum rockshelter	hearth in natural terrace profile			
2000-044-8	Khuzmum rockshelter	hearth in natural terrace profile			
2000-044-9	Khuzmum rockshelter	hearth in natural terrace profile			
2000-044-11	Khuzmum rockshelter	hearth in natural terrace profile			
2000-044-12	Khuzmum rockshelter	hearth in natural terrace profile			
2000-044-13	Khuzmum rockshelter	hearth in natural terrace profile			
2000-044-15	Khuzmum rockshelter	hearth in natural terrace profile			
2000-044-17	Khuzmum rockshelter	hearth in natural terrace profile			
2000-044-20	Khuzmum rockshelter	hearth in natural terrace profile			
2000-044-23	Khuzmum rockshelter	hearth in natural terrace profile			
2000-044-26	Khuzmum rockshelter	hearth in natural terrace profile			
2004-110-4-A-009 Bags 1 and 2	110 rockshelter	ashy layer in Quad A excavation		x	
1998-CS-2 2.3 m below top	Wādī Sanā Cave II	ashy layer in rockshelter natural profile	x	x	x
1998-CS-2 2.4 m below top	Wādī Sanā Cave II	ashy layer in rockshelter natural profile	x		x
1998-CS-2 0.25 m below top	Wādī Sanā Cave II	organic layer in rockshelter profile	x	x	
1998-FPIV	regional silt terraces	hearth in natural terrace profile		x	
1998-FPVI	regional silt terraces	hearth in natural terrace profile			
1998-hearth 8	regional silt terraces	hearth in natural terrace profile	x		
1998-hearth 9	regional silt terraces	hearth in natural terrace profile	x		
1998-hearth 13	regional silt terraces	hearth in natural terrace profile		x	
1998-hearth 14	regional silt terraces	hearth in natural terrace profile		x	
1998-000-1-A-003 Bag x	Gravel Bar Site	hearth from Quad A			
1998-000-1-B-003 Bag x	Gravel Bar Site				
1998-WS1 hearth 1	regional silt terraces	hearth in natural terrace profile		x	
1998-WS1 hearth 2	regional silt terraces	hearth in natural terrace profile		x	x
2004-110-6 Hx Bag 2	110 rockshelter	hearth in natural terrace profile	x	x	
2000-033-15-A Bag 3	regional silt terraces	hearth in natural terrace profile			
2004-154-1-H1 Bag 3	regional silt terraces	hearth in natural terrace profile	x	x	x

Sample #	Site or Area	Context	Seeds	Wood	Dung
2000-010-A-001 Bag 3	Gravel Bar Site	hearth sampled in silt terrace surface			
2004-080-4-H1 Bag 2	regional silt terraces	hearth in natural terrace profile	x	x	
2004-080-4-H2 Bag2	regional silt terraces	hearth in natural terrace profile		x	
2004-080-4-H3 Bag 2	regional silt terraces	hearth in natural terrace profile		x	x
2005-155-2 M9-002 Lot 3	Manayzah	sandy layer with charcoal	x	x	
2004-155-2 K9-Niv 2	Manayzah	upper layer in deep excavation probe	x	x	
2005-155-2 L11-B-005 Lot 6 Bags x and y	Manayzah	ashy sediment possibly dump from a hearth	x	x	x
2005-155-2 B-006 Lot 7	Manayzah	ashy brown laminated sediment, visually identified as compacted dung in excavation		x	x
2005-155-2 M11-B-007	Manayzah	upper layer of a hearth containing ashy and charred silty sediment	x		x
2005-155-2 M11-B-004 Lot 5	Manayzah	hearth or dump from a hearth		x	x
2005-155-2 M9-A-009 Lot 11	Manayzah	upper layer of a hearth containing ashy and charred silty sediment		x	
2004-155-2 K9-005/006	Manayzah	layer in deep excavation probe		x	x
2005 155-2 M10+M11-B-013	Manayzah	hearth or dump from a hearth		x	
2005-155-2-M8-A-012	Manayzah	hearth? circular ashy feature	x	x	
2004-155-2 K9-Niv 7	Manayzah	layer in deep excavation probe	x	x	x
2005-155-2 K9-009(pit)	Manayzah	pit fill cleaned from section in deep excavation probe	x	x	
2004-155-2 K9-008	Manayzah	layer in deep excavation probe	x	x	x
2004-155-2 K9-Niv 13	Manayzah	layer 13 in deep excavation probe	x	x	x
2004-155-2 K9-Foyer 1 (H1)	Manayzah	hearth 1 in layers 9 & 10	x	x	
2004-155-2 K9-014	Manayzah	top of hearth 2	x	x	
2004-155-2 K9-014 bottom	Manayzah	bottom of hearth 2	x	x	x
2004-155-2 K9-014/015	Manayzah	layer in deep excavation probe	x	x	x
2005-155-2 K9-017sup	Manayzah	ashy layer in deep excavation probe		x	
2005-155-2 K9-018	Manayzah	layer in deep excavation probe		x	
2005-151-1 H8 Bag 4	Kheshiya	hearth in natural terrace profile (gully)		x	
2005-151-1 H3 Bag 4	Kheshiya	hearth in natural terrace profile (gully)	x	x	
2005-151-1 H5	Kheshiya	hearth in natural terrace profile (gully)		x	
2005-151-1-H1 Bag 3	Kheshiya	hearth in natural terrace profile (gully)		x	
2005-151-1-H13 Bag 3	Kheshiya	hearth in natural terrace profile (gully)		x	
2005-151-1-H4 Bag 5	Kheshiya	hearth in natural terrace profile (gully)		x	
2005-151-1-H12 Bag 6	Kheshiya	hearth in natural terrace profile (gully)	x	x	
2004-151-1-H9 Bag 2	Kheshiya	hearth in natural terrace profile (gully)		x	
2004-151-1-H10 Bag 4	Kheshiya	hearth in natural terrace profile (gully)		x	
2004-151-1-H11 Bag 2	Kheshiya	hearth in natural terrace profile (gully)		x	
2005-151-1-A-017 Bag 6	Kheshiya platform	silty fill and debris over occuation layer		x	x
2005-151-1-A-018 Bag 2	Kheshiya platform	occupation layer under platform fill			
2005-151-1-C-004 Bags 3 and 4	Kheshiya	ashy layer		x	
2005-151-1-C-009 Bag 1	Kheshiya	ashy layer	x	x	

Table 3.4. RASA flotation samples 1998–2005. No samples were floated in 2008. *(continued)*

Sample #	Site or Area	Context	Seeds	Wood	Dung
2005-151-1-B-006 Bag 1	Kheshiya cattle skulls	sediments from upper ashy layer around cattle skulls		x	
2005-151-1-B/E-002 Lot 10 Bag 2	Kheshiya cattle skulls	sediments from upper ashy layer around cattle skulls		x	
2005-151-1-E-004 Lot 6 Bag x	Kheshiya cattle skulls	sediments from upper ashy layer outside skull ring		x	
2005-151-1-E-004 Lot 6 Bag 2	Kheshiya cattle skulls	sediments from upper ashy layer around cattle skulls	x	x	
2005-151-1-E-004 Lot 5 Bag x	Kheshiya cattle skulls	sediments from outside skull ring and adjacent to upper skulls		x	
2005-151-1-E-007 Lot 10 Bag 1	Kheshiya cattle skulls	ashy sediments associated with cattle skull ring	x	x	
2005-151-1-E-008 Lot 12 Bag 1	Kheshiya cattle skulls	ashy sediments associated with cattle skull ring		x	
2000-037-3-A-006 Bags 16 and 17	Khuzma platform	occupation layer under platform fill	x	x	
2000-037-3-A-009-010 Bag x	Khuzma platform	ashy occupation layer under platform fill		x	
2005-037-3-D-002 Bag x	Khuzma platform	ashy sediments overlying top of platform			
2005-037-3-B-006 Bag x	Khuzma platform	hearth in unexcavated occupation under platform fill			
2005-037-3-C-010 Bag 4	Khuzma platform	ashy occupation layer under platform fill			
2005-037-3-C-011 Bag 4	Khuzma platform	ashy occupation layer under platform fill			
2005-037-3-C-012 Bags 1 and 8	Khuzma platform	sediment from occupation around hearth			
2005-037-3-C-014 Bags 1 and 3	Khuzma platform	hearth in occupation layers under platform fill			
2005-037-3-C-014 Bag x	Khuzma platform	from outside hearth 014			
2005-037-3-E1/C1-014 Bag x	Khuzma platform	hearth (C-014)			
2005-037-3 E1+C 015 bag x	Khuzma platform	laminated silts under platform occupation			
2005-037-3-E1+ C1-016 Bag 1	Khuzma platform	ashy occupation layer under platform fill			
2005-037-3-E1+C-017 Bag 1	Khuzma platform	ashy occupation layer under platform fill			
2005-037-3-E1+C-018 Bag 4	Khuzma platform	laminated silts with occupation debris			
2004-009-1-T3 Bag 6	water management	sediments above stone lining of canal			
2004-009-1-T3 Bag 8	water management	sediments under buried canal		x	
2004-009-1-T2 Bag 8	water management	sediments in buried canal	x	x	
2004-W1-1-T1 Bag 4	water management	hearth near water management structures			
2004-W1-1-T2 Bag 4	water management	hearth near water management structures			
2004-W1-4-H2 Bag 3	water management	hearth near water management structures			
2004-W1-4-H3	water management	hearth near water management structure			
2004-W5-1-T2 Bag 8	water management	burned layer in section of check dam			
2004-W6-1-H1 Bag 4	water management	hearth near water management structures		x	
2004-W6-1-H2 Bag 3	water management	hearth near water management structures		x	
2004-W13-1-H1 Bag 2	water management	hearth near water management structures	x	x	
2004-W22-12-H1 Bag 2	water management	hearth near water management structures	x	x	
2004-W23-1-H1 Bag 2	water management	hearth near water management structures		x	
2004-W23-1-H2 Bag 2	water management	hearth near water management structures	x	x	
2004-W23-1-H3 Bag 2	water management	hearth near water management structures		x	

Table 3.5. List of modern Wādī Sanā vegetation (left) and analyzed archaeobotanical taxa (right) (after Kimiaie and McCorriston 2014:37, table 2).

Modern Woody Taxa	Archaeobotanical Taxa
Acacia hamulosa Benth.	*Acacia* sp.
Acacia mellifera (M. Vahl.) Benth.	
Acacia oerfota (Forssk) Schweinf.	*Acacia* sp.
Acacia ehrenbergiana Hayne	
Anogeissus bentii Baker.	*Anogeissus* sp.
Balanites aegyptiaca Del.	*Balanites* type
Boswellia sacra Flueck.	
Commiphora kataf Engl.	
Carissa edulis	Apocynaceae cf. *Carissa*
Cadaba heterotricha Stocks.	*Cadaba* sp., *Cadaba/Maerua* type
Calotropis procera (Ait.) Ait *f.*	
Commiphora gileadensis (L.) C.Chr	*Commiphora* sp.
Cyphostemma crinitum (Planch.) Desc.	*Cyphostemma* sp.
Delonix elata Gamble.	*Delonix* sp., *Acacia asak/Delonix* type
Ficus salicifolia Vahl.	*Ficus* sp.
Indigofera spp.	
Lycium shawii Roem. & Schult.	
Maerua crassifolia Forssk.	*Maerua* sp., *Cadaba/Maerua* type
Moringa peregrina Fiori.	
Phoenix dactylifera L.	*Phoenix* type
Salvadora persica (L.) Garcin	*Salvadora persica*
Tamarix aphylla (L.) Karst.	*Tamarix* sp.
Tamarix nilotica (Ehrenb.) Bunge.	*Tamarix* sp.
Ziziphus leucodermis (Baker) O. Schwartz	*Ziziphus* sp.

suggesting these as the source of seeds. Recent experimental investigations suggest that seeds greater than 2 mm in diameter are quite unlikely to survive digestion in small ruminants (Wallace and Charles 2013), so it may be that the slow-burning, high-reduction environment of dung mats charred and preserved incidental seed materials discarded at human camps—bedding, kindling, commensal rodent caches, and deliberately gathered foodstuffs. This interpretation augments an earlier published interpretation of these dung mats and their seed load as uniquely the by-product of domesticated animals' plant consumption.

Identifiable charred seeds come from plants still growing in the Wādī Sanā today. Given the strong taphonomic filters and low numbers, these taxa are best understood in qualitative context. The charred materials suggest that the same woody plant taxa have been available to humans and domestic browsers through the Early to Middle Holocene as constituents of a cover that was manipulated but not entirely overhauled by human intervention. Some herbaceous seeding plants have local continuity at least from the Middle Holocene to modern era. Chapter 9 contains a further discussion of the two rockshelters and plant taxa recovered from them.

Burned Surfaces

The sedimentary sequences naturally exposed along Wādī Sanā include a multitude of contexts in which burned surfaces (that is, fire-altered layers, often particularly rich in ash and charred material) are particularly informative in terms of paleovegetation. Burned surfaces were notably revealing in light of their geological, archaeological, and paleoecological contexts.

(1) Geological Context. In the course of geological studies, Eric Oches discovered a series of ashy layers embedded in the wadi silts. These layers were often clearly visible as dark, charcoal-flecked bands extending tens of meters in the natural sections created by recent down-cutting of the wadi channel. At times these layers sloped perceptibly, following an old land surface, and the archaeological survey team quickly learned to recognize their surface remains where erosion has uncovered sometimes extensive areas of ash-darkened surfaces—such as one 5,300 m^2 area at the base of a natural outcrop (Khuzma as-Shumlya) that attracted particularly intense human activity in the past. The visibility of such ashy surfaces and several distinctive criteria, such as dense charcoal flecks, burned shell inclusions, and the typical abrupt transition between the surface/layer and overlying and underlying deposits, meant that these distinctive remains of burning episodes could be readily differentiated from the darkened, organically enriched bands of paleosol or stable land surfaces that also characterize natural sedimentary profiles. We incorporated the enumeration (n = 123) and description of burned surfaces into our random, stratified, systematic survey design in 2000 and 2004. (See chapter 4 for details.)

Described by Oches in McCorriston et al. (2002:66), one 3.5 m high sediment profile (98-WS-3; figure 3.16) (E 337139, N 1744610 UTM Zone 39 North) shows a cyclical pattern of sedimentation followed by burning: An individual cycle begins with what we interpreted as aeolian silt accumulation, followed by a pulse of fluvial sand deposition. Above the sandy layer is a distinct burned horizon, identified visually (charcoal above reddened surfaces) in outcrop and appearing as a strong increase in magnetic susceptibility of the sediment. We identified at least three distinct cycles in the section. We hypothesize that the coarse sediment pulses indicate periods of flooding. Vegetation growth, followed by burning, was recorded in the sediments immediately above each sand-enriched zone.

Contiguous with the middle burned horizon in 98-WS-3, an ashy layer in a nearby profile (98-WS-2) (E 337233, N 1744559 UTM Zone 39 North) yielded charcoal with a radiocarbon age of 6851–6547 cal BP (OS16689). In our broader regional framework of fluvial geomorphology, it

Figure 3.16. RASA 1998 Wadi Section 3 after rough cleaning by Eric Oches. *Photograph by Joy McCorriston/RASA Project Archive.*

is clear that the burned surfaces visible today date to the Early to Middle Holocene period. Not every ashy layer can be assigned to an in situ burning event, but other profiles suggest that such events are widespread. For example, in SU125, underlying reddening and sharp transition between black ash and overlying yellow-brown silt are visual characteristics of three superimposed burned layers in one profile.

(2) Archaeological Context. Another important source of information on the burned layers comes from their archaeological context. There are examples of clear contiguity between man-made hearths and burned surfaces. For example, in 1998 Wadi Sections 1 and 2 (98-WS-1, 98-WS-2), hearths are cut from (Hearth 1998-2, OS16933, 6792–6561 cal BP) (figure 3.17) or lie on (Hearth 1998-1) burned surfaces, clearly indicating the presence of humans in the Wādī Sanā when burning took place. Another of many natural wadi profiles studied cuts through occupation layers and flood sediments that formed a terrace to the east of the rockshelter SU045-10 at Khuzma as-Shumlya (chapter 9). In profile (4 m segment published in McCorriston et al. 2002:73, figure 8), most of the detected hearths (n = 27) appear within two 0.20–0.30 m bands of ash- and charcoal-enriched deposits accumulated from human occupations of the adjacent shelters. In this case, the charcoal-enriched deposits probably formed partially through flooding and reworking of hearth deposits rather than in situ burning across a vegetated surface, but the deposits show a close association between human occupation and sediment history.

Figure 3.17. Hearth 1998-2 in natural wadi section, showing the distinct burned layer from which this bell-shaped pit containing a hearth was cut. *Photograph by Joy McCorriston/RASA Project Archive.*

We have argued elsewhere that the burned surfaces represent deliberate anthropogenic firing of vegetation (McCorriston et al. 2005:144), an idea that gains greater credence in light of the archaeological data for other human activities. The earliest water management structures along Wādī Sanā (chapter 13) are stratigraphically higher than and presumably later than burned surfaces. On the earlier side, recent techno-chronological analysis of stone toolmaking in Wādī Sanā suggests that virtually all the highly skilled techniques for producing hunting projectile points (javelins or arrowheads) were no longer practiced by 6500 cal BP (Crassard 2008), suggesting perhaps a shift in aesthetic, magical, and economic emphasis on hunting (McCorriston 2014; McCorriston et al. 2014). Neither ceramics nor sedentary occupation occurred, so we may infer that people remained mobile. Between hunting and agriculture, mobile people managed the Wādī Sanā landscape in part by burning off areas of dense vegetation.

(3) Paleoecological Context. There are good geobotanical reasons to suppose that the burned surfaces represent human-set fires. Charcoal analysis from hearths and occupation surfaces identified locally growing wood species selected by humans (primarily for firewood). The local species come from a flora adapted to (summer) monsoon precipitation; archaeobotanical research found no indication of plants adapted to the winter storm tracks of Mediterranean and continental climate types across northern latitudes.

The implications of a consistent, if formerly moister, climate regime are that regional analog vegetation stands can provide important insights into floristic composition, evolutionary ecology, and rangeland management. First, it is important to note that tropical vegetation of the Horn of Africa—including the arid highlands of Eritrea, southern Red Sea margins, and Dhofar (Oman)—lacks the evolutionary responses to regular natural fires that one finds in Mediterranean-type plant ecosystems, in which lightning strikes at the end of a dry season periodically set natural fires (e.g., Biswell 1974; Kruger 1984). In the latter, there are plants that cannot disperse seeds or regenerate without exposure to moderate fires; there are trees that bind nutrients in leaf litter until they are released as ash after a fire that destroys soil seed banks of other species; there are thick phloem layers that protect mature trees like cork oaks; there are species that die back to corms and rhizomes protected from moderate fires at the end of the dry season; and there are a large number of colonizing species after fire whose seeds survive brief surface burns (Mooney and Dunn 1970; Naveh 1975; Vogl et al. 1977; Wells 1969). The periodicity and suppression of fires can have dramatic composition effects on Mediterranean-type vegetation (Pignatti 1979; Trabaud 1981), as in all ecosystems, but firing is one of the salient evolutionary constraints in Mediterranean climates.

Different evolutionary forces, including water stress in arid environments and protection from ungulate and bovid browsers (that is, thorns, sap, latex, and toxins) shaped the adaptations of plants in Wādī Sanā (Al-Hubaishi and Müller-Hohenstein 1984:25–39; Grime 1977; Kürschner 1998). For example, *Ficus* (cf. *salicifolia*) exudes a latex that discourages browsers. *Maerua* sp. and *Cadaba* sp. have rings of included phloem at the margins of annual growth rings that may insulate plant vascular tissue from drought (cf. Baas 1986; Jagiella and Kürschner 1987). Short-lived annuals spring up at the first rain and have seeded by the dry months, and the fruit of *Citrullus colocynthus* is poisonous to livestock and humans. Until humans began setting fires—possibly no earlier than 7,000 years ago—there was insignificant incidence of lightning strikes across most of Arabia, where annual precipitation rarely exceeded 200 mm, vegetation depended as much on runoff as on rainfall, and—until the Early Holocene moist period—only localized areas produced sufficient cover to sustain wildfires.

There is much to learn from the historical ecology of anthropogenic burning in Africa, where humans may have practiced periodic firing to improve range for wild game or herded animals for 50,000 years. Burning is most effective, indeed should be used, to maintain grazing resources (perennial grasses) in semiarid and arid savanna with an *Acacia–Commiphora* association and perennial grassland cover (e.g., Pratt et al. 1966:371). Regular burning reduces woody cover while encouraging nutritious growth of the perennial grasses favored by cattle (Pratt and Gwynne 1977: 33; Wright 1974:7), but burning must be undertaken with sufficient frequency to remove seedlings (Tesfaye et al. 2004:360) while not generating a heat so incendiary from excess leaf litter and woody cover as to damage the roots of perennial grasses. Regular burning not only keeps down woody species undesirable for cattle but also enables smaller and more palatable species to persist when taller species get rank, unpalatable, and lower in nutritional value as they mature (Pratt and Gwynne 1977:60). A number of Arabian perennial grasses offer attractive graze; they include *Cymbopogon* spp. *Lasiurus sindicus*, *Pennisetum divisum*, *Stipagrostis drarii*, *S. plumosa*, *Panicum turgidum*, *Hyperrhenia hirta* and *Andropogon distachyos* (Assaeed 1997; Bokhari et al. 1990). Some of these, most notably tropical *Hyparrhenia* and *Andropogon*, respond well to deliberate regular firing (Naveh 1974:421; Pratt and Gwynne 1977:60). In Arabia today, most rangeland is too arid to generate sufficient cover for a burn, or else it is used for farmland. Nevertheless, cover was more extensive across Southern Arabia in the Early to Middle Holocene (Cremaschi and Negrino 2005:577; Kutzbach et al. 1996:625), and the formation of weak paleosols in Wādī Sanā sediments points to a formerly more extensive vegetative cover.

Among African pastoralists, the practice of annual burning to stimulate young growth and to reduce woody cover so that cattle can graze is well attested. The Nuer and other pastoralists deliberately set fire to dry pasturelands to stimulate new growth as soon as the grasslands have dried after the rainy season (e.g., Desta and Coppock 2002:450; Evans-Pritchard 1940:59–60, 83). Other groups migrate, or scatter and coalesce, to take advantage of seasonal surface water and vegetation flushes (Box 1968:391), a practice also known in Arabia (Janzen 1986:101; Lancaster and Lancaster 1999). Burning is also undertaken by the Turkana to reduce ticks that carry diseases (McCabe 2004:65). The factors underpinning vegetation and vegetative changes may be complex (Bollig and Schulte 1999; Ellis and Galvin 1994; Ellis and Swift: 454–55), so it is difficult to reconstruct a specific ecosystem dynamic from the evidence for burning alone. Nevertheless, burning was surely anthropogenic in the Middle Holocene Wādī Sanā and would have produced variable effects according to fire intensity, periodicity, and the adaptive characteristics of native grass species.

Hyrax Middens

Searches of the Wādī Sanā produced abundant rock hyrax (*Procavia capensis*) deposits near the ongoing Wādī Sanā excavations. Rock hyraxes maintain a limited range around the rocky crevices and caves they inhabit, and they defecate and urinate near these crevices. In arid habitats the urine crystalizes, cementing the contents together. If the urine-soaked deposits continue to be protected from precipitation by a rocky overhang or cave, they accumulate and can be preserved for thousands of years. Plant macrofossils from the middens represent a highly localized vegetation record, while fossil pollen represents vegetation over a far wider area for wind-borne pollen. Such middens potentially offer multiple paleorecords with the preservation of plant macrofossils, pollen, spores, bones, snails, stable isotopes, plant cuticles, arthropods, DNA, and other organic materials.

Twenty-four middens were collected within 1,500 m of Khuzmum over a period of two weeks (figure 3.18), and these fossils were so frequent in alcoves and caves (figures 3.19 and 3.20) in the limestone cliffs lining the wadi that many times that number could have been collected. Those collected are only a representative sample of many available for kilometers along middle Wādī Sanā. Thus far, 11 of these samples have been radiocarbon dated, using conventional radiocarbon dating on bulk samples of hyrax pellets. The resulting radiocarbon ages range from 415 to 5236 ^{14}C years BP (426 to 6047 cal BP; table 3.6), making this the oldest hyrax midden series yet found in the Middle East. More specific AMS radiocarbon dating of individual layers within middens is now under way.

Each deposit contains abundant plant macrofossils and pollen, with an assortment of insects, bones, and snails. Many of the plant macrofossils are well preserved, as illustrated through comparisons with modern specimens (figures 3.21, 3.22, 3.23, 3.24, and 3.25). Some, such as *Ziziphus leucodermis* and *Acacia hamulosa*, have been present throughout the series (table 3.7). *Acacia ehrenbergiana* (figure 3.22) was not found until the midden dated at 2159 years BP, in which it was also associated with a seed of *Citrullus colocynthis* (figure 3.25). Other interesting midden contents include arthropods and arachnids, such as an unidentified tick (figure 3.26).

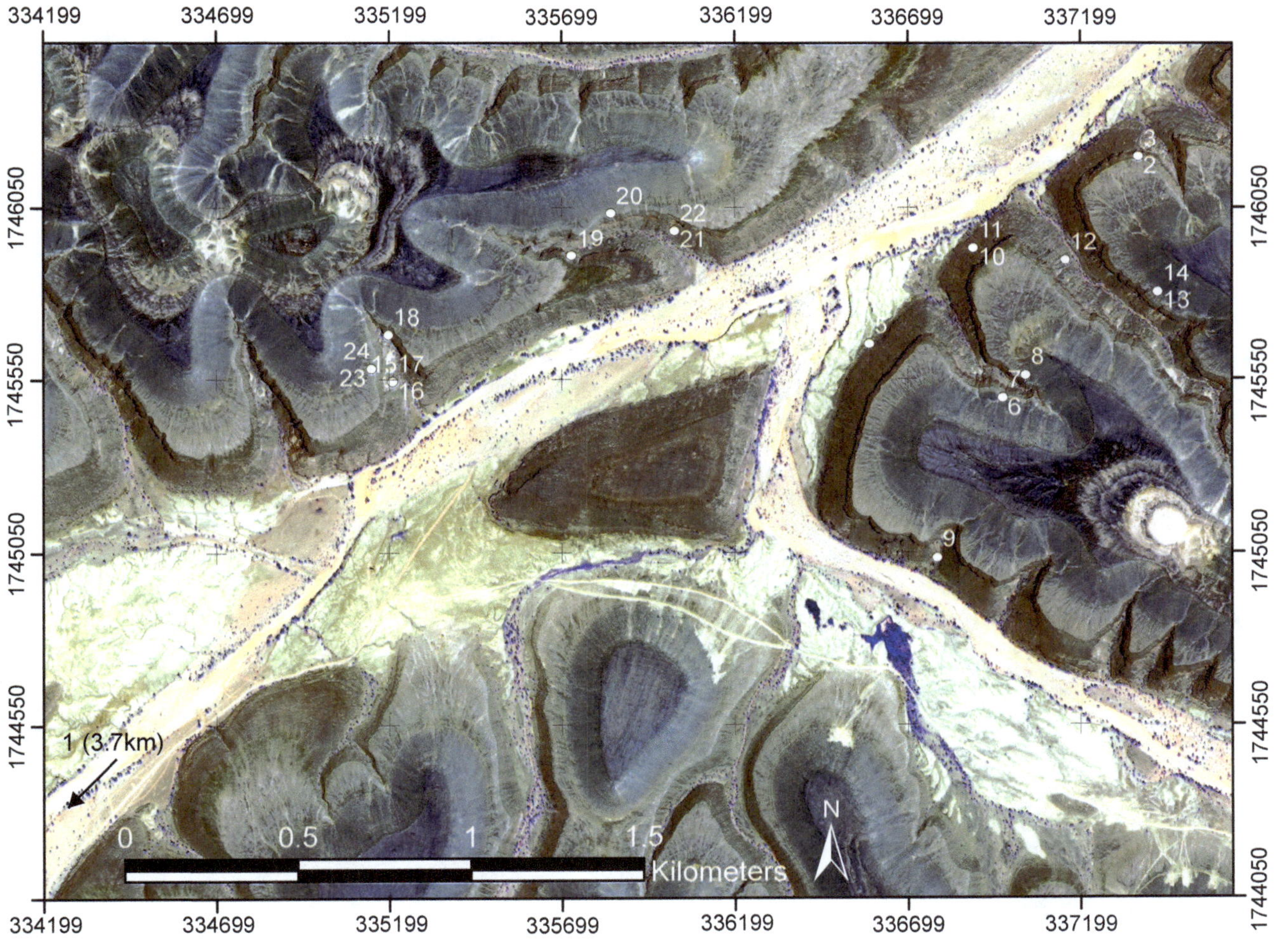

Figure 3.18. Map of hyrax middens collected from around Khuzmum (UTM Zone 39 North WGS1984).
Illustration by Michael Harrower.

Figure 3.19. Caves in the limestone cliffs lining the middle Wādī Sanā where ancient hyrax middens occur. *Photograph by Kenneth Cole.*

Figure 3.20. Nasser Al-ʿAlīy of the Ḥumūm bedouin group holding a packaged midden collection from cave in the background (left) and Muḥammad Al-Hijāzī (right). *Photograph by Kenneth Cole.*

Figure 3.21. Fossil seed of *Ziziphus leucodermis*. *Photograph by Kenneth Cole.*

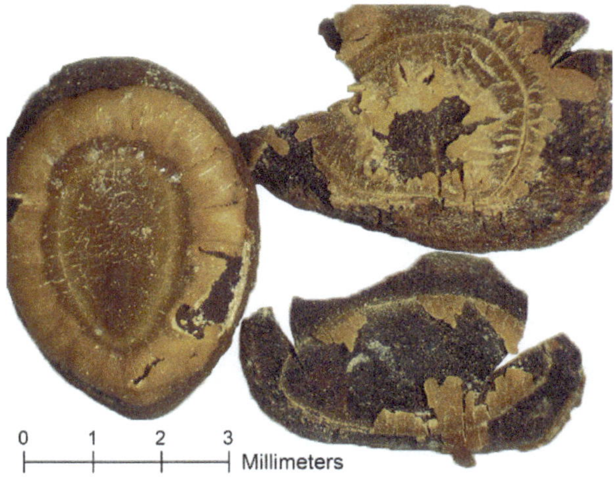

Figure 3.22. Fossil and modern seeds of *Acacia ehrenbergiana*. *Photograph by Kenneth Cole.*

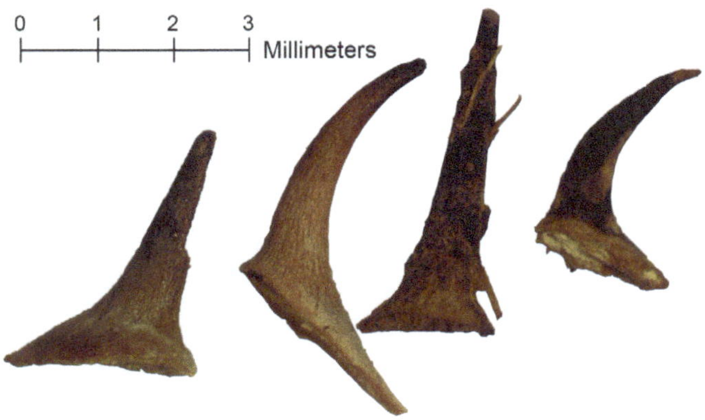

Figure 3.23. Fossil and modern spines of *Acacia hamulosa*. *Photograph by Kenneth Cole.*

Figure 3.24. Fossil and modern seed of *Pavonia subaphylla*. *Photograph by Kenneth Cole.*

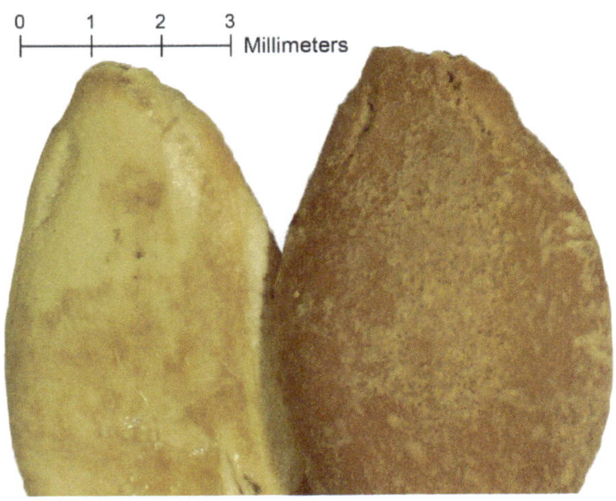

Figure 3.25. Fossil and modern seed of *Citrillus colocynthis*. *Photograph by Kenneth Cole.*

Figure 3.26. Closeup of the head of an unidentified fossil tick; length shown approximately 1.5 mm. *Photograph by Kenneth Cole.*

Table 3.6. Wādī Sanā hyrax midden locations (locations in UTM Zone 39 North WGS1984).

Sample Number	Location			Radiocarbon Data					Isotope Data				
	Elevation (m)	Northing	Easting	^{14}C Age	^{14}C SD	^{14}C Lab Number	Cal yr B.P.	Uncertainty 1SD	del-13C (‰)	%C	del-15N (‰)	%N	C/N
Wādī Sanā 1	750	1742017	331966	4590	90	GX24614	5258	203	-24.96	45.15	12.11	2.98	15.15
Wādī Sanā 2	725	1746201	337366										
Wādī Sanā 3	725	1746201	337366										
Wādī Sanā 4A	735	1746137	337722	4555	60	A11777	5187	131	-25.9				
Wādī Sanā 5	730	1745654	336587										
Wādī Sanā 6A	730	1745497	336974	415	40	A11778	426	90	-23.7				
Wādī Sanā 7A	730	1745497	336974	690	45	A11779	623	58	-24.2				
Wādī Sanā 8	730	1745620	336885	2159	45	AA39070	2184	119	-22.91	31.88	11.79	2.49	12.79
Wādī Sanā 9	730	1745006	336732										
Wādī Sanā 10	730	1745896	336977										
Wādī Sanā 11	730	1745896	336977	5236	55	AA38420	6047	127	-24.05	36.47	14.72	2.08	17.51
Wādī Sanā 12	730	1745895	337155										
Wādī Sanā 13	730	1745801	337422	4490	75	N/A	5166	124	-24.90	38.49	12.08	2.84	13.55
Wādī Sanā 14	730	1745801	337422	4425	115	A11780	5073	203	-25.6				
Wādī Sanā 15	735	1745545	335209										
Wādī Sanā 16	735	1745545	335209										
Wādī Sanā 17A	735	1745545	335209	4490	44	AA39071	5166	118	-23.89	35.53	14.20	2.55	13.92
Wādī Sanā 18	740	1745680	335196										
Wādī Sanā 19	730	1745909	335727										
Wādī Sanā 20	730	1746034	335840										
Wādī Sanā 21	735	1745983	336025										
Wādī Sanā 22	735	1745983	336025										
Wādī Sanā 23A1	730	1745583	335147										
Wādī Sanā 23A2	730	1745583	335147	4602	45	AA38421	5271	178	-24.20	33.52	15.94	2.14	15.64
Wādī Sanā 24A	730	1755583	335147	4230	50	A11052	4753	101	-21.33	35.42	15.58	3.76	9.41

Table 3.7. Wādī Sanā hyrax midden contents (age, δ ^{13}C, and most frequent botanical contents).

Midden Number	11	23A2	1	4	17A	13	14	24A	8	7A	6A
Radiocarbon Age (year BP)	5236±55	4602±45	4590±90	4555±60	4490±44	4490±75	4425±115	4230±50	2159±45	690±45	415±40
δ ^{13}C of Pellet Subsample	-24.05	-24.20	-24.96	-25.9	-23.89	-26.0	-25.6	-21.33	-22.91	-24.2	-23.7
Primary Identified Plant Macrofossils (number)											
Ziziphus leucodermis, seeds	72	2	28		12				42		
Ziziphus leucodermis, spines	1				1			2	25		
Acacia hamulosa, spines	445	197	25		397			45	147		
Acacia ehrenbergiana, seeds									127		
Acacia ehrenbergiana, spines									45		
Pavonia subaphylla, seed	1										
Citrullus colocynthis, seeds	1							1			
cf. *Capparis cartiliginea*, thorns		1	1						2		
Chloris barbata, flowers									7		
Total Distinctive Macrofossil Taxa	16	10	8		11			12	15		
Primary Identified Pollen (%)											
Leguminosae, (cf. *Cassia*)	14.3	12.1	40.8		38.3			48.2	12.5		
Acacia spp.	9.9	3.4	6.6		7.4			13.3	0		
Burseraceae	3.3	19.0	0		0			0	28.8		
Gramineae	8.8	13.8	17.1		12.8			0	5.0		
Chenopodiaceae/ *Amaranthus*	2.2	22.4	15.8		14.9			13.3	1.3		
cf. Capparadaceae	6.6	0	17.1		17.0			14.5	11.3		
cf. Zygophyllaceae	1.1	0	0		1.1			3.6	1.3		
Other Contents of Interest	*Procavia* incisor		rodent jaw					rodent jaw	linen cloth amber beads human phalange		

This Late Holocene appearance of *A. ehrenbergiana*, widespread throughout Yemen today, could indicate a return to moister climates following an extremely arid Late/Middle Holocene. When considered with the geophysical and archaeological results from this site that indicate a moist Early to Middle Holocene, these midden deposits suggest that the period of maximum Holocene aridity in the Southern Arabian highlands may have been between 5000 and 2500 cal BP (McCorriston et al. 2002). During this arid period, archaeological evidence also suggests, people abandoned the area. Rock hyrax is not found at this locality today, and the landscape is barren, possibly due to recent browsing by goats and camels.

We hope that many additional plant macrofossils will be identified in ongoing analyses of these deposits. Once unidentified specimens have been documented, they can be compared to multiple plant herbarium specimens collected in the area as part of this project (chapter 2).

Comparison of Hyrax Middens and Packrat Middens

The hyrax middens collected are remarkably similar in appearance and provenience to the fossil packrat middens that are very abundant throughout deserts of the American Southwest. They were first recognized by the plant ecologists Phillip Wells (Wells and Jorgensen 1964) and, somewhat later, Thomas Van Devender (1973), who discovered them to be rich sources of plant macrofossils. The middens contain plant parts identifiable to the species level, allowing the reconstruction of ancient plant assemblages from local sites. Wells preferred to examine individual horizontal strata within the deposit, recognizing 5 to 10 species, while Van Devender disaggregated thicker layers, yielding assemblages of up to 40 species from a single stratum. Records of fossil pollen and stable isotopes within the packrat middens were only studied much later than the plant macrofossils (Cole and Arundel 2005; Pendall et al. 1999; Thompson 1985).

More recent, fine-scale analyses of packrat middens, including all macrofossil specimens greater than 0.5 mm in size, reveal 30 or more identifiable species in samples of less than 200 g and less than 2 cm in thickness (Fisher et al. 2009). Although there are exceptions, packrat middens seem to be primarily deposited in brief periods of rapid accumulation. Multiple radiocarbon ages are usually indistinguishable from deposits as thick as 6 cm and lacking in internal layering. In contrast, hyrax middens are thought to be deposited over longer time intervals and to contain multiple thin layers (Chase et al. 2012). Multiple AMS radiocarbon determinations on plant macrofossils from different horizons within the Wādī Sanā middens are now being tested to explore this issue.

In contrast to packrat middens, hyrax middens were first studied by palynologists rather than plant ecologists. Pons and Quézel (1958) discovered deposits in the Haggar Mountains of the Sahara, while Fall et al. (1990) located a few at Petra, Jordan, and Scott (1990) began extensive studies in South Africa. The South African midden deposits have more recently been the source of isotopic paleoclimatic records (Chase 2013). Because the urine matrix of the middens contains extremely high concentrations of pollen and isotopes, only a very small sample, perhaps less than 20 g, is required. As a result, most hyrax midden studies have only analyzed small samples of the urine matrix, and if plant macrofossils were present, they were not reported.

Hyrax middens may not contain as diverse a macrofossil flora as packrat middens, as packrats are named for their prodigious collecting habits. A much larger sample than has been used for pollen, maybe 100 to 200 g, seems to be required for an adequate sample of plant macrofossils from a hyrax midden. Smaller samples expand the number of potential deposits and minimize contamination between layers of differing ages. However, in a locality like the Wādī Sanā, the number of potential deposits of large size is not a concern. Selecting deposits with obvious horizontal layers can be important. The stratigraphy of many middens is complex, and these should be avoided if possible.

The disparity between the plant macrofossil analyses on North American packrat middens and the palynological and isotopic studies conducted on hyrax middens may result from the primary focus of the investigators and the selection of middens to be studied as much as from the distinct characteristic of the deposits. While packrat middens are often collected as tabular slabs, rich in plant macrofossils that seem to originate from brief periods of accumulation, highly sampled hyrax middens are more often from composite thin layers of hyrax urine collecting through time at a single location. Both types of deposits can be produced by each animal, although the thick-layered urine deposits seem to be far rarer, at least for packrats.

A comparison between hyrax middens from Yemen and packrat middens from along the canyons of the Colorado River in the southwestern United States shows that the midden components are remarkably similar in proportion (figure 3.27). Each midden is cemented together by a water-soluble fraction, mainly consisting of urine averaging 75 percent of the mass for packrat middens (amberat) versus 70 percent for hyrax middens (hyraceum). This fraction also contains pollen and dust that fell on the midden. The mass of fecal pellets is also similar (12 percent versus 10 percent). The

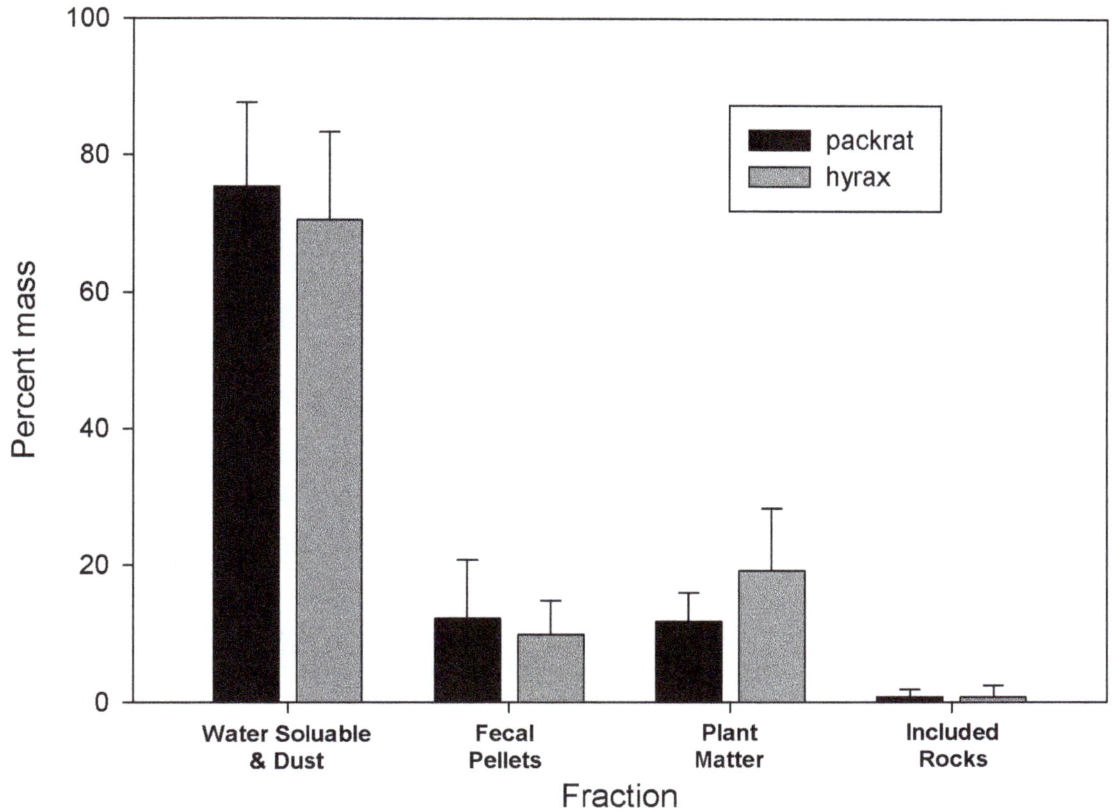

Figure 3.27. A comparison of the percent mass of primary components of packrat and hyrax middens. *Illustration by Kenneth Cole.*

packrat middens contained 12 percent of dried plant material matrix (> 0.5 mm), yielding numerous plant macrofossils. Although the mass of this fraction from the hyrax middens was higher (19 percent), much of the hyrax material is very small plant matter that may, or may not, ultimately be identifiable. Many larger plant macrofossils are present, but their identification remains difficult compared to the packrat middens that have been studied for 50-plus years, producing abundant knowledge and rich archives of comparative specimens. The hyrax middens should also ultimately produce a much wider array of species-specific macrofossils than can currently be identified.

The deposits are analogous to tiny archaeological sites. Experienced archaeologists can excavate a site, recognizing differing layers and cataloging their contents. A similar archaeological approach to what has been done with hyrax middens would be to collect a narrow core through the site and analyze only its fossil pollen and isotopes. Extreme caution and multiple AMS radiocarbon ages are needed to ensure the integrity of a macrofossil assemblage. But robust assemblages of plant species can produce valuable information on local paleoecological history.

Human Burial

One Wādī Sanā midden contained pieces of linen cloth (figure 3.28), hand-carved amber beads (figure 3.29), and a human finger bone (figure 3.30). A piece of the linen cloth was dated by an AMS radiocarbon date to 2159 ^{14}C years BP. The provenience of the midden suggested that it may have been constructed on top of a human burial already in the cave, and thus it likely postdates the cloth. The linen cloth pieces adhering to the bottom of the midden were visible when it was collected, although they were not recognized for what they were at that time. This cave burial may not have been unusual; at least one cave in the area appeared to be recently sealed with rocks stacked there by humans, and Nasser Al-ʿAlīy, our local bedouin guide, did not appear willing to investigate or excavate its entrance. We did not press the issue, as there were abundant nearby unsealed caves to investigate, but it is likely that there are more recent burials within the caves. One cave that was far too small for a camel to crawl into held a very pungent camel carcass, perhaps dating from decades ago (figures 3.31 and 3.32). So it appears that these caves have been, and may still be, used for burials of various kinds.

Figure 3.28. Linen cloth fabric found along the bottom of Wādī Sanā 8. *Photograph by Kenneth Cole.*

Figure 3.29. Small amber beads found in Wādī Sanā 8. *Photograph by Kenneth Cole.*

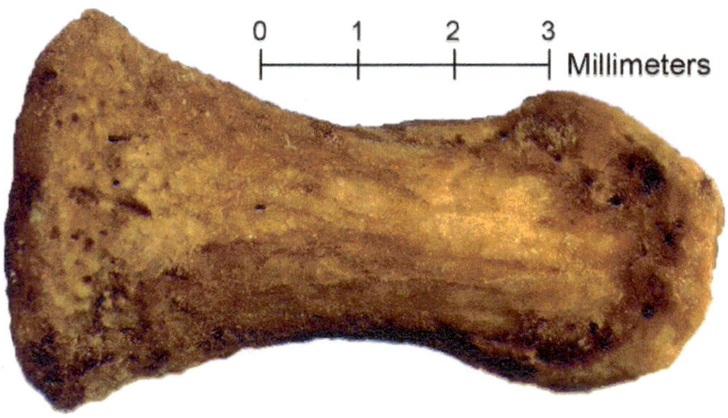

Figure 3.30. Human fingertip found in Wādī Sanā 8. *Photograph by Kenneth Cole.*

Conclusions

The data for geomorphology, paleohydrology and paleoecology come from disparate sources—physical formations and depositional sequences, pollen and macrobotanical remains, magnetic properties, and inclusions in silt sediments. These data manifest spatial and temporal discontinuities. Nonetheless, in combination such data provide an important local perspective on the long-term processes registered across Southern Arabia at coarser scales. In some instances, data are highly complementary—for example, where *Acacia ehrenbergiana* did not show up among wood charcoals (fuel choices) but did occur in hyrax middens as the manifestation of a Late Holocene return to moister climates.

In overview, the geomorphology, paleohydrology, and paleoecology of Wādī Sanā reflect a Late Pleistocene environment of high-energy activity across the Wādī Sanā that not only deposited stream-rolled gravels within the main channel but also likely inhibited much human activity and disrupted or destroyed most if not all Pleistocene sites in the lower elevations. The Early Holocene marked a calmer environment, with seasonal flooding of backwater marshlands pocked by stands of woody vegetation on higher ground. Landscapes generated through anthropogenic burning likely constructed favorable niches for grazing animals, whether hunted or herded or both. Contiguous stands of dry-season vegetation included reed beds and grassy mats that caught a spark and smoldered in its wake, providing ashy nutrients and stimulating new growth. By approximately 5100 cal BP, a major climatic shift widely influential across Southern Arabia precipitated an onset of channel incision and violent flooding (see chapter 18). Much of the former marshland was removed; other areas were stranded as silt terraces, without sufficient moisture to support the vegetation biomass of the Early Holocene. Lower biomass in Wādī Sanā in the Late Holocene period does not appear to accompany a significant loss in plant biodiversity, with most of the woody taxa represented throughout the paleoecological sequence. Above all, what we learned from the fine-grained study of Wādī Sanā is that the climatological emphasis on a dramatic pan-Arabian Middle to Late Holocene climatic shift masks subtle environmental changes, such as the seasonal waxing and waning of marshy grasslands, the aridification that constrained *Acacia ehrenbergiana*, and the stable spatial distributions of taxa such as *Delonix* and *Cyphostemma*.

Figure 3.31. A large hyrax midden covers a shelf in this small cave. The dark region of the foreground is where a small sample of the midden was broken off.. More of this midden was not sampled because a decomposing camel carcass lay on top of the midden to the left and the air was not breathable. *Photograph by Kenneth Cole.*

Figure 3.32. A midden collected from a ledge across the cave from figure 3.31: (a) a bisected view of the midden; (b) a closeup of the cross section through the midden; scale is in centimeters. Two horizontal laminae suggest that three layers may be present; AMS dating of specific layers is currently under way. Circular contents are hyrax fecal pellets; note two leaf macrofossils at bottom right. *Photographs by Kenneth Cole.*

References

Al-Hubaishi, Ahmed, and Klaus Müller-Hohenstein
1984 *An Introduction to the Vegetation of Yemen*. Deutsche Gesellschaft für Technische Zusammenarbeit, Eschborn, Germany.

Assaeed, Abdalaziz M.
1997 Estimation of Biomass and Utilization of Three Perennial Range Grasses in Saudi Arabia. *Journal of Arid Environments* 36:103–11.

Baas, Pieter
1986 Ecological Patterns in Xylem Anatomy. In *On the Economy of Plant Form and Function*, edited by Thomas J. Givnish, pp. 327–52. Cambridge University Press, Cambridge.

Berger, Jean-François, Jean-Paul Bravard, Louise Purdue, Anne Benoist, Michel Mouton, and Frank Braemer
2012 Rivers of the Hadramawt Watershed (Yemen) during the Holocene: Clues of Late Functioning. *Quaternary International* 266(0):142–61.

Beydoun, Ziad R.
1966 Eastern Aden Protectorate and Part of Dhufar. In *Geology of the Arabian Peninsula*, edited by R. W. Powers, L. F. Ramirez, C. D. Redmond, and E. L. Elberg Jr., pp. 1–49. Professional Paper 560H. U.S. Geological Survey, Washington, DC.

Birks, H. J. B., and Joy Deacon
1973 A Numerical Analysis of the Past and Present Flora of the British Isles. *New Phytologist* 72(4):877–902.

Biswell, Harold H.
1974 Effects of Fire on Chaparral. In *Fire and Ecosystems*, edited by T. T. Kozlowski and C. E. Ahlgren, pp. 321–64. Academic Press, New York.

Bokhari, Unab, Faisal Alyaeesh, and Mohmood Al-Noori
1990 Nutritional Characteristics of Important Desert Grasses in Saudi Arabia. *Journal of Arid Environments* 43:202–4.

Bollig, Michael, and Anje Schulte
1999 Environmental Change and Pastoral Perceptions: Degradation and Indigenous Knowledge in Two African Pastoral Communities. *Journal of Human Ecology* 27:493–514.

Bottema, Sytze
1991 Pollen Proxy Data from Southeastern Europe and the Near East. In *Evaluation of Climate Proxy Data in Relation to the European Holocene*, edited by Burkhard Franzel, Armand Pons, and Birgit Gläser, pp. 63–80. Fisher, Stuttgart.

Box, Thadis W.
1968 Range Resources in Somalia. *Journal of Range Management* 21:388–92.

Bronk Ramsey, C.
2009 Bayesian Analysis of Radiocarbon Dates. *Radiocarbon,* 51(1):337–60.

Chase, Brian M., Arnoud Boom, Andrew S. Carr, Michael E. Meadows, and Paula J. Reimere
2013 Holocene Climate Change in Southernmost South Africa: Rock Hyrax Middens Record Shifts in the Southern Westerlies. *Quaternary Science Reviews* 82:199–205.

Chase, Brian M., Louis Scott, Michael E. Meadows, Graciela Gil-Romera, Arnoud Boom, Andrew S. Carr, Paula J. Reimerg, Loïc Truc, Verushka Valsecchi, and Lynne J. Quick
2012 Rock Hyrax Middens: A Palaeoenvironmental Archive for Southern African Drylands. *Quaternary Science Reviews* 56:107–25.

Cole, Kenneth L., and Samantha T. Arundel
2005 Carbon 13 Isotopes from Fossil Packrat Pellets and Elevational Movements of Utah Agave Reveal Younger Dryas Cold Period in Grand Canyon, Arizona. *Geology* 33:713–16.

Crassard, Rémy
2008 *La préhistoire du Yémen. Diffusions et diversités locales, á travers l'étude d'industries lithiques du Hadramawt*. Archaeopress, Oxford.

Crassard, Rémy, Joy McCorriston, Eric Oches, ʿAbd Al-Aziz Bin ʿAqil, Julien Espagne, and Mohammad Sinnah
2006 Manayzah, Early to Mid-Holocene Occupations in Wādī Sanā (Hadramawt, Yemen). *Proceedings of the Seminar for Arabian Studies* 36:151–73.

Cremaschi, Mauro, and Fabio Negrino
2005 Evidence for an Abrupt Climatic Change at 8700 ^{14}C yr B.P. in Rockshelters and Caves of Gebel Qara (Dhofar-Oman): Palaeoenvironmental Implications. *Geoarchaeology* 20:559–79.

Davies, Caroline P.
2006 Holocene Paleoclimates of Southern Arabia from Lacustrine Deposits of the Dhamar Highlands, Yemen. *Quaternary Research* 66(3):454–64.

Davis, Margaret B., Ruth G. Shaw, and Julie R. Etterson
2005 Evolutionary Responses to Changing Climate. *Ecology* 86(7):1704–14.

Desta, Solomon, and D. L. Coppock
2002 Cattle Population Dynamics in the Southern Ethiopian Rangelands. *Journal of Range Management* 55:439–51.

Ellis, Jim, and Kathleen A. Galvin
1994 Climate Patterns and Land-Use Practices in the Dry Zones of Africa. *Bioscience* 44(5):340–49.

Ellis, James E., and David M. Swift

1988 Stability of African Pastoral Ecosystems: Alternate Paradigms and Implications for Development. *Journal of Range Management* 41:450–59.

Enzel, Yehouda, Yochanan Kushnir, and Jay Quade

2015 The Middle Holocene Climatic Records from Arabia: Reassessing Lacustrine Environments, Shift of ITCZ in Arabian Sea, and Impacts of the Southwest Indian and African Monsoons. *Global and Planetary Change* 129:69–91.

Evans-Pritchard, E. E.

1940 *The Nuer.* Oxford University Press, Oxford.

Fall, P. A., C. Lindquist, and S. Falconer

1990 Fossil Hyrax Middens from the Middle East: A Record of Paleovegetation and Human Disturbance. In *Packrat Middens: The Last 40,000 Years of Biotic Change*, edited by Julio L. Betancourt, Thomas R. Van Devender, and Paul S. Martin, pp. 408–27. University of Arizona Press, Tucson.

Fisher, J. F., K. L. Cole, and R. S. Anderson

2009 Using Packrat Middens to Assess Grazing Effects on Vegetation Change. *Journal of Arid Environments* 73:937–48.

Fleitmann, Dominik, Stephen J. Burns, Augusto Mangini, Manfred Mudelsee, Jan Kramers, Igor Villa, Ulrich Neff, Abdulkarim Al-Subbary, Annett Buettner, Dorothea Hippler, and Albert Matter

2007 Holocene ITCZ and Indian Monsoon Dynamics Recorded in Stalagmites from Oman and Yemen (Socotra). *Quaternary Science Reviews* 26(1–2):170–88.

Fleitmann, Dominik, Stephen J. Burns, Manfred Mudelsee, Ulrich Neff, Jan Kramers, Augusto Mangini, and Albert Matter

2003 Holocene Forcing of the Indian Monsoon Recorded in a Stalagmite from Southern Oman. *Science* 300(5626):1737–39.

Gasse, Françoise

2002 Diatom-Inferred Salinity and Carbonate Oxygen Isotopes in Holocene Waterbodies of the Western Sahara and Sahel (Africa). *Quaternary Science Reviews* 21(7):737–67.

Grime, J. P.

1977 Evidence for the Existence of Three Primary Strategies in Plants and Its Relevance to Ecological and Evolutionary Theory. *American Naturalist* 111:1169–94.

Harrower, Michael J.

2006 Environmental versus Social Parameters, Landscape, and the Origins of Irrigation in Southwest Arabia (Yemen). PhD dissertation, Department of Anthropology, The Ohio State University, Columbus.

Harrower, M., J. McCorriston, and E. A. Oches

2002 Mapping the Roots of Agriculture in Southern Arabia: The Application of Satellite Remote Sensing, Global Positioning System (GPS), and Geographic Information System (GIS) Technologies. *Archaeological Prospection* 9:35–42.

Harrower, Michael J., Eric A. Oches, and Joy McCorriston

2012 Hydro-Geospatial Analysis of Ancient Pastoral/Agro-Pastoral Landscapes along Wādī Sanā (Yemen). *Journal of Arid Environments* 86:131–38.

Hildebrandt, Anke, Mohammed Al Aufi, Mansoor Amerjeed, Mahaad Shammas, and Elfatih A. B. Eltahir

2007 Ecohydrology of a Seasonal Cloud Forest in Dhofar: 1. Field Experiment. *Water Resources Research* 43(10).

Hildebrandt, Anke, and E. A. B. Eltahir

2008 Using a Horizontal Precipitation Model to Investigate the Role of Turbulent Deposition in Survival of a Seasonal Cloud Forest in Dhofar. *Journal of Geophysical Research: Biogeosciences* 113(G4).

Hoorn, Carina, and Mauro Cremaschi

2004 Late Holocene Palaeoenvironmental History of Khawr Rawri and Khawr Al Balid (Dhofar, Sultanate of Oman). *Palaeogeography, Palaeoclimatology, Palaeoecology* 213(1–2):1–36.

Inizan, Marie-Louise, Anne-Marie Lézine, Bruno Marcolongo, Jean-François Saliège, Christian Robert, and Frédéric Werth

1997 Paléolacs et peuplements holocènes du Yémen: le Ramlat As-Sabat'ayn. *Paléorient* 23(2):137–49.

Jagiella, Christian, and Harald Kürschner

1987 *Atlas der Hölzer Saudi Arabiens*. Dr. Ludwig Reichert Verlag, Wiesbaden.

Janzen, Jörg

1986 *Nomads in the Sultanate of Oman: Tradition and Development in Dhofar*. Westview Press, Boulder.

Kimiaie, Masoumeh, and Joy McCorriston

2014 Climate, Human Palaeoecology, and the Use of Fuel in Wādī Sanā , Southern Yemen. *Vegetation History and Archaeobotany* 23:33–40.

Komex International Ltd.,

1997 Groundwater Resources Assessment, Hadramout–Masila Region, Republic of Yemen. Volume I Text. Unpublished report prepared for Ministry of Oil and Mineral Resources, Republic of Yemen, November 1997. Report No. KI94-3866 T18.

Kruger, F. J.

1984 Effects of Fire on Vegetation Structure and Dynamics. In *Ecological Effects of Fire in South African Ecosystems*, edited by Peter d. V. Booysen and Neil M. Tainton. pp. 219–31. Springer Verlag, Berlin.

Kürschner, Harald

1998 Biogeography and Introduction to Vegetation. In *Vegetation of the Arabian Peninsula*, edited by Shahina A. Ghazanfar and Martin Fisher, pp. 63–98. Kluwer Academic, Dordrecht.

Kürschner, Harald, Peter Hein, Norbert Kilian, and Mohammad Hubaishan

2004 The Hybantho durae-Anogeissetum dhofaricae ass. nova—Phytosociology, Structure and Ecology of an Endemic South Arabian Forest Community. *Phytocoenologia* 34:569–612.

Kutzbach, J. E., G. Bonan, J. Foley, and S. P. Harrison

1996 Vegetation and Soil Feedbacks on the Response of the African Monsoon to Orbital Forcing in the Early to Middle Holocene. *Nature* 384:623–26.

Kutzbach, J. E., X. Liu, Z. Liu, and G. Chen

2008 Simulation of the Evolutionary Response of Global Summer Monsoons to Orbital Forcing over the Past 280,000 Years. *Climate Dynamics* 30:567–79.

Lancaster, William, and Fidelity Lancaster

1999 Identities and Economies: Mountain and Coastal Ras al Khaimah. *Proceedings of the Seminar for Arabian Studies* 29:89–94.

Lézine, Anne-Marie

2009 Timing of Vegetation Changes at the End of the Holocene Humid Period in Desert Areas at the Northern Edge of the Atlantic and Indian Monsoon Systems. *Comptes rendus—Geoscience* 341(8–9):750–59.

Lézine, Anne-Marie, Christian Robert, Serge Cleuziou, Marie-Louise Inizan, Frank Braemer, Jean-François Saliège, Florence Sylvestre, Jean-Jacques Tiercelin, Rémy Crassard, Sophie Méry, Vincent Charpentier, and Tara Steimer-Herbet

2010 Climate Change and Human Occupation in the Southern Arabian Lowlands during the Last Deglaciation and the Holocene. *Global and Planetary Change* 72(4):412–28.

Lézine, Anne-Marie, Jean-François Saliège, Robert Mathieu, Thibaut-Louis Tagliatela, Sophie Méry, Vincent Charpentier, and Serge Cleuziou

2002 Mangroves of Oman during the Late Holocene: Climatic Implications and Impact on Human Settlements. *Vegetation History and Archaeobotany* 11(3):221–32.

Lézine, Anne-Marie, Jean-Jacques Tiercelin, Christian Robert, Jean-François Saliège, Serge Cleuziou, Marie-Louise Inizan, and Frank Braemer

2007 Centennial to Millennial-Scale Variability of the Indian Monsoon during the Early Holocene from a Sediment, Pollen and Isotope Record from the Desert of Yemen. *Palaeogeography, Palaeoclimatology, Palaeoecology* 243(3–4):235–49.

Marzin, C., and P. Braconnot

2009 Variations of Indian and African Monsoons Induced by Insolation Changes at 6 and 9.5 kyr BP. *Climate Dynamics* 33(2–3):215–31.

Marzin, C., P. Braconnot, and M. Kageyama

2013 Relative Impacts of Insolation Changes, Meltwater Fluxes, and Ice Sheets on African and Asian Monsoons during the Holocene. *Climate Dynamics* 37:2267–86.

Matter, Albert, Eike Neubert, Frank Preusser, Thomas Rosenberg, and Khalid Al-Wagdani 2015 Palaeo-Environmental Implications Derived from Lake and Sabkha Deposits of the Southern Rub' al-Khali, Saudi Arabia and Oman. *Quaternary International* 382:120–31.

McCabe, J. T.

2004 *Cattle Bring Us to Our Enemies: Turkana Economy, Politics, and Raiding in a Disequilibrium System.* University of Michigan Press, Ann Arbor.

McCorriston, Joy

2013 The Neolithic in Arabia: A View from the South. *Arabian Archaeology and Epigraphy* 24(1):68–72.

McCorriston, Joy, Michael J. Harrower, Louise A. Martin, and Eric A. Oches

2012 Cattle Cults of the Arabian Neolithic and Early Territorial Societies. *American Anthropologist* 114:45–63.

McCorriston, Joy, Michael J. Harrower, Eric A. Oches, ʿAbdalʿazīz Bin ʿAqīl, Catherine Heyne, AbdalBaset Noman, Rémy Crassard, Khalid BaDhofary, and Joshua Anderson

2005 Foraging Economies and Population in the Middle Holocene Highlands of Southern Yemen. *Proceedings of the Seminar for Arabian Studies* 35:143–54.

McCorriston, Joy, Michael Harrower, Tara Steimer, Kimberly D. Williams, Matthew Senn, Masʿūd Al Hādhari, Masʿūd Al Kathīrī, ʿAli Aḥmad Al Kathīrī, Jean-François Saliège, and Jennifer Everhart

2014 Monuments and Landscape of Mobile Pastoralists in Dhufar: The Arabian Human Social Dynamics (AHSD) Project 2009–2011. *Journal of Oman Studies* 19:117–43.

McCorriston, Joy, Eric A. Oches, Dawn Walter, and Kenneth Cole

2002 Holocene Paleoecology and Prehistory in Highland Southern Arabia. *Paléorient* 28(1):61–88.

Meister, J., M. A. Hubaishan, N. Kilian, and C. Oberprieler
2006 Temporal and Spatial Diversification of the Shrub Justicia areysiana Deflers (Acanthaceae) Endemic to the Monsoon Affected Coastal Mountains of the Southern Arabian Peninsula. *Plant Systematics and Evolution* 262(3):153–71.

Mooney, Harald A., and E. L. Dunn
1970 Convergent Evolution of Mediterranean-Climate Evergreen Sclerophyll Shrubs. *Evolution* 24:292–303.

Naveh, Zev
1974 Effects of Fire in the Mediterranean Region. In *Fire and Ecosystems*, edited by T. Kozlowski T. and C. E. Ahlgren, pp. 401–34. Academic Press, New York.
1975 The Evolutionary Significance of Fire in the Mediterranean Region. *Vegetatio* 29:199–208.

Oberprieler, Christoph, Jörg Meister, Christine Schneider, and Norbert Kilian
2009 Genetic Structure of Anogeissus dhofarica (Combretaceae) Populations Endemic to the Monsoonal Fog Oases of the Southern Arabian Peninsula. *Biological Journal of the Linnean Society* 97(1):40–51.

Parker, Adrian G., and Andrew S. Goudie
2008 Geomorphological and Palaeoenvironmental Investigations in the Southeastern Arabian Gulf Region and the Implication for the Archaeology of the Region. *Geomorphology* 101(3):458–70.

Parker, Adrian G., Andrew S. Goudie, Stephen Stokes, Kevin White, Martin J. Hodson, Michelle Manning, and Derek Kennet
2006 A Record of Holocene Climate Change from Lake Geochemical Analyses in Southeastern Arabia. *Quaternary Research* 66(3):465–76.

Pendall, E., J. Betancourt, and S. Leavitt
1999 Paleoclimatic Significance of δD and δ13C Values in Pinyon Pine Needles from Packrat Middens Spanning the Last 40,000 Years. *Palaeogeography, Palaeoclimatology, Palaeoecology* 147:53–72.

Pignatti, Sandro
1979 Plant Geographical and Morphological Evidences in the Evolution of the Mediterranean Flora (with Particular Reference to the Italian Representative). *Webbia* 34:243–55.

Pons, A., and P. Quézel
1958 Première remarques sur l'étude palynologique d'un guano fossile du Hoggar. *Comptes Rendus des Scéances de L'Académie des Sciences* 244:2290–92.

Pratt, D. J., P. J. Greenway, and M. D. Gwynne
1966 A Classification of East African Rangeland. *Journal of Applied Ecology* 3:369–82.

Pratt, D. J., and M. D. Gwynne
1977 *Rangeland Management and Ecology in East Africa*. Robert E. Krieger, Huntington, NY.

Reimer, P. J., E. Bard, A. Bayliss, J. W. Beck, P. G. Blackwell, C. Bronk Ramsey, P. M. Grootes, T. P. Guilderson, H. Haflidason, I. Hajdas, C. Hatte, T. J. Heaton, D. L. Hoffmann, A. G. Hogg, K. A. Hughen, K. F. Kaiser, B. Kromer, S. W. Manning, M. Niu, R. W. Reimer, R. D. A. Richards, E. M. Scott, J. R. Southon, R. A. Staff, C. S. M. Turney, and J. van der Plicht
2013 IntCal13 and Marine13 Radiocarbon Age Calibration Curves 0–50,000 Years Cal BP. *Radiocarbon* 55(4):1869–87.

Sale, J. B.
1980 The Ecology of the Mountain Region of Dhofar. In *The Scientific Results of the Oman Flora and Fauna Survey (Dhofar) 1977*, edited by S. N. Shaw-Reade, J. B. Sale, M. D. Gallagher, and R. H. Daly, pp. 25–54. Diwan of H.M. for Protocol, Sultanate of Oman.

Scott, L.
1990 Hyrax (Procaviidae) and Dassie Rat (Petromuridae) Middens in Palaeoenvironmental Studies in Africa. In *Packrat Middens: The Last 40,000 Years of Biotic Change*, edited by J. Julio L. Betancourt, Thomas R. Van Devender, and Paul S. Martin, pp. 398–407. University of Arizona Press, Tucson.

Thompson, R. S.
1985 Palynology and Neotoma Middens. In *The Late Quaternary Vegetation and Climates of the American Southwest*, edited by Bonnie F. Jacobs, Patricia L. Fall, and Owen K. Davis, pp. 89–112. AASP Contributions Series 16. American Association of Stratigraphic Palynologists, Dallas.

Trabaud, L.
1981 Man and Fire: Impacts on Mediterranean Vegetation. In *Mediterranean-Type Shrublands*, edited by Francesco di Castri, David W. Goodall, and Raymond L. Specht, pp. 523–38. Elsevier Scientific, Amsterdam.

Tesfaye, Getachew, Teketay Demel, Yoseph Assefa, and Masresha Fetene
2004 The Impact of Fire on the Soil Seed Bank and Regeneration of Harenna Forest, Southeastern Ethiopia. *Mountain Research and Development* 24(4):354–61.

Van Devender, T. R.
1973 Holocene Woodlands in Southwestern Deserts. *Science* 198:189–92.

Vogl, Richard J., P. Wayne, Keith L. White, and Kenneth L. Cole
1977 The Closed-Cone Pines and Cypresses. In *Terrestrial Vegetation of California*, edited by Michael G. Barbour and Jack Major, pp. 295–305. John Wiley and Sons, New York.

Wallace, Michael, and Michael Charles
2013 What Goes in Does Not Always Come Out: The Impact of the Ruminant Digestive System of Sheep on Plant Material, and Its Importance for the Interpretation of Dung-Derived Archaeobotanical Assemblages. *Environmental Archaeology* 18(1):18–30.

Wells, Philip V.
1969 The Relation between Mode of Reproduction and Extent of Speciation in Woody Genera of the California Chaparral. *Evolution* (23):264–67.

Wells, P. V., and C. D. Jorgensen
1964 Pleistocene Wood Rat Middens and Climate Change in the Mojave Desert: A Record of Juniper Woodlands. *Science* 143:1171–74.

Wright, Henry A.
1974 Range Burning. *Journal of Range Management* 27:5–11.

Wright, Henry E., and Neil Roberts
1993 Vegetational, Lake-Level, and Climatic History of the Near East and Southwest Asia. In *Global Climates since the Last Glacial Maximum*, edited by H. E. J. Wright, J. E. Kutzbach, T. Webb III, W. A. Ruddiman, F. A. Street-Perrott, and P. J. Bartlein, pp. 194–220. University of Minnesota Press, Minneapolis.

Part II

Archaeological Survey: Methods and Site Distributions

Chapter 4

History of the RASA Survey

Methodologies from Broad Exploration to Intensive Studies

Joy McCorriston and Michael J. Harrower

From the outset, the RASA archaeological survey sought to locate sites and settlements linked to the earliest introductions of domesticates and prehistoric landscapes of farming and pastoralism in Southern Arabia. In 1996 the General Organization for Antiquities, Museums, and Manuscripts, Yemen granted Joy McCorriston and Sheila McNally (University of Minnesota) a broad survey permit for Shabwa and Hadramawt Provinces. Our first visits were reconnaissance journeys, driving and walking through the Wādī Mayfaʿah, Wādī ʿAmāqīn, Wādī Ḥabbān, Wādī Ḥajar, and Wādī Fuwwah; the southern coastline; several routes across the Southern Jol; and Wadi Do'an, Wādī ʿAmd, Wādī al-ʿAyn, and Wādī 'Idim—the great southern tributaries to Wādī Ḥaḍramawt. Like our predecessors, we assumed that somewhere in the wilderness lay ancient settlements, long ago abandoned by farming folk and visited since only by mobile tribesmen like the bedouin guides we occasionally persuaded to help us. Even on first encounter, local people took no money and offered no threat, so we had wonderful help in accessing remote places, often guided by, dependent upon, and hosted by total strangers. Without their help, we would not have accomplished the broad explorations that led us to select Wādī Sanā and Wādī ʿIdim for intensive survey and excavation (figure 4.1).

Thus began a decade of exploratory and intensive survey, with an emphasis on the existing knowledge of local guides. In 1998 the first RASA team returned to the Southern Jol with a systematic plan for site discovery. Following on the promising results and generous sharing of information from a team from the Deutsches Archäologisches Institut, commissioned to conduct the Canadian Occidental (CANOXY-Yemen) oil pipeline survey (Vogt and Sedov 1994), the RASA crew explored widely between Wādī ʿIdim and Wādī Sanā. In 2000–2004, the RASA Project settled on Wādī Sanā for intensive coverage of randomly selected strips across the drainage system, supplemented with test excavations, assemblage analysis, and radiocarbon dating to establish a basic cultural-historical sequence. In 2000 Michael Harrower joined the team to assist with satellite imagery and GPS mapping (Harrower et al. 2002). In 2004 Rémy Crassard joined with a research interest in the spatio-temporal definition and distributions of formal lithic types and their cultural implications (Crassard 2007, 2008; Crassard and Bodu 2004). In 2005, while excavations continued at selected sites, Tara Steimer joined the RASA Project, bringing expertise in tombs and monuments of the third millennium BCE Bronze Age in Arabia. The team piloted a new survey of small-scale stone monuments, the so-called cairn survey, which led to new funding and research foci as the 2008 Arabian Human Social Dynamics (AHSD) Project (Bin 'Aqīl and McCorriston 2009; McCorriston et al. 2011). In this chapter, we recount the history of the RASA survey, including methods and their development over time. We review details of topic-specific surveys for lithics, water management, and cairns in chapter 5; a statistical analysis of survey results is presented in chapter 6.

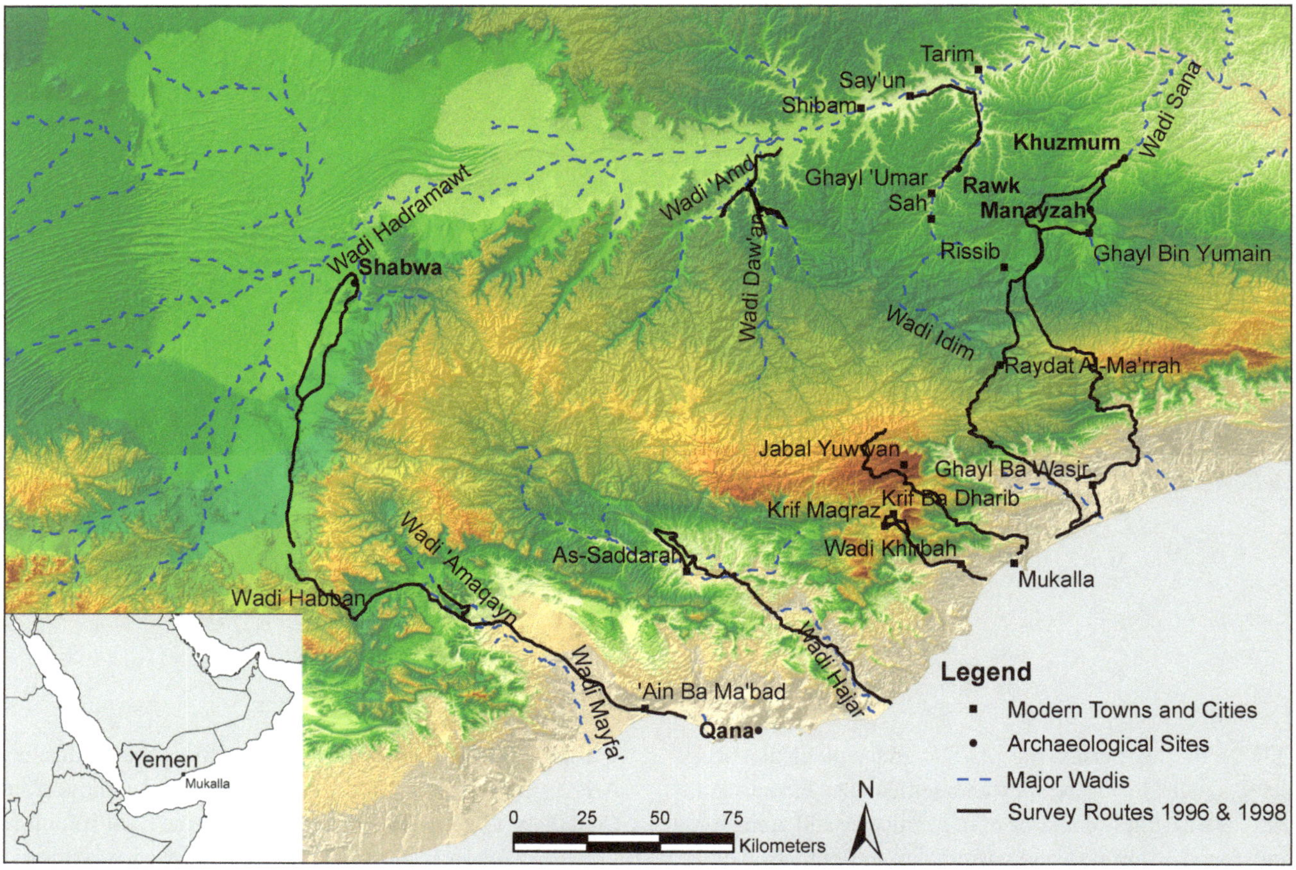

Figure 4.1. Southern Jol Exploratory survey in 1996 and 1998. Topography and hydrology based on Shuttle Radar Topography Mission (SRTM) version 4. *Illustration by Michael Harrower.*

Exploration (1996)

In winter 1996, the RASA team made an exploratory trip of a few weeks to visit well-known and excavated sites to explore archaeological ruins identified by local people. Most of these proved historical in date (glazed ceramics, graffiti and inscriptions, sculpted and worked architectural elements, and rectilinear architectural plans), while others were simply unknown, perched on bare rock, and offered little prospect for dating from surface indications, especially to our inexperienced eyes. A second trip included medieval archaeologist Ingrid Hehmeyer and an introduction to the renowned ʿAbdalʿazīz Bin ʿAqīl, widely admired among his countrymen for his experience, formal training abroad, and independent scholarship. "If you want to learn about Hadramawt," we were told, "go speak with Dr. ʿAbdalʿazīz."

Easier said than done, for neither my Arabic nor Ingrid's was strong. Yet ʿAbdalʿazīz offered a very gracious welcome, even if he might reasonably have had some reservations about the two Western women who appeared at his breezy sea-view office in the old Sultan's Palace in Mukalla. We had little introduction, inadequate language skills, and poor backgrounds in Hadramawt customs and history. But we had a vehicle; he had an unflagging professional curiosity; and just weeks previously a few tribesmen had dropped by the museum to report antiquities in the hinterlands above their village. Thus we found ourselves together on archaeological survey, relying on local informants to locate settlements and other sites in the vast and inaccessible terrain of southern Yemen's dissected limestone plateau.

Relying on local informants differs from the more systematic survey methods recommended to archaeology students since the 1970s, and certainly we strayed far from the rigorous plans I had laid in my faculty office at University of Minnesota. When we (Ingrid, Samīr, ʿAbdalʿazīz, and I) set out on Arabia's Southern Jol with a few very young tribal guides, we expected to climb a few hours to a place they knew well and start our survey there. We drove up to the last village on a dirt track past date plantings, flowing springs, and the dry gravel bed of a deep canyon, where the road ended. We were at the foot of the escarpment cliff at the head of Wādī Khirbah, one of the short, southward-draining wadis that segments southern Yemen's coastal plain. Women waited in the car while ʿAbdalʿazīz, seated on the stone curb of a house, spoke with the clustering men, asking who knew the route, how far was the rumored site, would they take us there? Craning, peering children increased in number, shutting out the circle of men so that we could see only dusty legs in cheap sandals. Whispers fluttered, while men's conversation rose and fell. I remember the oddest details, like the kindness of my new companion, ʿAbdalʿazīz, who absently caressed the head of an unkempt child picking at a discarded juice box with a filthy finger. A woman curiously watched from uphill. Meeting my gaze, she turned and disappeared, the bundle on her head pivoting as if upon a swivel. If we two women thought ourselves unobtrusive in the car, we fooled only ourselves. Suddenly we were departing with hasty triage. Carry the blanket, water, flashlight, film, food, GPS, and camera. Leave the car, driver, cash, equipment, and decisions. When one guide relieved me of my shoulder bag, I thought we had porters. When a steady grip on my elbow guided me along a fog-faced cliff, I realized we had hosts. Somewhere along the way I asked myself, "What would be worse: to climb all this way and find interesting sites too inaccessible for further study (we did, and they were) or to find nothing of interest after such effort?" (figure 4.2).

Figure 4.2. Jibal Yuwān. Hiking up the escarpment in 1996. *Photograph by Joy McCorriston.*

Our "two-hour" climb with young hosts from the Bani Hassan took two days to get us to ancient sites in the southern plateau, where no archaeologists had ventured previously. We struggled with holy Ramadan, semi-fasting (some did, some did not, but no one had sufficient food), and thirst. Many hours into our climb up the escarpment, we broke the fast with a rifle shot, water, and dates cached in sticky plastic that emerged from every pocket and tied-off shawl. Merriment turned to sobriety and prayer as a sundown fog swept suddenly cold up the canyon and onto the plateau. By moonlight we continued to a campsite beside a foamy pool edged with clotted camel dung and algal blooms. (We drank.) In the morning, we again began walking, sluggishly, warming slowly in single file, toting guns and bundles among the heavy dew and glowing lichens. It was low stone tumuli that we came to first, and cracked upon them were stelae carved in sparse style—men with beards, hands, belts, and straight daggers with pommel handles (McCorriston 2012). As elsewhere in the mountains, numerous small-scale stone monuments, such as wall tombs dated to the late third millennium BCE, lie along routes and promontories (McCorriston et al. 2011). Near Krīf Khyḍār, we passed many 4–9 m diameter tumuli with up to half a dozen upright anthropomorphic stelae, sometimes dated by the shape of daggers comparable to second-millennium metal types in Iran and Syria-Palestine (Moorey 1974; Newton and Zarins 2000; Philip 1995; Vogt 1997). As intriguing are the unknown mechanisms by which metal dagger styles found their way into these remote highlands 4,000 years ago, our immediate concerns were thirst and the treacherous footing on which one risked a turned ankle and a long wait for a camel (figures 4.3 and 4.4).

Every destination was another "hour" and then another. By the time we reached Krīf Maqrāẓ (Krif Magrad), we had been walking six hot hours without water. For us, the most immediate feature of the location was the *krif*, or artificially enhanced rainwater pool. This *krif*, we noted with quiet despair, was quite dry, and this circumstance precluded more than a brief visit to the adjacent settlement remains, with their conjoined curvilinear house

Figure 4.3. A *krif,* or seasonal pool fed by monsoon rains and artificially enhanced to capture runoff. Such sources provide seasonal access to upland basin pastureland. *Photograph by Catherine Heyne.*

Figure 4.4. Krīf Badrīb, in an upland basin near Jibal Yuwān. Anthropomorphic stelae surround a low stone tumulus. This unexcavated example probably is a Bronze Age tomb. *Photograph by Joy McCorriston.*

foundations, orthostat entrances, deep stone mortar, and thin scatter of ceramics and artifacts. Possibly this was the seasonal settlement of people whose dead lay in the stone tumuli some hours behind us.

The site at Krif Magrad lies on the southern gravel terrace along a shallow drainage of the upland basin near Jibal Yuwān, which is bounded by the principal fault escarpment to the south and by faulting (Wādī Ḥuṭī) to the north. Faulting and uplifting have created a basin atop the plateau and trapped sufficient sediment for limited runoff agriculture in select locations, which plausibly could have sustained a modest settlement of a few families for at least part of a year. In plan, the site resembles Bronze Age settlements of the northern Yemen highlands, with clustered, multiroom compounds, rooms up to 4 m in diameter, central stone pillar supports, and orthostat doorways (de Maigret 1990). Walls were of dry cobble masonry preserved to about half a meter in height, three to five courses high, and several courses thick. Internal hearths of thermally altered rock were constructed of smooth cobbles brought from a wadi bed. There were sparse surface ceramics. Surface lithics yielded only nondiagnostic debitage of multiple types of raw material. A 70 cm deep massive limestone mortar at the west end of the site suggested in situ processing, such as dehusking of plant food (figure 4.6).

Although there is no independent date for the Krif Magrad settlement, other settlements like it date to the Bronze Age—about 2500–1800 BCE (de Maigret 1990)—and the tumuli and wall tombs within a few kilometers probably date to the end of the third millennium BCE (McCorriston et al. 2011). Precipitation declined during the late fourth and third millennium BCE, reducing productive grasslands and pasturage for herders and narrowing the numbers and duration of rain-fed pools that made pasture accessible (Harrower et al. 2012). Small-scale agriculture would have been possible on the escarpment where check dams and sluices today retain and manage water runoff in field systems that may themselves date back many millennia (Harrower 2016:97–104).

But Krif Magrad and adjacent sites were a 12-hour walk from the nearest village and a good four-hour walk from the nearest car track, making further archaeological study there impractical. We returned after nightfall. Children lined our staggered entry to the village in Wadi Khirbah, from which our painful descent down the escarpment cliff had been watched all afternoon. Our arrival was a triumph. Village teens, all carrying automatic weapons to burnish their status as tribesmen, had safely guided two odd women and other strangers through the mountains. We broke Ramadan fast at Sulaymān's brother's tower house with one last, numbing climb up six flights to find our driver in one corner, already well fed and jesting with our gentle hosts.

Extensive Survey (1998)

Like any new archaeological research project initiated by foreigners in new terrain, RASA began with many questions and little understanding of the natural and cultural processes that formed the archaeological landscape. In the remote highlands of southern Hadramawt Province, where

Figure 4.5. Krif Magrad. Habitation site of about 10 circular enclosures, linked by an exterior wall and sheltered by a small cliff. This is probably a pastoral seasonal camp occupied in the Bronze and Iron Ages. *Photograph by Joy McCorriston.*

Figure 4.6. Stone mortar on the surface at Krif Magrad. (The use of stone as a pestle, as imagined here, is unconfirmed, but the mortar does bear clear traces of interior abrasion by undetermined material). *Photograph by Joy McCorriston.*

there had been almost no previous archaeological fieldwork, non-Yemeni team members faced a steep learning curve in adapting experience in the eastern Mediterranean to the mountainous terrain of Southern Arabia. Yemeni colleagues faced a different challenge—how to adapt to different field methods and different research goals from those imparted through joint Soviet–Yemeni projects at the huge urban centers of classical antiquity. In the forging of different perspectives and teaching one another, the team developed highly successful strategies for documenting and interpreting prehistoric Hadramawt.

To better understand site distributions and preservation processes in the arid uplands of Southern Arabia, the RASA Project conducted both extensive site discovery and intensive landscape survey. Until late-twentieth-century archaeological work in highland northern Yemen (de Maigret 1990; Edens et al. 2000; Ghaleb 1991; Lewis et al. 2010), Dhufar (Zarins 2001), and the RASA Project, there had been a puzzling dearth of settlement sites predating the appearance of complex centers and the fully complex societies of the Iron Age "caravan kingdoms" (Gunter 2005). Given the existence of large centers dependent on farming in late antiquity (e.g., Breton 1991; Schmidt 1988; Sedov and Griaznevich 1996), one would expect prior settlement and a development of irrigation technologies in the same areas. Conventional expectations that urban complexity emerged from the manipulation and elite appropriation of a farming surplus (Fried 1967:186; Wright 1977) suggest that there should have been prehistoric Arabian farming villages. Recent studies of sediments have also confirmed irrigation around state capitals, including Ma'rib, as early as the third millennium BCE, but without clear evidence of associated settlements (Kühn et al. 2010; Pietsch et al. 2010). This monograph summarizes our contributions to these important issues of long-term history, with specific focus in the Hadramawt Mountains.

Fully appreciating the contributions of the RASA archaeological surveys requires an understanding of preexisting challenges and gaps in prior archaeological research. One problem for archaeologists lies in locating ancient remains where itinerant farmers were committed to a lifestyle of mobile pastoralism. Another is that

Holocene land surfaces have changed. And only limited vehicle access is possible. Much of Southern Arabia in the late twentieth century remained nearly as inaccessible as it was in the late nineteenth century, but some major routes have been paved, and archaeological knowledge has developed accordingly along them. Archaeological explorations have emphasized major, easily recognizable, and relatively accessible sites in the deep Wādī Ḥaḍramawt itself, along with the margins of the desert interior, coastal plains, and major travel routes. These lowland areas can also be examined with aerial photography, in which features and sites undisturbed by recent farming sometimes show quite clearly (e.g., Gentelle 1991, 1998). Highland sites in southern Yemen, on the other hand, not only have remained relatively inaccessible but also suffer some of the visibility problems noted in other highlands of the Near East (Banning 1996), namely erosion and deposition in highly dissected terrain and the overlay of modern life in locations of constrained arable land and water resources, such as around highland springs.

In 1998 the RASA team conducted a vehicle survey across the upper and middle drainages of Wādī Sanā and Wādī ʿIdim, focusing on areas in which modern populations are low today but where sufficient water could have supported settlement in the past. Closely spaced field walking at 10 to 50 m transect intervals in modern, spring-fed oases such as Ghayl Bin Yumain, Ghayl ʿUmar, as-Ṣadārah, and Ghayl Bā Wazīr showed that date palm agriculture and the intensity of adjacent modern settlement destroy or bury surface traces of early agricultural settlement, if such existed. This combination of factors excludes much of the Southern Jol, and survey quickly focused on wadis and areas adjacent to cultivable sediments. Initial efforts built upon the 1993 CANOXY-Yemen oil pipeline survey carried out 70 km east of Wādī ʿIdim (Vogt and Sedov 1994). Site discovery was expanded to upstream tributaries of Wādī ʿIdim not documented as part of the earlier Russian archaeological survey in Wādī Ḥaḍramawt (Sedov 1996; Sedov and as-Saqqaf 1992) and to fill gaps in previous exploratory survey.

The RASA team used four-wheel-drive vehicles, Soviet 1:100,000-scale maps, and Landsat satellite imagery to select tributaries for complete coverage of wadi bottoms, lower terraces, and slopes. It proved impossible to acquire aerial photographs, even those of the 1930s, since access was restricted by security concerns. Vehicles dropped a team at the farthest upstream location reached by vehicle track. From this point, team members walked widely spaced intervals to return to the vehicle at the mouth. Team members focused on different aspects of the record, checking sediment sections for features in the profile, rockshelters, boulders that provide shade, and natural terrace surfaces. Because the team specifically targeted settlement discovery, individuals ranged as far as 500 m apart. In open terrain without vegetation, this assured visibility of all protrusive architecture, such as houses, structures, room clusters, and prominent platforms. Because the target of survey was settlement sites, not all fourth/third- and first-millennium high circular tombs (HCTs) were documented during this phase of fieldwork.

It became quickly apparent that there was little likelihood of finding archaeological settlement remains in the vicinity of modern villages, which tend to occur near the same water sources that would have been attractive in antiquity. For example, the modern houses in the village of Risib (E 294912, N 1708377)[1] cluster around a cistern and are built on gravel terraces and bedrock. The attraction for settlers is the broad expanse of silty soils across the shallow drainage, which has been bulldozed end to end to create retention bunds for runoff-water farming. This reconnaissance and others like it in villages of Sāh (E 270086, N 1724640), Sikdān (E 275226, N 1733357), Ghāyl Ba Wazīr (E 324529, N 1634809), Ghāyl Bin Yumain (E 324150, N 1720135), and Wādī Ḥarū (E 304625, N 1722793) allowed the team to assess preservation of sites and paleoecological records where water sources today support substantial population. Although occasional nondiagnostic chert flakes may be found in isolation in agricultural fields, the team found no concentrations of material culture suggestive of ancient occupations (figure 4.7).

In regions unaffected by modern agriculture—today accomplished by bulldozers scraping large bunded basins and by canalization for date palm irrigation—the team found good surface preservation and the material remains of ancient peoples. In the upper reaches of Wādī ʿIdim, north of the springs at Ghayl ʿUmar (E 270160, N 1733322), were the traces of ancient springs. These are manifest as tufa outcrops at the sides of the Tertiary wadi bed and isolated by later Holocene erosion of Early Holocene sediments that had accumulated around them. Lithic tools and chipping debris are scattered along the former margins of pools and oxbows in alluvium spread beside the former springs. On rock and gravel terraces upslope from tufa outcrops are the remains of high circular tombs and scattered, thinly stratified houses, with one dated occupation in the second millennium BCE (McCorriston 2000; chapter 16 this volume).

Earlier occupation seemed likely in the middle Wādī Sanā, where the 1993 pipeline survey had located remains

Figure 4.7. Farmland at Risib, Southern Jol. *Photograph by Joy McCorriston.*

of Early Holocene tool production at a site that suggested buried architectural remains. Therefore it was to Wādī Sanā, with its Arabian Neolithic architecture, that RASA directed further survey efforts. In 1998 the extensive survey team drove the middle Wādī Sanā from the military checkpoint (E 329121, 1741244 N) to the confluence of Wādī Sanā and Wādī as-Shumlyah. Holocene terraces bordering the modern wadi channel were explored on foot and by vehicle, as were lower bedrock and gravel terraces of the confluence area, known to local bedouin as Khuzma as-Shumlya (E 335980, N 1745365), as well as select small tributaries.

Throughout the regions covered by extensive survey—upper Wādī ʿIdim and selected small tributaries (*shiʿb*), middle Wādī Sanā, and the northern margins of Ghayl Bin Yumain—the RASA team registered visible sites and concentrations of archaeological remains. These records, while neither comprehensive nor all dated to period, serve as the basis for a broad typological categorization of material remains and provide broader insight into spatial distribution than afforded by the subsequent, more intensive and systematic survey of Wādī Sanā.

Mapping and Intensive Survey (2000–2005)

After two seasons of exploration (1996) and reconnaissance survey (1998) across the Southern Jol, it became clear that few settlements exist in the arid highlands and that detailed, systematic landscape survey—"siteless survey" or "distributional archaeology" (Dunnell and Dancy 1983; Ebert 1992) was particularly appropriate. Wide area coverage provides for better documentation of existing archaeological remains while also revealing natural and cultural site formation processes that bury, preserve, erode, scatter, concentrate, collapse, stratify, and differentially explain spatial patterning of material culture on current land surfaces today. Because these processes are rarely studied and poorly understood in South Arabian landscapes (cf. Fedele 1990; Wilkinson 1997, 1999, 2003), interdisciplinary landscape history is greatly beneficial to understanding the preservation and distribution of archaeological records in Hadramawt. Accordingly, the RASA Project adopted and adapted methods of nonsettlement documentation from Mediterranean surveys (Cherry et al. 1991; Terrenato and Ammerman 1996). Given Hadramawt's rugged and highly inaccessible terrain, these methods were far more appropriate than attempting "full-coverage regional" survey (Kowalewski 2008), which would have taken decades and yielded little, as most time would have been spent walking rocky upland areas devoid of water, with scant archaeological remains and no chance of encountering stratified sites suitable for excavation. Instead, RASA used stratified random and opportunistic sampling strategies to generate a representative sample of archaeological remains and landforms along Wādī Sanā while also targeting high-potential areas.

Throughout extensive survey, the RASA team struggled with inadequate maps and satellite imagery for local orientation, navigation, documenting topography, and site locations. Russian maps produced at 1:100,000 scale from aerial photography[2] and similarly scaled preprinted Landsat satellite image maps were helpful in planning routes and locating small tributaries,[3] but detailed documentation of the spatial relationships among sites requires higher spatial resolution. For the documentation of sites in Wādī Sanā, the team produced its own image maps in late 1999 from Landsat-5 imagery (Harrower et al. 2002; McCorriston and Harrower 2005).[4] While the 30 m spatial resolution of Landsat-5 allowed only relatively coarse mapping of landforms, it did offer substantial improvements in orienting and navigating the field team and led to more effective spatial understanding and comparisons of site locations.

The survey team began in 2000 with a detailed study of an area of interest, the landforms of middle Wādī Sanā around the confluence of Wādī Sanā and Wādī as-Shumlyah, and later expanded to a wider area up and down Wādī Sanā. Throughout our survey records, the designation "targeted" refers to the nonrandom choice of areas, including the confluence area around an inselberg known locally as Khuzma as-Shumlya. This location stands out

Figure 4.8. Khuzma as-Shumlya inselberg in the middle Wādī Sanā lies at the confluence of Wādī Shumlyah and Wādī Sanā. The surrounding silt terraces are fairly extensive and contain a remarkable array of sites, including open-air occupations adjacent to rockshelters, half-buried monuments, concentrations of animal bone, hearths, surface accumulations of chipped and ground stone, water management channels and dams, paleosol layers, and in situ burned vegetation. The ground-penetrating radar survey in the foreground (Tim Archer, Arrow Geophysics) produced no results that the team could confirm by subsequent testing. *Photograph by Joy McCorriston.*

as one of the very few inselbergs in the entire drainage, and indeed, such features are uncommon in the dendritic drainages of the Southern Jol. Exploratory survey had already noted important archaeological remains in this area. We endeavored to more comprehensively record them, including structures, artifact scatters, bone scatters, rockshelters, tombs, rock cairns, small-scale monuments, pits, burned layers within sediment profiles, water management structures, graffiti and rock art, trackways, and hearths, with specific attention paid to areas with stratified archaeological deposits. The archaeological record made this a promising locale at which to focus an intensive and multidisciplinary study of geomorphological processes, landscape history, paleoecology, and distributional archaeology. "Targeted" survey also refers to other survey units chosen from satellite imagery and from on-location decisions to investigate areas of particular interest—around a water trap, dried springs, or a particular branch of a catchment (figure 4.8).

The RASA team used geomorphological landform classes to structure archaeological survey. Because it was not immediately clear how natural and post-occupational processes had impacted cultural remains, it was imperative that cultural materials be documented in close and accurate association with the preservational environment—that is, the geomorphological landform class—on which they are found today (Gregory 2004:19; Terrenato and Ammerman 1996). The team defined a series of landforms that categorized the land cover of Yemen's Southern Jol. These seven landforms (plateau, bedrock slope, scree slope, bedrock terrace, gravel terrace, wadi silts, and wadi channel) include all the terrain of the Wādī Sanā watershed, regardless of underlying bedrock formations, which vary from south to north (chapter 3). These landform classes were mapped with 61 percent accuracy in unsupervised classification of Landsat imagery (Harrower et al. 2002) and can be readily identified along the 80 km extent of the main Wādī Sanā channel that formed the principal survey area (figure 4.9). Although satellite imagery landform classification accuracy was later improved to 67 percent using ASTER imagery (Harrower 2006:134–35; figure 4.10), our initial work with Landsat proved instructive in helping

Figure 4.9. Schematic cross section of middle Wādī Sanā that identifies landform classification categories. (For a color version of the background image, see chapter 1, figure 1.2.) *Photograph by Joy McCorriston. Illustration by Daniel Alvarez.*

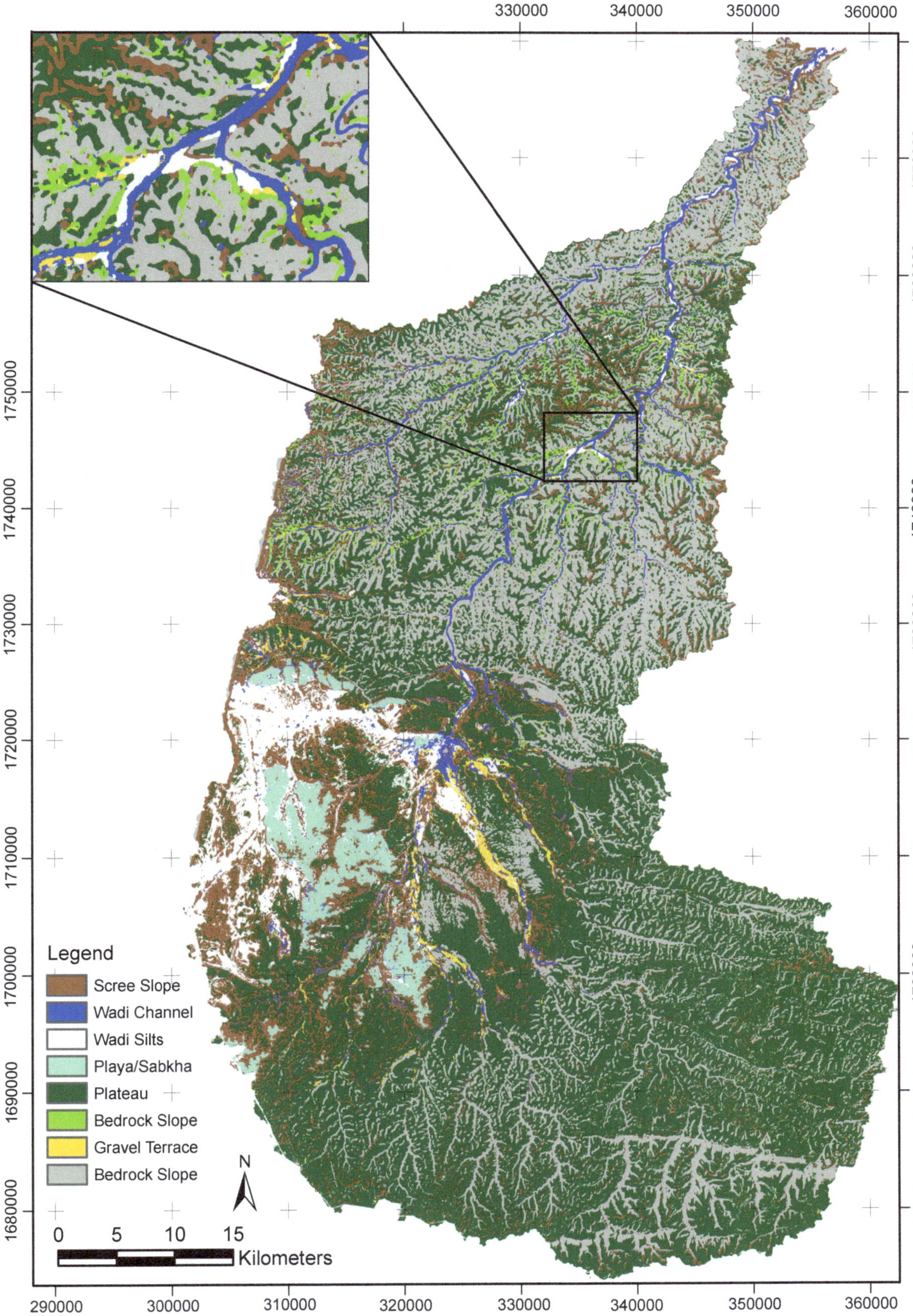

Figure 4.10. Landform classification map for the Wādī Sanā watershed produced by supervised classification of ASTER satellite imagery (Harrower 2006:134–35). Coordinates in UTM Zone 39 North WGS84. *Illustration by Michael Harrower.*

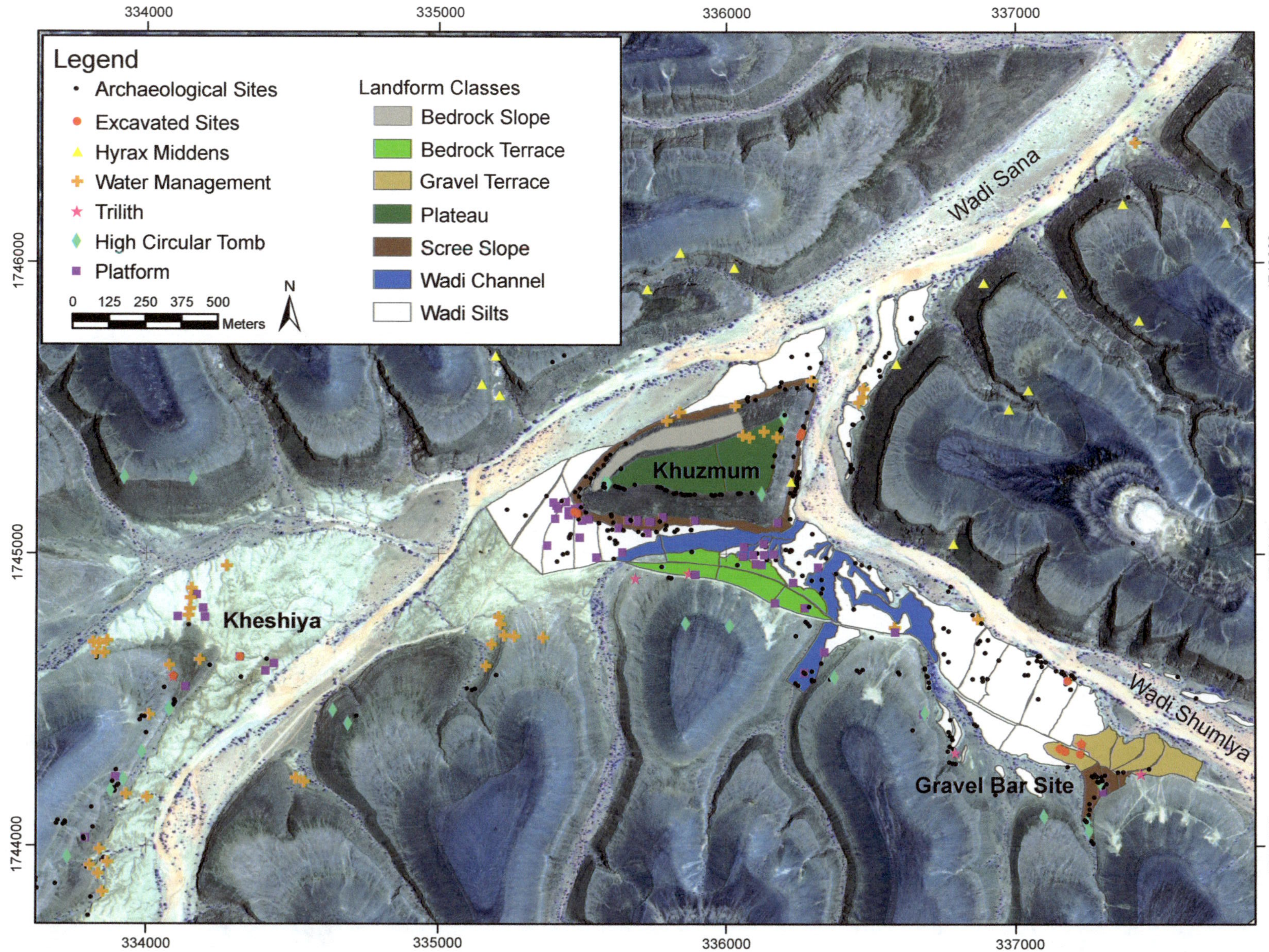

Figure 4.11 Map of the RASA geographic information system (GIS) for the Khuzma as-Shumlya area at the confluence of Wādī Sanā and Wādī as-Shumlyah (background QuickBird satellite imagery). Coordinates in UTM Zone 39 North WGS84. *Illustration by Michael Harrower.*

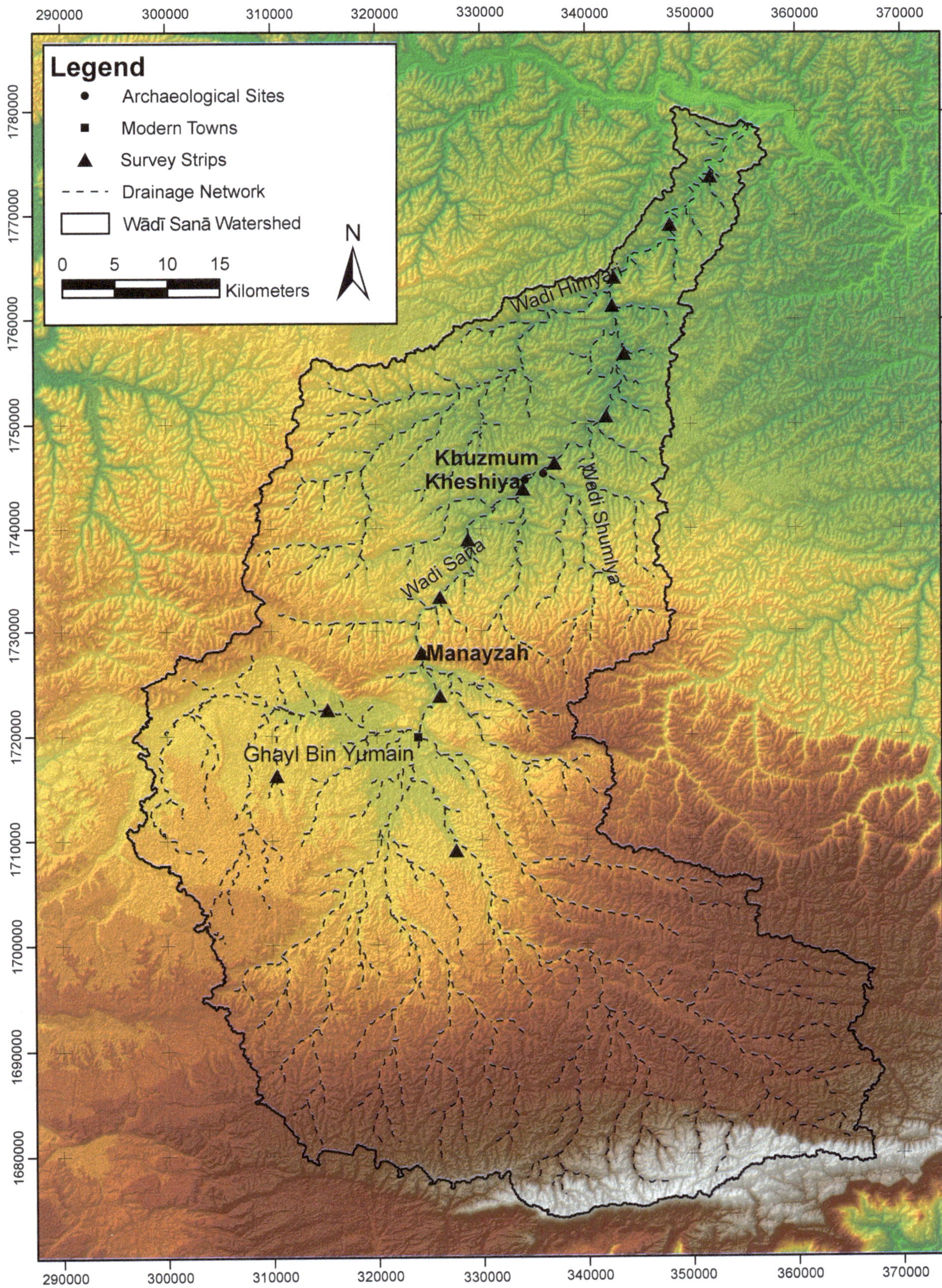

Figure 4.12 Map of the Wādī Sanā survey area with locations of survey transects indicated. Background imagery is ASTER GDEM topography with Wādī Sanā watershed boundaries and major drainages determined by ArcHydro. Coordinates in UTM Zone 39 North WGS84. *Illustration by Michael Harrower.*

us begin to examine spatial patterning of local landforms and alluvial geomorphology

Our first targeted area was the middle Wādī Sanā sediment terraces (wadi silts) and adjacent landforms. In 2000 the team completed full-coverage survey of 96 ha in the area known as Khuzma as-Shumlya, at the confluence of Wādī Sanā and Wādī as-Shumlyah. The RASA GIS in middle Wādī Sanā resembles a quilt of contiguous survey units radiating from the initial point of interest (figure 4.11). Before survey work commenced, a GPS operator mapped survey unit boundaries by walking the unit perimeter with a Trimble Pathfinder Pro XRS GPS backpack receiving Omnistar real-time differential correction.[5] Survey units never included more than one landform class, a rule that ensures that archaeological materials in the same survey unit are associated with one landform class and are thereby subject to comparable site formation processes. Such an approach is critical for interpreting patterns in survey results relative to landform (chapter 6). RASA survey unit boundaries sometimes were arbitrary; that is, the GPS operator selected a linear limit to the area a survey team could cover in working hours. In other cases, survey unit boundaries might also be fixed by the edge of a landform. Thus the plateau unit ends at a slope or the wadi channel unit ends at the beginning of the terrace. In practice the demarcation of a sharp, natural transition was frequently challenged by gradation of landforms: the interface of gravel terrace with wadi silts (alluvium) and wadi channels was often patchy and graduated across tens of meters. What looks like a neat patchwork of adjacent, sharply delineated survey units on our map reflects a myriad of field decisions and workday limits that defined the boundaries of survey units in the field.

Once survey units (each about 1 ha) were mapped, the field team walked transects spaced 10 m apart (using a compass to maintain straight lines) and recorded all visible archaeological remains. This spacing ensured that the team recorded all archaeological remains visible at 5 m. As on all archaeological surveys, visibility of the same materials varied across different landforms. Some smaller materials, such as lithics, stood out at 5 m against the generally bare surfaces of compacted Early Holocene silt terraces, but among the clastic rubble of a scree slope, they disappeared even at closer range. The survey records make no correction of this disparity (cf. Terranato and Ammerman 1996). Counts of sites are the tabulation of the remains actually encountered because differences in landform affected: (1) the unknown original disposition of human land use (the ultimate goal of survey); (2) the subsequent preservation of remains; and (3) the ability of archaeologists to detect them. Correcting Wādī Sanā survey counts of sites for the effects of these variables would be extremely difficult and would likely introduce more complexity and error than it would resolve.

To minimize the impact on archaeological sites and limit the amounts of material that would need to be analyzed and stored, the team collected a minimum of archaeological materials from surface finds, especially in the early years of survey. The goals were to record artifacts in the field and leave them in place so that future specialists might examine key archaeological assemblages, with research questions closely tied to collecting and with collection strategies tailored to research. (See chapter 5.) This decision is important in the context of a pan-Arabian impact on desert lithic scatters by nonprofessional archaeologists, who select arrowheads and other attractive tools from surface assemblages, leaving behind a nondiagnostic scatter with few traces of the typological forms knappers were trying to achieve. This is not a problem unique to Arabia, but its impacts are increasingly noted by Arabian heritage managers. Likewise, when local bedouin and military personnel see or learn of archaeologists collecting material, they impute monetary value to antiquities regardless of any protests to the contrary. This encourages looting and collecting. The RASA approach of minimal collection was approved by ʿAbdalʿazīz Bin ʿAqīl, regional director for the General Organization for Antiquities and Museums, Yemen. The relatively few collections made by RASA teams have been deposited in the Mukalla Museum, Hadramawt. (Recent damage to these collections and their dissociation from identifying records in the course of war suggest that objects left in the field may have survived better than some that were collected.)

During a second two-month field season (2004), the team expanded coverage of middle Wādī Sanā with stratified random survey. Full coverage survey of the entire length of Wādī Sanā would take decades. Furthermore, upland plateau areas away from the wadi are often very difficult to access and in our experience offer very sporadic archaeological remains, rarely preserved in stratigraphic context (chapters 7 and 14). We devised a compromise that would provide detailed understanding of site formation and preservation while revealing broader understanding of human activity along the wadi. The team completed intensive survey of 12 randomly selected 100 m wide survey strips (transects) that traversed the main channel of Wādī Sanā and extended to the plateau on either side (figure 4.12). Three more transects were also surveyed across the Ghayl Bin Yumain basin that forms the headwaters of Wādī Sanā (McCorriston et al. 2005). Extensive survey had shown that the deep profiles of sediment terraces are the most likely localities for site preservation and that sites, other than small lithic scatters, are visible from many meters away on the surface. Accordingly, the team targeted the main channel of Wādī Sanā for intensive survey, eschewing most of the surrounding plateau. Plateau and bedrock slope landforms contain no sediment, and therefore no possibility for stratified archaeological sites, and constitute 74.9 percent of the Wādī Sanā watershed (Harrower 2006:table 10.4). Restricting intensive survey within the Wādī Sanā main channel meant that our survey strips covered only a sampling of plateau and bedrock slope on either side of the wadi and had a vastly increased likelihood of encountering stratified and datable archaeological remains. The random strips were stratified by latitude, with one strip (100 m wide transect perpendicular to the wadi) every 0.04 degrees of latitude, so that the first strip, for example, fell between 15.740 and 15.780 at 15.767 N. Michael Harrower used a table of random numbers to select strip locations, used GPS to navigate to them, and coordinated survey at 10 m spacing (figure 4.12).

Conclusions

From wide exploration to more focused studies, the RASA archaeological survey adapted survey methods developed in very different regions to examine the ancient past of Yemen's Hadramawt Governorate. Initial investigations focused on the search for settlements that might yield early plant and animals domesticates. Research foci then broadened and expanded through time as detailed in subsequent chapters. Throughout the history of the RASA Project (1996–2008), major changes in available geospatial technologies transformed archaeological fieldwork. Approaches pioneered in the early years are now more routine to archaeological survey in Arabia and elsewhere. Nevertheless, what the RASA Project learned in fieldwork not only clarifies Yemen's ancient past but also is instructive for current and future methodology, practice, and analysis in archaeology.

In the late 1990s, early GPS receivers proved extremely valuable in navigating and documenting the location of archaeological sites, but with only wide-area, 1:100,000-scale maps, spatial relationships could be examined only from a coarse regional perspective. For fieldwork in 2000, RASA produced its own custom image maps, and landform classification maps, from Landsat imagery. These maps and real-time, sub-meter-accurate differential GPS greatly improved our ability to navigate, document, and understand ancient landscapes. By 2004 new techniques, including extraction of topographic digital elevation model (DEM) data from ASTER satellite imagery, and the availability of high-resolution QuickBird satellite imagery, in which small features like cairn tombs can be visually identified, propelled new lines of research, including investigations further described in chapter 5.

Notes

1 All coordinates in this chapter are listed in Universal Transverse Mercator coordinate system (UTM Zone 39 North WGS84).

2 With thanks to Burkhard Vogt and the German Archaeological Institute in Sana'a for permission to consult map holdings.

3 The RASA team thanks CANOXY Petroleum LLC (later Canadian NEXEN Petroleum Yemen) and especially contract geologist Kenneth Spraggs for providing access to and use of satellite image maps for the Southern Jol.

4 Maps made by Michael Harrower.

5 Although this configuration is designed to achieve sub-meter accuracy, we subsequently recognized that despite receiving the proper correction signal, the absolute accuracy of some of our data is less than expected.

References

Banning, Edward B.

1996 Highlands and Lowlands: Problems and Survey Frameworks for Rural Archaeology in the Near East. *Bulletin of the American Schools of Oriental Research* 301:25–45.

Bin ʿAqīl, ʿAbdalʿazīz J., and Joy McCorriston

2009 Convergences in the Ethnography, Semantics, and Archaeology of Prehistoric Small Scale Monument Types in Hadramawt (Southern Arabia). *Antiquity* 83:602–18.

Breton, Jean-Francois

1991 *Fouilles de Shabwa II: rapports preliminaires.* Syria 68. Geuthner, Paris.

Cherry, John F., Jack L. Davis, and E. Mantzourani

1991 *Landscape Archaeology as Long-Term History: Northern Keos in the Cycladic Islands from Earliest Settlement until Modern Times.* UCLA Institute of Archaeology, Los Angeles.

Crassard, Rémy

2007 Apport de la technologie lithique à la définition de la préhistoire du Hadruamawt, dans le contexte du Yémen et de l'Arabie du Sud. PhD dissertation, Département d'Anthropologie, Ethnologie et Préhistoire, Université de Paris 1 Panthéon-Sorbonne.

2008 *La préhistoire du Yémen. Diffusions et diversités locales, á travers l'étude d'industries lithiques du Hadramawt.* British Archaeological Reports 1842. Archaeopress, Oxford.

Crassard, Rémy, and Pierre Bodu

2004 Préhistoire du Ḥaḍramawt (Yémen): nouvelles perspectives. *Proceedings of the Seminar for Arabian Studies* 34:67–84.

de Maigret, Antonio (editor)

1990 *The Bronze Age Cultures of Khawlān at Tiyyāl and al-Hada (Republic of Yemen): A First General Report.* ISMEO, Rome.

Dunnell, Robert C., and William S. Dancey

1983 The Siteless Survey: A Regional Scale Data Collection Strategy. *Advances in Archaeological Method and Theory* 6:267–87.

Ebert, James L.

1992 *Distributional Archaeology.* University of New Mexico Press, Albuquerque.

Edens, Christopher, Tony J. Wilkinson, and Glynn Barrett

2000 Ḥammat al–Qā and the Roots of Urbanism in Southwest Asia. *Antiquity* 74:854–62.

Fedele, Francesco G.

1990 Man, Land, and Climate: Emerging Interactions from the Holocene of the Yemen Highlands. In *Man's Role in the Shaping of the Eastern Mediterranean Landscape*, edited by Sytze Bottema, Gerte Enjes-Nieborg, and Willem Van Zeist, pp. 31–42. A. A. Balkema, Rotterdam.

Fried, Morton H.

1967 *The Evolution of Political Society.* Random House, New York.

Gentelle, Pierre

1991 Les Irrigations Antiques a Shabwa. *Syria* 68:5–54.

1998 La nature et l'irrigation. In *Une vallée aride du Yémen antique: Le wādī Bayhān*, edited by Jean-François Breton, Jean-Charles Arramond, Brigitte Coque-Delhuille, and Pierre Gentelle, pp. 107–69. Éditions Recherche sur les Civilisations, Paris.

Ghaleb, Abdu O.

1991 *Agricultural Practices in Radman, Wādi Al-Jubah.* PhD dissertation, Department of Oriental Studies, University of Pennsylvania, Philadelphia.

Gregory, Timothy E.

2004 Less Is Better: The Quality of Ceramic Evidence from Archaeological Survey and Practical Proposals for Low-Impact Survey in a Mediterranean Context. In *Mediterranean Archaeological Landscapes: Current Issues*, edited by Effie Athanassopoulos and LuAnn Wandsnider, pp. 15–36. University of Pennsylvania Museum of Archaeology and Anthropology, Philadelphia.

Gunter, Ann C. (editor)

2005 *Caravan Kingdoms: Yemen and the Ancient Incense Trade.* Sackler Gallery, Smithsonian Institution, Washington, DC.

Harrower, Michael J.

2006 *Environmental versus Social Parameters, Landscape, and the Origins of Irrigation in Southwest Arabia (Yemen).* PhD dissertation, Department of Anthropology, The Ohio State University, Columbus.

2016 *Water Histories and Spatial Archaeology: Ancient Yemen and the American West.* Cambridge University Press, Cambridge.

Harrower, M., J. McCorriston, and E. A. Oches

2002 Mapping the Roots of Agriculture in Southern Arabia: The Application of Satellite Remote Sensing, Global Positioning System (GPS) and Geographic Information System (GIS) Technologies. *Archaeological Prospection* 9:35–42.

Harrower, Michael J., E. A. Oches, and J. McCorriston

2012 Hydro-Geospatial Analysis of Ancient Pastoral/Agro-Pastoral Landscapes along Wādī Sanā (Yemen). *Journal of Arid Environments* 86:131–38.

Kowalewski, Stephen A.

2008 Regional Settlement Pattern Studies. *Journal of Archaeological Research* 16(3):225–85.

Kühn, Peter, Dana Pietsch, and Iris Gerlach

2010 Archaeopedological Analyses around a Neolithic Hearth and the Beginning of Sabaean Irrigation in the Oasis of Ma'rib (Ramlat as-Sab'atayn, Yemen). *Journal of Archaeological Science* 37:1305–10.

Lewis, Krista, Lamya Khalidi, William Isenberger, and Ali Sanabani

2010 Mapping Maṣnaʿat Māryah: Using GIS to Reconstruct the Development of a Multi-Period Site in the Highlands of Yemen. *Proceedings of the Seminar for Arabian Studies* 40:213–26.

McCorriston, Joy

2000 Early Settlement in Hadramawt: Preliminary Report on Prehistoric Occupation at Shi'b Munayder. *Arabian Archaeology and Epigraphy* 11:129–53.

2012 Anthropomorphic Stelae from Hadramawt, Yemen. In *Proceedings of the Amman Table Ronde on Funerary Monuments and Rural Sanctuaries in the 4th and 3rd Millennia BC in Arabia and the Levant*, edited by Tara Steimer, pp. 159–66. British Archaeological Reports, International Series 2317. Institut Français d'Archéologie du Proche Orient, Amman.

McCorriston, Joy, and Michael J. Harrower

2005 Annales History, Geographic Information Systems and the Analysis of Landscape in Hadramawt, Yemen.

In *Temps et Espaces de L'Homme en Société: Analyses et modèles spatiaux en archéologie.XXVe Recontres Internationales d'Archéologie et d'Histoire d'Antibes 2004*, edited by Jean-François Berger, Fréderique Bertoncello, Frank Braemer, Gourguen Davtian, and Martin Gazenbeek, pp. 31–41. Editions ADPCA, Antibes.

McCorriston, Joy, Michael Harrower, Eric Oches, ʿAbdalʿazīz Bin ʿAqīl, Catherine Heyne, AbdalBaset Noman, Remy Crassard, Khalid BaDhofary, and Joshua Anderson

2005 Foraging Economies and Population in the Middle Holocene Highlands of Southern Yemen. *Proceedings of the Seminar for Arabian Studies* 35:143–54.

McCorriston, Joy, Tara Steimer-Herbet, Michael J. Harrower, Kimberly A. Williams, Jean-François Saliège, and ʻʿAbdalʿazīz Bin ʿAqīl

2011 Gazetteer of Small Scale Monuments in Prehistoric Hadramawt, Yemen: A Radiocarbon Chronology from RASA Project Research 1996–2008. *Arabian Archaeology and Epigraphy* 22:1–22.

Moorey, P. R. S.

1974 *Ancient Persian Bronzes in the Adam Collection.* Faber, London.

Newton, Lynne, and Juris Zarins

2000 Aspects of Bronze Age Art of Southern Arabia: The Pictorial Landscape and Its Relation to Economic and Socio-Political Status. *Arabian Archaeology and Epigraphy* 11(2):154–79.

Philip, Graham

1995 Warrior Burials in the Ancient Near-Eastern Bronze Age: The Evidence from Mesopotamia, Western Iran and Syria-Palestine. In *The Archaeology of Death in the Ancient Near East*, edited by Stuart Campbell and Anthony Green, pp. 140–54. Oxbow Books, Oxford.

Pietsch, Dana, Peter Kühn, Thomas Scholten, Ueli Brunner, Holger Hitgen, and Iris Gerlach

2010 Holocene Soils and Sediments around Ma'rib Oasis, Yemen: Further Sabaean Treasures? *Holocene* 20:785–99.

Schmidt, Jurgen

1988 The Sabaean Irrigation Economy of Marib. In *Yemen: 3,000 Years of Art and Civilization in Arabia Felix*, edited by Werner Daum, pp. 55–62. Pinguin Verlag, Innsbruck.

Sedov, Alexander V.

1996 Al-Guraf in the Wādī ʿIdim. Notes on an Archaeological Map of the Ḥaḍramawt, 2. *Arabian Archaeology and Epigraphy* 7:53–62.

Sedov, Alexander V., and Abdalrahman as-Saqqaf

1992 Stone Idols from Wādī 'Idim (Inner Hadramawt). In *Yemen studi archeologici, storici e filologici sull'Araba meridional*, edited by G. Gnoli and M. Poli, pp. 125–29. IsIAO, Rome.

Sedov, Alexander V., and P. A. Griaznevich (editors)

1996 *Raybun Settlement (1983–1987 Excavations)*. Preliminary Reports of the Soviet–Yemeni Joint Complex Expedition 2. Russian Academy, Moscow.

Terrenato, Nicola, and Albert J. Ammerman

1996 Visibility and Site Recovery in the Cecina Valley Survey, Italy. *Journal of Field Archaeology* 23(1):91–109.

Vogt, Burkhard

1997 La fin de la prehistoire au Hadramawt. In *Yemen au pays de la reine de Saba*, edited by Christian J. Robin and Burkhard Vogt, pp. 30–33. Institut du monde arabe—Flammarion, Paris.

Vogt, Burkhard, and Alexander V. Sedov

1994 *Surveys and Rescue Excavations in the Hadhramawt Governate, Republic of Yemen.* Manuscript on file. Deutches Archäologisches Institut, Sana'a.

Wilkinson, Tony J.

1997 Holocene Environments of the High Plateau, Yemen: Recent Geoarchaeological Investigations. *Geoarchaeology* 12(8):833–64.

1999 Settlement, Soil Erosion and Terraced Agriculture in Highland Yemen: A Preliminary Statement. *Proceedings of the Seminar for Arabian Studies* 29:183–92.

2003 The Organization of Settlement in Highland Yemen during the Bronze and Iron Ages. *Proceedings of the Seminar for Arabian Studies* 33:157–68.

Wright, Henry

1977 Recent Research on the Origin of the State. *Annual Review of Anthropology* 6:379–97.

Zarins, Juris

2001 *The Land of Incense. Archaeological Work in the Governate of Dhofar, Sultanate of Oman 1990–1995.* Ministry of Information, Sultanate of Oman, Project of the National Committee for the Supervision of Archaeological Survey in the Sultanate, Muscat.

Chapter 5
Topic-Specific Survey Approaches

Joy McCorriston, Michael J. Harrower, and Rémy Crassard

As RASA archaeological survey developed from wide-ranging exploration to more intensive studies focused on specific research questions, new topic-specific methods were devised to address particular aspects of Yemen's ancient past. In conjunction with general RASA survey, which recorded an extremely wide variety of remains from the Paleolithic to the Iron Age, topic-specific surveys were conducted to examine ancient stone tools, water management, and cairn tombs. These surveys generated substantive research outcomes, including a dissertation on the prehistory of Yemen through analysis of lithics (Crassard 2007), a dissertation on the origins of water management and irrigation in Yemen (Harrower 2006), a dissertation on pastoral landscapes revealed in phytolith assemblages (Buffington 2019) and a successor to RASA—the Arabian Human Social Dynamics (AHSD) Project, which focused on small-scale monuments including cairn tombs in Yemen and Oman from 2008 to 2010. (See chapter 14.) In this chapter we review the basic methodological approaches of lithic survey, water management survey, and cairn survey to lay the foundations for details presented in subsequent chapters.

Lithics Survey (2004)

Wādī Sanā offers a plentiful array of remains that document early human populations. In terms of artifacts, ceramics are rare, and lithics form the vast majority of the material record. Most sites along Wādī Sanā lie on the modern surface and in many cases have little or no stratigraphy. Such sites may be very old or very recent, and they often contain lithics from multiple periods or occupational episodes. Deflation and erosion contribute to mixtures of industries from different periods, which makes analysis challenging and requires careful attention to context.

In arid and hyper-arid environments of Hadramawt, surveys necessarily involve identifying lithic industries from surface deposits. In addition, the team targeted locations potentially protected and preserved from erosion to discover stratified archaeological remains and thereby refine the information from surface materials. A stratigraphic accumulation could occur in areas sheltered from prevailing winds at the top of the plateaus. Such sheltered areas also include huge limestone blocks or the base of cliffs providing protection from violent wadi flows at the foot of Hadramawt's deep gorges. Rockshelters and caves are ideal for the preservation of anthropogenic sediments, but it was rarely possible to discover deposits in this context. Albeit numerous in the limestone cliffs of Hadramawt, most habitable caves are still occupied by a modern seminomadic population. Archaeological survey and excavation are therefore difficult in caves and rockshelters, except when they are occasionally unoccupied. In such cases, it is possible to carry out soundings and test pits, but recurrent roof collapse and goat dung accumulations make it difficult to reach prehistoric layers, if any exist. Such overburden may have sealed archaeological levels in some caves, but most such deposits remain undiscovered.

In the case of surface sites, archaeologists are faced with a choice of collection method. Collecting artifacts from surface deposits presents advantages and disadvantages, which need to be defined before undertaking any action. Over multiple seasons, the RASA team employed several collection strategies in tandem with overall survey goals and team expertise, opting in 1998 to collect tools and diagnostic chipping products with a GPS record and site description. In 2000 the team included Dawn Walter Gagliano, a lithic analyst conducting her first fieldwork in the Middle East. In this season, the RASA team recorded lithic clusters with a brief site description and GPS point but eschewed large-scale collection of surface artifacts until an expert analyst (Rémy Crassard) with a clearly defined analytical approach joined the team as a permanent member.

Building upon preliminary studies of lithics in 1998 and 2000, in 2004 we adopted strategies and conducted survey specifically targeting lithics that utilized different collection strategies, depending on the material encountered.

In some cases we chose *not to collect anything* and to leave sites in their original state of discovery. This choice was motivated by the poorness or, conversely, the richness of the deposits and followed the general strategy adopted in 2000. Indeed, if the site's data were few, then by describing in detail the most relevant aspects, we chose to inventory the lithic material's characteristics directly on-site. The choice not to collect anything often occurred during the discovery of an element already known (typologically or technically) and lacking new information. The informative value of such a site type then lay in its presence and its geographic/spatial contribution to the distribution of associated pieces present on other sites. Conversely, if the site was particularly rich, whether quantitatively or technologically, we often chose not to make any preliminary brief collection but to leave it undisturbed until a research-oriented collection strategy could be devised.

Most commonly we utilized two main collection strategies: selective collection and systematic collection. The first strategy involved targeting specific types of materials to collect. Most often, analysts favor tool collection to obtain information on observed types and techniques. Archaeologists also pick up artifacts with a high yield of technological information (for example, cores and crested blades); such collecting may be done by sampling or by collecting the totality of a category. The second collection type, which we called systematic, consisted of collecting every visible piece on the surface, on the entire site or on a well-defined portion of the site. Systematic collection was particularly useful as a density measure and in generating statistically significant quantification.

Given the collections studied and the different levels of study applied, it is important to elaborate further the methodology we used in collecting the lithics.

We recognized three main types of sites:

- surface sites with low informative level
- surface sites with high informative level
- stratified sites

Surface sites with low informative level often required no more than a selective collection or no collection at all. Without a systematic collection, a qualitative analysis rather than a quantitative inventory was then carried out, building upon the technological observations made on the artifacts on-site. The objective was to document knapping techniques and methods observable within the assemblage. These observations were then compared with known knapping techniques and methods. Spatial location contributed to a broader quantitative and qualitative assessment of the distributions and densities of pieces over various terrain.

The *surface site with high informative level* combined distinctive additional features. These features may have been related to an unusual density and/or homogeneity in the lithic industries. A surface site with high informative level may have had one or more previously unknown knapping methods. In such cases, we applied a systematic collection method to all or parts of the site, using a surface grid system or by marking individual GPS points. Such collection had varying degrees of application: sometimes we decided to collect only tools; other times we sought elements characteristic of a flaking method.

Because *stratified sites* contributed chronological perspectives, such sites often provided the most information. Therefore, the search for stratified sites was a survey priority in Hadramawt. Once they were discovered, excavation and analysis also became high priorities. To collect all artifacts, lithic analysts carried out a systematic sieving of the sediments.

Thus these three types of sites merited study in different ways. Importantly, considering a surface site to have low informative level did not imply its irrelevance. Such sites contributed to the overall documentation of prehistoric activity. Each type of site was therefore important in addressing particular problems, such as understanding a technical knapping process, intrasite spatial analysis of occupation, or more simply serving as an element for future comparison with neighboring sites.

In general, the number and richness of surface lithic sites along Wādī Sanā was considerable. In total 173 lithic sites were recognized by RASA surveys (table 5.1). We studied, documented, and inventoried the vast majority of collected industries, and our collections have been deposited in the Mukalla Museum (Crassard 2007). One observation about surface lithic distributions called for a more extensive analysis of surface sites than might otherwise be undertaken. We observed that a clear majority of Lower/Middle Paleolithic pieces lie on the lower rock terraces directly overlooking the wadi bed. The upper terraces are mostly covered with Early/Middle Holocene sites, while the top of the plateau can be covered with flaked Levallois industries and/or Holocene industries. Within the wadis themselves, remnant silt accumulations also sometimes contain Holocene lithics. This book presents only a sample of the analyses conducted. We focus on the most informative Paleolithic sites (chapter 7), on the Terminal Pleistocene–Early Holocene stratified rockshelter occupation of Manayzah (chapter 8), and on the Khuzmum rockshelters and Gravel Bar Site (chapter 9).

Water Management Survey (2004)

In conjunction with RASA Project interest in early agriculture, the remains of water management and irrigation encountered along Wādī Sanā quickly become of particular interest. In 2004 Michael Harrower conducted survey specifically targeting water management along Wādī Sanā for PhD dissertation research (Harrower 2006). Surveys in 1998 and 2000 had identified approximately a dozen irrigation structures, but the sample of these structures was too small to generate reliable conclusions, so more targeted methods were needed to examine spatial patterning of ancient water use.

As further described in chapter 13, the search for water management sought to identify all possible remnants of water management, such as wells, cisterns, canals, channels, terraces, diversion structures, fields, dams, and check dams. Because water management structures are readily visible from a considerable distance (most commonly 50 to 100 m or more), the team relied on extensive coverage, including vehicular survey. Our approach involved rapid reconnaissance of areas identified in Landsat and ASTER satellite imagery, and surveyors' and locals' knowledge of potentially irrigable areas. When the team encountered areas with potential for preservation of ancient water management, it used two-way radios to survey at 30 m spacing. Where water management features were discovered, surveyors then expanded outward to cover the surrounding area and record all archaeological remains within a 200 m radius. This strategy aimed to assess spatial associations among water management and other ancient activities. By including data from areas subject to targeted and random survey by RASA (see chapter 4), Harrower (2006) ensured coverage of areas where irrigation might not have been expected. Assessments of local alluvial geomorphology were also used to help estimate the age of water management features, and small test excavations of water management were dug in five locales to examine subsurface remains and retrieve samples for radiocarbon and optically stimulated luminescence (OSL) dating (chapter 18).

Cairn Survey (2005)

Cairn tombs that often mark cliff lines throughout the Hadramawt are among the most visually prominent features for the archaeological record along Wādī Sanā. Since RASA from the outset focused on early plant and animals domesticates, field teams' attention initially concentrated on the search for stratified occupations. Tombs were only occasionally recorded when they happened to coincide with locations of interest. By the end of the 2004 field season, it had become increasingly apparent that cairn tombs deserved more focused attention. Additionally, in August 2004, RASA acquired 60 cm spatial resolution QuickBird satellite imagery in which many cairn tombs were readily visible, offering the possibility of developing methods to map them via imagery without physically visiting each tomb during survey. In 2005 the team conducted a pilot study of survey specifically targeting tombs and other small-scale monuments along Wādī Sanā. Groundtruthing was conducted to evaluate tombs (and possible tombs) identified on QuickBird imagery.[1] These efforts involved Tara Steimer-Herbet, whose expertise in South Arabian tombs greatly clarified the RASA team's understanding and identifications of different monument types.

From a very early stage it was clear that high-accuracy GPS mapping was important to RASA Project fieldwork. GPS accuracy became even more critical once high-resolution satellite imagery became a key part of our research. Until May 1, 2000, the U.S. Department of Defense intentionally introduced approximately 100 m of error into GPS signals using a method called selective availability (SA). This deliberate scrambling compounded error inherent to GPS, such as interference as GPS signals pass through the earth's atmosphere. To overcome these spatial inaccuracies in registering sites and landform areas, the team used a Trimble Pathfinder Pro XRS GPS backpack system with an Omnistar Virtual Base Station (VBS) subscription for real-time differential correction. This configuration was designed, under optimal conditions, to achieve sub-meter accuracy. Although we experienced somewhat

Table 5.1. Artifacts collected by lithics survey of Wādī Sanā in 2004. All collections deposited in the Mukalla Museum, Hadramawt.

	Coordinates		Cores			Products					Tools				Others				
Collection Bag	**UTM Northing**	**UTM Easting**	**Levallois Core**	**Core for Blades**	**Core for Flakes**	**Levallois Products**	**Blades**	**Retouched Flakes**	**Flakes**	**Obsidian**	**Tools**	**Bifaces**	**Arrowheads**	**Lithics Total**	**Beads**	**Ceramic**	**Observations**	**Estimated Period (Middle Paleolithic, general Holocene, Neolithic, undetermined)**	**Collection Strategy**
RASA-2004-000-1	1744314	337217							1					1				Holocene	selective
RASA-2004-045-1-A-1	1745413	336255							1			1		2				Neolithic	selective
RASA-2004-081-3-3	1743824	333813								1				1				Holocene	selective
RASA-2004-084-0-1	1743787	333362	2		5		1		2		1	2		13				Middle Paleolithic	selective
RASA-2004-084-2-1	1743787	333362	2											2				Middle Paleolithic	selective
RASA-2004-084-2-2	1743787	333362							3					3				Middle Paleolithic	selective
RASA-2004-084-2-2	1743787	333362	1											1				Middle Paleolithic	selective
RASA-2004-084-2-3	1743787	333362							1					1				Middle Paleolithic	selective
RASA-2004-090-1-1	1722562	315316							12					12				Holocene	selective
RASA-2004-092-0-1	1722418	315342									1			1				Holocene	selective
RASA-2004-092-1-2	1722418	315342												0			all natural	N/A	N/A
RASA-2004-092-1-3	1722418	315342			1				1					2				Holocene	selective
RASA-2004-092-1-4	1722418	315342						1	1					2				Holocene	selective
RASA-2004-092-1-5	1722418	315342					1							1				Holocene	selective
RASA-2004-093-0-2	1722722	315344							1					1				Middle Paleolithic	selective
RASA-2004-095-2-1	1722815	315287						1	1					2				Neolithic	selective
RASA-2004-095-2-2	1722842	315347			10				88					98				Neolithic	selective
RASA-2004-095-2-3	1722862	315255			2							1		3				Neolithic	selective
RASA-2004-096-1-1	1722833	315243			1		2		33		1			37				Middle Paleolithic	selective
RASA-2004-098-2-1	1708828	327367			2									2				Holocene	selective
RASA-2004-098-2-2	1708823	327366			1				3					4				Holocene	selective
RASA-2004-098-2-3	1708819	327367												0			all natural	N/A	N/A
RASA-2004-099-1-1	1708864	327431			1				3		1			5			tool: one fragment	Middle Paleolithic	selective
RASA-2004-103-2-1	1716178	310723							2					2				Neolithic	selective
RASA-2004-103-2-2	1716178	310723							2					2				Neolithic	selective
RASA-2004-103-2-3	1716178	310723										1		1				Neolithic	selective
RASA-2004-110-2-1	1739194	328521								1			1	2				Neolithic	selective
RASA-2004-110-2-2	1739194	328521					1							1				undetermined	selective

	Coordinates		Cores			Products					Tools				Others				
Collection Bag	**UTM Northing**	**UTM Easting**	**Levallois Core**	**Core for Blades**	**Core for Flakes**	**Levallois Products**	**Blades**	**Retouched Flakes**	**Flakes**	**Obsidian**	**Tools**	**Bifaces**	**Arrowheads**	**Lithics Total**	**Beads**	**Ceramic**	**Observations**	**Estimated Period (Middle Paleolithic, general Holocene, Neolithic, undetermined)**	**Collection Strategy**
RASA-2004-122-1-1	1756798	342344	3		6	1	1		1		2			14				Middle Paleolithic	selective
RASA-2004-124-1-1	1738945	328474	2		1	3	1							7				Middle Paleolithic	selective
RASA-2004-124-1-2	1738945	328474	7		1		5		1					14				Middle Paleolithic	selective
RASA-2004-124-1-3	1738945	328474	3		4	1	1		1					10				Middle Paleolithic	selective
RASA-2004-124-1-4	1738945	328474			2		1							3				Middle Paleolithic	selective
RASA-2004-134-1-1	1744542	334062										2		2				Neolithic	selective
RASA-2004-135-1-1	1743837	332491	5				2							7				Middle Paleolithic	selective
RASA-2004-136-1-1	1743345	332850			1									1				Middle Paleolithic	selective
RASA-2004-136-1-2	1743345	332850										1		1				Middle Paleolithic	selective
RASA-2004-136-1-3	1743345	332850	2				1				2			5				Middle Paleolithic	selective
RASA-2004-136-1-4	1743345	332850	3				1					1		5				Middle Paleolithic	selective
RASA-2004-141-1-1	1764658	342696	1			1	1							3				Middle Paleolithic	selective
RASA-2004-142-1-1	1764914	342467	4		3		3					3		13				Middle Paleolithic	selective
RASA-2004-149-1-1	1761201	341608	5								2			7				Middle Paleolithic	selective
RASA-2004-149-2-1	1761022	340863	12		3									15				Middle Paleolithic	selective
RASA-2004-151-1-02-1	1744647	334320							1					1				undetermined	selective
RASA-2004-153-1-1	1739460	330321	1			3	4		1		1			10				Middle Paleolithic	selective
RASA-2004-153-1-2	1739460	330321	2				1							3				Middle Paleolithic	selective
RASA-2004-155-2-1	1727877	324343										1		1				Neolithic	selective
RASA-2004-155-2-2	1727877	324343							3	2	2	1		8				Neolithic	selective
RASA-2004-155-2-3	1727877	324343			1					3	2	1		7				Neolithic	selective
RASA-2004-155-2-4	1727877	324343							3	15				18				Neolithic	selective
RASA-2004-155-2-6	1727877	324343								4		1		5				Neolithic	selective
RASA-2004-155-2-7	1727877	324343			1					1				2		2		Neolithic	selective
RASA-2004-155-2-9	1727877	324343							2	5		1		8				Neolithic	selective
RASA-2004-155-2-G9surf	1727877	324343							1					1				Neolithic	systematic
RASA-2004-155-2-H6surf	1727877	324343							4					4				Neolithic	systematic
RASA-2004-155-2-H7surf	1727877	324343							3	1				4				Neolithic	systematic

Table 5.1. Artifacts collected by lithics survey of Wādī Sanā in 2004. All collections deposited in the Mukalla Museum, Hadramawt. *(continued)*

	Coordinates		Cores			Products					Tools				Others				
Collection Bag	**UTM Northing**	**UTM Easting**	**Levallois Core**	**Core for Blades**	**Core for Flakes**	**Levallois Products**	**Blades**	**Retouched Flakes**	**Flakes**	**Obsidian**	**Tools**	**Bifaces**	**Arrowheads**	**Lithics Total**	**Beads**	**Ceramic**	**Observations**	**Estimated Period (Middle Paleolithic, general Holocene, Neolithic, undetermined)**	**Collection Strategy**
RASA-2004-155-2-H9surf	1727877	324343								2		1		3				Neolithic	systematic
RASA-2004-155-2-I4surf	1727877	324343							9	1				10				Neolithic	systematic
RASA-2004-155-2-I5surf	1727877	324343							12	5				17				Neolithic	systematic
RASA-2004-155-2-I6surf	1727877	324343											1	1				Neolithic	systematic
RASA-2004-155-2-I6surf	1727877	324343							7	1				8				Neolithic	systematic
RASA-2004-155-2-I7surf	1727877	324343							8	2				10				Neolithic	systematic
RASA-2004-155-2-I8surf	1727877	324343							8	1				9				Neolithic	systematic
RASA-2004-155-2-I9surf	1727877	324343							3		1			4				Neolithic	systematic
RASA-2004-155-2-J10surf	1727877	324343							2	4				6				Neolithic	systematic
RASA-2004-155-2-J4surf	1727877	324343							26	8				34			One copper?	Neolithic	systematic
RASA-2004-155-2-J5surf	1727877	324343			1				28	8	1			38				Neolithic	systematic
RASA-2004-155-2-J6surf	1727877	324343							26	7				33				Neolithic	systematic
RASA-2004-155-2-J7surf	1727877	324343							26	6				32				Neolithic	systematic
RASA-2004-155-2-J8surf	1727877	324343							7	3			1	11	1			Neolithic	systematic
RASA-2004-155-2-J9surf	1727877	324343							7	2				9				Neolithic	systematic
RASA-2004-155-2-K10surf	1727877	324343							26					26				Neolithic	systematic
RASA-2004-155-2-K11surf	1727877	324343							34					34				Neolithic	systematic
RASA-2004-155-2-K12surf	1727877	324343							27	1				28				Neolithic	systematic
RASA-2004-155-2-K4surf	1727877	324343							49	7				56				Neolithic	systematic
RASA-2004-155-2-K5surf	1727877	324343			2				47	11				60				Neolithic	systematic
RASA-2004-155-2-K-5surf	1727877	324343								4		1		5				Neolithic	systematic
RASA-2004-155-2-K6surf	1727877	324343							30	2				32				Neolithic	systematic
RASA-2004-155-2-K7surf	1727877	324343							76	10	1			87			tool: one fragment	Neolithic	systematic
RASA-2004-155-2-K8surf	1727877	324343							27	4	1			32			tool: one fragment	Neolithic	systematic
RASA-2004-155-2-L10surf	1727877	324343							76	2				78				Neolithic	systematic
RASA-2004-155-2-L11surf	1727877	324343							31					31		1		Neolithic	systemati
RASA-2004-155-2-L12surf	1727877	324343							8	1				9				Neolithic	systemati
RASA-2004-155-2-L6surf	1727877	324343							87	4				91				Neolithic	systemati

	Coordinates		Cores			Products					Tools				Others				
Collection Bag	**UTM Northing**	**UTM Easting**	**Levallois Core**	**Core for Blades**	**Core for Flakes**	**Levallois Products**	**Blades**	**Retouched Flakes**	**Flakes**	**Obsidian**	**Tools**	**Bifaces**	**Arrowheads**	**Lithics Total**	**Beads**	**Ceramic**	**Observations**	**Estimated Period (Middle Paleolithic, general Holocene, Neolithic, undetermined)**	**Collection Strategy**
RASA-2004-155-2-L8surf	1727877	324343							4	3			1	8				Neolithic	systematic
RASA-2004-155-2-L9surf	1727877	324343							82	3				85				Neolithic	systematic
RASA-2004-155-2-M10surf	1727877	324343							9					9				Neolithic	systematic
RASA-2004-155-2-M11surf	1727877	324343							31	1				32				Neolithic	systematic
RASA-2004-155-2-M13surf	1727877	324343							2					2				Neolithic	systematic
RASA-2004-155-2-M15surf	1727877	324343							2					2				Neolithic	systematic
RASA-2004-155-2-M6surf	1727877	324343							82	9	1			92				Neolithic	systematic
RASA-2004-155-2-M7surf	1727877	324343							38	8				46				Neolithic	systematic
RASA-2004-155-2-M8surf	1727877	324343							27	4				31				Neolithic	systematic
RASA-2004-155-2-N10surf	1727877	324343							20	4				24				Neolithic	systematic
RASA-2004-155-2-N11surf	1727877	324343							34	6				40				Neolithic	systematic
RASA-2004-155-2-N12surf	1727877	324343							62	3	1		1	67			arrowhead: one fragment	Neolithic	systematic
RASA-2004-155-2-N13surf	1727877	324343							13					13				Neolithic	systematic
RASA-2004-155-2-N13surf	1727877	324343							12	1				13				Neolithic	systematic
RASA-2004-155-2-N14surf	1727877	324343							13	1				14				Neolithic	systematic
RASA-2004-155-2-N15surf	1727877	324343							1					1				Neolithic	systematic
RASA-2004-155-2-N6surf	1727877	324343							28	5				33				Neolithic	systematic
RASA-2004-155-2-N7surf	1727877	324343							42	6				48				Neolithic	systematic
RASA-2004-155-2-N8surf	1727877	324343			1				60	3		1		65			BIF: one fragment	Neolithic	systematic
RASA-2004-155-2-N9surf	1727877	324343							32					32				Neolithic	systematic
RASA-2004-155-2-O10surf	1727877	324343							37	2				39				Neolithic	systematic
RASA-2004-155-2-O11surf	1727877	324343							11	2				13				Neolithic	systematic
RASA-2004-155-2-O12surf	1727877	324343							18	1	1			20				Neolithic	systematic
RASA-2004-155-2-O13surf	1727877	324343							16	2				18				Neolithic	systematic
RASA-2004-155-2-O14surf	1727877	324343			1			1	13	2				17	2			Neolithic	systematic
RASA-2004-155-2-O15surf	1727877	324343							5	1				6				Neolithic	systemati
RASA-2004-155-2-O6surf	1727877	324343							17	3				20				Neolithic	systemati
RASA-2004-155-2-O8surf	1727877	324343							22	4				26				Neolithic	systemati

Table 5.1. Artifacts collected by lithics survey of Wādī Sanā in 2004. All collections deposited in the Mukalla Museum, Hadramawt. *(continued)*

	Coordinates		Cores			Products					Tools				Others				
Collection Bag	**UTM Northing**	**UTM Easting**	**Levallois Core**	**Core for Blades**	**Core for Flakes**	**Levallois Products**	**Blades**	**Retouched Flakes**	**Flakes**	**Obsidian**	**Tools**	**Bifaces**	**Arrowheads**	**Lithics Total**	**Beads**	**Ceramic**	**Observations**	**Estimated Period (Middle Paleolithic, general Holocene, Neolithic, undetermined)**	**Collection Strategy**
RASA-2004-155-2-P10surf	1727877	324343							13	2				15				Neolithic	systematic
RASA-2004-155-2-P11surf	1727877	324343							12	1		2		15			BIF: two fragments	Neolithic	systematic
RASA-2004-155-2-P12surf	1727877	324343							10					10				Neolithic	systematic
RASA-2004-155-2-P13surf	1727877	324343							11	1				12				Neolithic	systematic
RASA-2004-155-2-P15surf	1727877	324343							10	2				12				Neolithic	systematic
RASA-2004-155-2-P6surf	1727877	324343							19	4				23				Neolithic	systematic
RASA-2004-155-2-P7surf	1727877	324343							11	2				13				Neolithic	systematic
RASA-2004-155-2-P8surf	1727877	324343							18	5				23				Neolithic	systematic
RASA-2004-155-2-P9surf	1727877	324343			1				47	8				56				Neolithic	systematic
RASA-2004-155-2-S6surf	1727877	324343										1		1				Neolithic	systematic
RASA-2004-155-2-T11surf	1727877	324343								1				1				Neolithic	systematic
RASA-2004-155-2-Y6surf	1727877	324343										1		1			BIF: one fragment	Neolithic	systematic
RASA-2004-156-1-1	1723719	325560							1					1				undetermined	selective
RASA-2004-165-1-1	1745846	334582	4				1		1			3		9				Middle Paleolithic	selective
RASA-2004-166-1-1	1745995	335026	4				1		2			1		8				Middle Paleolithic	selective
RASA-2004-166-1-2	1745995	335026	5				6		1			1		13				Middle Paleolithic	selective
RASA-2004-166-1-3	1745995	335026	2	3					4					9				Middle Paleolithic	selective
RASA-2004-166-1-4	1745995	335026	17	7			12		1		2			39				Middle Paleolithic	selective
RASA-2004-167-1-1	1745039	337402	1				11		1					13				Middle Paleolithic	selective
RASA-2004-167-2-1	1745226	337485					8		8					16				Middle Paleolithic	selective
RASA-2004-168-1-1	1745254	337944	8		1									9				Middle Paleolithic	selective
RASA-2004-000-1 (GBS surf)	1744325	337164							2			4		6				Neolithic	selective
RASA-2004-W3-6-1	1742592	329661								2				2				Holocene	selective
RASA-2004-W5-2-1	1743421	333769							2					2				Holocene	selective
RASA-2004-W5-2-3	1743421	333769							3					3				Holocene	selective
RASA-2004-W5-3A-1	1743457	333812									1			1				Holocene	selective
RASA-2004-W5-3A-2	1743461	333773						1						1				Holocene	selective
RASA-2004-W9-2-10	1740751	330357					1							1				Holocene	selective

	Coordinates		Cores			Products					Tools				Others				
Collection Bag	**UTM Northing**	**UTM Easting**	**Levallois Core**	**Core for Blades**	**Core for Flakes**	**Levallois Products**	**Blades**	**Retouched Flakes**	**Flakes**	**Obsidian**	**Tools**	**Bifaces**	**Arrowheads**	**Lithics Total**	**Beads**	**Ceramic**	**Observations**	**Estimated Period (Middle Paleolithic, general Holocene, Neolithic, undetermined)**	**Collection Strategy**
RASA-2004-W9-2-2	1740733	330349											1	1				Neolithic	selective
RASA-2004-W9-2-3	1740734	330349									1			1				Holocene	selective
RASA-2004-W9-2-4	1740735	330348										1		1				Neolithic	selective
RASA-2004-W9-2-5	1740734	330354					1							1				Holocene	selective
RASA-2004-W9-2-6	1740740	330356											1	1				Neolithic	selective
RASA-2004-W9-2-8	1740745	330364									1			1				Holocene	selective
RASA-2004-W9-2-9	1740746	330356							1					1				Holocene	selective
RASA-2004-W9-4-1	1740670	330182											1	1				Neolithic	selective
RASA-2004-W9-4-10	1740687	330172									1			1				Holocene	selective
RASA-2004-W9-4-11	1740694	330175							1					1				Holocene	selective
RASA-2004-W9-4-12	1740711	330176											1	1				Neolithic	selective
RASA-2004-W9-4-13	1740796	330141	1											1				Holocene	selective
RASA-2004-W9-4-14	1740789	330152												0				Holocene	selective
RASA-2004-W9-4-15	1740751	330175							1					1				Holocene	selective
RASA-2004-W9-4-2	1740672	330182			1									1				Holocene	selective
RASA-2004-W9-4-3	1740672	330180	1											1				Holocene	selective
RASA-2004-W9-4-4	1740669	330185			1									1				Holocene	selective
RASA-2004-W9-4-5	1740664	330174			1									1				Holocene	selective
RASA-2004-W9-4-6	1740651	330184	1											1				Holocene	selective
RASA-2004-W9-4-7	1740649	330187											1	1				Neolithic	selective
RASA-2004-W9-4-8	1740671	330165							1					1				Holocene	selective
RASA-2004-W9-4-9	1740680	330170			1									1				Holocene	selective
			100	10	61	9	68	5	1946	249	30	38	11	**2527**	3	3	Totals		

less accuracy during fieldwork, probably because of the lack of Omnistar ground stations in Yemen, our equipment was sufficient to ensure mapping of archaeological remains according to landforms. However, as we began identifying dozens and then hundreds of cairn tombs on QuickBird imagery, it immediately became clear that matching the precise locations on the ground with locations marked on imagery was essential for evaluating our ability accurately to identify cairn tombs. So the 2008 field season marked a major improvement in GPS accuracy when we began using a Trimble 5700 Base Station and a Trimble 5700 Rover in a post-processed kinematic (PPK) configuration that achieved less than 15 cm accuracy.[2] Further details on small-scale monument survey in 2005 and related research in 2008 are reported in chapter 14.

Conclusions

The RASA Project began with general wide-ranging exploration but over time concentrated its attention and involved a number of focused studies on topics including lithic technologies, water management, and cairn tombs. Different topics and different aims required different methods, but together the collection of information presented in subsequent chapters builds a broad, holistic landscape history of ancient Hadramawt.

Notes

1 Maps made by Nisha Patel under the supervision of Michael Harrower.

2 Equipment borrowed from the Department of Civil, Environmental, and Geodetic Engineering, The Ohio State University.

References

Buffington, Abigail F.

2019 Using Phytolith Assemblages to Detect a Pastoral Landscapein Neolithic Wādi Ṣanā, Yemen. PhD dissertation, Department of Anthropology, The Ohio State University, Columbus.

Crassard, Rémy

2007 Apport de la technologie lithique à la définition de la préhistoire du Hadramawt, dans le contexte du Yémen et de l'Arabie du Sud. PhD dissertation, Département d'Anthropologie, Ethnologie et Préhistoire Université de Paris 1-Panthéon-Sorbonne.

Harrower, Michael J.

2006 Environmental versus social parameters, landscape, and the origins of irrigation in Southwest Arabia (Yemen). PhD dissertation, Department of Anthropology, The Ohio State University, Columbus.

Chapter 6

Survey Results and Landforms

A Statistical Analysis

Joy McCorriston

From 2000 to 2005, RASA Project survey generated a valuable dataset on archaeological remains organized by survey units defined by landform type. In addition to the wide range of analyses and results cataloged in subsequent chapters, this dataset offers opportunities for future statistical analysis. This chapter presents a statistical analysis of RASA survey data to address the following questions: Are sites more plentiful on targeted versus randomly selected survey units? Did the RASA survey focus too much on the wadi silts landform class? How are different types of sites and site densities associated with different landform classes?

Archaeological Site Types

From reconnaissance and extensive surveys in earliest seasons, the RASA team in Wādī Sanā developed a catalog of site types that included all archaeological remains encountered. This classification included 10 categories (table 6.1) that served (with minor modifications) the RASA survey field seasons of 2000, 2004, and 2005 (including the lithics survey, water management survey, and cairn survey discussed in chapter 5).

Post Survey Reassessment of Site Types

To understand the distribution of remains across different landform categories and to discern chronological differences in human landscape use in Wādī Sanā, it has been necessary to recategorize some of the raw survey records. The site categories described above were successful in two ways: One, they were readily applicable in the field, with minimal training needed to recognize and differentiate among them. Two, they offered a readily quantifiable assessment of the distribution of archaeological remains throughout Wādī Sanā. The categories served as standards that could be used by all members of the team across several languages and skill sets. Lengthy discussions provided a basis for translation of RASA survey recording forms into Arabic by Yemeni archaeologist and ethnographer ʿAbdalʿazīz Bin ʿAqīl. In subsequent seasons, Yemeni colleagues took notes in Arabic, which were then translated into English through the same collaborative process. Preliminary studies using raw data from the site categories described above found no significant difference in the distribution of structures around the (targeted) Khuzma as-Shumlya and elsewhere in the (randomly selected) Wādī Sanā survey strips (McCorriston et al. 2005).

Nevertheless, subsequent excavations and linguistic study (Bin ʿAqil and McCorriston 2009; McCorriston et al. 2011; Steimer-Herbet 2004) have taught us much about the original construction techniques and chronologies of small-scale stone monuments. With greater experience at recognizing the construction techniques, purposes, and chronologies of cairns and structures, it made sense to reassign some of the original survey records in consultation with detailed notes, sketches, and photographs made in the field. A reexamination of all survey records showed that sites can be securely (re)assigned to the categories listed below. Where preservation was too poor to recognize a monument/structure category, the record was

Table 6.1. Site categories used in RASA survey, 2000–2004.

Site Category	Description
Artifact Cluster	An artifact cluster was an unusually dense spatial distribution of artifacts—usually chipped stone. Artifact clusters were never defined quantitatively but were registered in the context of the survey unit. Thus a plateau survey unit with a continuous area of thousands of accumulated flakes on desert reg (these are fairly common conditions on plateaus overlooking wadis) would have one artifact cluster registration. A sediment terrace survey unit with two or three small clusters of 5 to 10 chipped flakes each might have three artifact clusters registered. The project did not record the spatial location but did count each isolated artifact encountered on survey. Thus one or two marine shells, which are clearly rare and anthropogenic introductions in this highland valley, would also receive a designation of artifact cluster. The team did make detailed records in instances where artifacts (usually chipped stone) were unusually dense for that particular survey unit. There was no quantified and absolute density measure applied uniformly to all survey units. The purpose of documenting remains in association with identifiable geomorphological landforms is to develop analytical understandings of the archaeological record, including the natural and cultural site formation processes that account for the current distributions of archaeological remains. Therefore the survey team did not make field decisions about the primary context of artifact clusters or whether materials had been used together in the past. A typical artifact cluster is the overlay of thermally altered rock (TAR) that may include the remnants of dismantled or eroded hearths. The original hearths from which these TAR clusters derive could not be enumerated, but notable densities of TAR were recorded on survey forms. In a few instances, such as the Gravel Bar Site (SU 000-001) and others documented in detail (Crassard 2008), extensive collections of diagnostic chipped stone, flakes, and other artifacts were made for such purposes. An artifact cluster form usually served to document artifacts the team did not collect.
Burned Layer	In the course of survey and geological studies, the team discovered a series of burned sedimentary layers in the alluvial deposits left by Early Holocene flooding. These burned layers were often clearly visible as dark, charcoal-flecked bands extending tens of meters in the natural sections created by recent down-cutting of the wadi channel (McCorriston et al. 2002). At times these bands sloped perceptibly, following an old land surface, and the archaeological survey team quickly learned to recognize their surface remains where erosion has uncovered sometimes extensive areas of burned surfaces—such as one 5,300 m^2 area at the base of Khuzma as-Shumlya. The visibility of burned surfaces and several distinctive criteria such as dense charcoal flecks, burned shell inclusions, and the typical abrupt transition between burned surface or layer and overlying and underlying deposits meant that these distinctive remains of burning episodes could be readily differentiated from the darkened, organically enriched bands of paleosol or stable land surfaces that also characterize natural sedimentary profiles. The archaeologists quickly learned to recognize burned layers and incorporate their enumeration and description into survey registers in 1998, 2000, and 2004.
Cairn/Tomb	Cairns are artificial piles of stone. In Arabia many cairns are the remains of tombs of various ages and types, which may or may not retain readily recognizable remnants of a deliberate construction plan, chamber, or burials. Such features may be relatively well preserved under an outer collapse of stone, as subsequent studies have shown (McCorriston et al. 2011). For the purposes of RASA survey, the circular piles of stone that later proved to be Arabian high circular tombs were documented as cairns alongside other artificial piles of stone, including modern or reused corbelled pens for kid goats, and stone tumuli that also may or may not contain burials. As with artifact clusters, the field team did not approach documentation of cairns with an a priori understanding of the site formation processes, natural and cultural, that resulted in the remains visible today. Nor did the team assign cairn categories based on survey unit landform: such associations became apparent as the outcome of intensive survey.
Hearth	Hearths were visible in several forms. There were buried hearths in sediment terraces, the wadi silt category of landforms. These were apparent when RASA surveyors walked along the many sediment profiles revealed through natural gullying through the silt terraces. Counts of buried hearths in survey records are partly a function of the degree that a sediment terrace has in cross-cutting erosional gullies and partly reflect the intensity and frequency of human encampments. Hearths also occur on the surface, often ringed by large cobbles and small boulders and filled with stream-rolled gravel. Such hearths are built and used today, and there are identical examples built and used two millennia ago. The RASA team recorded all hearths that could be enumerated regardless of whether these were at modern bedouin camps at rockshelters or occurred in uninhabited places. Relatively dense concentrations of TAR on surfaces are likely the residues of eroded, deflated, or deconstructed hearths whose retaining ring of large cobbles had been removed.

Site Category	Description
Rockart	Rockart includes a range of graffiti, petroglyphs, and the rare inscription found on rock surfaces. In most cases these occurred on the smooth faces of bedrock, in situ. In cases where stones of structures bore graffiti, rockart was documented as part of the documentation of structures. The survey team included Yemeni colleagues trained in Old South Arabian epigraphy so that graffiti in Arabic and Old South Arabian languages were documented by experts. The survey team photographed and drew all rockart.
Rockshelter	The porous limestone formations of the Southern Jol naturally form overhangs, rockshelters, clefts, and very rarely true caves with constricted entrances and enlarged chambers. In most cases, the intermittent and powerful flows of water unabsorbed by soils maintain bedrock floors and preclude long-term preservation of archaeological remains in sedimentary sequences. There are exceptions, especially where Early Holocene flooding in the Wādī Sanā deposited silt terraces against overhanging rock. In some cases such overhangs have collapsed, capping and protecting Early Holocene archaeological sequences under rockfall (for example, SU 044). In other cases, erosion has stripped the surfaces of sediments unprotected by overhangs, leaving stranded sediment sequences in rockshelters high in the cliffs (for example, 1998 CS-1 and 1998 CS-2, 2004 SU 155-2). In Wādī Sanā , rockshelters were also the location from which RASA team recovered Holocene-era fossil hyrax middens (Cole in McCorriston et al. 2002), but no pre-Holocene archaeological deposits were found. Whether they contained archaeological remains or not, rockshelters were enumerated in association with adjacent survey units. Where traces of human activity occurred—whether sooty stains on the ceiling, charred goat dung mats on rock floors, modern goat pens and hearths near the entrance, or clearly ancient petroglyphs and graffiti—rockshelters and associated remains were documented in detail.
Structure	This category constitutes a large and variable range of constructions, and the documentation of structures as a single category of records recognizes archaeologists' inability during survey to differentiate among a wide variety of constructions that stem from many periods and reflect use and reuse sequences. While cairns (see entry above) and water management structures (see entry below) constitute two recognizable categories of construction, others are less easy to differentiate in the field. Subsequent excavations and research have greatly contributed to refinement of this category (Bin ʿAqil and McCorriston 2009; Harrower 2006 ; McCorriston et al. 2002, 2011, 2012; Williams et al. 2014). Where the team encountered more than five transported stones together that could not be assigned to an alternative category (cairn, water management structure, hearth), these were registered together as a structure, and any details of construction technique, association, or subsequent use and modification were described. As with artifact clusters, the team did not attempt to make field decisions about the original form of destroyed structures—a discipline that proved valuable when comparisons between structures and the landforms on which they occurred later demonstrated that natural site formation (or deformation) processes on different landforms created very different remnants from what had originally started as identical constructions.
Trackway	The team made no attempt to differentiate between modern and ancient trackways, and registry of trackways includes smoothed paths across bedrock, which may be ancient animal trails over modern silt and gravel surfaces, old roadbeds, or tracks left by modern vehicles. As with other features, trackways are subject to differential preservation on different landforms and topography. They serve as a crude index of human and herd animal activity in any given survey unit and only rarely are of intrinsic archaeological value as constructed features or long-term landscape paths.
Water Management Structure	Archaeological survey in 2000, 2004, and 2005 designated water management structures where constructed features controlled the flow of water through diversions, dams, and canals. As with other archaeological features, there were differences in preservation as the outcome of different landforms on which they were constructed. The eroded remains of structures originally constructed as polygon-shaped monuments or shelters were in some cases aligned or scattered by the massive force of *sayl* (floodwater) episodes, and the survey teams learned to differentiate these residues from in situ archaeological water management. Water management structures proved especially challenging to date and in 2005 were the focus of independent studies (chapter 13).
Other	Any feature that could not be described in the preceding categories received the designation "other." Many offered no possibility for dating, such as small rock piles (typical of land boundary markers in the post-1991 unification of the Yemens and subsequent privatization of lands in formerly communist South Yemen). Clearings and dry-stone crude shelters roofed to retain baby goats were also likely modern. Finally, bedrock mortars were also recorded under this category.

Table 6.2. Site categories reassigned post-survey for analytical purposes.

Site Category (reassigned)	Description (post-survey)
Firepit/Hearth	Hearths in Wādī Sanā were constructed in one of three ways: They were dug as steep pits lined with small slabs of limestone (8000–6900 BP); later they were shallow, unlined scoops filled with smooth gravel from wadi beds (6900–4800 BP); finally, hearths were constructed on the surface by filling a ring of large cobbles with smooth gravel (post-4500 BP) (Kimiae and McCorriston 2013). To be included in this category, hearths had to be countable, so unstructured scatters of thermally altered rock are in the lithic cluster category (see entry below).
Burned Surface	Burned surfaces were directly dated to 6000 BP in only one instance (McCorriston et al. 2002), but they appear only in the Early/Middle Holocene silt terraces—this is of course the only viable preservation environment for burned surfaces. The radiocarbon date is consistent with relative dates apparent in stratigraphic relationships of burned surfaces to older sediments and to younger archaeological remains in and on the silt terraces of middle Wādī Sanā. People might have continued to burn vegetation, but in Middle/Late Holocene climatic conditions, there seems little likelihood that a spatially continuous vegetative growth was present to sustain an anthropogenic burn (Harrower et al. 2012) .
Lithic Cluster	In reassessment of survey data, it makes sense to quantify lithic clusters, including thermally altered rock (TAR) and chipped stone, separately from the records of other types of artifacts. Published dissertation research of Rémy Crassard (2008) included the most significant surface collections of knapped stone from RASA survey and constitutes a comprehensive report of this material.
Bone Cluster	These include a wide numeric range of bones, from one buried mandible eroding from Early Holocene sediments to an unquantified number eroding and on the surface. While it was not always possible to establish in the field whether a bone on the surface was a remain of recent or ancient activity, it became clear that bone embedded in and eroding from Early Holocene silt terraces conveyed significant information, even as survey observations of bone not subsequently collected or analyzed (McCorriston et al. 2012). Survey records were recoded to identify from which survey units bones were recorded, separating bone from other artifact clusters.
Other Artifact	Use of this category differentiated records of bone and lithics from less common finds, such as heavy stone mortars, bedrock mortars, marine shell ornaments, a few ceramics, and a basalt incense burner.
Water Management Structure	Unchanged from the original field category, this category did not have any reassigned observations. Excavated features and stratigraphic relationships suggest that these first appeared after 5300 BP and continued to be used through the Middle Holocene (Harrower 2008).
Trilith	Clearly documented in field records as a subtype of cairns, this distinctive monument type occurs widely across Southern and Eastern Arabia (Al Shahri 1991; De Cardi et al. 1977; Zarins 2001). With alternating tripod and stela arrangements of three and one stone supported in low, gravel platform elements, these elongated monuments also include sets of four small boulders in a square in front of each platform element. Before these are hearths, also in a line. These hearths have consistently yielded dates of 2300–1700 BP and are recorded and classified as part of the trilith, not as individual hearths (see entry above) or *madhbaḥ* (see entry below).
High Circular Tomb (HCT)	Recent studies by Tara Steimer-Herbet (2004) and others (Braemer et al. 2001; Crassard and Hitgen 2007; De Maigret et al. 2005) clearly associate this type with Bronze Age construction (5300–4000 BP). A hollow chamber was created by setting large unworked limestone blocks or slabs in a semicircular or circular shape, capped by corbelled pavers and a final capstone. Outer walls were well constructed of dry-laid facing blocks, also unworked, and the space between chamber and facing was packed with rubble and chinking stones to make a wall about 1 m in thickness. Systematic documentation and excavations reveal that the modern condition of HCTs is a function of whether they were later quarried for the supply of facing stones, dismantled for robbing burials, subject to heavy erosion on a slope, or reused, as was often the case from 3000 to 1500 BP (McCorriston et al. 2011). Survey records from 2000 and 2004 documented a number of cairn subtypes—conical, conical with tail, pillbox—that are now recognizably HCTs and thus reclassified.

Site Category (reassigned)	Description (post-survey)
Madhbaḥ	A type of hearth specifically for cooking sacrifices at large gatherings was the *madhbaḥ* (see entry below), which could be differentiated from smaller roasting hearths by its size (greater than 1 m across) and by its construction on a prominent artificial basal platform. Dated examples fall within the range of living memory (McCorriston et al. 2011), aligning with oral accounts of their use (Bin ʿAqil and McCorriston 2009). Whereas the hearths in front of triliths may once have had a similar communal purpose, the survey categorized as *madhbaḥ* only obvious examples unassociated with triliths. Most features recategorized as *madhbaḥ* had been described and drawn as unknown structures in original survey records.
Islamic Grave	Islamic graves have included upright markers for nearly 1,400 years (Halevy 2007). The regional tradition expressed in Wādī Sanā of head and foot markers for males and three uprights (a third on the belly) for women differs from urban conventions (one for males, two for women). Islamic graves are never set in the immediate vicinity of prehistoric monuments, nor do they reuse prehistoric tombs. Local pastoralists assert that this is by deliberate avoidance. The RASA team recorded Islamic graves and noted the additions of new individuals in subsequent seasons. Islamic graves often cluster, with burials demarcated by a ring of large stones and a cap of cobbles, into which *shuwāhid*, or tombstones, are set. The team never excavated or attempted to excavate any burials locally identified as Islamic graves; ethnographic research by Bin ʿAqil did establish that isolated graves with Islamic markers might be rare cases of Islamic appropriation of preexisting monuments (Bin ʿAqil and McCorriston 2009), possibly containing pre-Islamic ancestors. Although this is widely supported by circumstantial evidence across Hadramawt, it would be inappropriate and unwelcome to probe these tombs with archaeological excavations (McCorriston 2011). Cairns originally categorized as witness graves in survey records were reassigned to this Islamic graves category.
Miscellaneous Monument	Within this category are several distinctive and readily recognized monument types, including dolmens and wall tombs. Encountered across a broad sector of the Near East and widely studied (e.g., Braemer et al. 2003, Steimer-Herbet 2004; Steimer-Herbet et al. 2006, Eddy and Wendorf 1998), these stone monuments sometimes contain burials and often were robbed in antiquity. The survey team in Wādī Sanā encountered so few of these in the course of systematic survey that their rare instances contribute nothing to statistical analysis. Subsequent excavation tested dolmens and wall tombs across the Southern Jol (McCorriston et al. 2011), providing clear chronological and structural information. From the 2000 and 2004 survey records, records of box cairns were reassigned as wall tombs and dolmens.
Rockart/Inscription	There were no changes or recategorization of the records of rockart and inscriptions. Those graffiti of sufficient length and clarity to be read have been sent to Alessia Prieti at Centre Nationale de Recherche Scientifique of France.
Platform	Platforms are small-scale stone monuments built of upright limestone slabs as a perimeter and then filled with cobbles, boulders, slabs, and rubble. Excavations show they were occupied and abandoned before being filled. They date to the late Neolithic, about 7000–6500 BP (McCorriston et al. 2002, 2011, 2012). Often one finds one or several standing stones, sometimes worked on the top, situated outside and a few meters from the platform. Many of the platforms in Wādī Sanā are in poor condition, but the excavation and documentation of better preserved examples have made it possible to reclassify a number of structures, including damaged ones, as remnants of Neolithic platforms.
Rockshelter	There have been no changes to the original survey classifications of rockshelters.
Indeterminate	This category includes most stone rings, many stone emplacements, damaged structures, scattered hearth remains, and stone piles.
Trackway	With one exception, trackways documented in survey were in current use, without any way to recognize or assign chronological depth to their use. Almost all were animal and human footpaths used by bedouin herders. Trackway counts may prove a useful index of modern activities. They provide no a priori index to ancient activities and are therefore not included in a dataset of archaeological features.
Modern Activity	Where remains were clearly marked as modern or associated with modern occupation on the survey records, they were recategorized as modern activities. A common example is the case of goat pens constructed of dry-wall undressed stone in the vicinity of rockshelters inhabited by Wādī Sanā's current Ḥumūm bedouin.

categorized as "indeterminate." This category included most stone rings, many stone emplacements, damaged structures, scattered hearth remains, and stone piles (table 6.2).

Quantitative Data Analysis

The RASA team collected data from 166 survey units in randomly selected and targeted locations. Post hoc removal of nonarchaeological features (modern activities; trackways) and reclassification of structures into categories of known features, such as platforms (Neolithic), triliths, and high circular tombs (HCTs), has refined the dataset, allowing for the exclusion of indeterminate remains and modern activities from analysis of the distributions of archaeological remains across different landform classes.

Summary Statistics

At first glance, Wādī Sanā is a landscape everywhere visited by humans, with hearths or fire pits as the only archaeological site category identified on or buried in all landform classes. (See chapters 3 and 4 for descriptions and maps of RASA landform classes: plateau, bedrock slope, scree slope, gravel terrace, bedrock terrace, wadi channel, and wadi silt.) A tabulation of total sites of each category shows that indeterminate remains ($n = 831$) are the most numerous of the 2,265 archaeological records registered in the survey, comprising 37 percent of total. If indeterminate remains are excluded from further consideration, the most numerous remains are in the fire pits and hearths category ($n = 803$). These comprise more than half (56 percent) of the identifiable sites ($n = 1,434$) in the survey. Lithic clusters, the second most prolific site category ($n = 173$; 12 percent), occur everywhere except on gravel terraces and bedrock slopes.

Table 6.3. Summary statistics of reassigned site categories from Wādī Sanā survey (2000–2004).

Site Category (reassigned)	n	%
Firepit/Hearth	805	56
Burned Surface	123	9
Lithic Cluster	173	12
Bone Cluster	16	1
Other Artifact	57	4
Water Management Structure	42	3
Trilith	4	> 1
High Circular Tomb	39	3
Madhbaḥ	11	1
Islamic Grave	60	4
Miscellaneous Monument	6	> 1
Rockart	13	1
Platform	51	4
Rockshelter	34	2
Totals	**1,434**	**100**

Other clear associations appear between landforms and some site categories. Gravel terraces and bedrock slopes are the only landform classes on which triliths occur. On the other hand, water management structures do not occur, or at least are not now found, on gravel terraces and bedrock slopes. High circular tombs are on plateau and scree slopes, which are also the highest-elevation landforms in Wādī Sanā. Dolmens and wall tombs are so rare that only a few were encountered, and they total less than 1 percent of all survey sites (table 6.3, table 6.4, figure 6.1).

Figure 6.1. Percentages of identifiable sites from the Wādī Sanā survey (2000–2004), excluding categories "indeterminate," "trackway," and "modern activities" (n = 1,434). *Illustration by Joy McCorriston.*

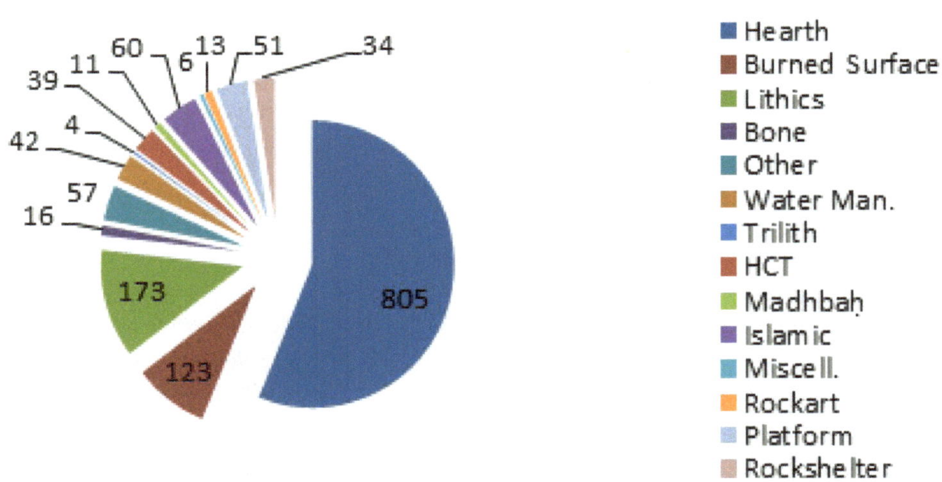

Table 6.4. Observed sites from 2000–2004 systematic survey in Wādī Sanā. Survey conditions: 3 = random; 4 = targeted. Landform class: 1 = wadi channel; 2 = wadi silts; 3 = gravel terrace; 4 = bedrock terrace; 5 = bedrock slope; 6 = scree slope; 7 = plateau.

Survey Unit #	Survey Conditions (random; targeted)	Landform Class	Area (ha)	Firepit/Hearth	Burned Surface	Lithic Cluster	Bone Cluster	Other	Water Management	Trilith	HCT	*Madhbaḥ*	Islamic Grave	Miscellaneous Monument	Rockart/Petroglyph	Platform Structure	Rockshelters	Indeterminate	Trackway	Modern Activities	Total Sites	Sites/ha	Random or Targeted Mean Sites/ha	Landform Class Mean Sites/ha	Landform Class Percentage of Total Area	Landform Class Percentage of Total Sites
020	T	1	2.278	0	0	0	0	0	3	0	0	1	1	0	0	0	0	0	2	0	5	11.4				
021	T	1	2.201	2	0	0	0	0	0	0	0	0	0	0	0	0	0	5	0	0	2	4.4				
022	T	1	1.54	1	0	0	0	1	0	0	0	0	0	0	0	1	0	9	2	0	3	4.62				
027	T	1	0.268	0	0	0	0	0	0	0	0	0	0	0	0	0	0	0	0	0	0	0				
028	T	1	0.421	3	0	0	0	1	0	0	0	0	0	0	0	0	0	0	0	0	4	1.68				
032	T	1	2.568	0	0	0	0	0	1	0	0	0	0	0	0	0	0	5	0	0	1	2.57				
036	T	1	0.718	0	0	0	0	0	0	0	0	0	0	0	0	0	0	0	0	0	0	0				
094	T	1	0.45	0	0	0	0	0	0	0	0	0	0	0	0	0	0	0	1	0	0	0				
100	T	1	1.83	8	0	0	0	0	0	0	0	0	0	0	0	0	0	3	6	0	8	14.6				
101	T	1	1.73	1	0	0	0	0	0	0	0	0	0	0	0	0	0	0	2	0	1	1.73				
105	T	1	1.54	7	0	2	0	2	0	0	0	0	0	0	0	0	0	0	4	0	11	16.9	5.27			
058	R	1	4.589	0	0	0	0	0	0	0	0	0	0	0	0	0	0	0	2	0	0	0				
064	R	1	3.913	0	0	0	0	0	0	0	0	0	0	0	0	0	0	0	0	0	0	0				
067	R	1	0.534	0	0	0	0	0	2	0	0	0	0	0	0	0	0	0	0	0	2	1.07				
079	R	1	0.92	2	0	0	0	0	0	0	0	0	0	0	0	0	0	0	2	0	2	1.84				
081	R	1	0.99	3	0	1	0	0	0	0	0	0	0	0	0	0	1	4	2	0	5	4.95				
112	R	1	1.92	1	0	0	0	0	0	0	0	0	0	0	0	0	3	5	4	0	4	7.68				
117	R	1	0.89	0	0	3	0	1	1	0	0	0	0	0	0	0	0	0	4	0	5	4.45				
120	R	1	1.92	1	0	0	1	0	0	0	0	0	0	0	0	0	1	0	1	1	3	5.76				
126	R	1	1.75	0	0	0	0	0	0	0	0	0	0	0	0	0	0	0	2	0	0	0				
129	R	1	2.3	1	0	0	0	0	0	0	0	0	0	0	0	0	0	0	2	1	1	2.3				
138	R	1	1.23	0	0	0	0	0	0	0	0	0	0	0	0	0	0	0	2	0	0	0				
139	R	1	2	0	0	0	0	0	0	0	0	0	0	0	0	0	0	0	1	0	0	0				
143	R	1	1.09	6	0	1	0	0	0	0	0	0	0	0	0	0	0	3	1	2	7	7.63				
144	R	1	1.22	1	0	0	0	0	0	0	0	0	0	0	0	0	0	0	1	0	1	1.22				
155	R	1	1.45	10	0	3	0	0	0	0	0	0	0	0	0	0	1	0	4	6	14	20.3				
159	R	1	4.33	6	0	0	0	0	0	0	0	0	0	0	0	0	0	0	2	0	6	26	5.20	1.39	0.19	0.06
008	T	2	2.236	0	0	0	0	0	0	0	0	0	0	0	0	0	0	3	3	0	0	0				
009	T	2	2.371	3	0	0	3	1	1	0	0	0	0	0	0	0	0	2	0	0	8	19				
010	T	2	1.649	8	0	0	0	0	0	0	0	0	0	0	0	0	0	0	0	0	8	13.2				
011	T	2	3.602	11	9	4	0	0	0	0	0	0	0	0	0	0	0	6	4	0	24	86.4				
012	T	2	2.011	2	11	0	0	1	0	0	0	0	0	0	0	0	0	3	2	0	14	28.2				
013	T	2	3.606	10	11	0	1	0	1	0	0	0	0	1	0	0	0	3	3	0	24	86.5				
014	T	2	1.297	1	3	0	0	0	0	0	0	0	0	0	0	0	0	2	2	0	4	5.19				
015	T	2	1.997	10	1	0	0	0	2	0	0	0	0	0	0	1	0	4	1	0	14	28				
016	T	2	0.594	23	0	1	0	0	0	0	0	0	0	0	0	1	0	7	0	0	25	14.9				
017	T	2	2.444	36	0	4	0	0	1	0	0	1	0	0	0	3	0	9	0	0	45	110				
018	T	2	0.478	5	2	0	0	0	0	0	0	0	0	0	0	0	0	2	0	0	7	3.35				
019	T	2	0.438	1	2	0	0	0	0	0	0	0	0	0	0	0	0	0	0	0	3	1.31				
024	T	2	0.357	22	0	3	0	0	0	0	0	0	0	0	0	4	0	0	0	0	29	10.4				
025	T	2	0.363	0	0	0	0	0	0	0	0	0	0	0	0	2	0	0	0	0	2	0.73				
026	T	2	0.191	2	1	0	0	0	0	0	0	0	0	0	0	1	0	0	0	0	4	0.76				
033	T	2	2.451	24	4	2	3	1	0	0	0	0	0	0	1	2	0	19	3	0	37	90.7				
037	T	2	1.463	2	0	1	0	0	0	0	0	0	0	0	0	5	0	8	0	0	8	11.7				
038	T	2	2.291	17	1	0	1	0	0	0	0	0	0	3	0	6	0	9	1	0	28	64.1				
039	T	2	2.316	27	3	0	1	0	0	0	0	0	0	0	0	3	0	4	0	0	34	78.7				
040	T	2	1.095	21	0	0	0	0	0	0	0	0	0	0	0	0	0	3	0	0	21	23				
041	T	2	2.934	17	2	0	0	0	0	0	0	0	0	0	0	0	0	2	2	0	19	55.7				
042	T	2	2.447	7	4	0	0	0	1	0	0	0	0	0	0	0	0	3	1	3	12	29.4				
043	T	2	2.873	55	3	2	0	1	0	0	0	0	40	0	0	0	0	6	0	0	101	290				
044	T	2	0.571	27	4	0	0	0	0	0	0	0	0	0	1	0	0	7	0	0	32	18.3				
046	T	2	0.878	10	3	0	0	0	0	0	0	0	0	0	0	0	0	16	0	0	13	11.4				
047	T	2	2.114	8	2	0	0	0	3	0	0	0	0	0	0	0	0	8	0	0	13	27.5				
048	T	2	1.633	4	3	1	0	0	0	0	0	0	0	0	0	0	0	2	0	0	8	13.1				
049	T	2	0.129	2	0	0	0	0	0	0	0	0	0	0	0	0	0	0	0	0	2	0.26				
050	T	2	0.542	2	0	0	0	0	0	0	0	0	0	0	0	0	0	2	0	0	2	1.08				
051	T	2	1.17	7	0	0	0	0	0	0	0	0	0	0	0	0	0	26	1	0	7	8.19				
052	T	2	2.73	1	1	0	0	0	0	0	0	0	0	0	0	2	0	3	5	0	4	10.9				
053	T	2	2.99	2	6	1	0	0	0	0	0	0	0	0	1	0	0	0	5	0	10	29.9				
076	T	2	1.117	13	5	0	0	0	0	0	0	0	0	0	0	0	0	0	1	0	18	20.1				
089	T	2	0.33	1	0	1	0	1	1	0	0	0	0	0	0	0	0	1	0	1	4	1.32				
090	T	2	0.54	0	0	0	1	0	2	0	0	0	0	0	0	0	0	0	1	0	3	1.62				
091	T	2	0.64	2	0	1	0	0	0	0	0	0	0	0	0	0	0	1	1	0	3	1.92				
092	T	2	2.53	1	0	2	1	0	6	0	0	0	0	0	0	0	0	1	2	0	10	25.3				
103	T	2	0.88	3	0	2	0	0	0	0	0	0	0	0	0	0	0	1	3	0	5	4.4				
110	T	2	0.56	9	1	2	0	0	0	0	0	1	16	0	0	0	2	17	0	2	31	17.4				
111	T	2	1.39	8	5	14	0	0	0	0	0	0	0	0	0	0	0	2	0	0	27	37.5				
151	T	2	2.26	65	1	19	3	19	0	0	0	0	0	0	0	3	0	15	1	0	110	249				
152	T	2	3.7	67	0	4	0	1	2	0	0	0	0	0	0	3	0	8	1	0	77	285				
154	T	2	0.005	1	0	0	0	0	0	0	0	0	0	0	0	0	0	0	0	0	1	0.01	42.21			
059	R	2	0.132	1	0	0	0	0	0	0	0	0	0	0	0	0	0	0	0	0	1	0.13				

Table 6.4. Observed sites from 2000–2004 systematic survey in Wādī Sanā. Survey conditions: 3 = random; 4 = targeted. Landform class: 1 = wadi channel; 2 = wadi silts; 3 = gravel terrace; 4 = bedrock terrace; 5 = bedrock slope; 6 = scree slope; 7 = plateau. *(continued)*

Survey Unit #	Survey Conditions (random; targeted)	Landform Class	Area (ha)	Firepit/Hearth	Burned Surface	Lithic Cluster	Bone Cluster	Other	Water Management	Trilith	HCT	*Madhbaḥ*	Islamic Grave	Miscellaneous Monument	Rockart/Petroglyph	Platform Structure	Rockshelters	Indeterminate	Trackway	Modern Activities	Total Sites	Sites/ha	Random or Targeted Mean Sites/ha	Landform Class Mean Sites/ha	Landform Class Percentage of Total Area	Landform Class Percentage of Total Sites
065	R	2	0.176	2	0	0	0	0	0	0	0	0	0	0	0	0	0	2	1	0	2	0.35				
066	R	2	0.536	14	2	0	0	0	0	0	0	0	0	0	0	0	0	1	0	0	16	8.58				
086	R	2	0.42	0	0	1	0	0	0	0	0	0	0	0	0	0	1	0	0	0	2	0.84				
080	R	2	3.15	27	8	0	0	0	0	0	0	0	0	0	0	0	0	8	0	0	35	110				
119	R	2	1.04	2	0	0	0	1	0	0	0	0	0	0	0	0	0	0	0	0	3	3.12				
125	R	2	2.89	8	21	1	1	7	0	0	0	0	0	0	0	0	0	11	0	1	38	110				
127	R	2	0.69	1	0	0	0	0	1	0	0	0	0	0	0	0	0	0	2	0	2	1.38				
130	R	2	0.48	3	2	2	0	13	1	0	0	0	0	0	0	0	0	3	0	0	21	10.1				
140	R	2	0.28	0	0	0	0	0	0	0	0	0	0	0	0	0	0	0	0	0	0	0				
147	R	2	0.5	1	0	0	0	0	0	0	0	0	0	0	0	0	0	0	0	1	1	0.5	22.28	2.00	0.32	0.68
000	T	3	0.916	3	0	1	0	0	0	0	0	0	0	0	3	0	0	6	1	0	7	6.41				
004	T	3	1.022	0	0	0	0	0	0	0	0	0	0	0	0	0	0	8	2	1	0	0				
005	T	3	0.893	1	0	0	0	0	0	0	0	0	0	0	0	0	0	2	1	1	1	0.89				
006	T	3	1.34	1	0	0	0	0	0	1	0	0	0	0	0	0	0	0	3	0	2	2.68				
007	T	3	1.408	2	1	0	0	0	0	1	0	0	0	0	0	0	0	1	1	0	4	5.63				
102	T	3	1.55	0	0	0	0	0	0	0	0	0	0	0	0	0	0	0	3	0	0	0				
150	T	3	0.09	2	0	0	0	1	0	0	0	0	0	0	0	3	0	2	0	3	6	0.54	2.31			
073	R	3	0.733	4	0	0	0	0	0	0	0	1	1	0	0	0	0	38	2	0	6	4.4				
145	R	3	1.24	6	0	0	0	0	0	0	0	0	0	0	0	0	0	0	1	4	6	7.44				
148	R	3	0.39	1	0	0	0	0	0	0	0	0	0	0	0	0	0	0	2	0	1	0.39				
160	R	3	0.3	1	0	0	0	0	0	0	0	0	0	0	0	0	0	1	2	0	1	0.3	3.13	3.33	0.04	0.02
082	R	4	0.8	0	0	2	0	0	0	0	0	0	0	0	0	0	0	0	2	0	2	1.6				
085	R	4	1.8	0	0	0	0	0	0	0	0	0	0	0	0	0	0	0	3	0	0	0				
133	R	4	0.21	3	0	5	0	1	0	0	0	0	0	0	0	0	1	2	0	0	10	2.1				
158	R	4	0.6	1	0	28	0	0	0	0	0	0	0	0	0	0	0	10	6	0	29	17.4	5.28			
023	T	4	0.434	3	0	0	0	0	0	0	0	0	0	0	0	0	0	0	2	0	3	1.3				
029	T	4	0.852	2	0	1	0	0	0	0	0	0	0	0	0	0	0	0	4	0	3	2.56				
030	T	4	0.769	2	0	1	0	1	0	0	0	0	0	0	0	0	0	2	1	0	4	3.08				
031	T	4	1.383	1	0	0	0	0	0	0	0	0	0	0	0	0	0	2	9	1	1	1.38				
034	T	4	1.039	0	0	0	0	0	0	1	0	0	0	0	1	1	0	3	3	0	3	3.12				
035	T	4	0.935	1	0	1	0	0	0	0	0	0	0	0	0	2	0	0	3	0	4	3.74				
104	T	4	1.77	5	0	1	0	0	0	0	0	0	1	0	0	0	0	4	15	1	7	12.4				
134	T	4	8	18	0	1	0	0	4	1	0	0	0	1	0	1	0	14	14	1	26	208				
077	T	4	1.666	0	0	0	0	0	0	0	1	0	0	0	1	1	0	0	0	0	3	5	26.73	1.80	0.08	0.07
054	T	5	3.135	0	0	0	0	0	0	0	0	0	0	0	0	0	0	0	0	0	0	0				
106	T	5	2.25	2	0	0	0	0	0	0	0	0	0	0	0	0	0	0	0	0	2	4.5	2.25			
062	R	5	0.979	0	0	0	0	0	0	0	0	0	0	0	0	0	2	0	0	0	2	1.96				
063	R	5	0.047	2	0	0	0	0	0	0	0	0	0	0	0	0	1	0	0	12	3	0.14				
071	R	5	1.353	0	0	0	0	0	0	0	0	0	0	0	0	0	0	0	0	0	0	0				
074	R	5	0.126	2	0	0	0	0	0	0	0	0	0	0	0	0	3	1	0	5	5	0.63				
083	R	5	1.88	0	0	0	0	0	0	0	0	0	0	0	0	0	0	0	2	0	0	0				
116	R	5	0.73	0	0	0	0	0	0	0	0	0	0	0	0	0	0	0	5	0	0	0				
118	R	5	0.98	0	0	1	0	0	0	0	0	0	0	0	0	0	0	0	3	0	1	0.98				
128	R	5	0.2	1	0	0	0	0	0	0	0	0	0	0	0	0	0	0	1	1	1	0.2				
137	R	5	1.08	0	0	0	0	0	0	0	0	0	0	0	0	0	0	1	0	0	0	0				
146	R	5	0.88	0	0	0	0	0	0	0	0	0	0	0	0	0	1	0	1	0	1	0.88				
157	R	5	2.34	0	0	0	0	0	0	0	0	0	0	0	0	0	0	0	3	0	0	0				
088	R	5	0.49	0	0	0	0	0	0	0	0	0	0	0	0	0	0	0	0	0	0	0	0.40	0.00	0.10	0.01
001	T	6	0.409	0	0	0	0	0	0	0	0	0	0	1	0	0	0	4	4	0	1	0.41				
002	T	6	0.136	0	0	0	0	0	0	0	0	0	0	0	0	0	0	1	4	0	0	0				
003	T	6	0.799	5	0	0	0	0	0	0	0	0	0	0	1	0	0	11	1	0	6	4.79				
045	T	6	0.507	8	0	2	0	0	0	0	0	0	1	0	2	0	0	13	0	2	13	6.59				
056	T	6	1.717	18	1	1	0	0	3	0	0	0	0	0	2	0	13	12	0	5	38	65.2				
057	T	6	2.274	11	0	1	0	0	0	0	1	2	0	0	0	6	0	28	2	0	21	47.8				
093	T	6	0.93	18	0	1	0	1	0	0	0	2	0	0	0	0	1	32	6	2	23	21.4				
095	T	6	0.67	5	0	1	0	0	0	0	1	0	0	0	0	0	0	2	7	0	7	4.69				
099	T	6	1.4	0	0	1	0	0	0	0	0	0	0	0	0	0	0	0	0	0	1	1.4	16.92			
108	R	6	0.88	0	0	0	0	0	0	0	0	0	0	0	0	0	0	0	1	0	0	0				
113	R	6	0.33	1	0	0	0	0	0	0	0	0	0	0	0	0	2	0	2	0	3	0.99				
131	R	6	0.82	0	0	0	0	0	0	0	0	0	0	0	0	0	1	0	0	1	1	0.82				
132	R	6	0.41	0	0	0	0	0	0	0	0	0	0	0	0	0	0	0	0	0	0	0				
161	R	6	0.79	0	0	0	0	0	0	0	0	0	0	0	0	0	0	0	0	0	0	0	0.36	0.00	0.05	0.08
055	T	7	4.523	2	0	0	0	0	0	0	10	0	0	0	0	0	0	308	1	18	12	54.3				
078	T	7	4.084	6	0	0	0	0	3	0	12	0	0	0	0	0	0	26	4	4	21	85.8				
096	T	7	0.24	0	0	1	0	0	0	0	1	0	0	0	0	0	0	0	2	0	2	0.48				
097	T	7	0.19	2	0	1	0	0	0	0	0	0	0	0	0	0	0	12	1	2	3	0.57				
098	T	7	0.15	0	0	1	0	0	0	0	0	3	0	0	0	0	0	5	2	5	4	0.6				
109	T	7	1.52	1	0	1	0	0	0	0	0	0	0	0	0	0	0	0	3	1	2	3.04				
124	T	7	0.62	0	0	1	0	0	0	0	0	0	0	0	0	0	0	0	0	0	1	0.62				
135	T	7	4.68	0	0	1	0	0	0	0	0	0	0	0	0	0	0	0	0	0	1	4.68				
136	T	7	12	0	0	1	0	0	0	0	0	0	0	0	0	0	0	0	0	0	1	12				

Survey Unit #	Survey Conditions (random; targeted)	Landform Class	Area (ha)	Firepit/Hearth	Burned Surface	Lithic Cluster	Bone Cluster	Other	Water Management	Trilith	HCT	*Madhbah*	Islamic Grave	Miscellaneous Monument	Rockart/Petroglyph	Platform Structure	Rockshelters	Indeterminate	Trackway	Modern Activities	Total Sites	Sites/ha	Random or Targeted Mean Sites/ha	Landform Class Mean Sites/ha	Landform Class Percentage of Total Area	Landform Class Percentage of Total Sites
141	T	7	0.33	0	0	1	0	0	0	0	0	0	0	0	0	0	0	0	0	0	1	0.33				
142	T	7	0.5	0	0	1	0	0	0	0	0	0	0	0	0	0	0	0	0	0	1	0.5				
149	T	7	8	0	0	2	0	0	0	0	0	0	0	0	0	0	0	0	0	0	2	16				
153	T	7	2	0	0	1	0	0	0	0	0	0	0	0	0	0	0	0	0	0	1	2				
165	T	7	1.5	0	0	0	0	0	0	0	0	0	0	0	0	0	0	0	0	0	0	0				
166	T	7	2	0	0	1	0	0	0	0	0	0	0	0	0	0	0	0	0	0	1	2				
167	T	7	2	0	0	2	0	0	0	0	0	0	0	0	0	0	0	0	0	0	2	4				
168	T	7	0.8	0	0	1	0	0	0	0	0	0	0	0	0	0	0	0	0	0	1	0.8				
114	T	7	0.52	0	0	0	0	0	0	0	0	0	0	0	0	0	0	0	0	0	0	0	10.43			
061	R	7	0.691	0	0	0	0	0	0	0	0	0	0	0	0	0	0	0	0	0	0	0				
068	R	7	0.575	1	0	0	0	0	0	0	0	0	0	0	0	0	0	0	0	0	1	0.58				
069	R	7	0.931	0	0	0	0	0	0	0	0	0	0	0	0	0	0	0	2	0	0	0				
070	R	7	0.964	3	0	0	0	0	0	0	0	0	0	0	0	0	0	0	1	0	3	2.89				
072	R	7	0.417	0	0	0	0	0	0	0	10	0	0	0	0	0	0	0	1	0	10	4.17				
075	R	7	0.663	0	0	0	0	0	2	0	0	0	0	0	0	0	0	0	1	0	2	1.33				
084	R	7	2.04	2	0	4	0	0	1	0	0	0	0	0	0	0	0	1	4	0	7	14.3				
087	R	7	1.56	0	0	0	0	0	0	0	0	0	0	0	0	0	0	11	0	0	0	0				
107	R	7	2.93	0	0	8	0	0	0	0	0	0	0	0	0	0	0	1	4	0	8	23.4				
115	R	7	2.49	1	0	1	0	0	0	0	1	0	0	0	0	0	0	3	7	0	3	7.47				
121	R	7	1.42	0	0	1	0	0	0	0	1	0	0	0	0	0	0	0	0	0	2	2.84				
122	R	7	0.88	0	0	1	0	1	0	0	0	0	0	0	0	0	0	0	3	1	2	1.76				
123	R	7	0.81	0	0	0	0	0	0	0	0	0	0	0	0	0	0	0	0	0	0	0				
156	R	7	0.36	1	0	14	0	1	0	0	0	0	0	0	0	0	0	0	4	3	16	5.76				
162	R	7	2.11	7	0	1	0	0	0	0	1	0	0	0	0	0	0	0	3	0	9	19	5.57	4.27	0.26	0.08
Totals			**248.3**	**805**	**123**	**173**	**16**	**57**	**42**	**4**	**39**	**11**	**60**	**6**	**13**	**51**	**34**	**831**	**258**	**92**	**1434**					

The Effect of Random and Targeted Survey Methods

Before one can discern further patterns that result from chronological and spatial differences in human landscape use in Wādī Sanā, it is critical to consider the outcome of survey methodologies—random and targeted. By employing random, stratified survey methods, the RASA team generated results that can be used to assess the observed distributions of archaeological remains in targeted areas. Were these distributions richer in areas that drew the attention of archaeologists because of a visible density of protrusive archaeological remains? In light of the publications and interpretations already highlighting the Khuzma as-Shumlya region of middle Wādī Sanā (e.g., McCorriston 2011; McCorriston et al. 2012), it is interesting to ask whether the area is especially rich in archaeological sites (McCorriston et al. 2005).

A traditional survey approach has long been to lay out quadrats or transects of fixed size so that the remains from different survey units can be readily compared (Banning 1985; Plog et al. 1978). The RASA survey's innovative adoption of GPS to record the perimeter of each irregular survey unit allows a precise calculation of its area while maintaining the integrity of a single landform within each survey unit. Contiguous survey units also alleviated the modifiable area unit problem (Openshaw 1983) inherent in interval sampling and provided total coverage of significant portions of the Khuzma as-Shumlya region. By generating a simple distribution of sites per hectare, one can standardize the numbers of survey sites in survey units of different types, allowing comparison across survey units in different locations and different landform classes. This approach overestimates rare sites (for example, triliths), so the analysis that follows focuses on the more common sites, such as hearths, whose representation is more reliable. In a subsequent multivariate discriminant assessment of the distributions of different site categories, we use survey area size as a constraining, explanatory variable.

To determine whether the results from random and targeted surveys differed, one can consider the density of archaeological remains encountered in each survey method as an index for whether there were more sites encountered by targeted methods focused on locations of interest (where the team noted highly visible sites). At first glance it would seem so, with larger mean site densities from almost all landform classes in the targeted survey. For these calculations, trackways and modern activities were excluded. These mean densities were derived by calculating site densities for each survey unit and then calculating a mean site density from the random survey

Table 6.5. Sites per hectare means from random and targeted survey units used in F-test.

Landform Class	Sites/ha Random	Sites/ha Targeted
Wadi Channel	6.2505	9.19718182
Wadi Silts	27.66981818	50.7576977
Gravel Terrace	10.1705	4.74328571
Bedrock Terrace	6.88	40.7847778
Bedrock Slope	0.499583333	2.25
Scree Slope	0.362	31.6446667
Plateau	7.5402	93.8865556

units for each landform class and a mean site density from the targeted of each landform class. I used a two-tailed probability F-test of the (null) hypothesis that there was no statistically significant difference between the numbers of sites per hectare in random versus targeted survey units datasets. The results of the F-test (F = 0.00715914; *ndf* = 6) failed to reject the null hypothesis at the 95 percent confidence level (table 6.5).

Although this result indicated no significant difference between site densities of random versus targeted datasets, there may be significant differences in the densities of identified sites per hectare *within* landform classes across random and targeted samples. Perhaps specific landform classes produced statistically different means from random and targeted samples, but were these differences masked in the combined datasets from all landform classes? To explore this further, one can compare random and targeted survey units from the same landform class, selecting wadi silts (alluvial terraces) because this landform class contains many sites of chronological and analytical interest (for example, burned surfaces, platforms, and bone clusters). The uneven numbers of survey units per landform class in random and targeted samples make the F-test inappropriate for this approach. Furthermore, the random and targeted survey units are drawn from the same population (wadi silts), violating one of the test assumptions.

The Mann-Whitney U test is appropriate for samples of different sizes and for samples not normally distributed. It evaluates differences in medians (not variance or means). For this test of significant difference in the medians of site distributions in random and targeted survey units, I excluded the indeterminate, trackway, and modern activities site categories. The exclusion of indeterminate sites reduces identifiable site numbers but refines the data, so that ancient activities are clearly those that are represented by site counts (n = 1,434). The test considers one landform category—wadi silts—making it unlikely that differential preservation plays an important role in the numbers of identifiable observed sites in random and targeted survey units (unlike the F-test, which included all landform classes). The medians of sites per hectare for randomly selected wadi silt and targeted wadi silt survey unit groups were 2.25 and 17.82, respectively. I performed a Mann-Whitney U test to evaluate the difference in the responses of sites per hectare across random and targeted groups. I found no significant difference between population medians at the 95 percent confidence interval (W = 183.0; Z = -9.97; p = 0.0585, adjusted for ties; r = 1.383). Although very close to the threshold, this is still a failure to reject the null hypothesis at the 95 percent confidence interval at which there is no statistically significant difference in the median density of sites observed through random and targeted survey methods. This finding is consistent with earlier studies of the field data before they were reassigned to the current site categories (McCorriston et al. 2005) (table 6.6).

Did the RASA survey focus too much on particular landform classes? From the outset, it was clear in the field that wadi silts offered a preservation environment (sedimentation) unlike any other landform class, and there were sites like burned surfaces and Type II hearths buried in wadi silts encountered on no other landform. This led the team to target wadi silts, especially in the area of middle Wādī Sanā, and the inclusion of these in the Mann-Whitney U test shows that there were high-median site densities that nearly led us to reject the null hypothesis. So the initial field hunch that the Khuzmum silt terraces in middle Wādī Sanā were distinguished by unusually high densities of sites was technically unsupported—but only just.

Wadi silt terraces were not the only landform class examined, but they made up about one-third (31.62 percent) of the total—random and targeted—surveyed area (n = 248 ha). If the team had equally covered all seven landforms, the expected percentage of wadi silts would be 14.29. A 2 x 7 chi-square test established that the actual (observed) survey outcomes were not significantly different from random factors in landform class selection ($\chi 2$ = 5.23482, df = 6, α = 0.05). Targeting large numbers of wadi silt survey units with seemingly dense remains did not produce a dataset differentiated from strictly random survey. The results of all our surveys can therefore be treated as a representative sample of landscape use in antiquity (table 6.7).

Table 6.6. Wadi silt (silt terraces) survey units: distributions of sites per hectare across random and targeted Mann-Whitney U test values.

Wadi Silts Survey Units, Sites/ha.	
Random	**Targeted**
0.132	18.968
0.352	13.192
8.576	86.448
0.84	28.154
110.25	86.544
3.12	5.188
109.82	27.958
1.38	14.85
10.08	109.98
0.5	3.346
	1.314
	10.353
	0.726
	0.764
	90.687
	11.704
	64.148
	78.744
	22.995
	55.746
	29.364
	290.173
	18.272
	11.414
	27.482
	13.064
	0.258
	1.084
	8.19
	10.92
	29.9
	20.106
	1.32
	1.62
	1.92
	25.3
	4.4
	17.36
	37.53
	248.6
	284.9
	0.005

Site Distributions and Correlations: Methods

How variable are site classes and site densities across landform classes? Since the results above indicate no obvious statistically significant differences between random and targeted survey, all survey units were used to address problems of chronological and spatial distributions of the remains of human landscape use. Using the percentages of each landform class area as an expected value for the percentage of total sites found in that landform, I designed a 2 x 7 chi-square test to see whether the survey (observed) percentages of total sites (including indeterminate) in each landform deviated significantly from even distribution of sites across all landform categories. ($\chi2 = 8.74981$, $df = 6$, $\alpha = 0.05$). The test failed to disprove a null hypothesis, which supports the idea that people using all landforms over time left a ubiquitously dense distribution of remains (table 6.8).

The major components of the tally of sites, fire pits/hearths, and indeterminate remains are likely influencing these outcomes. By removing all survey sites except those of recognizable age—that is, removing fire pits/hearths and indeterminate remains — the sample of sites is greatly reduced ($n = 629$). The sample of included sites is no longer normally distributed and cannot be transformed to a normal distribution (too many zeros). Nonparametric methods, such as correspondence analysis and canonical correspondence analysis, are appropriate approaches to explore relationships between site types and landform categories (ter Braak and Šmilauer 2002).

Once I removed fire pits/hearths and indeterminate sites as response variables, many survey units contained no other observed sites and were therefore eliminated from further analysis, leaving site distributions in 111 survey units. Rarer sites, also of greatest interest for their cultural-historical significance, assume a more robust proportion in this dataset, with representation across a higher percentage of survey units (ubiquity). Those site categories that remain rare (< 6 percent ubiquity) do not contribute significantly to the variance between sample assemblages (the sites within survey units) and have been dropped from further consideration. (These categories are trilith, *madhbaḥ*, Islamic grave, and miscellaneous monument). Site categories further considered include burned surfaces, lithic clusters, bone clusters, other artifacts, water management, HCT, rockart, platform, and rockshelter sites.

Correspondence analysis (CA) and canonical correspondence analysis (CCA) reported here were performed in the CANOCO 4.5 statistical package, designed for use with ecological data (which typically return many zero values and are appropriate for multivariate, nonparametric pattern recognition and significance testing). Like much

Table 6.7. Landform class areas in Wādī Sanā survey for chi-square test values.

	Landform Class Percentages of Total Surveyed Area						
	Wadi Channel	Wadi Silt	Gravel Terrrace	Bedrock Terrrace	Bedrock Slope	Scree Slope	Plateau
Expected	14.285710	14.285710	14.285710	14.285710	14.285710	14.285710	14.285710
Observed	18.765331	31.620730	3.980232	8.15943482	9.68823529	4.862311	25.978242

Table 6.8. Wādī Sanā site distributions used as chi-square test values.

	Site Distributions as Percentages in Total Surveyed Area						
	Wadi Channel	Wadi Silt	Gravel Terrrace	Bedrock Terrrace	Bedrock Slope	Scree Slope	Plateau
Expected	18.765331	31.620730	3.980232	8.15943482	9.68823529	4.862311	25.978242
Observed	5.253863	53.0684327	4.06181015	5.82781457	0.750552	9.580574	21.456954

ecological data, the Wādī Sanā survey dataset is unimodal (the relationship between independent and response variables is not linear because site location is a response to an optimum combination of environmental variables—landform class is just one—and can be described by a quadratic equation). In a general paradigm of human behavioral ecology, it is not surprising that unimodal ecological models best describe the artifact outcomes of accumulated human behavioral choices. Unfortunately, the sample sizes of most site types of interest with cultural-historical specificity (for example, triliths, dolmens, and wall tombs) are too small to be statistically meaningful in such analyses.

Site Distributions and Correlations: Results

Visualization helps in presenting and reading the results of CA and CCA. A common presentation is the scattergram (CA), which can be read using the centroid principle (ter Braak and Šmilauer 2002:414–16). In the scattergram of site categories, each point represents the center (in two dimensions) of the three-dimensional distribution of survey units (samples) containing that site category. Figure 6.2 shows a clustering of samples (lower left quadrat) that contain site categories (bone, burned surfaces, platforms, water management) preserved on wadi silts, suggesting that the classification of landforms indeed reflects meaningful variables in the observed distribution of sites. The first (horizontal) and second (vertical) axes of ordination have the greatest explanatory power in the distribution of samples (but the distance of centroids from the origin is not a direct measure of separation). Rockshelters are separated from other site types on the first axis of ordination, possibly because they are found on slopes, where other site preservation tends to be poor. On the second axis, HCT separates from platforms, water management, bone accumulations, and burned surfaces. This differentiation may also be a factor of landform and preservation, as well as human behaviors in antiquity.

Where presenting the results of an analysis (CCA) that evaluates the strength of environmental variables (here landform classes and area of survey unit are the ones considered), a biplot rather than a scattergram is used, and the biplot rule is used in interpretation. The biplot rule is used to connect an indicator or constraining point with the origin point (the crossing point of ordination axes). Samples (here site classes) can be assessed by perpendicular projection to this line; the distances from the origin point at which projected connections cross the species-origin line reflect that indicator (here landform, area, survey method) frequency in the sample (Lepš and Šmilauer 2003:figure 10–11; ter Braak 1994; ter Braak and Verdonschot 1995:270–73).

This analysis focused on examining the influence of landform class on the observed distribution of site categories. With CA it was possible to explore patterning in the site categories with the expectation that landform class would provide post hoc interpretive explanation for obvious clustering. Other factors also affected human choice (optimizing a combination of environmental-economic, social, and ideological variables, some of which are unknowable and unknown to archaeologists), so there

Figure 6.2. CA scattergram of site categories. Note a cluster of site categories preserved in wadi silts in the lower left quadrat of the diagram. These are clearly separated from rockshelters on the first axis of ordination and separated from rockshelters, HCTs, and lithic clusters on the second axis. HCT = high circular tomb; Rocksh = rockshelter; Lithic = lithic scatter; Platfo = platform structures; WaterM = water management sites; BSurf = burned surfaces; Bone = bone concentration. *Illustration by Joy McCorriston.*

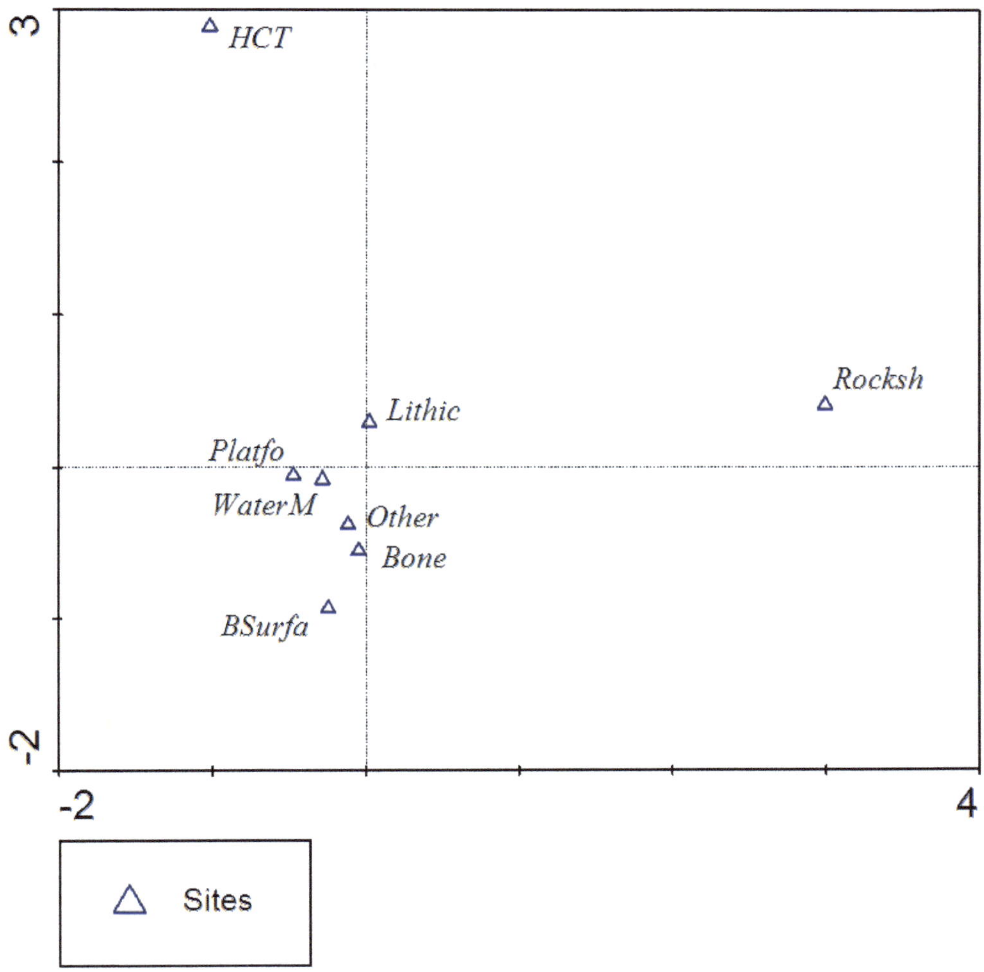

remains a question as to how significant landform classes are in the distributions of site categories. CCA was used primarily to identify which variables—landform classes (nominal variables, conventionally shown as points), area of landform surveyed (ordinal variable, exhibited by an arrow), and random or targeted survey unit selection (nominal)—had statistically significant power to explain variability in observed sites.

Canonical Correspondence Analysis

The differing approaches of correspondence analysis and canonical correspondence analysis provide complementary results, shown here in figure 6.2 and figure 6.3. CCA has constrained the axes of ordination, so the analytical results explain site variability in terms of the indicator (landform, area, survey method) variables. The resulting CCA biplot in figure 6.3 presents the site scores when constrained by all variables of landform classes, area, and survey method. Since random and targeted are collinear, targeted has been removed as a variable. Several indicator variables do not significantly explain sample variance (within 95 percent confidence). These are scree slope ($p = 0.0609$), wadi channel ($p = 0.0689$), area ($p = 0.5255$), gravel terrace ($p = 0.6733$), and bedrock terrace (not tested to improve fit; minimal variance in model). It is important to recognize that the biplot is based on only the first two axes of ordination, which in this case explain 19.4 percent of variance in site distributions and only 76.3 percent of the variance in fitted site data (site–indicator relations). This is not unusual for abundance data, and there are still useful implications to be drawn from the ordination diagram (ter Braak and Šmilauer 2002:123) (figure 6.3, table 6.9).

Figure 6.3. CCA biplot with site classes and constraining variables. Nominal variables (landform classes and survey method) are represented by points, while the ordinal survey unit area is a linear projection. This diagram is interpreted by the biplot rule: The farther along an imaginary line (for nominal variables; the arrow serves for ordinal ones) there intersects a perpendicular line from the sample centroid (site types), the more important is the constraining variable in explaining the site type. The actual distance of a site type centroid perpendicular to a vector is irrelevant. For example, plateau is much more important in explaining the site class HCT than in explaining rockart. The distance from origin is important regardless of on which side of the line through the origin a site class lies. Thus plateau is roughly equally important in explaining lithics and (negatively) bone. BEDSLO = bedrock slope; WCHAN = wadi channel; SCRSLO = scree slope; RANDOM = random survey method; WSILTS = wadi silt terrace; GRATER = gravel terrace; BEDTER = bedrock terrace; PLATEA = plateau; AREA = area of survey unit. *Illustration by Joy McCorriston.*

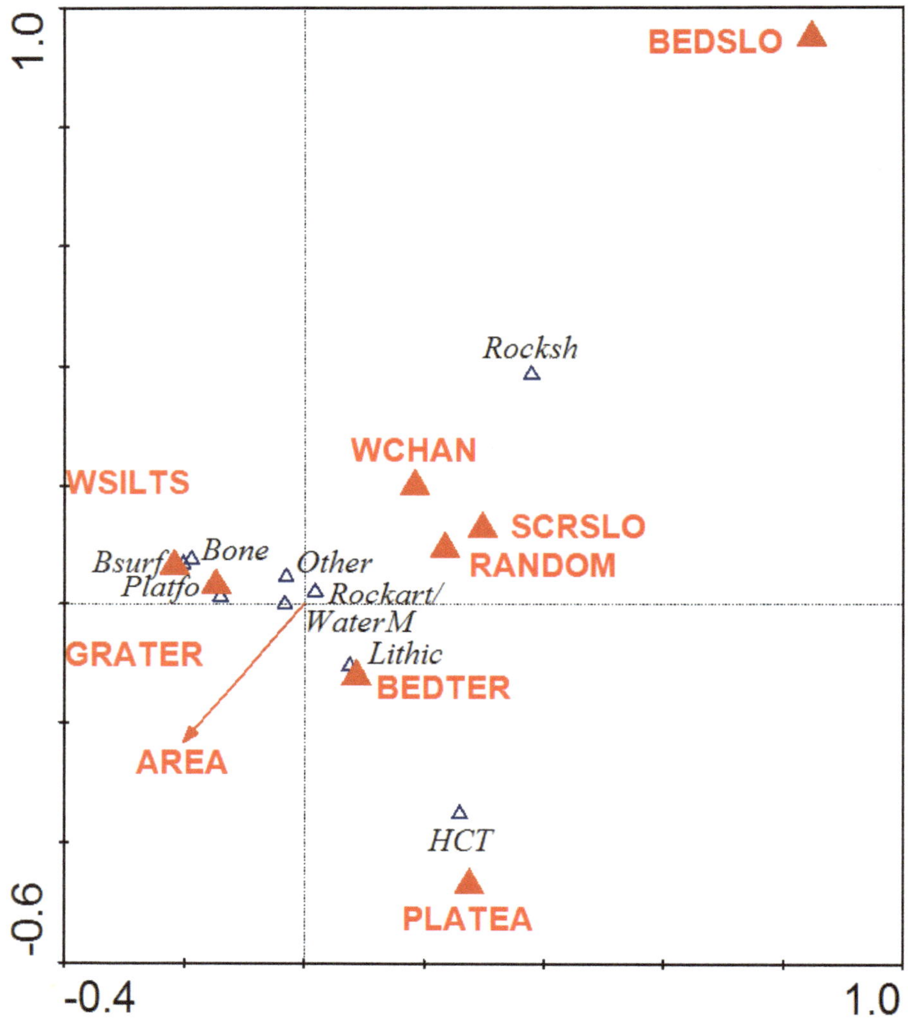

Table 6.9. CCA summary results from Wādī Sanā survey data. All four eigenvalues are canonical and correspond to axes constrained by the landform class, area, and survey method variables.

Axes	1	2	3	4
Eigenvalues	0.385	0.317	0.125	0.047
Site–Landform Correlations	0.779	0.708	0.54	0.318
Cumulative % Variance of Site Data	10.7	19.4	22.9	24.2
Cumulative % Variance of Site–Landform Relation	41.9	76.3	89.9	95
Sum of All Eigenvalues				3.611
Sum of All Canonical Eigenvalues				0.919

From this biplot, which is interpreted using the biplot rule (Lepš and Šmilauer 2003:figure 10–11), one can appreciate how sample composition (sites in survey units) varies by landform class and survey method. There is a significant difference between wadi silts, plateau, and bedrock slope survey units. If one draws an imaginary line from wadi silts through the origin and another from plateau through the origin, then the points for bone and bsurf (burned surfaces), when connected by imaginary perpendicular offsets to the imaginary lines, fall further from the origin than the point for water management (biplot rule above). Thus survey units from wadi silts differ from plateau survey units in bone and bsurf, but they have similar representation as rockart and rockshelters. Platforms, connected closer to the origin, are somewhat less strongly associated with wadi silts than are bone and burned surfaces. The position of HCT on the biplot shows that these high circular tombs are strongly associated with plateau. HCT is roughly equidistant from the origin along the projected imaginary lines of bedrock slope, random strategy, and wadi silts, suggesting that these have similar and statistically significant low association with HCT.

Through a forward stepwise model, CCA shows that some variables, notably wadi silts ($p = 0.001$), plateau ($p = 0.001$), bedrock slope ($p = 0.001$), and random strategy ($p = 0.002$), had significant explanatory power for the variability in site composition of survey units in Wādī Sanā. The use of Monte Carlo simulation with 1,000 permutations allows the site data to be randomly shuffled for 1,000 trials, against which the variance in the original dataset was compared. This approach assesses whether the composition of sites in the survey units can be explained solely through random principles or if there exist other underlying factors that must account for the variance in site composition. In the cases of wadi silts, plateau, bedrock slope, and random strategy nominal classes, p-values given above show statistically significant variance in site composition. Interestingly, survey unit area, an ordinal variable, does not significantly explain survey unit variance in site composition. In other words, it does not matter that some survey units were differently sized in our survey method.

Correspondence Analysis and the Types of Sites on Landforms

Environmental (constraining) variables have been coded on a CA scattergram plotting survey units to explore variability in site compositions of different survey units (samples). (For image clarity, the graphing program [CanoDraw] suppresses samples that are redundant.) In figure 6.4, properly interpreted using the centroid principle, ordination axes 1 and 2 account for 40.9 percent of the variance in site data (all four axes account for 71.1 percent). The first axis separates plateau and wadi silts from other landforms; the result mirrors the outcome of CCA in which these are two variables with significant explanatory power in site locations. The second axis of ordination differentiates wadi silts from plateau, but the arch effect suggests that this axis does not express new information in the distribution (ter Braak 1995:105). Figure 6.5 shows the results of the same CA ordination and should be viewed alongside figure 6.4. In the case of figure 6.5, pie diagrams visualize the composition of individual samples (survey units). Sample composition indicates that platforms, HCT, burned surfaces, and water management largely drive the ordination results. Because the plotted location of each survey unit is the same in figure 6.4 and figure 6.5, one can readily glance between them to assess the composition of sites in survey units on different landforms (figures 6.4 and 6.5).

Conclusions

Through systematic survey and statistical analysis, the RASA team reconstructed the structure of the human landscape through time and the integration of artifact and structural remnants into landscape processes. Survey methods were effective. Survey unit area had no significant effect on the observed distributions of sites across landform classes, probably because ancient people's use of particular landforms, the unknown variables of human decision-making, and the preservation of archaeological materials in those environments were far greater determinants of the situation of sites than the effort and coverage of the team looking for them. Random survey strategy accounted for only a minimal amount of site distribution variance (11 percent within the constraints of indicator variables). We conclude from our studies of different landforms that our survey results offer strong lessons for future survey and the representation of sites across landforms and important insight into human behavior in ancient Hadramawt.

The results of archaeological survey demonstrate that strong factors of association were at play in the construction and siting of small-scale stone monuments. People built HCT high on the plateau, away from direct contact with (but still visible from) waterways and grazing areas. While this observation is not new, it is for the first time quantifiably demonstrable through the analysis of random survey results. Anecdotally it is also apparent that HCT are visible from the wadi channel and lower terraces. The locations of HCT are in strong contrast with the locations of a very different monument type, platforms, which people built at lower elevations on terraces of silt, bedrock, and gravel near to the modern wadi channel and near to the best pasture and long-term shelter.

Figure 6.4. CA Scattergram of survey units coded for landform classes. *Illustration by Joy McCorriston.*

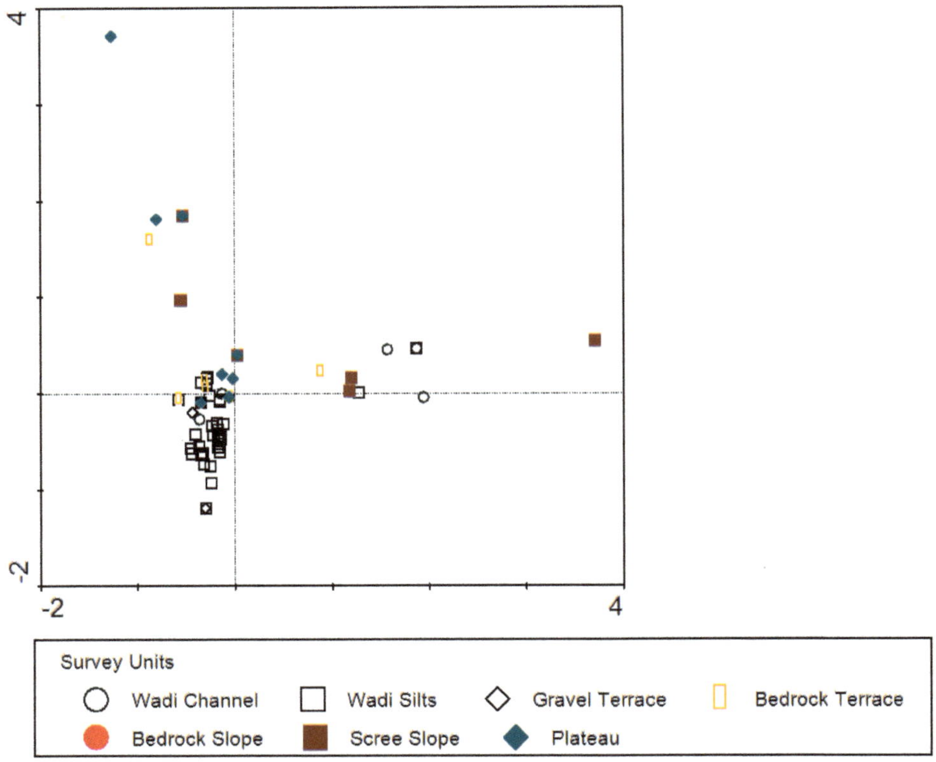

When the RASA team began survey in 1998, the distinctive Neolithic platform monuments, many with associated worked standing stones, had never been recognized or categorized as such. Chapters 10–12 report the excavation and analytical work that went into documenting, dating, and interpreting these remarkable memorials of ancient gatherings and feasts. These monuments appear throughout Wādī Sanā, but the intensive study of wadi silts at the Khuzma as-Shumlya confluence led the team to notice and return to a particular example, at Shiʿb Khishiyah. An erosional gully in the wadi silt terrace had exposed vertically embedded cattle skulls in the natural section. As the survey team gently collected teeth and bone crumbling from the section, more skulls appeared. Survey results show us that platform monuments mark areas where there once occurred a suite of activities—open camp occupations, gatherings, feasts, ceremonies, and collective building activities—all close to the richest resources of vegetation and water (and, one must presume, game and prey).

Unfortunately, the sample sizes of many other site types of interest (for example, triliths, dolmens, and wall tombs) are too small to be statistically significant. But these site types are culturally significant as readily identifiable, dated cultural markers. Survey observations show that triliths, dolmens, and wall tombs occur on lower terraces in the Wādī Sanā and tributaries (Wādī Ḥimayrī, Wādī as-Shumlyah, and Wādī Washaʿah); that they are infrequent; and that, in the case of triliths and wall tombs, they are clustered. Relative stratigraphy and geomorphological indications suggest that triliths postdate wall tombs (Trilith SU134-3 was built with slabs most likely robbed from Wall Tombs SU134-6 and SU134-2) and that they also occur with some frequency near preexisting sites, such as wall tombs, HCT, or platforms. Dolmens freestanding on rock or gravel terraces offer little dating potential, but the discovery of eroded fragments of what was most likely a former dolmen embedded in wadi silts suggests that these predate many of the other infrequent monument types (McCorriston et al. 2011:8–9). Once the team had ascertained a need to target relatively rare, culturally significant remains, we dedicated a survey and excavation season to this effort in 2008 (piloted in 2005; see chapter 5). The results of this season appear elsewhere (McCorriston et al. 2011) and in chapter 14.

Archaeological survey over five seasons integrated exploratory, systematic, random, and targeted methods to produce an integrated regional and highly detailed local understanding of human landscape use and preservation processes. This work would have been impossible without the multidisciplinary perspectives afforded by working

Figure 6.5. CA Scattergram of survey units as pies of site types. *Illustration by Joy McCorriston.*

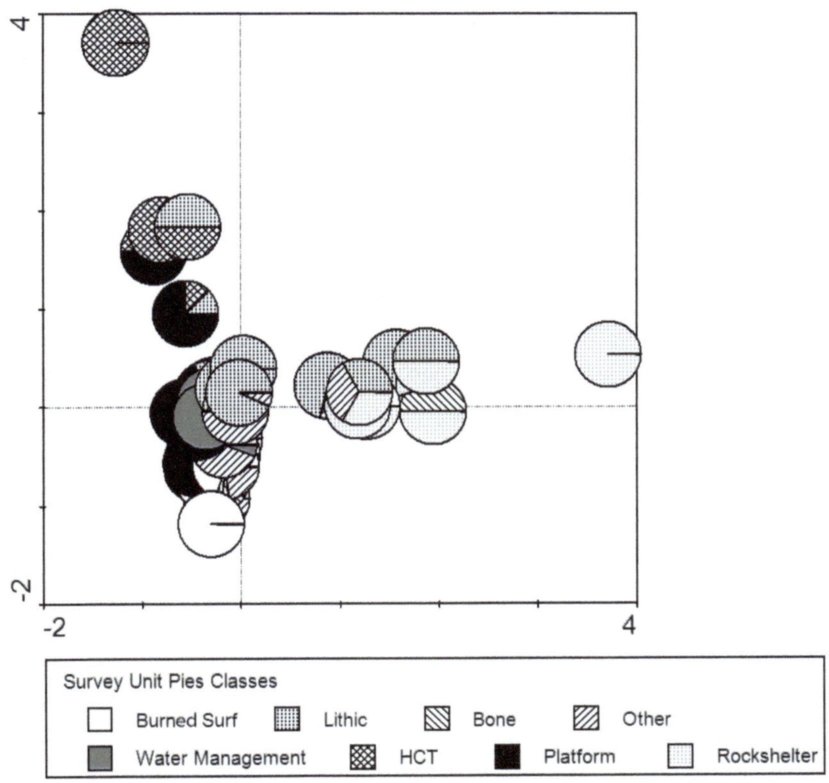

alongside geologist-paleoecologist Rick Oches and his students and Hadrami ethnographers ʿAbdalʿazīz Bin ʿ Aqīl and Ietha Al-ʿmari. In the course of survey, important examples of well-preserved sites became targets for intensive study and test excavations. In the following chapters we discuss the archaeological collections, small-scale excavations, and analytical results from RASA fieldwork and subsequent research from 1998 to 2008. We have excavated and provide radiocarbon ages on rockshelters, open-air sites, hearths, platforms, HCT, other small-scale stone monuments, a sanctuary, stone houses in Wādī ʿIdim, and water management structures. These sites have subsequently contributed valuable data for culture history and human activities in ancient Southern Arabia. A holistic and synthetic interpretation of the human past in the Southern Jol appears in the final chapter.

References

Al-Shaḥri, ʿAli Aḥmad Mahash

1991 Recent Epigraphic Discoveries in Dhofar. *Proceedings of the Seminar for Arabian Studies* 21:173–91.

Banning, Edward B.

1985 Pastoral and Agricultural Land Use in the Wadi Ziqlab Survey, 1981. PhD dissertation, Department of Anthropology, University of Toronto.

Bin ʿAqil, Abdalaziz J., and Joy McCorriston

2009 Convergences in the Ethnography, Semantics, and Archaeology of Prehistoric Small Scale Monument Types in Hadramawt (Southern Arabia). *Antiquity* 83:602–18.

Braemer, Frank, Serge Cleuziou, Luc Buchet, Jean-François Saliège, and Hervé Guy

2001 Le bronze ancien du Ramlat as-Sabatayn (Yémen). Deux nécropoles de la première moitié du IIIe millénaire à la bordure du désert: Jebel Jidrān et Jebel Ruwaiq. *Paléorient* 27:21-44.

Braemer, Frank, Serge Cleuziou, and Tara Steimer-Herbet

2003 Dolmen-like Structures: Some Unusual Funerary Monuments in Yemen. *Proceedings of the Seminar for Arabian Studies* 33:169–82.

Crassard, Rémy

2008 *La Préhistoire du Yémen. Diffusions et diversités locales, à travers l'étude d'industries lithiques du Hadramawt.* British Archaeological Reports S1842. Archaeopress, Oxford.

Crassard, Rémy, and Holger Hitgen

2007 From Safer to Balhaf: Rescue Excavations along the Yemen LNG Pipeline Route. *Proceedings of the Seminar for Arabian Studies* 37:1–17.

De Cardi, Beatrice, B. Doe, and S. P. Roskams

1977 Excavation and Survey in the Sharqiyah, Oman 1976. *Journal of Oman Studies* 3:17–33.

de Maigret, Alessandro, and Sabina Antonini

2005 *South Arabian Necropolises: Italian Excavations at Al-Makhdarah and Kharibat al-Ahjur (Republic of Yemen)*. Istituto Italiano per l'Africat e L'Oriente, Rome.

Eddy, Frank W., and Fred Wendorf

1998 Prehistoric Pastoral Nomads in the Sinai. *Sahara* 10:7–20.

Halevi, Leor

2007 Muhammad's Grave: Death Rites and the Making of Islamic Society. Columbia University Press, New York.

Harrower, M. J.

2006 Environmental versus Social Parameters, Landscape, and the Origins of Irrigation in Southwest Arabia (Yemen). PhD Dissertation, Department of Anthropology, The Ohio State University, Columbus.

2008 Mapping and Dating Incipient Irrigation in Wādī Sanā, Hadramawt (Yemen). *Proceedings of the Seminar for Arabian Studies* 38:187–202.

Harrower, Michael J., Eric A. Oches, and Joy McCorriston

2012 Hydro-Geospatial Analysis of Ancient Pastoral/Agro-Pastoral Landscapes along Wādī Sanā (Yemen). *Journal of Arid Environments* 86:131–38.

Kimiaie, Masoumeh, and Joy McCorriston

2014 Climate, Human Palaeoecology, and the Use of Fuel in Wādī Sanā , Southern Yemen. *Vegetation History and Archaeobotany* 23:33–40.

Lepš, Jan, and Petr Šmilauer

2003 *Multivariate Analysis of Ecological Data Using Canoco*. Cambridge University Press, Cambridge.

McCorriston J., E. A. Oches, D. E. Walter, and K. L. Cole

2002 Holocene Paleoecology and Prehistory in Highland Southern Arabia. *Paléorient* 28(1):61–88.

McCorriston, Joy

2011 *Pilgrimage and Household in the Ancient Near East*. Cambridge University Press, Cambridge.

McCorriston, Joy, Michael J. Harrower, Louise A. Martin, and Eric A. Oches

2012 Cattle Cults of the Arabian Neolithic and Early Territorial Societies. *American Anthropologist* 114:45–63.

McCorriston, Joy, Michael J. Harrower, Eric A. Oches, ʿAbdalʿazīz Bin ʿAqīl, Catherine Heyne, AbdalBset Noman, Rémy Crassard, Khalid BaDhofary, and Joshua Anderson

2005 Foraging Economies and Population in the Middle Holocene Highlands of Southern Yemen. *Proceedings of the Seminar for Arabian Studies* 35:143–54.

McCorriston, Joy, Tara Steimer-Herbet, Michael J. Harrower, Kimberly A. Williams, Jean-François Saliège, and ʿAbdalʿazīz Bin ʿAqīl

2011 Gazetteer of Small Scale Monuments in Prehistoric Hadramawt, Yemen: A Radiocarbon Chronology from RASA Project Research 1996–2008. *Arabian Archaeology and Epigraphy* (22):1–22.

Openshaw, Stan

1983 *The Modifiable Areal Unit Problem*. Geo Books, Norwich, UK.

Plog, Steven, Fred Plog, and W. Wait

1978 Decision Making in Modern Surveys. In *Advances in Archaeological Method and Theory*, vol. 1, edited by Michael B. Schiffer, pp. 384–421. Academic Press, New York.

Steimer-Herbet, Tara

2004 *Classification des sepultures a superstructures lithiques dans le Levant et l'Arabie meridionale (IVe et IIIe millenaires avant J.-C.)*. British Archaeological Reports, International Series 1246. Archaeopress, Oxford.

Steimer-Herbet, Tara, Frank Braemer, and Gourguen Davtian

2006 Pastoralists' Tombs and Settlement Patterns in Wādī Wash'ah during the Bronze Age (Ḥaḍramawt, Yemen). *Proceedings of the Seminar for Arabian Studies* 36:257–65.

ter Braak, Cajo J. F.

1994 Canonical Community Ordination. Part I, Basic Theory and Linear Methods. *Ecoscience* 1:127–40.

1995 Ordination. In *Data Analysis in Community and Landscape Ecology*, edited by R. H. G. Jongman, C. J. F. ter Braak, and O. F. R. van Tongeren, pp. 91–173. Cambridge University Press, Cambridge.

ter Braak, Cajo J. F., and Petr Šmilauer

2002 *CANOCO Reference Manual and CanoDraw for Windows User's Guide: Software for Canonical Community Ordination (Version 4.5)*. Microcomputer Power, Ithaca, NY.

ter Braak, Cajo J. F., and Piet F. M. Verdonschot

1995 Canonical Correspondence Analysis and Related Multivariate Methods in Aquatic Ecology. *Aquatic Ecology* 57:255–89.

Williams, K. D., T. Steimer-Herbet, L. Gregoricka, J.-F. Saliège, and J. McCorriston

2014 Bioarchaeological Analyses of 3rd Millennium BC High Circular Tower Tombs from the Arabian Human Social Dynamics (AHSD) Project in Dhofar, Oman. *Journal of Oman Studies* 18: 153–73.

Zaris, J.

2001 *The Land of Incense: Archaeological Work in the Governorate of Dhofar, Sultanate of Oman, 1990–1995*. Sultan Qaboos University Publications.

Part III

Pleistocene to Early Holocene: Hunter-Foragers and the Introduction of Domesticates

Chapter 7

Middle Paleolithic Populations of Wādī Sanā

Rémy Crassard

For years, prehistorians considered the Arabian Peninsula an archaeological no-man's-land. The terms *cul-de-sac*, *elephant graveyard*, and *tabula rasa* were used to describe the impoverished, if not entirely absent, archaeological remains in all periods of prehistory (Crassard 2008). There are few sites; almost all of them are missing stratigraphy and therefore are nondatable. Moreover there are virtually no bone remains even for the most recent periods of prehistory, and until recently (Groucutt et al. 2018) there were no human fossils from earlier periods. So many inconvenient archaeological challenges for a territory as large as one-third of the continental United States! Fortunately this view has gradually changed through deep commitment and the tenacity of a few pioneers. The shift in perception is radical, so that Arabia is today one of the central places in debates on the origins and dispersals of early human groups.

There existed several waves of colonization by early hominids outside the African cradle (e.g., Antón and Swisher 2004; Bar-Yosef 1987; Bar-Yosef and Belfer-Cohen 2001; Dennell 2003; Dennell and Roebroeks 2005; Trinkaus 2005), including at least one migration into Arabia. Distant ancestors of *Homo sapiens*, these hominids thereby dispersed into most of the regions of our planet. In Africa, the first forms of *Homo sapiens* (also known as anatomically modern humans, amHs, our species) appear about 200,000 to 150,000 years ago, probably through the evolution of local forms of African *Homo erectus* (or *H. ergaster*). Expansion of *Homo sapiens* within and outside Africa is of particular research interest; the processes involved are critical if one seeks to understand the history of peopling the earth by our species, which has today reached near total hegemony. The Middle Stone Age period, when the first *Homo sapiens* appeared in East Africa, is contemporaneous to the Middle Paleolithic in the Arabian Peninsula. Distinctive lithic productions (stone tools and characteristic debitage) of these early humans are regularly found in both Africa and Arabia.

The Discovery of Levallois Industries in Hadramawt

The analysis of lithic industries from Wādī Sanā is part of a larger study covering the central and eastern part of the province of Hadramawt, where Middle Paleolithic materials occur. In addition to Wādī Sanā, areas farther north, such as the region of Wādī Washaʿah and tributaries of Wādī Masilah near the village of As-Ṣūm, have been explored (Crassard 2007, 2008, 2009; Crassard and Thiébaut 2011). I and my colleagues found that the tops of plateaus overlooking the wadis in Hadramawt yield rich lithic production remains. Wādī Sanā is a good example. The disappointing aspect is that all these sites occur on the surface without any possibility of dating elements independent of typo-technology. If found, a stratified context could potentially provide independent dating of lithic elements. What follows describes surface collections in the Wādī Sanā.

Apart from rare bifaces resembling Acheulian types and possibly dated to the Lower Paleolithic, the knapping operational schemes (or *chaines opératoires*) relate to the Levallois concept, which is generally dated to the Middle Paleolithic. These Levallois-like schemes are typologically and technologically among the oldest in the Hadramawt. The state of weathering (or patina) on the artifacts' surfaces is also particularly developed and much more so than the weathering on industries independently dated to the Early and Middle Holocene.

Mainly known from Africa and Eurasia, the Levallois concept (Boëda 1994) of knapping was in use for 400,000 to 500,000 years. Levallois appears from the Terminal Acheulean–type industries in Africa; the evidence in Western Europe shows Levallois-type knapping among Middle Acheulean–type assemblages since the end of Marine Isotopic Stage (MIS) 10 (about 400,000 to 300,000 years ago; Tuffreau 2004). Researchers generally place Eurasian Levallois industries in the Middle Paleolithic (300,000 to 30,000 years ago) from MIS 8. The presence of Levallois technology traditionally signals Upper Pleistocene industries across Africa and the Middle East.

Discovery Context

The first researchers to explore the Hadramawt region mentioned the use of Levallois technology there. Gertrude Caton-Thompson (1953) was the first to identify a potential human presence during the Pleistocene in South Arabia. Subsequent researchers also noting Levallois included Gus van Beek (Van Beek et al. 1963), Marie-Louise Inizan (Inizan 1989; Inizan and Ortlieb 1987), Hizri Amirkhanov (1991, 1994) and Paul Zimmerman (2000).

During field operations, the RASA team documented 12 surface sites with characteristic elements of Levallois technology (table 7.1). Except in rare cases where lithic scatters were relatively abundant and homogenous (with possible refitting on site), Levallois technology occurred as discrete presences within low-density surface assemblages. Most of the time, discovered pieces showed a poor state of conservation, with heavily weathered artifacts showing rounded edges and *arises* as well as exfoliated surfaces. Only a few sites yielded less-weathered industries with knapping scars perfectly readable. All artifacts seen on surface sites were made of local Eocene/Oligocene high-quality, fine-grained chert. This material derives from outcrops common directly on the top of the plateau where the sites were found. For all sites with Levallois industries, I made a selective collection with a special focus on cores. These cores form the basis of a technological study that was as complete as possible. Because only the last stages of exploitation—the discarded cores—are visible, information obtained from these cores alone represents a truncated view of the knapping sequence. RASA data provide significant value in documenting the final phases of the Levallois operational schemes encountered in Hadramawt.

Table 7.1. RASA sites with Levallois artifacts.

Sites	UTM Easting	UTM Northing	Number of Analyzed Cores (this study)	Number of Collected Cores
RASA 2004-84-0	333362	1743787	3	3
RASA 2004-84-2	333362	1743787	3	3
RASA 2004-124-1	328474	1738945	7	12
RASA 2004-135-1	328474	1738945	1	5
RASA 2004-136-1	332491	1743837	1	2
RASA 2004-141-1	332850	1743345	1	1
RASA 2004-149-1	342696	1764658	4	5
RASA 2004-149-2	341608	1761201	8	12
RASA 2004-153-1	340863	1761022	2	3
RASA 2004-165-1	330321	1739460	2	4
RASA 2004-166-1	334582	1745846	4	11
RASA 2004168-1	335026	1745995	1	1
Totals			**37**	**62**

The search for Levallois products (for example, preferential, recurrent, or preparation flakes) was also a priority. Therefore I adapted my collection method to maximize the information these pieces would hold. As the only material markers of Middle Paleolithic hominids in Arabia, Levallois products are key to understanding Arabia's population history and its larger role in global population and human species dynamics. Research in Arabia has yielded only one hominid fossil, so knapped-stone products are the only way to understand ancient hominids. To provide chrono-stratigraphic data, the RASA surveys implemented an intensive search for rockshelters and caves with stratified deposits, unfortunately with none found from the Middle Paleolithic. Regular explorations on top of the limestone plateaus have routinely led to the discovery of Levallois sites, making Wādī Sanā one of the first places to yield critical evidence for a better understanding of Levallois variability in South Arabia.

Methods and Definitions

Where my analysis could completely read the scars of a Levallois core, that core has a systematic illustration. Along with diacritic schemes, these illustrations constitute a good basis for the analysis of the assemblages (figures 7.1–7.10).

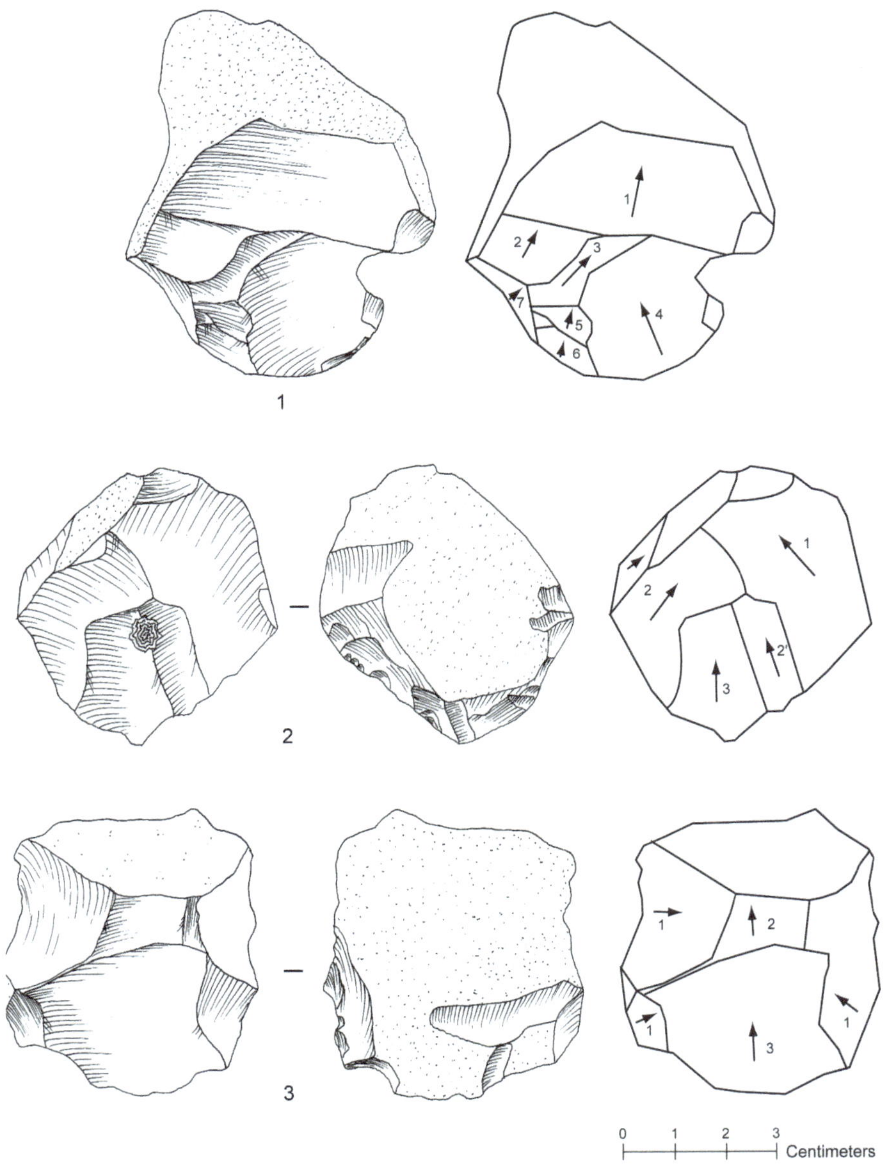

Figure 7.1. Levallois cores from RASA-2004-84-0. (1) RASA-2004-84-0 Core 1; (2) RASA-2004-84-0 Core 2; (3) RASA-2004-84-0 Core 3. *Illustration by Rémy Crassard.*

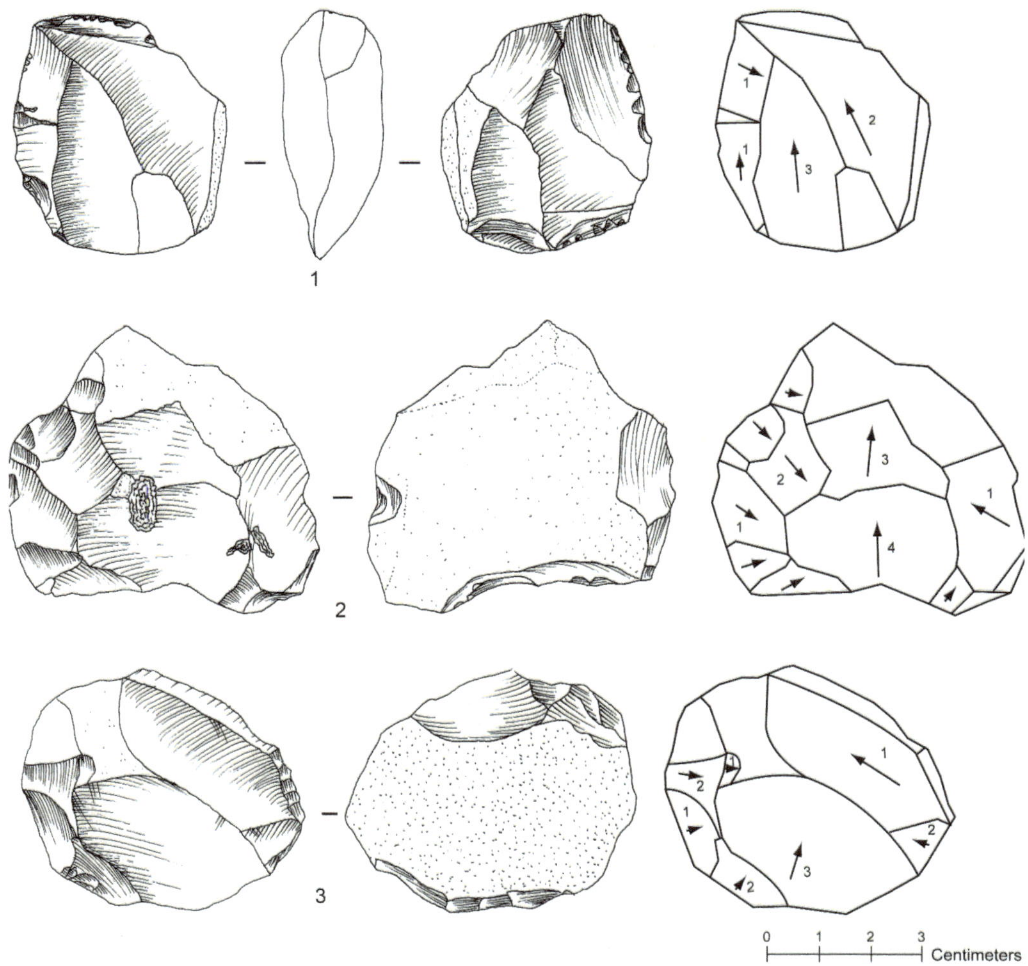

Figure 7.2. Levallois cores from RASA-2004-84-2. (1) RASA-2004-84-2 Core 1; (2) RASA-2004-84-2 Core 2; (3) RASA-2004-84-2 Core 3. *Illustration by Rémy Crassard.*

The Levallois concept implies a specific preparation of the core: the knapper shaped the block to obtain two intersecting convex surfaces. One is the striking platform's plan; the other is the production (or Levallois) surface. The knapper prepared the core to present convexities that would determine the shape and size of the future Levallois flakes(s). Two main classes of methods are identified in Wādī Sanā: the preferential Levallois flake methods, where only one flake is extracted from the core surface after preparation, and recurrent methods, in which several successive flakes are struck from the core surface, without the need to prepare it again. If the successive flakes converge toward the center of the core, it is a Centripetal Recurrent Levallois scheme. If the successive flakes are parallel and extend from a single striking platform, it is a Unidirectional Recurrent Levallois scheme. If the successive flakes are parallel and are based on two, opposed, striking platforms, it is a Bidirectional Recurrent Levallois scheme. Convergent unidirectional, and bidirectional are other methods that allow a knapper to obtain the characteristic triangular flakes called Levallois points.

This analysis of the Middle Paleolithic cores from RASA surveys relied on principles of dynamic evolution of the flaking. Such analysis encompasses predetermination characteristics (Boëda 1994). This approach implies that the analyst can reconstruct the prehistoric knapper's individualized conceptualization of Levallois operational schemes. The Preferential Method cores were grouped into several types of preferential products (quadrangular, oval, or trapezoidal flakes, Levallois points). On the other hand, Recurrent Method cores were grouped by modality of flaking surface management: unidirectional (parallel), bidirectional, and centripetal.

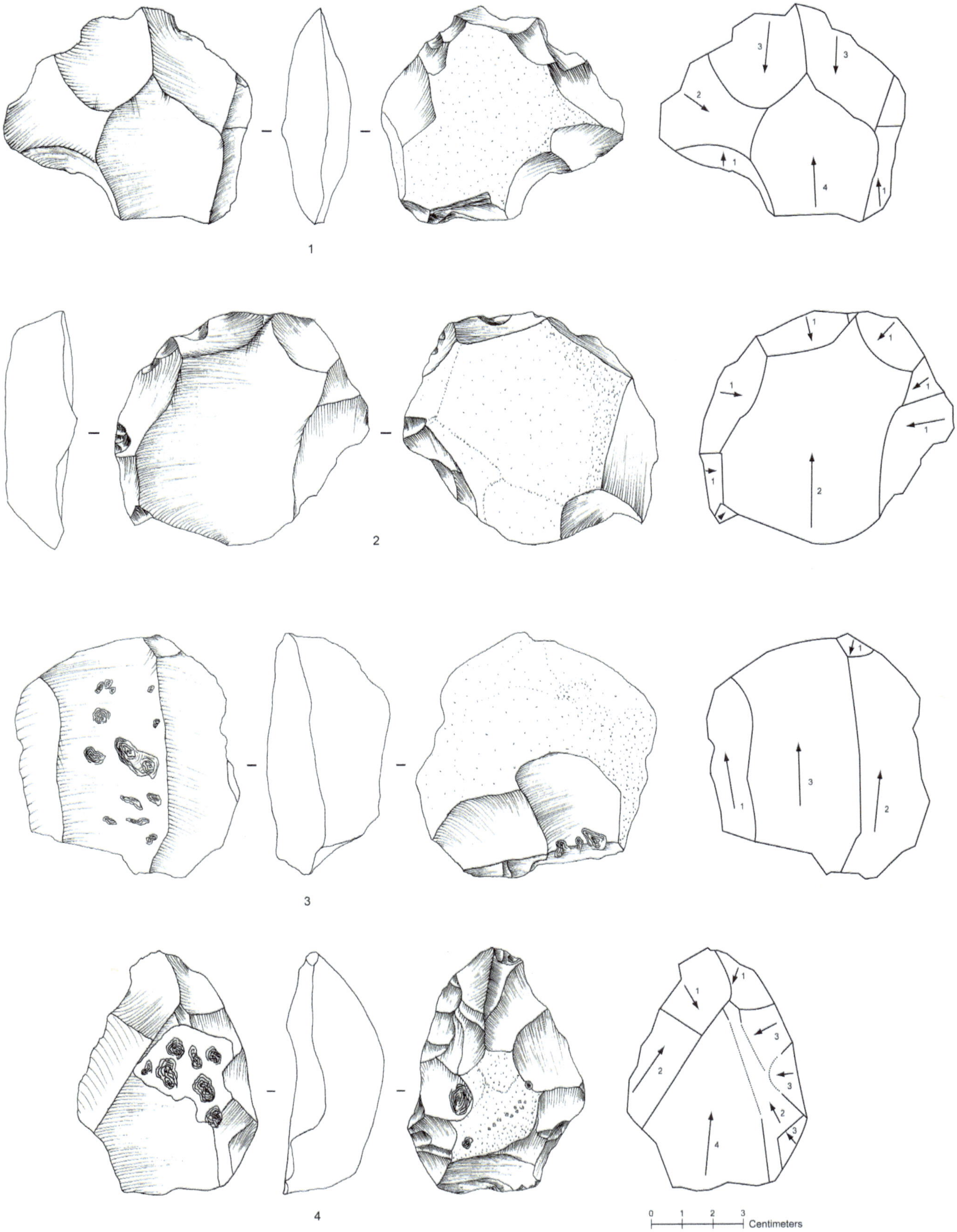

Figure 7.3. Levallois cores from RASA-2004-124-1. (1) RASA-2004-124-1 Core 1; (2) RASA-2004-124-1 Core 2; (3) RASA-2004-124-1 Core 3; (4) RASA-2004-124-1 Core 4. *Illustration by Rémy Crassard.*

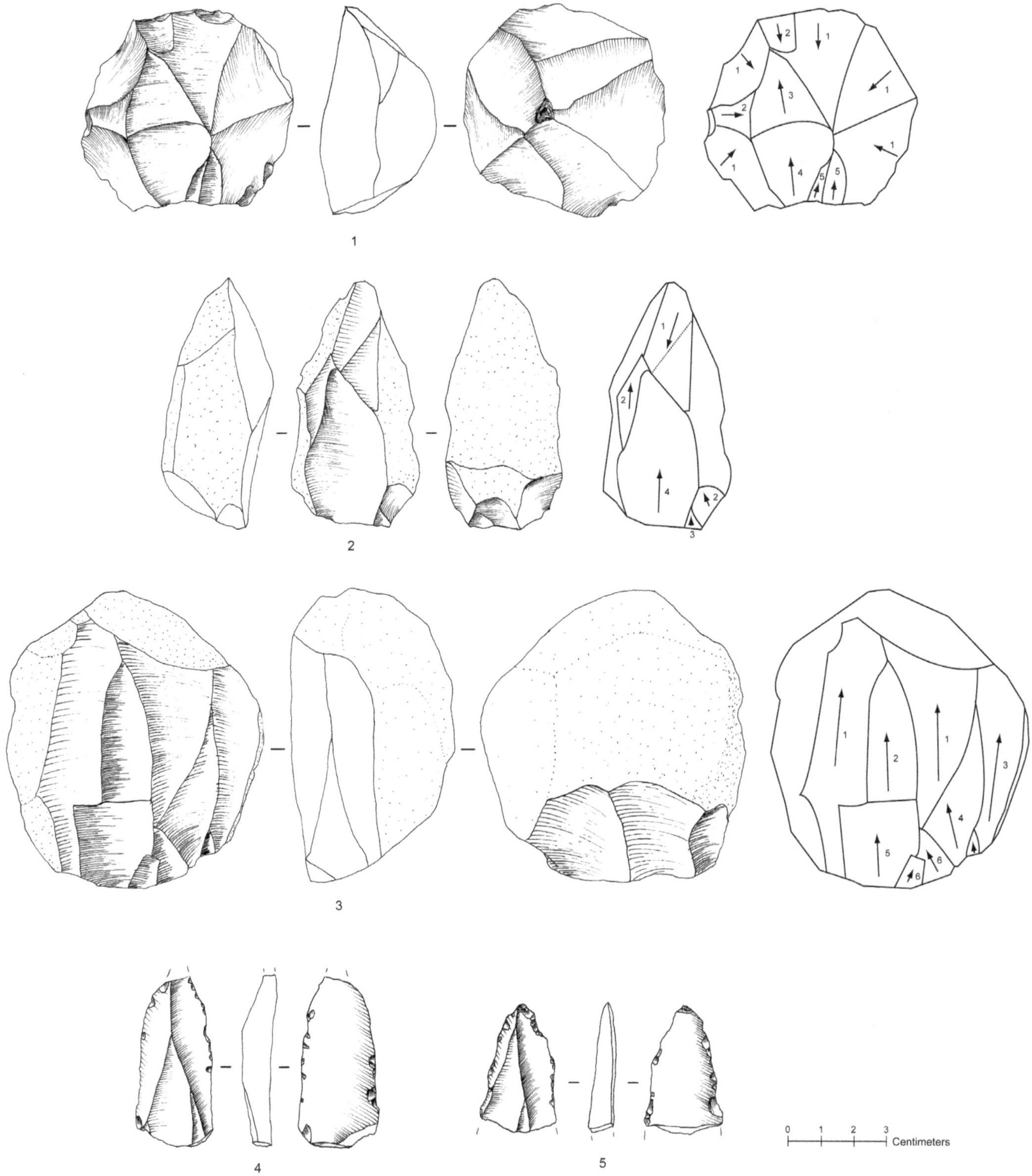

Figure 7.4. Levallois cores and products from RASA-2004-124-1. (1) RASA-2004-124-1 Core 5; (2) RASA-2004-124-1 Core 6; (3) RASA-2004-124-1 Core 7; (4) and (5) fragmentary Levallois triangular flakes from RASA-2004-124-1. *Illustration by Rémy Crassard.*

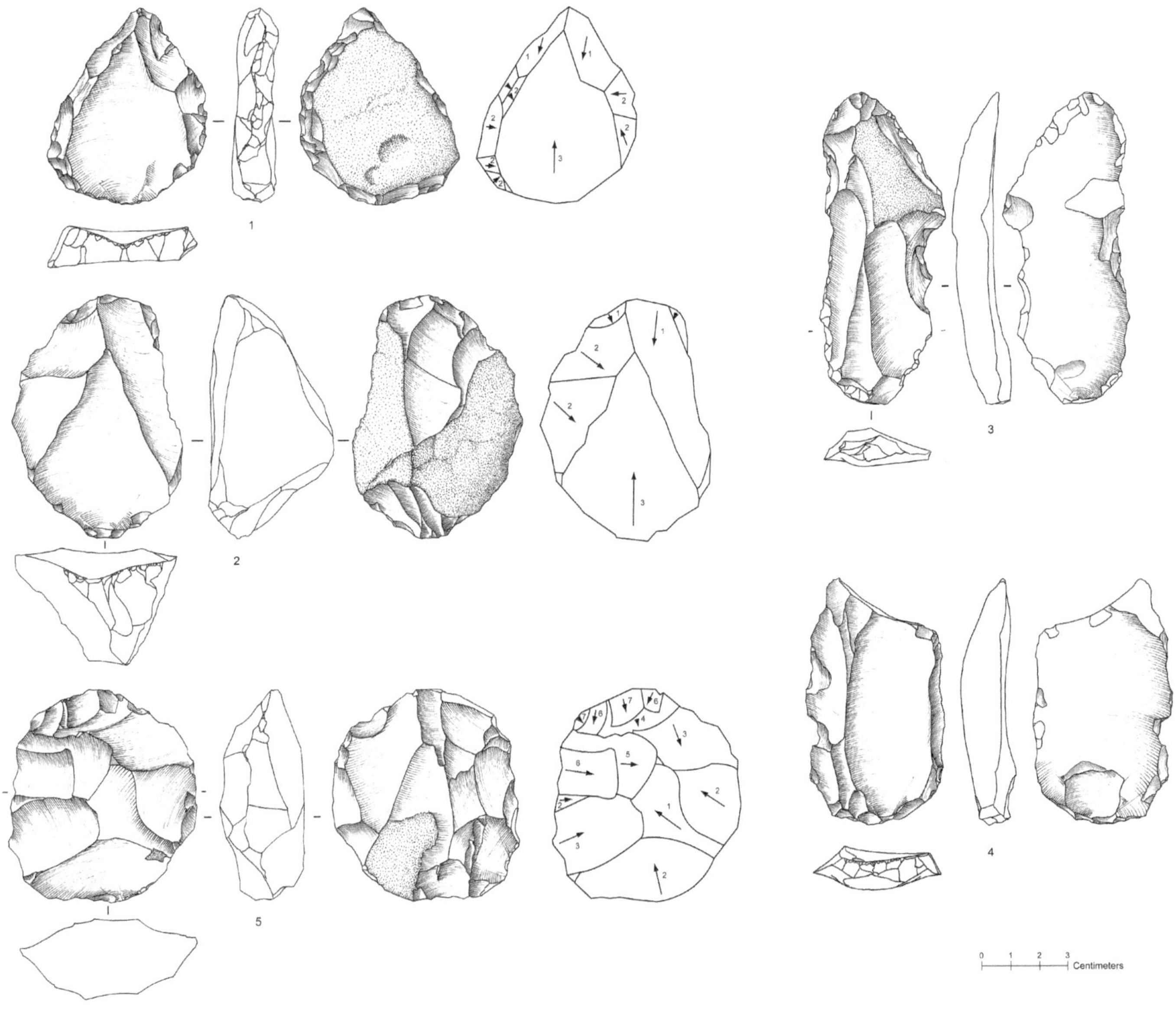

Figure 7.5. Levallois cores from RASA-2004-135-1 and RASA-2004-136-1, and Levallois cores and products from RASA-2004-141-1. (1) RASA-2004-135-1 Core 1; (2) RASA-2004-136-1 Core 1; (3) and (4) Levallois debitage from RASA-2004-141-1; (5) RASA-2004-141-1 Core 1. *Illustration by Julien Espagne and Rémy Crassard.*

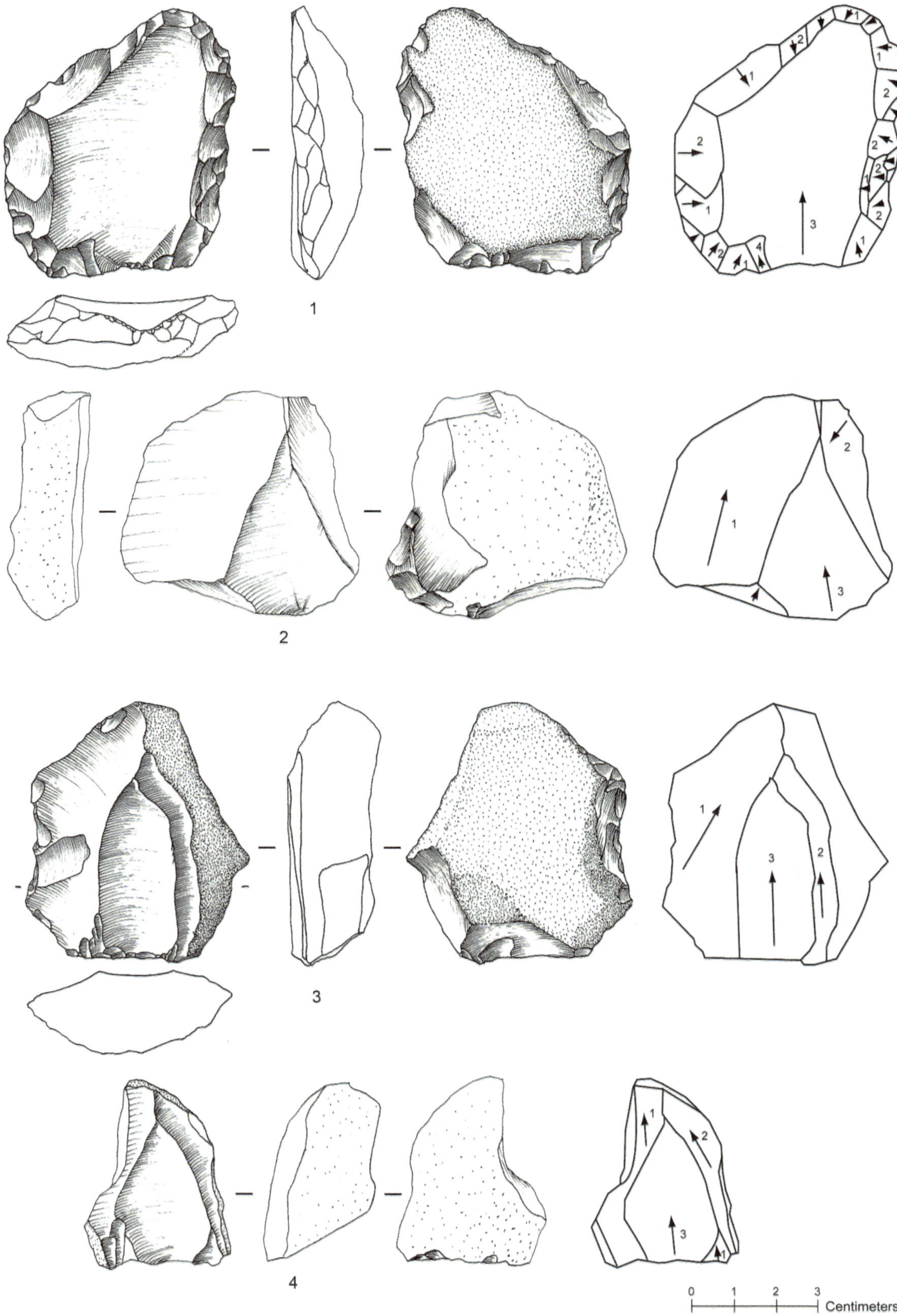

Figure 7.6. Levallois cores from RASA-2004-149-1. (1) RASA-2004-149-1 Core 1; (2) RASA-2004-149-1 Core 2; (3) RASA-2004-149-1 Core 3; (4) RASA-2004-149-1 Core 4. *Illustration by Julien Espagne and Rémy Crassard.*

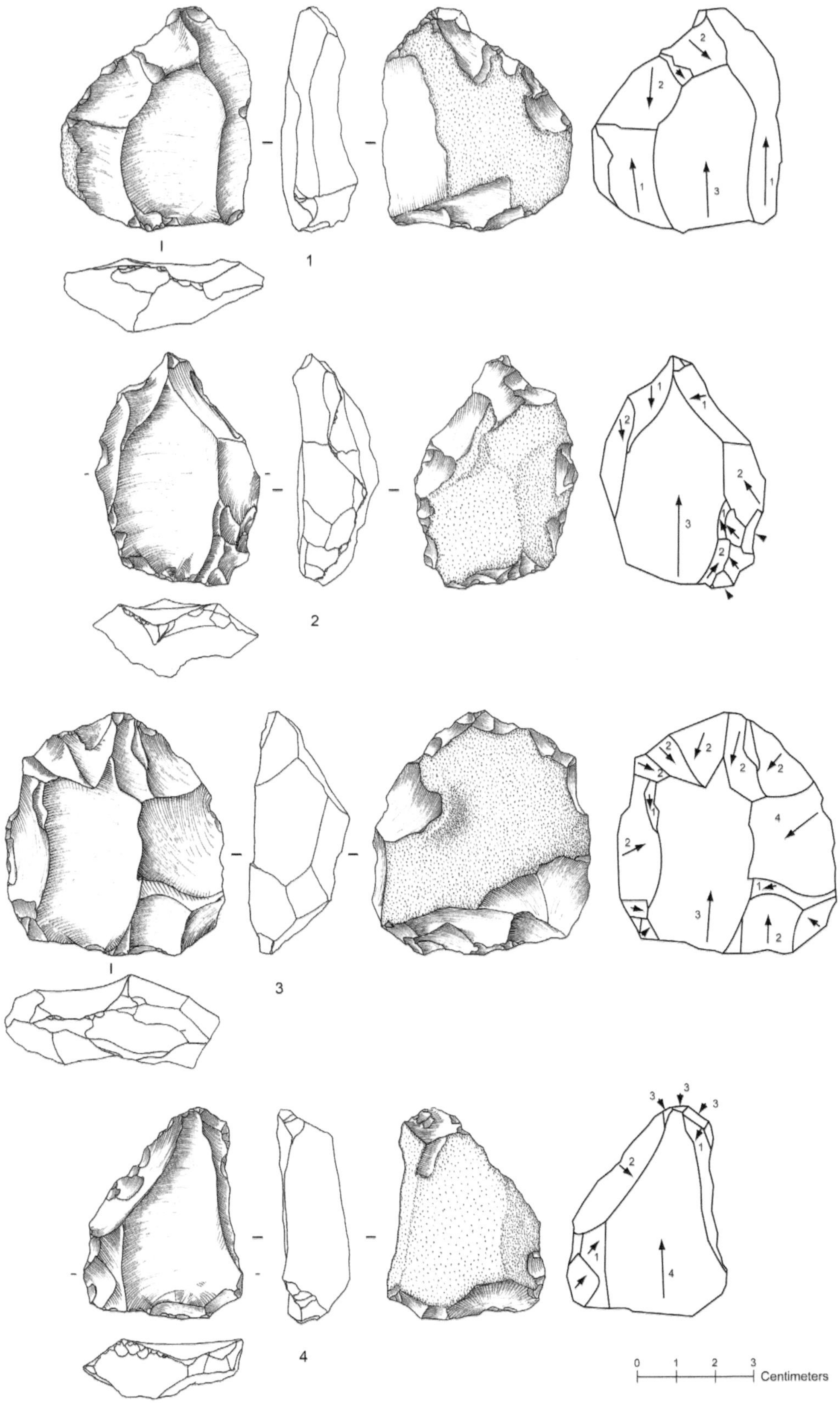

Figure 7.7. Levallois cores from RASA-2004-149-2. (1) RASA-2004-149-2 Core 1; (2) RASA-2004-149-2 Core 2; (3) RASA-2004-149-2 Core 3; (4) RASA-2004-149-2 Core 4. *Drawing by Julien Espagne and Rémy Crassard.*

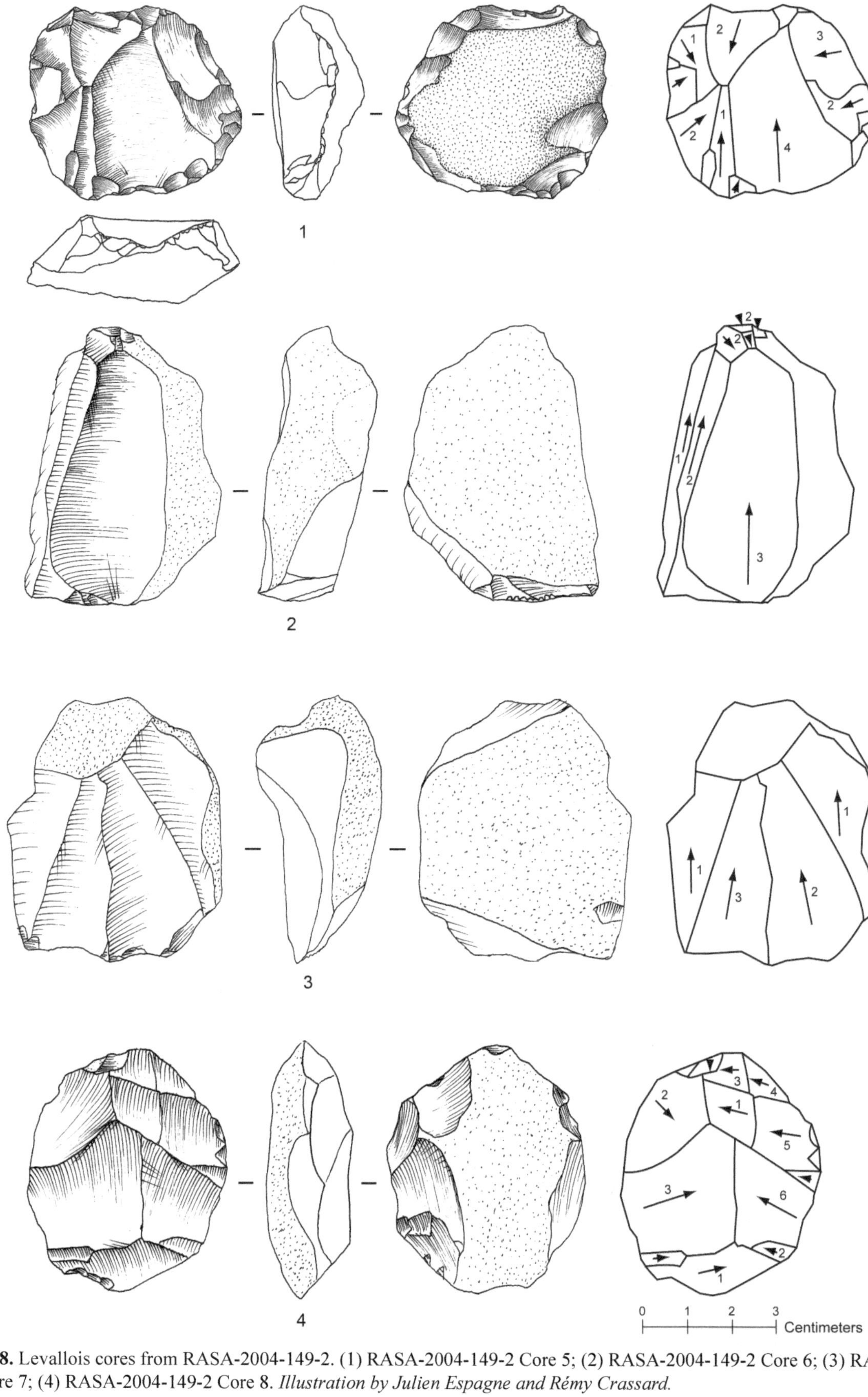

Figure 7.8. Levallois cores from RASA-2004-149-2. (1) RASA-2004-149-2 Core 5; (2) RASA-2004-149-2 Core 6; (3) RASA-2004-149-2 Core 7; (4) RASA-2004-149-2 Core 8. *Illustration by Julien Espagne and Rémy Crassard.*

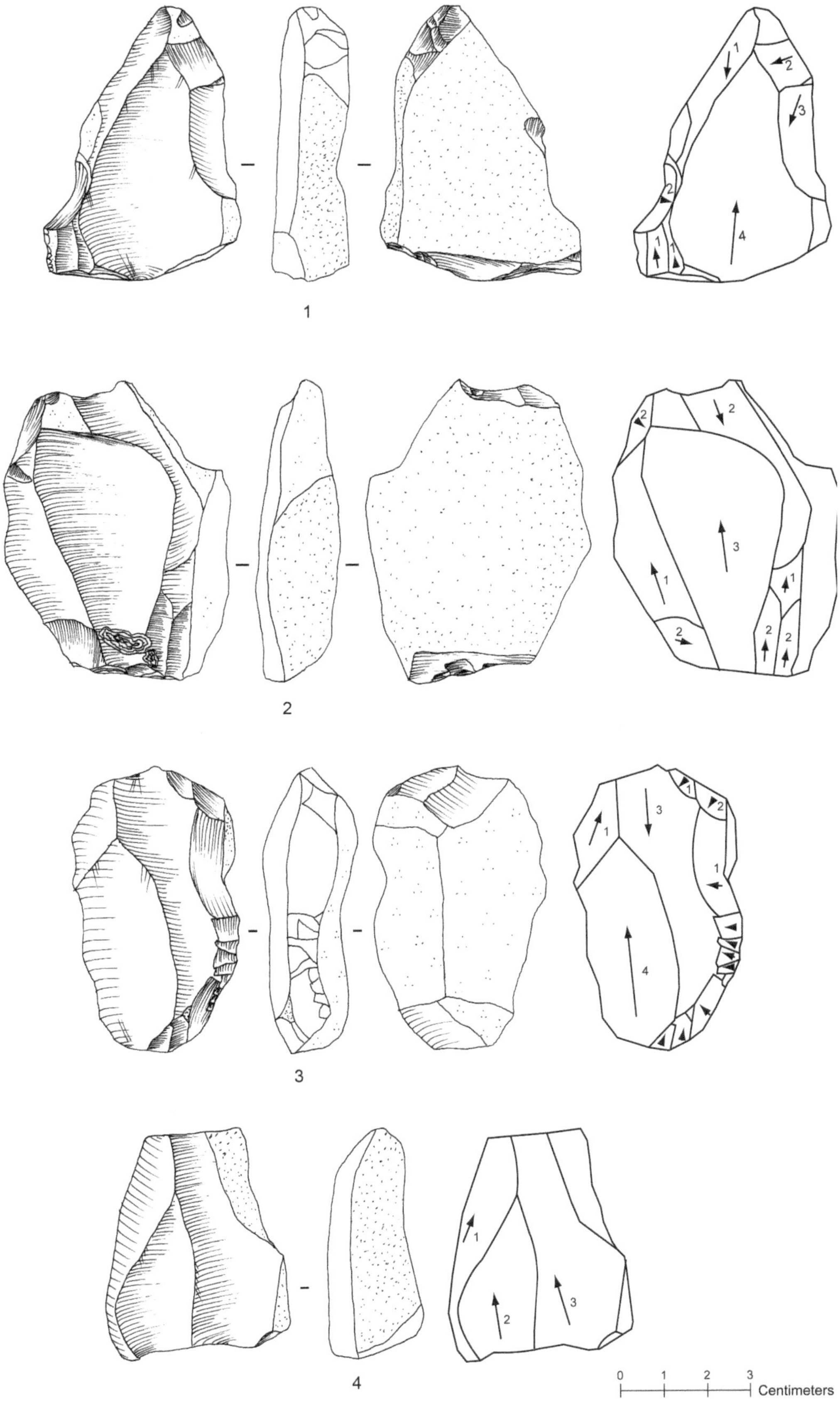

Figure 7.9. Levallois cores from RASA-2004-153-1 and RASA-2004-165-1. (1) RASA-2004-153-1 Core 1; (2) RASA-2004-153-1 Core 2; (3) RASA-2004-165-1 Core 1; (4) RASA-2004-165-1 Core 2. *Illustration by Rémy Crassard.*

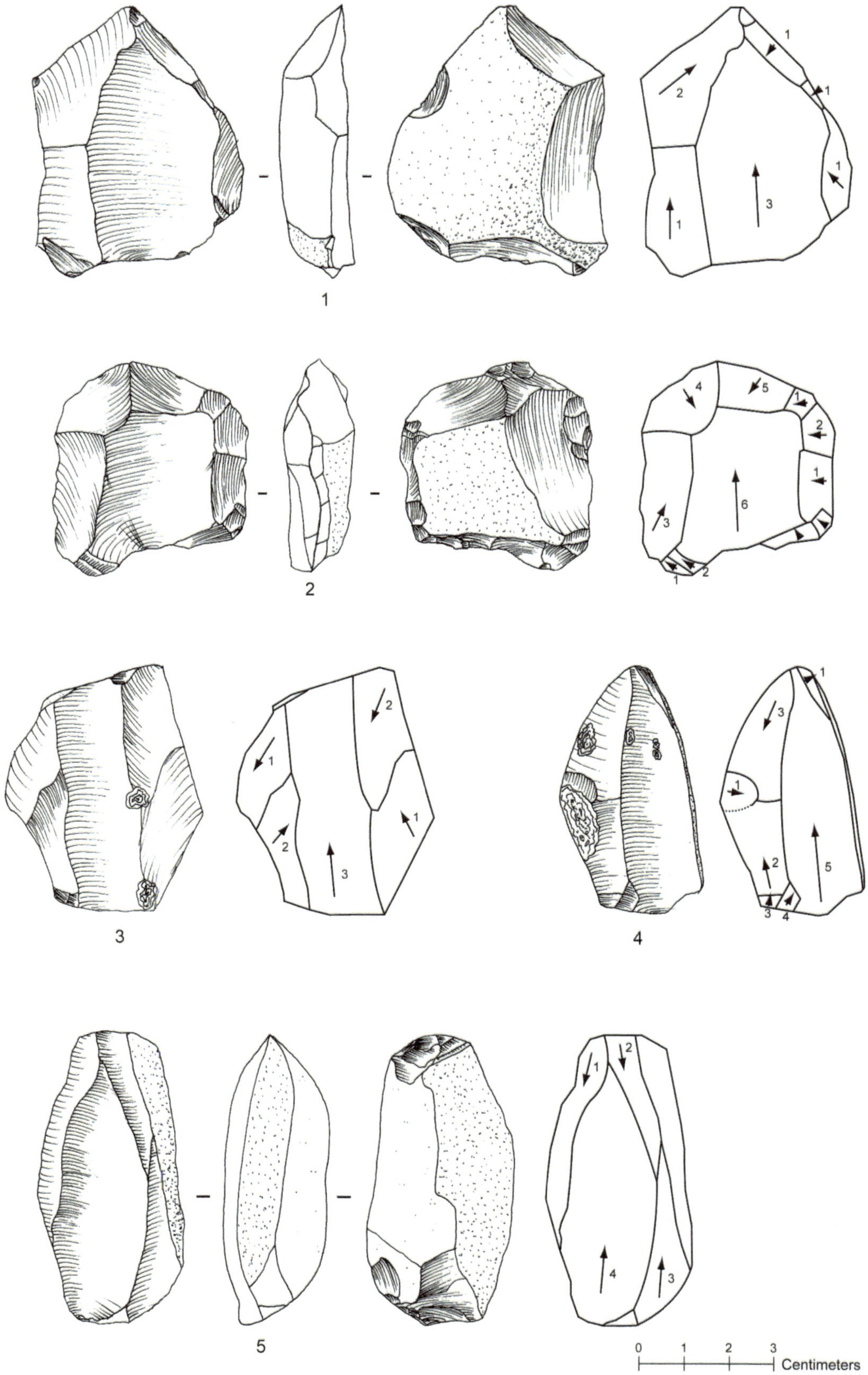

Figure 7.10. Levallois cores from RASA-2004-166-1 and RASA-2004-168-1. (1) RASA-2004-166-1 Core 1; (2) RASA-2004-166-1 Core 2; (3) RASA-2004-166-1 Core 3; (4) RASA-2004-166-1 Core 4; (5) RASA-2004-168-1 Core 1. *Illustration by Rémy Crassard.*

Analyses of Levallois Industries from Wādī Sanā

Two Levallois Methods and Three Technological Groups from Wādī Sanā

The technological analysis focused on a total of 37 cores. Two cores were found isolated, while 35 are from 10 assemblages of at least two cores. All sites lie in Wādī Sanā.

By studying the final stages of visible preparation on the Levallois cores, the author could differentiate three distinct groups, A, B, and C (Crassard 2009). A number of operational schemes have been highlighted for the entire study area (figure 7.11). These operational schemes reveal the variability within Group A and within Group B. In sum, the study of the cores from Wādī Sanā distinguishes two Levallois methods: one for Preferential Levallois flakes (Groups A and B) and the other for Centripetal Recurrent surfaces (Group C).

Group A is characterized by:

- Production of a preferential flake with centripetal preparation (Schemes A1 and A2)
- Production of a preferential flake with bilateral preparation (Scheme A3)

Group B is characterized by:

- Production of preferential triangular flakes with convergent, unidirectional preparation (Scheme B1). The "constructed" version with opposed preparations could be related to production of a "Nubian" type (Crassard and Hilbert 2013; Guichard and Guichard 1965; Usik et al. 2013).
- Production of preferential triangular flakes with distal divergent preparation from a secondary striking platform opposed to the striking platform of the preferential flake (Scheme B3). This scheme is often related to the Nubian production (Crassard and Hilbert 2013; Guichard and Guichard 1965; Usik et al. 2013).
- Production of preferential triangular flakes (or proper Levallois points) with crossed preparation (Scheme B4)
- Production of preferential triangular flakes (or proper Levallois points) with bidirectional preparation (Scheme B5)

Group C is characterized by:

- Centripetal Recurrent Levallois production (Scheme C)

Levallois production in Wādī Sanā has three main characteristics. First is the production of unique preferential Levallois flakes. A second characteristic is the predominance of modalities striving to achieve triangular flakes, whether Levallois "classical" points or "constructed" points. Third, one can detect a certain complexity of knappers' technical behavior, which is mainly observed in the production of constructed points, with installation or reinstallation of surface convexities at various stages of the flaking.

Thus, within a same Levallois method, variability is evident. At the same time, note that some operating schemes absent in the Wādī Sanā assemblages are known elsewhere in Eurasia. Thus the laminar Levallois and Unidirectional or Bidirectional Recurrent Levallois schemes are methods that have not yet appeared in Wādī Sanā.

Origins and Development of Middle Paleolithic Populations in Southern Arabia

Arabia has a special place in the investigation of our origins: it is located near East Africa, the very place where paleontologists have found the first *Homo sapiens*. The Arabian Peninsula, and especially Yemen, is potentially one of the first lands outside of Africa where modern humans trod. Geneticists first tried to find evidence of crossing through Arabia in the great journey that would lead these early modern humans from South Asia to the borders of the far reaches of Oceania (about 60,000 years ago) and to Western Europe (about 40,000 years ago) before finally reaching the Americas (15,000 years ago or more). By calculating the rhythms of genetic modifications, geneticists have generated predictive models to suggest dates for the Out-of-Africa dispersals about 70,000 years ago. Some prehistorians (e.g., Mellars et al. 2013) hypothesized a late Out-of-Africa event and a rapid colonization of the rest of the Eurasian lands by modern humans around 60,000–50,000 years ago, mainly by following the coasts to reach Oceania.

Meanwhile, increased fieldwork in Arabia has pushed forward archaeological research on the problems raised by a Middle Paleolithic period (see references in Groucutt and Petraglia 2012). The human fossils are still tantalizing and may be found, but it is the material productions that have to date greatly intrigued archaeologists, to the point of contradicting estimations made by geneticists. These lithic products occur on sites even older than the 70,000 years estimation for out-of-Africa. This inconsistency with geneticists' molecular clocks stirs debate: How could modern humans deposit tools in Arabia during a time period that is earlier than the out-of-Africa genetic models say they should be there? Or does Levallois production show activity prior to or contemporaneous with the first *Homo sapiens* in Arabia? Might stone tools alone tell us about the people who made them?

The answers to these questions are unclear, even with the recent discovery of an 85,000-year-old human finger bone in Saudi Arabia (Groucutt et al. 2018). It is still

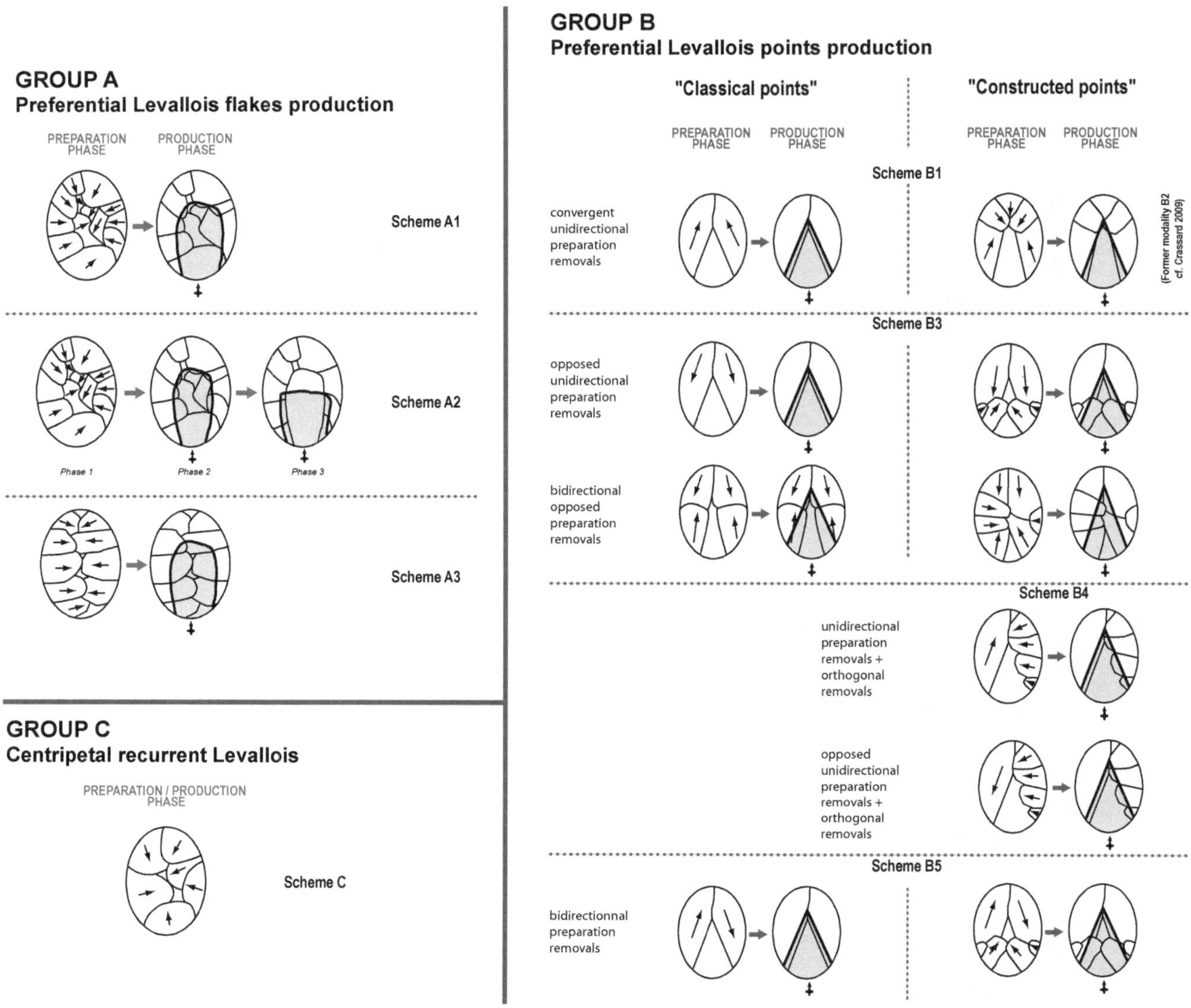

Figure 7.11. Theoretical representations of Levallois production schemes from Wādī Sanā (based on a regional comparative study in central Hadramawt; see Crassard 2009; Crassard and Thiébaut 2011). *Illustration by Rémy Crassard.*

difficult to associate a lithic production with a particular species of hominid. What is certain is evidence that ancient hominids occupied parts of Yemen and Arabia in general. Furthermore, these occupations occurred during very different climatic phases that were not necessarily optimal for hominids. Drier periods are one example. Some of these Arabian lithic productions—such as the Nubian production, which is a very specific preferential Levallois production—resemble ones made by the first modern humans in Africa. Until its discovery in Arabia, this Levallois method had never been found outside Egypt, Sudan, Eritrea, Ethiopia, and Somalia. Only one of the sites that show a Nubian type of production in the Dhofar (southern Oman) could be dated to slightly over 100,000 years ago (Rose et al. 2011; Usik et al. 2013). At this time, Nubian technology was still very much present in East Africa.

To return to an earlier problem, what can lithic industries tell us about human groups? First, we must not forget the lack in Arabia of hominid fossils dated earlier than a few thousand years ago. The passage of prehistoric human groups is evident from their material productions, with mostly stone as the only evidence, well preserved as tools such as flakes, blades, and points. These lithic products are very important for understanding hominid stylistic and technological behaviors. Critical information comes from diagnosing multiple modalities for obtaining tools and some types of flakes or blades. Modalities in knapping directly reveal the methods hominids invested in their conceptions of knapping and show the analyst multiple conceptualizations of forms and cultural traditions.

In this context, three recent discoveries in Arabia revive the groundwork for a major debate on the origin and dispersals of modern humans. First, in the United Arab Emirates is the Jebel Faya site (Armitage et al. 2011), which has been dated by optically stimulated luminescence (OSL). Around 125,000 years old, the oldest archaeological level yielded a lithic industry interpreted as being close to African MSA industries, with the presence of Levallois cores and Levallois flakes, as well as blades and bifacial pieces. This interpretation has skeptics among archaeologists who see equally important similarities with contemporary sites in the Mediterranean Levant, such as the Skhul site in Israel.

A second discovery is the presence of the Nubian technology in Arabia, first found in the south of the Arabian Peninsula and more recently in central and northern Saudi Arabia (Crassard and Hilbert 2013; Hilbert et al. 2016). Its presence has often been interpreted as clear evidence of a link (not yet well defined) between the known techno-complexes in East Africa and Arabia. In the absence of more sites that are well dated and that have yielded sufficient evidence, it is still difficult to conclude definitively that such a link implies a specific human population or culture group.

Finally, a series of discoveries in northern Saudi Arabia, especially in the Jubbah Basin, is of primary importance to understanding the potential population dispersals to and from the north. Many sites are dated to a period between 130,000 and 75,000 years ago (Petraglia et al. 2012), corresponding with more humid climate phases in the history of the peninsula. And the Levallois stone tools from this time range are quite similar, both among themselves and with those known to date to the same period in Africa and the Levant on sites that have yielded remains of modern humans. Nevertheless, it is premature to directly and clearly link northern Saudi Arabia with other regions so distant, especially because dating of lithic sites, when even available, is sometimes controversial. The future of research is therefore eminently linked to the discovery of well-dated sites, yielding stratified industries as comparable bases for study.

The use of the Levallois concept is evident in diverse regions of Yemen. The main occurrences, mostly known from surface sites, are:

- Aden region (Whalen and Pease 1991; Whalen and Schatte 1997): Wadi Shahar and Wadi Ghadin. Cores with Centripetal Recurrent Levallois production are present.
- Jibal Ṣafīr region (Ramlah as-Sabʿatayn central desert): Wādī Hirāb and Wādī Sadbā (Cleuziou et al. 1992)
- Shabwa region: Khushum Tuhayfah at the edge of Wadi Thib, Wadi Muqqah, and Hayd al-Ghalib (Inizan 1989; Inizan and Ortlieb 1987)
- Western Hadramawt region: Wādī Jirdān (YLNG-012 site) (Crassard and Hilbert 2006; Crassard and Hitgen 2007)
- Say'ūn region (central Hadramawt): Wādī al-Gabr (al-Gabr 1 site) and Wādī Ḥajar (Amirkhanov 1994:218); Wādī Bin ʿAlī (Zimmerman 2000)
- Eastern Hadramawt: Wādī Washaʿah and Wādī al-Khūn region (Crassard and Bodu 2004); Wādī Sanā and Wādī as-Shumlyah (Crassard 2004)
- Wādī Surdūd region (Shiʿbāt Diḥyah sites SD1 and SD2, dated by OSL to about 55,000 years ago; Delagnes et al. 2012, 2013)

Combined with the absence of stratified contexts (except for SD1/SD2), the absence of detailed technological study for the vast majority of discovered pieces precludes further comparisons with the Middle Paleolithic material from Wādī Sanā.

As noted, the geneticists' hypothesis that dates an out-of-Africa migration by *Homo sapiens* around 70,000 years ago is unconfirmed by archaeological data. Also interesting is that archaeologists' prior assumption of a single out-of-Africa migration around 60,000 to 50,000 years ago along the coasts is probably not supported by recent discoveries. Lithic evidence suggests that at least one out-of-Africa migration likely occurred earlier. If it left only lithic production and no genetic trace, the evidence nevertheless has very important consequences for understanding the evolution of our global genetic heritage. Arabia therefore now occupies a prominent place in research, as many answers to these questions are buried there.

Conclusions

The analysis of the Levallois cores presented here remains limited due to the lack of available data in the earlier phases of Levallois surface preparation. For the most part, the only documented phases are the final stages of production of Levallois preferential flakes or recurrent flakes. Technological study is also limited by the total absence of stratified contexts for the Middle Paleolithic in Wādī Sanā. There is no basis on which to propose dates for the assemblages described here.

Notwithstanding these limitations, this study describes distinct knapping methods and knappers' objectives (preferential flaking or not). It is still unknown if these operating schemes are also chronologically distinct or whether these schemes show different traditions across time. It remains for future studies to address how these knapping methods may conform to strong cultural constraints. Attested on many surface sites across the Hadramawt region and more widely all over Arabia, Levallois technology ultimately presents significant variation in the methods used in its production. To further address the significant questions tied to this material production, archaeologists must generate more comparisons across surface collections and an increased corpus of well-dated data.

References

Amirkhanov, Hizri

1991 *Paleolit yuga Aravii* (The Paleolithic in South Arabia). Scientific World, Moscow.

1994 Research on the Palaeolithic and Neolithic of Hadramaut and Mahra. *Arabian Archaeology and Epigraphy* 5:217–28.

Antón, Susan C., and Carl C. Swisher

2004 Early Dispersals of *Homo* from Africa. *Annual Review of Anthropology* 33:271–96.

Armitage, Simon J., Sabah A. Jasim, Anthony E. Marks, Adrian G. Parker, Vitaly I. Usik, and Hans-Peter Uerpmann

2011 The Southern Route "Out Of Africa": Evidence for an Early Expansion of Modern Humans into Arabia. *Science 331*(6016):453–56.

Bar-Yosef, Ofer

1987 Pleistocene Connexions between Africa and Southwest Asia: An Archaeological Perspective. *African Archaeological Review* 5:29–38.

Bar-Yosef, Ofer, and A. Belfer-Cohen

2001 From Africa to Eurasia: Early Dispersals. *Quaternary International* 75(1):19–28.

Boëda, Eric

1994 *Le concept Levallois: variabilité des méthodes*. Monographie du CRA 9. CNRS Editions, Paris.

Caton-Thompson, Gertrude

1953 Some Palaeoliths from South Arabia. *Proceedings of the Prehistoric Society* 19:189–218.

Cleuziou, Serge, Marie-Louise Inizan, and Bruno Marcolongo

1992 Le peuplement pré- et protohistorique du système fluviatile fossile du Jawf-Hadramawt au Yémen (d'après l'interprétation d'images satellite, de photographies aériennes et de prospections). *Paléorient* 18(2):5–28.

Crassard, Rémy

2004 *Prehistory in Wādī Sanā : The Lithic Industries, Roots of Agriculture in South Arabia Preliminary Report 2004*. Preliminary report, unpublished, filed in the Department of Anthropology, The Ohio State University, Columbus.

2007 Apport de la technologie lithique à la définition de la préhistoire du Hadramawt, dans le contexte du Yémen et de l'Arabie du Sud. PhD dissertation, Departement d'Anthropologie, Ethnologie, et Préhistoire, Université de Paris 1 Panthéon-Sorbonne.

2008 *La Préhistoire du Yémen. Diffusions et diversités locales, à travers l'étude d'industries lithiques du Hadramawt*. British Archaeological Reports, International Series 1842. Archaeopress, Oxford.

2009 Middle Paleolithic in Arabia: The View from the Hadramawt Region, Yemen. In *The Evolution of Human Populations in Arabia: Paleoenvironments, Prehistory and Genetics*, edited by Michael D. Petraglia and Jeffrey I. Rose, pp. 151–68. Springer, Dordrecht.

Crassard, Rémy, and Pierre Bodu

2004 Préhistoire du Hadramawt (Yémen): nouvelles perspectives. *Proceedings of the Seminar for Arabian Studies* 34:67–84.

Crassard, Rémy, and Yamandú H. Hilbert

2006 *Final Report: "Bronze Age Tombs" Project (YLNG-BAT 2006); Archaeological Excavations, Rapport interne Yemen LNG*. Centre des Etudes Français Arabes à Ṣanā'a/Deutches Archäologisches Institut, Sana'a.

2013 A Nubian Complex Site from Central Arabia: Implications for Levallois Taxonomy and Human Dispersals during the Upper Pleistocene. *PLoS ONE* 8(7): https://doi.org/10.1371/journal.pone.0069221.

Crassard, Rémy, and Holger Hitgen

2007 From Sāfer to Bālhāf: Rescue Excavations along the Yemen LNG Pipeline Route. *Proceedings of the Seminar for Arabia Studies* 37:43–59.

Crassard, Rémy, and Céline Thiebaut

2011 Levallois Points Production from Eastern Yemen and Some Comparisons with Assemblages from East-Africa, Europe and the Levant. In *The Lower and Middle Palaeolithic in the Middle East and Neighboring Regions,* edited by Jean-Marie Le Tensorer, Reto Jagher, and Marcel Otte, pp.131–42. ERAUL, Liège, Belgium.

Delagnes, Anne, Rémy Crassard, Pascal Bertan, and Luca Sitzia

2013 Cultural and Human Dynamics in Southern Arabia at the End of the Middle Palaeolithic. *Quaternary International* 300:234–43.

Delagnes, Anne, Chantal Tribolo, Pascal Bertran, Michel Brenet, Rémy Crassard, Jacques Jaubert, Lamya Khalidi, Norbert Mercier, Sébastien Nomade, Stéphane Peigné, Luca Sitzia, Jean-François Tournepiche, Mohammad Al-Halibi, Ahmad Al-Mosabi, and Roberto Macchiarelli

2012 Inland Human Settlement in Southern Arabia 55,000 Years Ago: New Evidence from the Wadi Surdud Middle Paleolithic Site Complex, Western Yemen. *Journal of Human Evolution* 63:452–74.

Dennell, Robin

2003 Dispersal and Colonisation, Long and Short Chronologies: How Continuous Is the Early Pleistocene Record for Hominids outside East Africa? *Journal of Human Evolution* 45(6):421–40.

Dennell, Robin, and Wil Roebroeks

2005 An Asian Perspective on Early Human Dispersal from Africa. *Nature* 438:1099–104.

Groucutt, Huw S., Rainer Grün, Iyad S. A. Zalmout, Nick A. Drake, Simon J. Armitage, Ian Candy, Richard Clark-Wilson, Julien Louys, Paul S. Breeze, Mathieu Duval, Laura T. Buck, Tracy L. Kivell, Emma Pomeroy, Nicholas B. Stephens, Jay T. Stock, Mathew Stewart, Gilbert J. Price, Leslie Kinsley, Wing Wai Sung, Abdullah Alsharekh, Abdulaziz Al-Omari, Muhammad Zahir, Abdullah M. Memesh, Ammar J. Abdulshakoor, Abdu M. Al-Masari, Ahmed A. Bahameem, Khaled S. M. Al Murayyi, Badr Zahrani, Eleanor M. L. Scerri, and Michael D. Petraglia

2018 *Homo sapiens* in Arabia by 85,000 Years Ago. *Nature Ecology and Evolution* 2:800–809.

Groucutt, Huw S., and Michael D. Petraglia

2012 The Prehistory of the Arabian Peninsula: Deserts, Dispersals, and Demography. *Evolutionary Anthropology* 21(3):113–25.

Guichard, Jean, and Genevieve Guichard

1965 The Early and Middle Palaeolithic of Nubia: A Preliminary Report. In *Contributions to the Prehistory of Nubia,* edited by Fred Wendorf, pp. 57–116. Southern Methodist Press, Dallas.

Hilbert, Yamandú H., Rémy Crassard, Guillaume Charloux, and Romolo Loreto

2016 Nubian Technology in Northern Arabia: Impact on Interregional Variability of Middle Paleolithic Industries. *Quaternary International* 435(A):77–93.

Inizan, Marie-Louise

1989 Premiers éléments de Préhistoire dans la région de Shabwa (R. D. P. du Yémen). *Raydan* 5:71–77.

Inizan, Marie-Louise, and Luc Ortlieb

1987 Préhistoire dans la region de Shabwa au Yémen du sud (R. D. P. Yémen). *Paléorient* 13(1):5–22.

Mellars, Paul, Kevin C. Goric, Martin Carre, Pedro A. Soares, and Martin B. Richards

2013 Genetic and Archaeological Perspectives on the Initial Modern Human Colonization of Southern Asia. *Proceedings of the National Academy of Sciences* 110(26):10699–704.

Petraglia, Michael D., Abdullah Alsharekh, Paul Breeze, Chris Clarkson, Rémy Crassard, Nick A. Drake, Huw S. Groucutt, Richard Jennings, Adrian G. Parker, Ash Parton, Richard G. Roberts, Ceri Shipton, Carney Matheson, Abdulaziz al-Omari, and Margaret-Ashley Veall

2012 Hominin Dispersal into the Nefud Desert and Middle Palaeolithic Settlement along the Jubbah Palaeolake, Northern Arabia. *PLoS ONE* 7(11):e49840.

Rose, Jeffrey I., Vitaly I. Usik, Anthony E. Marks, Yamandú H. Hilbert, Christopher S. Galletti, Ash Parton, Jean Marie Geiling, Viktor Černý, Mike W. Morley, and Richard G. Roberts

2011 The Nubian Complex of Dhofar, Oman: An African Middle Stone Age Industry in Southern Arabia. *PLoS ONE* 6(11):e28239.

Trinkaus, Erik

2005 Early Modern Humans. *Annual Review of Anthropology* 34:207–30.

Tuffreau, Alain

2004 *L'Acheuléen, de* l'Homo erectus *à l'homme de Néandertal*. La Maison des Roches Editeur, Paris.

Usik, Vitaly I., Jeffrey Ian Rose, Y. H. Hilbert, P. Van Peer, and A. E. Marks

2013 Nubian Complex Reduction Strategies in Dhofar, Southern Oman. *Quaternary International* 300:244–66.

Van Beek, Gus W., Glen H. Cole, and Albert Jamme

1963 An Archaeological Reconnaissance in Hadramaut: A Preliminary Report. *Annual Report of the Smithsonian Institution* 1963:521–45.

Whalen, Norman M., and David W. Pease

1991 Archaeological Survey in Southwest Yemen. *Paléorient* 17(2):127–31.

Whalen, Norman M., and K. E. Schatte

1997 Pleistocene Sites in Southern Yemen. *Arabian Archaeology and Epigraphy* 8:1–10.

Zimmerman, Paul

2000 Middle Hadramawt Archaeological Survey: Preliminary Results, October 1999 Season. *Yemen Update* 42:27–29.

Chapter 8

Manayzah: A Terminal Pleistocene-Early Holocene Rockshelter Occupation

Rémy Crassard, Joy McCorriston, Louise Martin, and Thomas S. Dye

Discovered in 2004 in the upper channel of Wādī Sanā, the stratified Neolithic site of Manayzah (E 324343, N 1727876 UTM Zone 39 North) is an important benchmark in the study of Holocene occupation in the Hadramawt and more broadly in the Arabian Peninsula. The Manayzah rockshelter revealed both substantial stratigraphy and an extensive surface accumulation of archaeological remains, which makes this site the only one of this magnitude in the region (Crassard et al. 2006). Presented here, two brief seasons of excavation and subsequent analysis already demonstrate its status as a reference site.

The place-name Manayzah properly refers to the system of small natural cisterns adjacent to the rockshelter rather than the rockshelter itself (figure 8.1). These cisterns collect rainwater runoff. If an ancient spring fed this system, it has not left tufa deposits like those in Wādī ʿIdim (chapter 2). Furthermore, there is no evidence for the presence of a modern source. Stagnant water nowadays makes this place nonetheless a favorable location in a region of arid climate. The presence of water most certainly was a decisive factor for the installation of prehistoric populations in the vicinity of the rockshelter. Located at the foot of an abrupt cliff 40 to 50 m in height, the site offers good protection against morning sunlight while the cliff face absorbs afternoon rays and radiates them by night, making the shelter an even more attractive location. The site's surface, which produced many archaeological remains, lies on a slope of approximately 9 degrees by a small mound descending from the entrance of a narrow and shallow cave. It would seem that the cave was once higher and larger than it is today, for the accumulation of sediments on its floor has blocked the lower part of the cavity. The entire terrace is topographically higher than the bed of the Wādī Sanā and nowadays beyond the ordinary reach of the main wadi flow. In antiquity, particularly powerful flooding occurred through a narrow canyon at this location, where the canyon constricts northward-flowing waters arriving from the large Ghayl Bin Yumain basin.

Discovery of the Manayzah Rockshelter

Manayzah is one of the sites discovered and tested by the RASA Project during systematic surveys along transects perpendicular to the course of the Wādī Sanā. Arbitrarily set to 100 m north–south in width, these transects were selected with stratified random sampling of latitudes between the mouth of the wadi channel at Wādī Ḥaḍramawt and its source in the region of Ghayl bin Yumain.

Our discovery was not without its tensions. A multilingual, multicultural research team desert camping 40 km from the nearest shower does have interpersonal frictions. Sometimes there is not enough food, or too much eggplant; someone takes your last cigarette, or you can't agree on strategy. If directing such a team is like herding cats, then the addition in 2004 of French prehistorians added new pathways—and new skills and perspectives.

Figure 8.1. View of the Manayzah rockshelter and terrace (left) with *qīr* water catchment in the cliff (right), as seen from the northwest. *Photograph by Rémy Crassard.*

Without the alternate trainings and deep convictions we brought together, there would have been fewer debates over method and interpretation, many held under the raw bulb and dying insects of our nightly commons under the stars. And the team would never have excavated at Manayzah. Indeed, we never would have found this remarkable site in the first place.

We had lively methodological differences. With only modest protest and frank incredulity, the French helped the Americans clamber over slopes and gullies in our stratified systematic survey of transects across Wādī Sanā (chapters 4 and 6). They echoed the disbelief of our Yemeni colleagues, who preferred to go where they already could see or knew sites to exist. The empty spaces in between didn't concern them—why look where you know nothing will be found?

So there were low expectations as the team plodded across one of our high-energy streambed locations in the narrow canyon of the upper Wādī Sanā. With massive rounded cobbles on a watercourse scoured out of soaring limestone cliffs, what hope could there be for any site older than the last flash flood? What a futile waste of precious time! At the far side against the cliff face was a narrow gravel terrace at the base of an ancient seep. Here Rémy dropped to his knees in ecstasy—the soft surface was littered with thousands of delicate flakes and aborted tools, the workshop site of ancient chert knappers, whose stratified debris promised an entirely new window into the prehistory of Hadramawt. It took American methods to find the site, French ones to recognize and dig it, and Yemenis to protect and archive its treasures. The al-ʿAlīy bedouin still know its location and quietly omit its mention to those who venture into Wādī Sanā. Someday it will be possible to continue the excavations at Manayzah and answer questions raised during our first two seasons of excavation. But for now, this chapter and the publications it summarizes stand as the definitive report on the site.

Taphonomic Preservation and Formation of Manayzah

It is puzzling that this site exists at all. It survived the massive flooding evident in the scoured canyon walls. Floods removed the Early Holocene sediment overburden, of which only a few high traces remain in the overhangs of the cliff face. Our hypothesis is that the unique local conditions of water catchment contributed to the site's preservation. During excavations, archaeologists noted that Manayzah's surface was covered with a crust of slightly hardened sand, the result of the calcification of sediments. When it rains, or rained more frequently in antiquity, rainwater flows along and across the adjacent limestone cliff, then spreads over the site, producing a precipitate as it evaporates. The calcified crust certainly played an important part in the preservation of delicate archaeological layers by preventing deflation during the later Holocene, a time when the erosion of fine loams of the Wādī Sanā region was widespread.

The Early Holocene accumulation and preservation of the site under thick beds of silt remains poorly understood, and we draw here in part on the geomorphology of the regional system (chapter 3) for explanation. Excavators noted thin, interleaving layers of orange-brown sandy silt with dark brown, charcoal-flecked sediment. It is our hypothesis that the Manayzah site formed as a seasonal streamside encampment (dark layers) as annual floodwaters receded, only to be gently inundated and coated with a fresh deposit of silt (orange-brown layers) when the narrow canyon flooded. Later the volume of floodwaters gradually increased, expanding the silty deposition without removing underlying archaeological sediments. In the narrow canyon, it must have been a delicate balance between flood volume and velocity to accumulate silt beds over the site. We know they were there: the stranded silts high in the canyon walls attest to the towering height of now vanished, regionally attested flood deposits, scoured out by Late Holocene down-cutting (chapter 3).

Methods

Rémy Crassard directed Manayzah fieldwork as a part of his dissertation analysis of Paleolithic deep prehistory of the Hadramawt (Crassard 2008; Crassard and Bodu 2004). The Manayzah site's surface (figure 8.2), at the foot of a small rockshelter, was densely carpeted with knapped and worked chert pieces: flakes, projectile points, and many other tools of chert, as well as obsidian flakes, bladelets, and tools. The quantity of archaeological items found prompted the team quickly to establish a grid of 1 m² units (alphabetical axis from east to west and numerical axis from north to south).

In 2004 the team opened a sounding of 1 m² in K9 (figure 8.3). With the discovery of many lithics of varied types and whose manufacture made use of many different raw materials, this first test sounding immediately demonstrated the site's strong potential. Fifteen archaeological layers clearly appeared along the 50 cm height of sounding cuts, with a succession of occupations, as well as structures like perfectly preserved hearths (figure 8.4). Confronted with this exceptional occurrence of prehistoric remains, the RASA team planned further excavation for the month of February 2005. The open area method was favored to study the spatial distribution of remains. In tandem, the resumption of excavation in K9 enabled the team to reach the deepest of lower layers.

Figure 8.2. Gridded collection of surface lithics and objects at Manayzah. Left to right: Rémy Crassard, Mohammad Sinnah. At rear: Nasser Al-ʿAlīy. *Photograph by Joy McCorriston.*

Figure 8.3. Overview of Manayzah excavations at the end of 2005 season. Note K9 sounding to the far left. *Photograph by Rémy Crassard.*

Figure 8.4. Manayzah grid square K9 near end of excavation in 2005. East section viewed, looking east. Scale is 20 cm. *Photograph by Rémy Crassard.*

Systematic Surface Collection and Excavation Strategy

In 2004 excavators made an initial study of the surface and a deep probe in a 1 x 1 m square, K9. In tandem with excavation, Rémy Crassard and Julien Espagne conducted systematic collecting and sampling of the surface over almost all the site (136 m^2; figure 8.5). Each piece of chert or obsidian was collected, recorded, and registered by square meter to assess spatial distribution. The entire surface lithic material is composed of 2,462 pieces. Spatial analysis of such an assemblage, even surface material, provides useful information, particularly on the degree of disturbance during the most recent occupations. Animal droppings and stone alignments show that Manayzah was recently and perhaps recurrently used as a bedouin camp. Analysis of the site's entire surface eventually should be finalized by finishing the sampling across the unsampled grid squares. What the process has revealed thus far is the quite homogeneous character of lithic industries on the surface, with only one particular area distinct from the rest of the site in its concentration in lithic remains (figure 8.6). It is nevertheless overly ambitious to detect clear working areas from such a surface accumulation, as a comparison with the spatial distribution of obsidian remains shows no such concentration and does not suggest an area specialized in work on this volcanic glass.

After reopening the stratigraphic probe in K9 in 2005, excavators decided to extend the excavation southward to follow stratigraphy already evident in the profile of Square K9 (figure 8.7). An initial sector of 2 x 2 m, which included Squares L8, L9, M8, and M9, was called Quadrat A (Quad A), in concord with the recording method of the RASA Project. To obtain better visibility of the spatial distribution of archaeological remains, excavators opened another sector, Quadrat B (Quad B), including Squares L10, L11, M10, and M11, to the east of Quadrat A. Excavators also opened two trenches to establish the entire stratigraphic sequence, from the uppermost layers in the cave from a stratigraphic viewpoint toward the foot of the deep sounding in K9. Hence Quadrat C (Quad C) includes Squares I14, J14, and K14, and Quadrat D

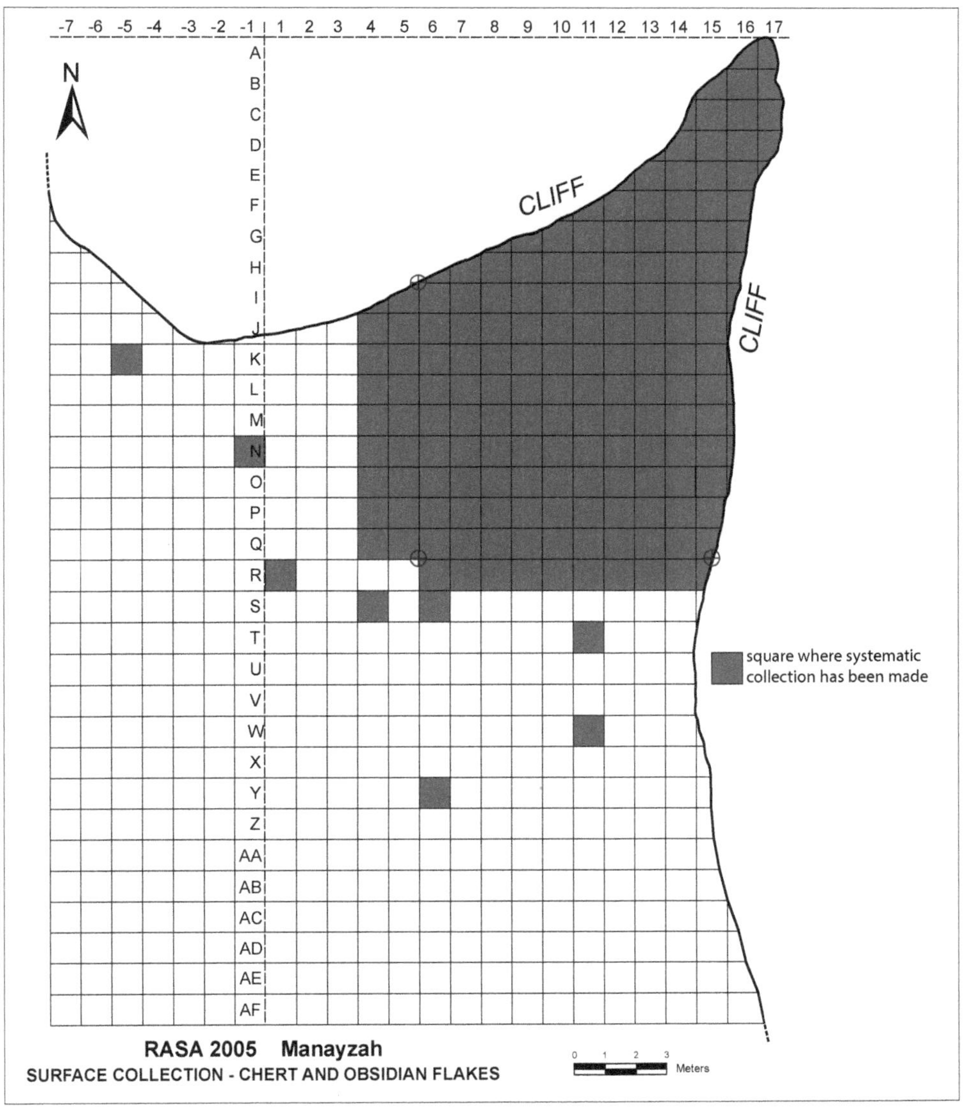

Figure 8.5. Manayzah terrace surface collection showing the site grid and where complete collection was made. *Illustration by Rémy Crassard.*

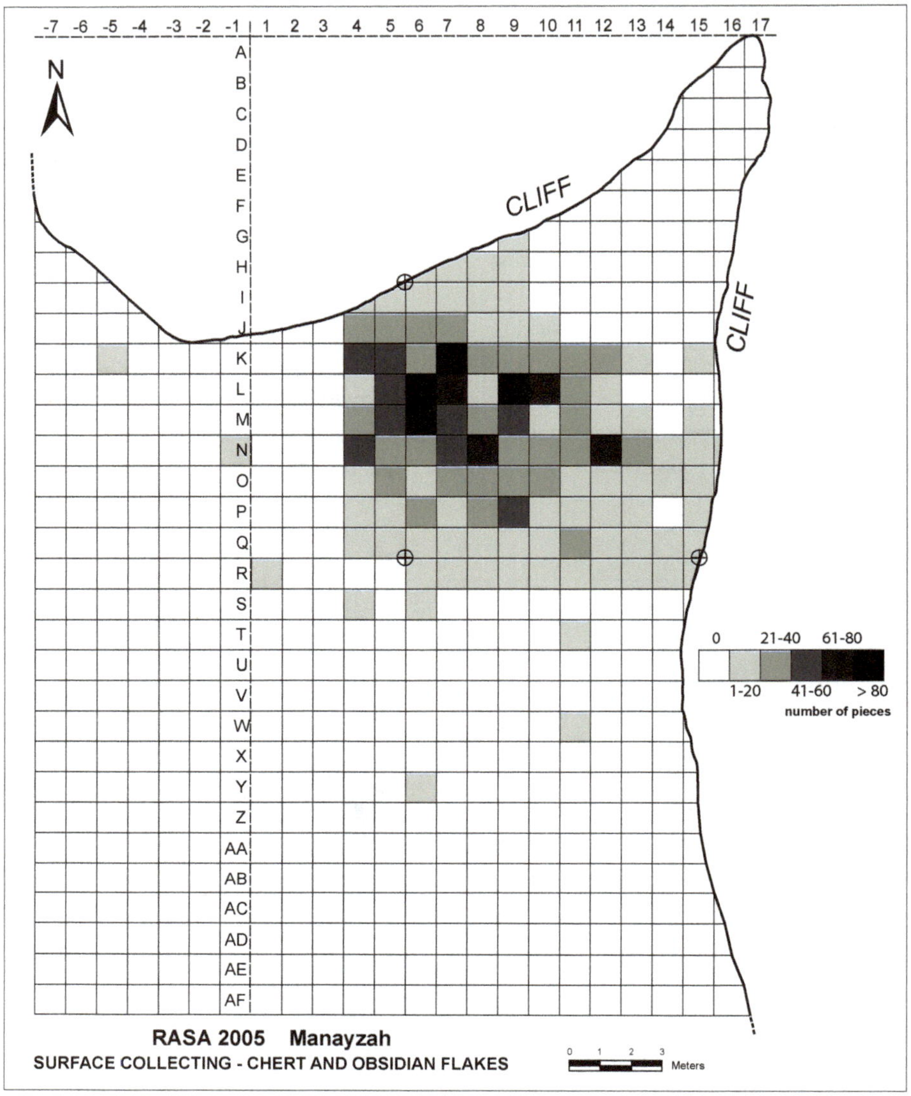

Figure 8.6. Manayzah terrace surface collection showing (a) density of chert and obsidian artifacts and (b) density of obsidian artifacts. Surface concentrations were significant in deciding where to locate excavations. *Illustration by Rémy Crassard.*

(Quad D) links B and C via a trench of 0.6 m width through the northern halves of Squares L12, L13, and L14.

The excavation was halted in Quadrats A and B after reaching the surface of a layer of hardened sand (Layer 132) and in C and D after arriving at a layer of fine yellowish sand (Layer 125), which could be followed in the two quadrats and in part of Quadrat B, thus establishing a stratigraphic connection among all sectors of the excavation. In Square K9, at a depth of 2.20 m below the highest layer of Quadrat D, no sterile base or bedrock was yet reached. Using a sieve with a regular 0.2 cm mesh, systematic sieving was practiced for all excavated sediments. These methodological and stratigraphic approaches permitted both a diachronic study (through the resumption of the chrono-stratigraphic study in K9) and a synchronic analysis, removing occupation layers one after another across Quadrats A and B.

Results: Stratigraphy, Dates, Flora, Fauna, and Small Finds

Stratigraphy

For a site largely made of wind-blown sand and sandy and loamy accumulations of loamy origin, enriched by organic materials, Manayzah holds a particularly important stratigraphy. Because taphonomic processes such as deflation and erosion eliminate most occupational remains, it is extremely

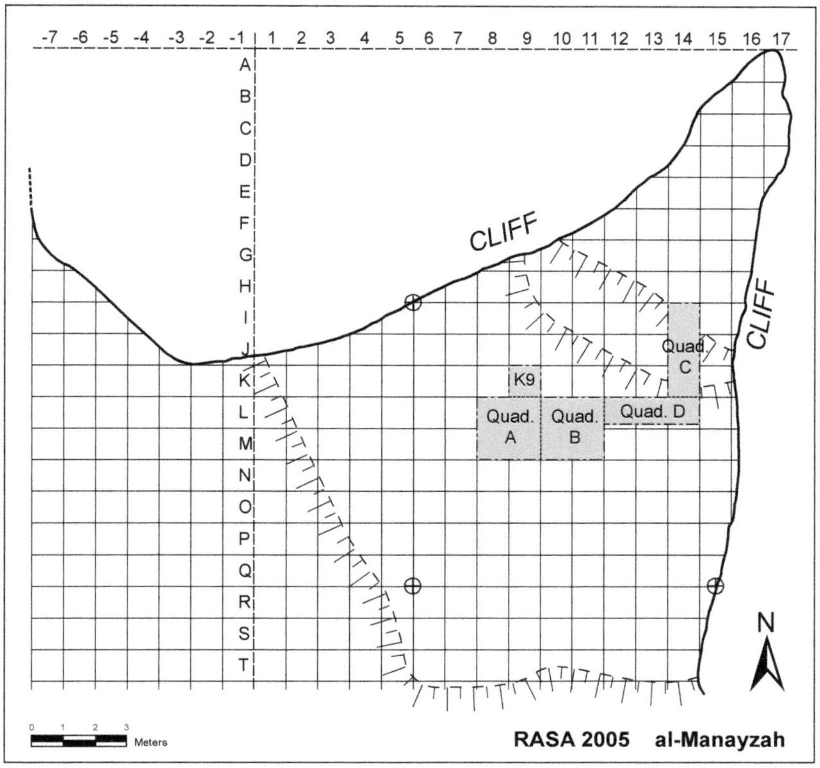

Figure 8.7. Manayzah excavation quadrats and grid plan. *Illustration by Rémy Crassard.*

rare in the Arabian Peninsula to discover sites of the Early to Middle Holocene that are well preserved. At Manayzah, a stratigraphy some 2.20 m deep yielded a relatively important Early to Middle Holocene regional chronology. Since bedrock has not been reached in K9, the probe square with deepest excavation, the base of this accumulation of sediment is still unexplored; taphonomic conditions that preserved the lower layers of the site remain a matter of conjecture.

Approximately 60 stratigraphic layers were identified (figure 8.8, table 8.1), and each was sampled for eventual phytolith analysis. Many of the upper layers contain distinctive phytolith assemblages (Buffington et al. 2017). Unfortunately, samples from the lower layers in K9 were destroyed in a flooding episode in basement storage at the University of South Florida. The thin Layer 143 in Quadrats A and B was partly exposed over a large surface. Layer 143 is an occupation level, clearly associated with structures: a pit, at least three constructed fireplaces, and a posthole.

Though the stratigraphy is clearly a good source of information on depositional episodes, certain layers are poorly preserved in the form of very thin accumulations. These circumstances at times make it impossible to differentiate layers during excavation. It was nevertheless possible to identify discrete chert workshops. For instance, the remains of micro-debitage, probably from the retouching of obsidian blanks, were apparent in the uppermost part of K9 Layer 13 on a flat surface interpreted as an activity area for retouching.

Surface collection has moreover contributed an important assemblage of diagnostic tools. Objects found on the surface are typologically similar to those found in the excavated layers closest to the surface (upper layers). Their shared attributes suggest that surface disturbance and erosion made little impact on spatial distribution—a hypothesis that should be tested with spatial statistics and by comparing the composition of surface assemblages and those located directly under the surface. Such studies will only be robust when the remains originating from the site's entire surface are sampled and the uppermost layers are more extensively excavated. While surface items include mostly chert and obsidian flakes, the few tools found on the surface are rich in information. Tools show traces specific to techniques known in the Early to Middle Holocene (pressure retouch, bifacial manufacture, and so on). A dagger made in jasper-like chert (figure 8.9) shows *en écharpe* retouch made by pressure flaking. Projectile points (figure 8.10), typologically akin to examples found on the excavations of Khuzmum SU45-1A, belong to a Neolithic cultural style.

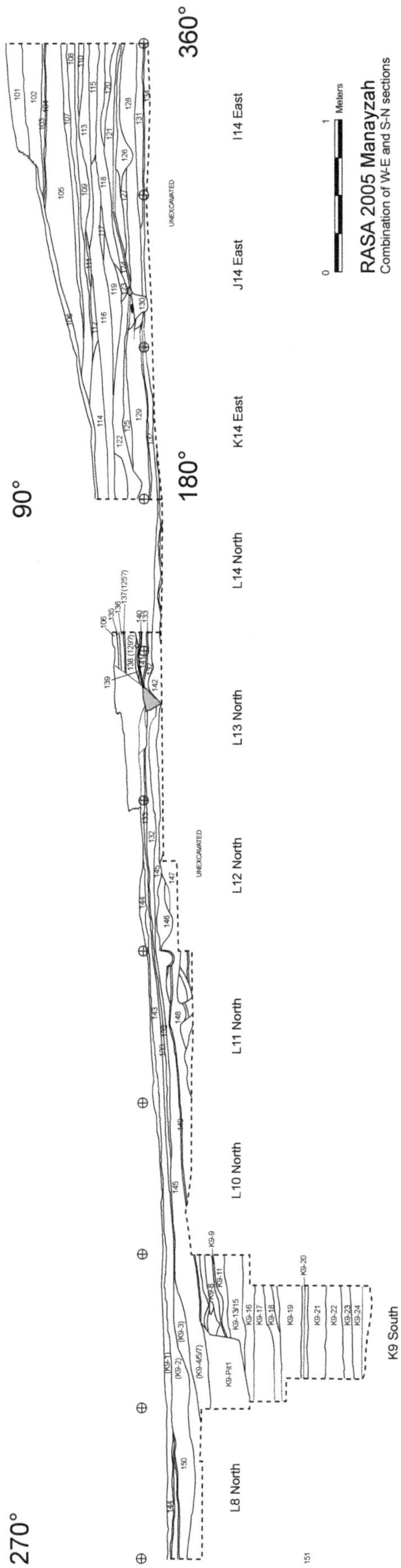

Figure 8.8. Manayzah composite section at the close of excavations 2005. Combines west–east and south–north sections. Bedrock (gravel terrace) not reached in any excavated quadrant. *Illustration by Rémy Crassard.*

Table 8.1. Stratigraphic layers at Manayzah.

Stratigraphic Layers from Section Analysis (as seen in chapter 8, figure 8, "Composite Section"), at End of Excavation, 2005					
Present in Quadrat	**Layer**	**Under**	**Over**	**Equals**	**Description**
C	101	(surface)	102	144, K9-1, all 000	Highly compacted sandy silt, light brown, many animal dungs (goat) with burned areas.
C	102	106	103		Aeolian stratified sand, light yellow, sterile.
C	103	106, 102	104		Sand, light grey, with burned flat limestone fragments (1 to 5 cm).
C	104	105	103		Compacted silt, light yellow, sterile, porous and light in weight.
C	105	106, 104	107, 108		Compacted sandy silt, grey, porous, micro wood charcoals and calcite nodules.
C, D	106	(surface)	102, 103, 105, 108, 114, 135		Surface layer, sand, yellow-grey, with animal dungs (goat).
C	107	105	108		Silty sand, light rust-colored, calcite micro-nodules are numerous, bioturbated (roots), sterile.
C	108	107, 105	109, 112, 114		Sandy silt, light grey, ashy, rare calcite micro-inclusions, sterile.
C	109	108	110, 111, 112, 113		Silty sand, grey-brown, many wood charcoal micro-inclusions and calcite micro-inclusions.
C	110	109	113		Sandy ashy silt, grey-blue, many wood charcoal fragments (infra-millimetric), calcite nodules.
C	111	109	112, 113, 116		Very fine and very ashy silt, light grey (pure ash?).
C	112	108, 109, 111	114, 116		Sandy silt, compacted, light grey, few calcite nodules.
C	113	109, 110, 111	115, 116		Sandy ashy silt, light grey-blue, many calcite nodules.
C	114	108, 112	116		Very fine silt, very compacted, light yellow, rare calcite nodules.
C	115	113	116, 118		Same as 113, darker and fewer calcite nodules.
C	116	111, 112, 113, 114, 115	117, 118, 119, 122		Stratified silty sand, dark grey, very rich in wood charcoals, calcite nodules.
C	117	116	118, 119		Ashy pocket, light grey, calcite nodules, black micro-layer of burned sediments at the base.
C	118	115, 116, 117, 119	120, 121		Clayey silt, compacted, beige, calcite micro-nodules (probably same as 122).
C	119	116, 117	118, 122, 123		Same as 116, grey-beige, few wood charcoals.
C	120	118	121		Silty sand, very ashy, light grey-blue, many calcite nodules, few wood charcoals.
C	121	118, 120, 123	124		Sand, dark brown, many wood charcoals (infra-millimetric).
C	122	116, 119, 123	125, 129		Clayey silt, compacted, dark beige, rare calcite nodules (probably same as 118).
C	123	119	118, 122, 124		Sandy silt, dark grey, slightly ashy, few wood charcoals, calcite nodules.
C	124	121	126, 127		Sandy silt, brown-grey, few calcite nodules.
C, D?	125	122	129	137?	Silt, yellow-white, compacted, calcite nodules (more yellow to the south).
C	126	124	127, 128		Sandy silt, very ashy, many wood charcoals, infra-centimetric and infra-millimetric, many burned calcite, burned lithics.
C	127	124, 126	128		Silty sand, dark grey, wood charcoals.
C	128	126, 127	131, 134		Silt, highly ashy, compacted, grey-white, wood charcoal nodules, infra millimetric, less compacted in the lower part.
C, D?	129	122, 125	133	138?	Silty sand, grey-brown, many calcite nodules.
C	130	125	134		Silty sand, dark grey, many calcite nodules, bones.
C	131	128	134		Sandy silt, dark grey-brown, slightly stratified, slightly porous, rare wood charcoals.
A, B, C, D	132	133	134, 142		Sandy silt, compacted, light yellow, few rare calcite nodules.

Table 8.1. Stratigraphic layers at Manayzah. *(continued)*

Present in Quadrat	Layer	Under	Over	Equals	Description
A, B, D	133	129, 141	132	151, K9-3	Sand, brown-black, few calcite nodules.
C	134	132	142		Silty sand, ashy, grey, calcite nodules.
D	135	106	136		Silty sand, dark grey, rare calcite nodules.
D	136	135	137, 125?		Sandy silt, brown, rare calcite micro-nodules, wood charcoals, infra-millimetric.
D	137	136	138, 129?	125?	Silty sand, dark yellow, rare calcite micro-nodules.
D	138	137, 125?	139, 140	129?	Silty sand, dark brown–grey, slightly compacted, rare calcite nodules.
D	139	138, 129?	140		Silty sand, ashy, light grey.
D	140	138, 129?, 139	141		Silty sand, dark brown, slightly compacted, calcite micro-nodules.
D	141	140	133	129?	Silty sand, grey-beige, many wood charcoals, slightly ashy, few calcite nodules.
D	142	132, 134	147		Silty sand, stratified, brown-grey-blue, ashy (probably two distinct layers).
B, D	143	144	133	K9-2, 150	Silty sand, brown-yellow, calcite micro-nodules, rare wood charcoals.
A, D	144	(surface)	143, 150	K9-1	Silty sand, beige-grey, animal dung (goat), surface.
B, D	145	132, 133, 151, K9-3	146, 149	K9-4, K9-5, K9-7	Silty sand, dark brown, calcite micro nodules, many wood charcoals, infra-millimetric.
D	146	145	147, 148		Silty sand, light grey, very slightly compacted, ashy, very numerous wood charcoals infra millimetric.
D	147	145, 146	(not excavated)	148?	Silt, dark brown, homogeneous, rare wood charcoal micro-nodules and rare calcite micro-nodules.
B	148	149	(not excavated)	147?	Same as 145, light brown.
B	149	145	148		Fine silt layer, compacted, dark yellow with micro-holes ("cheese layer").
A	150	144	151	143, K9-2	Sandy silt, yellow, porous, aeolian.
A	151	150	K9-4, K9-5, K9-7	133, K9-3	Sandy silt, grey-beige, compacted, calcite micro-nodules, wood charcoal micro-nodules.
Stratigraphic Layers in K9 Square (from east–west section analysis, as seen in chapter 8, figure 8, "Composite Section"), Excavated in 2004 and 2005					
Square	**Layer**	**Under**	**Over**	**Equals**	**Description**
K9	K9-1	(surface)	K9-2	144, 101, all 000	Silty sand, beige-grey, animal dung (goat).
K9	K9-2	K9-1	K9-3	143, 150	Sandy silt, yellow, porous, aeolian, micro-nodules of wood charcoals.
K9	K9-3	K9-2	K9-4	133, 151	Sandy silt, grey-beige, porous, rare calcite micro-nodules.
K9	K9-4	K9-3	K9-5	145	Silty sand, calcified, slightly compacted, grey-white, small calcite nodules.
K9	K9-5	K9-4	K9-6	145	Sand, compacted, grey-brown, small limestone nodules, small wood charcoal nodules.
K9	K9-6	K9-5	K9-7		Sand, compacted, grey, calcite micro-nodules (not visible in east–west section).
K9	K9-7	K9-6	K9-Pit1	145	Silty sand, dark beige, slightly compacted, small wood charcoal nodules.
K9	K9-Pit1	K9-7	K9-8, K9-9, K9-11, K9-13, K9-15, K9-16		Hollow structure, only partially excavated (in corner of test excavation square).
K9	K9-8	K9-Pit1	K9-9		Sand, slightly compacted, dark grey, small calcite nodules, many wood charcoals.
K9	K9-9	K9-8	K9-11		Sand, grey-black, very rich in wood charcoals pluri-millimetric and pluri-centimetric (= H1 or Hearth 1).

Present in Quadrat	Layer	Under	Over	Equals	Description
K9	K9-10	K9-9	K9-11		Same as K9-9, slightly lighter in color (= H1') (not visible in east–west section).
K9	K9-11	K9-9	K9-13		Sand, porous, grey, wood charcoal micro-nodules, many calcite nodules.
K9	K9-12	K9-11	K9-13		Same as K9-11, with rare calcite nodules (not visible in east–west section).
K9	K9-13	K9-11	K9-14		Sand, yellow, highly compacted.
K9	K9-14	K9-13	K9-15		Sand, grey, very slightly compacted, wood charcoal micro-nodules (not visible in east–west section).
K9	K9-15	K9-13	K9-16		Sand, yellow, very slightly compacted (same as K9-13?).
K9	K9-16	K9-15	K9-17		Sandy silt, dark yellow–grey, many burned stones on top.
K9	K9-17	K9-16	K9-18		Silt, very ashy, brown-grey, many wood charcoals and limestone rocks (small and medium in size).
K9	K9-18	K9-17	K9-19		Silty sand, coarse grain, beige-grey, some micro-limestone fragments, many wood charcoals, pluri-centimetric (manual sampling for anthracology and radiocarbon dating).
K9	K9-19	K9-18	K9-20		Sand, brown-grey, many small limestone fragments, calcite micro-nodules.
K9	K9-20	K9-19	K9-21		Sand, compacted, light brown, many limestone exfoliated rocks (small and medium in size) found horizontally.
K9	K9-21	K9-20	K9-22		Silty sand, brown, few calcite micro-nodules, many small limestone rocks, flat horizontal limestone rocks on top.
K9	K9-22	K9-21	K9-23		Sandy silt, yellow, slightly compacted with calcite nodules, many small and medium limestone rock horizontally disposed on top.
K9	K9-23	K9-22	K9-24		Same as K9-22 but beige.
K9	K9-24	K9-23	(not excavated)		Sandy silt, dark beige, lighter in color at the bottom, high density of limestone rock (small and medium in size), very dense in calcite (slabs).

Stratigraphic Description of Excavated Layers and Features in 2005 (extension of excavation from the tested K9 square in 2004. Loci are described for each square or quadrat)

Quadrat	Square	Locus	Under	Over	Equals	Description
A	L8	000	(surface)	001	101, 144, K9-1, all other 000	Surface layer.
A	L9	000	(surface)	001	101, 144, K9-1, all other 000	Surface layer.
A	M8	000	(surface)	001	101, 144, K9-1, all other 000	Surface layer.
A	M9	000	(surface)	001	101, 144, K9-1, all other 000	Surface layer.
A	L8	001	000	002	all 001 in Quad A	Sand, beige-grey, very loose.
A	L9	001	000	002	all 001 in Quad A	Sand, beige-grey, very loose.
A	M8	001	000	002	all 001 in Quad A	Sand, beige-grey, very loose.
A	M9	001	000	002	all 001 in Quad A	Sand, beige-grey, very loose.
A	L8	002	001	003	all 002 in Quad A	Sand, grey-brown, ashy, slightly compact with calcite micro-nodules, few wood charcoal fragments.
A	L9	002	001	003, 004	all 002 in Quad A	Sand, grey-brown, ashy, slightly compact with calcite micro-nodules, few wood charcoal fragments.
A	M8	002	001	003	all 002 in Quad A	Sand, grey-brown, ashy, slightly compact with calcite micro-nodules, few wood charcoal fragments.
A	M9	002	001	003	all 002 in Quad A	Sand, grey-brown, ashy, slightly compact with calcite micro-nodules, few wood charcoal fragments

Table 8.1. Stratigraphic layers at Manayzah. *(continued)*

Quadrat	Square	Locus	Under	Over	Equals	Description
A	L8, L9, M8, M9	003	all 002 in Quad A	004	005	Silt, yellow-beige, compacted, very few artifacts.
A	L8, L9, M8, M9	004	005, 003, 002	008		Sandy silt, dark grey, loose, ashy, with wood charcoal micro-nodules, rich in bones (fragment of polished bone?); difficult to distinguish if this is a lower phase of 002.
A	L8, L9, M8, M9	005	002	003 (top)	004	Silt, yellow-beige, same as 003 (top), first interpreted as a layer going down 004.
A	L8	006	from 002	003		Square-shaped hollow structure (15 x 15 cm), should come from 002 layer through 003, sand grey-brown, loose, ashy with wood charcoals. Three vertical stones in the hole and one big bone found at vertical along border of the hole.
A	L8	007	within 003			bioturbation, no artifact, silt, compact, brown with calcite nodules.
A	L8, L9, M8, M9	008	004	009		Sand compact, dark green, ashy with charcoals; burned stone surface with no organization.
A	M9	009-10		009-11		Discrete concentration of rocks (on edge) inside a clear pit edge. Probably a hearth. Compacted clayey silt, almost pinky (color: 75YR6/4).
A	M9	009-11	009-10			Very distinct change in sediment matrix to black ashy and very rich in charcoal around rocks in pit. Compact sorted silt (color: 10YR3/1).
A	M9	009-12	009-11	009-13		Grey-black ashy sandy silt, with burned stones. This is a new layer of burned stones in a big hearth.
A	M9	009-13	009-12	009-14		Grey-black ashy sandy silt, with burned stones: this is a new layer of burned stones in a big hearth.
A	M9	009-14	009-13			Grey-black ashy sandy silt, with burned stones. This is the last layer of burned stones in a big hearth.
A	L8, L9, M8, M9	010		149	003 in Quad B	Grey sandy silt with charcoals; probably part of 004; 010 is directly above Layer 149 as described in the general section. Easy to follow in all of Quad 1 and in the western part of Quad B.
A	L8, L9	011	010	K9-Pit1?		Very compacted layer of grey-white calcified silt. No clear limit found at the end of the 2005 excavation season.
B	L10	000	(surface)	001	101, 144, K9-1, all other 000	Surface layer.
B	L11	000	(surface)	001	101, 144, K9-1, all other 000	Surface layer.
B	M10	000	(surface)	001	101, 144, K9-1, all other 000	Surface layer.
B	M11	000	(surface)	001	101, 144, K9-1, all other 000	Surface layer.
B	L10, L11, M10, M11	001	000	002	all 001 in Quad A	Sand, beige-grey, very loose, charcoal and calcite micro-nodules.

Quadrat	Square	Locus	Under	Over	Equals	Description
B	L10, L11, M10, M11	002	001			Heterogeneous loose and compacted sands, general color brown-grey with inclusion of yellow, orange, dark grey, green. Presence of ash, charcoal, roots, and several bioturbations.
B	L10, L11, M10, M11	003	002	149	010 in Quad A	Sandy silt, grey, micro-nodules of calcite, charcoals, infra-millimetric, compact.
B	M11	004			dug in 003 (contemporary with 007?)	Little area of burned stones with ashy grey-black charcoaly silt in a small depression dug in 003.
B	L11	005	003			Ashy and loose layer with burned stones; a possible area of dump from a hearth; many burned stones but no organization. Very loose dark brown silt.
B	L10, L11, M10, M11	006	002			Sandy silt; distinct fine brown material, probably overlying other exposed areas of Quad B. Some of the ashy-grey lenses in horizontal chunks, as if the remains of goat dung layers burned.
B	L10	002-8	002	006		
B	M11	007-9		007	dug in 003	First layer of sterile yellow silt sealing a big hearth (same type as Quad A 009). This hearth seems to be dug in 003.
B	M11	007-10	007-9	007-10		Ashy and highly charcoaly black-grey sandy silt with many burned stones (> 10–15 cm).
B	M11	007-11	007-10			New layer of burned stones directly under 007-10 in the big hearth. Grey-black ashy sandy silt.
B	L10, L11, M10, M11	008	003	011		Ashy sandy grey layer of silt (where 005 may have been dug?).
B	L10, L11	009	006	008		Brown-beige sandy silt with flat and thin limestone fragments and ash, slightly compact.
B	L10, L11	010		008	012?	Silty sand, yellow-pink in north of L10 and M10, with ash. Natural depression.
B	L10, L11, M10, M11	011	008			Silty sand, slightly compacted, grey-beige with many charcoals, infra-millimeter and infra-centimeter.
B	L10, L11	012			010?	Sandy silt, dark brown, slightly compacted.
B	east M10, west M11	013			dug in 011	Possible hearth or ash midden (or dump), sandy silt, grey-beige, with ash and charcoals, > 1–2 cm.
B	L11 north	014			in 012	Sterile white pure ash with three rocks (west), porous, slightly compacted.
C	I14, J14, K14	000	(surface)	001		Surface layer, no artifact.
C	I14	001	000	002		First layer under surface in I14 (the higher square in Quad C). Sandy silt, brown-black, goat dungs, no artifacts.
C	I14, J14, K14	002	001	003		Pure sand, yellow-beige, aeolian, no artifacts.

Table 8.1. Stratigraphic layers at Manayzah. *(continued)*

Quadrat	Square	Locus	Under	Over	Equals	Description
C	I14	003	002	004		Indurated silt surface with burned stones, yellow, very compact.
C	I14, J14, K14	004	003	005		Aeolian, yellow-grey, sterile, and very compacted silts.
C	I14, J14, K14	005	004	006	107	Silty sand, light rust-colored, calcite micro-nodules are numerous, bioturbated (roots), sterile.
C	I14, J14, K14	006	005	007	108	Sandy silt, light grey, ashy, rare calcite micro-inclusions, sterile.
C	I14, J14	007	006	008		Blue-grey compact silts with ash, charcoal, and insect burrows.
C	I14, J14, K14	008	007	009		Arbitrary layer, 20 cm thick. Micro-layers almost impossible to follow. Mostly brown and dark grey silt and ashy silt.
C	I14, J14, K14	009	008	010		Arbitrary layer, 20 cm thick. Micro-layers almost impossible to follow. Silt and ashy silt. The goal is to reach Locus 010 and to have continuity with Quad D.
C	I14, J14, K14	010	009		003 in Quad D, 010 in Quad C, 132	Sandy silt, compacted, light yellow, few rare calcite nodules.
D	half north of three squares: L12, L13, L14	000		001		Surface layer.
D	half north of three squares: L12, L13, L14	001	000	002		Very loose yellow-grey-brown silt, slightly compacted on surface.
D	half north of three squares: L12, L13, L14	002	001	003		Two layers, hard to individualize, excavated as the same layer to reach the yellow layer below; easier to follow; brown-yellow to brown-black, very loose silt.
D	half north of three squares: L12, L13, L14	003	002	004	010 in Quad C, 003 in Quad D, 132	Sandy silt, very loose in west and more compacted in east, light yellow to yellow-grey, few rare calcite nodules.
D	half north of three squares: L12, L13, L14	004	003	005		Slightly compacted layer of dark brown–grey sandy silt with calcite micro-nodule and numerous charcoals.
D	half north of three squares: L12, L13, L14	005	004	006		Ashy compacted light grey sandy silt, maybe a pit or ash dump, full of charcoals.
D	half north of three squares: L12, L13, L14	006	005	(not excavated)	007	Homogeneous dark brown silts with very few charcoals.
D	quarter northwest of L12	007	005	(not excavated)	006	Same as 006, but only in L12 to have a good continuity in section with Quad B.

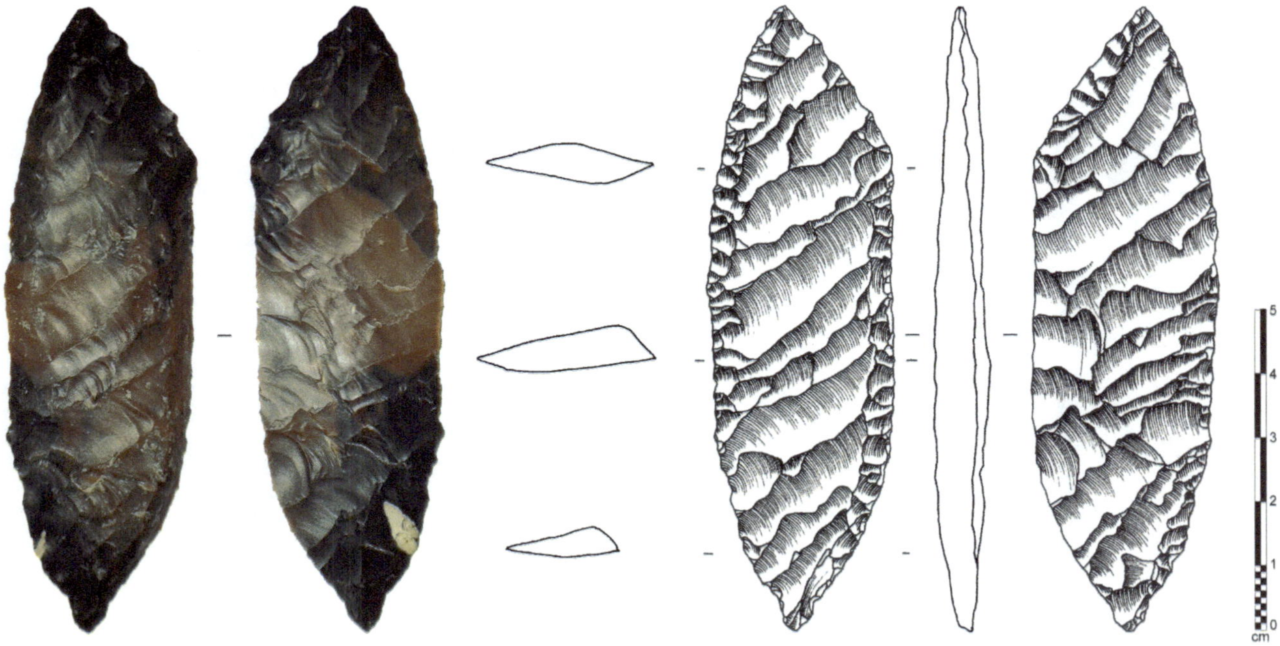

Figure 8.9. Dagger of jasper-like chert from Manayzah (surface). *Photograph by Rémy Crassard. Drawing by Julien Espagne.*

Chronometric Dating

A set of radiocarbon ages is available for a few of Manayzah's archaeological layers. These ages are highly precise, with a margin of about a century or two. Manayzah's stratigraphic sequence, within which samples were situated, provides good prior knowledge about the parameters for radiocarbon ages, further constraining their ranges (chapter 18). The results allow archaeologists chronologically to situate some of the various tool industries found. Table 8.2 presents the radiocarbon ages from Manayzah. For each sample, excavators collected charcoal from one branch, avoiding the inherent error in pooling fragments from multiple species and perhaps multiple death dates for each.

The posterior means of a Bayesian analysis of five radiocarbon ages from the site and a broader model of siltation chronology in the Wādī Sanā indicate an overall occupation span at Manayzah of about 1,300 years (chapter 18). Using the outer ranges of posterior distributions, the radiocarbon ages span the Early to Middle Holocene, from a beginning around 9550 cal BP. The posterior distributions extend from the middle of the tenth millennium to the middle of the eighth millennium cal BP, and the beginning and end medians span 1,288 years. The difference between a dated event associated with Layer 17 of Square K9 (early eighth millennium cal BP) and a dated event associated with Layer 20 in the same square (early ninth millennium cal BP) suggests the existence of a chronological hiatus within the occupational span. This hiatus also appears in posterior distributions (chapter 18) as a gap of 250 years or more. This gap is certainly a cultural and stratigraphic one, noticed during excavations and during the study of the material remains from lower and upper layers.

Some of the Structures Exposed

Four combustion structures (figure 8.11) and many ashy areas (perhaps for refuse from fireplaces or remains of discrete hearths) were excavated. These deposits demonstrate a dense prehistoric occupation of the site. Two constructed fire pits belong to Type I within a broader Wādī Sanā typology (Kimiaie and McCorriston 2014) and could represent a component of material culture (technical tradition) belonging to a specific group at a given time. These Type I circular hearths measure between 0.5 m and 0.6 m in diameter and are 0.2 to 0.25 m deep. They are dug hollows and intentionally filled with clastic limestone slabs averaging 0.10–0.20 m in length. The two fire pits are only 3.5 m distant one from one another. They were dug from the same occupation level (Layer 149). Other fire pits (Type II) found in stratigraphic context are generally constructed with a simple, unlined, hollow dug into a sedimentary matrix and almost always contain the remains of charcoal. Another hearth (K9 Hearth 1) was made of a bed of flat stones laid on a surface in the shape of a full circle (equals Type III; see Kimiaie and McCorriston 2014).

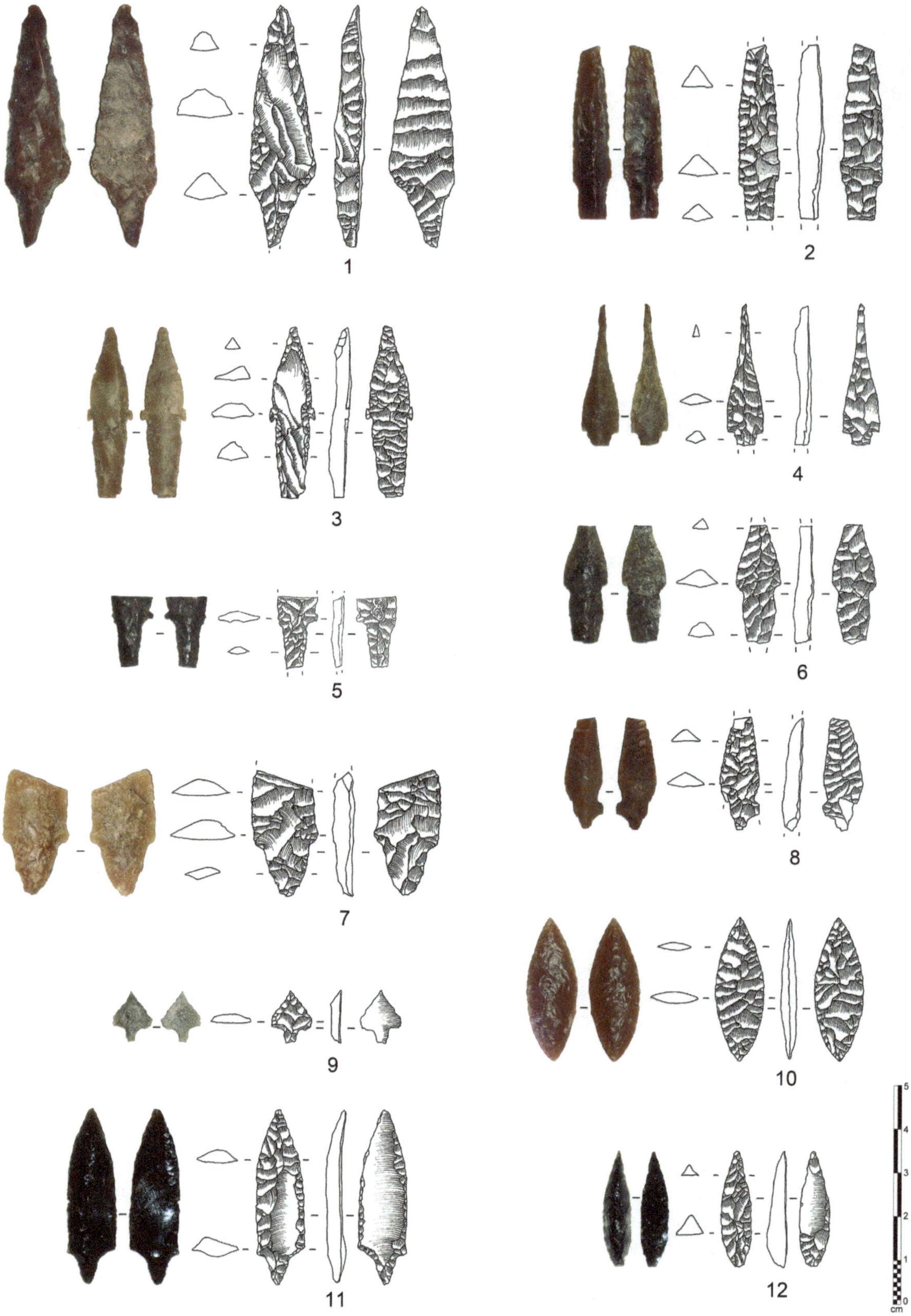

Figure 8.10. Arrowheads from Manayzah: (1) from M10, Locus 011; (2) from K9, Locus 016; (3) from L10, Locus 011; (4) from K9, Locus 017 top; (5) from K9, Hearth 1 (high flot); (6) from K9, Locus 016; (7) from K9, Locus 018; (8) from K9, Locus 016; (9) from M8, Locus 001; (10) from M8, Locus 004; (11) in obsidian, from L9, Locus 002; (12) in obsidian, from L8, Locus 001. *Photographs by Rémy Crassard. Drawings by Julien Espagne.*

Table 8.2. Radiocarbon ages from Manayzah. For further information on Bayesian posterior distributions, consult chapter 18.

Lab #	Material	Grid	Quad	Locus	Lot	Context	^{14}C Year BP	Cal BP (OxCal 4)	Bayesian Posterior Distribution (chapter 18)	Event Posterior Distribution (chapter 18)
AA66684	cf. *Tamarix* sp. charcoal piece	I14	C	009	10	Arbitrary layer, 20 cm deep in compact ashy silts with charcoal, bone, and lithic inclusions. Younger than AA57570.	6981 +/- 51	7932–7679	7644, 250	7879–7360
AA66683	*Acacia hamulosa/ mellifera* charcoal piece	L9	A	010	15	Grey, compact sediment overlying good surface. Contains ash, charcoal, and burned stone. Younger than AA59570. Provides indirect date on use of fluting technique.	6987 +/- 57	7935–7695	7764, 74	7919–7636
AA59570	charcoal fragments, all *Ziziphus leucoderma*	K9 north half Hearth 1		Layer 9		Hearth in stratigraphic context in test pit K; first indirect relative date on lithics; stratigraphically younger than AA66685, AA66686.	6902 +/- 41	7834–7665	7822, 72	7963–7692
AA66685	10 charcoal fragments, at least five *Ziziphus* sp.	K9		Layer 17 upper		Brown-grey ashy sediment layer with dense charcoal, containing significant new lithic technology: arrowhead, "concorde plane"; stratigraphically younger than AA6686; stratigraphically older than AA59570.	7133 +/- 51	8035–7845	7979, 98	8157–7813
AA66686	three *Acacia* sp. charcoal fragments	K9		Layer 20		Lowest excavated layer to date in which occurs datable charcoal; lower lithics differ in patination; stratigraphically under AA66685, AA57570.	8072 +/- 79	9254–8653	8975, 342	9550–8404

Figure 8.11. Hearth constructed as a shallow pit in an occupation surface, Manayzah Quad A, Grid M9, Locus 009 (in the process of excavation). Scale is 20 cm. *Photograph by Rémy Crassard.*

Part of a pit (K9 Pit 1) was recorded in K9 during the 2004 season. It appears to measure 1.5 m in length and is at least 0.30 m deep. Although it was only partially excavated, the small fraction exposed in K9 yielded a large quantity of animal bones. It was also very rich in charcoal, with pieces of carbonized wood still in place.

Two postholes were very clearly identified from two different occupation surfaces. The circular postholes (0.10 m in diameter and 0.10 m deep) contained small chock stones placed vertically, lining the postholes. These suggest the presence of wooden structures. Given the incomplete excavation of these occupational surfaces, it is impossible at this time to define any structure plans or confirm their existence.

Faunal Remains

Apart from the abundant and particularly varied presence of lithic industries and archaeological structures in the entire stratigraphy, the faunal remains of Manayzah are one of the few—and therefore one of the richest—assemblages of the Early to Middle Holocene of South Arabia. They provide (together with lithic industries, archaeological structures, plant remains, and other finds potentially pointing to a cultural tradition) elements of information on the socioeconomic structure, thus contributing to a better definition of Yemen's Neolithic. Bone assemblages, exclusively comprising animal remains, are present in just about all the layers above Layer 20 in K9 (table 8.3). Some bone specimens were worked, including polished pieces. Prior publication (Martin et al. 2009) contains a complete study of the few identifiable faunal remains, including identification details of fully domesticated and imported sheep in K9 Layer 16 and domesticated cattle (Quad C-009) higher in the stratigraphy. Attempts to extract DNA from the cattle bone have been unsuccessful. The results of Bayesian analysis now place the domesticated sheep around 8000 cal BP and the domesticated cattle some 400 years later, based on posterior means. (See chapter 18 for details.) Stratigraphically earlier finds of *Bos* in the Manayzah sequence could also be from domesticated animals, but the bones recovered offer no definitive markers of such status. The assemblage of wild and domesticated fauna points to an encampment of hunter-pastoralists, accessing cattle, sheep, and perhaps goats alongside wild ungulate prey.

Archaeobotanical Remains

During excavation, Manayzah appeared to be very rich in organic materials, including bone, charcoal, burned animal droppings, and visible charred seeds of *Ziziphus* sp. Upper deposits of Quadrat C contained a wide range of colors with grey, black, and whitish ash and yellow-brown silt mixes. In the K9 sounding, excavators also noted numerous dark grey and black ashy deposits, some clearly delineated as combustion features and others across the entire excavated surface (figure 8.4). The most likely source of this dramatic coloration range is plant ash and burned dung. Such deposits may contain seeds, depending on the firing temperatures, available oxygen during combustion, and the original source

Table 8.3. Identifiable faunal remains from Manayzah. Adopted from Martin et al. 2009:274–75, figure 4.

Taxon	Element	Portion	Fusion	Condition	Quadrat	Locus	Analytical Phase
Bos	radius	proximal	fused	fair	C	9	middle
Bos	metacarpal	distal	fused	fair	C	9	middle
Bos	calcaneum	fragment	?	burned	C	9	middle
Bos	sesamoid	complete		fair	A	4	middle
Bos	phalanx 1	complete	unfused	burned	C	9	middle
Bos	phalanx 2	distal	unfused	fair	K9	posthole 1	middle
Bos	phalanx 3	complete	fused	burned	C	9	middle
Bos	horncore	fragment		fair	A	1	middle
Bos	tooth	fragment		fair	C	9	middle
Bos?	thoracic vert	fragment		fair	K9	14	middle
***Bos* Total**							**10**
gazelle	calcaneum	complete	fused	fair	A	2	middle
gazelle	cuneiform	complete		burned	B	2	middle
gazelle	metatarsal	proximal	fused	fair	B	6	middle
gazelle	metatarsal	distal	fused	fair	L9	2	middle
gazelle	phalanx 1	distal		burned	D	2	middle
gazelle	phalanx 1	distal		fair	K9	18	middle
gazelle	phalanx 2	distal		burned	C	10	middle
gazelle	phalanx 3	complete	fused	burned	C	9	middle
gazelle	patella	complete		fair	K9	14	middle
gazelle	max M1/M2	crown		fair	L8	10	middle
gazelle	max M1/M2	crown		fair	A	4	middle
gazelle	mandible	fragment		fair	K9	14	middle
gazelle	mand P4	crown		fair	K9	13	middle
gazelle	mand P2	complete		fair	A	4	middle
gazelle	tooth	fragment		fair	K9	8	middle
gazelle	tooth	fragment		fair	K9	7	middle
gazelle	tooth	fragment		fair	A	11	middle
gazelle	tooth	fragment		fair	A	2	middle
Gazelle Total							**18**
caprine	metacarpal	distal		fair	C	7	upper
caprine	phalanx 1	distal		fair	C	7	upper
caprine	scapula	proximal		fair	K9	7	middle
caprine	scaphoid	complete		fair	D	2	middle
caprine	semi-lunaire	complete		fair	D	2	middle
caprine	pyramidal	complete		fair	D	2	middle
caprine	semi-lunaire	complete		fair	K9	16	middle
caprine	capetum-trapezoid	complete		fair	C	9	middle

Table 8.3. Identifiable faunal remains from Manayzah. Adopted from Martin et al. 2009:274–75, figure 4. *(continued)*

Taxon	Element	Portion	Fusion	Condition	Quadrat	Locus	Analytical Phase
caprine	metacarpal	distal	fusing	fair	B	10	middle
caprine	femur	proximal	unfused	fair	K9	posthole 1	middle
caprine	max M1/M2	complete		fair	B	11	middle
caprine	atlas	cranial		fair	B	10	middle
caprine	phalanx 1	proximal	fused	fair	B	3	middle
caprine	maxilla	complete		fair	K9	pit 1	middle
caprine	tooth	fragment		fair	K9	16	middle
Caprine Total							**15**
Capra	phalanx 3	fragment	fused	fair	C	7	upper
Capra	astragalus	complete		burned	K9	7	middle
Capra	metacarpal	distal	fusing	burned	D	5	middle
Capra	phalanx 3	complete	fused	fair	K9	16	middle
***Capra* Total**							**4**
Ovis	astragalus	complete		fair	K9	16	middle
***Ovis* Total**							**1**
Total							**48**

of organic material (Canti 2003; Miller and Smart 1984). In permanent settlement sites, plant macroremains and wood charcoals may be abundant; in ash mounds of burned cattle dung, they may be very scarce (Fuller et al. 2004:117).

The source of Manayzah's combusted material was a priori unknown. To obtain the best available range of archaeobotanical data from Manayzah, excavators collected fragments of wood charcoal and a burned seed by hand and also bagged 20 samples of sediment for flotation. Flotation samples ranged from 0.1 to 16.75 liters of sediment (average is 6 liters), with a volume determined in the field by the visible richness of charred plant remains and the overall volume of an integral stratigraphic unit. As with other sites in Wādī Sanā, sampling was strategic and targeted: excavators selected deposits based on visual evidence of charcoal, likely primary context for in situ burning (for example, a hearth), geomorphological-taphonomic context, and a good preservation environment buried at depth. Available water resources precluded a random or stratified systematic flotation sampling strategy (chapter 3).

Where excavators saw charcoal, they collected 20 separate specimens presumed to originate from 20 separate branches or wood sources. To collect specimens most likely to conform to this assumption, excavators chose pieces separated from one another by sediment and individually wrapped each in foil to preclude post-excavation fragmentation and mixing. Analysts recognize that one dead tree could furnish all the wood for an in situ burning event like a hearth. Where diversity in species exists, such a collection method should better capture it.

Excavators also sampled sediments for eventual analysis of phytoliths and spherulites. This sampling targeted each occupation level in section and required small sediment collections of 0.25 to 0.60 liters. Unfortunately, the K9 sequence of phytolith samples was lost to flooding damage in the University of South Florida's basement storage. Current phytolith studies of upper strata of the site show promising assemblage differences (Buffington et al. 2017) and may enhance the inferences drawn from macroremains and wood charcoal. Spherulite studies of upper deposits may help determine the source and activities that deposited concentrated ash at Manayzah (Canti 1998).

To evaluate recovery rates, team member Catherine Heyne added 50 count of modern poppy seeds to several sediment bags before flotation (Wagner 1989). In sorting the light fraction flot, Heyne, McCorriston, and students recovered poppy seeds in a variable range, from a mere 7 to 36 count. Consequently, we recognize that recovery

Table 8.4. Flotation samples from Manayzah, organized with oldest deposits at bottom of table and youngest at the top.

Sample ID	FS #	Sediment Volume (l)	Flotation Volume (ml)	Sorted Volume (ml)	Organics/ Liter	Seed Count	Seeds/Liter	Shell	Wood	Charred Dung	Calcified Sediment	LF Lithics	HF Lithics
M9-002 Lot 3	1395	5	75	75	15.0	2	0.4		x		x		
K9-Niv 2	1411	5.5	58.4	58.4	10.6	4	0.7		x			x	
L11-B-005 Lot 6	1404	10.75	168	168	15.6	0	0.0		x	x			
L11-B-005 Lot 6	1405	8.75	116	98.5	13.3	1	0.1		x		x		
all B-006 Lot 7	1406	9.75	186	138.13	19.1	0	0.0		x	x			
M11-B-007	1399	15	242	242	16.1	4	0.3	x		x	x		
M11-B-004 Lot 5	1403	6	63	63	10.5	0	0.0		x	x	x		x
M9-A-009 Lot 11	1396	16.75	168	74	10.0	0	0.0		x				x
K9-005/006	1415	6.5	214.7	214.7	33.0	0	0.0		x	x			x
M10+M11-B-013	1401	5.5	56	56	10.2	0	0.0		x				x
M8-A-012	1398	3	5.3	5.3	1.8	3	1.0		x				x
K9-Niv 7	1409	6	63.5	63.5	10.6	0	0.0		x	x		x	
K9-009 (pit)	1393	3.5	60	60	17.1	0	0.0		x		x		
K9-008	1412	5	145.7	145.7	29.1	4	0.8		x	x		x	x
K9-Niv 13	1417	5	78	78	15.6	15	3.0	x	x	x		x	
K9-Foyer 1 (H1)	1413	7	75.6	75.6	10.8	2	0.3		x				x
K9-Niv 11	1498	not flot	–	–	–	1	–						
K9-014	1416	0.1	2.5	2.5	25.0	4	40.0		x	x			x
K9-014 bottom	1414	0.2	0.5	0.5	2.5	12	60.0		x	x			
K9-014/015	1410	1.5	50.5	50.5	33.7	3	2.0		x	x		x	x
K9-017sup	1391	2.25	6.5	6.5	2.9	0	0.0		x				
K9-018	1394	2	20	20	10.0	0	0.0		x				x
Average Volume		**5.96**			**Total # Seeds**	**55**							
Total Volume		**125.05**											
# of Samples with Seeds Present						12		2	18	10	4	4	10
Ubiquity						57.14		9.52	82	45	18	18	45

Table 8.5. Seed, achene, and nutlet taxa from Manayzah flotation samples.

Flotation Sample #	1391	1393	1394	1395	1396	1398	1399	1401	1403	1404	1405	1406
Portulaca sp.	0	0	0	0	0	0	0	0	0	0	0	0
Aerva javanica	0	0	0	0	0	0	0	0	0	0	0	0
Ziziphus sp.	0	0	0	0	0	0	0	0	0	0	0	0
small legume	0	0	0	0	0	0	0	0	0	0	0	0
Cleome sp.	0	0	0	0	0	0	0	0	0	0	0	0
Eragrostis cf. *ciliaris*	0	0	0	0	0	0	0	0	0	0	0	0
cf. *Polypogon*	0	0	0	0	0	1	0	0	0	0	0	0
cf. *Solanum*	0	0	0	0	0	0	0	0	0	0	1	0
cf. *Heliotropium*	0	0	0	0	0	0	0	0	0	0	0	0
Compositae cf. *Centaurea*	0	0	0	0	0	0	0	0	0	0	0	0
grass culm	0	0	0	0	0	0	0	0	0	0	0	0
grass caryopsis	0	0	0	0	0	0	0	0	0	0	0	0
cf. *Schoenoplectus*	0	0	0	0	0	0	0	0	0	0	0	0
cf. Cucurbitaceae	0	0	0	0	0	0	0	0	0	0	0	0
unidentified charred seeds	0	0	0	2	0	2	4	0	0	0	0	0
Seed Count	**0**	**0**	**0**	**2**	**0**	**3**	**4**	**0**	**0**	**0**	**1**	**0**

rates of seeds also varied. Some sediment samples were so rich in ash and decaying organics (dung) that they were "hydrophobic," dispersing across the surface of the flotation tank. Much of the charcoal visibly shattered in contact with water, making analysis of hand-picked charcoals all the more important. Heyne sorted heavy fraction residues in the field. Heavy fraction residues contained fragments of consolidated sediment with rich charcoal content and also contained charcoals so impregnated by crystalline precipitates (possibly gypsum) that no one could identify the original cellular structures. From heavy fractions Heyne also recovered bone, soil gastropods, and micro-debris from chert (table 8.4).

Manayzah samples were sorted under 6–10x magnification using Leitz MZ-5 and MZ-12 microscopes. Joy McCorriston finalized identifications of macrobotanical material using modern reference collections made largely by Catherine Heyne in Yemen, with McCorriston's confirmed identifications of voucher specimens. McCorriston, Abigail Buffington, and Masoumeh Kimiaie sorted and identified wood charcoal specimens using a compound Leica forensic microscope of 50–400x. Wood collections from Yemen provided modern charcoals for reference, supplemented by reference manuals. From small flotation assemblages and hand-picked anthracology collections, we identified all fragments larger than 2 mm; from larger collections, we usually identified the largest 20 fragments. In some cases these larger fragments showed poor species richness (all of one type). In such instances, we continued to select and identify up to 100 fragments.

The overall numbers of seeds recovered from Manayzah are extremely low, with most seeds from one hearth (K9 Layer 14) and adjacent sediments (K9 Layer 13) (table 8.5). In addition to few seeds, light fraction flot returned glassy nodules, soil gastropods, ostracods, and chunks of calcined sediment. Modern dung—or at least dung that had not fully charred or decayed—appeared in some deeper contexts, and other soil fauna disturbances probably introduced modern seeds as deep as K9 Layer 14. Some ostracods show signs of burning, especially the darkened ostracods in M9-A-002 and M8-A-012.

Flotation recovered only 55 seeds from 125 liters of flotation-processed sediment. An extremely weak Pearson's correlation coefficient ($R^2 = 0.06$) suggests that recovery rate of seeds is independent of the volume of sediment floated; flotation of larger sediment samples would not have yielded a substantially greater overall richness of seeds from the Manayzah deposits. Nor were there substantially higher seed counts in FS no. 1391 (K9 Layer 17 upper), from which analysts recovered the highest number ($n = 36$) of poppy seeds in the calibrated assessment of seed recovery. Of the other sediment samples to which poppy seeds were added, FS no. 1414 (K9 Layer 14 bottom) had one of the highest ancient seed counts ($n = 12$)

Flotation Sample #	1409	1410	1411	1412	1413	1414	1415	1416	1417	1498	**Total**
Portulaca sp.	0	0	0	0	0	0	0	0	1	0	**1**
Aerva javanica	0	0	1	0	2	0	0	0	3	0	**6**
Ziziphus sp.	0	0	0	0	0	0	0	0	0	1	**1**
small legume	0	0	0	0	0	1	0	0	0	0	**1**
Cleome sp.	0	0	0	0	0	4	0	0	1	0	**5**
Eragrostis cf. *ciliaris*	0	0	2	0	0	0	0	0	0	0	**2**
cf. *Polypogon*	0	0	0	0	0	0	0	0	0	0	**1**
cf. *Solanum*	0	0	0	0	0	0	0	1	0	0	**2**
cf. *Heliotropium*	0	0	0	0	0	0	0	1	0	0	**1**
Compositae cf. *Centaurea*	0	0	0	0	0	1	0	0	0	0	**1**
grass culm	0	0	0	2	0	1	0	0	0	0	**3**
grass caryopsis	0	0	0	0	0	0	0	0	1	0	**1**
cf. *Schoenoplectus*	0	1	0	0	0	0	0	0	3	0	**4**
cf. Cucurbitaceae	0	0	0	0	0	0	0	0	4	0	**4**
unidentified charred seeds	0	2	1	2	x	6	0	2	2	0	**23**
Seed Count	**0**	**3**	**4**	**4**	**2**	**12**	**0**	**4**	**15**	**1**	**55**

and the highest ancient seed density (60 seeds per liter), despite the loss of many poppy seeds (n = 9 recovered) during flotation. The highest seed counts and the majority of seeds came from two contexts in the deeper deposits of the site (K9 Layer 13, K9 Layer 14 bottom). These are deposits that generally lack the calcined sediment nodules that trap buoyant air pockets and largely dominate the light fraction flot of many upper deposits. Although the sample size is small for statistically significant conclusions, the results of flotation suggest a contrast in seed density between contexts from the lowest layers of the site and those in the upper.

Another characteristic of the charred seeds is that they are minute and diverse, with few repetitions of taxa. All charred seeds, achenes, and nutlet shell fragments measure less than 1 mm, and most less than 0.5 mm. Most samples contained wood charcoal; the largest fragments of charcoal from the site survived as hand-picked specimens.

Hearth F9, Layer 14 contains relatively more seeds than other deposits, but charcoal nonetheless dominates the assemblage. None of the seeds came from *Ficus* sp., the only source for the identified wood charcoal from this hearth (Kimiaie and McCorriston 2014:36, table 1) (table 8.6). While half the seeds from this hearth could not be identified, all were minute. As in many deposits, fragments of charred dung are also present, another potential source of seeds (Miller and Smart 1984).

Identifiable charcoal fragments from Manayzah are less numerous than excavators had assumed by observing black flecks in soil. In microscopic view, many items proved to be decaying organic matter (modern roots that had penetrated deep), burned stone or fragments of grey daub, and charcoal so heavily infused with crystalline gypsum precipitates that the fragments sank in flotation and were recovered in the heavy fraction. Because the cell structure is warped and obscured, these and similar fragments saved and foil-wrapped in excavation proved impossible to identify. Albeit as small fragments and in small numbers, wood charcoals do occur in 82 percent of the flotation samples, twice as ubiquitous as charred dung (41 percent) and more ubiquitous than seeds (57 percent). Dung occurred with greater ubiquity if one considers the browned, incompletely charred materials that may represent modern dung carried down through insect activity, of which there was ample evidence.

The source of Manayzah dung fragments is not clear, although the deep ash in upper layers suggests that the residues originated from penning animals on site. The digestive systems of cattle and sheep, the two unequivocally domesticated fauna represented in the bone assemblage from Manayzah, suggest that few seeds would survive in dung (Wallace and Charles 2013). Ash mounds deliberately compiled of burned cattle dung by villagers in the southern Indian Neolithic yielded almost no identifiable

Table 8.6. Wood charcoal identified from Manayzah.

	Sample #	Weight (g)	Volume (ml)	# Pieces (> 2 mm)	Weight of Identified Pieces	*Acacia* spp.	*Anogeissus* sp.	*Cadaba* spp.	*Commiphora* spp.	*Ficus* spp.	*Maerua* sp.	*Ziziphus* spp.
M9 002 Lot 3	1395	N/A	N/A	N/A	N/A							–
K9 Niv-2	1411	0.018	0.2	4	0.018	–	–	–	–	–	–	0.015
L11 B 005 Lot 6	1404 and 1405	0.353	N/A (> 0.4)	38	0.25	0.07	0.019	–	–	–	0.039	0.122
all Quad B 006 Lot 7	1406	0.012	0.2	8	0.013	0.002	–	–	–	–	–	0.011
M11 B 007	1399	0.129	N/A	17	0.077	0.024	0.021	–	–	0.032	–	–
Quad B 008 Lot 12	1499	5.208	11.3	41	0.168	–	0.088	–	–	–	–	0.08
M9 A 009 Lot 11	1396	0.71	8	36	0.084	0.047	–	–	0.037	–	–	–
K9 005/006	1415	0.005	0.2	2 (1 > 1 mm)	0.003	–	–	–	–	0.003	–	–
M10 + M11 B 013	1401	0.521	1.2	19	0.046	–	0.001	–	–	0.02	–	0.025
M8 A 012	1398	0.225	0.2	6	0	–	–	–	–	–	–	
K9 Niv-7	1409	0.014	0.2	5	0.016	–	0.008	–	–	–	–	0.008
K9 009 pit	1393	1.84	9.1	103	0.268	–	–	–	–	–	–	0.233
K9 008 N 1/2	1412	0.186	1	33	0.09	–	–	–	–	0.028	–	0.062
K9 Niv-13 (= 008 south half)	1417	0.015	0.2	5	0.016	–	–	–	–	0.016	–	–
K9 foyer (H1)	1413	0.222	–	20	–	–	0.061	–	–	–	–	–
K9 014	1416	0.128	–	20	0.076	–	–	–	–	0.076	–	–
K9 014 bottom	1414	0.004	0.2	2 (1 > 1 mm)	0.004	–	–	–	–	0.004	–	–
K9 014/015	1410	0.013	0.2	7	0.01	–	–	–	–	0.008	–	0.002
K9 017 sup	1391	0.052g	0.2	1	0.052	–	–	–	–	–	–	0.052
K9 018	1394	16.519	15+	111	1.072	0.25		0.102		0.576	0.103	0.041
Total Weights					**2.263**	**0.393**	**0.198**	**0.102**	**0.037**	**0.763**	**0.142**	**0.651**
# Samples with Species Present						5	6	1	1	9	2	11
Ubiquity						25	30	5	5	45	10	55

plant material, suggesting that the digestive tracts of cattle and the open-air firing of dung destroy most seeds (Fuller et al. 2004). While taphonomic conditions and general environmental circumstances of southern Indian ash mounds may differ, seed antiquity and survival are comparable. In a considerably later period, medieval burned cattle dung from Zabid, Yemen, also contained few or no seeds. In Zabid's cattle dung only hard-shelled seeds survived, like *Ziziphus* and *Acacia* (the latter erroneously reported as *Convolvulus*) (McCorriston and Johnson 1998). Ashy upper deposits probably came from slow-burning dung fires, either as deliberate camp cleanings or lit by hearth sparks or brush fires. Future spherulite and phytolith studies may clarify their origins (Canti 1998). A pilot study of phytoliths from C5 Layer 13 is unusually rich in phytoliths and especially in multicell phytoliths and occluded carbon, consistent with the burned dung mat hypothesis (Buffington et al. 2017:39).

The rare incidence of seeds in sediments and the small overall numbers preclude significant quantitative analysis, but this assemblage can provide some qualitative insights. Manayzah's seed assemblage suggests no deliberate collection and processing of seeds for human food. The rare surviving seeds are minute and come from a diverse array of taxa (even those unidentifiable). Small seeds like chia and teff may provide human food, but they require processing like grinding and cooking to access sufficient nutrition. One *Eragrostis* caryopsis may be related to a teff-like grass, but there is no pattern to suggest routine collection and processing in which significant numbers of seeds were lost and burned. No processing tools such as grinding stones, mortars, or sickles occurred in the durable assemblage from Manayzah. And there were no domesticates present in the charred plant remains. Of the other identifiable seeds, a single *Ziziphus* drupe (hand collected in excavation) represents an edible fruit, but no other examples—whether intact or as diagnostic shell—appeared in flotation.

From the wood charcoals identified, it is clear that a pattern of wood use seen in Manayzah hearths is evident in other layers at the site. *Ziziphus* and *Ficus* wood are the most frequently encountered, with *Anogeissus* and finally *Acacia* as other frequent species (figure 8.12). There is a wider availability of other woody taxa in Wādī Sanā today and probably also in the past; we infer that Manayzah's inhabitants selected particular woods useful as dense, long-burning fuels (*Acacia, Anogeissus,* and *Ziziphus*) and perhaps for smoke qualities (*Ficus*) (Kimiaie and McCorriston 2014). *Acacia* is the more common and widespread of these taxa in Wādī Sanā today, and this

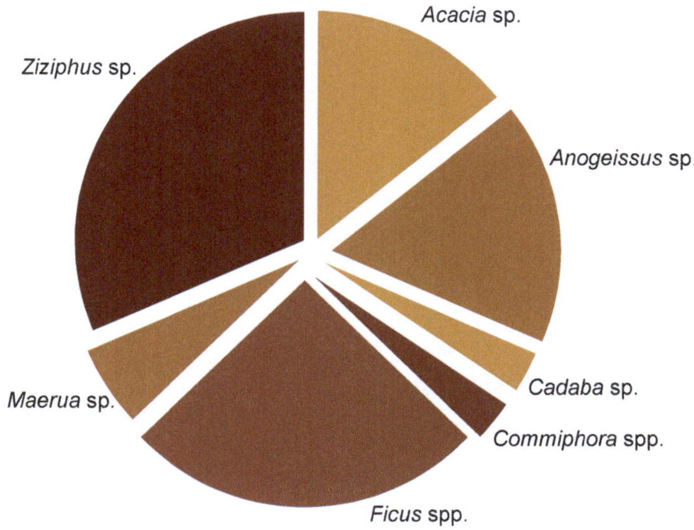

Figure 8.12. Percentages of wood charcoal taxa at Manayzah (by weight). *Illustration by Joy McCorriston.*

modern distribution may be in part an outcome of the substantial climatic and environmental changes that have taken place in Wādī Sanā since the Early Holocene occupations at Manayzah.

In chapter 3, table 3.5 summarizes the presence in archaeological context of many taxa found growing in Wādī Sanā today, and Manayzah's identifiable plant remains come from species found locally. Seeds may have been introduced to the K9 Layer 14 hearth via kindling (dry grass, dry annual plants). Their presence hints that dry plants bearing seeds were available in the site environment and that occupation, at least when the lower hearth in K9 Layer 14 was last in use, took place in winter months after the rainy season, flood subsidence, and fruiting of seasonal annuals. *Polypogon*, *Portulaca*, and *Schoenoplectus* are wetland-loving plants that may have thrived in the vicinity of Manayzah's spring.

Beads and Ornaments

By excavating across an occupation surface, excavators recovered multiple objects associated with its abandonment. These included elements of body decoration (figure 8.13). Manayzah excavations yielded six beads: four on the site's surface and two in a stratified context. People manufactured these beads of stone and shells. Some of the very small shell beads were made by perforation and polishing. Fragmentary seashell (a small cowrie) found in the oldest layer (K9 Layer 20) is also a very likely relic of body or clothing decoration. K9 Layer 8 also yielded a fragment of a stone pendant. It consists of a flattened

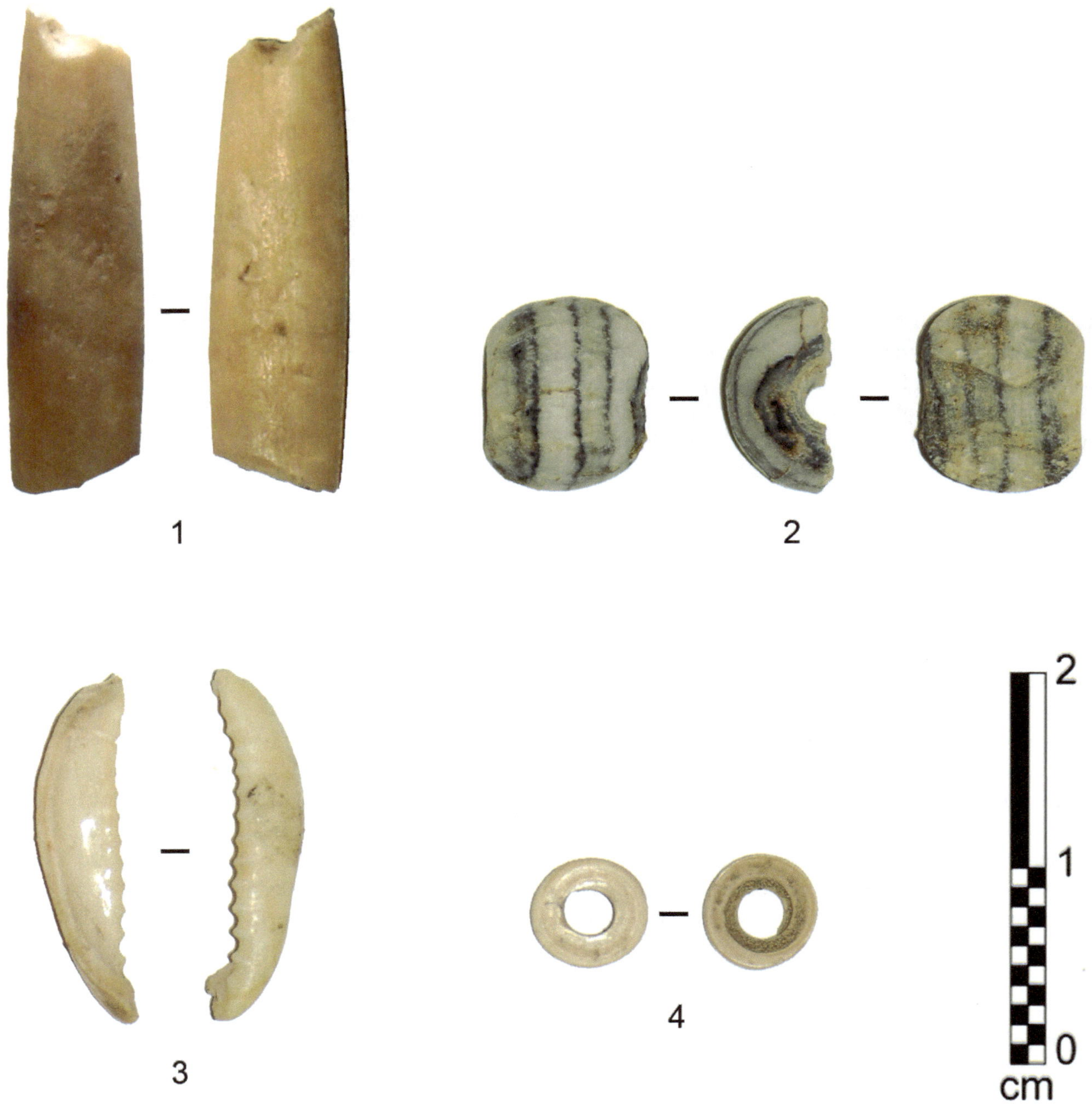

Figure 8.13. Ornaments from Manayzah: (1) drilled stone pendant from K9-008; (2) biconically drilled bead of metamorphic stone from L2 surface; (3) (marine) cowrie shell from K9-20inf; (4) *Conus* shell bead from L9-001. *Photographs by Rémy Crassard.*

and polished cylinder made of undetermined yellowish stone—perhaps alabaster or calcite—broken at the two extremities. Of the entire pendant, only 2.5 cm survives. In its proximal part is the trace of half a perforation.

Results: Lithic Industries

Most archaeological layers contain dense deposits of tools, products, and debris from chert and obsidian knapping (table 8.7). These remains are represented in each layer, across almost all the site's surface to the lowest layer excavated in Square K9. There were 7,525 knapped lithic pieces found: 5,063 in stratigraphic context and 2,462 on the surface. Most of the flakes from the Manayzah assemblage come from bifacial manufacturing or retouch operations (for example, bifacial with pressure flaking, *en écharpe* bifacial shaping)

Raw Materials

Knappers used many raw materials (figure 8.14). By examining both the excavated material at Manayzah and the Wādī Sanā regional geology, the excavators identified two local main sources. First, cobbles in the streambed of wadis were a close and easily accessible source. At such locations, the Manayzah flintknappers found an almost unlimited means of supplying themselves with varied materials. Raw materials originating from wadi beds in the Southern Jol (plateau) are highly diversified—in their majority chert (including high-quality flint) with also chalcedony and jasper. All were used because of their good knapping qualities, particularly for pressure flaking, which gives excellent results on fine-grained rocks. The summit of plateaus overlooking the Wādī Sanā provides another good source of raw material. Very good-quality chert (and flint) is available in the exposed Eocene and Oligocene beds. During excavation at Manayzah, we found fragments of thin tabular chert blocks. These have orange-like or pinkish cortex on blocks that were sometimes worked and at other times not worked. Finally, Manayzah knappers also extensively used obsidian, evident by a high ubiquity index in the recorded stratigraphy. In principle, this volcanic glass is extrinsic to the Hadramawt region, since the closest documented sources of obsidian are in the highlands of western Yemen, some 600 km away.

Thermal treatment could improve the quality of rocks used. Thanks to the presence of a characteristic luster, analysts could identify some examples of preknapping thermal treatment applied to the archaeological material found at Manayzah.

Figure 8.14. Local raw materials represented in Manayzah lithic industries: (1) streambed cobbles; (2 and 4) plateau source; (3) tabular flint band eroding from limestone beds. *Photographs by Rémy Crassard.*

Table 8.7. Manayzah surface and excavation finds inventory. This table compiles three datasets: (1) materials from excavation; (2) surface collection by grid square; (3) general surface collections prior to site gridding—that is, as areas of SU155-002.

(1) Material from Excavations						Lithics							
						Products				Tools			Cores
Date	Quad	Square	Locus	Lot	Bag	Chert Flakes	Obsidian Flakes and Bladelets	Channel Flakes	Other	Arrowheads	Bifaces	Misc. Tools	Cores for Flakes
2/27/05	A	M9	001	2	1	74	8					1	
2/10/05	A	L8	001	2	1	49	6			1			
2/10/05	A	M8	001	2	1	62		1		2			
2/10/05	A	L9	001	2	1	93	7	2			1		
2/11/05	A	M9	002	3	1	87	9	1	1				
2/12/05	A	L9	002	3	1	184	13			1	1	1	
2/11/05	A	M8	002	3	1	90	5	1		1			
2/12/05	A	L8	002	3	1	126	8	1					1
2/12/05	A	ALL	003	4	1	37	4						
2/13/05	A	ALL	004	5	3								
2/12/05	A	ALL	004	5	1	97	4	1		1	1		
2/13/05	A	ALL	005	6	1	129	7	2				1	
2/23/05	A	M9	009	13	1	3							
2/23/05	A	M9	009	14	1	1	1						
2/23/05	A	M9	009	11	1	3	1						
2/22/05	A	M9	009	10	1	11	3						
3/3/05	A	L8	010	15	1	146	5	2					
2/28/05	A	M8	010	15	1	71		1					
2/27/05	A	L9	010	15	1	45	5				1		
2/27/05	A	M9	010	15	1	37	2						
2/27/05	A	L8	011	16	1	20							
2/27/05	A	L9	011	16	1	26	1	1			2		
2/15/05	B	ALL	001	2	1	142	14	2		1			
2/15/05	B	ALL	002	3	1	480	15	1	2	2	3		
2/19/05	B	L10/L11	003	4	1	89	5	1			2		
2/17/05	B	ALL	003	4	1	31	3						1
2/19/05	B	L11	005	6	1	56	4				1		
2/16/05	B	L10	006	7	1	23							
2/20/05	B	ALL	008	12	1	79	6		1		1		
2/19/05	B	M11	008	12	3								
2/20/05	B	L10/L11	009	13	1	50	1	1			1		
2/27/05	B	M10	003	4	1	5							
2/20/05	B	L10/M10	010	14	1	23	3						
2/21/05	B	L10	010	14	1	23	2			1	2		
2/21/05	B	ALL	011	15	1	58	18			2		1	
2/21/05	B	L11	012	16	1	24	4				1		
2/22/05	C	ALL	008	9	1	6	4						
3/2/05	C	ALL	009	10	1	92	12				2		
2/16/05	C	ALL	007	8	1	21	2						
3/3/05	C	ALL	010	11	1	25	5				1		
2/27/05	C	ALL	008	9	3								
3/1/05	C	ALL	009	10	3								
2/28/05	D	L12/L13	001	2	1	101	2	1					

	Other Than Lithics					
Totals	**Worked Bones**	**Beads**	**Shells**	**Other Stones**	**Ceramic**	**Observations**
83						one scraper
56						one obsidian arrowhead (drawn)
65						one fragment fluted arrowhead (drawn) + one micro-arrowhead on flake (drawn)
103		1				one possible preform biface
98						one burin-like flake from a trihedral or plano-convex tool
200						one entire fluted arrowhead from base (drawn) + one flake bifacially shaped from an old flake (double patina) + scraper with very fine direct retouches on a big flake
97						one broken plano-convex arrowhead (drawn)
136						one core for flakes in obsidian
41						
	1					one fragment of polished bone?
104						one biface with remaining cortex on both extremities (drawn) + one entire foliated shaped arrowhead (drawn)
139						one retouched bifacial thinning flake as a side scraper and notch
3						
2						
4						
14						
153						
72						
51						one preform biface with remaining cortex on both sides
39						
20						
30						two fragments preform plano-convex biface
159						one broken arrowhead, plano-convex biface (drawn)
503						two fragments plano-convex biface arrowhead + three fragments preform biface + one possible greenstone polished ax + one burin-like flake from a trihedral or plano-convex tool
97						one bifacial piece (in L10) with fine *en écharpe* shaping (drawn)
35						one flat + T14 core for flakes on obsidian
61						one very fragmentary piece of biface
23						
87						one possible fragment of crested blade + one possible biface broken during fluting operation (drawn)
				1		one fossil brought from the plateau
53						one preform biface with typical remaining cortex in both extremities (drawn)
5						
26						
28						one very little piece of arrowhead tang + one fragment very precisely shaped biface (drawn) + one fragment foliated tool
79						two entire arrowheads (all drawn) + one retouched flake
29						one fragment preform biface
10						
106						one big and very finely shaped fluted bifacial pieces (broken) + one preform biface
23						
31						one possible fragment biface preform
				1		one fragment polished sandstone (grindstone)
				1		one fragment polished sandstone (grindstone)
104						

Table 8.7. Manayzah surface and excavation finds inventory. This table compiles three datasets: (1) materials from excavation; (2) surface collection by grid square; (3) general surface collections prior to site gridding—that is, as areas of SU155-002. *(continued)*

(1) Material from Excavations						**Lithics**							
						Products				**Tools**			**Cores**
Date	**Quad**	**Square**	**Locus**	**Lot**	**Bag**	**Chert Flakes**	**Obsidian Flakes and Bladelets**	**Channel Flakes**	**Other**	**Arrowheads**	**Bifaces**	**Misc. Tools**	**Cores for Flakes**
2/28/05	D	ALL	003	4	1	9	2						
3/1/05	D	ALL	004	5	1	9	1				1		
3/1/05	D	ALL	005	6	1	35	7						
3/2/05	D	ALL	006	7	1	10							
3/2/05	D	ALL	007	8	1	16	3			1			1
02/2004	-	K6	1	-		91	15			1			
02/2004	-	K9 North	1	-		162	8				1		
02/2004	-	K9 North	2	-		212	6					1	
02/2004	-	K9 North	3 + 4	-		56	4						
02/2004	-	K9 North	5 + 6	-		2	1						
02/2004	-	K9 North	7	-		53	2				1		
02/2004	-	K9 North	8	-		34	5						
02/2004	-	K9 North	9 + 10 = H1	-		1				1			
02/2004	-	K9 North	11 + 12	-		13	2						
02/2004	-	K9 North	13	-		9	2						
02/2004	-	K9 North	14	-		1							
02/2004	-	K9 North	15	-		3							
02/2004	-	K9 South	1	-		187	6						
02/2004	-	K9 South	2 + 3	-		245	10				1		
02/2004	-	K9 South	4	-		9							
02/2004	-	K9 South	5	-		2				1			
02/2004	-	K9 South	5/6 limit	-		11	2						
02/2004	-	K9 South	6	-		17	1						
02/2004	-	K9 South	7	-		31							
02/2004	-	K9 South	8	-		65	4						
02/2004	-	K9 South	9 = H1	-									
02/2004	-	K9 South	10 = H1'	-		5	2						
02/2004	-	K9 South	11	-		27	11						
02/2004	-	K9 South	12	-		2	18						
02/2004	-	K9 South	14 + 15	-		13	1						
02/2004	-	K9 South	P1	-		92	8					1	1
02/2004	-	K9 South	PH1	-		6							
3/1/05	-	K9	016	-	1	64	4			6	1		
3/1/05	-	K9	017	-	1	64				2		1	1
3/1/05	-	K9	018	-	1	134				1			1
3/2/05	-	K9	019	-	1	15							
3/3/05	-	K9	020up	-	1	7							
3/3/05	-	K9	020down	-	1	5							
3/3/05	-	K9	020down	-	2								
3/3/05	-	K9	021	-	1	11							1
3/3/05	-	K9	022	-	1	16							1
3/3/05	-	K9	023	-	1	13							
3/4/05	-	K9	023	-	2								
3/4/05	-	K9	024	-	1	12							
Totals						**4653**	**322**	**19**	**4**	**25**	**25**	**7**	**8**

	Other Than Lithics					
Totals	**Worked Bones**	**Beads**	**Shells**	**Other Stones**	**Ceramic**	**Observations**
11						
11						one small fragment biface, broken in fire
42						
10						
21						one small arrowhead (?) drawn + one residual core in obsidian
107						one fragment arrowhead
171						one fragment biface
219						one retouched flake
60						
3						
56						one fragment biface
39			1			
2						
15						
11			3			
1						
3						
193						
256						one fragment biface
9						
3						one fragment arrowhead
13						stratigraphic interface (real layer in itself?)
18						
31						
69		1				
						Hearth 1 (= H1)
7						Hearth 1' (= H1')
38			1			
20						
14						
102						one retouched flake; Pit 1 (= P1)
6			1			Posthole 1 (= PH1)
75						three broken plano-convex arrowheads (drawn) + 1 broken trihedral arrowhead (drawn) + two almost entire arrowheads (drawn) + one burin-like broken fragmented biface
68						one entire arrowhead found at bottom of ashy area with charcoals from 017 up to U46 (drawn) + one arrowhead tang + one fragment retouched flake + 1 core for flakes
136						one broken biface arrowhead with two small barbs (drawn) + one unidirectional core with single striking platform in bad-quality chert
15						one flake with typical Levallois prepared butt removed in continuity with other flake
7						
5						laminar tendency, decalcified lithics
			1			one fragment cowry shell, very probably used as a bead
12						decalcified lithics; one core with strong Levallois flake debitage modality (drawn)
17						decalcified lithics; laminar tendency; one core with strong Levallois flake debitage modality (drawn)
13						laminar tendency; decalcified lithics
			1			one apex of marine shell (maybe *Conus* sp. or *Strombus* sp.)
12				1		laminar tendency; decalcified lithics + one fossils (sea star?)
5063	**1**	**2**	**8**	**4**	**0**	

Table 8.7. Manayzah surface and excavation finds inventory. This table compiles three datasets: (1) materials from excavation; (2) surface collection by grid square; (3) general surface collections prior to site gridding—that is, as areas of SU155-002. *(continued)*

(2) Surface Collection by Square						Lithics							
						Products				Tools			Cores
Date	Quad	Square	Locus	Lot	Bag	Chert Flakes	Obsidian Flakes and Bladelets	Channel Flakes	Other	Arrowheads	Bifaces	Misc. Tools	Cores for Flakes
02/2005	-	G9	000	-	1	1							
02/2005	-	H6	000	-	1	4							
02/2005	-	H7	000	-	1	3	1						
02/2005	-	H8	000	-	1	9	3						
02/2005	-	H9	000	-	1		2				1		
02/2005	-	I4	000	-	1	9	1						
02/2005	-	I5	000	-	1	12	5						
02/2005	-	I6	000	-	1	7	1			1			
02/2005	-	I7	000	-	1	8	2						
02/2005	-	I8	000	-	1	8	1						
02/2005	-	I9	000	-	1	3						1	
02/2005	-	J4	000	-	1	26	8						
02/2005	-	J5	000	-	1	28	8					1	1
02/2005	-	J6	000	-	1	26	7						
02/2005	-	J7	000	-	1	26	6						
02/2005	-	J8	000	-	1	7	3			1			
02/2005	-	J9	000	-	1	7	2						
02/2005	-	J10	000	-	1	2	4						
02/2005	-	K5	000	-	1		4				1		
02/2005	-	K4	000	-	1	49	7						
02/2005	-	K5	000	-	1	47	11						2
02/2005	-	K6	000	-	1	30	2						
02/2005	-	K7	000	-	1	76	10					1	
02/2005	-	K8	000	-	1	27	4					1	
02/2005	-	K10	000	-	1	26							
02/2005	-	K11	000	-	1	34							
02/2005	-	K12	000	-	1	27	1						
02/2005	-	K13	000	-	1	1							
02/2005	-	K15	000	-	1							1	
02/2005	-	L2	000	-	1								
02/2005	-	L4	000	-	1	9			1				
02/2005	-	L5	000	-	1	42	11						
02/2005	-	L6	000	-	1	87	4						
02/2005	-	L7	000	-	1	57	5						
02/2005	-	L8	000	-	1	4	3			1			
02/2005	-	L9	000	-	1	82	3						
02/2005	-	L10	000	-	1	76	2						
02/2005	-	L11	000	-	1	31							
02/2005	-	L12	000	-	1	8	1						
02/2005	-	M4	000	-	1	28	3						
02/2005	-	M5	000	-	1	38	3						
02/2005	-	M6	000	-	1	82	9					1	
02/2005	-	M7	000	-	1	38	8						
02/2005	-	M8	000	-	1	27	4						
02/2005	-	M9	000	-	1	41	1					1	1
02/2005	-	M10	000	-	1	9							

	Other Than Lithics					
Totals	**Worked Bones**	**Beads**	**Shells**	**Other Stones**	**Ceramic**	**Observations**
1						
4						
4						
12						
3						
10						
17						
9						one preform for fluted arrowhead (double flute from both extremities, drawn)
10						
9						
4						
34						
38						
33						
32						
11		1				
9						
6						
5						
56						
60						
32						
87						one tool fragment
32						one tool fragment
26						
34						
28						
1						
1						one fragment small end scraper on flake
		1				one-half bead in polished stone
10						one possible proximal-medial of crested blade (?) or accident on biface
53						
91						
62						
8						
85						
78						
31					1	
9						
31						
41						flakes, including big bifacial shaping flakes; very thin, soft percussion
92						
46						
31						
44						one retouched flake
9						

Table 8.7. Manayzah surface and excavation finds inventory. This table compiles three datasets: (1) materials from excavation; (2) surface collection by grid square; (3) general surface collections prior to site gridding—that is, as areas of SU155-002. *(continued)*

(2) Surface Collection by Square						Lithics							
						Products				Tools			Cores
Date	Quad	Square	Locus	Lot	Bag	Chert Flakes	Obsidian Flakes and Bladelets	Channel Flakes	Other	Arrowheads	Bifaces	Misc. Tools	Cores for Flakes
02/2005	-	M15	000	-	1	2							
02/2005	-	N1	000	-	1							1	
02/2005	-	N4	000	-	1	39	4						1
02/2005	-	N5	000	-	1	21	2						
02/2005	-	N6	000	-	1	28	5						
02/2005	-	N7	000	-	1	42	6						
02/2005	-	N8	000	-	1	60	3				1		1
02/2005	-	N9	000	-	1	32							
02/2005	-	N10	000	-	1	20	4						
02/2005	-	N11	000	-	1	34	6						
02/2005	-	N12	000	-	1	63	3	1		1		2	
02/2005	-	N13	000	-	1	25	1						
02/2005	-	N14	000	-	1	13	1						
02/2005	-	N15	000	-	1	1							
02/2005	-	N16	000	-	1							1	
02/2005	-	O4	000	-	1	7							
02/2005	-	O5	000	-	1	25							
02/2005	-	O6	000	-	1	17	3						
02/2005	-	O7	000	-	1	38	1						
02/2005	-	O8	000	-	1	22	4						
02/2005	-	O9	000	-	1	20	6						
02/2005	-	O10	000	-	1	37	2						
02/2005	-	O11	000	-	1	11	2						
02/2005	-	O12	000	-	1	18	1					1	
02/2005	-	O13	000	-	1	16	2						
02/2005	-	O14	000	-	1	13	2					1	1
02/2005	-	O15	000	-	1	5	1						
02/2005	-	P4	000	-	1	3	1						
02/2005	-	P5	000	-	1	18	1						
02/2005	-	P6	000	-	1	19	4						
02/2005	-	P7	000	-	1	11	2						
02/2005	-	P8	000	-	1	18	5						
02/2005	-	P9	000	-	1	47	8						1
02/2005	-	P10	000	-	1	13	2						
02/2005	-	P11	000	-	1	12	1				2		
02/2005	-	P12	000	-	1	10							
02/2005	-	P13	000	-	1	11	1						
02/2005	-	P15	000	-	1	10	2						
02/2005	-	Q4	000	-	1	8	2						
02/2005	-	Q5	000	-	1	16				1			
02/2005	-	Q6	000	-	1	8	3			1			
02/2005	-	Q7	000	-	1	6	6						1
02/2005	-	Q8	000	-	1	9	1	1					
02/2005	-	Q9	000	-	1	3							
02/2005	-	Q10	000	-	1	5	1						
02/2005	-	Q11	000	-	1	25	1						
02/2005	-	Q12	000	-	1	7	2						

	Other Than Lithics					
Totals	Worked Bones	Beads	Shells	Other Stones	Ceramic	Observations
2						
1						one big end scraper on flake
44						one fragment
23						
33						
48						
65						one fragment biface
32						
24						
40						
70						one fragment arrowhead + one retouched flake, one triangles obsidian, one medial of long channel flake
26						
14						
1						
1						one small end scraper on flake
7						
25						
20						
39						
26						
26						
39						
13						
20						
18						
17		2				one retouched flake
6						
4						
19						
23						
13						
23						
56						
15						
15						two fragments biface
10						
12						
12						
10						
17						one fluted arrowhead (from tip, drawn)
12						one preform tang; one obsidian with very strong abraded butt
13						one core fragment or biface?
11					1	one small medial of channel flake + one fragment ceramic
3						
6						
26						
9						

Table 8.7. Manayzah surface and excavation finds inventory. This table compiles three datasets: (1) materials from excavation; (2) surface collection by grid square; (3) general surface collections prior to site gridding—that is, as areas of SU155-002. *(continued)*

(2) Surface Collection by Square						Lithics							
						Products				Tools			Cores
Date	Quad	Square	Locus	Lot	Bag	Chert Flakes	Obsidian Flakes and Bladelets	Channel Flakes	Other	Arrowheads	Bifaces	Misc. Tools	Cores for Flakes
02/2005	-	Q15	000	-	1	4							
02/2005	-	R1	000	-	1						1		
02/2005	-	R6	000	-	1	1	2						
02/2005	-	R7	000	-	1	2							
02/2005	-	R8	000	-	1	2	1						
02/2005	-	R9	000	-	1	2							
02/2005	-	R10	000	-	1	7							
02/2005	-	R11	000	-	1	8							
02/2005	-	R12	000	-	1	4	1						
02/2005	-	R13	000	-	1					1			
02/2005	-	R14	000	-	1	7							
02/2005	-	R15	000	-	1	1							
02/2005	-	S4	000	-	1					1			
02/2005	-	S6	000	-	1						1		
02/2005	-	T11	000	-	1		1						
02/2005	-	W11	000	-	1							1	
02/2005	-	Y6	000	-	1						2		
Totals						**2109**	**258**	**2**	**1**	**7**	**10**	**14**	**9**

(3) Surface Collection by Area (first collection)						Lithics							
						Products				Tools			Cores
Date	Quad	Square	Locus	Lot	Bag	Chert Flakes	Obsidian Flakes and Bladelets	Channel Flakes	Other	Arrowheads	Bifaces	Misc. Tools	Cores for Flakes
2004	-	155-2-1	000	-	1						1		
2004	-	155-2-2	000	-	1	3	2				1	2	
2004	-	155-2-3	000	-	1		3				1	2	1
2004	-	155-2-4	000	-	1	3	15						
2004	-	155-2-6	000	-	1		4				1		
2004	-	155-2-7	000	-	1		1						1
2004	-	155-2-8	000	-	1		2			1			
2004	-	155-2-9	000	-	1	2	5				1		
Totals						**8**	**32**	**0**	**0**	**1**	**5**	**4**	**2**

Synthesis of Inventories	Lithics							
	Products				Tools			Cores
Totals	Chert Flakes	Obsidian Flakes and Bladelets	Channel Flakes	Other	Arrowheads	Bifaces	Misc. Tools	Cores for Flakes
From Excavations (1)	4653	322	19	4	25	25	7	8
Surface Collection by Square (2)	2109	258	2	1	7	10	14	9
Surface Collection b Area (3)	8	32	0	0	1	5	4	2
	6770	**612**	**21**	**5**	**33**	**40**	**25**	**19**

	Other Than Lithics					
Totals	Worked Bones	Beads	Shells	Other Stones	Ceramic	Observations
4						add one fragment tested plaquette
1						one biface preform
3						
2						
3						
2						
7						add one fragment tested plaquette (from wadi bed)
8						
5						
1						one obsidian arrowhead, small and thick
7						
1						
1						one arrowhead with barbs and tang; plano-convex with trihedral tendency; notching of barbs in face; tang off-sided
1						
1						
1						one backed bladelet in obsidian
2						one biface fragment on plaquette; brown-grey chert with cortex remaining of brown-orange color + one other biface fragment
2410	**0**	**4**	**0**	**0**	**2**	

	Other Than Lithics					
Totals	Worked Bones	Beads	Shells	Other Stones	Ceramic	Observations
1						
8						
7						
18						
5						
2						
3						
8						
52	**0**	**0**	**0**	**0**	**0**	

	Other Than Lithics					
Totals	Worked Bones	Beads	Shells	Other Stones	Ceramic	
5063	1	2	8	4	0	
2410	0	4	0	0	2	
52	0	0	0	0	0	**Total Other**
7525	**1**	**6**	**8**	**4**	**2**	**21**

Fluting at Manayzah and Its Remarkable "Extra-Americas" Context

The fluting method is frequently attested at Manayzah. This method, initially described by Donald Crabtree (1966), whose experimental knapping defined its technical aspects, is well attested over a long period in the Americas: in the Arctic region (as evident from paleo-Eskimo points) through to Patagonia, as well as in North America (with Paleo-Indian points like the Folsom and Clovis points dated 11,500/10,500 cal BP). Fluting consists of refining a bifacial piece by removing a long and flat flake (a channel flake). A channel flake scar (the flute) appears along the central axis of a small bifacial piece and originates from one of the biface's extremities. Fluted tools and channel flakes are therefore easily recognizable and thus can be associated with this method. There is no possibility that these pieces could have been created accidentally. Southern Arabia is the first place outside the Americas where archaeologists identify an absolutely incontestable independent invention of this technology.

This method was already identified in Yemen from pieces discovered on the surface of sites (Charpentier 2003; Charpentier and Inizan 2002). All these fluted items found before the Manayzah excavations were correctly interpreted as projectile points and show the removal of the channel flake (the characteristic—and diagnostic—waste product following a fluting "strike") from the apical extremity (the tip). Manayzah is the first discovery in Yemen of fluted pieces made from the basal extremity of the bifacial tool (figure 8.15:3). In a unique example from Manayzah, fluting occurred from both extremities (figure 8.15:1). Some of the fluted tools could be preforms of projectile points, abandoned after a knapping accident (such as a break due to failed percussion; figure 8.15:2) or following a cultural or personal choice, like an insufficiently satisfying bifacial shape). The discovery of a considerable number (21 in total, including 19 in stratigraphic context and 2 on the surface) of channel flake fragments (proximal, medial, distal) points to an in situ production of points with fluting (figure

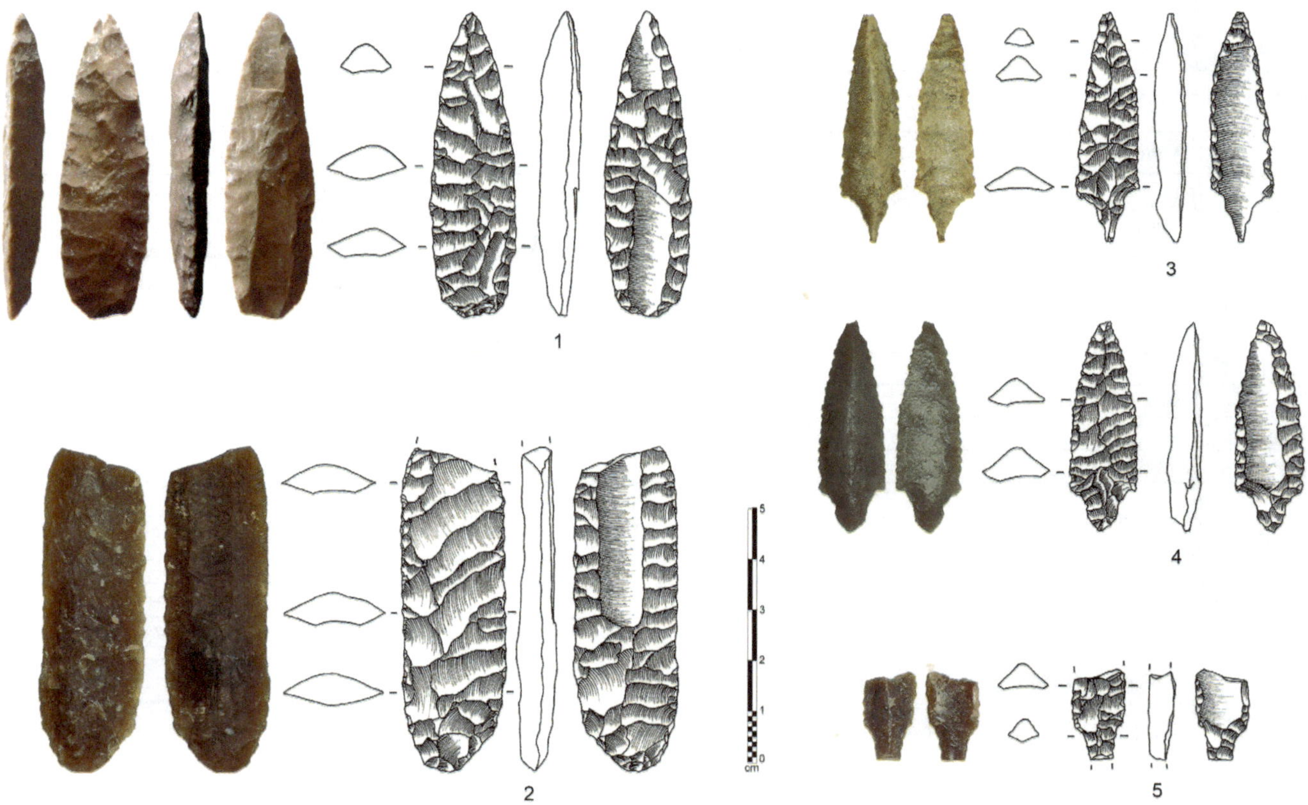

Figure 8.15. Fluted pieces from Manayzah: (1) bifluted piece from I6, Locus 000; (2) fluted piece abandoned after breakage from Quad. C, Locus 009; (3) complete arrowhead fluted from the base from L9, Locus 002; (4) complete arrowhead fluted from the tip from Q5, Locus 000; (5) fragmentary arrowhead from the tip from M8, Locus 001. *Drawings by Julian Espagne. Photographs by Rémy Crassard.*

8.16). Given the difficulty in making a fluted point and the know-how implied by the fluting method and all the different steps leading to a successful flute, the channel flakes indicate the presence of very experienced flintknappers, perhaps even quasi-specialized "master" stone tool manufacturers.

Fluting *Chaine Opératoire* at Manayzah: How and Why

The detachment technique or techniques used in fluting at Manayzah are still difficult to reconstruct. It remains unclear whether fluting was done with pressure flaking or by direct or indirect percussion. Observation of the platforms of proximal parts of channel flakes indicates frequent *en éperon* preparation, and the bulb is particularly well outlined and marked. This tends to favor a hypothesis of pressure flaking. The systematic softening (strong abrading) of the platforms, sometimes to the point of polishing, strengthens this interpretation. Such preparations were used to avoid shattering the striking platform during extraction using this technique. Plausibly, some knappers thermally treated some bifacial tools before fluting. Thermal treatment would be a precious aid in extracting elongated channel flakes. The presence of a shiny surface detected on channel flakes suggests as much, but we cannot clearly confirm thermal treatment without microscopic analyses. (It is currently impossible to export the archaeological material.) Despite these observations, it is still difficult to ascertain the technique employed for fluting; direct percussion (particularly with an organic and/or mineral soft hammer) and indirect percussion should remain viable hypothetical manufacturing techniques tested in future analysis, especially those with experiments and reproduction.

There are three main stages of the operational scheme of fluting at Manayzah (figure 8.17).

First, knappers chose raw materials (strictly local) focused on naturally globular chert blocks (cobbles at the bottom of wadi beds), but the use of fine tabular flint, whose source lies at the top of Hadramawt's plateaus, is also proven. Nothing at Manayzah or in the rest of Yemen shows obsidian use in fluting operations. A small bifacial tool with an asymmetrical biconvex section was manufactured in a good-quality, fine-grained material. The knapper used a soft hammer percussion technique for retouching and pressure retouch, often *en écharpe*. Thus produced, the bifacial piece can have an almost trihedral cross section, and most complete fluted projectile points from Manayzah show a trihedral cross section with a flatter lower face and a medial ridge along the upper face.

Second, knappers carried out fluting from one of the bifacial piece's ends, along the axis of its lower face. The knapper had to undertake substantial effort to soften the base to avoid crushing it with the subsequent use of an indenting tool or hammer. In all likelihood, the extraction technique occurred by applying pressure, but percussion is also plausible. Examples of fluted projectile points from the apical (the future tip) part are more numerous, with variability at this stage. Fluting can take place from the point (or future point), from the base, or from both extremities of an oval or almond-shaped bifacial piece. The knapper determines where to originate a flute. His or her interest may be in the function of the fluted form or in demonstrating a technical ability to achieve a channel along the entire bifacial piece.

Third, the fluted bifacial piece was retouched (bifacially shaped by pressure) to obtain a plano-convex cross section with a trihedral tendency or a strictly trihedral cross section. Retouch also introduces a slight tang. As the point emerges in this finishing or refining stage, the fluted face is not retouched or is only slightly retouched.

To conclude, there is relatively low variability noticeable in the *chaine opératoire*. One variant is that a bifacial piece used for fluting is not necessarily already pointed. Pointing can take place during the third stage during the final retouching of the piece.

Fluting *Chaine Opératoire* at Manayzah: A Bifluted Point, Indicating an Attempt at Extended Channel Flaking

Discovered on the site's surface (Square I6; figure 8.15:1), one example of a bifacial point offers particularly valuable evidence for reconstructing the *chaine opératoire* for fluted tools described above. This item has a plano-convex cross section and unusual fluting from both extremities of the lower face. Within the context of a rich cohort of fluted pieces from Manayzah, it especially clarifies the knapper's aims, and an extended contextual analysis below interprets this point as the preform for a fluted arrowhead:

First, the stone toolmaker manufactured a small, almond-shaped bifacial tool with a biconvex, asymmetrical cross section. He or she first used a semi-abrupt retouch on the upper face, then a shallow retouch on the lower face, always employing first a soft percussion tool, then applying pressure.

The flintknapper then proceeded to fluting, starting with the lower face and the apical extremity. The flute extraction technique remains unclear. If by pressure, then the knapper used an indenter—probably a shoulder

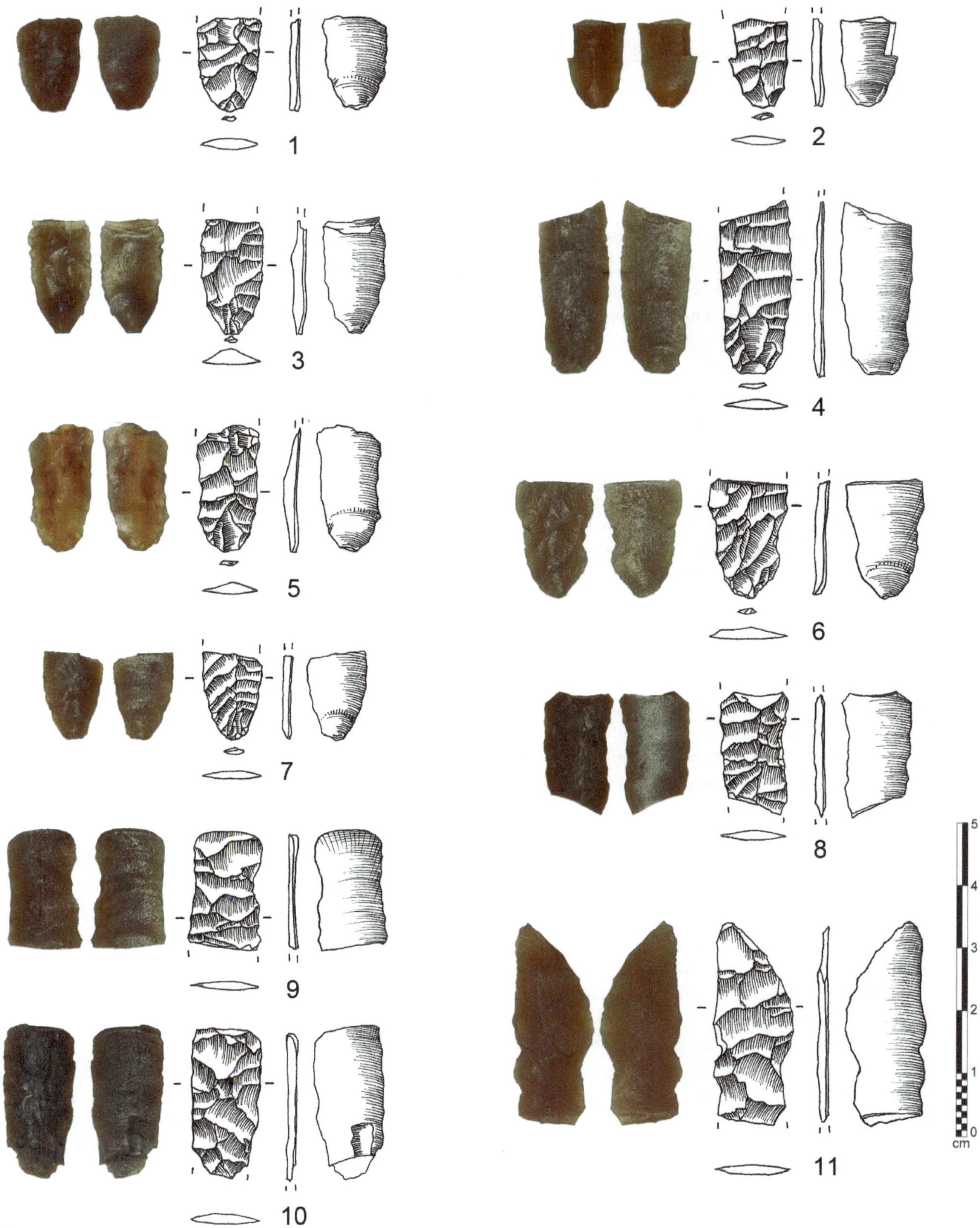

Figure 8.16. Fragmentary channel flakes from Manayzah: (1) proximal from L9, Locus 011; (2) proximal from L9, Locus 011; (3) proximal from Quad A, Locus 005; (4) proximal from Quad B, Locus 001; (5) proximal from M8, Locus 002; (6) proximal from M8, Locus 001; (7) proximal from M8, Locus 010; (8) medial from M8, Locus 002; (9) distal from L12, Locus 001; (10) distal from Quad B, Locus 001; (11) distal from L9, Locus 001. *Drawings by Julien Espagne. Photographs by Rémy Crassard.*

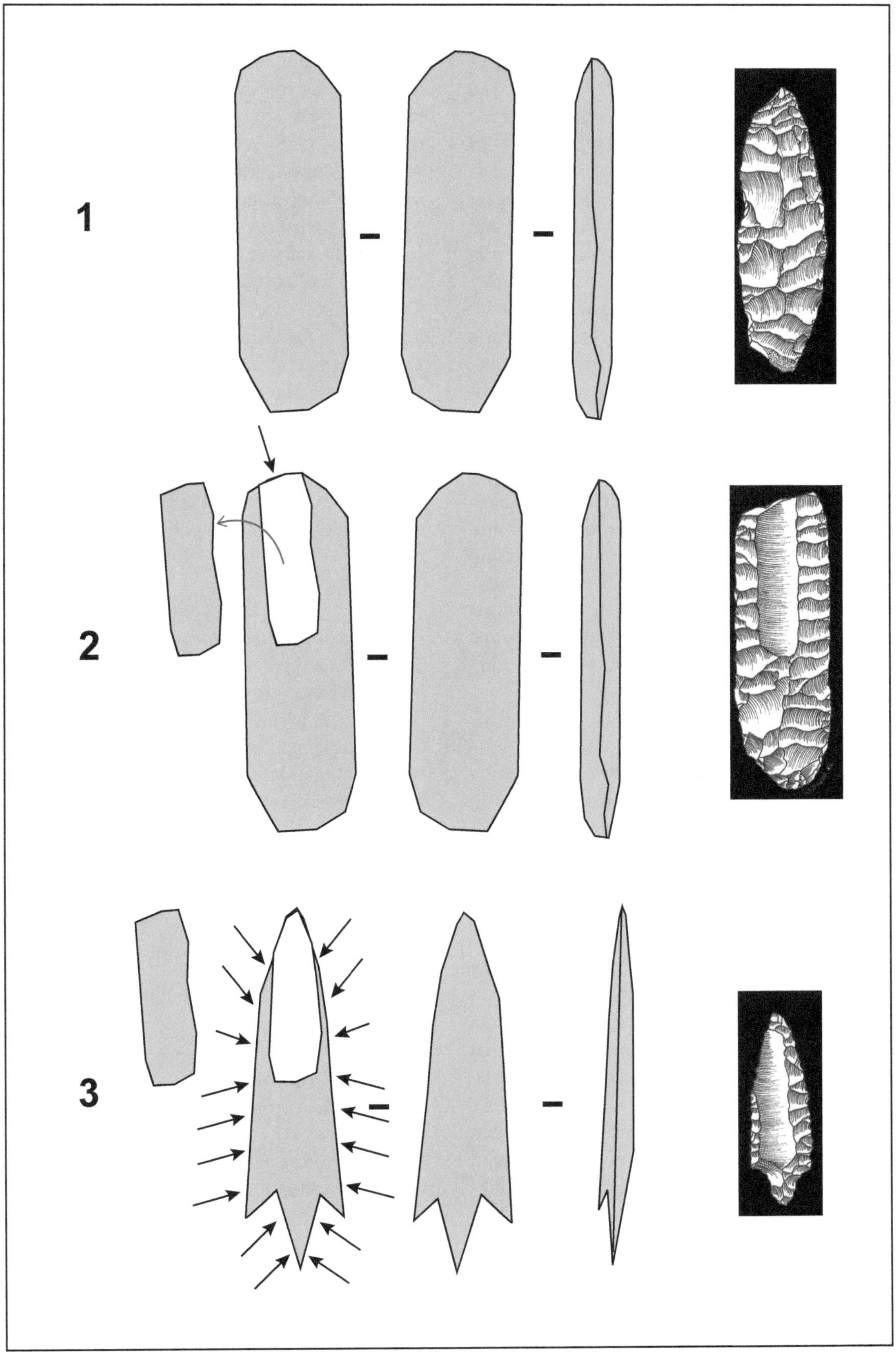

Figure 8.17. Three main stages of the operational scheme/*chaine opératoire* of fluting at Manayzah. *Illustration by Rémy Crassard.*

crutch, allowing one to obtain more substantial pressure force than that provided by a simple hand. This first operation of fluting failed to satisfy the knapper, whose mental schema clearly demanded a longer fluted channel. The flute channel ended in a step fracture. Because his or her goal was a complete channel flaking along a fair part of the piece's length, the toolmaker decided to reproduce fluting from the base of the small bifacial piece, again on the lower face. The same failure occurred again, and the toolmaker abandoned what we interpret as the preliminary form of a plano-convex fluted arrowhead. Our hypothesis of preform discard is also based on characteristics of fluted projectile points, which usually show a tang and are less thick. This small, bifluted bifacial item was not retouched after fluting. Therefore it seems logical to classify it as a discarded item.

This piece indicates that the toolmaker's preliminary aim was to obtain fluting along the entire surface of the arrowhead, and he or she therefore attempted to have enough channel flaking length. Either the tool's functionality or a specific techno-cultural constraint can explain why this was the knapper's aim. Even without clarifying the reasons for fluting, this particular tool—with its reconstructed *chaine opératoire* and its final unattained goal—offers important insight into the general fluting *chaine opératoire* in Southern Arabia.

Why Fluting in Southern Arabia?

The aim and relevance of fluting in this Yemeni setting warrants explanation. If fluting marks improvements to the point's functions, how did these improvements happen? The occurrence of fluting in North America has suggested many interpretations (e.g., Whittaker 1994): it made hafting easier, it demonstrated the toolmaker's skill, or it introduced a "blood groove" that would have made hunted prey bleed faster and tire more quickly, reducing the chase. These interpretations have had varying popularity; that favoring hafting is now preferred. Unlike the majority of those from Southern Arabia, American fluted points systematically show fluting beginning from the tool's base, which does favor a functional interpretation.

In the case of fluting at Manayzah, it is clear that such points were used as projectile points. The fluting technique itself makes the weapon lighter and offers significant kinetic advantages. We have demonstrated that toolmakers sought to obtain fluting removals of certain, preconceived length. They evidently abandoned tools whose fluting length was unsatisfactory, even in the case of a preformed tool otherwise perfectly suited for final retouch and use.

In most cases, the fluting operation originated from the projectile point's apical extremity. In such cases, the fluted area cannot serve to facilitate hafting. A fluted area located at the top of projectile points cannot really be explained from a functional viewpoint, calling into question functionality in hafting as an aim for all Southern Arabian fluting. Perhaps one should rather consider the action of fluting as proof of the aptitude or technical skill of highly specialized knapping "masters." Indeed, the preparation and action of fluting requires considerable practical and abstract cultural knowledge. Moreover, fluting risks wasting fastidious preliminary work expended in the manufacture of a complex bifacial piece. Removing one or several channel flakes not absolutely necessary for the hafting process could easily break an item previously obtained with much effort. And archaeologists find preforms that broke after or during the channel flake removal. In Southern Arabia, fluting may indicate highly specialized knappers' choices determined by a particularly salient cultural or stylistic template.

Obsidian Industries

Above the K9 Layer 20/Layer 21 stratigraphic interface, obsidian is ubiquitous in each stratified layer. The interface marks a typological and technological disruption in lithics. Even the very nature of the material used by knappers changes at this stratigraphic juncture. At Manayzah a greenish-black obsidian of unidentified source was worked alongside various types of chert. Obsidian flaking used percussion, and retouched tools are rare. (Only two such obsidian tools have been discovered, apart from projectile points.) In South Arabia, Manayzah is one of the only archaeological examples in the Early to Middle Holocene where the remains of obsidian knapping occur in stratigraphic context. Another Hadramawt example is the HDOR 419 site, with stratigraphy yielding some obsidian remains (Crassard 2008; see also Amirkhanov 1994 for other occurrences).

Obsidian Projectile Points

Three obsidian projectile points were found at Manayzah (Grid Square L9, Locus 002; Grid Square L8, Locus 001; Grid Square R13, surface; figure 8.10:11–12). They were manufactured on a flaked blank—either a flake or a bladelet. The first arrowhead reflects the use of a laminar blank, with a slightly curved profile, while the second was manufactured on a thick flake. The arrowhead thus obtained is plano-convex. Obsidian exhibited no example of fluting, whether on projectile points, on bifacial pieces, or as channel flakes.

Obsidian Bladelets

Manayzah's knappers worked obsidian to produce bladelets and flakes (figure 8.18). Even with problematic interpretation of some platforms and bulbs on bladelets, there is no obvious evidence of *débitage* using the pressure technique. As with the procurement of channel flakes, obsidian bladelets systematically show platforms that were subjected to substantial softening (through strong abrading). The bulbs of bladelets are also frequently prominent, yet the edges of blanks are insufficiently regular to conclude that pressure flaking normally occurred. Moreover, the negatives of previous removals do not show regular parallel ridges, a characteristic that also prevents an interpretation of pressure-flaked, standardized bladelets. This *chaine opératoire* of precision blade production would yield a characteristic obsidian core, of which no example was found. Instead, the sole obsidian core recovered (figure 8.18:8) in all likelihood did yield bladelets; but this example shows none of the flake scars reflecting the regular and standardized removals usually associated with cores for so-called classic bladelets, such as, for instance, "bullet-cores" (e.g., Wilke 1996). The Manayzah obsidian core reveals unidirectional flaking of lengthened and non-standardized flakes. Manayzah's obsidian bladelets were likely obtained by percussion.

The intended use of bladelet production is still unknown. Only two backed (or used on one edge) micro-bladelets (figure 8.18:10–11) and a triangle (figure 8.18:9) are tools processed from bladelets. From Square N12, this triangle is a piece made of a small obsidian flake or of a bladelet fragment (with a possible truncation). It has the shape of an elongated isosceles triangle, with obverse and very short retouch along both main edges and also along one-half of the base (a possible partial retouch on truncation).

Obsidian in Stratigraphy

The remains of obsidian represent a little more than 8 percent of the recovered lithics by count (614 flakes, bladelets, and obsidian tools from a total of 7,725 pieces; table 8.7). The average size of obsidian flakes and bladelets is generally around 0.5–2.0 cm, which suggests that raw materials were not plentiful. By this calculation, obsidian was the rarest of raw materials at the site, almost certainly because obsidian rocks came from a source outside Wādī Sanā.

Within the stratigraphic superimposition of assemblages, the presence and absence of obsidian demonstrates changes in the material supply through time. From the surface to K9 Layer 16, one finds different assemblages left by the production of fine and elongated obsidian blanks. In K9 Layer 16 and lower, older layers, obsidian disappears from lithic assemblages. Not only is obsidian absent, but the older layers exclusively contain chert of lesser quality than in the upper layers. These differences indicate different raw material supplies associated with different production strategies. An occupation of the site by populations with less elaborate technical traditions and alternate raw materials preceded Manayzah's upper occupation. While the upper technologies are typical of populations of the Early to Middle Holocene, one cannot determine with current evidence whether the earliest occupation is pre-Holocene.

To summarize, obsidian at Manayzah was specially employed in the production of blanks (flakes and bladelets). Prehistoric toolmakers used obsidian differently from chert, for chert and obsidian clearly differed in technological and functional purposes.

Projectile Points

Mainly pressure-retouched, chert and obsidian projectile points are present in a great variety of forms. Thirty-two forms have been registered—twenty-five in the stratigraphy and seven on the surface of the site. Because excavations at Manayzah are unfinished, typological studies are not definitive at this time. Nonetheless, some preliminary observations can be suggested.

Four types of arrowhead were defined at Manayzah:

- Manayzah Type 1 (figure 8.19). This type seems to be one of the most recent; (rare) examples occurred in the surface level or directly beneath it. These examples are tanged points on a flaked blank. The upper face underwent general retouch, and the tang was subjected to bifacial retouch.
- Manayzah Type 2 (figure 8.19). A bifacial point—rarely trifacial at the apical extremity—with a trihedral section and with a slightly off-sided and not very salient tang.
- Manayzah Type 3 (figure 8.19). A bifacial tanged point, trihedral in section at the apical extremity and mostly with a plano-convex section ("plano-convex with a trihedral tendency") in the basal-medial part. Some very small barbs, sometimes appearing as small spurs, were clearly manufactured. A very slight off-side of the tang can occur (Manayzah Type 3A; figure 8.25A). A subtype was isolated when a fluting removal was made in the lower face (Manayzah Type 3B; figure 8.25B), from the apical or basal part or from both extremities.
- Manayzah Type 4 (figure 8.19). A foliated bifacial point with a symmetrical or subsymmetrical section, void of any barb or tang, with *en écharpe* retouch. Examples belonging to this type are very infrequent.

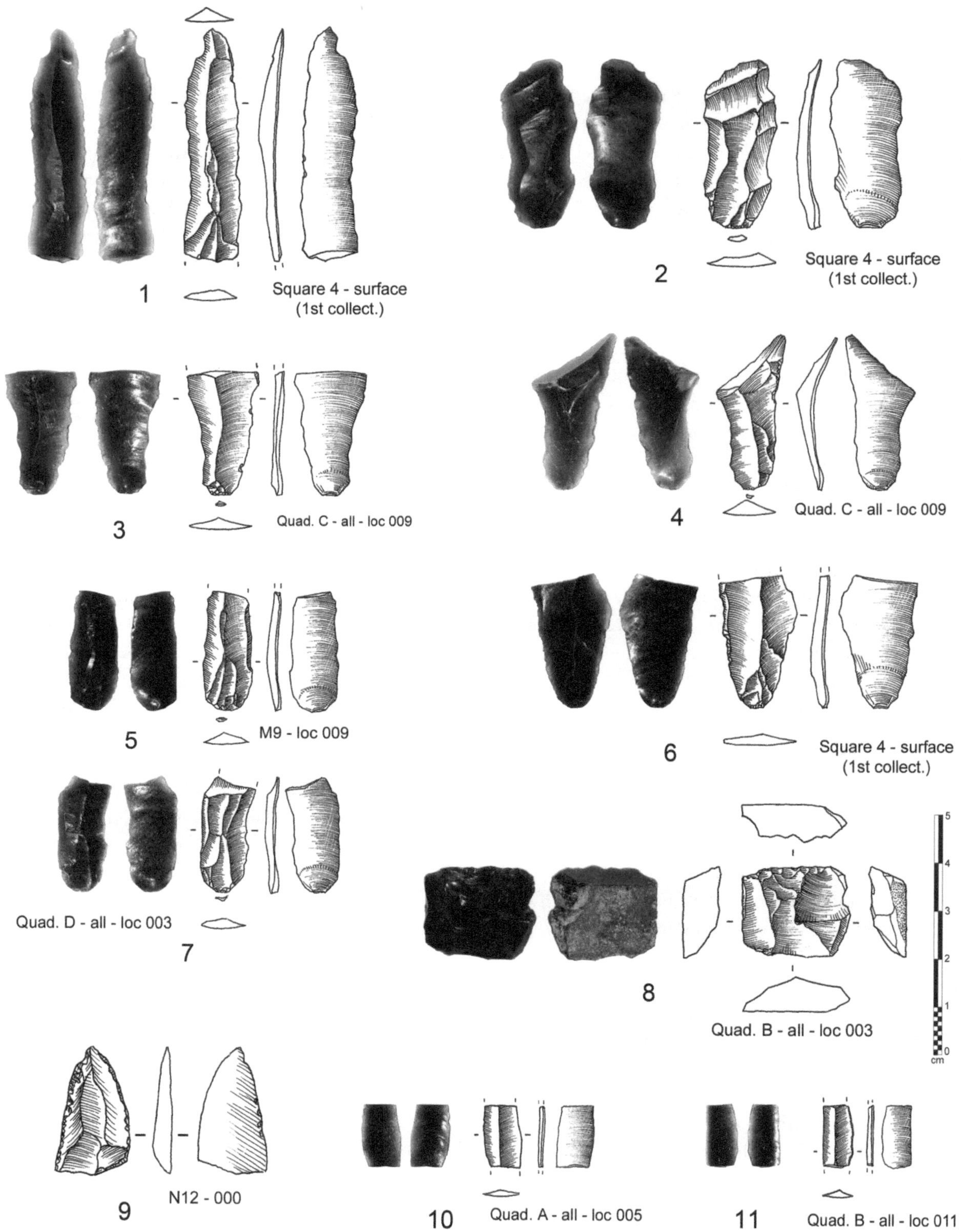

Figure 8.18. Obsidian pieces from Manayzah: (1) medio-distal of long bladelet from Square 4 surface (first collection); (2) flake from Square 4 surface (first collection); (3) proximo-medial of elongated flake from Quad C, Locus 009; (4) flake from Quad C, Locus 009; (5) proximo-medial of bladelet from M9, Locus 009; (6) proximo-medial of flake from Square 4 surface (first collection); (7) proximo-medial of bladelet from Quad D, Locus 003; (8) small bidirectional core from Quad B, Locus 003; (9) retouched triangle from N12, Locus 000; (10) medial of bladelet with traces of use on one edge from Quad A, Locus 005; (11) medial of bladelet with traces of use on one edge from Quad B, Locus 011. *Drawings by Julien Espagne and Rémy Crassard. Photographs by Rémy Crassard*

Manayzah Type 1 is seldom present; it appears only with obsidian projectile points exclusively manufactured on a flaked blank (figure 8.10:11–12) and a single instance occurred of an arrowhead on a flaked chert blank. This chert example was achieved by making a micro-point with very slightly salient barbs and tang (figure 8.10:9).

Points with a trihedral section (the trihedral points) are numerous at Manayzah (figure 8.10:1–2). A first type (Manayzah 2) can be distinguished by the regularity of the trihedral section all along the length of the piece. A tang is most often only slightly shaped, showing a slight off-siding.

Another type (Manayzah 3) differs from the previous one by the delimiting of very small barbs and sometimes even small spurs (figure 8.10:3, 5, 8). The section of the piece is once more trihedral, with often a tendency to be plano-convex in its basal-medial part. There is a tang at times showing a slight off-siding. Within this type, the fluted points (Manayzah 3B) represent a subtype. Many occurrences of fluted points show the same characteristics of Type 3, with an additional removal of a channel flake on the ventral face, the plane face, or the least convex face (Figure 8.15:3–5).

Finally, a last type (Manayzah 4) is characterized by a foliated form. These bifacial points show a symmetrical or subsymmetrical section (figure 8.10:10).

Other Retouched Tools

Apart from finished projectile points, tools number 65. The bifacial industry is the most prominent technology in the tool assemblage: 40 complete or fragmentary bifacial pieces, at different stages of their manufacture, were discovered (figure 8.20). Expertly worked in brown-beige jasper, a dagger is the most technically accomplished biface example (figure 8.9). Using pressure, particularly intricate and careful retouch of this piece was carried out by long *en écharpe* removals. Almost all remaining bifacial pieces probably belong within a set of rough outlines of projectile points. Some examples were clearly bifacial pieces not intended for further shaping as projectile points, and other pieces were preforms for projectile points—fluted bifacial pieces clearly belong to the latter.

Two fragments of polished axes were also present in the assemblage on the site's surface (figure 8.21). Due to their surface context, it is impossible to assume these fragmentary tools are coeval with other surface finds, such as fluted points. The first fragment is part of the cutting edge of an axe, manufactured in a greenish rock. The second fragment is the middle part of an axe in grey stone, perhaps chert. TThis discovery is one of the rare examples from a secure archaeological context in Yemen, even as the two fragments of Manayzah are surface finds. The polishing of axes is a little-known technological phenomenon in Yemen (for example, in the western highlands in Wadi Dhar; Kallweit 1996:201–3, plates 13–15; in the Ramlah as-Sabʿatayn at the site of al-Hawa; Inizan et al. 1998:143. See also examples from Oman at the RH-6 site [Biagi 1999] and on the island of Maṣīrah [Oman]; Charpentier et al. 2012). Polished ax technology remains undated. Examples of this shaping technique occur at surface sites attributed to the Early to Middle Holocene. Manayzah's axes seem to confirm that preliminary chronological attribution.

Both on the surface and in the earliest layers of the stratigraphy, excavators recovered small scrapers (figure 8.22). These tools are thumbnail end-scrapers, showing very regular, obverse, and abrupt retouch, finely done on the edge. The techno-cultural relevance of these pieces remains to be confirmed, but the context of a few similar examples discovered outside Manayzah tends to suggest that these scrapers have an important typo-chronological value (see, for instance, a scraper found in HDOR 561; Crassard 2008). They seem to belong to a specific technical and stylistic tradition, as similar scrapers were found at al-Quwīd (Cleuziou et al. 1992:figure 5/8–9), at the Sa'ada region at Jabal al-Makrūg (Inizan and Rachad 2007:68–69, figures 33–34), and in the Wādī Dahr (Kallweit 1996:203–5, plates 15–17), as well as at Mundafan (southwest Saudi Arabia; Crassard et al. 2013).

Lithic Industries Prior to K9 Layers 19 and 20

In the lowest excavation of a 0.5 x 1.0 m area of K9, excavators noted a clear stratigraphic interface between Layers 19 and 20. While Layer 19 is a layer of a grey-brown silty sand with small calcite nodules and tiny gravels, lower Layer 20 (below Layer 19) is an indurated, light brown, silty sand layer showing numerous thin, flat limestone clasts that are horizontally lying on the top of the layer. If confirmed as a site-wide phenomenon, this interface could be the result of a hiatus in the perennial character of occupation at Manayzah. Or it might mark an environmental change that altered the nature of sedimentation. Whatever the explanation for its formation, this deep stratigraphic entity seems to represent a crucial transition—even a *disruption*—in the site's history. Furthermore, analysts noted a drastic change in technical traditions between K9 Layers 19 and 20. The use of various different raw materials is also clear.

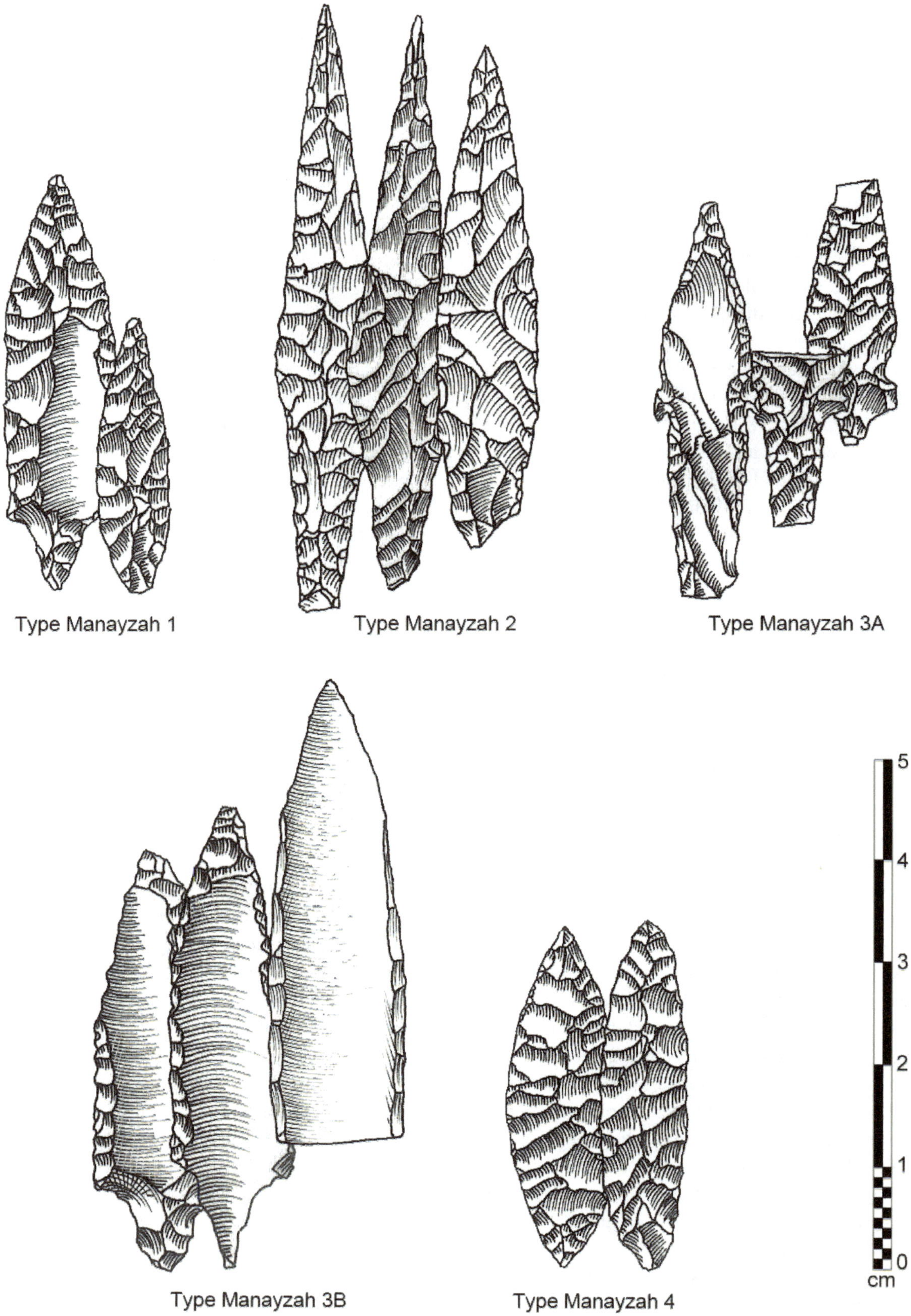

Figure 8.19. Manayzah types. *Drawings by Julian Espagne. Illustration by Rémy Crassard.*

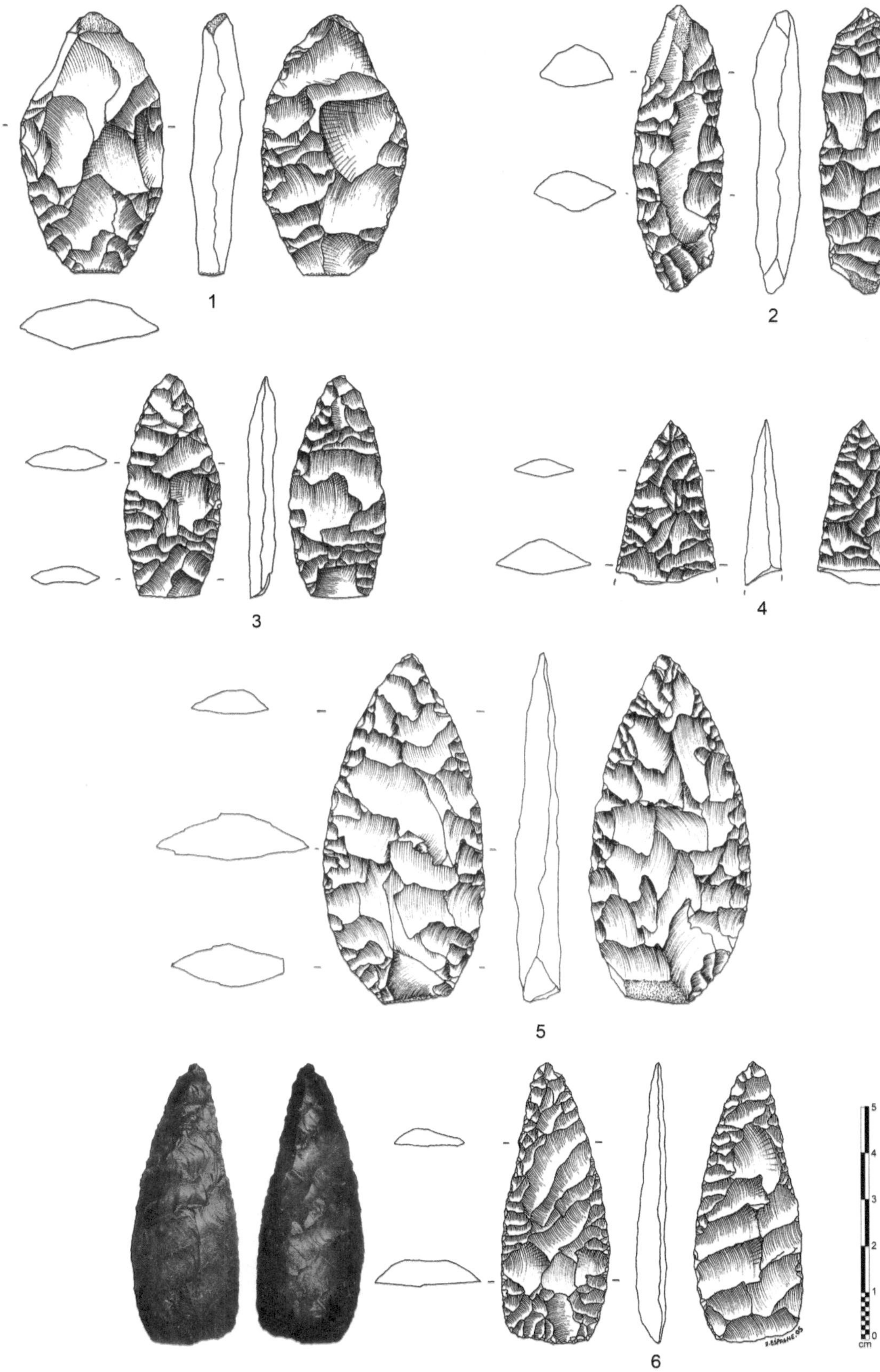

Figure 8.20. Bifacial pieces from Manayzah: (1) from Quad A, Locus 004; (2) from L10/L11, Locus 009; (3) from Quad B, Locus 008; (4) from L10, Locus 010; (5) from H9, Locus 000; (6) from L10, Locus 003. *Photograph by Rémy Crassard. Drawing by Julien Espagne.*

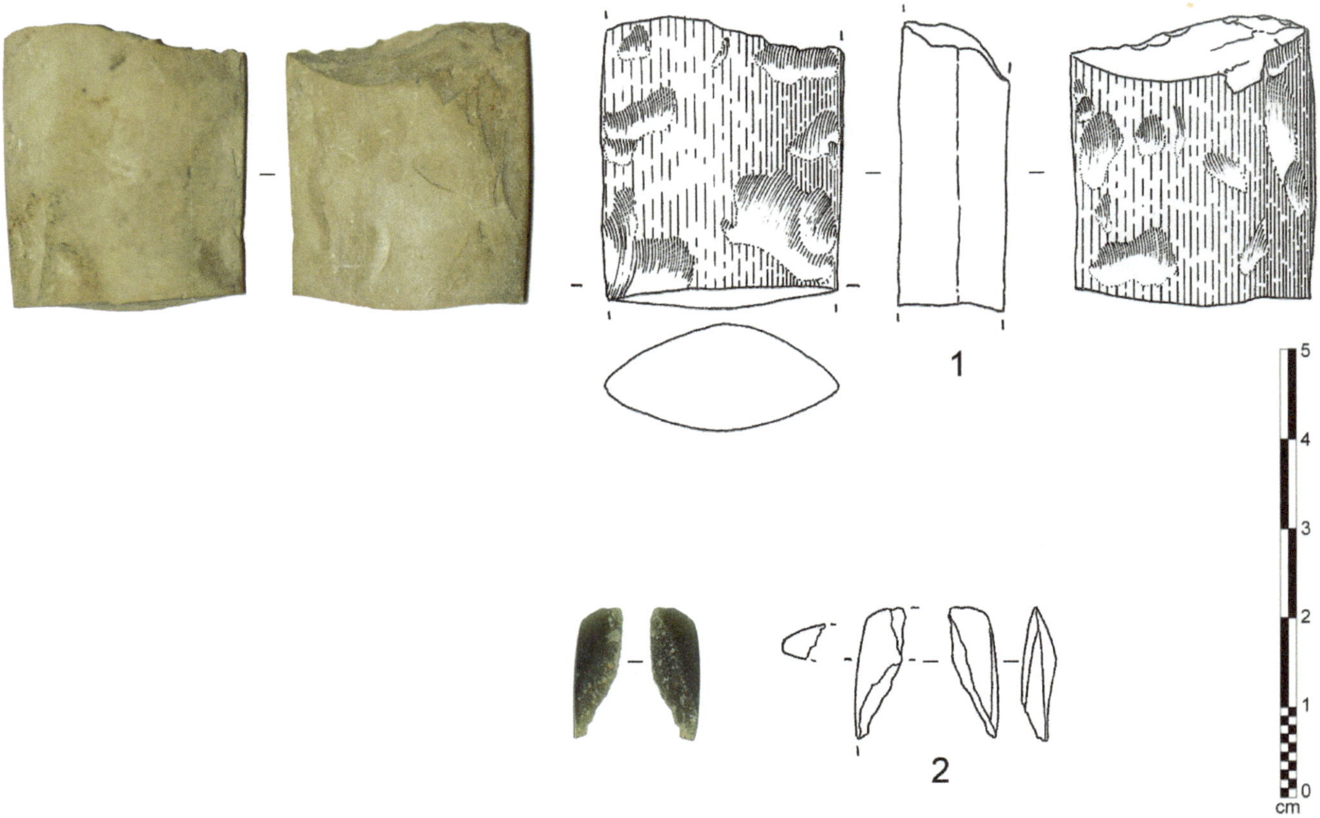

Figure 8.21. Polished axes from Manayzah: (1) distal part made of grey stone from R15, Locus 000; (2) edge fragment made of greenish jaspoid stone from I5, Locus 000. *Drawing by Julien Espagne. Photographs by Rémy Crassard.*

Initial observations on the corpus of lithics below K9 Layer 19 indicate that the oldest occupants used knapping techniques very distinct from those of their successors, whose already-described assemblages occur in overlying deposits. The older techniques are expressed in a very different lithic assemblage than that of recent archaeological layers. Most older pieces are longer, probably because the toolmakers sought to produce flaked blanks (*débitage* as elongated flakes) rather than bifacially shaped tools (*façonnage*, as seen in later layers). Moreover, a core with recurrent centripetal flaking (figure 8.23:2) attests a production of flakes following a pattern absent during the Holocene in Yemen. Found in one of the oldest known layers at Manayzah (K9 Layer 22), this core recalls a mode of *débitage* akin to the Levallois technology. Differences between this core and those known in the Middle Paleolithic in the Wādī Washaʿah and Wādī Sanā lie first in much smaller dimensions than those usually found in Hadramawt. Next, the knapper's management of the Levallois surface employs less concern with convexity. These oldest layers are not chronometrically dated. The presence of this core in the stratigraphy need not imply a chronological Middle Paleolithic occupation at Manayzah.

The raw material also differed significantly from materials encountered in the upper layers. In addition to cherts, coarser-grained rocks were knapped and worked; this range of material is absent from more recent layers. In the lowest layers, the siliceous content of chert has leached, and the stone is therefore very light in weight. Desilicified flakes all show a white or grey patina. The latter reflects erosive action due to water percolation. Finally, the use of obsidian is completely absent from layers located under the K9 Layer 19/Layer 20 interface, an absence that implies a technological break in the assemblages deposited above and below this stratigraphic interface.

Albeit from a limited sounding, this excavated older lithic assemblage heralds an important stage in regional prehistoric research. It could mark the discovery of industries of the Early Holocene or even the Late Pleistocene in well-stratified context. The resumption of excavations

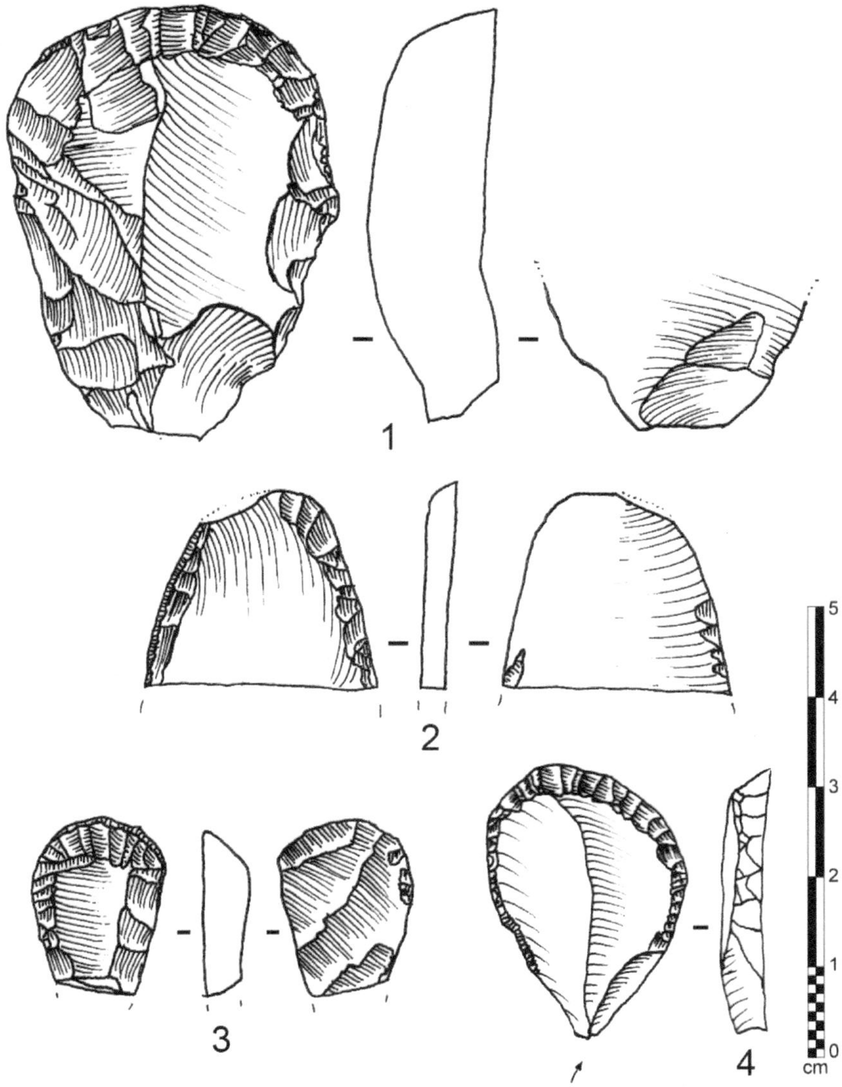

Figure 8.22. Scrapers from Manayzah: (1) large semicircular scraper on thick flake from N1, Locus 000; (2) fragmentary thin scraper from L9, Locus 002; (3) small thumbnail scraper on bifacial blank from N16, Locus 000; (4) semicircular end scraper on flake from Square 2, surface (first collection). *Drawing by Rémy Crassard.*

with an emphasis on chronometric dating of clearly associated organic materials will allow archaeologists more clearly to answer the new questions raised by the presence of these distinct industries. From the Manayzah stratigraphy, their relative chronological status is clearly antecedent to the well-known Early to Middle Holocene industries.

Results: The Chronological Framework of Lithic Industries

A Bayesian analysis of the radiocarbon ages from Manayzah and other sites provides a chronological framework for lithic industries and other aspects of Manayzah occupation (chapter 18). The earliest dated event related to Manayzah occupation begins in the (2-sigma posterior distribution) range of 9550–8404 cal BP. The end of radiocarbon-dated occupation is in the (2-sigma) range of 7879–7360 cal BP. The dated events suggest a duration of occupation between 525 and 2,190 years (2-sigma range), with 1,288 years between the distribution means (figure 8.24, table 8.2). Remember that excavators obtained radiocarbon ages from sediments near the surface, making the end of dated occupation close to the real end of occupation (as preserved). By contrast, the earliest deposits at Manayzah have no associated radiocarbon ages; occupation began earlier than the earliest dated event with a radiocarbon age estimate.

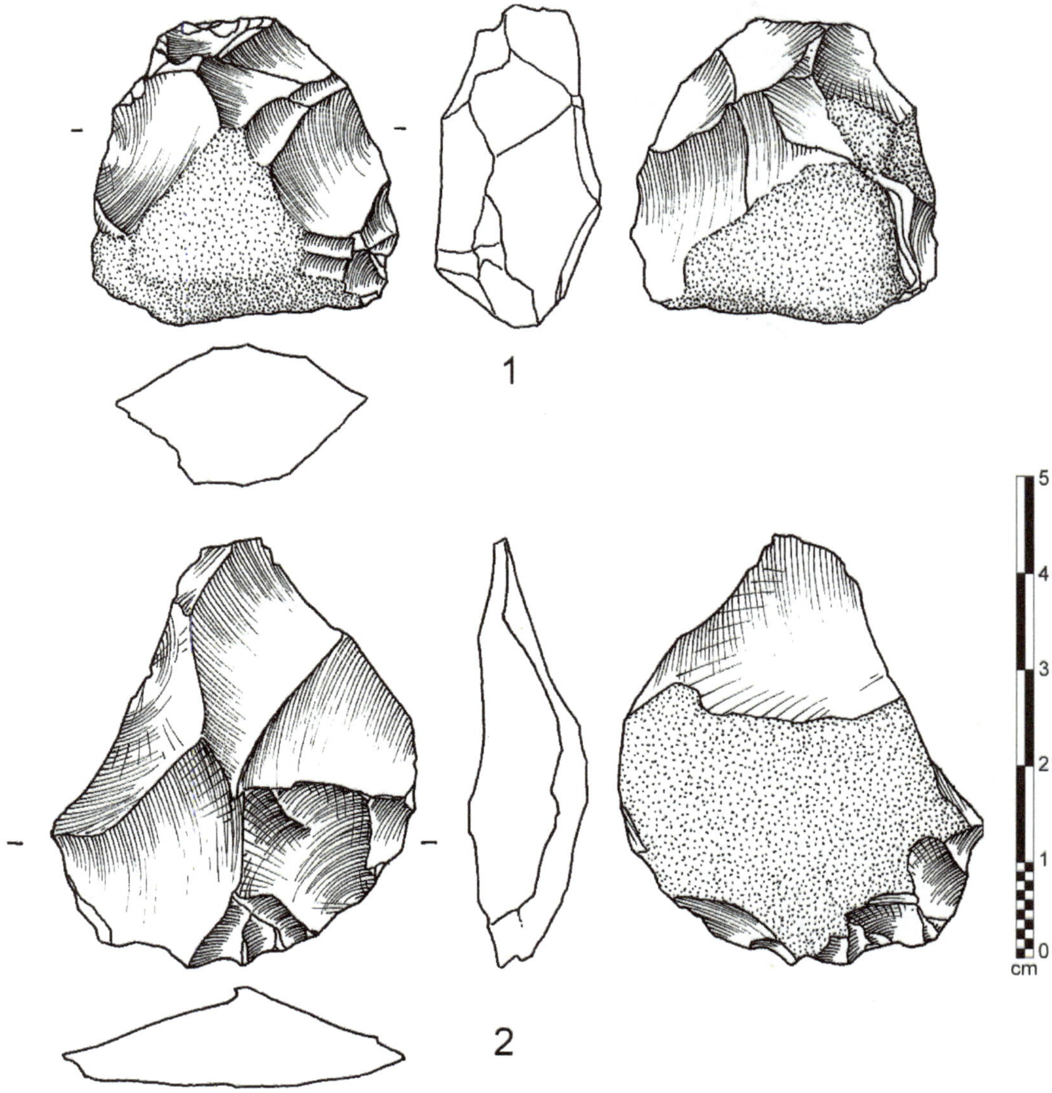

Figure 8.23. Cores from Manayzah's lower layers: (1) multidirectional core for flakes from K9, Locus 021; (2) recurrent centripetal core from K9, Locus 022. *Drawing by Julien Espagne.*

Complete and fragmentary fluted points emerged from stratigraphic context, allowing excavators to suggest a relative and an absolute date for fluting. Channel flakes occur in strata between K9 Layer 11 and the surface. Fluted elements appear in layers located above—and therefore more recent than—a combustion structure (K9 Layer 9, Hearth 1). From this combustion structure, *Ziziphus leucoderma* charcoal fragments yielded a radiocarbon age (AA59570) of 6902 ± 41 years BP (7834–7665 cal BP). In the stratigraphy, channel flakes are present in layers that are older by one or two centuries (but not more, according to the dating of K9 Layer 17, AA66685, 8035–7845 cal BP). The fluting method can therefore be attributed to the first quarter, or more broadly to the first half of the eighth millennium cal BP (sixth millennium BCE). Surface finds make it impossible to exclude that this technical tradition lasted until more recent times.

The beginnings of fluting are now well dated. Moreover, tools such as the probable bifacial preforms of fluted points and the associated typo-technological points were manufactured according to this technical process and therefore belong to the same chronological framework. The projectile points closely associated with fluted points are trihedral points (of Manayzah Types 2, 3A, and 3B). They appear to be contemporary or older than the use of the fluting method. Within the same radiocarbon and

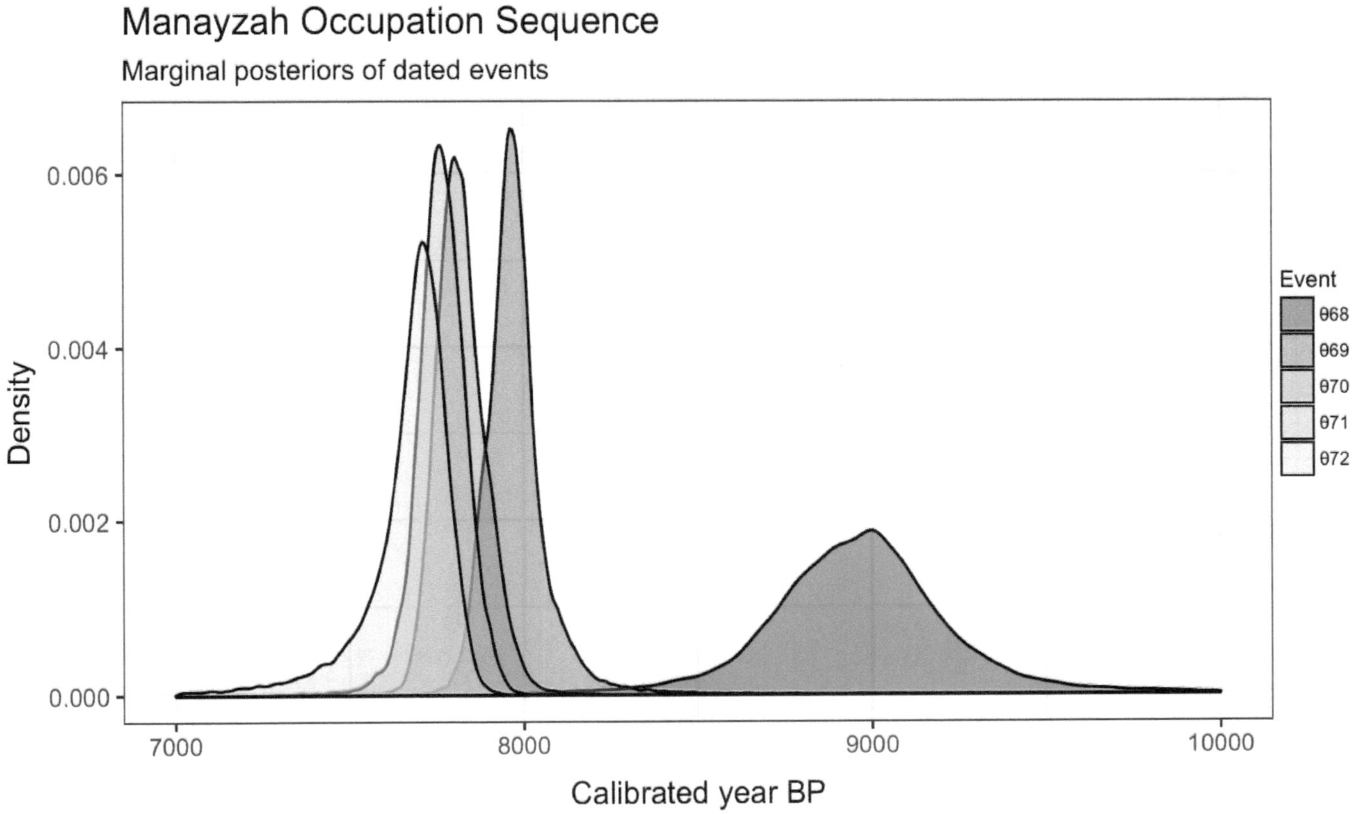

Figure 8.24. Bayesian distribution of posterior distributions at Manayzah. *Illustration by Thomas Dye.*

stratigraphic constraints, they can be related to the end of the ninth millennium BP and the first half of the eighth millennium cal BP. Some bifacial projectile points possess a tang whose section is slightly curved, here described as slightly off-sided. This typological peculiarity could be typical of the end of the ninth millennium cal BP, since one finds them in association with trihedral points in the same layers and older than the use of the fluting technique. These two types of projectile points were noticed among the surface material collected during the previous campaigns in Hadramawt, particularly at the Gravel Bar Site (McCorriston et al. 2002) and also at the stratified site of Khuzmum SU45-1A (chapter 9).

The two other projectile points known at Manayzah (Manayzah Types 1A and 1B, and Manayzah Type 4) apparently coexisted with the previous types in the recent phases of the stratigraphy—that is, until the early eighth millennium cal BP. If one considers that surface material has been only slightly disturbed because it is the same material as that found in the first layers of the stratigraphy, one can therefore—albeit cautiously—date the ax-polishing technique to the first half of the eighth millennium cal BP. A similar argument applies to the small thumbnail-shaped end-scrapers with a carefully manufactured edge.

The dates of layers located under the K9 Layer 19/Layer 20 interface are earlier, around the end of the tenth millennium BP, yet this seems still quite late for an industry of the Levallois tradition. Because of limited excavated exposure, the data are few from the bottom of Manayzah's stratigraphy. Whether in terms of lithic industries or in the knowledge of the site's full stratigraphic depth, future excavation promises greater resolution; our date for the site's earliest occupations is still a hypothesis. Despite the gaps in our information, Manayzah undoubtedly remains a key site and a unique chrono-stratigraphic sequence for understanding the broader regional chronology of Yemen. Figure 8.25 displays the chrono-typology suggested for the projectile points and points found at Manayzah.

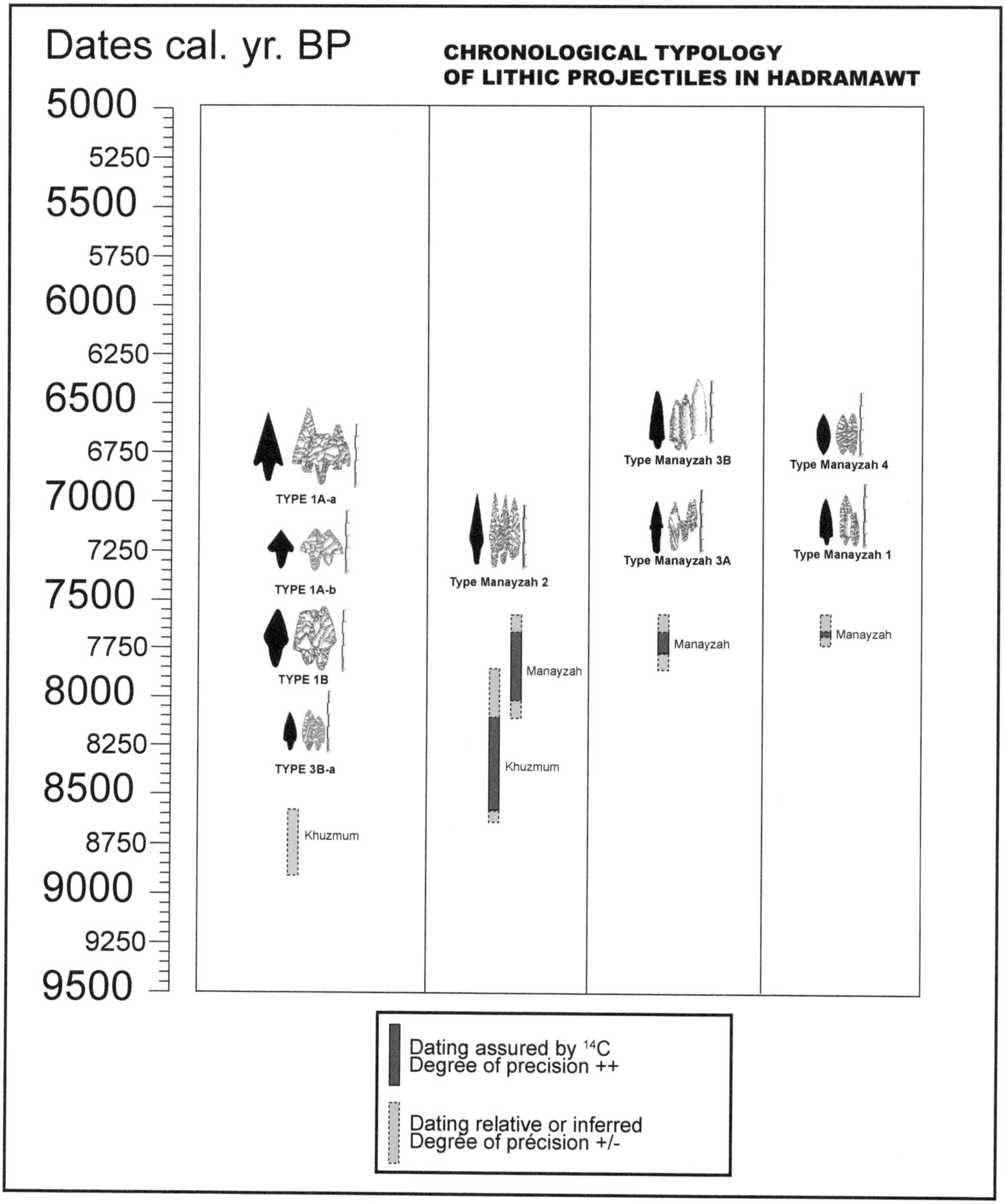

Figure 8.25. Multiple types of arrowheads relatively dated through stratigraphic series (adapted from Crassard 2008:144, figure 147). *Drawings by Julien Espagne and Rémy Crassard.*

Conclusions

Excavations at Manayzah ended too soon, cut short by political and military conditions in Yemen that ended RASA fieldwork. No excavations have been possible from 2008 to the date of this publication, and it is not clear when archaeological field research may resume. Future excavation at Manayzah should take into account the following observations from the 2004 and 2005 seasons and the future potential to which they point. First, horizontal excavation across multiple grid squares remains very difficult. Isolating each fine layer is nearly impossible, complicated by dramatic color changes within the same depositional event, such as the black-grey-white ash of in situ burning of dung mats. Such challenges make it difficult to follow surfaces, particularly occupation surfaces. These surfaces could clearly be identified during excavation, despite the fine matrix, which in places has a powdery consistency that prevented excavators from definitely isolating the finest occupational laminations. The aforementioned occupational level Layer 149 (visible in the sections of Quadrats A, B, C, and D) offers attractive prospects for site activity areas that can be better defined with the extension of excavation across the site's entire surface, within the 1 x 1 m grid for strict excavation control. Furthermore, the low-energy deposition environment enhanced the preservation in situ of stone tool knapping areas associated with archaeological structures, other objects of daily life, and refuse in primary discard context.

Second, an imperative is surely to extend a 1 m wide trench from K9 toward the south to augment with a greater sample size the already stratified and tightly dated reference corpus of diagnostic tool assemblages. This sequence is critical in the techno-typological seriation of Hadramawt's prehistoric material culture (Crassard 2008) and that of Arabia's more broadly (e.g., Charpentier 2008).

Third, to reach the base of the stratigraphy is also an aim of paramount importance. The variation in knapping techniques observed in the currently known stratigraphy suggests that deeper layers and broader excavation of them will substantially increase technological knowledge of stone tool production in the Early Holocene and perhaps even in older periods. The site offers every promise that it contains stratified and organic remains associated with knapping techniques not previously dated in a relative or chronometric fashion.

Fieldwork carried out elsewhere in the Wādī Sanā and broader geomorphological analyses (chapter 3) make it possible for archaeologists to situate Manayzah within a regional ecological and cultural context. Future excavations may revise some of our conclusions, and the site has the potential to substantially augment the conclusions we have been able to draw from its rich material and structural remains that are clearly stratified and datable. Nevertheless, the two seasons of excavation revealed a key sequence in the chronology and characterization of the Arabian Early Neolithic.

Manayzah is one of the few known shelters that preserve traces of human occupation as early as the Early and Middle Holocene in the upper and middle part of the wadi. The occupations at Manayzah and other shelters (chapter 9) indicate that human groups probably adopted bovid, sheep, and goat pastoralism as an adaptation to available pasture and fodder provided by the Wādī Sanā. A stratigraphy of intercalated flood and occupation deposits suggests seasonal occupation when rising floodwaters would not threaten livestock and humans trapped in the canyon. Pastoral populations even today occupy rockshelters in the region, but the water resources have seriously dwindled since the Middle Holocene. Wādī Sanā can no longer support cattle—there is no grass and insufficient water. Pastoralists in the Wādī Sanā today never cultivate—if indeed ever they did—domesticated plants. Evidence from Manayzah provides key indicators of the beginnings of cattle pastoralism, the earliest domesticated cattle and sheep yet found in Southern Arabia, and a rich detail of the technological strategies and cultural affinities of Wādī Sanā's indigenous herders.

Manayzah provides a strong potential for socioeconomic studies focusing on the Neolithic lifestyle in South Arabia, a lifestyle still poorly known. The site's stratigraphy has survived the passing of millennia largely intact, with numerous in situ lithic assemblages, remains attesting to important technical and typological variety. These are clearly associated with plant and faunal remains. There was a substantial series of occupations, characterized by multiple built and coeval combustion structures, likely domestic structures (postholes), and various archaeological features, including at least one pit that was partly excavated in Square K9. The site's location in the vicinity of perennial water and seasonal vegetation flushes certainly attracted hunter-foragers, pastoral populations, and possibly even proto-agrarian populations. The presence of hunting tools, associated with still-preserved animal droppings and with the remains of domesticated and wild fauna, make it all the more urgent that excavators return someday to Manayzah to explore Neolithic strategies at the intersection of hunting, herding, and landscape modification.

Manayzah was certainly a privileged location for specialized chert and obsidian toolmakers and craftsmen. Fluting evidence from the stratigraphy allowed excavators to date this toolmaking method with precision, thanks to an initial corpus of consistent radiocarbon ages. These chronometric dates supported the construction of a still-incomplete chrono-typology of particular types of stone projectile points. Manayzah's contribution to Hadramawt's prehistory is incomparable, notwithstanding the lingering questions about basal stratigraphy. Therefore, there is much left to learn at Manayzah about Early to Middle Holocene human occupations and about the poorly known earlier populations. Future work at the site will surely extend excavations and penetrate the deepest layers to conclude the analysis of lithics as well as richer study of faunal and plant remains.

References

Amirkhanov, H.

1994 Research on the Palaeolithic and Neolithic of Hadramaut and Mahra. *Arabian Archaeology and Epigraphy* 5:217–28.

Biagi, P.

1999 Excavations at Shell-Midden of RH-6 1986–1988, Muscat, Sultanate of Oman. *al-rāfidān: Journal of Western Asiatic Studies* 20:54–84.

Buffington, A. F., Michael J. Harrower, Joy McCorriston, and E. A. Oches

2017 A Niche Construction Approach to Vegetation Community Development in the South-West Arabian Neolithic: Preliminary Results. *Proceedings of the Seminar for Arabian Studies* 47:31–42.

Canti, M. G.

1998 The Micromorphological Identification of Faecal Spherulites from Archaeological and Modern Materials. *Journal of Archaeological Science* 25(5):435–44.

2003 Aspects of the Chemical and Microscopic Characteristics of Plant Ashes Found in Archaeological Soils. *Catena* 54:339–61.

Charpentier V.

2003 From the Gulf to Hadramawt: Fluting and Plunging Processes in Arabia. In *Archaeology of the United Arab Emirates: First International Conference on Emirates Archaeology*, edited by Daniel T. Potts, Hasan Al Naboodah, and Peter Hellyer, pp. 66–71. Trident Press, London.

2008 Hunter-Gatherers of the "Empty Quarter of the Early Holocene" to the Last Neolithic Societies: Chronology of the Late Prehistory of South-Eastern Arabia (8000–3100 BC). *Proceedings of the Seminar for Arabian Studies* 38:59–82.

Charpentier, V., J.-F. Berger, R. Crassard, M. Lacaze, and G. Davtian

2012 Prehistory and Palaeo-Geography of the Coastal Fringes of the Wahiba Sands and Bar al-Hikman, Sultanate of Oman. *Proceedings of the Seminar for Arabian Studies* 42:57–78.

Charpentier V., and M. L. Inizan

2002 Fluting in the Old World: The Neolithic Projectile Points of Arabia. *Lithic Technology* 27(1):39–46.

Cleuziou S., M.-L. Inizan, and B. Marcolongo

1992 Le peuplement pré- et protohistorique du système fluviatile fossile du Jawf-Hadramawt au Yémen (d'après l'interprétation d'images satellite, de photographies aériennes et de prospections). *Paléorient* 18(2):5, 28.

Crabtree D. E.

1966 A Stoneworker's Approach to Analyzing and Replicating the Lindenmeier Folsom. *Tebiwa* 9(1):3–39.

Crassard R.

2008 *La Préhistoire du Yémen. Diffusions et diversités locales, à travers l'étude d'industries lithiques du Hadramawt.* British Archaeological Reports S1842. Archaeopress, Oxford.

Crassard, R., and P. Bodu

2004 Préhistoire du Hadramawt (Yémen): nouvelles perspectives. *Proceedings of the Seminar for Arabian Studies* 34:67–84.

Crassard R., J. McCorriston, E. A. Oches, 'A. Bin 'Aqil, J. Espagne, and M. Sinnah

2006 Manayzah, Early to Mid-Holocene Occupations in Wādī Ṣanā (Ḥaḍramawt, Yemen). *Proceedings of the Seminar for Arabian Studies* 36:151–73.

Crassard, R., M. D. Petraglia, A. G. Parker, A. Parton, R. G. Roberts, Z. Jacobs, 'A. Alsharekh, A. Al-Omari, P. Breeze, N. A. Drake, H. S. Groucutt, R. Jennings, E. Régagnon, and C. Shipton

2013 Beyond the Levant: First Evidence of a Pre-Pottery Neolithic Incursion into the Nefud Desert, Saudi Arabia. *PLoS ONE* 8:e68061.

Fuller, D., R. Korisettar, P. C. Venkatasubbaiah, and M. K. Jones

2004 Early Plant Domestications in Southern India: Some Preliminary Archaeobotanical Results. *Vegetation History and Archaeobotany* 13:115–29.

Inizan M.-L., A.-M. Lézine, B. Marcolongo, J.-F. Saliège, C. Robert, and F. Werth

1998 Paléolacs et peuplements holocènes du Yémen: le Ramlat as-Sab'atayn. *Paléorient* 23(2):137–49.

Inizan M.-L., and M. Rachad

2007 *Art rupestre et peuplements préhistoriques au Yémen.* CEFAS, Sana'a.

Kallweit H.

1996 Neolithische und Bronzezeitliche Besiedlung im Wadi Dhahr, Republik Jemen. Eine Untersuchung auf der Basis von Geländebegehungen und Sondagen. PhD dissertation, Urgeschichtliche Archäologie, Frühgeschichtliche Archäologie und Archäologie des Mittelalters, Freiburg University, Freiburg im Breisgau.

Kimiaie M., and J. McCorriston
2014 Climate, Human Palaeoecology and the Use of Fuel in Wādī Sanā , Southern Yemen. *Vegetation History and Archaeobotany* 23(1):33–40.

Martin L., J. McCorriston, and R. Crassard
2009 Early Arabian Pastoralism at Manayzah in Wādī Sanā, Hadramawt. *Proceedings of the Seminar for Arabian Studies* 39:271–82.

McCorriston J., E. A. Oches, D. E. Walter, and K. L. Cole
2002 Holocene Paleoecology and Prehistory in Highland Southern Arabia. *Paléorient* 28(1):61–88.

McCorriston, J., and Z. Johnson
1998 Agriculture and Animal Husbandry at Zayidid Zabid, Yemen. *Proceedings of the Seminar for Arabian Studies* (28):175–88.

Miller, N. F., and T. L. Smart
1984 Intentional Burning of Dung as Fuel: A Mechanism for the Incorporation of Charred Seeds into the Archeological Record. *Journal of Ethnobiology* 4:15–28.

Wagner, G.
1989 Comparability among Recovery Techniques. In *Current Paleoethnobotany*, edited by Christine A. Hastorf and Virginia S. Popper, pp. 17–35. University of Chicago Press, Chicago.

Wallace, M., and M. Charles
2013 What Goes in Does Not Always Come Out: The Impact of the Ruminant Digestive System of Sheep on Plant Material, and Its Importance for the Interpretation of Dung-Derived Archaeobotanical Assemblages. *Environmental Archaeology* 18(1):18–30.

Whittaker J. C.
1994 *Flintknapping, Making and Understanding Stone Tools.* University of Texas Press, Austin.

Wilke P. J.
1996 Bullet-Shaped Microblade Cores of the Near Eastern Neolithic: Experimental Replicative Studies. In *Neolithic Chipped Stone Industries of the Fertile Crescent and Their Contemporaries in Adjacent Regions*, edited by Hans Georg Gebel, Stefan Karol Kozłowski, and Lorraine Copeland, pp. 289–310. Ex Oriente, Berlin.

Chapter 9

Early Holocene Forager Encampments
Rockshelters and the Gravel Bar Site

Joy McCorriston, Rémy Crassard, Dawn Walter, and Louise Martin

The middle Wādī Sanā has a number of low-elevation rockshelters that open onto the gravel and silt terraces of the current wadi bed or can be reached from the lowest bedrock shelf. Many are in current use by bedouin; most show signs of some previous occupation. Scorched ceilings, adjacent small pens and dry-stone curtain walls, scattered thermally altered rock, graffiti, and abandoned debris occur frequently. Much of this can be difficult to date and indeed may accrue over many visits and periods. Archaeological survey showed that rockshelters were important habitation sites, especially in the absence of other housing from most periods. To better understand the dates of rockshelter occupations and the relationships of these human occupations with the environmental history of Wādī Sanā, the team conducted excavations at several sites.

Through survey, the team recognized several surface concentrations of chert debitage and diagnostic lithic projectile points associated with rockshelter occupations. Manayzah was one such site (chapter 8), but other sites also offered the promise of stratigraphic and spatial associations that would provide multiple dates for stone tools and technologies. At the beginning of RASA research, there were neither regional nor local chrono-stratigraphic sequences with which artifacts could be compared. As some of the project's primary questions were cultural-historical and chronological, RASA chose promising surfaces and sections that would yield stratified radiocarbon ages associated with cultural remains.

Beyond mere dating, RASA has pressed questions about the occupation of Wādī Sanā across different environmental conditions. Stratified rockshelters provide a view of the permanency, duration, repetition, timing, and gaps in occupation, contributing to a broader understanding of long-term land use and landscape patterns in Wādī Sanā.

To the rockshelter excavations and testing reported in this chapter, we add the important Wādī Sanā Gravel Bar Site (GBS), an open-air site initially documented by the CANOXY pipeline survey (Vogt and Sedov 1994) and revisited numerous times (with repeated amateur and professional archaeological collecting). Despite the impressive surface accumulation of tools and knapping debris, there are no stratified integral cultural remains. The results of test excavations and detailed analysis of the surface and its artifacts nevertheless are important indicators of open-air strategies in ancient Wādī Sanā and have been included here to complement the results of rockshelter and rockshelter terrace excavations.

Khuzmum Rockshelters

Where Wādī as-Shumlyah joins Wādī Sanā, Early Holocene floodwaters slackened to deposit alluvial silts across the Tertiary channel. When this happened, the bedrock Khuzmah as-Shumlyah inselberg at the confluence was a real island, perhaps for weeks at a time (figure 9.1). As the floodwaters receded, the Khuzma served as a marshland landmark, with an incomparable view of the hunting grounds and pastureland around it. The Late

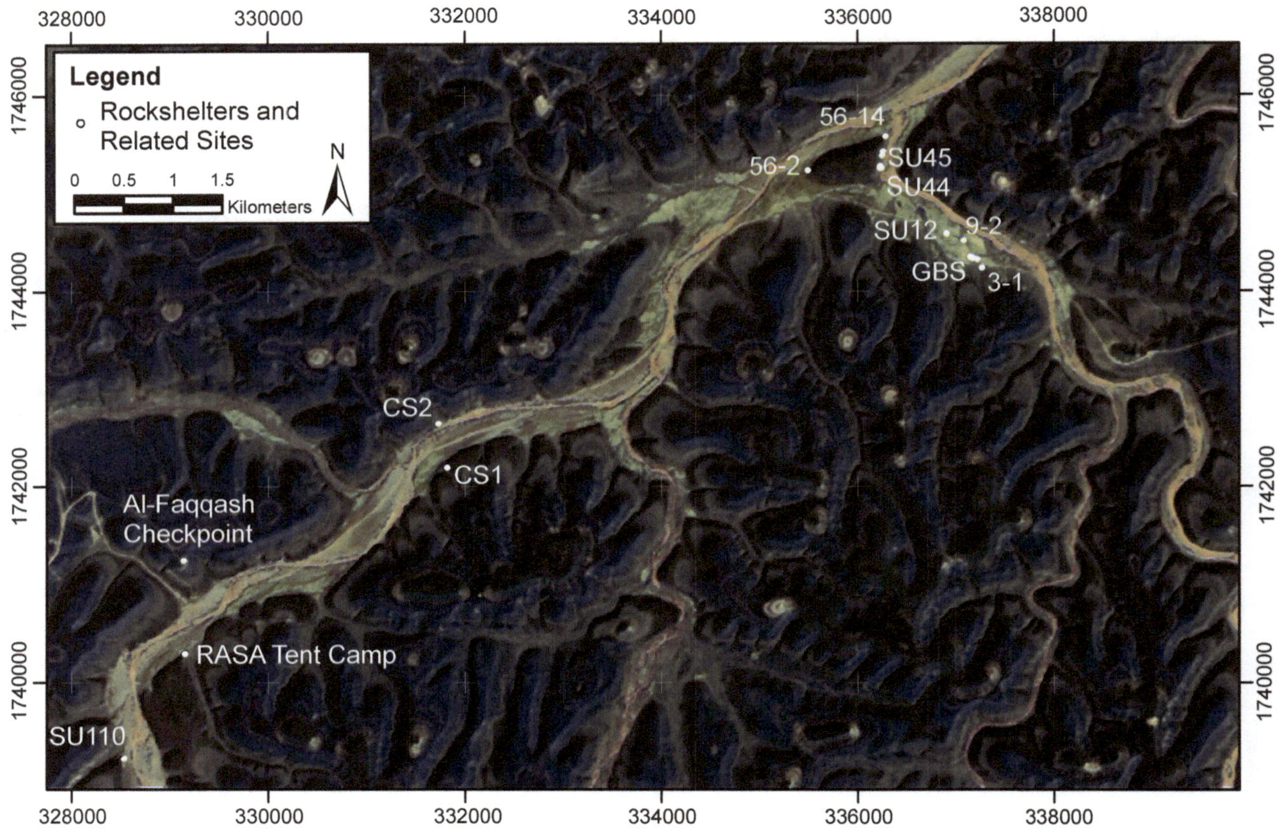

Figure 9.1. Map of Khuzma as-Shumlya and Wādī Sanā confluence with Khuzmum rockshelters (SU 45 series) and other rockshelter sites. *Illustration by Michael Harrower.*

Holocene channel sliced through silt beds and gravel substrates to create abrupt truncation of the terraces of the eastward-facing rockshelters. This natural process has revealed a deep stratigraphic sequence of repeated terrace occupations interlayered with evidence of disturbance and alluvial deposition. Today, modern bedouin intermittently use the rockshelters and the present-day terrace surface.

SU44—A Natural Profile with Radiocarbon Ages

Paleoecologist Rick Oches first recognized the distinctive sequence at Survey Unit (SU) 44, a naturally cut profile with multiple hearths (25 count) in section in front of rockshelter SU45-10 (figure 9.2). The rockshelters are named after a field reading ("Khuzmum") of a word pecked onto a patinated limestone bedrock face (SU45-7), located several meters north of the biggest rockshelter, SU45-10 (figure 9.3). Among images of camels and ibex were words and names, including "Place of Ḥuzma" (ḤS³MM) in Ancient South Arabian writing. Other graffiti were less readily recognized, but according to Alessia Prioletta, who specializes in Ancient South Arabian philology, their lettering appears to be consistent with two undeciphered Dhofari scripts. If these graffiti are about 2,000 years old (as such scripts' association with camels, date harvests, and ships in Dhofar would seem to imply), then the graffiti and rock images plausibly date the stable rock face of Khuzma as-Shumlya. A more detailed discussion of graffiti and rock art from the Late Holocene occupations appears in chapter 17. Much of the sediment terrace is protected from surface erosion by clastic rockfall, and it is apparent that much of this must have tumbled before people pecked images and text into the face. Lithics and hearths in underlying sediments date to an occupation more than 8,000 years old.

The team used a combination of approaches to document the long, east-facing Holocene terrace and rockshelters. These approaches included profile documentation and sampling of the natural profile, test excavations (see below), GPS registration of rockshelters, and documentation of human modifications of bedrock (graffiti; chapter 17). Some of the clastic overburden includes massive boulders nearly 1 m deep over the profile. A large horizontal terrace

Figure 9.2. Khuzmum rockshelter with a natural profile through the silt terrace. The entire profile was designated SU44, with multiple site numbers for individual hearths and visible ashy occupation layers. The Khuzmum rockshelters themselves received separate survey numbers. The SU045-10 rockshelter is pictured here (center), with a solitary figure standing on the terrace just left of the natural profile in Figure 6.3. *Photograph by Joy McCorriston.*

exposure in front of the SU45-10 rockshelter was beyond the resources of the RASA Project, but 25 hearths in the natural profile were measured and sampled for flotation and radiocarbon dating, and the team produced careful geological notes and an archaeological drawing of that profile.

Most hearths ($n = 18$) were within one of two dark ashy layers traceable for more than 20 m along the profile, and many hearths ($n = 9$) had retouched lithic debris and broken tools or blanks made of chert within them (figure 9.4). While it is clear that the hearths were an anthropogenic source of charcoal and ash that contributed to the discoloration of the sediments in which they were buried, field observations of the sediments documented a sorted layering of fine gravel, sand, and silt, deposited by overbank flooding of varying intensities. These repeated events built the sediment terrace in front of the Khuzmum rockshelters.

Flotation samples taken from the natural profile were a few liters per hearth (SU44-3, SU44-4, SU44-9, SU44-13, SU44-20), and lithic counts in table 9.1 tally these samples, not the full contents of hearths. Two charcoal samples from hearths were radiocarbon dated, with the lowest dated hearth (SU45-20, AA38545) containing rich ash and tiny retouch flakes from 8384–8066 cal BP. No debitage or tools were recovered from the upper hearth, which was about 1 m below the surface and dated a millennium later, to 7418–7173 cal BP (SU45-25, AA38546). There were hearths stratigraphically positioned between these dated examples.

These hearths and their depositional context attest to intermittent episodes of occupation and overbank flooding. The predominant record of human occupation is preserved from two periods when the surface was relatively stable (two ashy layers, SU44-28 and SU44-29, visible in profile). These occupations were partially eroded by flooding, mixing anthropogenic charcoal with sedimentary deposits. Violent floods may have truncated some deposits, effacing traces of other occupational episodes. Given its location within the marshy floodplain of the confluence area, the

Figure 9.3. ʿAbdālʿazīz Bin ʿAqīl photographing SU45-7 rockart and graffiti panel. *Photograph by Jennifer Everhart.*

rockshelter area offered attractive and recurrent temporary shelter that was nonetheless too vulnerable to flooding to provide a permanent home base.

SU45—Test Excavation

Archaeological excavation at a nearby rockshelter offered an important opportunity to clarify chronological issues, probe the integrity of a promising occupation site, and collect samples for radiocarbon analysis, lithic analysis, bones, and flotation. Rockshelter SU45-1 lies northward, along the Khuzma as-Shumlya's eastern face. Although rockfall has deposited massive boulders on the terrace surface, there were surface areas at the wadi edge where abundant lithic tools and debris appeared atop an accessible natural profile several meters deep. While horizontal exposure was limited by rockfall, two narrow 1 x 3 m test excavations (SU45-1-A and SU45-1-B) were excavated in 2000 by Dawn Walter and ʿAbdalBaset Nʿoman. The first trench (SU45-1-A) was located along a natural profile, while the other was situated on the terrace surface abutting the back wall of the rockshelter (SU45-1-B).

The rockshelter walls and large boulders obstructed GPS and sightlines, making mapping challenging. A datum point consisting of a brass screw cemented to bedrock, referred to as Main SU45 Khuzma Datum, was established at E 336247.86, N 1745385.85 (UTM Zone 39 North WGS84), elevation 681.38m MSL (EGM96), as measured on March 4, 2000, with a Trimble Pathfinder Pro XRS GPS using Omnistar real-time correction (elevation value 662.22 m HAE). Excavations at SU45-1-A used Sub-Datum A, located with tapes (E 7.80 m, N 37.19 m, and 2.06 m below Main SU45 Khuzma Datum).

Excavation followed the natural stratigraphic sequence (figure 9.5). All tools and lithic debris from the surface were collected before the removal of a layer of high-energy accumulation of clastic cobbles and pebbles, likely generated through rockfall off the rockshelter ceilings and Khuzma as-Shumlya's eastern face. Excavators sieved all deposits through 0.5 cm screens, retrieving bone and chipped stone. Using an Ankara-style flotation tank built for the project by CANOXY engineers, Catherine Heyne and Joy McCorriston processed flotation samples taken

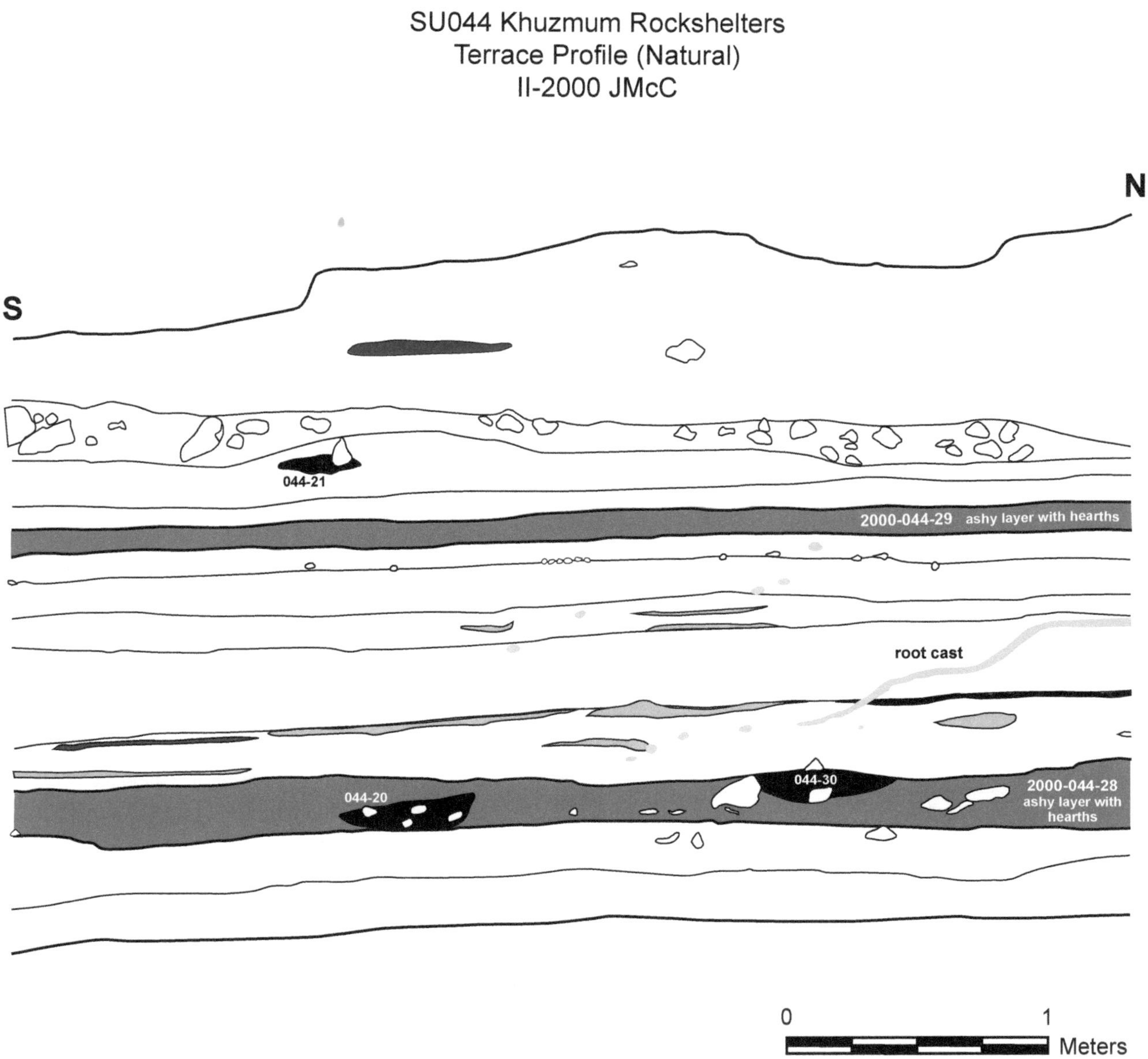

Figure 9.4. Natural profile of the Khuzmum rockshelter terrace in front of SU45-10. *Illustration by Joy McCorriston and Jarrod Burks.*

strategically from deposits rich in ash and concentrations of lithics or contained within a clear hearth. The taphonomic context of low-energy floodplain deposition is not appropriate for comprehensive sampling of all deposits. Charcoal could accumulate from multiple sources and reflect a wide range of human activities.

With almost 2 m of stratigraphy, Test Excavation A was the more productive of the two trenches. Surface cobbles were embedded in a matrix of loose sandy silt, still containing lithic material. Underneath was a more compact layer of sandy silt and natural clasts (Loci 001, 002, and 003), still rich with stone tools, blanks, and knapping debris. This overlay the uppermost of two stratified ashy layers (Loci 004, 005, and 006) containing hearths, thermally altered limestone clasts, bone, and a relatively dense accumulation of lithic tools and debris. A radiocarbon age on charcoal from the upper layer (AA38543) of 8370–8047 cal BP is the same age range as the lower ashy layer in Profile SU44. Separated from the upper ashy layer by only a few centimeters of silt (Loci 007 and 008), the lower ashy layer (Locus 009) (figure 9.6) included a number of hearths, one of which yielded a radiocarbon age some hundreds of years earlier (AA38548, 8723–8365 cal BP). This layer was also dense with lithics from a wide variety of chert types and colors. Some cherts were burned, whether intentionally or in the hearths, and the knapping debris was unrolled and unsorted.

Table 9.1. Hearths from SU44 associated with ashy layers and lithic contents. Data from field observations and heavy fractions of flotation samples.

Hearth	Within Ashy Layer	Tools	Blanks	Retouch Flakes
044-003	044-029	0	0	0
044-004	--	0	0	0
044-005	--	0	0	0
044-006	--	0	0	1
044-007	--	0	0	0
044-008	044-028	0	0	7
044-009	044-028	1	0	4
044-010	044-028	0	0	0
044-011	044-028	0	0	14
044-012	044-029	0	0	4
044-013	044-028	1	0	21
044-014	--	0	0	0
044-015	044-028	0	1	5
044-016	044-028	0	0	0
044-017	044-029	0	0	2
044-018	044-028	0	0	0
044-019	044-028	0	0	0
044-020	044-028	0	0	11
044-021	--	0	0	0
044-022	--	0	0	0
044-023	--	0	0	0
044-024	044-028	0	0	0
044-025	044-029	0	0	0
044-026	--	0	0	0
044-027	044-029	0	0	0
044-030	044-028	0	0	0

Underlying layers contained less ash. A layer of lighter-colored compacted silt with gravel inclusions underlay the ashy occupations and contained lithic tools and debris. Albeit in diminishing quantities, lithics also occurred in an underlying silt layer (Loci 012–020), which was also a compact sediment with unsorted clastic inclusions. Excavations through a compact silty sand at the base of the trench (Loci 021–023) uncovered no cultural material (figure 9.7).

The sequence suggests reiterative visits to the site, with in situ stone tool production and use. Surface disturbances, indicated by nearby abandoned goat pens (SU44-1), modern tin cans, and flooding of the rockshelter area, may have contributed to mixing earlier lithic material with later clastic rockfall. There may have been truncations of the stratigraphic deposits through flooding and erosion. The most concentrated occupational evidence dates to a 700-year period of the seventh millennium BCE. The radiocarbon ages from Khuzmum rockshelters are the first stratified association of dated organic material in situ with Neolithic projectile points (McCorriston et al. 2002) and anchor a chrono-typological series for Hadramawt (Crassard 2008). Previously, these were known from surface finds (di Mario 1989; Edens 1982) and a similar profile in the Ḥabarūt oasis on the Yemen side (Amirkhanov 1997:figures 24.2 and 26.7; Zeuner 1954).

After excavating only 10 cm of sterile rocky fill without any artifacts, the 1 x 3 m excavation (SU45-1-B) reached immovable boulders from rockfall on the bedrock floor of the terrace. This brief probe established that roof collapse postdates the terrace strata containing lithics and that sedimentation under the modern shelter overhang is very shallow.

Despite flotation efforts targeting hearths and ashy deposits, few identifiable seeds and animal bones were recovered from the limited soundings at Khuzma as-Shumlya rockshelters. Because the hearths are small, so are samples (3–10 liters in volume), which have yielded several small wild legumes (*Crotalaria* cf. *oocarpa*) and Compositae family (*Hochstettera/Pulicaria*-type) seeds that show the possibility of preservation in open fires. The available inventory is too small to suggest roasted food use of these or other wild plant species, and they may as easily stem from local brush kindling or from animal dung in hunter-gatherers' camps.

Animal bones were often charred. The very fragmented and very small excavated collection of Khuzmum Rockshelter bones includes only two identifiable specimens, both from the contexts containing dated charcoals. Both bones are from caprines (table 9.2).

The lithics recovered from SU45-1-A were initially studied by Dawn Walter in 2000, then stored in the Mukalla Museum in Hadramawt, Yemen. In 2004 Rémy Crassard reexamined the tools. Initiated by Dawn Walter and updated by Rémy Crassard, the analysis of lithic artifacts included an assessment of all pieces and their classes as types and debitage (Walter) and a study of technical aspects using bifaces and projectile points (Crassard). Because Crassard selected tools from the

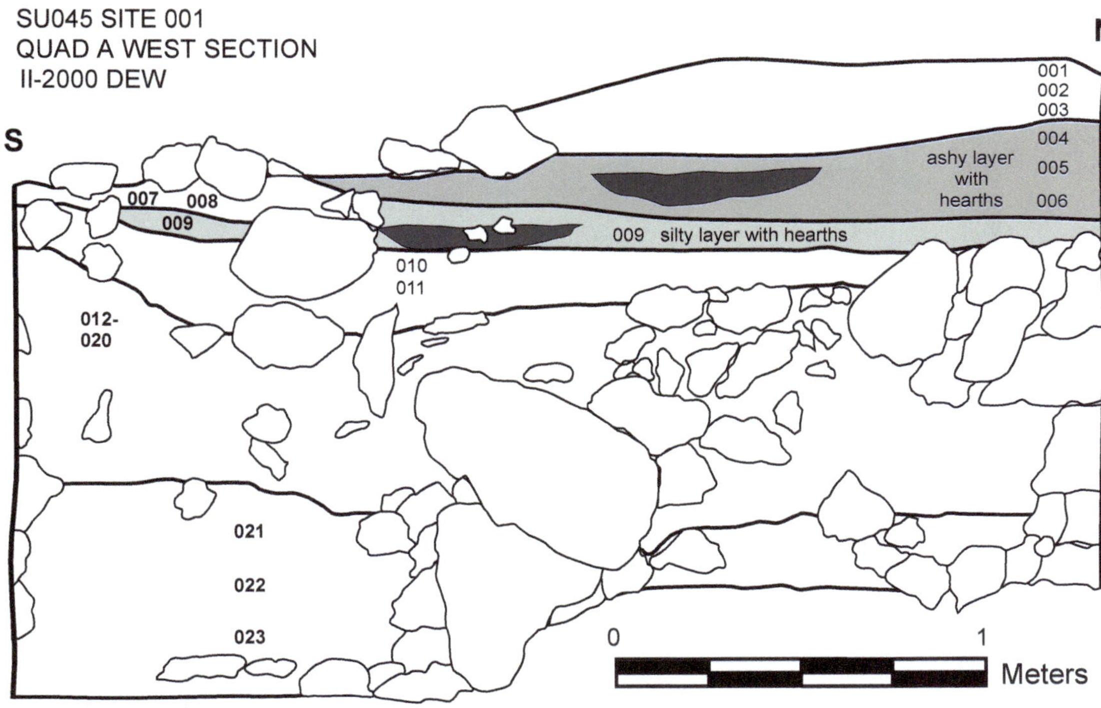

Figure 9.5. East-facing west section of the SU45-1-A excavation in 2000. *Drawing by Dawn Walter, Jarod Burks, and Clara Hickman.*

Figure 9.6. Hearths visible south and west of terrace edge under excavation in SU45-001-A. *Photograph by Joy McCorriston.*

bags stored by archaeological find layer, his counts did not always match Walter's counts, which are presented here. Due to current conditions in Yemen, no further analysis of the artifacts stored in Mukalla is possible in the immediate future. The assemblage recovered from the Khuzmum rockshelters excavation includes 105 diagnostic tools and 4,682 pieces of manufacturing debitage. Tools include 56 fragmentary or complete projectile points (including trihedral points), 11 fragmentary or complete bifacial pieces in various stages of manufacture (including foliates), scrapers, borers, and retouched flakes. The artifact classes identified in the rockshelter assemblage are described below.

The tool assemblage contains a high frequency of projectile points (figures 9.8 and 9.9). Sixteen percent (n = 9) of the projectile points are complete, while 61 percent (n = 34) of the fragmentary projectile points are complete enough to be separated into categories. The remaining items (n = 13) have been identified as undetermined fragments of projectile points but could perhaps be the bases/tips of other tool types. Because their morphological characteristics are similar to the complete arrowheads in the assemblage, these 13 items have been included in the arrowhead category.

The high number of generally broken projectile points indicates that the Khuzmum rockshelters site was frequently visited by local hunters as a location to stop

Figure 9.7. Khuzmum rockshelter SU45-1-A: final photo of excavated terrace profile. Note hearths in upper layers overlying sterile clastic talus and smooth stream boulders at base. *Photograph by Joy McCorriston.*

between different hunting places and replenish their tool kits with locally available raw material. The typological and morphological characteristics of the projectile points indicate that they were used for hunting purposes. The distal ends of the projectiles discovered at the site were most probably broken during production or use, while the proximal ends were probably replaced on hafts when suitable raw material or time allowed for retooling.

The projectile point assemblage shows some degree of variability, with in situ conservation of typological styles over centuries. Arrowheads that were found in the most recent layers (Loci 000, 002, 003, 004, 005, 007, and 009) are the most numerous within the corpus (31 out of 56, representing 55.4 percent of the total) (table 9.4). They were made by bifacial, sometimes trifacial retouch. They are systematically constructed with a trihedral section or with a plano-convex section with a tendency to become trihedral at the tip. They also have a long tang, most of the time slightly shifted in comparison with the general rectilinear profile of the projectile point. Rarer examples show a similar kind of shift of the barbs. In some of the rather old layers (Loci 009, 011, and 012), arrowheads are made on flaked blanks (flakes, laminar flakes, or blades?). These points are not numerous (only 3, maybe 4, out of 56, representing 5.4 or 7.1 percent of the total of projectile points). Only the dorsal face of the flaked blank is retouched, by pressure removals, most of the time *en écharpe*. The tang is systematically made by bifacial retouch, until the obtainment of a slight shift of the tang in comparison with the linear profile of the piece. As observed with the previous type, barbs can also be slightly shifted. Finally, in the oldest layers (Loci 011 and 018), two examples of bifacial arrowheads with barbs and tangs of a different type were found. The section is symmetric. This type is well-known from many surface sites in Hadramawt (Crassard 2008) and elsewhere in South and Central Arabia (e.g., Crassard 2008; Crassard et al. 2013; Edens 1982), as well as in another stratified site in Hadramawt (HDOR 561, Sounding 3, Layer 3; Crassard 2008).

Table 9.2. Identified faunal remains from Khuzmum rockshelter excavations, SU45-1 Quad A. Identifications by Louise Martin.

Bag Number	Element	Part	Taxa	Side	Additional Information	Context	^{14}C Date on Charcoal
2000-045--1-A-006 Bag 15	lower incisor	fragment	caprine	left	left half; about 1 cm	Early Neolithic rockshelter occupation	8150 cal BP
2000-045-1-A-009 Bag 33	distal phalanx	complete	caprine	left	burned/singed: black on the caudal and lateral sides	Early Neolithic hearth	8550 cal BP

Table 9.3. Lithic debitage recovered from Khuzmum rockshelter excavations, SU45-1 Trench A. Dawn Walter identified blanks, cores, finishing flakes (BFF), thinning/shaping flakes (BTF), flakes of indeterminate reduction (FIR), biface initial reduction flakes (BIRF, IRF), and shatter (chert that shows no evidence of being humanly struck but may nonetheless be a waste product from a knapping episode).

		BFF		BTF		BTF/BFF		FIR		IBTF		IRF		Shatter		Total	
Level		#	%	#	%	#	%	#	%	#	%	#	%	#	%	#	%
1	surface	5	0.6	54	2.2	21	3.2	13	6.6	4	1.9	33	11.1	5	4.7	**135**	**2.9**
2	loose scree	1	0.13	67	2.8	11	1.7	7	3.6	2	0.9	14	4.7	2	1.9	**104**	**2.2**
3	ash	315	40.8	815	33.5	255	38.3	40	20.3	43	20.4	71	24	9	8.4	**1548**	**33.1**
4	compact silt and scree	227	29.4	585	24	86	12.9	65	33	38	18	64	21.6	21	19.6	**1086**	**23.2**
5	compact silt with gravel	216	27.9	639	26.3	160	24.1	50	25.4	22	10.4	79	26.7	33	30.8	**1199**	**25.6**
6	very compact silt with scree	9	1.2	233	9.6	132	19.8	19	9.6	87	41.2	29	9.8	36	33.6	**545**	**11.6**
7	compact silty sand	0	0	40	1.6	0	0	3	1.5	15	7.1	6	2	1	0.9	**65**	**1.4**
	Total	**773**	**16.5**	**2433**	**52**	**665**	**14.2**	**197**	**4.2**	**211**	**4.5**	**296**	**6.3**	**107**	**2.3**	**4682**	**100**

The types of arrowheads found at Khuzmum 45-1-A are:

- **Type Khuzmum 1.** Most recent and most numerous: bifacial or trifacial point with a trihedral section, with a long tang, slightly shifted (Khuzmum 1A); sometimes with a slight shift of barbs (Khuzmum 1B).
- **Type Khuzmum 2.** Older than Type Khuzmum 1; not often represented: point with a plano-convex section on a flaked blank with a tang; regular retouch, covering the total superior face and bifacial retouch of the tang (Khuzmum 2A); sometimes bifacial retouch of the tip (Khuzmum 2B).
- **Type Khuzmum 3.** Older than Khuzmum 1 and Khuzmum 2; very few examples known: flat bifacial projectile point with barbs and tang, with symmetric or subsymmetric section.

The typological variability of the projectile points found at Khuzmum 45-1-A shows three types of points that can be placed in a relative chronological order, based

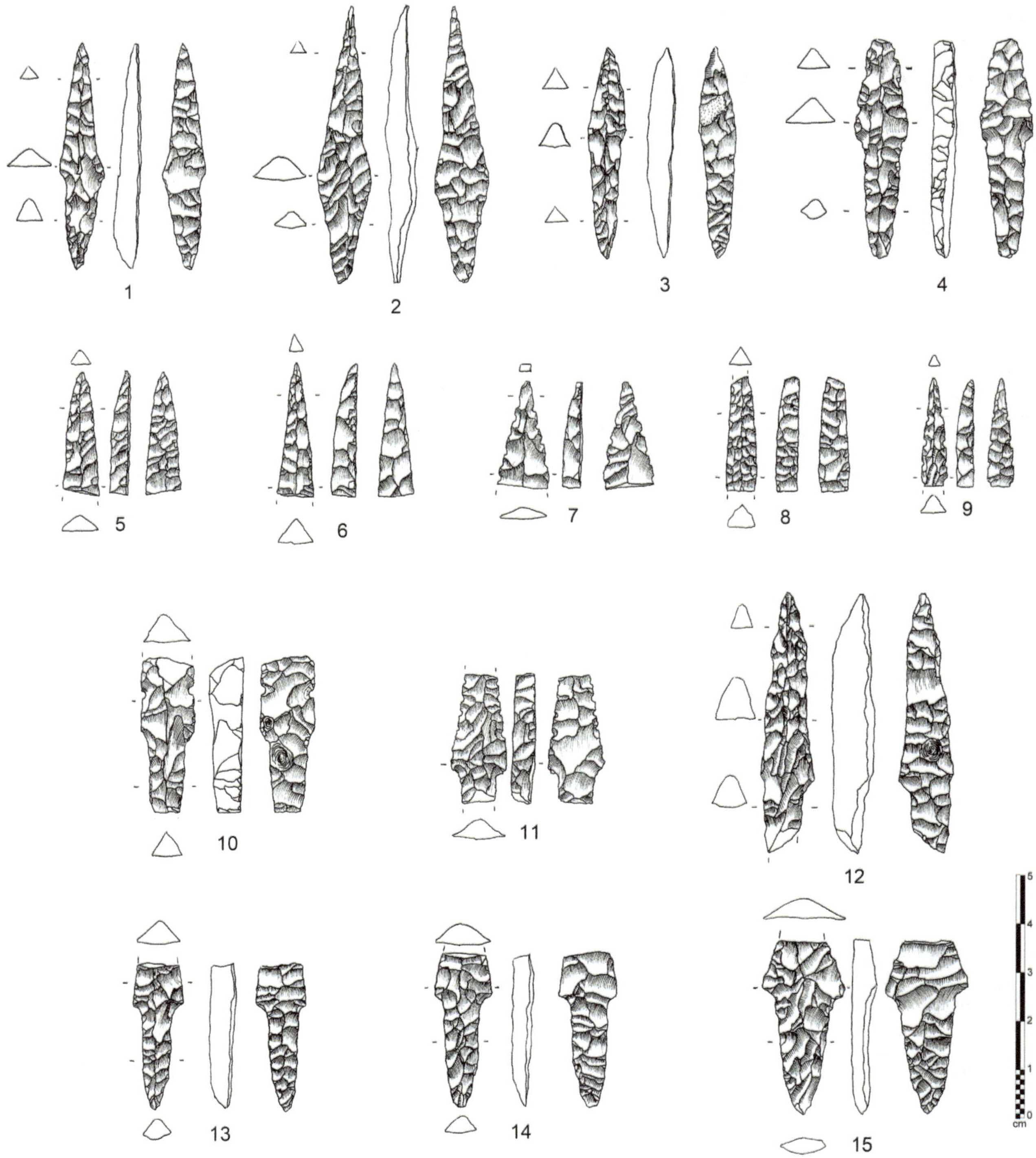

Figure 9.8. Khuzmum rockshelter SU45-1-A projectile points. By locus: (1) Locus 016; (2) Locus 012; (3 and 9) Locus 011; (4 and 10) Locus 010; (5, 6, 13 and 15) Locus 005; (7 and 8) surface; (11 and 14) Locus 003; (12) Locus 007. *Drawings by Julien Espagne. Illustration by Rémy Crassard.*

on the excellent stratigraphic information from Manayzah (chapter 8) and the stratigraphic sequence at Khuzmum 45-1-A. (See chapter 8, figure 8.25.) Both sequences have multiple radiocarbon dates in context with points. The trihedral points seem to reveal a younger technical tradition than the one used for making the bifacial points with barbs, tang, and symmetric section. This latter tradition could then be dated to the beginning of the ninth

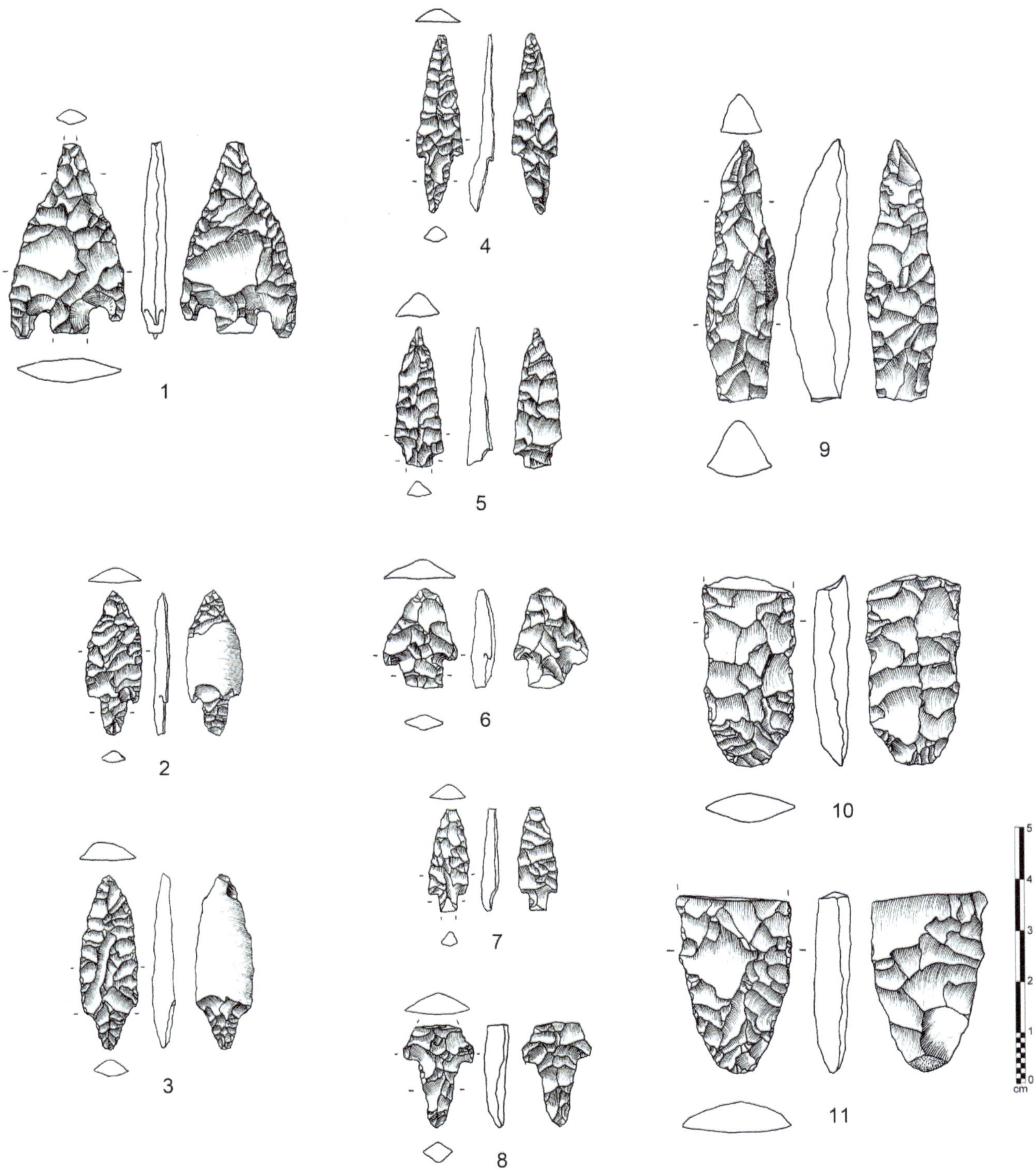

Figure 9.9. Khuzmum rockshelter SU45-1-A points and bifaces. By locus: (1, 4, 8, and 10) Locus 003; (2, 5, and 6) Locus 005; (3) Locus 007; (7 and 11) surface; (9) Locus 018. *Drawings by Julien Espagne, Illustration by Rémy Crassard.*

millennium cal BP, even if the accuracy of such a dating will need many more dates and discoveries. The trihedral point type is dated to the second half of the ninth millennium cal BP, confirming similar dating in regional sites (such as Manayzah, chapter 8, or HDOR 410 and HDOR 419; Crassard 2008, 2009).

The variability of projectile point styles indicates a diachronic occupation of the site most probably by different

Table 9.4. Projectile points recovered from Khuzmum rockshelter excavations, SU45-1 Trench A, studied by Rémy Crassard. Measurement is in millimeters.

Piece No.	Locus	Lot	Bag	Conservation							Typology					Section					Dimensions (mm)				Observations
				Complete	Tang Only	Proximal	Proximal-medial	Medial	Medial-distal	Distal	Barbed and Tanged Bifacial	Simple Tang Bifacial	Trihedral	On a Flaked Blank	Undetermined	Plano-convex	Asymmetric	Symmetric	Trihedral	Undetermined	Total Length	Maximum Width	Tang Length	Maximum Thickness	
1	012	50		1							1					1					28	11	6	2	
2	011	48		1								1				1					33	11	6	3	
3	011	48					1				1								1		28	9	2	5	
4	011	48		1								1								1	39	27		10	
5	011	48						1											1		13	7		3	burned
6	010	41		1							1					1					35	9	11	3	
7	010	41		1								1							1		56	11	14	5	trihedral tip
8	010	41					1				1									1	19	14	12	4	
9	009	35	1				1				1							1			35	12	16	5	
10	009	35	2							1			1						1		20	7		4	
11	009	35	3					1					1						1		15	14		7	
12	009	35	5							1				1		1					32	17		5	
13	005	13	8							1			1						1		20	6		4	
14	005	3	7							1			1						1		22	17		4	
15	005	13	8					1							1	1					26	12		4	burned
16	005	flot								1			1						1		16	6		3	
17	005	13	1				1								1		1				36	12		6	
18	005	13	2				1						1						1		30	10	20	5	
19	005	13	9						1				1						1		27	6		5	
20	005	13	4					1					1						1		26	11		4	four bifacial notches on edges
21	005	13	6							1			1						1		22	5		4	four bifacial notches on edges
22	005						1					1				1					35	17	21	5	
23	005	13	8					1					1						1		18	11		4	maybe on fluted blank?
24	005	flot						1					1						1		11	6		4	
25	005	flot				1									1		1				15	3		5	
26	005	flot						1					1						1		12	7		5	
28	003	8	11							1					1	1					22	10		3	
29	003	8	9	1									1						1		42	8		5	

Piece No.	Locus	Lot	Bag	Conservation							Typology					Section					Dimensions (mm)				Observations
				Complete	Tang Only	Proximal	Proximal-medial	Medial	Medial-distal	Distal	Barbed and Tanged Bifacial	Simple Tang Bifacial	Trihedral	On a Flaked Blank	Undetermined	Plano-convex	Asymmetric	Symmetric	Trihedral	Undetermined	Total Length	Maximum Width	Tang Length	Maximum Thickness	
31	003	8	10						1				1						1		51	14		12	
32	003	8	1					1					1						1		22	5		5	
33	003	8	10							1					1	1					24	7		5	
34	003	8	4					1					1						1		23	5		5	trifacial retouch
35	003	8	5					1					1						1		13	6		4	
36	003	8								1			1						1		16	5		5	trifacial retouch
37	000	surface								1					1			1			22	17		3	
38	000	surface								1					1			1			13	6		3	
39	002	4								1			1						1		20	6		4	
40	004	10						1					1						1		17	9		6	trifacial retouch
41	004	10								1			1						1		24	7		3	
42	007	21	1	1									1						1		46	9	16	5	two small lateral “ears”
43	007	21	2				1					1					1				30	13		7	
44	007	21	3							1			1						1		25	7		3	
45	007	21								1					1	1					12	9		3	
46	015	56				1									1	1					12	9		3	
47	015	56			1										1			1			12	7		2	
48	014	52		1							1					1					19	7		3	
49	006	17	5							1			1						1		16	5		4	
50	006	17	2					1					1						1		22	8		5	two notches on edges
51	006	17	1				1						1						1		42	11		6	two small lateral “ears”
52	006	17	3		1										1		1				19	7		3	
53	006	17	10					1			1		1						1		15	10		5	
54	018	64		1							1							1			37	22	4	5	
55	001	22					1				1					1					32	10		5	
56	001	2						1							1			1			6	6		2	
Totals				**9**	**2**	**2**	**10**	**15**	**2**	**16**	**9**	**5**	**28**	**1**	**13**	**13**	**4**	**6**	**31**	**2**					

Table 9.5. Bifacial pieces recovered from Khuzmum rockshelter excavations, SU45-1 Trench A, studied by Rémy Crassard. Measurement is in millimeters; $n = 11$. Note that all shaping is balanced on upper face and unbalanced on lower face.

Locus	Lot	Bag	Conservation					Section			Dimensions (mm)			Observations
			Complete	Basal	Baso-medial	Apical	Undetermined	Plano-convex	Asymmetric	Symmetric	Total Length	Maximum Width	Maximum Thickness	
009	35	4		1					1		25	38	6	foliate
005	13	4	1					1			52	18	6	foliate, irregular retouch
005	13	12				1		1			42	20	11	
005	flot						1	1			18	10	5	
003	8	3			1				1		44	31	11	burned
003	8	2				1			1		27	25	8	
000	surface		1					1			69	27	14	
000	surface		1						1		41	32	12	
016	60			1						1	32	35	5	foliate
004	10					1		1			49	22	17	
007	21					1		1			47	24	9	
Totals			**3**	**2**	**1**	**4**	**1**	**6**	**4**	**1**				

human groups with cultural traditions that differentiate through time. The analysis of the rockshelter lithic material identified 11 bifacially worked pieces, fragmentary and complete, other than arrowheads. They show a shaping (*façonnage*) that is highly standardized, as almost all (10 out of 11) bifacial pieces have an asymmetric or plano-convex section. The retouch on each face is also systematically unbalanced, meaning that retouch (by soft percussion and pressure) is longer from one edge. This shows that the bifacial pieces are preforms for arrowheads and confirms the specialized aspect of the knapping activities at Khuzmum 45-1-A. They show plano-convex preforms in preliminary stages of production, maybe even before creation of the trihedral section that is characteristic of many arrowheads found across the stratigraphic accumulation. Dimensions of the complete bifacial pieces confirm this conclusion (table 9.5).

The remaining tools in the assemblage, especially the retouched flakes, scrapers, borers, and "blade-like" flakes, appear to be ad hoc tools, manufactured for a specific task and immediately discarded. There does not appear to be any preparation of other formal tools or any evidence of curation of specific tools. The few "blade-like" flakes were identified as such due to the fragmentary nature of the artifacts. Very little information can be gleaned from these items.

Dawn Walter's analysis applied a *chaine opératoire* approach to the lithic collection (flakes and cores). Rémy Crassard studied projectile points and bifacial pieces only. A lithic *chaine opératoire* considers the technical organization expressed through raw material selection; tool manufacturing, use, breakage, and sharpening; and discard trajectories. If each component of this sequence refers to a specific project or projects, this analysis provides insight into task-oriented behavior. Debitage categories used in this analysis are based on classification schemes widely employed (Bordes 1961; Tixier et al. 1980) and separate cores and shatter fragments from flakes, which can be further subdivided into classes within a reduction sequence. Walter's analysis identified blanks, cores, finishing flakes (BFF), thinning/shaping flakes (BTF), flakes of indeterminate reduction (FIR), biface initial reduction flakes (BIRF, IRF), and shatter (pieces of chert that show no evidence of being humanly struck but may nonetheless be waste products from knapping episodes (table 9.3). In addition, badly broken flakes that were unidentifiable in terms of their reduction sequence were classified as shatter.

The high numbers of biface thinning and finishing flakes indicate that tool manufacture and maintenance were occurring in this area. A moderate frequency of biface finishing flakes as micro-debitage (< 1–2 mm) indicates local resharpening of points. The rockshelter was repeatedly used as a location to refashion hunting tools in an area that provided ample raw material. The hunters' camps appear to have been good places for producing and retooling points, since few other stages of production were identified and there was a dearth of other tools.

Other Wādī Sanā Rockshelters

SU110—Test Excavation

The east-facing, huge rockshelter SU110-4 (E 328469, N 1739242 UTM Zone 39 North WGS84) is located along Wādī Sanā approximately 11 km upstream (south) of Khuzma as-Shumlya. It has a sloping rock floor that opens onto a relict silt terrace, which is bisected by erosion through the shelter itself, possible because of an unusual hole in the back roof that provides access to the plateau above. The Holocene silt terrace at the northern and southern ends of the rockshelter opening has been retained by rockfall, and modern bedouin occupations in the shelter have created thick dung mats that prevent surface erosion of any underlying archaeological deposits (figure 9.10). A surface scatter of chert knapping debris, points, thermally altered rock, intact roasting hearths, goat pens, and intact installations is evidence that multiple occupations have taken place. There is no rock art or graffiti. What seem to be Islamic graves are marked by stone cairns on the surface of the terrace, and at least one *madhbaḥ* (SU110-3) of unknown age is present. In the silt terrace sections, several buried hearths and an ashy layer (SU110-5) are visible at a depth of 1–2 m (figure 9.11).

Figure 9.10. Rockshelter SU110-4 viewed from the east, with silt terrace preserved to the right of parked vehicles. *Photograph by Joy McCorriston.*

Figure 9.11. Michael Harrower (right) indicating location of Hearth SU110-6, from which radiocarbon sample (AA60240) was taken. Trench A appears between two chaining pins to the left of the spade, below ʿAbdallah Nasser Ṣarām (standing, upper left). *Photograph by Joy McCorriston.*

Brief test excavations at SU110-4 Trench A cut through the ashy layer of Hearth SU110-5 and recovered no lithics. A radiocarbon sample yielded an age determination (6177–5751 cal BP; AA60243) for this layer, which probably formed much the way the ashy deposits formed in comparable terraces at Khuzmum (figure 9.12). There is a clear stratigraphic contiguity between this layer and a buried hearth, SU110-6, with a radiocarbon age 5997–5754 cal BP (AA60240) on charcoal removed from its section. This hearth also contained unidentified and uncharred animal bone and was constructed, like the buried hearths at the Khuzmum rockshelters, of wadi cobbles and pebbles in a shallow scoop in the ancient surface. Further test excavations extending Trench A recovered no diagnostic chipped stone in stratified deposits. Stratified occupation at the SU110-4 rockshelter appears to have occurred after the disuse of Neolithic knapping technologies.

CS1 and CS2—Records of Alluvial Deposition

Downstream roughly 5 km from SU110, a pair of shallow rockshelters face each other across the main channel of Wādī Sanā (98-CS1 E 331815, N 1742198 and 98-CS2 E 331730, N 1742646). Both rockshelters today preserve high relicts of silt and sandy strata built during the last of the aggradation in the main Wādī Sanā channel (chapter 3, figure 3.12). Wadi erosion has removed these last deposits elsewhere, exposing a section of the remnant protected by the overhang (figure 9.13). In Wādī Sanā CS1, wood from a basal ashy layer produced a radiocarbon age of 5568–5067 cal BP (OS16958), while a similar layer in Wādī Sanā CS2 dates to 5651–5327 cal BP (OS18691). Because they have been sheltered from surface erosion, these layers provide a terminus post quem for widespread fluvial incision and sediment erosion (see chapter 18 for details of radiocarbon dating).

Beyond their paleoenvironmental significance, there is archaeological evidence to consider, primarily anthropogenic assemblages of charred plant remains. Occupations at both CS-1 and CS-2 produced ashy layers with a moderate density of charred plant material. Flotation samples of several liters each were taken from the exposed sections. Analysis yielded dead wood (termite-ridden precharring) from acacia and tamarisk trees, and charred seeds from *Ziziphus leucoderma* as well as from fruiting herbs such as *Cleome* sp., *Helianthemum*/Malvaceae, *Corchorus trilocularis*, *Cucumis figarei*, *Crotalaria oocarpa*, *Chenopodium* cf. *murale*, and other annuals (McCorriston et al. 2002) (figure 9.14, table 9.6). The density of identifiable material and the presence of intact pellets of caprine-size animal dung suggest that these deposits were formed by slow-burning dung mats of the kind one can recognize today. Modern bedouin who reoccupy a rockshelter will deliberately fire the compacted residues of herded animals kept by the previous occupants—this practice sterilizes a pest-ridden shelter. Slow-burning, high-reduction firing of dung mats may have charred and preserved incidental seed materials such as bedding, kindling, commensal rodent caches, and deliberate human-gathered foodstuffs. Even so, the quantities of macroremains are very small, so that interpretations remain limited. The presence of caprine-size dung accords with our evidence of a late-ninth- to early-eighth-millennium BP introduction of sheep in Wādī Sanā (chapters 8 and 18). By the middle eighth millennium BP, sheep and goats had been introduced into foraging-based economies across Arabia, where they continued to play an important role (Magee 2014; Uerpmann et al. 2000).

Figure 9.13. Wādī Sanā CS-2 showing Middle Holocene sediments protected from erosion. Banded sediments in cave are about 2.5 m thick. *Photograph by Joy McCorriston.*

Figure 9.12. Trench A in silt terraces in front of SU110-4. Photo faces west. The ashy layer SU110-5, from which radiocarbon sample AA60243 was taken, is visible at the base of the right-hand chaining pin and extends across the back section of the stepped trench behind the chaining pins. *Photograph by Rémy Crassard.*

SU56—Food Processing

On the north side of the Khuzmah as-Shumlyah inselberg are several shallow rockshelters, some with graffiti to be described in chapter 17. Like many rockshelters in Wādī Sanā, SU56-13 has a floor of exposed bedrock without any deposited sediment. But there are nonetheless traces of human use. On a rock ledge over SU56-13 are three oblong bedrock mortars about 0.40 m long and about 0.08

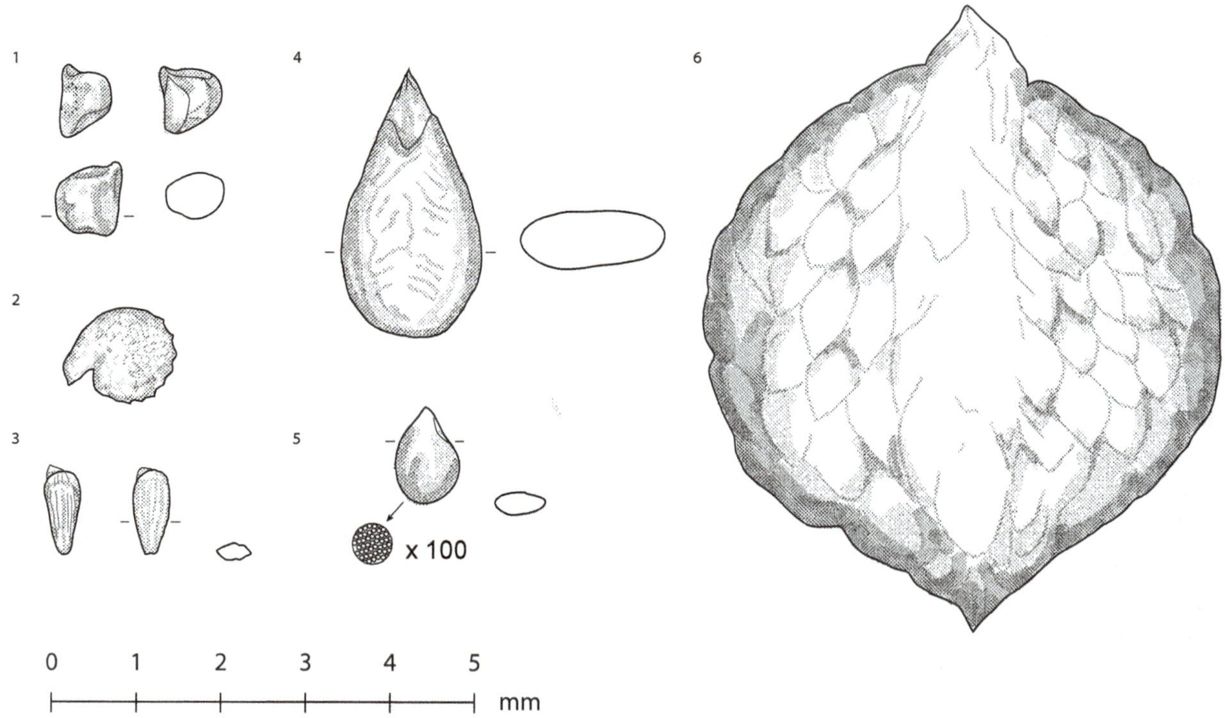

Figure 9.14. Charred seeds from Wādī Sanā rockshelter 98-CS-2: (1) *Corchorus* cf. *trilocularis*; (2) *Cleome* cf. *scaposa*; (3) *Pulicaria* sp.; (4) *Cucumis sp.* (*figarei/prophetarium*); (5) *Solanum* sp.; (6) *Ziziphus leucoderma*. *Drawings by Joy McCorriston. Illustration by Clara Hickman.*

m deep. These were the only bedrock mortars recorded by the survey. They are artificially smoothed within and whitened by grinding, producing a contrast with weathered bedrock around them. There is no clear indication what they were used for—grinding pigments, polishing wooden objects, preparing plant foods, pounding a paste for infants and toothless consumers, or perhaps to release bark tannins for leatherworking. They occur near the rockshelters with the densest and most recurrent evidence of reoccupations in Wādī Sanā (figure 9.15). Whenever they first came into use, such mortars must have added to the local resources at Khuzmum, attracting returning visitors who also used the rockshelters and constructed check dams and upslope *shrūj* features for diverting water, presumably for cultivation. The rockshelters afforded potential food storage sites. Perhaps a limited cultivation by mobile peoples may be implied by this constellation of shelter, water management, and processing installations.

There were several surface finds of freestanding limestone mortars in middle Wādī Sanā, including one from SU12. One mortar collected in 1998 is stored in the Mukalla Museum; others were photographed and left where encountered in the field, on the surfaces of silt terraces. These mortars are each about 0.3 m deep and conical in form (figure

Figure 9.15. Bedrock mortars adjacent to Khuzmum rockshelter SU45-8. *Photograph by Joy McCorriston.*

Table 9.6. Charred plant remains from dung mats in Wādī Sanā (after McCorriston et al. 2002:80, table 4).

	1998 CS-2, Meters below Surface		
Taxon/Type	**FS 1122 0.25 m**	**FS 1120 2.4 m**	**FS 1121 2.3 m**
Amaranthus/Chenopodium		74	56
Chenopodium sp.		5	
Cleome cf. *scaposa*	1	2	
Cleome cf. *ambylocarpa*			1
Cleome sp.		2	1
unknown 1 (Cruciferae?)		5	2
Crotalaria sp.		1	
Ziziphus spina-christi/leucoderma		3	8
Ziziphus leucoderma			4
Corchorus cf. *trilocularis*		27	7
unknown 4 (drupe)		3	
unknown 6			1
Solanum sp.		3	
Cucumis sp. *(figarei/prophetarium)*		22	10
Pulicaria sp.			1
parenchymous fragments		+	+
monocotyledonous stem	1		
animal dung (goat)		+++	++
wood	+	+	

9.16). No elongate stone pestles were found; these are more portable than mortars, and anyhow wooden pestles may have been more effective. Such mortars are useful for cracking and husking and are scarcely portable. There is no clear evidence by which they may be dated; nor need their shaping and use be confined to any particular period. Additionally on the surface of SU9-2, the RASA team documented a limestone grinding slab with a circular mano. Foods in Wādī Sanā that might be processed in such mortars include small seeds, cattail (*Typha*) and Cyperaceae roots (in marshy ground), pods from *Acacia*, dried fruits of *dom* (*Ziziphus spina-christi, Z. leucoderma*), and grass inflorescences.

Figure 9.16. Surface find of a freestanding limestone mortar in Wādī Sanā. *Photograph by Catherine Heyne.*

The Gravel Bar Site (GBS)

Upstream along Wādī as-Shumlyah from its confluence with Wādī Sanā, the modern dirt road slices through several stone monuments and bypasses others, including a trilith built on silts (SU6-1). Among the fill forming the platform bases of its trilith uprights are smooth wadi cobbles and Neolithic projectile points, signaling that the construction materials surely came from a nearby gravel bar deposited in violent overbank flooding many thousands of years previously (figure 9.17). It was surely the visible trilith monument that caused petroleum prospectors and archaeologists to stop here, but the adjacent gravel bar provided a great photographic viewpoint—and there are hundreds of knapped chert tools, blanks, and associated debris underfoot. From here Neolithic knappers could see the Khuzmum inselberg while perched on a rocky highland

Figure 9.17. Overview of Gravel Bar Site from the north. Note alignment of dark uprights (Trilith SU6-1) in middle ground next to the road. GBS rises on the far side of the road, with a figure and tripod at the left end and two vehicles parked among the foliage behind. *Photograph by Joy McCorriston.*

Figure 9.18. Overview of the surface of the GBS. Note the alignment of limestone manuported uprights in left foreground. These were a target of early excavations in 1998. *Photograph by Joy McCorriston.*

Figure 9.19. GBS Trench A after excavation to sterile silt surface. Note limestone blocks embedded in underlying silts. *Photograph by Joy McCorriston.*

within marshy ground. Figure 1.2 (chapter 1) also shows the surrounding area, looking up the Wādī as-Shumlyah toward the GBS (SU0). The site may have been a perfect hunting blind surrounded by marsh and a great chert source of stream cobbles on high ground.

Test Excavations

The surface of the Gravel Bar Site had a dense scatter of chert knapping debris and tools, several hearths, and large limestone slabs protruding from the gravel (figure 9.18). Given the high density of cultural material, three test excavations, Trenches A (at the west end of the GBS), B (central), and C (east end), were concluded in 1998.

Trench A (E 337147, N 1744333) extended as a 3 x 3 m square with a depth of 28 cm. Under the surface, excavators encountered no lithics or other artifacts. The underlying matrix is rounded gravel and small cobbles, which have filled in around and probably displaced massive limestone slabs (figure 9.19). There is a disconformity between the smooth gravel and the underlying silt, which was very compact and sterile. Massive flooding, which laid down stream-rolled gravel, must have removed some upper strata of silt; flooding nonetheless lacks the velocity to have carried (and placed upright) limestone slabs without removing underlying silt. Some limestone slabs were embedded in the silt; others were aligned with flat surfaces tipping westward downstream. Stream gravel is sorted in several layers, with the heaviest gravel at the bottom, and the gravel layers settled around tilted limestone blocks.

This sequence was repeated in Trench B (4 x 4 m) (E 337164, N 1744325) (figure 9.20) and Trench C (3 m north–south x 1 m east–west) (E 337216, N 1744314), where large limestone blocks were embedded in sterile gravel under a cap of surface lithics. No cultural material occurred below the surface except the limestone blocks as manuports to the site.

It is clear that violent floods shifted large limestone slabs—some more than 1 m in length—that once belonged to a human construction on the Holocene silt terrace. Humans built an unknown something there; geomorphologist Eric Oches argues that no natural force could carry such limestone slabs across intact silt (figure 9.21). The nearest source of slabs is tabular bedrock more than 100 m distant across a subsidiary channel several meters deep. Subsequent violent flooding, plausibly a single episode, deposited a bed of sterile cobble and pebble-size stream gravel that has capped silts and prevented their further erosion. But had the flow also carried limestone slabs, no silt base could have remained intact

Figure 9.20. GBS Trench B at close of excavations in 1998. *Photograph by Joy McCorriston.*

beneath it. Subsequent erosion of the Holocene silts has left the gravel at higher elevation than surrounding silt. Thus the Gravel Bar Site is today a terrace where in antiquity it fanned across a flatland or filled a channel.

Why did humans drag limestone blocks to build on the Holocene silts? Was this once a water management structure? A monumental platform? Something else? Despite initial published interpretations (McCorriston and Oches 2001; McCorriston et al. 2002), it remains difficult to tell for certain. Limestone blocks could not have been transported by stream flow, but they aligned according to the direction of flow. Patination of the limestone blocks is impressive where they protrude above the gravel, suggesting a very long period of partial burial in gravel. The lithic knapping debris caps the site and includes Neolithic pieces roughly comparable to stratified lithics at the Khuzmum rockshelters and Manayzah and dated to the eighth millennium cal BP at those sites (radiocarbon ages on associated materials). Surface collections and study clearly show that these lithics accumulated in place after flooding destroyed prior structures.

Surface Collections

Surface collections from the GPS have picked the site almost clean of most tools and preforms. Only the RASA-collected material is archived in the Mukalla Museum in Hadramawt. Archaeologists conducting a contracted survey in 1993 to identify archaeological sites prior to an oil pipeline construction first identified the GBS. Their collections are held at the German Archaeological Institute in Sana'a. As these selections—presumably of end-stage knapping products or tools with a formal shape—have never been published, it is difficult to know the purpose and method of their collection. In 1996 McCorriston visited the site with Burkhard Vogt of the German Archaeological Institute in Sana'a but removed no new material. Some artifacts have since been taken by amateur visitors, and the quantities of artifacts on the GBS surface diminished over the years of RASA fieldwork. McCorriston suspects that most were removed as illegal souvenirs by oil company subcontractors and security forces. In 1998 a handful of end-stage knapping products (projectile points) and a hammerstone were collected at the GBS as a nonsystematic grab sample during RASA extensive survey (chapter

4). These were supplemented by all tools and debitage from specific, systematically sampled locales on the GBS surface (see below). These also are archived in the Mukalla Museum and form part of the collection analyzed first by Dawn Walter (Walter et al. 2000) and by Rémy Crassard in 2004 (see below). Finally, in 2004 Rémy Crassard and Julien Espagne removed all visible remaining blanks, tools, and diagnostic debitage fragments that could be used to reconstruct bifacial production and projectile point typology. This 2004 study was done to expand the inventory of diagnostic pieces and to compare them with identified types already becoming known and dated in stratigraphic sequences nearby, at Manayzah (chapter 8) and at Khuzmum rockshelter SU45-1. The purpose was to date lithic production at the GBS if possible and to expand regional knowledge of typology and ancient knapping.

With 1998 excavation results from three test pits indicating no stratified deposits of intact cultural material, the team initiated a detailed and systematic surface examination to determine whether the lithic debris on the surface was the result of in situ knapping or whether it had been rolled or eroded. If in situ, it would seal and potentially offer a relative date for the human-transported limestone blocks below, whatever their original purpose.

In 1998 RASA mapped the GBS surface by establishing a main north–south transect over the highest elevation, with perpendicular east–west transects (A through I) crossing at every 5 m interval. Along these east–west transects at 5 m intervals, Zack Johnson laid a 1 x 1 m drawing frame or cross tapes centered on the 5 m mark (figure 9.22). Within this sample area, he noted all tools, blanks, and knapping debris, assigning to each a relative value of battering on a scale of 1 ("unworn") to 5 ("very badly worn"). At each sample area where sediment existed, Johnson used a household sieve (mesh size about 1 mm) to sift for tiny micro-debitage and retouch flakes, thereby capturing the full size range of debris. For each lithic, Johnson recorded the amount of wear. The study intended to assess the taphonomy of the surface scatter and to explore whether and how natural sorting had rearranged lithic debris after its deposition, which must have occurred after a violent flood destroyed an existing structure.

Figure 9.21. GBS Trench C (erroneously labeled Area B in image). Note the limestone block exceeding 1 m in length. Area C is at the highest elevation of the GBS, and here one finds the largest limestone blocks, transported by humans for an unknown purpose. *Photograph by Joy McCorriston.*

Figure 9.22. GBS surface survey. Left: Pieter Vlah (foreground) and Zachariah Johnson (rear) lay out survey grid across the GBS. Right: Zack Johnson collects and sieves for artifacts at regular intervals on the surface. *Photographs by Joy McCorriston.*

Results of the 1998 Study

Johnson found that along the crest of the site there existed little sediment in many places, but there were tools, debitage, and tiny thin chips from retouching, often all "unworn" (Stage 1). Even under 10x magnification, thin retouching flakes show no sign of the battering they would receive if rolled into place during the deposition of water-sorted gravel. (A comparative sample is the thin retouching flakes picked from the heavy fraction of flotation samples from hearths at SU44; these show microscopic chipping from being stirred in a flotation tank even though they were sealed in a primary context for thousands of years.) There were very few natural clasts on the GBS; gravel came from stream deposits. Lithics at the lower elevations of the site did tend to be more worn (Stages 3–5)—perhaps because they had rolled downhill or because erosion of other materials had abraded them in place over millennia (figure 9.23). Everywhere that Johnson could employ sieving, he found few or no lithics in the upper centimeters below surface, confirming the excavation results that indicated only surface lithics. Debris is present from all stages of processing, including tested cobbles, cortical flakes, discarded cores, broken blanks, secondary flakes, retouching flakes, a fluting flake from a near-finished tool, and broken and discarded tools (figure 9.24). The site appears to be an in situ knapping floor, perhaps reused many times by transient knappers. The occasional obsidian flake attests to long-distance circulation of material.

Lithic Technology

A first analysis by Dawn Walter focused on a *chaine opératoire* inventory, comparable to that at Khuzmum rockshelter SU45-1-A. Walter identified a sample of 64 tools, 1,048 pieces of manufacturing debitage, and

115 pieces of shatter recovered from Shumlyah GBS in 1998. Tools include 24 bifacial foliates (broad and narrow), 20 trifacially worked tools, 11 trihedral drills/rods, six borers, two side scrapers, and one retouched flake. Approximately 80 percent of the bifacially and trifacially worked tools (trifacial foliates and trihedral drills) are fragmentary, while a high percentage of the broken items appear to be basal/proximal portions of the implements (Walter et al. 2000:figure 12).

A second inventory made by Crassard in 2004 focused on analyzing the bifacial production and the projectile point types, including now the pieces he and Julien Espagne collected. Within the previously inventoried corpus, 33 complete or fragmentary bifacial pieces (table 9.7) and 22 complete or fragmentary arrowheads that were retouched by pressure were counted (table 9.8).

Debitage

Based on Walter's analysis, the manufacturing debitage consists of flakes from all stages of bifacial reduction, from large corticated initial reduction flakes to small bifacial thinning flakes associated with retouch (table 9.9). Bifacial thinning flakes (BTF) dominate the debitage assemblage, representing 56.6 percent of collected pieces by count. It is highly likely that many of these flakes derive from the reduction of biface preforms or the sharpening of existing tools, both carried onto the site from elsewhere. Approximately 20 percent of the debitage flakes retain cortical material and are initial reduction/primary flakes, suggesting that initial-stage manufacture used previously unworked pieces of raw material transported to this location. Later-stage finishing, retouching, and sharpening also occurred, but no inference can be made about their

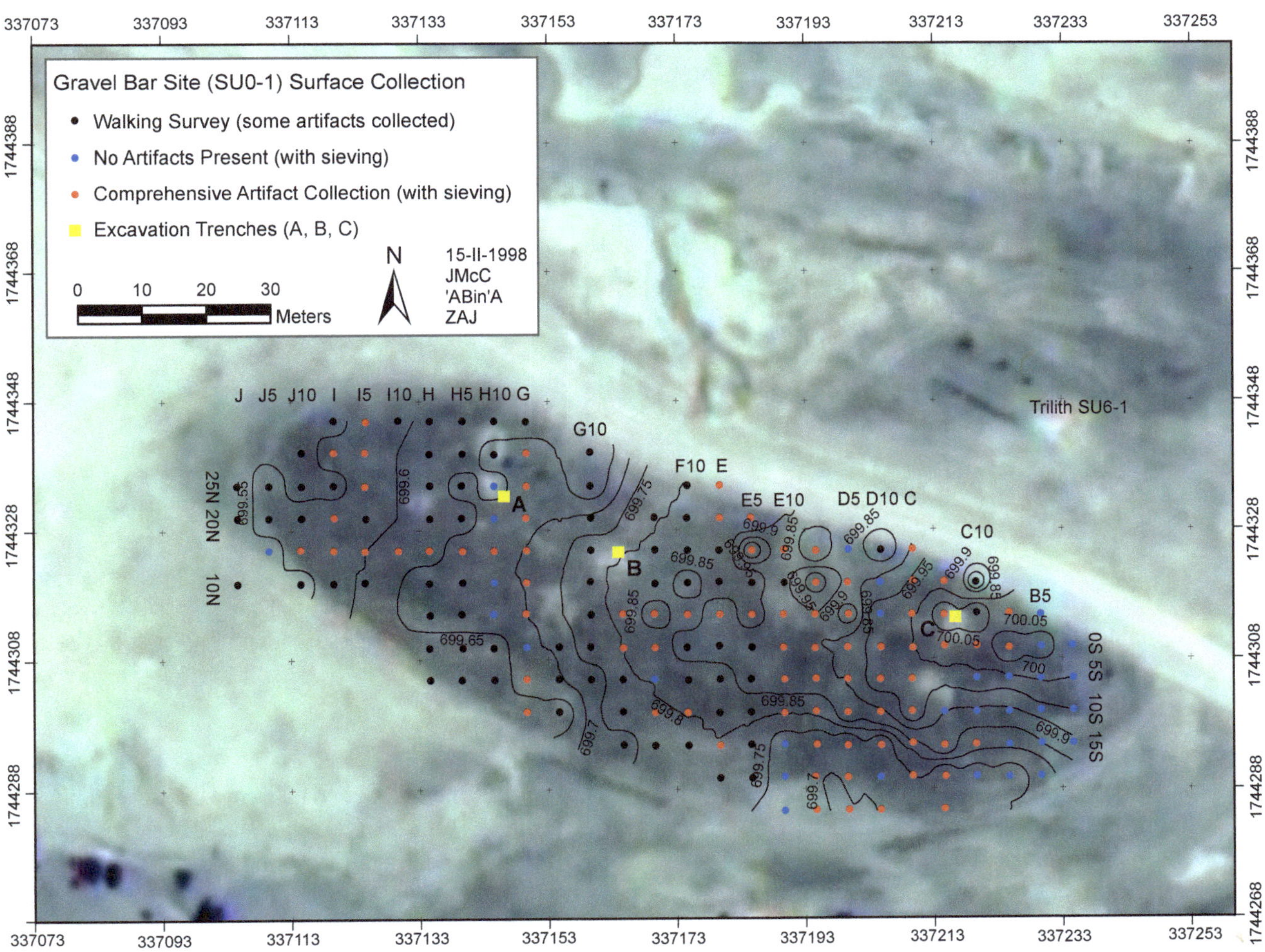

Figure 9.23. Map of Zachariah Johnson's GBS surface collection showing sample areas where micro-artifacts were recovered by sieving and artifact collections were made. Field map with a Brunton compass by ʿAbdalʿazīz Bin ʿAqīl and Joy McCorriston (UTM Zone 39 North, MSL EGM96). *Illustration by Michael Harrower.*

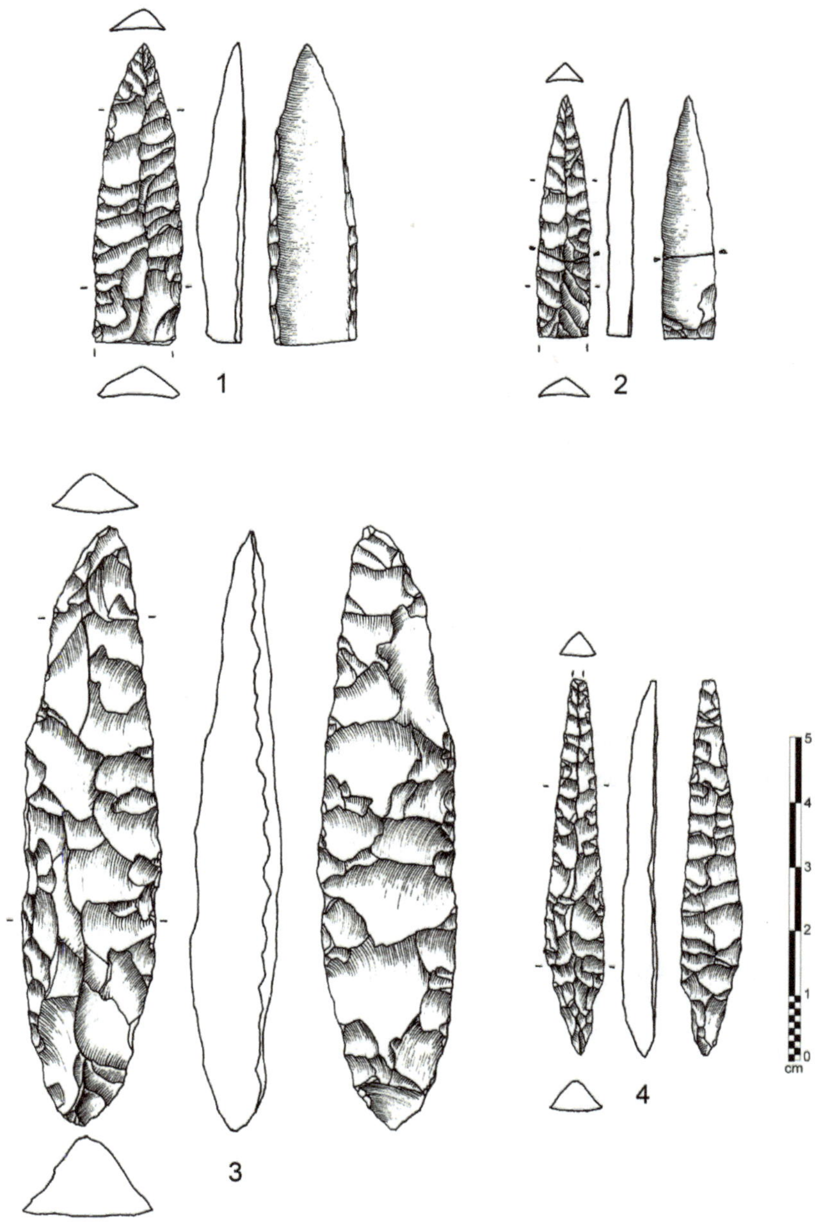

Figure 9.24. Selected surface finds of tools from Gravel Bar Site SU0-1: (1 and 4) general surface collection; (2) mapped surface collection point B4-O2; (3) D5-5S. (Refer to figure 9.23 for locations.) Collections remain in the Mukalla Museum. *Drawings by Julien Espagne. Illustration by Rémy Crassard.*

frequency because sampling strategies favoring larger items can account for the paucity of chips and micro-debitage. Although there is evidence for late-stage reduction, such as finishing flakes, there is little evidence for the later stages, such as retouching, platform preparation, sharpening, or maintenance. It seems that Neolithic hunters shaped many, if not most, of the tools from the vantage of a gravel bar overlooking the Wādī as-Shumlyah tributary to Wādī Sanā.

Projectile Points

The projectile points studied by Rémy Crassard (table 9.8) have a high stylistic homogeneity. They clearly show a standardized production and probably specialized production by one or more skilled knappers. The majority of these tools (81.8 percent) have a trihedral section. Other types of sections are rather plano-convex (13.6 percent) or asymmetric (4.5 percent). Their presence, combined with the absence of pieces with symmetric section, shows the evident search

Table 9.7. Bifacial pieces recovered from the Gravel Bar Site and analyzed by Rémy Crassard. Measurement is in millimeters; $n = 32$. Grid location refers to figure 9.23, surface collection of the Gravel Bar Site.

Piece No.	Grid Location (all surface)	Conservation						Section		Dimensions (mm)			Observations
		Complete	Proximal-medial	Medial	Medial-distal	Distal	Undetermined	Plano-convex	Asymmetric	Total Length	Maximum Width	Maximum Thickness	
1	**B4**					1		1		35	23	9	
2	**B4**					1		1		23	24	9	
3	**B1**			1				1			26	12	
4	**B3**			1					1	28	19	10	
5	**B3**			1				1		16	10	8	
6	**undetermined**	1						1		86	34	22	oval type, cortical proximal (base)
7	**undetermined**					1		1		32	22	10	on flake distal; cortex on tip (distal)
8	**undetermined**	1						1		45	28	11	almond type
9	**undetermined**	1							1	53	18	7	almond type; bipoint (projectile?)
10	**undetermined**			1				1		38	23	12	retouch on proximal breakage
11	**undetermined**						1	1				15	
12	**undetermined**			1				1		39	13	7	projectile?
13	**undetermined**					1		1		34	20	10	
14	**undetermined**					1		1		28	17	10	
15	**E5-N5**			1				1		28	23	17	
16	**E5-5N**			1				1		44	18	11	
17	**E5-20N**				1			1		51	22	13	narrow oval type
18	**E10-10S**				1			1		47	19	10	narrow oval type
19	**F-5N**	1						1		74	25	16	long oval type; small extremity fragment missing
20	**G-20N**			1				1		40	19	11	
21	**J10-15N**				1			1		40	14	7	
22	**I5-15N**				1			1		45	26	11	
23	**D10-15S**			1					1	32	20	8	
24	**D-10N**	1							1	73	15	11	probably projectile preform
25	**D5-5S**		1					1		38	28	10	
26	**D5-0S**		1						1	40	15	8	
27	**D5-0S**				1			1		35	20	10	
28	**D5-5N**					1		1		39	27	11	
29	**D5-5N**				1			1		45	19	7	
30	**D-10S**					1		1		30	12	5	projectile?
31	**D5-0S**					1		1		30	23	9	
32	**D5-0S**					1		1		27	14	7	projectile?
	Totals	**5**	**2**	**9**	**6**	**9**	**1**	**27**	**5**				

Table 9.8. Projectile points recovered from the Gravel Bar Site and analyzed by Rémy Crassard. Measurement is in mm. n=22. Grid location refers to Figure 9.23: surface collection of the Gravel Bar Site.

Piece No.	Grid Location or Trench/Locus (Xn-000)	Conservation					Typology				Dimensions (mm)			Observations
		Complete	Proximal-medial	Medial	Medial-proximal	Proximal	Undetermined	Bifacial with shifted tang	Trihedral	Fluted trihedral	Total Length	Maximum Width	Maximum Thickness	
1	indeterminate				1					1	45	13	6	fluting of ventral face, from the tip
2	indeterminate	1							1		57	9	4	
3	C2-001		1						1		26	8	5	
4	A3-009			1					1		23	8	4	
5	B4-002				1					1	36	7	3	two refitting parts
6	B4-002					1			1		21	7	3	
7	B4-002			1					1		24	14	5	
8	B4-002			1			1				21	11	6	
9	B3-002				1				1		27	6	4	
10	B3-002				1				1		44	14	6	cortical tip (proximal)
11	B4-001					1			1		13	5	3	
12	B4-001			1					1		14	8	4	burned
13	B4-000					1			1		13	6	4	
14	B3-000			1					1		16	9	5	
15	H10-15N			1					1		11	5	4	
16	D5-5S					1			1		16	6	4	
17	C-5S			1					1		41	10	5	
18	G-15N					1	1				16	9	4	
19	F5-5N			1			1				20	10	5	tang?
20	H5-15N					1	1				18	12	5	
21	D-5N	1						1			48	9	3	
22	D5-5S			1					1		13	5	5	
	Totals	**2**	**1**	**9**	**4**	**6**	**4**	**1**	**15**	**2**				

for trihedral section or for plano-convex section with a trihedral tendency. The very low number of other types of tools discovered, except the arrowheads and the bifacial pieces, confirms the specialization of the knapping activities.

Type GBS 1

From this assemblage, one type of arrowhead can be individualized, as well as two subtypes:

Type GBS 1 is a bifacial or rarely trifacial point with trihedral section and with a small tang, bipointed (GBS

Table 9.9. Lithic debitage recovered from the GBS and tabulated by Dawn Walter. The counts include bifacial finishing flakes (BFF), thinning/shaping flakes (BTF), flakes of indeterminate reduction (FIR), initial reduction flakes (IRF), and shatter (chert that shows no evidence of being humanly struck but may nonetheless be a waste product from a knapping episode).

Level		BFF		BTF		FIR		IRF		Shatter		Total	
#		%	#	%	#	%	#	%	#	%	#	%	
0	**Surface**	20	2	657	57	209	18	159	14	115	9	**1160**	**100**

1A). It's sometimes fluted along the ventral face, from the tip (GBS 1B). Type GBS 1A is close to Type Khuzmum 1A, but without a shifted tang.

As observed at Khuzmum SU45-1-A, the bifacial industry at GBS represented by 32 pieces is standardized to the search for a plano-convex section (table 9.7, cf. table 9.5). The bifacial pieces show a large proportion of this type of section (84.4 percent), while pieces with an asymmetrical section complete the rest of the corpus (15.6 percent). Most of the pieces with an asymmetrical section show a tendency toward plano-convexity. The reduction stages can be seen preceding the shaping of a flat face and a convex one. This technique is characterized by management of the removals, typical of the preliminary phases of retouching. These bifacial pieces at GBS can then be interpreted without much doubt as preforms for projectile points, abandoned at stages when their section was not trihedral enough, as the trihedral points might have been the final goal of the knappers. As observed at Khuzmum, the morphometric analysis of the complete pieces, and especially of the thicknesses, confirms this interpretation.

GBS is relatively close (1.4 km) to Khuzmum SU45-1-A and presents considerable similarity in some types of observed materials, particularly trihedral projectile points. The trihedral characteristic of the sections from many arrowheads is actually a strong technical and stylistic aspect. The material from the younger layers at Khuzmum SU45-1-A, which revealed almost only bifacial points with trihedral sections, is dated to the middle of the ninth millennium BP. The typological concordance between trihedral industries from Khuzmum and GBS is nevertheless imperfect. The earliest ones show a frequent shifted tang. Moreover, the fluting technique observed at GBS (Type GBS 1B) is well dated from Manayzah to the beginning of the eighth millennium cal BP. It seems then, after comparisons with Khuzmum and Manayzah, that the trihedral industry from GBS could be dated between the end of the ninth millennium and the beginning of the eighth millennium cal BP. Finally, these first analyses of lithics seem to reveal a specialized feature of the GBS site, oriented to massive production of quasi-exclusively trihedral points. This distinguishes the GBS site from the Khuzmum rockshelters or Manayzah, where the variability of the industries and the chronological depth indicate dwelling sites or hunting stops.

In the course of survey (chapter 4), the RASA team also documented surface find spots and collected lithic tools, some of which have been represented here (figures 9.25 and 9.26). These artifacts in surface context mostly have parallels at Manayzah and the Khuzmum rockshelters (for example, figure 9.25:3, 7, 8; figure 9.26). An exceptional type (Crassard's Hadramawt Type 3A; cf. Crassard 2008:144) (figure 9.25:1, 2, 4, 5, 6) has been found only on the surface in Wādī Sanā.

Conclusions

Through excavations and testing at multiple rockshelters, several important general observations can be made. First, these locations have ensured the preservation of deposits critical for establishing important chrono-stratigraphic sequences. Wādī Sanā rockshelters, including Manayzah (chapter 8), provide a fundamental record of the cultural-history sequence of southern Yemen. An important surface accumulation like the Gravel Bar Site requires material from stratified sites for ages and associations. Second, the multiple locations tested by the RASA team offer a glimpse of occupational history of Wādī Sanā and regions like it in the Southern Jol. There are too few radiocarbon ages to estimate continuous occupation (see chapter 18), but there are radiocarbon ages showing rockshelter use from the late ninth millennium cal BP, with subsequent occupations dated to the eighth, seventh, and sixth millennia cal BP. Some of these radiocarbon ages indicate repeated reuse of the same shelters. After 5,000 years ago, erosional processes in Wādī Sanā may have prevented the aggradation of cultural deposits on rockshelter terraces and scoured the rock floors of *abris*. Nevertheless, there are shelters occupied by modern goat herders, who form thick dung mats and burn them; build and reuse pens, curtain walls, and hearths; and blacken the ceilings of shelters with their soot.

The Place of Khuzma still retains its name from ancient times, graven on the bedrock above a campsite that

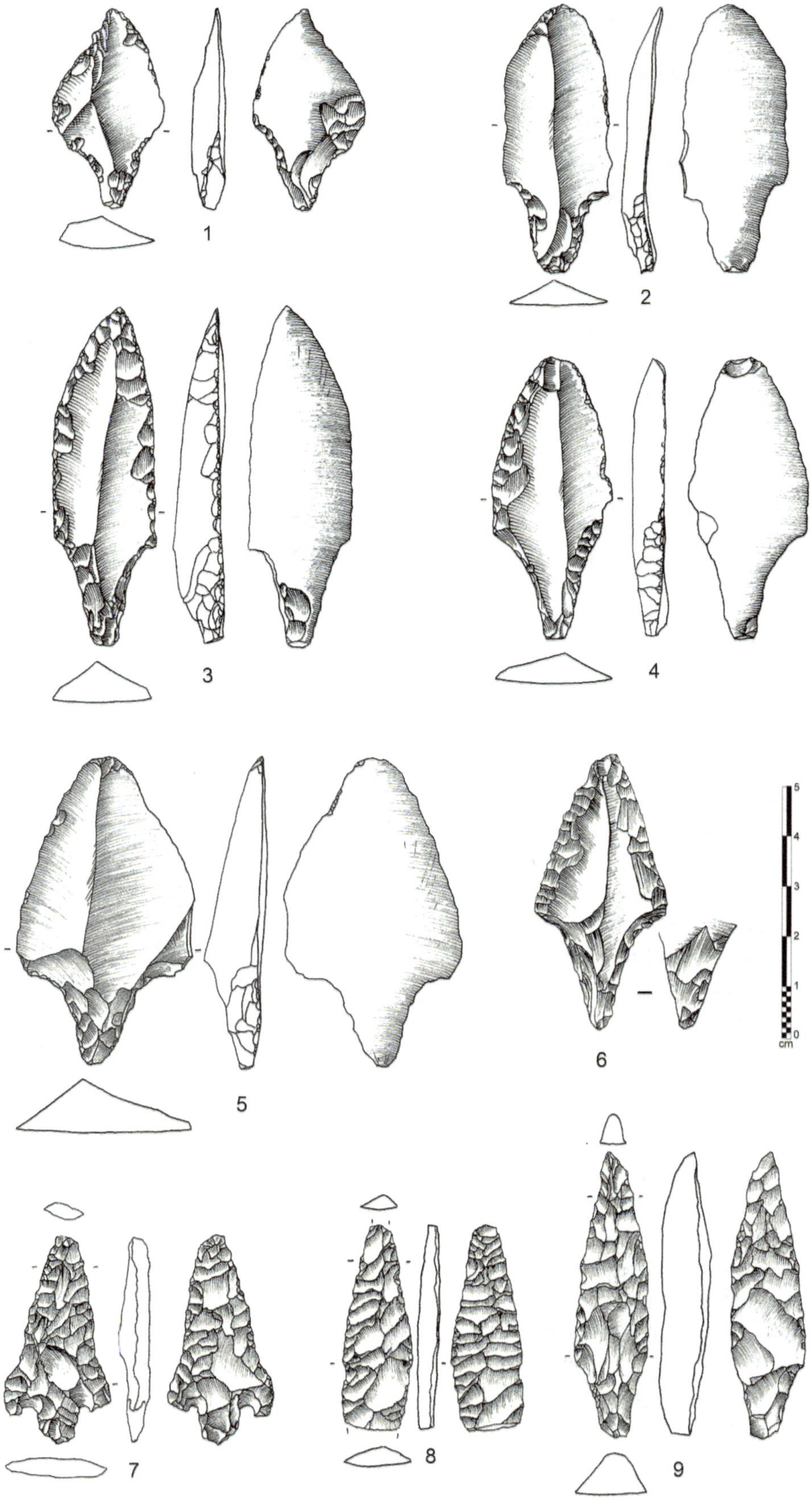

Figure 9.25. Surface finds from Wādī Sanā survey. Projectile points: (1 and 6) W1-0; (2) SU48-1; (3) SU30-1; (4) SU53-1; (5) SU76-1; (7) SU56-x; (8) SU56-13. *Drawings by Julien Espagne. Illustration by Rémy Crassard.*

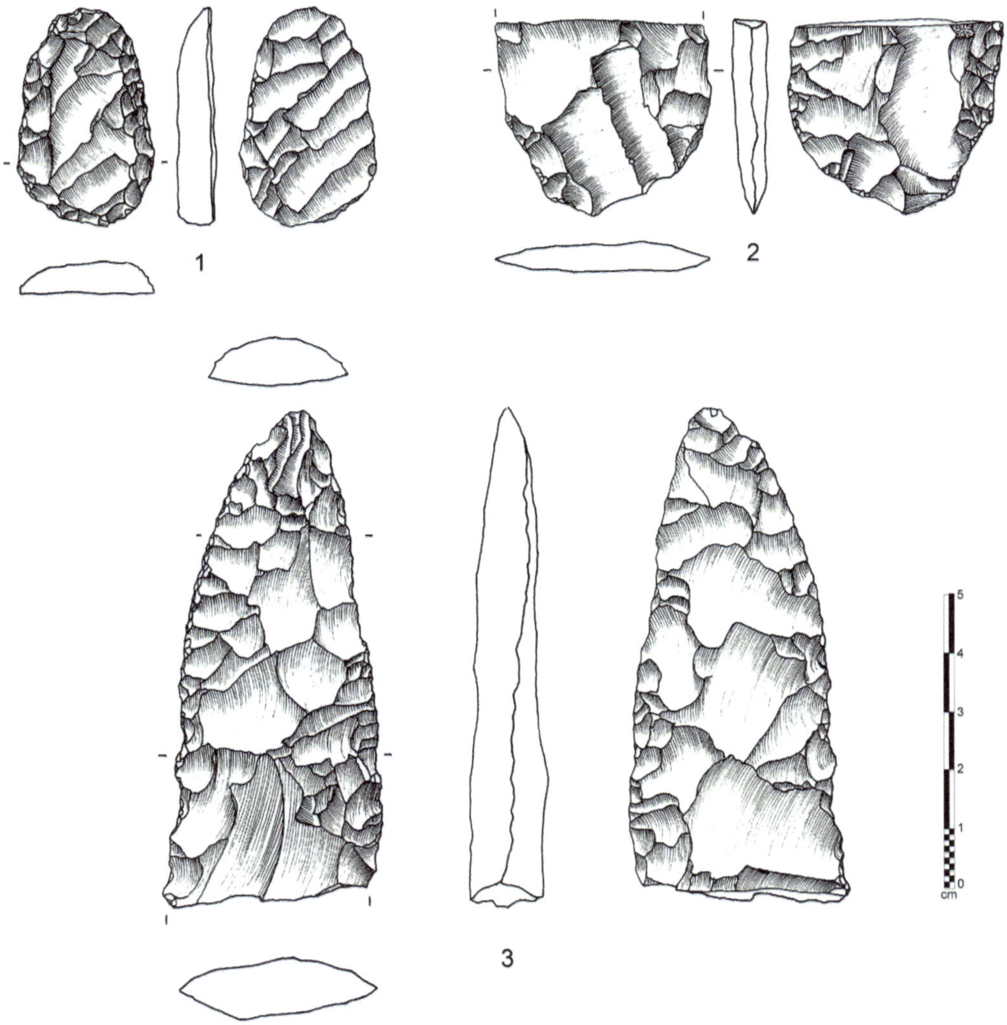

Figure 9.26. Surface finds from Wādī Sanā survey. Bifaces: (1) SU33-s; (2) SU16-40; (3) SU76-1. *Drawings by Julien Espagne. Illustration by Rémy Crassard.*

was already ancient, even when the lettering was fresh. Excavations at rockshelters have provided an important glimpse into material life of the past, with evidence of ancient hunting camps, technologies, herding, roasting meat, processing plant foods, storage of long-vanished goods, shelter and habitation, and long-range transport or exchange manifest in rare marine shells and obsidian fragments. What is more difficult to grasp in archaeological interpretation is the social practice and meanings such habits sustained. Even as one recognizes the material reoccupation and reuse of rockshelters in a landscape shaped by people and changing with shifts in climate, erosion, and deposition, there exists a faint echo of ancient practice in herder traditions today.

References

Amirkhanov, Hizri A.

1997 *The Neolithic and Post-Neolithic of the Hadramaut and Mahra.* Scientific World, Moscow.

Bordes, F.

1961 *Typologie du Paléolithique ancien et moyen.* Delmas, Bordeaux.

Crassard, Rémy

2008 *La préhistoire du Yémen. Diffusions et diversités locales, á travers l'étude d'industries lithiques du Hadramawt.* British Archaeological Reports, International Series 1842. Archaeopress, Oxford.

2009 Modalities and Characteristics of Human Occupations in Yemen during the Early/Mid-Holocene. *Comptes Rendus de l'Académie des Sciences: Geoscience* 341:713–25.

Crassard R., M. D. Petraglia, N. Drake, P. Breeze, B. Gratuze, A. Alsharekh, M. Arbach, H. S. Groucutt, L. Khalidi, N. Michelsen, C. J. Robin, and J. Schiettecatte
2013 Middle Paleolithic and Neolithic Occupations on the Mundafan Lakeshore, Empty Quarter, Saudi Arabia: Implications for Climate Change and Human Dispersals. *PLoS ONE* 8(7):e69665.

di Mario, F.
1989 The Western al-Rub' al Khali "Neolithic": New Data from the Ramlat Sab'atayn. *Annali Istituto Universario Orientale Napoli* 49:109–48.

Edens, Christopher
1982 The Rub al-Khali "Neolithic" Revisited: The View from Naqdan. In *Araby the Blest*, edited by Daniel T. Potts, pp. 15–43. Carston Niehbur Institute of Ancient Near Eastern Studies 7. Museum Tusculanum Press, Copenhagen.

Magee, Peter
2014 *The Archaeology of Prehistoric Arabia: Adaptation and Social Formation from the Neolithic to the Iron Age*. Cambridge University Press, Cambridge.

McCorriston, Joy, and Eric Oches
2001 News and Notes: Two Early Holocene Check Dams from Southern Arabia. *Antiquity* 75(290):675–76.

McCorriston, Joy, Eric A. Oches, Dawn Walter, and Kenneth Cole
2002 Holocene Paleoecology and Prehistory in Highland Southern Arabia. *Paléorient* 28(1):61–88.

Tixier, J., M.-L. Inizan, H. Roche, and M. Dauvois
1980 *Préhistoire de la pierre taillée, Terminologie et technologie*. Cercle de recherche et d'études préhistoriques, Antibes.

Uerpmann, Margarethe, Hans-Peter Uerpmann, and Sabah A. Jasim
2000 Stone Age Nomadism in SE Arabia: Paleoeconomic Considerations on the Neolithic Site of Al-Buhais 18 in the Emirate of Sharjah, U.A.E. *Proceedings of the Seminar for Arabian Studies* 30:229–34.

Vogt, Burkhard, and Alexander V. Sedov
1994 *Surveys and Rescue Excavations in the Hadhramawt Governate, Republic of Yemen*. Manuscript on file, Deutches Archäologisches Institut, Sana'a.

Walter, Dawn, Joy McCorriston, and Eric A. Oches
2000 Shumlyah GBS: An Arabian Bifacial Tradition Assemblage from Hadramawt Province, Yemen. *Neolithics* 2-3/00:12–14.

Zeuner, F. E.
1954 Neolithic Sites from the Rub' al Khali, Southern Arabia. *Man* 54:133–36.

Part IV

Middle Holocene: A Pastoralist Landscape

Chapter 10

Excavations at the Kheshiya Cattle Skull Ring and Neolithic Monument

Joy McCorriston

Discovery of the antiquity and functions of Neolithic monuments in Hadramawt has been one of the major contributions of RASA research to Arabian prehistory, and the presence and activities of Neolithic herding peoples have major implications for the subsequent development of Arabian societies. While summaries of RASA research on Neolithic monuments and its implications have been published elsewhere (Henton et al. 2014; McCorriston 2011, 2013a, 2013b, McCorriston et al. 2011, 2012), this report offers a comprehensive account of the excavations, taphonomic circumstances (this chapter), and finds (chapter 11), with critical documentation of figures and tables. Neolithic monuments are, it turns out, widely distributed in Southern Arabia (Braemer et al. 2003; McCorriston et al. 2014), but hitherto they have been poorly investigated.

Discovery and Distribution of Neolithic Monuments

Neolithic monuments were apparent from the first reconnaissance visits to Wādī Sanā, but none had documented their true antiquity and functions. In 1996 Burkhardt Vogt introduced McCorriston to Wādī Sanā and to sites identified by Vogt and Alexander Sedov during a Canadian Nexen pipeline mitigation survey. Vogt and Sedov (1994) included structures built of large blocks on or in low-lying terraces as components of a "Hadramawt Megalithic Complex." A number of collapsed structures lie near lithic surface scatters that Vogt and Sedov recognized as Neolithic scatters because of the distinctive projectile point types, then designated as the Arabian Bifacial Tradition, or ABT (Edens 1982). Among the visited sites were stone structures protruding from Early Holocene silt terraces adjacent to the modern Wādī Sanā channel (figure 10.1) as well as a short alignment (up to seven stones) of uprights on a gravel terrace in the Wādī Washaʿah, an upstream tributary just north of Ghayl Bin Yumain.

RASA survey and test excavations taught us that these ruins, some very poorly preserved, once belonged to a Middle Holocene tradition of building monumental platforms, subcircular, trapezoidal, or D-shaped in plan, with worked standing stones erected in front. As noted in chapter 6, platforms are small-scale stone monuments built of upright limestone slabs (or occasionally rounded boulders) that form a perimeter that is then filled with cobbles, boulders, or slabs, sometimes in combination with pebbles, chipped stone debitage, charcoal, and bone. Preservation of such monuments varies greatly, with near-perfect conditions of some on bedrock terraces or buried in silts (figure 10.2) and near-total erosion of others (figure 10.3).

Low-lying platformed structures appear throughout the middle and upper Wādī Sanā drainages and exhibit strong spatial association with wadi silt terraces known to have been richly vegetated in the Early and Middle Holocene (figure 10.4). Geomorphological and paleoecological reconstructions have moreover shown that the sedimentary wadi bottoms and broad sedimentary basins of the upper drainage were the only richly vegetated zones north of the

Figure 10.1. Neolithic monuments SU37-3 (left) and 38-7 (right) in foreground of the Khuzma as-Shumlya area, with Wādī Sanā channel edge vegetation in background (south). Aḥmad Nagī shown as scale. *Photograph by Joy McCorriston.*

Figure 10.2. Natural erosion cut through SU33-18, showing perimeter of upright limestone blocks and platform fill preserved in aggraded sediments. *Photograph by Joy McCorriston.*

Figure 10.3. Deflation and erosion of Middle Holocene silt terraces was impeded by the protection of limestone blocks and cobbles. Now scattered over a relict knoll in the silts, these stones likely once formed a Neolithic monument (SU26-1) at the confluence of Wādī Sanā and Wādī as-Shumlyah. *Photograph by Joy McCorriston.*

Southern Jol escarpment; plateau and rocky slopes with scant cover today also held no soil and therefore had scant cover in the Early Holocene. The RASA team encountered Neolithic platforms only in the Wādī Sanā watershed. While there was probably a wider distribution across the upper drainages of Wādī ʻIdim and certainly across Southern Arabia, RASA survey concentrated on Wādī Sanā and surely missed platforms elsewhere.

French archaeological survey in the Wādī Washaʻah, which drains the Northern Jol toward the desert interior, documented platform structures constructed in the same plan and fashion as Neolithic platforms of Wādī Sanā, and these monuments were moreover constructed at lower elevations than the Bronze Age tombs and in landscape contexts comparable to platforms in Wādī Sanā (Braemer et al. 2003; Steimer-Herbet, personal communication 2010). Farther away, comparable platforms that date to the Neolithic with absolute and relative chronological indicators appear in Dhofar, where they also conform to the locational pattern in Wādī Sanā: they appear inland of the escarpment in upper and middle drainages and are locally situated on lower terraces near good grazing and water resources (McCorriston et al. 2014).

Taphonomy and Spatial Distribution

The variable condition of standing monuments today led to a confusing first classification system based on present-day plan and features that, in retrospect, proved to be the outcomes of variable preservation. Only through a combination of intensive survey (chapters 4 and 6), analysis of the landform type and its role in preservation (chapter 2), and excavations have the conformity in construction technique, the original plan, and the widespread occurrence of such monuments become apparent. It has been helpful to rely on geomorphological and paleoenvironmental studies, which show that Neolithic monuments sit on lower slopes and terraces undisturbed since the Middle Holocene or are embedded in or eroding from the Early Holocene silt terraces. They are in locations that have not endured substantial inundation since the Middle Holocene and lie beyond the normal flood zone of modern stream discharge. A stone monument on bedrock offers poor stratigraphic potential for dating in the course of archaeological survey. Early Holocene silt terraces with monuments buried by alluviation (which ceased around 5,000 years ago) provided a terminus ante quem for construction (chapter 18).

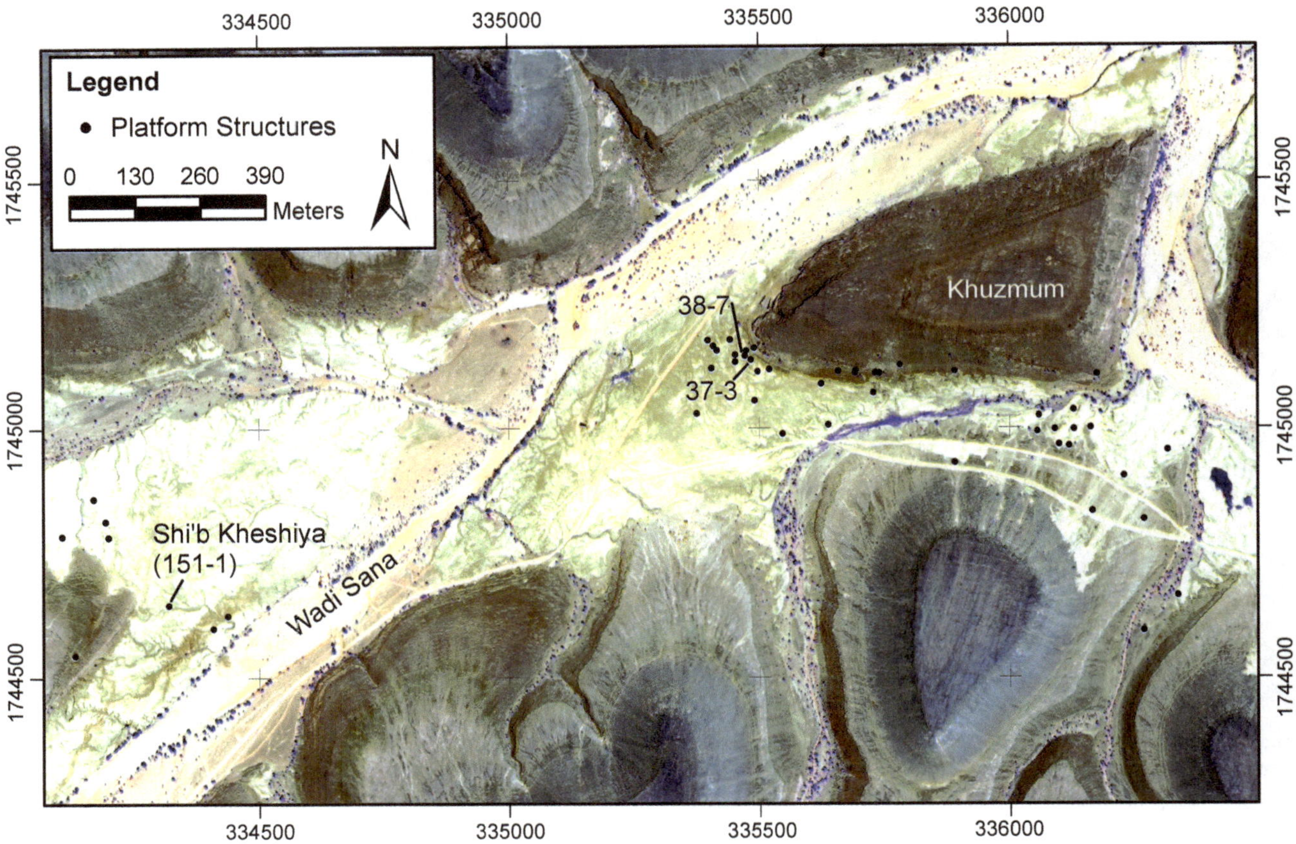

Figure 10.4. Map of SU151-1 and other Neolithic platforms near Khuzma as-Shumlya along the middle Wādī Sanā. *Illustration by Michael Harrower.*

Chronology and Distribution in Time

Monuments buried in silt terraces also can be associated with stratified artifacts and environmental proxies. All stratified Neolithic monuments lie near or at the top of the silt accumulations, and despite extensive examination of cut-banks and gullies, there are no cases of deeply buried monuments (greater than 0.40 m depth). Optically stimulated luminescence (OSL) dates supplemented by radiometric dating of charcoals show the silts to have been deposited between 13,000 and 5,000 years ago (chapter 18), so Neolithic monuments appear near the end of this time frame. While many of the platforms in Wādī Sanā are in poor condition, the excavation and documentation of better-preserved examples have made it possible to reclassify a number of structures, including extremely damaged ones, as remnants of Neolithic platforms. Excavations have shown platforms to have been occupied and abandoned before being filled and to date to the Late Neolithic, about 7000–6500 cal BP (see also McCorriston et al. 2002, 2011, 2012).

Shiʿb Kheshiya: The Cattle Skull Ring, Site SU151-1

In the middle Wādī Sanā, at its confluence with the Wādī as-Shumlyah tributary, the most distinctive landscape feature is Khuzma as-Shumlya, a bedrock inselberg isolated by ancient stream channels. Still locally called after the "Khuzmum" ("Place of Khuzma" in Old South Arabian) graffiti on its eastern rock face (chapter 17), the flat-topped feature commands a wide view and today marks important bifurcating routes north–south through Wādī Sanā and east–west across the Southern Jol. Smugglers say the eastward road goes "all the way to Oman," even as it climbs out of the Wādī as-Shumlyah tributary a few kilometers east of Khuzma as-Shumlya. Khuzma as-Shumlya is surrounded at its base by alluvial infill preserving diverse archaeological remains, including some of the earliest stratified camp sites in Southern Arabia (Crassard 2008; Martin et al. 2009; McCorriston 2006; chapter 9 this volume), numerous meat-grilling hearths, and early water management structures (Harrower 2008, chapter 13 this volume). Nearby are high circular tombs from the Bronze Age, and Iron Age trilith monuments (Bin ʻAqil and McCorriston 2009). Within Holocene alluvial terraces and on adjacent low-elevation bedrock outcrops sit the remains of more than 50 Neolithic stone platforms, some visibly associated with carefully shaped standing stones placed in front. Close study of natural sections also suggests an association between stone platforms and medium-large animal bone. Among the best preserved of these associations is the Neolithic platform and cattle skull ring at Shiʿb Kheshiya.

Discovery of Shiʿb Kheshiya

While surveying Early Holocene silt terraces southwest of Khuzma as-Shumlya in 2004, RASA Project team member Catherine Heyne noted large animal crania with maxillary teeth embedded in the section of a small gully south of Shiʿb Kheshiya (a minor tributary to Wādī Sanā) (figure 10.5). Only a few meters away was a D-shaped perimeter of upright tabular limestone slabs protruding through the alluvial terrace, along with the tip of a worked slab upright (figure 10.6). The slab-ring interior appeared to be filled with sediment, but subsequent excavations would uncover a stone platform made by deliberately filling the perimeter of limestone uprights with flat-lying limestone slabs and angular cobbles. With a quick sketch of the gully section, Heyne documented 12 stratified hearths, some of which yielded dates on charcoal as early as 7207–6839 cal BP (AA69754, 6,097 ± 39; H2 in figure 10.19). Exposed cranial fragments and teeth about 0.30 m below surface fell away readily, exposing behind them more crania and teeth in the same vertical orientation.

Methods: Excavations at Shiʿb Kheshiya

Recognizing that further gouging of a natural section impacted the stratigraphic integrity of archaeological deposits, the RASA team dedicated a season of fieldwork to excavations in 2005. Michael Harrower established a Main Datum on the nearest bedrock west of the site at E 334219.36, N 1744618.78 (UTM Zone 39 North WGS84), elevation 695.57 m MSL (EGM96) as measured with a Trimble Pathfinder Pro XRS GPS system using Omnistar real-time correction (elevation value 676.43 m HAE). Sub-Datum A was located at the base of a wooden post anchored in the silt terrace near the excavation trenches. Most elevations were obtained with a stadia rod and dumpy level from Sub-Datum A and referenced back to the Main Datum. The excavation trench was oriented approximately north–south, east–west by compass and was laid out with measuring tapes. As with most RASA small-scale excavations, we established no site grid, preferring

Figure 10.5. Cattle skulls visible in a natural gully section southeast of Neolithic monument SU151-1. Note the dark, ashy occupation at the nasal tip of the skulls and a lighter (upper) ashy layer at the level of eye sockets. *Photograph by Joy McCorriston.*

Figure 10.6. At SU151-1, the only limestone slab with signs of working was a standing stone southeast of the Neolithic platform (and outside the skull ring). *Photograph by Joy McCorriston.*

opportunistically to situate and extend small test trenches to recover stratigraphic and chronological information, probe architectural details, and maximize recovery of stratified finds and objects. Architectural outlines visible from the surface of platforms and monuments made this an efficient strategy.

Excavation followed natural stratigraphic divisions, so a locus is normally a stratigraphically constrained unit of sediment within one excavation trench or quad (for example, SU151-1 B, Locus 005). A lot is a subdivision of a locus imposed arbitrarily for various reasons—for example, when the locus includes a large volume of sediment; when excavators return after a hiatus; when a stratigraphic unit such as a hearth includes sediment of different colors, textures, or inclusions; or when excavators choose to impose spatial control with arbitrary subdivisions. Further control was maintained by sequential bag numbers within lots or a whole locus—one for bone, one for flotation samples, one for lithics, one per uncommon artifact, one for handpicked charcoal, such as a sample for radiocarbon dating.

All sediments were sieved through 0.5 cm mesh. Excavators sampled hearths, occupation surfaces, ashy deposits, and other charcoal-bearing sediments for flotation. Flotation sample sizes varied from 0.5 to 10 liters according to the availability of deposits.

Bone received somewhat unorthodox treatment. Whereas excavators normally would hurry to free and remove exposed bone from surrounding sediments, the circumstances at Shiʿb Kheshiya demanded a different treatment. First of all, bone was highly friable, and it offered less resistance than the concrete-hard surrounding silts. To remove it intact, excavators had to dampen the silty matrix, rendering it soft and easily disengaged from embedded bone, which also unavoidably had to be wetted numerous times in this process. Second, excavations revealed that bone had been used as architectural elements in a structure. Each skull received its own lot number within Locus 009, and from the first realization that the cattle ring represented, in addition to its archaeozoological significance, architectural integrity, excavators treated the cattle ring as a construction and excavated the sediments in Quads D, E, and B/E by recognizing their stratigraphic continuity with Quad B. Fragile bone, now rewetted many times, needed to remain in situ, freed from sediment protection, while the architectural plan was exposed. The team worked furiously to limit this exposure time. ʿAbdalKarīm Al-Burkānī developed painful inflammation of connective tissue in both arms, which he insisted on wrapping to continue incessant and repetitive picking to free bone alongside his colleagues, who were equally fatigued and equally dedicated. On his own initiative, Nasser Al-ʿAlīy hiked 10 km to the half-exposed structure and slept beside it to guard it from tampering when a flash flood separated the RASA team camp from the site for several crucial days. Once the structural configuration of bone elements was documented, each skull was freed, with its sediment fill intact inside the brain cavity. The skulls were wrapped in cotton bandages and secured within a plaster jacket. Despite concerns that damp plaster would affect the bone, subsequent analysis (chapter 11) showed this to be an effective stabilization for seven hours' transport over unpaved trackways to the Mukalla Museum and a two-year wait for analysis (chapter 11).

Results: Excavations at Shiʿb Kheshiya

These methods uncovered both the monumental Neolithic platform and the adjacent faunal remains at Shiʿb Kheshiya, revealing important evidence of Neolithic activities and constructions in the Early Holocene—namely, a ring of cattle skulls (chapter 11) subsequently marked by a stone platform and at least one standing stone (table 10.1). What follows is a description of each excavation quadrat, any extensions, and the stratigraphic associations of the stone monument and cattle skull ring.

Quad A: The Stone Platform and Interior Deposits

Quad A was laid out 2 m north–south x 1.90 m west–east in the north end of Structure SU151-1 (figure 10.7). The excavators designated as Quad A the area entirely within the protrusive limestone slab uprights. These uprights were 0.45 to 0.55 m high, ranged in thickness from 0.03 to 0.55 m, and converged to a corner at the north end of the D-shaped structure. The perimeter of the slabs was clearly visible, protruding from the surface, which was a 0.10–0.12 m thick, compact, yellowish–pale brown clayey silt

Figure 10.7. SU151-1 from the northeast at the beginning of excavation. The Neolithic stone platform was visible only at the edge of a natural gully as a few upright limestone blocks at its south end. The gully cut through ashy layers and hearths, here visible in the south section. Quads A and B here are each 1 m in width. *Photograph by Catherine Heyne.*

Table 10.1. Stratigraphic sequences for SU151-1, concordance with figures 10.9, 10.11, and 10.13.

Trench	Locus	Under	Over	Equals	Description
A	001	--	002, 003	010, ①	Compact, yellowish–pale brown clayey silt (alluvium), surface
	002	001		003, ②	Large flat-lying limestone boulders, cobbles, and pebbles in compact yellow silty matrix
	003	001	005	002	Compact yellow silt
	004	003	006	005	Small stones within compact yellow silty matrix
	006	004	007	005	Yellow silt
	008	007	009	--	Compact yellow silt
	009	008, 014	015	--	Silt and clay with few clastic cobbles and pebbles, bottom layer of rock fill
	010	--	002, 003, 011	001	Compact, yellowish–pale brown clayey silt (alluvium), surface
	011	010	012	002, 003	Compact yellow silt matrix with limestone inclusions
	012	011	013	--	Compact yellow silt
	013	012	014	--	Compact silty matrix with unsorted flat-lying limestone boulders, pebbles, and cobbles
	014	013	009	008	Pale brown–yellow silt with unsorted limestone cobbles
	015	009	016	--	Compact yellow-brown clayey silt with few unsorted limestone cobbles and pebbles and charcoal flecks
	016	015	017	--	Compact yellow clayey silt with charcoal flecks.
	017	016	018	-- ③	Brown-yellow silty matrix with sorted angular pebbles and charcoal inclusions, rootlets
	018	017	019	-- ④	Pale brown–yellow loam with inclusions of charcoal and ash, terrestrial shells, sorted angular limestone cobbles and pebbles
	019	018	020	-- ⑤	Hard yellow silt
	020	019	unexcavated	019	Hard yellow silty loam with rodent and insect burrows
C	000	surface	001/002		Very pale brown compact surface
	001	000	003	002	Light brownish–grey matrix with sorted clay peds, 1–2 cm in diameter, and a few terrestrial shells, bone
	003	001, 002	004, 005		Light brown sediment with charcoal, bone, and burned terrestrial shell inclusions
	004	003	006, 007	⑥	Very compact fine soil with a few pieces of charcoal and bone, in east part of excavation area only
	005	003, 004	006	②	Ashy sediment with numerous large land snails, lithics.
	006	004, 005	007, 008		Very pale brown, hard silt with charcoal and bone inclusions
	007	006	008		Hard silt with charcoal inclusions
	008	007	009		Pale brown silt and clay with some sand, ashy with charcoal and lithics
	009	008	011		Pale brown compact clayey silt in northeast area of excavation.
	010	007			Compact dark grey silt in southwest area of excavation, with lithic inclusion
	011	009	012, 013		Hard pale brown silt with bone and charcoal inclusions.
	012	011	013		Hard pale brown silt with charcoal and lithic in northwest area of excavation
	013	011, 012	014, 017		Hard pale brown silt with charcoal, lithic, and bone.
	014	013	015	③	Hard pale brown silt with charcoal inclusion, tree root, or insect burrow
	015	014	016	④upper	Brown hard silt with charcoal and bone as a depression in the east area of excavation
	016	015		④lower	Sterile pale brown silt
	017	013	014		Observed tree root penetrating matrix, filled with loose pale brown silt
	018	016	019		Hard pale brown silt appeared as a surface in eastern area of excavation, upon which bone was lying; charcoal, lithics also present.
	019	018			Loose pale brown silt filling a tree root
	020	014	021	015	Hard pale brown silt with charcoal inclusions in west area of excavation
	021	020	022	016, ⑤	Hard pale brown silt with charcoal and lithic inclusions in west area of excavation
	022	021	unexcavated	018, ⑤	Hard pale brown silt with charcoal and bone inclusions in west area of excavation; two possible postholes exposed at base of excavation.

Trench	Locus	Under	Over	Equals	Description
B	000	--	001	--	Compact
	001	000	002	①	Loose to compact very pale brown clayey silt
	002	001	003	②	Hard very pale brown silt
	003	B002, D002	004		Light yellowish–brown hard silty clay with a surface of small angular pebbles and a few small cobbles, bone, lithic, and thermally altered rock inclusions. Surface abuts exterior of limestone uprights of platform. Two flat-lying limestone slabs lie on this surface.
	004	003	005	③	Light grey compact layer of clayey silt with gravel inclusions as well as charcoal, distinct cluster of bone, and a few lithics. Very clayey regions adjacent to bone characteristic of decomposed tissue staining.
	005	004	006		Slightly darker, light grey, compact layer of silt with inclusions of bone, charcoal, a few lithics, and an obsidian fragment; bone clusters now readily differentiated into lots of different skulls (Locus 009, Lots 11, 12, 13, 14, 15).
	006	005	007		Light brownish–gray matrix with silt and sand and distinctly darker patches across the excavation area; inclusions of bone lots (Locus 009), charred *dom* seeds, charcoal, one piece of hair.
	007	006	008/009	D003, ④	Pale brown ashy layer of clayey silty sediment rich in charcoal inclusions as well as bone, thermally altered rock pebbles, and half a marine cowrie shell; Locus 009, Lots 16, 17, and 18 now also exposed; Locus 007 abuts the exterior of the limestone uprights of platform.
	008	006	007	④	Pale brown ashy area outside and to the west of cattle skull ring
	009	003, 004, 005, 006, 007	010	⑧, D009, E009	Cattle skulls, Lots 9–41; embedded in a silty matrix with pebble-size thermally altered rock, ash, and charcoal.
	010	009	unexcavated	E010	Pale brown silty matrix
	011	009	unexcavated	E011	Dark ashy matrix with thermally altered rock, small cobbles, and pebbles; a hearth.
D	000	--	001	B000, surface	Surface
	001	surface	002	B001, A001	Very pale brown compact silty sediment; hard peds; well sorted.
	002	001	B003	B002	Light gray hard silty clayey sediment over a likely surface (with bone, lithics, and rock lying flat over B003)
	excavation of D continued as B/E following locus numbering of B				
E	001	surface	002		Very pale brown compact silty sediment, occasional lithic and bone fragment
	002	001	003	B005	Arbitrary removal of hard light grey silt sediments to expose cattle skulls (Lots 3 and 10)
	003	002	004	B007	Pale brown ashy layer of clayey-silty sediment rich in charcoal inclusions as well as bone, a few lithics; Lot 7.
	004	001	005, B009	002 & 003	Arbitrary removal of very pale brown ashy sediment to expose cattle skulls outside and east of cattle skull ring
	005	004	unexcavated	B009	Temporary designation for cattle skulls in E; later labeled as B009.
	006	surface	005, B009	001, 002, 003	Arbitrary removal of very pale brown compact and hard light grey silt sediments to expose and remove cattle skulls
	007	006	008	B007, 004	Hard pale brown ashy silt with bone inclusions
	008	007	unexcavated	B007-010	Hard silt between ashy layers outside cattle skull ring
	009	007	010	B009	Cattle skulls in E
	010	009	unexcavated	B010	Pale brown silty matrix with many thermally altered rock cobble and pebble inclusions and charcoal inclusions around and slightly under cattle skulls
	011	010	unexcavated	B011	Dark ashy matrix with thermally altered rock, small cobbles, and pebbles; a hearth.

Figure 10.8. SU151-1 Quad A with limestone slab fill. *Photograph by Joy McCorriston.*

(alluvium). Final measurements of the structure perimeter were 2.54 m north–south and 2.02 m east–west.

Excavation revealed an underlying layer of large flat-lying limestone boulders up to 1 m long; these had been placed within the upright perimeter and showed no sign of tipping, as might be expected had they accumulated as a collapse event (figure 10.8). Cobbles and pebbles were also present as chinking material, particularly near the center of the platform, where slabs fitted against the perimeter did not meet smoothly. In excavation, a few fragments of nondiagnostic animal bone (shaft fragments) were recovered from the silt fill between the slabs. The structure contained no burial in the two-thirds excavated.

Beneath the flat-lying slab fill was a 0.05 m thick layer of clay patches. These clay patches probably accumulated as multiple accretions of super fine-grained aeolian deposits that were subsequently flooded; thereby they were water sorted so that the finest particles rested on top. The patches then dried to form mud cracks. Clay deposits and loamy silt were flecked with charcoal and a fragment of bone, plus two (nonhuman) molars. The clay deposits and underlying loamy silt probably represent episodes of post-occupational flooding before final infill of the platform (figure 10.9). The deep impressions of flat-lying slabs in the structure center strongly suggest that there was infilling when the interior ground surface was soft, as after a flood, and it is likely that the weight and perhaps violence of placing flat-lying slabs disturbed underlying wet layers.

These layers overlay a 0.05–0.12 m deep compact loam surface mottled with charcoal flecks, bone fragments, an obsidian chip, and distinctly burned areas (scattered burned pieces of clay), with concentrations of charcoal flecks in the surface against the eastern and western limestone uprights. There areas appeared to be informal hearths, subsequently disturbed by post-occupation flooding. Near these hearth areas were pieces of thermally altered rock and a few charred drupe pits (*Ziziphus spina-christi* or *Z. leucoderma*). A localized campfire had burned the interior face of two adjacent limestone uprights. However faintly, this surface sloping

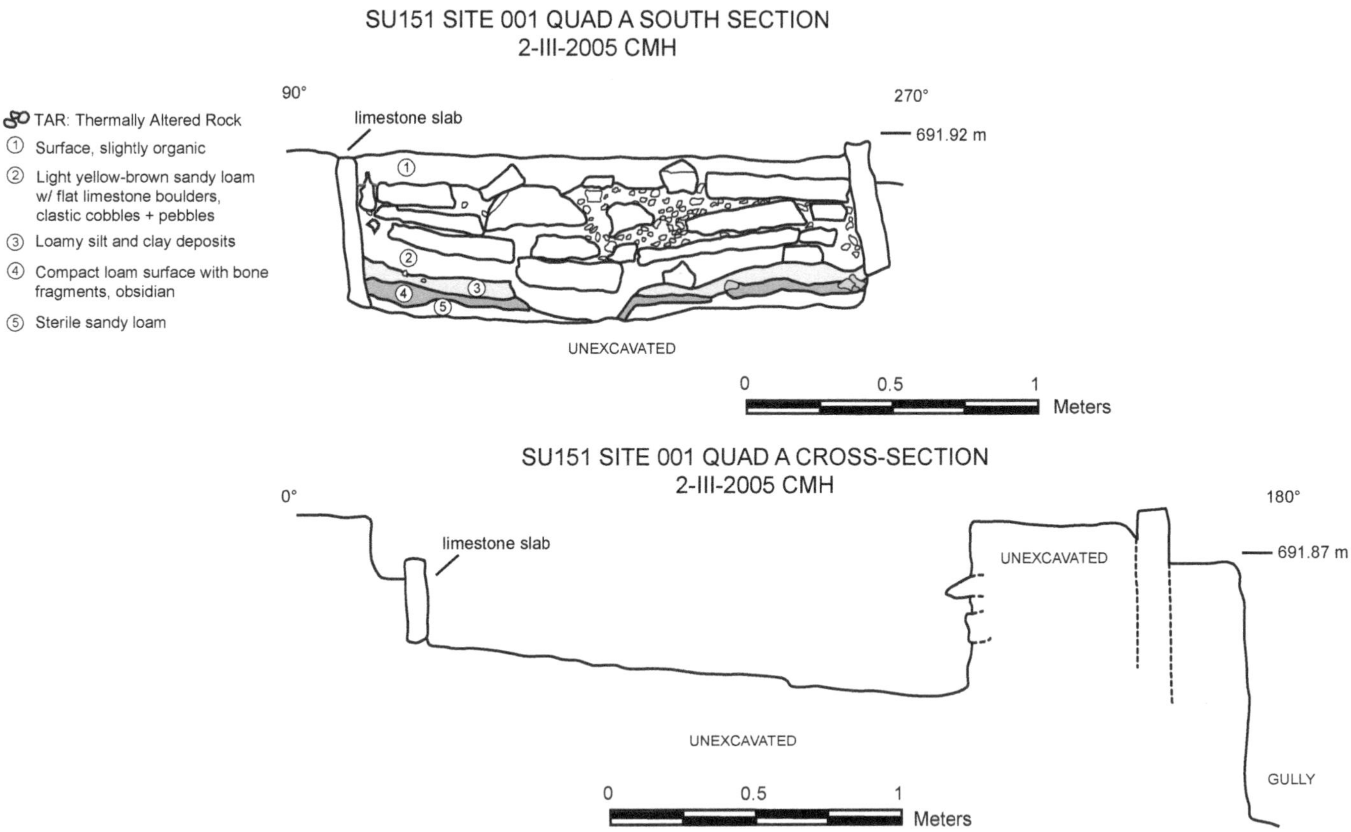

Figure 10.9. SU151-1 Quad A south section showing fill and underlying occupation levels. *Drawing by Catherine Heyne and Petra Creamer.*

down into the center of the structure (shape of the pit construction?) shows signs of in situ campfires and occupation, perhaps briefly while the platform was under construction (figure 10.10).

Across Quad A, excavators removed the compact occupation surface. Below lies a largely unexcavated layer of sterile sandy loam with rodent disturbances. This layer also underlies the limestone slab uprights forming the platform perimeter and is the layer into which the original platform was set. The west half of Quad A was excavated another 0.10 m in depth. Apart from rodent holes, there was no detectable disturbance (such as a burial pit) underlying the occupation floor.

From excavations of the interior (Quad A), it is clear that the D-shaped structure was a slab-filled platform constructed by excavating a slight pit in the ground surface to retain the outer bottoms of slabs. Some of these slabs were set on point, a construction technique also noted among Neolithic stone platforms nearby. D-shaped construction presumably began with two upright slabs at angles to one another to support them, and the rest of the slabs were set upright, sometimes with overlap to support adjacent slabs. Once set upright and supported on their outer sides by the pit edge, some uprights were chocked on the interior with blocky boulders to hold them upright. At this stage, the floor of the pit was occupied, accruing charcoal, fire areas, and limited debris like charred seeds, bone, and an obsidian flake. The limestone uprights were deliberately filled with four layers of flat-lying stone slabs with chinking material and cobble fill. The longest of these slabs measured 1.75 m, suggesting at least six to eight adult bearers to transport it across the silt terrace, based on the RASA team's experience in moving a similarly sized slab at the Gravel Bar Site (Trench C) in 1998. The entire monument was subsequently buried by alluvial sedimentation (chapters 2 and 18) so that the modern surface revealed only the tops of the upright slabs.

Quad C: Platform Exterior

Quad C was laid out west of the stone platform 151-1 and extended 1.8 m east–west x 1.4 m north–south (at

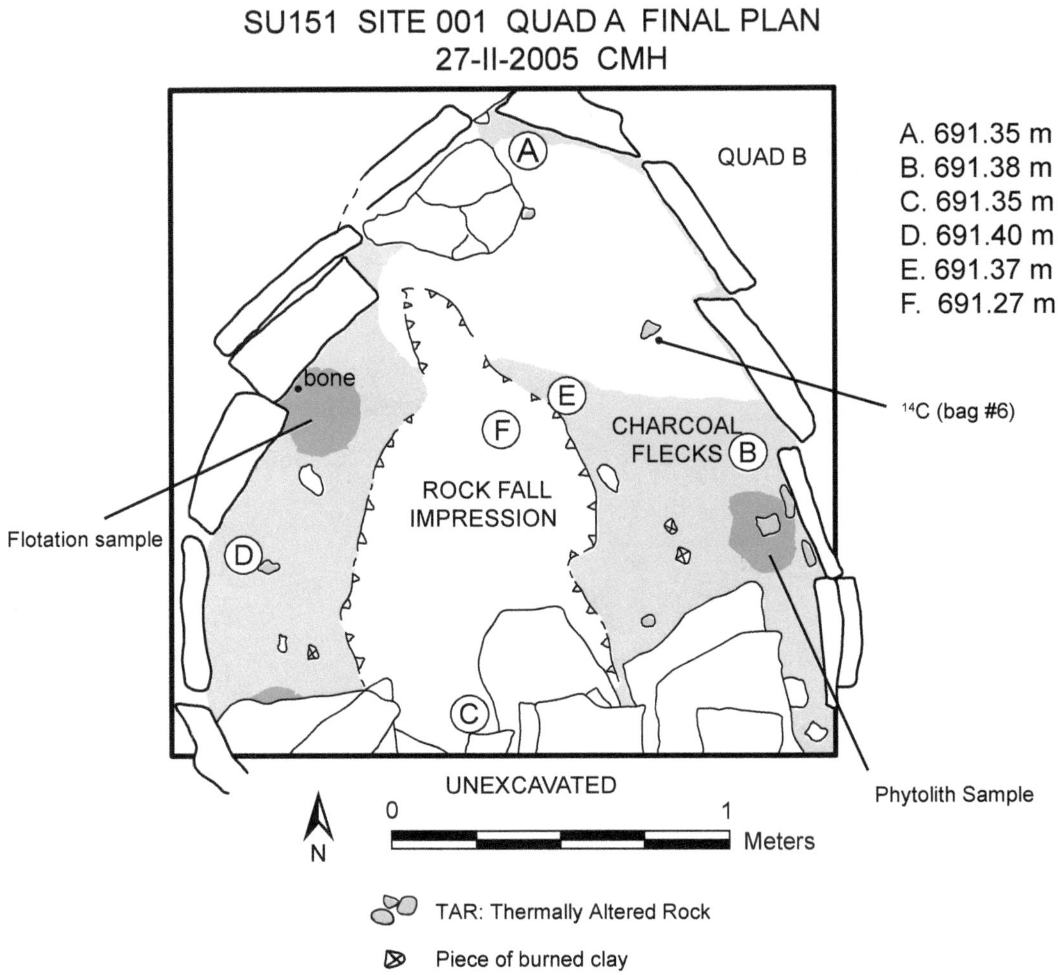

Figure 10.10. SU151-1 Quad A, plan of Locus 018 occupation level inside the D-shaped perimeter of limestone uprights. *Drawing by Catherine Heyne and Petra Creamer*.

the west end). Quad C was rectangular and formed a three-sided trench excavated to 0.55 m depth into the natural gully section (roughly east–west) west of the stone platform. Its purpose was to clarify stratigraphic context outside the stone platform structure. In excavation, a number of layers that did not appear distinctly in section were defined, as was true also for narrow vertical fill identified as stake holes (up to 0.06 m diameter) or tree roots.

A 0.20 m deep layer of light yellow–brown silty alluvium overlay the entire area. Below this surface layer in the eastern end (nearest the stone platform) was a distinct pocket of very compact brownish-grey organic sediment flecked with charcoal and bone. This pocket (Locus 015) cut into and likely marked a disturbance (near the platform) through a layer of yellowish pale brown sediment (Locus 005) that contained charcoal flecks, several pieces of thermally altered rock, and many large land snails (especially abundant nearest the platform).

Underlying this silt alluvium was a darker (brown-grey) compact organic loam (Loci 017 and 014), the first in which a root or stake hole 0.40 m deep was evident. This overlay a fine-grained compact gray-brown layer (Loci 020, 021, and 022), cut into at the east end by a disturbance of mottled backfill with yellowish-brown and light gray components (Loci 015, 016, 018, 019; figure 10.11).

An interpretation drawn from Quad C is that construction of the platform did disturb preexisting ashy sediments that had formed after the formation of the cattle skull ring. The disturbance (Loci 015, 016, 018, 019) may well be a foundation trench backfill, from when limestone slab uprights were set in a slight pit and backfilled outside to support the outer edges. It should

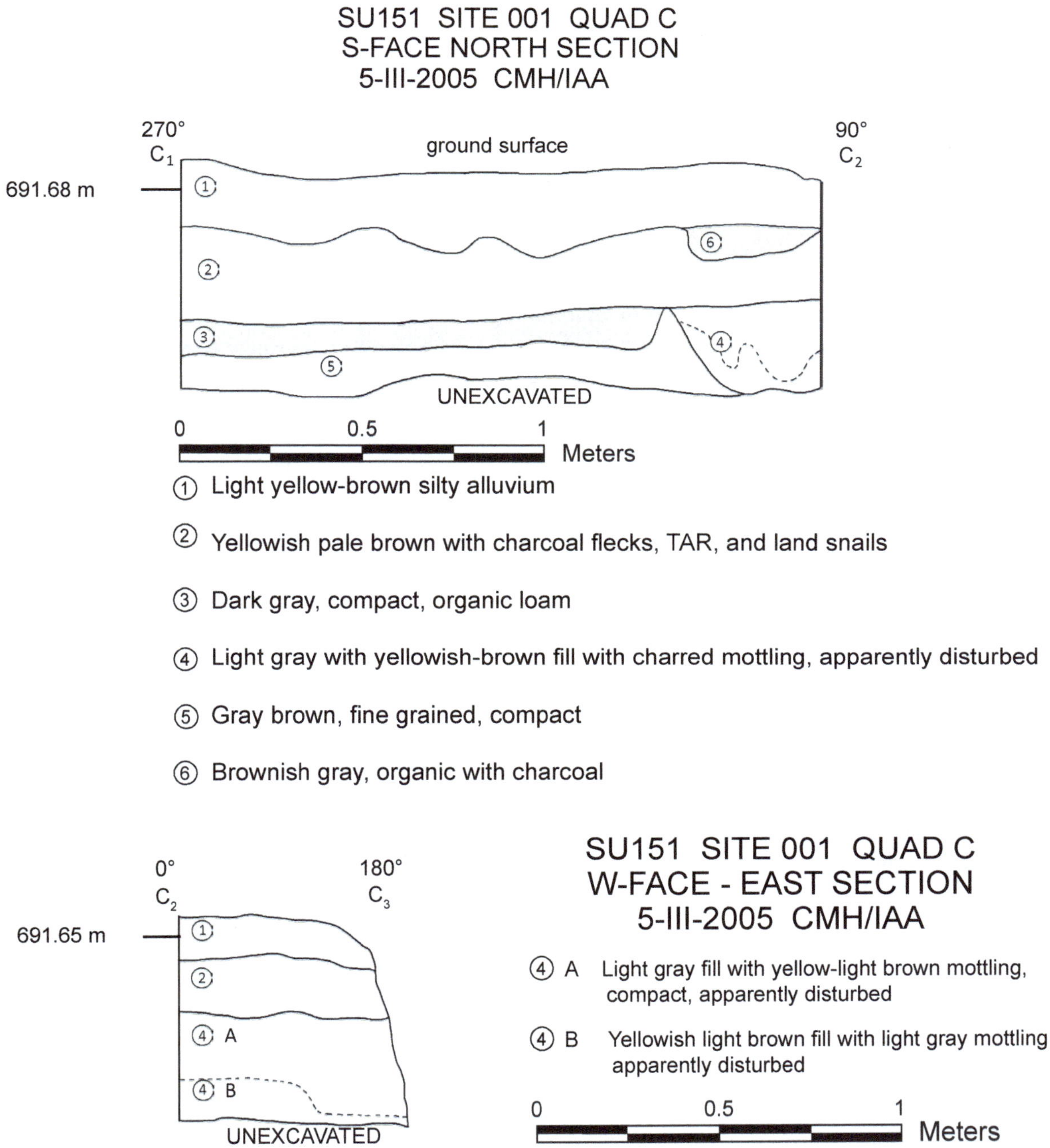

Figure 10.11. SU151-1 Quad C, north section showing disturbances (foundation trench) along the east side abutting the Neolithic platform exterior. *Drawing by Catherine Heyne, Ietha Al-ʿmari, and Petra Creamer.*

be noted that the foundation trench seen in this section does not match the gully profile on the east side of the stone platform (see below), where the upper, ashy, charcoal-flecked deposits (Quad B, Locus 007) that postdate the cattle skull ring abut the outer pavers of the stone platform, postdating it. Possibly because the ground was uneven, the builders excavated a pit cut more sharply to the east, where preexisting layers run up against the outside of slab uprights (set against the side of the pit), and more shallow in the west, where a foundation trench was filled against the outside of upright slabs to support them. Or the uneven ground may have meant virtually no cutting on the east side but leveling activities that left a low cut into rising ground on the west side. Although there were flecks of charcoal, bone, a bead, and few chipped stone flakes, artifact densities in Quad C were too slight to suggest sustained activity areas. Rather, the layers show discontinuous sedimentation through alluviation episodes disturbing nearby hearths, bone deposits, and other cultural residues.

Quads B, D, and E: The Cattle Skull Ring

Excavators opened Quad B to the east of and abutting the stone platform (Quad A). At its layout, Quad B measured 2.0 m west–east and 1.0 m north–south. During the course of excavation and with the notable discovery of bone deposits described below, excavators decided three times to extend the area, resulting in Quads D, E, and B/E (figure 10.12). To the north, a second 2.0 m west–east and 1.0 m north–south area was designated Quad D, and to the east, a 2.0 x 2.0 m extension became Quad E. Finally Quad B/E extended to the north of B and E, with a 3.5 m span east–west and a southern boundary defined by the edge of a natural gully. In total, this contiguous area to the east of the stone platform measured about 4.5 m (at the north) by 2.25 m (at the east). Excavations in this area revealed a stratigraphic sequence with significant human activity linked to the stone platform.

As in Quads A and C, a layer of yellowish–pale brown clayey silt overlay the entire area to a depth of 0.09 m. This current ground surface was less compact and showed darkened organic enrichment and soil peds, unlike the underlying 10 to 14 centimeters, which were also yellowish–pale brown clayey silt but considerably more compact (Layers ① and ② in Quad B/E, south section) (figure 10.13).

In the western end of the B–D–E area appeared two flat-lying unworked limestone slabs, (figure 10.14) both measuring less than 0.50 m in length. These slabs lay on a compact surface defined by a layer of small gravel, about 0.26 m below surface. The compaction and clear change to darkened sediment with charcoal inclusions clearly identified the underlying layer as an occupational accumulation (Layer ③ in Quad B/E, south section), richer in clay content than the overlying clayey silt and gravel lens. With extended excavation trench areas and multiple controlled stratigraphic probes, the distinct yellowish–pale brown compact layer with charcoal flecking appeared across the entire area, built up around an extraordinary deposit of bone (figures 10.15, 10.16).

Also built up around the bone deposit (discussion below) was a pale brown silty sediment rich in grey organic matter (likely decayed charcoal) and charcoal flecks (Quad B, Locus 007), Layer ④ in Quad B/E, south section. This layer contained pieces of thermally altered rock showing as grey limestone clasts, which likely derived, like the grey ash and charcoal, from the destruction of multiple hearths set on and in a paleosurface and visible throughout the sections of the natural gully. In addition, there were a very few nondiagnostic chert flakes from the ashy deposit. In contrast with the human disturbance sequence in Quad C, the upper ashy layer of Quad B (Locus 007) abuts the outer edge of the limestone uprights defining the monument.

The Quad B/E south section represents a view after cattle bone was lifted in excavation, but the presence of bone is well represented by disturbances (Layer ⑥) around and through layers underlying Locus 007. With horns interlocking and frontals facing inward to form an oval ring, the skulls of more than 40 head of cattle were inserted into (formerly) soft ground (figures 10.17 and 10.18). These sediments included a yellow–pale brown, very compact silty clay layer (Layer ⑦ in Quad B/E, south section) overlying a lower ashy layer (⑧, Locus 009) with thermally altered limestone clasts and charcoal flecks. Excavators noted a 2 to 6 mm wide, ultrafine, dark, and greasy clay around the contours of many skulls.

A radiocarbon date from a hearth underlying this layer (Quad B/E, Locus 010) indicates a terminus post quem for the cattle skull ring between 7201 and 6722 cal BP (AA69755, 6010 ± 69). The upper ashy layer in Quad D (Locus 003 = Quad B, Locus 007) also yielded a radiocarbon date on charcoal, indicating deposition of the skulls before 6457–6263 cal BP (AA66861, 5514 ± 48).

It appears that the ashy layer accumulated through overbank flooding and human revisitation events that churned and redeposited in and around the cattle skull ring the anthropogenic debris from nearby hearths. The stone monument was subsequently constructed into a sharp pit cut on its eastern side, removing ashy layers formed after the cattle skull ring. The disturbance of anthropogenically enriched sediments by a foundation pit on the west side of the monument (see Quad C, above) and the construction method of the monument itself indicate pit-cutting for monument construction. To build the monument, limestone slabs were erected by sinking a pointed end into soft sediment. This construction method indicates the need for some shallow pit or trench supporting the upright slabs, which would not balance themselves on their tips on flat ground. Over time, the entire cattle skull ring and most of the associated platform were covered by alluvial overbank flooding, resulting in the Holocene silt terraces on which slow soil formation processes have been at work to enrich upper centimeters with organic material.

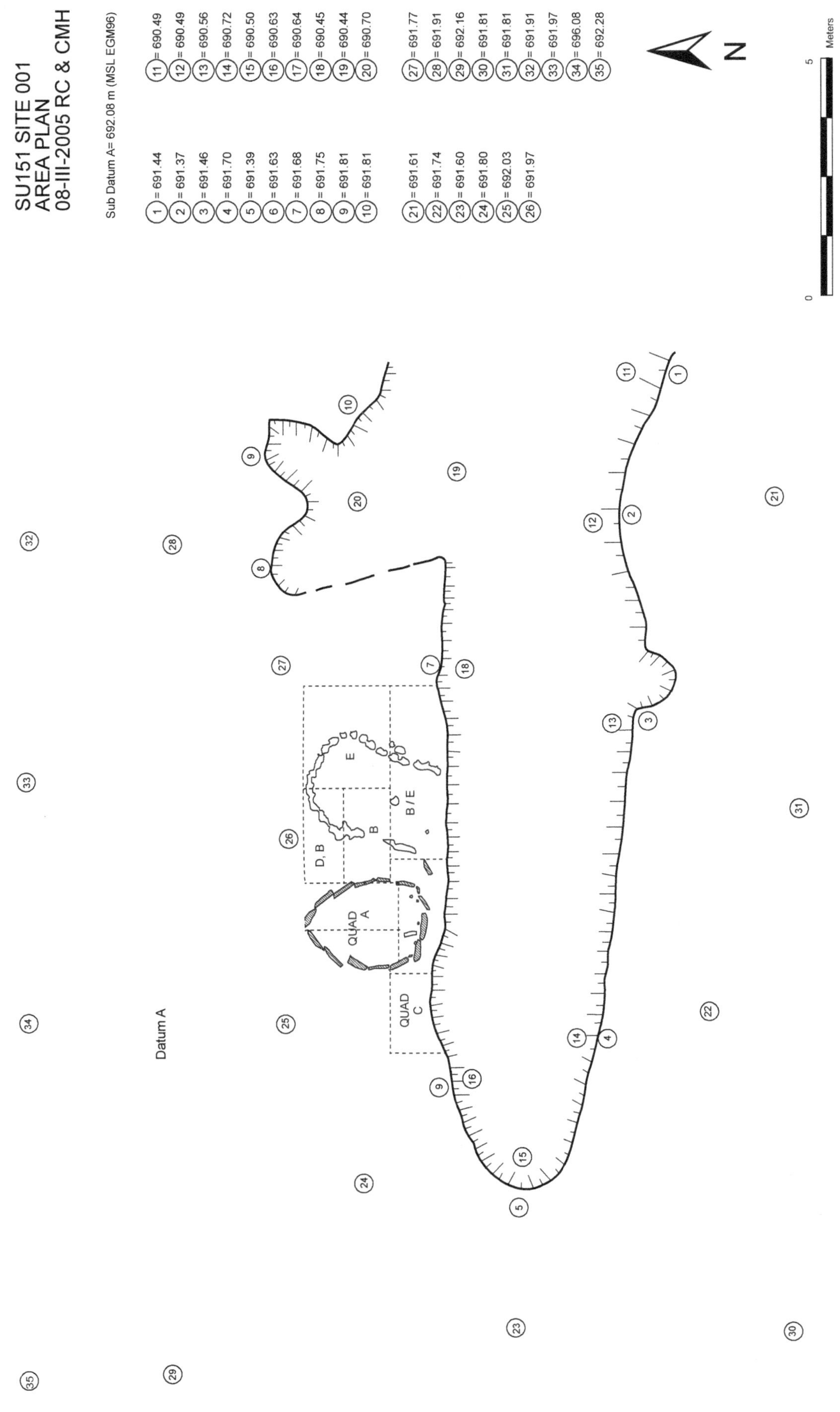

Figure 10.12. SU151-1 plan. *Drawing by Catherine Heyne and Clara Hickman.*

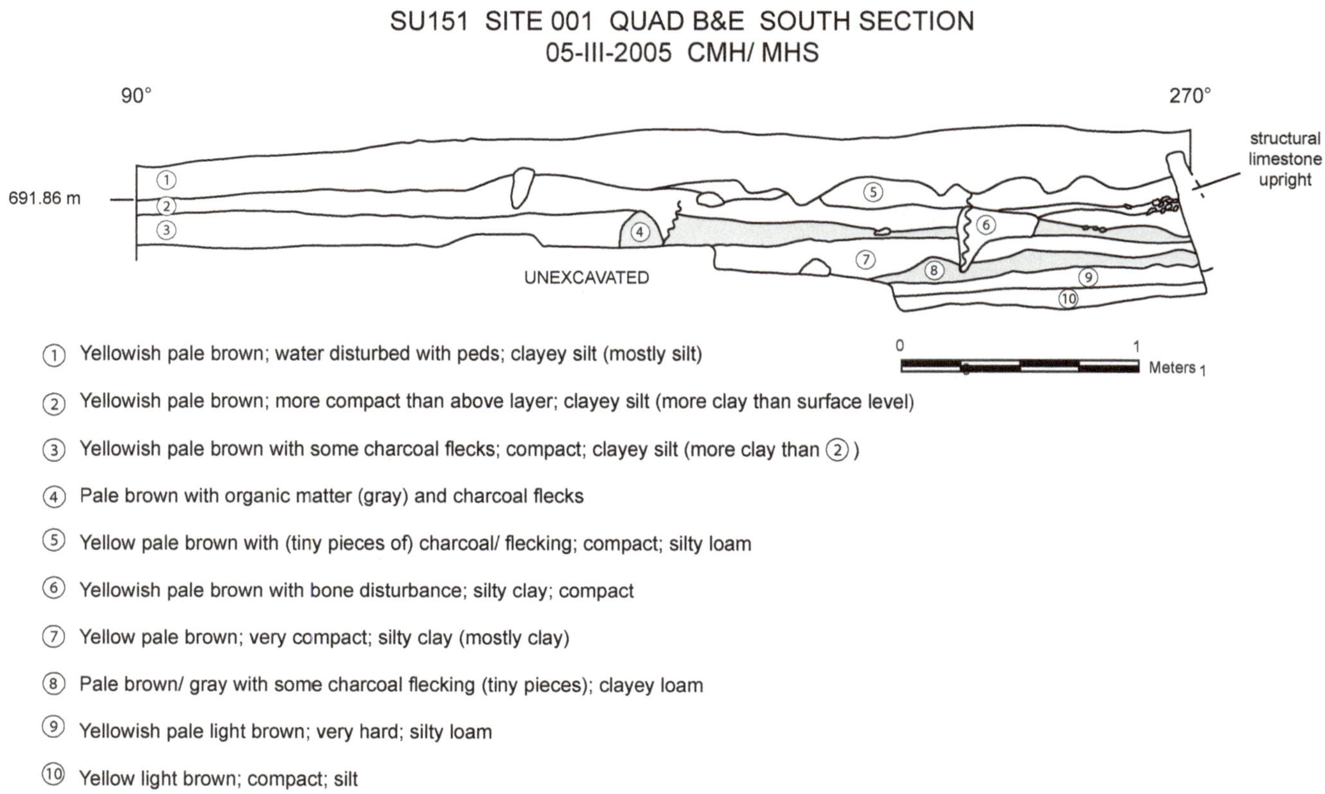

Figure 10.13. SU 151-1 Quads B and E, south section. *Drawing by Catherine Heyne, Mohammad Sinnah, and Petra Creamer.*

Figure 10.14. SU151-1 Quads A (foreground) and B. Limestone slabs probably broken from the platform perimeter lie on a sediment surface higher than the underlying cattle skull ring. *Photograph by Joy McCorriston.*

Figure 10.15. SU151-1 Quad B (foreground) with emergent cluster of bone (below figure 10.14). *Photograph by Joy McCorriston.*

Figure 10.16. SU151-1 Quad B: close view of emergent cluster of bone showing sweep of horns in the upper left corner, lower right corner (faint), and upper center. *Photograph by Joy McCorriston.*

Figure 10.17. SU151-1 Quads B, D, E, and B/E. *Photograph by Michael Harrower.*

The Gully Section

Cleaning and analysis of the natural gully section (figure 10.19) that defined the southernmost limit of the excavation also provided information, including a clear definition of extensive upper and lower ashy layers and 13 hearths that could be clearly viewed in section. Hearth 3 was clearly contiguous with the lower ashy layer, as was the hearth in Quad B, Locus 010. The ashy disturbances contained charcoal and burned limestone pebbles and were contiguous with many of these hearths. Ashy layers were also the deposits from which were recovered the few stone chips and flakes of human manufacture (table 10.2).

The gully section also revealed a lower sequence of activities not probed in excavation. To the southeast of the stone platform and set into deeper levels was a pair of large limestone uprights, one measuring nearly 0.75 m in height. The top of one of these, visible from the modern surface, was worked by pecking to produce a smooth top. Other large slabs lay in the sandy floor of the gully, evidently fallen through erosion. A hearth (Hearth 2) situated at the base of the silty sediments in the gully section and 0.5 m below the cattle skull ring is stratigraphically earlier than the upright slabs. A radiocarbon dating on charcoal from Hearth 2 yielded a date not many years earlier at 7207–6839 cal BP (AA69754, 6069 ± 48).

In consideration of the radiocarbon dates and the preservation of fragile bone that could not survive long exposed, it seems that episodes of fairly rapid sedimentation must have occurred between human visits and revisits to the site. The presence of many hearths detectable in gully section and many others on the eroding surfaces of the silt terrace indicates that there were many and repeated instances of warming and cooking at the site, and these visible manifestations surely are an indication of many other hearths buried in sediment. Some hearths were contemporaneous with the ashy deposits and stratigraphic context of the cattle ring. The site had several phases of monument construction, including an earlier stone emplacement, the installation of the cattle skull ring, and, thereafter, construction of the stone platform.

Conclusions: Regional Neolithic Activities at Shiʿb Kheshiya

The Kheshiya cattle skull ring site sequence developed beside a slow-moving stream, as previously published (McCorriston et al. 2012:51–52) (figure 10.20). For the purpose of incorporating regional and local paleoecological studies (chapter 3) with the site's archaeological records presented above, we review below the site's long-term development. This conclusion is important regional context for the subsequent analysis of the cattle skull bone assemblage (chapter 11).

Nearby geomorphological studies on the same Holocene silt terraces as the site showed a distinct facies traceable as a sinuous feature approximately 30 m wide and 500 m long. Exposures in gullies bisecting this deposit revealed a U-shaped paleochannel morphology surrounded by coarser reddish-brown, sandy silts, infilled with laminated, gray, organic, fine sandy silt. Charcoal samples from the top and base of this paleochannel infilling yielded radiocarbon ages of 6208 ± 61 cal BP (AA59761, 5402 ± 42) and 6815 ± 87 cal BP (AA599762, 5970 ± 72). The dated feature represents a shallow, quiet-water wetland environment, an abandoned channel on the floodplain of the main Wādī Sanā fluvial channel. The proximity of similar-age archaeological features (hearths and ashy layers at Shiʿb Kheshiya) indicates that people occupied the wetland margins and would have exploited the channel as a source of freshwater, especially in the winter months, when the risk of flooding would be lowest and the availability of water in vernal pools on the uplands was very limited.

The archaeological sequence shows repeated reuse of the area. First, many hearths were constructed as shallow, cobble-filled depressions on open ground. These hearths, dated between 7208 and 6722 cal BP (AA69754, AA69755, AA59571), contain thermally altered limestone clasts and fine-grained, black, ashy sediment with charcoal. Hearths are the most numerous archaeological features in Wādī Sanā, and some contain burned bone, including from cattle, probably from grilling meat as the bedouin do today.

Figure 10.18. SU 151-1 Quads A (right) and B (left) with the cattle skull ring under excavation and the Neolithic platform sectioned. *Photograph by Joy McCorriston.*

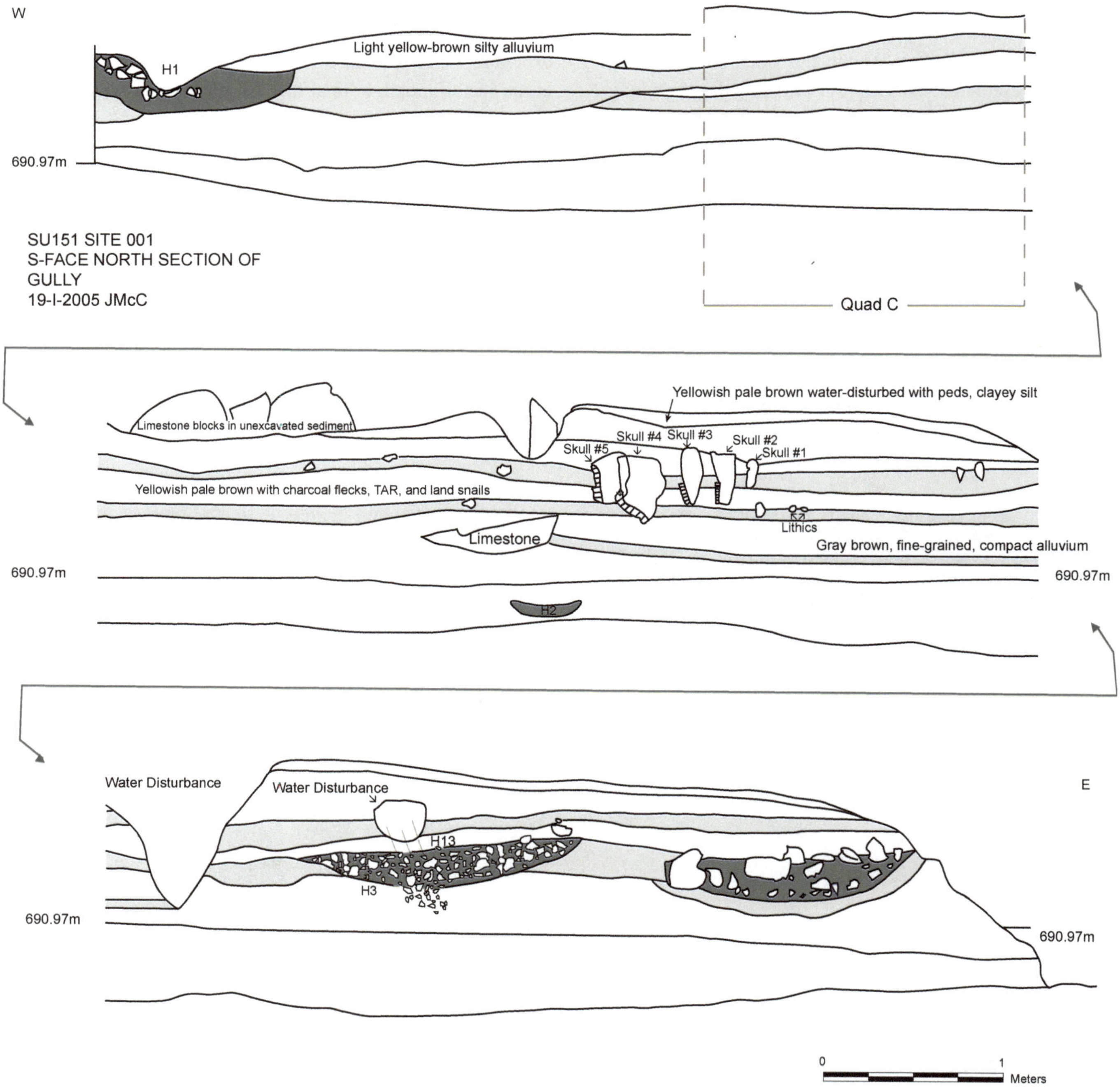

Figure 10.19. North section of a natural gully that cuts the south edge of SU151-1. Quad C was excavated from the gully section, so layers in Quad C have here been reconstructed from its northern section. *Drawing by Ramzi Ladeh, Joy McCorriston, Matt Indrutz, and Clara Hickman.*

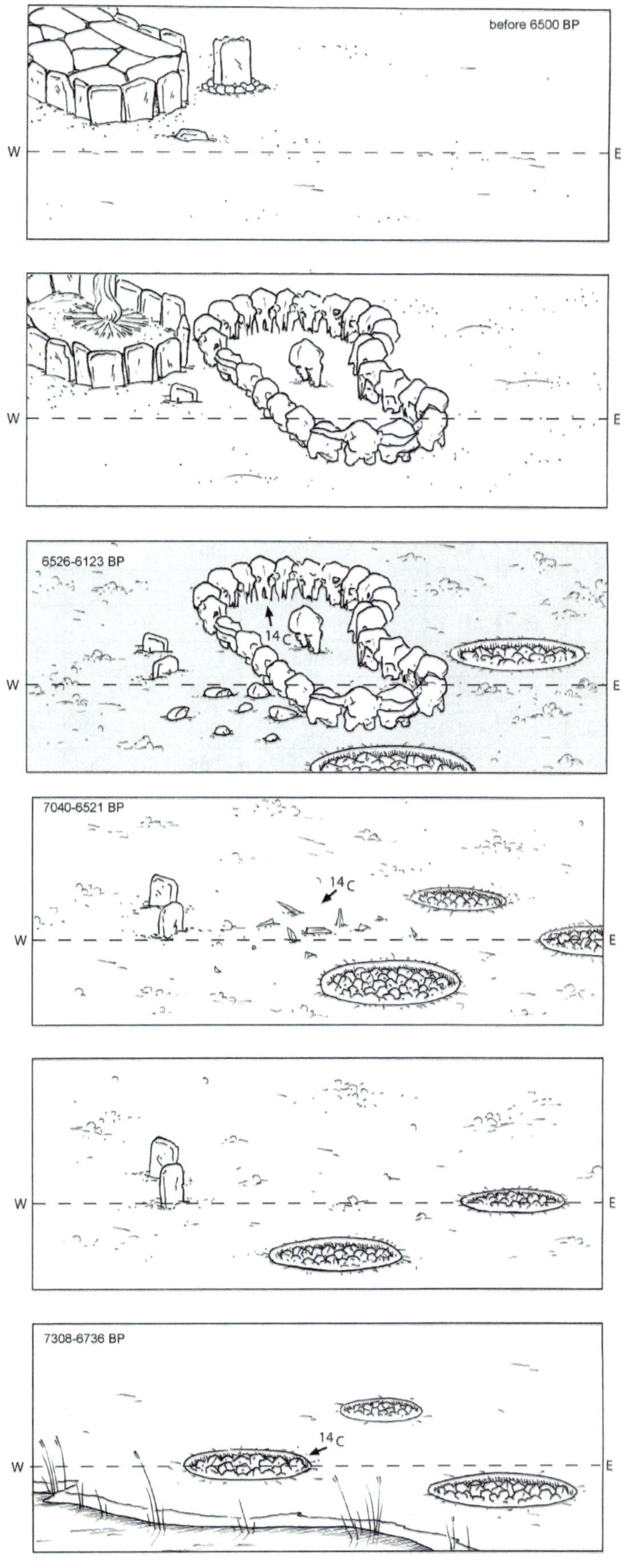

Figure 10.20. Schematic development of SU151-1, all within the geomorphological aggradation of silt terraces (chapter 3). Reproduced from McCorriston et al. 2012:52–53, figure 4. Radiocarbon ages here are from Bayesian analyses (chapter 18).

Table 10.2. Lithic artifacts from surface and excavations at SU151-1. Analysis by Joy McCorriston.

							Products				Tools		
Date	**Quad**	**Site**	**Location (cm) from Sub-Datum A**	**Locus**	**Lot**	**Bag #**	**Flakes**	**Primary Flakes**	**Obsidian Flakes**	**Other**	**Utilized Flakes**	**Arrowheads**	**Misc. Tools**
2/23/05	A	151-1		017	17	5			1				
2/23/05	A	151-1		018	18	4	2			12			
2/12/05	B	151-1		002	3	3	1						
2/15/05	B	151-1		003	3	6	1			3	3		
2/12/05	B	151-1		003	3	4				1			
2/13/05	B	151-1		003	3	3	1						
2/12/05	B	151-1		003	3	1		1					
2/23/05	B	151-1		004	4	10	1			1			
2/27/05	B	151-1		005	5	3	2		1	2			
3/21/05	B	151-1		006	6	8	1						
2/28/05	B	151-1		006	6	4				1			
3/12/05	B	151-1		007		1	4			5			
3/15/05	B	151-1		007	1					2			
3/19/05	B	151-1		007	1			2		3			
2/16/05	B	151-1	in section			1				1			
3/17/05	B	151-1					2						
2/12/05	C	151-1		002	1	2				6			
2/10/05	C	151-1		005	4	1							
2/14/05	C	151-1		008	7	2	1						
2/15/05	C	151-1		010	9	1	1						
2/16/05	C	151-1		012	11	2	1						
2/16/05	C	151-1		013	12	1	5			6			
3/13/05	C	151-1		018	3	3		1					
3/15/05	C	151-1		021		1		1					
2/23/05	C	151-1	in section			1	1	1					
3/18/05	E/B	151-1		009	21		2			2			
3/9/05	H5	151-1	wadi cut profile			4	1						
2/9/05		151-1	103.29N/102.52E	000	1	3	1						
2/8/05		151-1	106.2N/105.63E	000	1	1							
2/7/05		151-1	102.10N/102.62E	000	1	4							
2/18/05		151-1		004		8	2						
3/6/05		151-1	wadi cut profile	H12		4	1						
		151-1	profile	H12		6		1		1			
		151-1		H3						9			
3/5/05		151-1	wadi cut profile	H4		4	1						
2/19/05		151-1	94.1N/106.3E	000		1	8	2		5			
3/9/05		151-1	west wadi cut profile			1				1			

Cores				
Cores for Flakes	Core Trim Fragments	Lithics Total	Weight (all bag contents)	Observations
		1	0.2 g	Obsidian chip; flaked but an intact flake?
1		15	20.0 g	
		1	0.2 g	
		7	25.4 g	
		1	10.7 g	
		1	3.3 g	
		1	2.5 g	One graver/borer on a flake? Bifacial trimming fragment looks utilized?
	1	3	16.6 g	
	2	7	39.4 g	
		1	6.10 g	
1	2	4	9.5 g	
		9	13.5 g	
		2	8.6 g	
		5	14.6 g	
		1	1.9 g	
		2	3.5 g	
		6	4.8 g	
1		1	21.7 g	
		1	3.4 g	
		1	0.4 g	
		1	0.5 g	
	2	13	43.9 g	
		1	2.1 g	Core trim fragment? Primary
		1	3.0 g	
		2	10.7 g	
		4	4.3 g	
		1	0.0 g	Heavy fraction
		1	2.5 g	primary chunks and chips (three with cortex)
1		1	36.5 g	broken
	1	1	13.4 g	
		2	1.4 g	
	1	2	4.6 g	
		2	0.1 g	one thin trimming flake
		9	1.0 g	
		1	1.8 g	Heavy fraction; chunks and chips, shatter
	1	16	57.4 g	broken
		1	2.7 g	

Most hearths could be linked stratigraphically with one of two episodes of human activity, which deposited also a few chipped stone flakes, dark ash, and charcoal enrichment in two layers. The lower ashy layer (Quad B, Locus 009) developed alongside at least one standing stone with a finely dressed (stone-pecked) top. Another upright remains visible in gully section, and several lie on the gully floor, suggesting that an early stone monument—perhaps the focus of human activity—has been removed through natural erosion.

After about 500 years, represented by sterile, sandy-silty alluvial sedimentation, more than 40 cattle skulls were pressed nose-down and facing inward in an oval ring around a central skull. (Erosion destroyed an unknown number of additional skulls.) Neolithic celebrants inserted the skulls into soft mud, with rocks wedged to help stabilize the skulls in upright position. The interlocking horns once protruded visibly above ground. Evidence that the skulls were buried relatively quickly after the death of the animals also comes from well-preserved, delicate nasal bones, which break and disintegrate quickly if unprotected. Breakage and overlapping of the (more durable) unfused frontal sutures were caused by the weight of the skulls (and horns) pushing down into fresh mud. Like a house of cards, the skulls supported one another, implying that the ring was constructed simultaneously, not accretively.

Further analysis of the excavated bone may be found in chapter 11 (also Henton et al. 2014). What is clear from skull placement is that the cattle skull ring formed an architectural installation created as one event, in which some fully fleshed, some partially skinned skulls were pushed into soft ground. Excavators counted 33 skulls plus at least 5 lost to erosion and accept that the ancient arc truncated by the modern gully was most likely 45 to 50 skulls. The occipital parts stood aboveground with horns and horncores, while ashy sediments from human activities and flooding dispersed hearth charcoals around the level of the eye sockets. The mandibles were removed before burial, and their fate, along with that of other postcranial elements, is unknown. The skull ring had a deliberate 50 cm gap in the southwest. A central and the largest cattle skull faced east–northeast, outward and away from the worked standing stone still visible outside the ring.

Subsequent flooding deposited alluvium mixed with ash, burying the bone ring almost completely. Atop the alluvium, likely in another season or two while horns and frontal bone still showed, people placed stone slabs with a lens of pebbles that may have once served to chock or stabilize them. Whether these slabs once actually stood is unknowable, but throughout this sequence, the original standing stones, with their bases deep in accrued sediments, could still be seen next to the cattle skull ring.

The adjacent stone platform was built last, perhaps to commemorate the cattle skull ring or the event it signified, and itself has a complex sequence, requiring the labor of several strong individuals. First, a leveling event or shallow pit excavation carved a recess against the bank to the west. Upright limestone slabs forming the perimeter needed one another for support, as each supported the next slab, beginning with a pair at the north end set to form an obtuse angle. A blocky boulder was placed at the interior at this point to retain these key uprights as anchors for the subsequent perimeter-forming uprights, overlapped domino-fashion for mutual support. Limestone slabs were often sunk with a corner or point as the lowest part.

Nor was this stone ring immediately filled to create a platform. Despite lack of a doorway, it was briefly occupied, leaving the traces of several ephemeral fires against the inner sides of uprights and a compact interior floor. Occupants could easily have stepped in over one or several low uprights, which presented a modest impediment to animals. No sign of roofing survives, and best surmise is that if roofed at all, the ring of upright slabs would have supported a brush, reed, and mud superstructure. Clay deposits from flooding suggest that at least one season passed between construction of the slab ring and the infilling of its occupied interior to make a platform. Based on the labor required to excavate and remove heavy stone slabs, at least several strong people cooperated to lift them over and into the standing slab ring, laying slabs horizontally against the interior of uprights. This infill achieved the construction of a platform, which contained no burial or indeed any notable material culture. Subsequently, both skull ring and platform were buried quickly by seasonal overbank flooding associated with the Middle Holocene fluvial regime. Throughout the entire sequence described, including construction of the skull ring, the earlier standing stone with worked top would have been visible. We conclude that the worked standing stone provided a meaningful reminder of events past and marked a landscape place for further commemorative practices, where people could gather, sacrifice animals, and feast.

References

Bin ʻAqil, Abdalaziz J., and Joy McCorriston
2009 Convergences in the Ethnography, Semantics, and Archaeology of Prehistoric Small Scale Monument Types in Hadramawt (Southern Arabia). *Antiquity* 83:602–18.

Braemer, Frank, Serge Cleuziou, and Tara Steimer
2003 Dolmen-like Structures: Some Unusual Funerary Monuments in Yemen. *Proceedings of the Seminar for Arabian Studies* 33:169–82.

Crassard, Rémy

2008 *La préhistoire du Yémen. Diffusions et diversités locales, á travers l'étude d'industries lithiques du Hadramawt*. British Archaeological Reports 1842. Archaeopress, Oxford.

Edens, Christopher

1982 Towards a Definition of the Western ar-Rub' al-Khali "Neolithic." *Atlal* 6(1402):109–24.

Harrower, M.

2008 Mapping and Dating Incipient Irrigation in Wādī Sanā, Hadramawt (Yemen). *Proceedings of the Seminar for Arabian Studies* 38:187–202.

Henton, Elizabeth, Joy McCorriston, Louise L. Martin, and Eric A. Oches

2014 Seasonal Aggregation and Ritual Slaughter: Isotopic and Dental Microwear Evidence for Cattle Herder Mobility in the Arabian Neolithic. *Journal of Anthropological Archaeology* 33:119–31.

Martin, Louise, Joy McCorriston, and Rémy Crassard

2009 Early Arabian Pastoralism at Manayzah in Wādī Sanā, Ḥaḍramawt. *Proceedings of the Seminar for Arabian Studies* 39:271–82.

McCorriston, Joy

2006 Roots of Agriculture (RASA) Project 2005: A Season of Excavation and Survey in Wādī Sanā , Hadramawt. *Bulletin of the American Institute for Yemeni Studies* 47:23–28.

2011 *Pilgrimage and Household in the Ancient Near East*. Cambridge University Press, Cambridge.

2013a Pastoralism and Pilgrimage: Ibn Khaldūn's Bayt-State Model and the Rise of Arabian Kingdoms. *Current Anthropology* 54:607–40.

2013b The Neolithic in Arabia: A View from the South. *Arabian Archaeology and Epigraphy* 24:68–72.

McCorriston, Joy, Michael J. Harrower, Louise Martin, and Eric Oches

2012 Cattle Cults of the Arabian Neolithic and Early Territorial Societies. *American Anthropologist* 114(1):45–63.

McCorriston, Joy, Michael J. Harrower, Tara Steimer, Kimberly D. Williams, Matthew Senn, Mas'ūd Al Hādhari, Mas'ūd Al Kathīrī, 'Ali Aḥmad Al Kathīrī, Jean-François Saliege, and Jennifer L. Everhart

2014 Monuments and Landscape of Mobile Pastoralists in Dhofar: The Arabian Human Social Dynamics (AHSD) Project, 2009–2011. *Journal of Oman Studies* 18:117–43.

McCorriston, J., Eric Oches, Dawn Walter, and Kenneth Cole

2002 Holocene Paleoecology and Prehistory in Highland Southern Arabia. *Paléorient* 28(1):61–88.

McCorriston, Joy, Tara Steimer-Herbet, Michael J. Harrower, Kimberly Williams, Jean-François Saliège, and 'Abdalaziz Bin 'Aqīl

2011 Gazetteer of Small Scale Monuments in Prehistoric Hadramawt, Yemen: A Radiocarbon Chronology from RASA Project Research 1996–2008. *Arabian Archaeology and Epigraphy* 22:1–22.

Vogt, Burkhard, and Alexander V. Sedov

1994 *Surveys and Rescue Excavations in the Hadhramawt Governate, Republic of Yemen.* Manuscript on file. Deutches Archäologisches Institut, Sana'a.

Page 272–273: Imaginative reconstruction of the Kheshiya cattle skull ring by Judith Dobie, based on archaeological, zooarchaeological, and geoarchaeological evidence. This scene imagines the phase between 6526 and 6123 cal. yr. BP (see Figure 10.20).

Chapter 11

The Kheshiya Cattle Skull Ring
Zooarchaeological Analyses

Louise Martin with a contribution by Joe Roe

On December 31, 2005, Louise Martin, Lisa Usman, and Joy McCorriston settled on a hard floor in a sparse hotel in Mukalla to watch Pakistan ring in the New Year a few hours to the east. Toddler Jojo slept a cherubic sleep propped up by all the available pillows, having exhausted all episodes of *Balamory*. During the day, Louise and Lisa unwrapped 6,000-year-old cattle skulls and cleaned them for photographs, measurements, and curation. To say the conservation lab was improvised would overly gloss a battered room with rigged lighting and peeling floors. But the onshore breeze fills the Mukalla Museum, there's a five-star overlook of the brilliant sea, and you could get a rock lobster dinner for two dollars in those days. ʿAbdalʿazīz Bin ʿAqīl left us only for the morning of Eid al-Fitr, working through his holiday and the final Ramadan vigil. He and Joy kept Jojo busy so that his mother, Louise, could measure the frontal bones and wear patterns on cattle molars. This chapter is the outcome of her analysis, supported by Lisa's clever conservation solutions and Joe Roe's statistical skills in the comparison with East African cattle.

Domestic Cattle in Arabia and the Nature of Herding

The 2004 discovery of the site of Kheshiya SU151-1 in the highland Southern Jol region of Yemen was a gift not only to prehistorians and historians of the Arabian Peninsula and beyond but also to those interested in human–animal relations in the Neolithic. The oval installation constructed of at least 40 partially buried cattle skulls, adjacent to a similar-shaped stone "platform" structure, was unique at the time of discovery and remains so at the time of writing. It is immediately clear that the cattle skulls are no normal faunal assemblage made up of discard from everyday food consumption. Indeed, the site is a monument rather than an occupation location (McCorriston 2011; chapter 10 this volume), with the cattle skulls forming a central part of monument construction. The sample of cattle skulls retrieved from the site provides an as-yet-unchallenged insight into the nature of cattle, and human–cattle relations, in the Neolithic of Southern Arabia. This chapter focuses on analysis of the skulls themselves and the zooarchaeological information they yield, both at a site and local level, and also at a broader regional level, where they contribute to discussion on the appearance of early domestic cattle in Arabia and their role in subsistence and ritual.

Zooarchaeological analyses of the cattle skulls contribute to three main research spheres. First, despite their fragmented state, the skulls allow assessment of cattle cranial morphology, which has implications for species assignation of the cattle from which they derived. Morphometrics also inform on the domestic/wild status of the animals. Second, assessment of the animals' ages at death and their sex distributions within the sample allow discussion of the selection of animals for the cull and subsequent skull monument construction. Third, bone surface modification data and skull breakage patterns contribute toward our understanding of how skulls were prepared and installed as part of the construction of the Kheshiya monument.

The discovery, excavation, stratigraphy, and associated finds of the site—known also as the Kheshiya cattle skull ring—are detailed in chapter 10. To recap briefly, in 2004 the RASA survey team found several large mammal skulls eroding out of a gully in the Shiʿb Kheshiya. The team returned in 2005 to excavate the site to which these skulls belonged. By this time the skulls had been identified as cattle. The team first excavated the small (2 x 2.5 m) semisubterranean stone structure to reveal details of its construction and use. Intriguingly, the structure did not emerge as either occupational or a tomb but was classified as a "platform" of slabs with deliberate infill. Chapter 10 details how the occupational deposit cut by the platform extends externally, containing burned rock and flakes of chipped stone. It was into this deposit that the 40 or so cattle skulls were placed. Their placement formed an oval similar in size to the adjacent platform. The skulls faced frontals inward, maxillary teeth outward. Horns and horncores, which did not survive, would have risen aboveground, pointing upward and inward toward the center of the oval. Considering the close placement of some skulls, many horns would have interlocked. Mandibles had been removed before burial. There is good evidence (chapter 10) that the skulls were pushed into soft marshy sediment (seemingly in winter) all at one time, without a sequence of deposition. Only after the cattle skull ring was installed was the stone platform constructed. The skull ring sits approximately 1.5 m southeast of the stone structure.

Several publications have discussed preliminary zooarchaeological results of the cattle skulls from the ring (e.g., Henton et al. 2014; McCorriston 2011; McCorriston and Martin 2009; McCorriston et al. 2012); some zooarchaeological details have been refined as a result of the analyses presented in this chapter.

It is widely accepted that Arabia did not witness indigenous domestication of local plants and animals; instead, the area received domesticates from elsewhere (see chapter 1 this volume; Boivin and Fuller 2009). By the time of the Kheshiya cattle ring construction in the mid-seventh millennium cal BP, we have the following picture of domestic cattle in the peninsula.

Domestic cattle are seen in the Persian Gulf area from the seventh millennium cal BP, where they appear alongside a much larger assemblage of domestic sheep and goats at the site of Jabal al-Buhais 18 in Sharjah, UAE. Excavators interpreted the site as a station within a mobile herding system rather than a location of year-round habitation (Uerpmann and Uerpmann 2008:127–31). Late-seventh-millennium cal BP Ras al-Hamra 6 in northern Oman yielded a faunal assemblage that also included small numbers of domestic cattle and caprines, and by the sixth millennium cal BP, Ras al-Hamra 5 provides substantial samples of the same package of domesticates, interpreted as being for meat production rather than secondary products (Uerpmann and Uerpmann 2003). Domestic sheep, goats, and cattle are mentioned from the site of H3 in Kuwait (Beech and Glover 2005:99), which may indicate an earlier regional appearance of animal domesticates than previous evidence has suggested, considering the eighth-millennium cal BP dating of the main occupation of the site (Carter and Crawford 2003). It is clear that by the seventh millennium cal BP at the latest, the three major animal domesticates (cattle, sheep, and goats) formed a dominant part of subsistence systems in the Gulf region, with caprine herding the dominant activity.

A slightly different picture has emerged from the southwest of the Arabian Peninsula. The rockshelter site of Manayzah in the Southern Jol mountains of Yemen provides the earliest evidence of domestic cattle, sheep, and possibly goats in Southern Arabia, dated to the early eighth millennium cal BP (Martin et al. 2009; McCorriston and Martin 2009; chapter 8 this volume). The small sample hints at a mixed herding economy, alongside equally important gazelle hunting. Domestic cattle and caprines were also found at highland Wādī ath-Thayyilah 3, dating to the seventh millennium cal BP (along with Jibal Qutrān and Najd al-Abyad) (Fedele 2008). The slight dominance of cattle at Wādī ath-Thayyilah 3, together with architectural evidence, has led to interpretations of a Neolithic village-based cattle-herding economy (alongside caprines), which by 5,000 years ago gives way to caprine-dominant subsistence.

The north of the Arabian Peninsula is less well researched, but small samples of very likely domestic cattle have been found at Jebel Oraf 2 in the Nefud region in late seventh millennium cal BP, among a series of open-air hearths, strongly suggestive of mobile cattle pastoralist activity (Guagnin et al. 2017).

Despite evidence for wild cattle (*Bos primigenius*) in Arabia into the Holocene (see review in McCorriston and Martin 2009), finds of wild cattle bones are few and far between, and archaeological consensus reasonably holds that domestic livestock was introduced. Debate continues as to whether Neolithization, with domestic cattle included, represented a mobile pastoralist expansion from the Levant in the north (e.g., Drechsler 2007; Uerpmann et al. 2000) or whether evidence points more to an indigenous development of specialized cattle pastoralism (Cleuzio and Tosi 1998; McCorriston et al. 2012), albeit on introduced stock.

The Kheshiya cattle assemblage, which provides morphological and metrical evidence of cattle species, status, and size, contributes to these debates. The age and sex data that the Kheshiya skulls yield also allow comment on the herding system from which the cattle derived.

Domestic Cattle Possibilities

Southern Arabian Neolithic domestic cattle could potentially derive from three broad sources: domestic European cattle (*Bos taurus*), or taurines; Asian zebu cattle (*Bos indicus*), or indicines; or early African domesticates (taurines, sometimes referred to as *Bos africanus*). Cattle domestication evidence—both genetic and osteological—sees frequent revision in the literature, but a brief summary shows which species and types of cattle should be considered in attempts to identify the species and status of the Kheshiya skulls.

Bos taurus shows good evidence for domestication from the wild aurochs (*Bos primigenius*) in southeast Anatolia and the northern Levant region, where the process appears to have begun by the mid-eleventh millennium cal BP, with morphologically distinct domestic cattle identifiable (Peters et al. 2005). There is genetic support for a single taurine domestication event (Decker et al. 2014; Magee et al. 2014). Domestic cattle spread to western Anatolia and the Aegean by the end of the tenth millennium cal BP (Arbuckle et al. 2014) and to the southern Levant (Horwitz and Ducos 2006) and southern Europe by the late ninth millennium cal BP. We can thus assume they were present in a southern Levantine context for possible dispersal south by this time.

Bos indicus underwent separate domestication on the Indian subcontinent, with some suggestion that they were under cultural control by the late tenth millennium cal BP in the northwest of the region (Patel and Meadow 2017). They had spread to the Middle East region by the end of the sixth millennium cal BP (Chen et al. 2009; Matthews 2002) and are first witnessed in Africa at a 2,000-year-old site in Kenya (Hanotte et al. 2002; MacHugh et al. 1997; Marshall 2000). Genetic evidence suggests that zebu/taurine introgression occurred once zebu reached Africa. While there is no evidence that zebu cattle were in regions bordering South Arabia by the time of the Kheshiya occupation, the evidence for seafaring activity and maritime exchange in the Gulf region from the late eighth millennium cal BP (summarized in Boivin and Fuller 2009) means that the presence of *Bos indicus* needs consideration at Kheshiya.

African cattle are more complex. Genetic evidence supports the idea of a separate domestication of *Bos taurus* in Africa (e.g., Decker et al. 2014; Magee et al. 2014), perhaps from local aurochs (as Grigson [2000] predicted from her morphometrical zooarchaeological study) but also possibly from a hybridization of incoming Near Eastern *Bos taurus* with the resident wild African auroch population (Magee et al. 2014). Most later Egyptian cattle breeds seem descended from founder domestic herds from the Near East (Olivieri 2015), although one haplogroup might stem from a more southerly *Bos primigenius* ancestor, and only in far later millennia do zebu arrive and hybridize. While the zooarchaeological evidence for Early Neolithic domestic cattle in northeastern Africa is controversial (see Stock and Gifford-Gonzales 2013), there is evidence that domestic cattle were present in East Africa at least by the middle of the seventh millennium cal BP (Marshall and Hildebrand 2002).

As McCorriston states in chapter 1, highlighting the aims of the RASA Research Project, Southern Arabia is at the crossroads between the Near East, East Africa, and South Asia—a factor that makes it germane to questions of domestic cattle dispersals and introductions. We cannot entirely dismiss the possibility of any introduced cattle interbreeding with remnant indigenous aurochs in Arabia. For example, Park et al. (2015) have found genetic introgression between domestic European/Near Eastern cattle introduced to the United Kingdom and indigenous British aurochs. That said, populations of *Bos primigenius* in Arabia were probably thin on the ground by the Neolithic.

The Kheshiya Cattle Skulls: What Was Found?

The Kheshiya skull ring originally consisted of about 40 cattle skulls forming the installed monument; an exact count was impossible because part of the site had eroded into a small gully immediately to the south, taking several skulls with it (see chapter 10, figure 10.17). Extremely careful excavation and lifting of the remaining in situ skulls yielded 35 cattle crania sufficiently intact to allow recording of zooarchaeological data to various degrees, depending on states of preservation. The "skull" technically includes all the separate bones of the cranium plus the mandible (lower jaw). The Kheshiya skulls were buried without associated mandibles but we retain the term *skull* here for ease of use. A single additional piece of animal bone was retrieved from excavations within the skull ring; this was identified as a fragment of cattle mandible, although it could not be associated with any individual skull. The site produced no other fragments of animal bone.

Each of the 35 cattle skulls and the single mandible fragment was allocated a 'lot' number upon excavation, and these same numbers were used for zooarchaeological recording and analysis. The terms *lot number* and *skull number* are used interchangeably in this chapter. Numbers range from 2 to 42 (although 6 to 10 are not used), with Lot 35 referring to the cattle mandible. Figure 11.1 shows the location within the monument of each of the 35 skulls excavated and analyzed. We can observe how the cattle skulls were placed close together, often touching, with frontals (tops of skulls) facing inward, nasals pointing down (chapter 10, figure 10.17). It is notable, too, that the oval ring of skulls has a gap in the western curve, maybe serving as an "entrance"; that the ring seems to show part of a second row of skulls immediately behind the first on the east side; and that there is a single skull placed in the center. The placement alone raises questions, which I return to later in discussion.

Methods: Identification, Morphology, Age, and Taphonomy

Original observation and data recording of the Kheshiya cattle skulls were undertaken in January 2006 in the Mukalla Museum, southern Yemen, where the collection was housed. A small team consisting of RASA project director Joy McCorriston, conservator Lisa Usman, zooarchaeologist Louise Martin, and ʿAbdalʿazīz Bin ʿAqīl, director of the General Organization for Antiquities and Museums, Hadramawt Province, worked together to achieve the four aims of the study trip:

1. To clean each skull to the level where zooarchaeological study could be undertaken

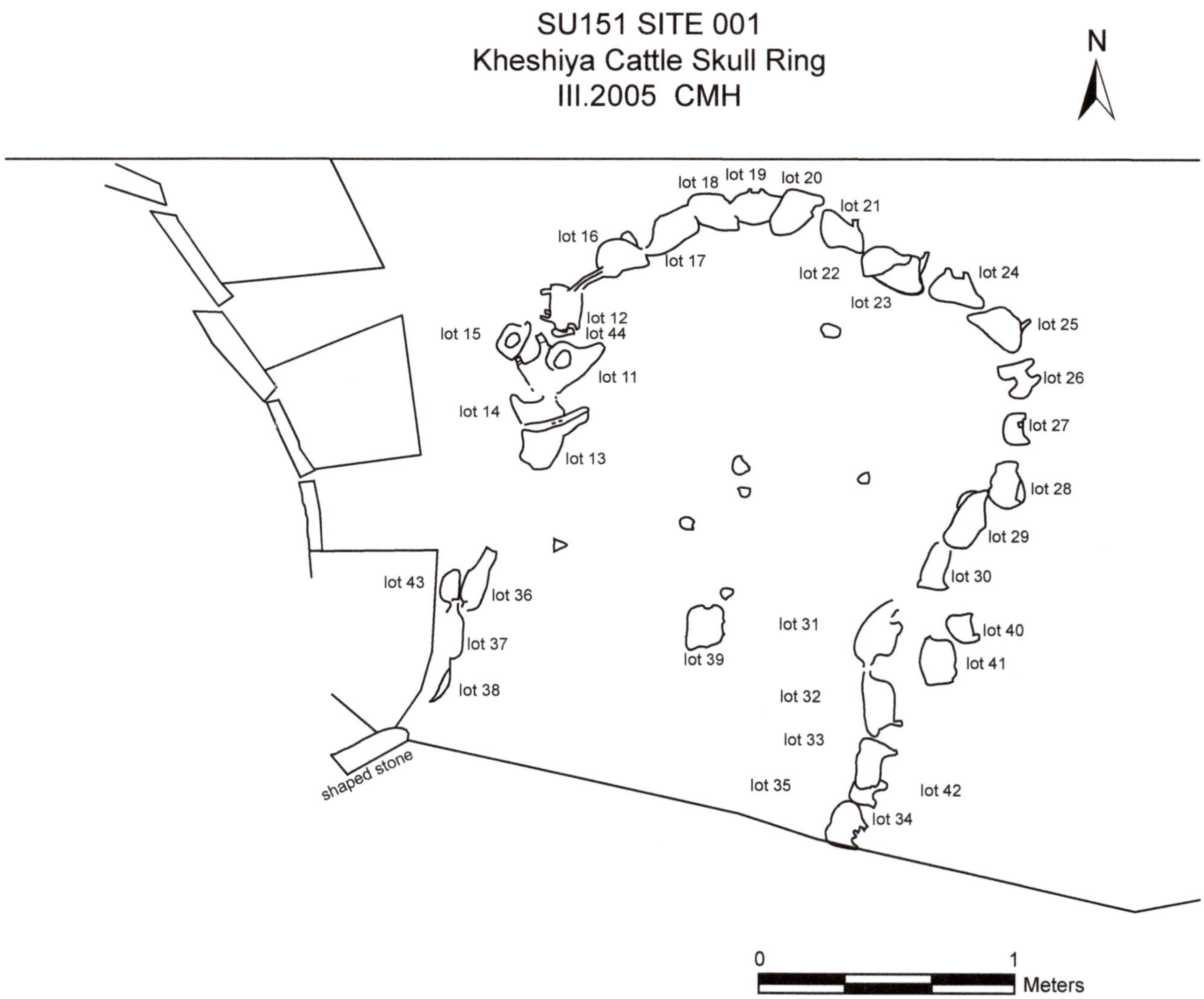

Figure 11.1. Plan of the Kheshiya monument showing the location of the 35 numbered cattle skulls described and discussed in chapter 11. *Drawing by Catherine Heyne, Illustration by Clara Hickman.*

2. To undertake detailed zooarchaeological recording related to the taxonomic status of the skulls, size and status of the cattle, and age of death of the animals; also to document any evidence for human modification and treatment of the skulls, or natural taphonomic processes
3. To create full documentation of the skull assemblage by means of a data and photographic archive
4. To stabilize and pack the skull assemblage for long-term storage in the Mukalla Museum

The director and staff of the Mukalla Museum kindly allowed the team to set up a temporary laboratory for the study within the museum. Appendix 11.1 provides an account of the process of unpacking, cleaning, and conservation of the skulls, explaining decisions taken to keep the skulls as intact and stable as possible to maximize data collection, given time and equipment constraints.

One main challenge to observation and recording was the fragile condition of the skulls. Despite excellent in-field excavation, lifting, first aid, and packing of the skulls in spring 2005 (chapter 10; see appendix 11.1), bone texture on all skulls was invariably dry, brittle, and prone to breakage and collapse, undoubtedly the result of millennia of seasonal changes in temperature and wetness/dryness of the Kheshiya burial environment, leaving bone leached out and very fragile. Furthermore, the skulls appear to have been originally only partially buried, with the posterior portion (all areas distal to the palatines/frontals, including horncores) exposed aboveground, with a high likelihood of subsequent repeated burial and reexposure. Horncores had therefore not survived beyond the occasional horncore base, and posterior parts of skulls were in very poor condition, if present at all.

On initial assessment, it became clear that the skulls were held together only by the fine silts laid down internally in their crania, so in the interests of both time and keeping skulls intact, we decided not to remove these internal deposits, or indeed any deposits in and around the skulls that was supporting bone in place. Deposit was removed selectively only from cranial areas providing the most useful zooarchaeological information, such as the maxillary tooth rows, palatines, frontals, lacrimals, and orbits. Throughout observation and recording, we avoided overhandling the skulls to reduce breakage. We studied the maxillary dentition first and then the palatine area, with skulls resting on their frontals, before turning skulls over to clean and make observations on the top part of the cranium.

Zooarchaeological Data Recording

A series of six cattle skull recording forms were developed specifically for recording the zooarchaeological data captured from the Kheshiya skulls. (Examples of Forms 1–4 are shown in appendix 11.2.)

Form 1, Basic: with fields for describing overall condition, bone surface weathering stages, presence/absence of burning, and any bone surface modifications, plus a table for scoring which skull parts were present

Form 2, Morphology: for recording nonmetrical morphological skull traits

Form 3, Aging Data: for recording dental eruption and wear stages, horncore texture, and cranial bone fusion

Form 4, Measurements on Cranium of *Bos*: for recording metrics taken on the cranium and maxillary dentition

Form 5: Image template line drawings of *Bos* cranium from von den Driesch 1976 (pp. 29–30:figure 8a, dorsal view, figure 8b, nuchal view, dorsal view, and nuchal view) for shading Kheshiya skull part presence/survival

Form 6: Image template line drawings of *Bos* cranium from von den Driesch 1976 (pp. 29–30:figure 8c, left side view, figure 8d, basal view) for shading Kheshiya skull part presence/survival

Identification

A range of large bovids potentially inhabited Southern Arabia during the Early and Middle Holocene, and care was taken to check taxonomic identification of each of the Kheshiya skulls against other possibilities. Because there are few comparative zooarchaeological datasets and the current wildlife is much diminished from earlier diversity, predicting the range of the Early Holocene native fauna of the region is challenging. Of the medium-size bovids (60–200 kg), the Arabian oryx (*Oryx leucoryx*) is likely to have been widespread in the past. Although there is no direct evidence that the addax (*Addax nasomaculatus*) or hartebeest (*Alcelaphus buselaphus*) ever inhabited Southern Arabia, their grassland/semidesert habitats—known from neighboring East and North Africa—mean that a wider distribution cannot be ruled out (McCorriston and Martin 2009:240–41). This may also hold true for the kudu (*Tragelaphus imberbis*). Of the larger bovids (above 200 kg), the African buffalo (*Syncerus caffer*) required consideration because of its known wide habitat range, even though it has not been directly evidenced in Arabia, apart from in prehistoric rockart (Rachad 2007).

Close observation of the morphology of the Kheshiya skulls, particularly their dentitions, allowed for these alternatives to be discounted, and all skulls were

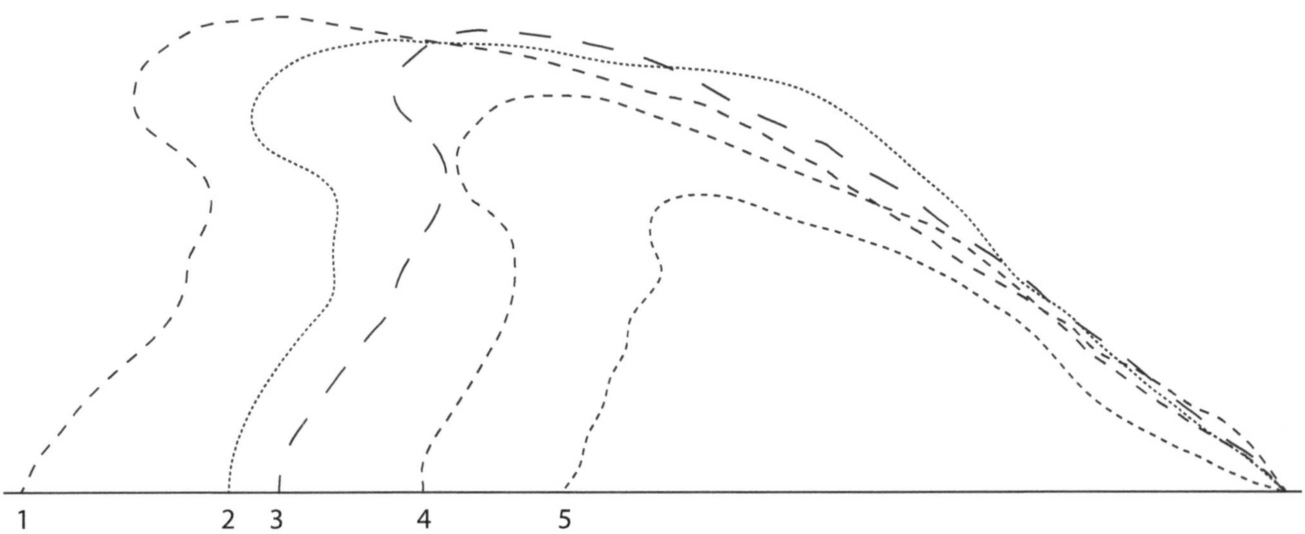

Figure 11.2. Cattle sagittal profile shapes for *Bos indicus* (after Grigson 1980:18, figure 11) showing adult forms (1–4) and a younger shape (5). Adults display a convex frontal profile, concave occipital area, and intercornual ridge facing upward and backward. *Drawing by Clara Hickman.*

identified as cattle (*Bos* sp). Whether the cattle represented were wild aurochs (*Bos primigenius*—evidenced in the Pleistocene and Early Holocene of Southern Arabia; see above) or herded domesticates, and if the latter, whether they were likely to belong to a domestic European cattle type (*Bos taurus*), Asia zebu (*Bos indicus*), or African cattle type, can be approached only through morphological and metrical analyses (described below).

Morphology: Nonmetrical Traits

Where possible, morphological features of the skulls were scored in an attempt to use these nonmetrical traits to assess the species of cattle present at Kheshiya and to gauge morphological variation between the individuals represented in the skull circle. Bear in mind that morphological variation may relate to taxonomic status (for example, taurine or zebu cattle), sexual dimorphism (male or female), wild or domestic status, and age.

Morphological skull features were recorded following criteria described by Grigson (1976, 1980), whose detailed studies of the craniology of four *Bos* species to assess their taxonomic relationships are exceptionally useful zooarchaeological aids (see also Grigson 1974, 1975, 1978). Grigson directly compares morphological criteria across *Bos* taxa, including taurines and zebu cattle, with line drawings highlighting the most useful distinguishing features. Ten separate traits were considered for the Kheshiya skulls to assess whether they had more taurine or zebu features. Unfortunately, preservation did not allow consistent observation/scoring of many of the traits, which are described below:

The sagittal profile was recorded where possible following Grigson's (1976, 1980) criteria. Grigson finds that the "the sagittal profile of *Bos indicus* differs very significantly in almost all of the skulls examined from that of *Bos taurus*" (1980:18), with *Bos indicus* displaying a convex frontal and concave occipital, with the intercornual ridge directed upward and backward (figure 11.2). *Bos taurus* has a flatter frontal profile, with either rounded or pointed intercornual ridges (figure 11.3). Only two Kheshiya skulls were complete enough for full sagittal profiles to be taken: Skulls 18 and 25. This was done using dental wax, heated in water, molded onto the skull in the sagittal plain (method follows Grigson 1976:115), and left to harden. The wax was then removed and the shape was drawn onto tracing paper. Grigson considers this the most important difference between *indicus* and *taurus* (1980: 30). Another four skulls gave an indication of sagittal shape but did not allow full profile.

The orbital rim is also considered a good criterion for *indicus*/*taurus* separation, with Grigson (1980:23) finding this feature in all *Bos indicus* she observed to be flat (see also Grigson 1976:123, figure 8), as opposed to having a sharp rim in *Bos taurus* (figure 11.4). Because it protrudes from the skull, the orbital rim of the Kheshiya assemblage often is damaged and was observable in only one specimen (Skull 41).

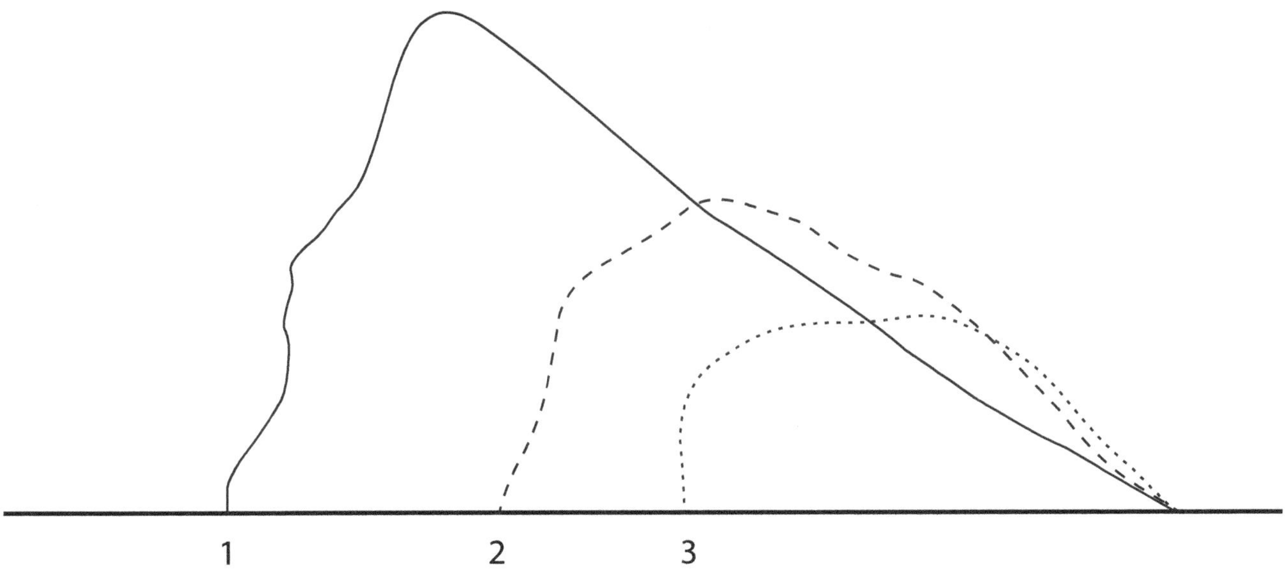

Figure 11.3. Cattle sagittal profile shapes for *Bos taurus* (after Grigson 1976:figure 5) showing the adult form (1) and profiles for younger animals (2, 3). The adult form displays a flat frontal profile with a rounded intercornual ridge. *Drawing by Clara Hickman.*

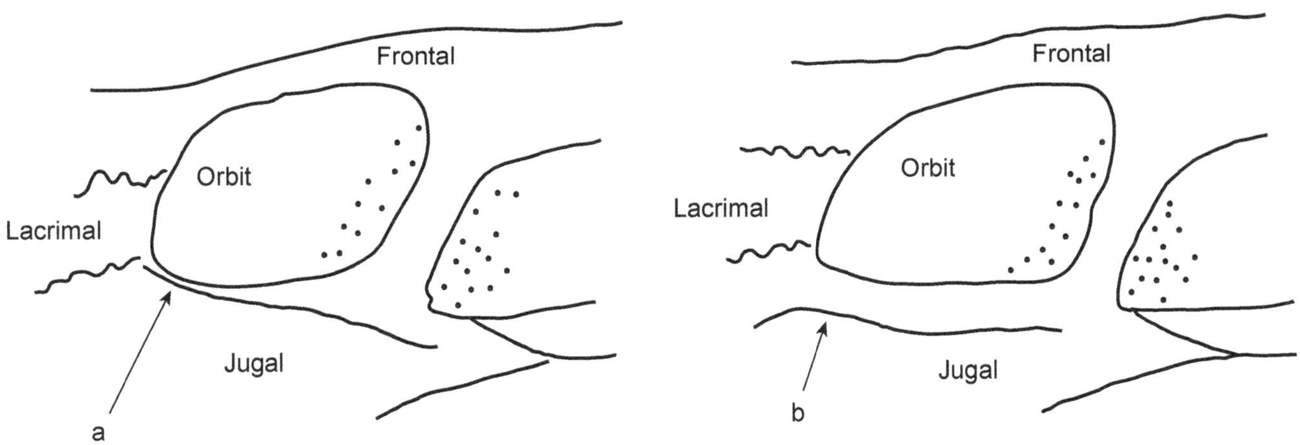

Figure 11.4. Forms of the cattle orbital rim. Left: *Bos taurus* with a sharp rim (a) (after Grigson 1976:123, figure 8). Right: *Bos indicus* form displaying a flat rim (b) (after Grigson 1980:23). *Drawing by Clara Hickman.*

The shape of the nasal–frontal suture was also often difficult to observe on the Kheshiya specimens, due to the frequent forward slumping of the skulls and bone breakage in this area. Grigson (1980:24, figure 21) shows that a simple inverted V shape tends to characterize *Bos indicus,* while *Bos taurus* often shows a more complex M shape (figure 11.5). This is an insecure separation criterion, however, with Grigson (1980:23) reporting exceptions in specimens of both species examined.

Grigson also found the frontal–lacrimal and lacrimal–jugal sutures to differ between the specimens of taurines and indicines she studied (Grigson 1980:23), with the frontal–lacrimal suture bowed downward in all taurines and the lacrimal–jugal suture bowed correspondingly upward (figure 11.6). In most but not all *Bos indicus* skulls, both sutures were straight. These sutures were observable in 14 of the Kheshiya skulls.

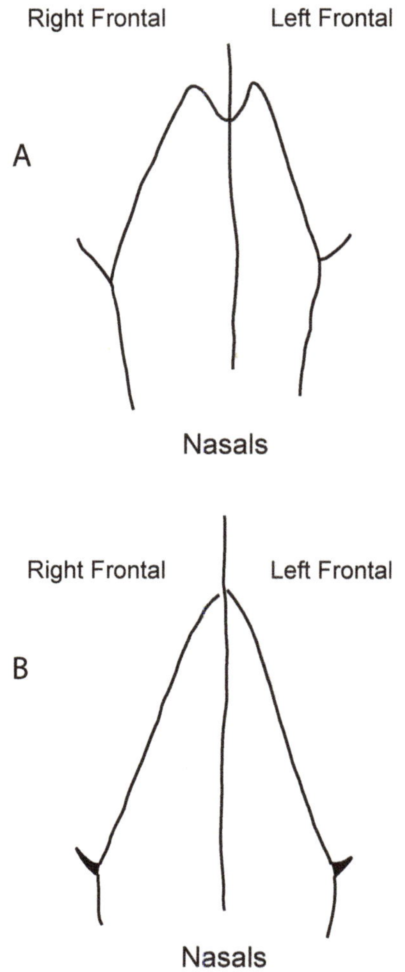

Figure 11.5. Forms of the nasal–frontal suture in cattle, following Grigson 1976, 1980. Top: The complex M shape seen in *Bos taurus* (A). Bottom: The simple inverted V shape (B) that tends to characterize *Bos indicus* (after Grigson 1976:125, Figure 11). *Drawing by Clara Hickman.*

Recording of the frontal profiles was attempted for 18 of the Kheshiya specimens. This shape should be observed between the horncores and viewed from above, but since many of the Kheshiya skulls did not have the horncore area surviving, the profile was often taken some centimeters anterior to this point. Grigson shows various frontal shapes of taurine breeds (1976:126, 1980:25) alongside a typical zebu profile between the horncores (figure 11.7). Taurines appear quite variable and can have convex or flat profiles, or rise in a boss, while *Bos indicus* is characterized by concave profiles (figure 11.7, profile 5). Grigson notes that the frontal profile is not a firm criterion, but it is fairly diagnostic (1980).

The intercornual ridge can also be distinctive between taurine and zebu cattle (Grigson 1976:128, 1980:26) (figure 11.8). Shapes 1–6 in figure 11.8 were all observed in taurine skulls, while 7 and 8 tended to be found in *Bos indicus*, although variations existed (Grigson 1980:26). These ridge forms were scored for the Kheshiya skulls in seven cases, where observation was possible.

Regarding horns and horncores, Grigson (1978, 1980:27–28) finds the overall shape and direction of these quite distinctive between taurines and *indicus*, with separation possible on the majority of the cattle skulls studied. While there is much breed-, age-, and sex-related variation, all *Bos indicus* skulls have horns that point upward and slope backward from their bases, unlike taurines, whose horncores leave the skull in an outward direction, whatever morphology the rest of the horn takes (figure 11.9). There is little difference between the actual shape of the horncore bases between the two species (Grigson 1980:28), so this was not recorded.

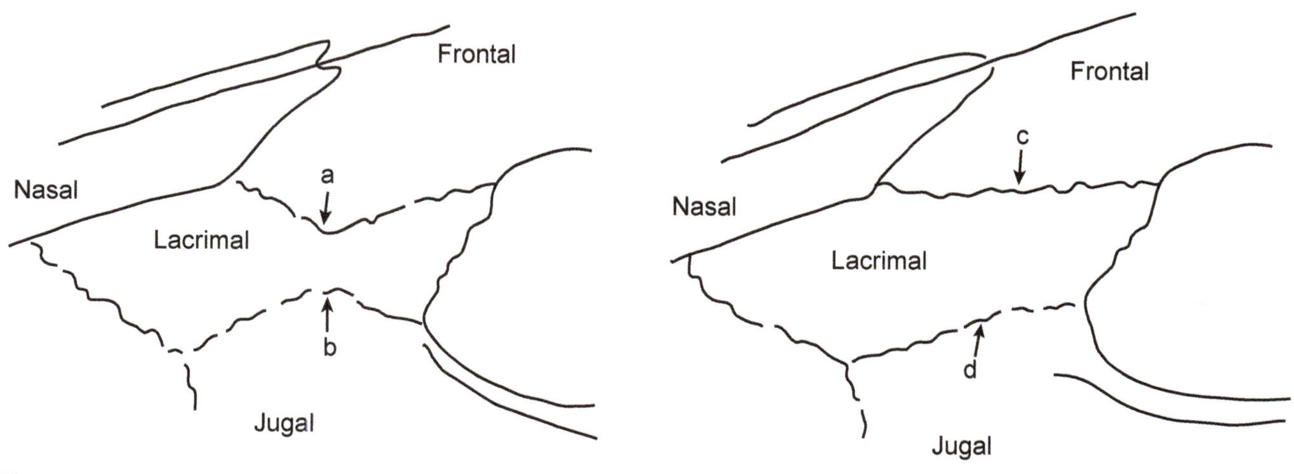

Figure 11.6. Forms of frontal–lacrimal and lacrimal–jugal cattle sutures that Grigson (1980:23) observed as tending to differ between taurines and indicines, with *Bos taurus* showing the frontal–lacrimal bowing downward (a) and lacrimal–jugal bowing upward (b), contrasting the relatively straight sutures (c) and (d) in *Bos indicus* (after Grigson 1976:124, Figure 9). *Drawing by Clara Hickman.*

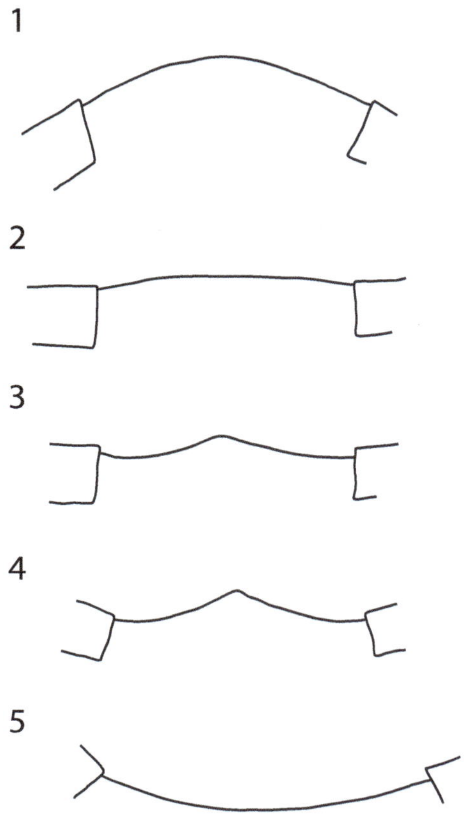

Figure 11.7. Forms of cattle frontal profiles (between the horns) observed by Grigson, with 1 to 4 showing various frontal shapes of taurine breeds, ranging from convex or flat profiles to those that rise in a rounded or pointed boss, alongside a typical zebu profile (5), which is concave. After Grigson 1976:126, 1980:25. *Drawing by Clara Hickman.*

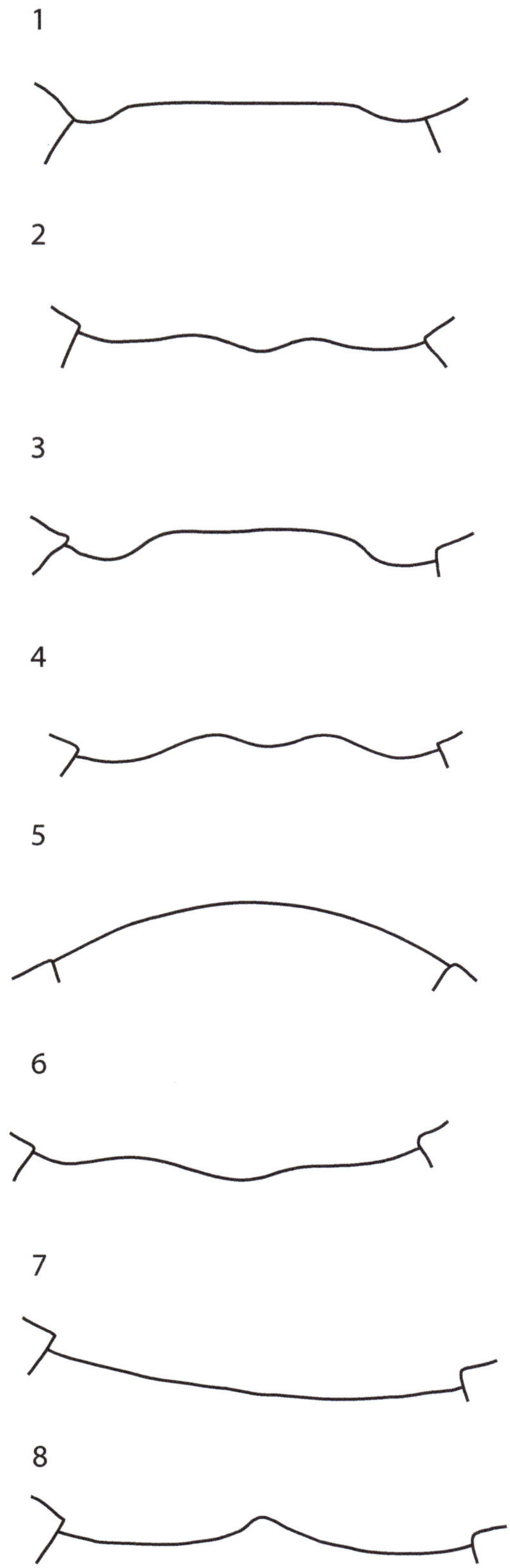

Figure 11.8. Forms of the intercornual ridge in taurine and zebu cattle, showing shapes 1 to 6 observed by Grigson in *Bos taurus* and shapes 7 and 8 tending to characterize *Bos indicus* (after Grigson 1976:128, 1980:26). *Drawing by Clara Hickman.*

Finally, the morphology of the posterior wings of the skull palate was an area that commonly survived in the Kheshiya skulls, so the shape was recorded following Grigson's observations (1976:126, figure 12, 1980:24–25, figure 22). There is much age- and sex-related variation in this character, but in general Grigson finds two morphologies: straight and broad, and convex and narrow (figure 11.10). The former is characteristic of adult taurines, and the latter is recorded for younger and female taurines and the few *indicus* skulls Grigson managed to study (1980:24).

In sum, of the 10 criteria described above, the most reliable for separating *Bos taurus* and *Bos indicus* appear to be the sagittal profile and the shape and direction of the horncores as they leave the skull. Grigson also found the orbital rim, frontal profiles, and intercornual ridges to be good discriminating criteria, if slightly less secure. Facial suture shapes seem less reliable as diagnostic criteria, and the morphology of the posterior wings of the palate is

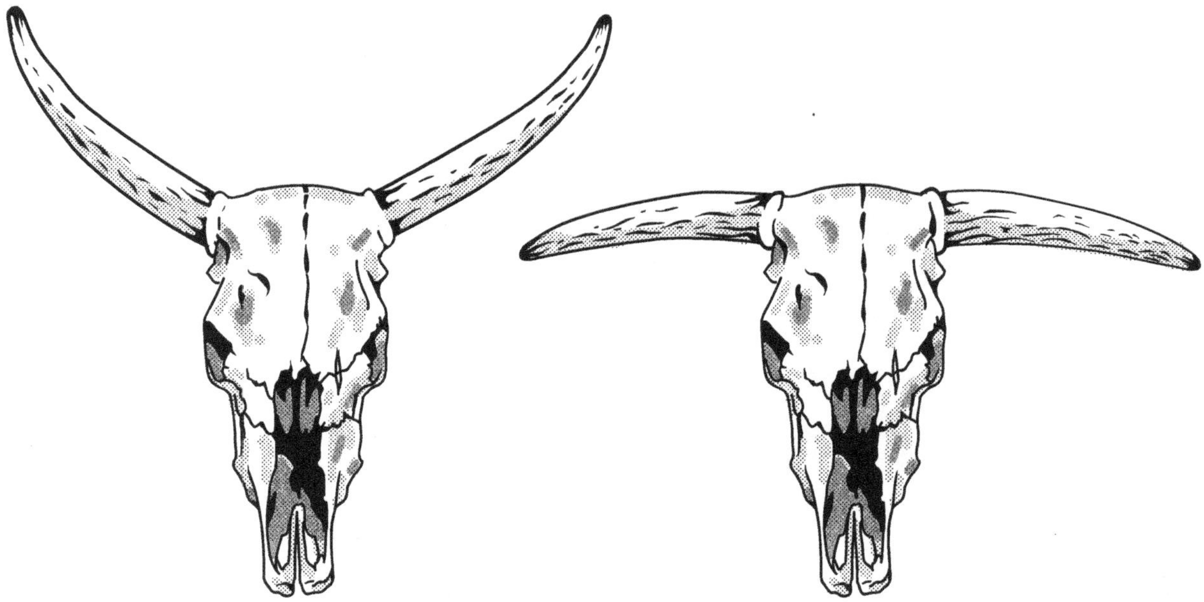

Figure 11.9. Left: horncore shape and direction of *Bos indicus*, showing skull with horns pointing upward and sloping backward from their bases. Right: *Bos taurus* skull showing horncores leaving the skull in an outward direction, whatever morphology the rest of the horn takes. *Drawing by Clara Hickman.*

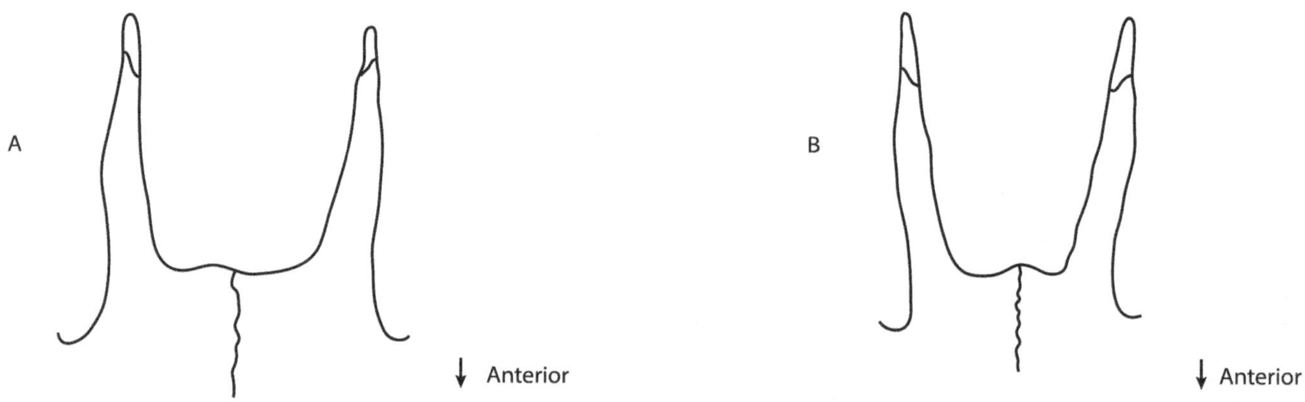

Figure 11.10. Two forms of the posterior wings of the palate in cattle. Left: The straight-sided and broader morphology of adult *Bos taurus* (A); right: the more convex-sided and narrow form (B), characteristic of younger and female taurines and noted in some *Bos indicus* skulls (after Grigson 1976:126, figure 12, 1980:24). *Drawing by Clara Hickman.*

highly variable according to age and sex, as well as species; hence it is less useful.

While Grigson's criteria give us extremely useful scoring systems for the nonmetrical traits of cattle skulls, we need to bear in mind that they were devised from studies of modern cattle, far distant from the Kheshiya cattle population(s) in both time and space. In her study, Grigson drew on a sample of about 24 modern *Bos indicus*, mostly from India, and a larger sample of *Bos taurus* skulls of various breeds, all from Britain. Geographical and temporal variation in cattle skull morphology should therefore be expected when interpreting criteria.

Biometrics

There are two reasons for undertaking biometrical analyses of the Kheshiya skulls: first, to explore the overall size of the skulls and implications for the wild/domestic status of the cattle (although their context and date strongly suggest domesticates); and second, metrical analysis allows for interpretation of sexual dimorphism and sex ratios of the skulls. Did the skulls belong to males, females, or both? All cranial metrics were taken where possible, even though only some are used in comparative analyses.

Measurements follow standards set by von den Driesch (1976). For cranial, tooth row, and horncore measurement, von den Driesch's Codes 1–47 were used (von den Driesch 1976:27–30, figure 8a–d) (see appendix 11.4). For individual teeth, standards for measuring ruminant teeth were followed (von den Driesch 1976:57) but applied to the maxillary dentitions at Kheshiya (that is, length [L] and breadth [B] of the molar teeth and P4 were added for completeness). All dentitions in the maxilla were erupted adult teeth. Tooth rows were measured along the alveoli on the buccal side, and individual tooth measurements were taken at the biting surface, following von den Driesch's standards. To capitalize on the fact that the dental arcades are the most intact areas of most skulls, three additional measurements were devised and taken. Measurements were taken between the left and right sides of the maxillary dentitions, at the alveolus on the lingual side, in the following locations: LM1: P2-P2 internal least breadth; LM2: M1-M1 internal least breadth; LM3: M3-M3 internal least breadth. All measurements were taken using a vernier caliper to 0.1 mm. Appendix 11.4 shows the full set of resulting osteometric data. Individual measurements marked as "estimated" were taken where bone surfaces were slightly eroded or fragmented but the dimension was still clear; these measurements were considered sufficiently accurate to be used in analyses. Measurements marked as "highly estimated" are less reliable and were not included in metrical analyses.

DNA

Three cattle maxillary teeth from separate Kheshiya skulls were sampled for preliminary testing for aDNA preservation, with unsuccessful results (appendix 11.3). Collagen preservation is likely to be poor in the Kheshiya environment; future studies might target the petrous temporal, which is proven to give better results than dentition (Hansen et al. 2017).

Aging Data: Dental Eruption and Wear, Skull Suture Closure, and Horncore Texture

Aging data were collected primarily through assessment of dental eruption and wear stages. The system described by Grant (1982) for cattle mandibular cheek teeth was adapted for the Kheshiya skulls. It included only their maxillary teeth (with no associated mandibles/mandibular dentition) because schemes for recording maxillary tooth wear stages were not found in the literature. Maxillary teeth have clear differences in morphology, proportions, and size compared to their mandibular counterparts, which needs to be taken into account in analysis, particularly when mandible wear scores (MWSs) are used to estimate age (or age stage) at death; rates of occlusal attrition through Grant's stages (1982:92, figure 1) are likely to vary between upper and lower dentitions.

For the Kheshiya skulls, the Grant tooth wear system was applied to both left and right maxillary cheek teeth. Teeth had sometimes fallen out and were missing; a single tooth had also been selected and removed in-field for sampling (normally M3 or M2). Thus the aim of recording both sides of dentition was, first, to check for asymmetrical wear and, second, to maximize chances of having a full set of molars with wear stages. (See "Results: Dental Aging," below.)

The coming into wear of the accessory pillar, which sits between the anterior and posterior cusps of cattle molars, has been proposed as a useful additional criterion for separating adult from older cattle (Halstead 1985). It was not used as a separate criterion in the current study (accessory pillar wear is included within Grant's original 1982 stages) since the pillar is now considered too variable in size to accurately reflect increased wear/age (Jones and Sadler 2012a:11). An alternative method for assessing age of death in older cattle that uses "the position of the cement/enamel junction and the root arch in relation to the alveolar border in molar teeth" (Jones and Sadler 2012a) was not published at the time of studying the Kheshiya assemblage, and so it was not recorded. It has proven difficult to observe these criteria from the Kheshiya cattle skull photographic archive because the maxillary alveolar border is frequently damaged. Hence the Jones/Sadler early to middle aging method (intended for mandibular dentitions anyway) was not attempted.

One further aspect of dental wear recorded for the Kheshiya assemblage was the movement and wearing of the distal end of the P4 into the mesial end of the M1, which sometimes occurred to a great extent where the P4 had wedged itself into the M1, forming a continuous occlusal surface. The aim was to view this alongside the dental anomalies (described below) that mainly appeared to result from premolar maleruption.

Skull suture closure also was recorded for various skull parts (see "Results: Cranial Suture Fusion," below), although poor visibility of sutures due to skull fragmentation and adherence of deposits limited observations. Areas most often visible and recordable were the medial–palatine suture and the maxilla–lacrimal–zygomatic sutures, all of which are relatively late fusing, plus the frontal halves, which are earlier fusing, and the basioccipital area, which is even earlier fusing, following Grigson's summary of data on cattle suture closure timings (Grigson 1982:20, appendix 1). Data tend to give broad age ranges for suture closure (ranging from one to five-plus years for adult animals), and there certainly will be much variation between cattle breeds across wide geographical areas and historical time frames. Nevertheless, by analyzing relative suture closure times, an analyst can assign very approximate age-at-death ranges. These ranges can be considered alongside ranges from dental data (best accuracy for early to middle age).

The skull recording form allowed for the documentation of horncore texture, useful to assess relative age (following Armitage 1982). Firsthand study of the skulls, however, found that in the few instances where horncore bases were preserved, none of the horncores survived beyond the first couple of centimeters from the skull, so this aspect of recording was abandoned.

Dental Pathologies and Anomalies

Two types of dental anomalies were observed and recorded. (See "Results: Dental Anomalies," below.) The first is tooth rotations—when a single tooth has erupted at an angle to the main line of the cheek tooth row but remained in occlusion. The instances of this in the Kheshiya assemblage were all observed with premolars, hinting at a link with maleruption. Tooth rotations were recorded, with details about which tooth and side of the jaw was affected and also the approximate angle of rotation, taking the buccal edge of the tooth row as a rough curved line and estimating the angle (in degrees) by which the rotated tooth diverged from that line. The second, very rare anomaly recorded was the absence of a particular tooth during the life of an animal.

Condition of Skulls, Breakage, and Treatment

The initial aim was to record which cranial parts were present and absent for each cattle skull to assess whether the skulls had seen any modifications prior to deposition. Cattle skulls used in installations in other Neolithic and later contexts from the broader Middle East/Anatolia/North African area often are not complete. At Neolithic Çatalhöyük in central Anatolia, for example, horns often were removed for separate installation, or anterior portions of the skull were removed to create the well-known bucrania (a term itself implying modification) that were built into domestic walls (Mellaart 1967; Russell and Martin 2005; Twiss and Russell 2009), with similar practices seen at nearby Early Neolithic Boncuklu (Baird et al. 2016:figure 6). The thousands of domestic cattle skulls buried at the Kerma necropolis in northern Sudan see varied modification through the fourth millennium cal BP, with an earlier practice of creating a bucranium of frontals, horncores, and nasals for deposition, while later examples had nasals removed too (Chaix 2007:173–75). The forms for the Kheshiya skulls, therefore, aimed at recording skull parts present (appendix 11.2; see Forms 1, 5, and 6).

It became apparent on lab examination that any skull parts absent had been broken not because of pre-depositional human modification but through post-depositional processes, either from the extreme burial environment (wetting/drying of the matrix; surface exposure) or the challenges of excavation and lifting. While retrieval was excellent, fragile skull parts often were broken off and fragmented, but they were carefully collected nevertheless. Skull part presence was recorded on the template diagrams (appendix 11.2, Forms 5 and 6) as a record of preservation; the process of close examination led to interesting observations about patterns of damage and breakage.

Bone Weathering

Bone surface weathering was recorded using Behrensmeyer's (1978) weathering stages, which define bone surface weathering in subaerial/surface contexts, with the aim—following further controlled experiments—of determining periods of time between bone deposition and eventually burial. This is relevant to the current study because there is a strong chance that rear parts of the Kheshiya skulls (including horns and horncores) remained aboveground, and examination of differential weathering patterns might aid understanding of the extent of original burial. Thus general weathering stages were noted across the skull frontals, since this is largest flat area of bone visible; where other cranial areas differed in weathering stage, this too was recorded.

Behrensmeyer's seminal article (1978) defined six weathering stages, which range from bone appearing fresh and still greasy (Stage 0), through increasing stages of surface cracking and exfoliation (Stages 1 and 2), to deeper cracks opening in bone (Stages 3 and 4), until bone eventually falls apart (Stage 5). There has been considerable subsequent discussion about whether stages can be usefully related to periods of time that bone has been left exposed aboveground before burial (e.g., Lyman and Fox 1989). Studies conclude that so many factors are at play—such as variations in bone size, element morphology, the microenvironment of burial, and temperatures and moisture (even before differences in prediscard treatment of bone is considered)—that weathering stages cannot be employed to read even approximate lengths of time between bone deposition and subsequent burial. It is now also acknowledged that bone weathering does not stop with burial, although it probably slows down, depending on the stability of the burial environment (Lyman and Fox 1989). Bone buried in deposits that continue to experience variation in moisture and temperature is likely to continue the weathering process, particularly if burial is shallow.

Beherensmeyer's scheme dealt with postcranial elements rather than skulls, and other observations (on human skulls) suggest that cranial bone weathers differently from long bones (e.g., Ross and Cunningham 2011:132, table 3)—for example, they exhibit surface pitting rather than cracking. Therefore, a slightly modified set of descriptors for the weathering stages was developed for use with the Kheshiya skulls (as described in "Results: Weathering," below).

Burning and Cut Marks

Skulls were examined for any signs of burning, heating, or charring, although none were found. Any cut marks were recorded by placement, number, and morphology.

Photographic Archive

Once cleaned, each skull was photographed from *six* angles (dorsal views, basal views, both lateral views, nasal and nuchal views), plus detailed close-ups of dentition, important morphological diagnostic features, and cut marks.

Results: Origin, Age, and Taphonomy

Data relating to the cattle skull analysis are here presented and discussed, focusing first on the cattle themselves, in terms of the species to which the skulls belonged, their size, status, ages at death, and sex balance of the cull. Second, data relating to the condition, treatment, and modification of skulls are assessed, with the aim of unraveling the cultural and natural processes that affected them prior to burial, during burial itself, and post-depositionally.

Table 11.1 summarizes the analyses to which each of the 35 skulls contributes information. The 35 skulls vary widely in terms of how much data they provide for various analyses, with many contributing to most areas of analyses. Other poorly preserved specimens, however, such as Skull 15, which had no surviving dentition and was too fragmentary to be measured, contribute less, although this skull could be assessed for weathering and cranial fusion data.

To summarize, table 11.1 shows that all 35 skulls could be assessed for bone surface weathering data, and some cranial fusion data could be assessed on each skull. Most skulls (n = 33) provided information on morphological traits and metrics, and there is a good sample of dental early to middle aging data (n = 32) for reconstructing ages at death. Cut marks were recorded on very few skulls (n = 4). Perhaps there were few because the other skulls never had any signs of butchery or preparation, or, more likely, because the fragmentary condition of many skulls obscured the visibility of cut marks. Dental anomalies, however, could potentially be observed on all 34 skulls because the dentition survived well, but in fact they are visible in only eight skulls, which is likely to be a roughly accurate frequency. Table 11.1 thus gives an indication of how representative the following discussions of results are in relation to the assemblage as a whole.

Taurine or Zebu Cattle?

The "Methods" section above describes how each cattle skull was scored where possible for nonmetrical morphological traits, with the aim of using these criteria to determine whether the skulls belonged to *Bos taurus or Bos indicus*. Appendix 11.5 shows the resulting data and comments for each skull recorded for 10 nonmetrical traits. It is notable how many traits were not assessable due to poor preservation. Very few skulls had posterior parts surviving, meaning that horncore shape and direction were visible in only a few cases, and sagittal profiles could not be taken in most. Orbits and facial sutures, too, suffered badly from breakage. Of the 35 potentially assessable skulls, the proportion of traits scored was relatively low, with the exception of the shape of the posterior wings of the palate, protected by its more internal skull location, which survived well.

Information in appendix 11.5 is summarized in table 11.2, which shows how many skulls exhibited morphological characteristics of either taurine or indicine cattle, following Grigson's (1976, 1980) criteria (see "Methods" section), or had more questionable criteria (leading to the "taurine?" and "*indicus*?" assignations). Where criteria were present but ambiguous, skulls were recorded as "indeterminate." The bottom row of table 11.2 shows how many of the 35 skulls could be scored for each trait; the right-hand column shows how many traits in the total Kheshiya skull assemblage could be counted as taurine, taurine?, *indicus*, *indicus*?, and indeterminate. Table 11.3 also summarizes the information by skull, showing how many morphological characteristics interpreted as taurine or taurine? and *indicus* or *indicus*? each skull exhibited.

An initial view of table 11.2 seems to suggest the presence of both taurine and *indicus* morphological traits in the Kheshiya skull assemblage, with more of the former in the totals. Further consideration needs to be given, however, to the reliability of each of the criteria—which Grigson notes as being variable in usefulness—particularly since 14 of the 35 assessable skulls display characteristics of both species within an individual skull (table 11.3).

Grigson found the most important criterion for separating taurine from indicine skulls to be the sagittal profile (1980:30), with all her study specimens being reliably separable using this skull shape. Kheshiya Skulls 18 and 25 allowed sagittal profile shapes to be taken (figures

Table 11.1. Table summarizing which zooarchaeological analyses each of the 35 Kheshiya cattle skulls provides data for, showing totals in each case (not including Lot 35, which is a mandible fragment). The right-hand column also shows which skulls provided tooth samples exported to University College London for our dental microwear and isotope study (Henton et al. 2014).

Skull/ Lot Number	Weathering	Posterior–Anterior Slumping	Morphological Traits	Cranial Metrics	Dental Metrics	Cranial Fusion	Dental Aging	Dental Anomalies	Cut Marks	Henton Isotope Study	Henton Microwear Study
2	x		x	x	x	x	x				
3	x	x	x	x	x	x	x	x	x		
4	x			x	x	x	x				
5	x	x	x	x	x	x	x				
11	x	x	x	x	x	x	x	x		x	x
12	x		x	x	x	x	x				x
13	x		x	x	x	x	x			x	x
14	x	x	x	x	x	x	x				x
15	x					x					
16	x	x	x	x	x	x	x				x
17	x	x	x	x	x	x	x			x	x
18	x	x	x	x	x	x	x			x	x
19	x		x		x	x	x			x	x
20	x	x	x	x	x	x	x			x	x
21	x		x	x	x	x	x			x	x
22	x	x	x	x	x	x	x			x	x
23	x		x	x	x	x	x			x	x
24	x	x	x	x	x	x				x	x
25	x	x	x	x	x	x	x				x
26	x		x	x	x	x	x				x
27	x	x	x	x	x	x	x			x	x
28	x	x	x	x	x	x	x	x		x	x
29	x	x	x	x	x	x	x	x			x
30	x	x	x	x	x	x	x				
31	x	x	x	x	x	x	x				x
32	x	x	x	x	x	x	x		x	x	x
33	x	x	x	x	x	x	x		x	x	x
34	x		x	x	x	x	x	x		x	x
36	x	x	x	x	x	x	x				x
37	x	x	x	x	x	x	x		x		x
38	x		x	x	x	x	x	x			x
39	x		x	x	x	x	x			x	
40	x		x	x		x					x
41	x	x	x	x	x	x	x	x		x	x
42	x		x	x	x	x	x	x			x
Total (35)	**35**	**21**	**33**	**33**	**33**	**35**	**32**	**8**	**4**	**17**	**28**

Table 11.2. A summary of information in appendix 11.5, showing the number of Kheshiya skulls that exhibit morphological characteristic of either taurine or indicine cattle, following Grigson's 1976 and 1980 criteria (see "Methods" section). Some exhibited questionable criteria leading to the taurine? and *indicus*? assignations; skulls with ambiguous criteria are listed as indeterminate. The bottom row shows how many of the 35 skulls could be scored for each trait; the right-hand column shows how many traits in the total Kheshiya skull assemblage can be assigned to taurine, taurine?, *indicus*, *indicus*?, and indeterminate.

Type	**Sagittal Profile**	**Orbital Rim**	**Nasal–Frontal Suture**	**Frontal–Lachrymal Suture**	**Lachrymal–Jugal suture**	**Frontal Profile**	**Intercornual Ridge**	**Horncore Shape**	**Horncore Direction**	**Shape of Posterior End of Palate**	**Total Traits, All Skulls**	**Total Most Reliable Criteria (sagittal profile, horncore shape and direction)**
taurine	2	0	1	0	0	14	2	4	1	8	**32**	**7**
taurine?	4	0	2	0	1	4	2	0	1	0	**14**	**5**
indicus	0	0	0	0	8	0	2	0	0	8	**18**	**0**
indicus?	0	1	5	1	5	0	0	0	2	3	**17**	**2**
intermediate	0	0	0	0	1	0	1	0	0	11	**13**	**0**
Total Assessable: 35	**6**	**1**	**8**	**1**	**15**	**18**	7	**4**	**4**	**30**		

Table 11.3. A summary of information in appendix 11.5, showing how many morphological traits interpreted as taurine/taurine? and *indicus*/*indicus*? each of the Kheshiya skull exhibits (following criteria of Grigson 1976, 1980; see "Methods" section, figures 11.2–11.10).

Skull/Lot Number	Taurine/Taurine? All Traits	*Indicus*/*Indicus*? All Traits
2	1	0
3	2	0
4	0	0
5	1	0
11	2	1
12	2	1
13	1	2
14	0	1
15	0	0
16	1	1
17	0	4
18	2	2
19	0	1
20	2	0
21	4	0
22	2	0
23	3	3
24	1	1
25	3	1
26	0	0
27	0	0
28	2	1
29	3	0
30	4	1
31	1	0
32	0	1
33	0	2
34	2	0
35		
36	2	3
37	1	1
38	0	0
39	1	1
40	0	1
41	2	3
42	1	1

11.11 and 11.12), and both show flat frontals characteristic of taurine cattle rather than the convex frontals of zebu. Figure 11.13 also shows a photograph of Skull 25 with the characteristic *Bos taurus* frontal profile, although the posterior end of skull is missing. Intercornual ridges do not protrude upward and backward as they would in zebu (Figure 11.2) but are more rounded, which is consistent with a taurine interpretation. Four other Kheshiya skulls (21, 23, 30, and 36) had enough of their frontals and occipitals surviving to gauge the sagittal profile shape, even if they were not sufficiently complete to be drawn. All four match the taurine profile. On the basis of this most distinctive and reliable criterion, therefore, no skulls are zebu-like, although the sample size is small (6 out of 35).

Horncore morphology is also considered reliable, with Grigson (1978, 1980:27–28) finding that the majority of taurines and zebu/indicines could be separated using their shapes and directions. As described above, horncores, horncore bases, and posterior parts of skulls preserved terribly at Kheshiya, probably because they protruded aboveground and were exposed at least initially after skulls were buried. Horncore shape and direction could be gauged from only seven skulls, and in each case from the small broken remains of horncore bases. Four specimens exhibited taurine horncore base shapes (Skulls 22, 23, 25, and 29) (figure 11.14); a further two skulls (21 and 41) showed horncores leaving the skull outward from the frontals, characteristic of taurines. Another two skulls were scored as having horncore bases with horns appearing to angle more backward, as they would in *indicus* (Skulls 17, 23), although in both cases this was noted as questionable because of the highly fragmented state of the skulls (figure 11.15). Considering that one of these tentative *indicus* shapes (Skull 23) was also recorded as having taurine-shaped horncores, identification seems inconsistent and the evidence is perhaps not strong enough to be sure of a presence of *indicus*-shaped horns. It seems prudent to conclude that while several examples have characteristic taurine horns (with three of the same skulls—21, 23, and 25—also having taurine-like sagittal profiles), two skulls have more backward-sloping horncores, which may hint at *indicus* shape or may simply represent taurine variation. It is worth mentioning here that some photographs of the skull ring upon excavation allow observations that laboratory study did not, where very fragile areas of horncore base were in some cases still supported by pillars of sediment deposits and horncore shape could be traced. Figure 11.16, for example, shows Skulls 13, 14, 11, and 12, with indications of horncores leaving their skulls outward from the frontals, as they would in taurines.

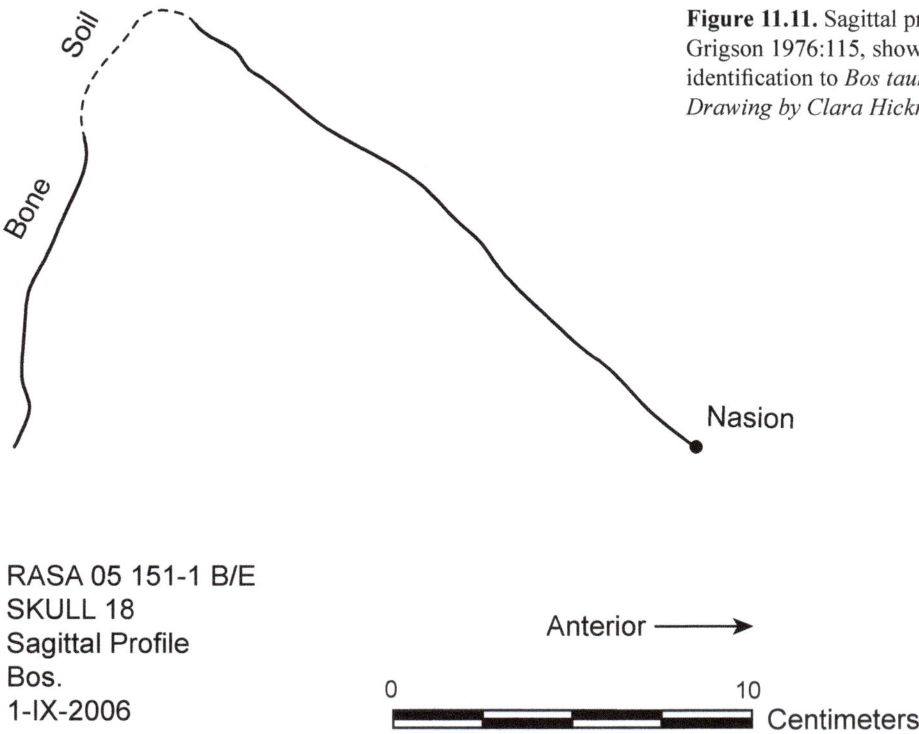

Figure 11.11. Sagittal profile of Kheshiya Skull 18, following Grigson 1976:115, showing flat frontal profile, consistent with identification to *Bos taurus* (compare to figures 11.2 and 11.3). *Drawing by Clara Hickman.*

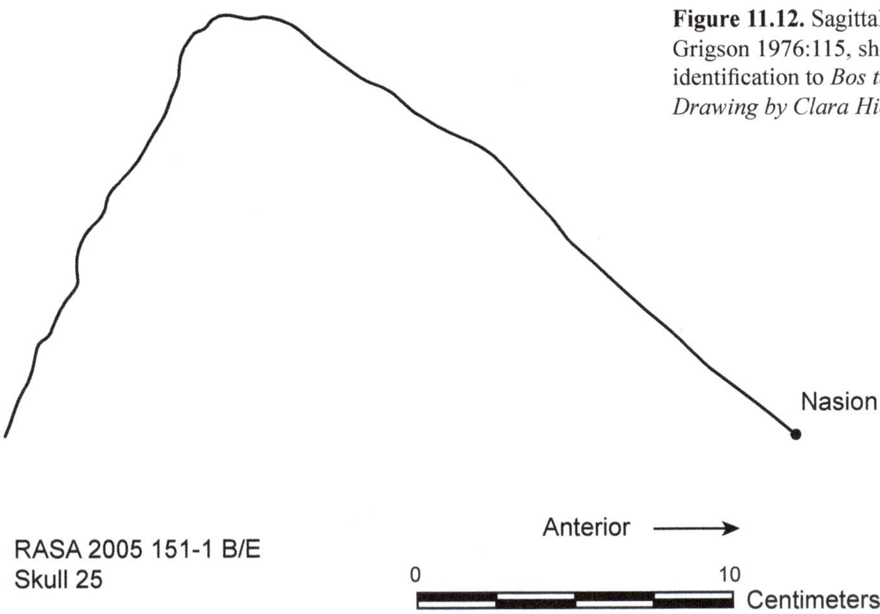

Figure 11.12. Sagittal profile of Kheshiya Skull 25, following Grigson 1976:115, showing flat frontal profile, consistent with identification to *Bos taurus* (compare to figures 11.2 and 11.3). *Drawing by Clara Hickman.*

The single orbital rim (zygomatic) from the Kheshiya skulls that allowed recording of its shape is flat (figure 11.17, Skull 41), which fits better with Grigson's (1980:23) description of *Bos indicus* rims than with sharp taurine forms. Grigson considers this a good diagnostic criterion.

Moving to less consistently secure criteria, Grigson states that the frontal profiles viewed from above are, with some exceptions, fairly diagnostic. Of the 18 Kheshiya skulls for which frontal profiles could be recorded, 15 are *taurus* shaped and none are *indicus* shaped. The intercornual ridge—also considered less reliable—was recorded for seven Kheshiya skulls, of which four showed taurine shapes while two (Skulls 17 and 23) had a slight boss in the center (figure 11.8, Shape 8), suggestive of the *Bos*

Figure 11.13. Skull 25, lateral view, showing the flat sagittal profile characteristic of *Bos taurus* (also shown in Figure 11.12). *Photograph by Louise Martin and Lisa Usman, digitally enhanced by Stuart Laidlaw.*

Figure 11.14. Skull 23, superior view, showing highly fragmented posterior/occipital end, with hint of horncore base, and horncores leaving frontals in an outward direction, characteristic of *Bos taurus*. *Photograph by Louise Martin and Lisa Usman, digitally enhanced by Stuart Laidlaw.*

Figure 11.15. Skull 17, superior view, showing highly fragmented posterior/occipital end, with hint of horncore base, showing horncores leaving skull slightly angled backward, although questionable. *Photograph by Louise Martin and Lisa Usman, digitally enhanced by Stuart Laidlaw.*

Figure 11.16. Part of the Kheshiya cattle skull ring upon excavation in 2005, showing Skulls 13, 14, 11, and 12, with hints of horncores leaving their skulls in an outward direction from the frontals, characteristic of *Bos taurus*. *Photograph by Michael Harrower.*

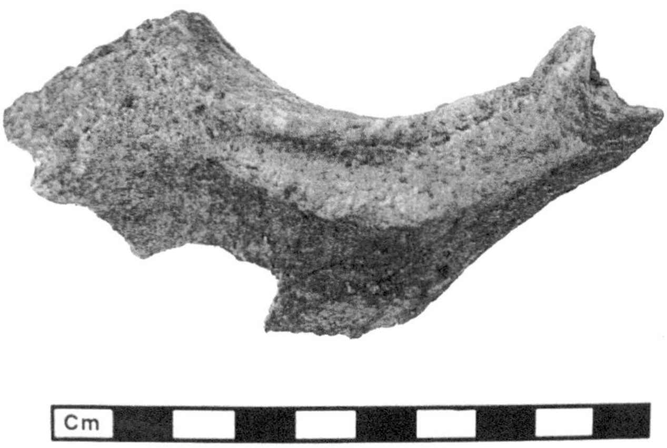

Figure 11.17. Orbital rim of Skull 41. The flat rim (compare with Figure 11.4) is considered diagnostic of *Bos indicus* (Grigson 1980:23). *Photograph by Louise Martin and Lisa Usman, digitally enhanced by Stuart Laidlaw.*

indicus morphologies recorded by Grigson (1976:128, 1980:26). Interestingly, these are the same two skulls observed above to have backward-angled horncores, as *indicus* does.

Bos facial suture shapes are less reliable indicators of species (Grigson 1980:23–24), with nasal–frontal suture morphology showing wide variation between taurines and indicines but with some characteristic forms (figure 11.5). Eight nasal–frontal sutures were scored for the Kheshiya skulls, with three being taurine-like and five with the simpler V shape of *indicus*, although these were all recorded as questionable due to much breakage in this skull area.

One frontal–lacrimal suture and 15 lacrimal–jugal sutures were assessed for shape, and all except two appeared straight rather than bowed, which Grigson (1980:23) tends to associate with *Bos indicus* rather than *taurus* (figure 11.6). This is puzzling since in several cases these morphologies were observed on skulls where more reliable criteria (for example, sagittal profiles) identified them as taurines. Suture criteria, therefore, seem unreliable, perhaps because in most cases the Kheshiya facial sutures were unfused and thus would not have been fully formed.

Finally, while the posterior wings of the palate often survived and could be observed and scored, they have quite variable morphologies, with some seeming to fit one or the other of Grigson's two observed shapes (figure 11.10); 10 others have intermediate forms, and some have shapes quite different from those described by Grigson. This criterion, therefore, was considered to have high variability—as Grigson noted—and was deemed unreliable.

In sum, this discussion of nonmetrical traits suggests that some criteria are more useful than others in the attempt to identify the Kheshiya cattle skulls to the broad species level of either *Bos taurus* or *Bos indicus*. Returning to the summary of traits in table 11.2, the extreme right column shows skull assignations based on sagittal profiles and horncore shapes alone—traits considered most reliable by Grigson. We can see that seven skulls appear strongly taurine, five questionably taurine, and two questionably indicine. As shown in table 11.4, when the least reliable criteria are excluded (facial suture morphology and the form of the wings of the palate), 19 of the 35 assessable Kheshiya skulls have traits that fit only a *Bos taurus* assignation, two skulls display both taurine and indicine morphologies (Skulls 23 and 41), a single skull (17) has two traits considered indicine, and none is taurine.

Table 11.4. Summary of how many morphological traits are indicative of *Bos taurus* and *Bos indicus* for each skull, excluding the least reliable criteria (facial suture morphology; form of the wings of the palate). Skulls still exhibiting *indicus* traits are highlighted.

Skull/Lot Number	Taurine Excluding Sutures and Wings of Palate	*Indicus* Excluding Sutures and Wings of Palate
2	0	0
3	1	0
4	0	0
5	1	0
11	0	0
12	1	0
13	1	0
14	0	0
15	0	0
16	1	0
17	0	2
18	2	0
19	0	0
20	1	0
21	3	0
22	2	0
23	2	2
24	1	0
25	3	0
26	0	0
27	0	0
28	1	0
29	2	0
30	3	0
31	1	0
32	0	0
33	0	0
34	1	0
35		
36	2	0
37	1	0
38	0	0
39	1	0
40		
41	2	1
42	0	0

It seems reasonable to view the evidence as suggesting that the Kheshiya skulls belonged to taurine cattle with some variation in horncore and posterior skull shape, which might be considered more zebu-like. It is difficult to make a strong case that any skulls are unambiguously identifiable as zebu, given that preservation, particularly of horncores and diagnostic skull areas, is poor and that traits exhibit ambiguity (for example, table 11.4). We should also acknowledge that this exercise in assessing morphological traits is based on a single published system (Grigson 1980) that draws on modern British cattle breeds and modern primarily Indian zebu stock. While extremely rigorous, Grigson's system was never intended to cover global and temporal variation in cattle morphology, and Southern Arabian Neolithic cattle are distant from Grigson's study samples both temporally and geographically.

This section therefore provides transparency about how the Kheshiya skulls were assessed and clearly documents details of their morphological traits for future users.

Cattle Skull Size

Although measurements were taken wherever possible on all skulls, the high degrees of fragmentation meant that most fragile bone areas and extremities could not be measured. Appendix 11.4 gives the full set of measurements for each skull, including those estimated. As described in the "Methods" section above, most follow von den Driesch (1976) and use her numerical codes.

Maxillary dentitions were the most intact and measurable areas of skulls with the length of cheek tooth row (Measurement 20), length of molar row (Measurement 21), and length of premolar row (Measurement 22) being possible to capture or estimate on at least 30 skulls of the total 35 assessed. Other dimensions that provided good samples are von den Driesch's Measurement 32, least frontal breadth ($n = 18$), and Measurements 4, 35, and 38, which provide about 20 data points each (appendix 11.4).

Size Comparison with Prehistoric Cattle in Arabia, Egypt, the Levant, and Anatolia

Although there is evidence for *Bos primigenius* in the Arabian Peninsula in the Neolithic, it is reasonably assumed that the Kheshiya skulls derived from herded domesticates, most likely European taurines, *Bos taurus*. Given that they are among the earlier domestic cattle in Arabia, it would be interesting to see how their size compares with regional wild cattle.

Since most *Bos* finds are postcranial, one difficulty is finding samples of *Bos primigenius* skull measurements for comparison. Where cranial finds survive, they tend to be mandibular or mandibular tooth fragments. Therefore, these measurements are available rather than maxillary measurements. No comparisons were found from the Arabian Peninsula. From Upper Egypt, a measurement of a Late Paleolithic *Bos primigenius* maxillary molar length (von den Driesch's Measurement 21) shows it to be far larger than any at Kheshiya (with the Egyptian specimen having a length of 95 mm compared to the largest at Kheshiya measuring 87.6 mm) (Baker and Gautier 1997). Linseele (2004) indeed finds that the Pleistocene African aurochs are as large as their European counterparts, although they grew smaller into the Holocene. In the Levant, where they occur more commonly at prehistoric sites, *Bos primigenius* remains consist mostly of long bones and trunk elements, with cranial portions often highly fragmented and not measurable (for example, at PPNB Kfar Hahoresh PPNB; Horwitz and Goring-Morris 2004). Even in the northern Levant/Euphrates Valley, where PPNB sites show cattle skulls and horns in installations, alongside evidence for local cattle domestication, published skull metrics are very few. At Middle/Late PPNB Halula, for example, measurements are given only for isolated maxillary molars (M1-M3) (Seguí 1999).

Turning to central Anatolia, where cattle bucrania installations from Late PPNB Çatalhöyük and nearby Boncuklu Höyük are believed to belong to *Bos primigenius*, the skulls appear vastly larger than those from Kheshiya. For Measurement 32 (least frontal breadth), Çatalhöyük has one skull measuring 320 mm (Russell et al. 2013, skull from 4040, Hodder Phase G, circa 9000–7500 cal BP), while the Kheshiya skulls range from 147 to 197.5 mm for the same dimension. A *Bos primigenius* skull from Boncuklu (Building 4, west skull) measures 250 mm across its least frontal breadth (Baird et al. 2016). Given both that domesticates are smaller than wild counterparts and that a north–south size cline is likely at play (with northerly examples

Figure 11.18. Maxillary tooth row lengths (von den Driesch Measurement 20) expressed in millimeters; Kheshiya cattle measurements (from appendix 4) compared to those of Danish *Bos primigenius* males and females (data from Degerbøl and Fredskild 1970:85, table 9). *Illustration by Louise Martin.*

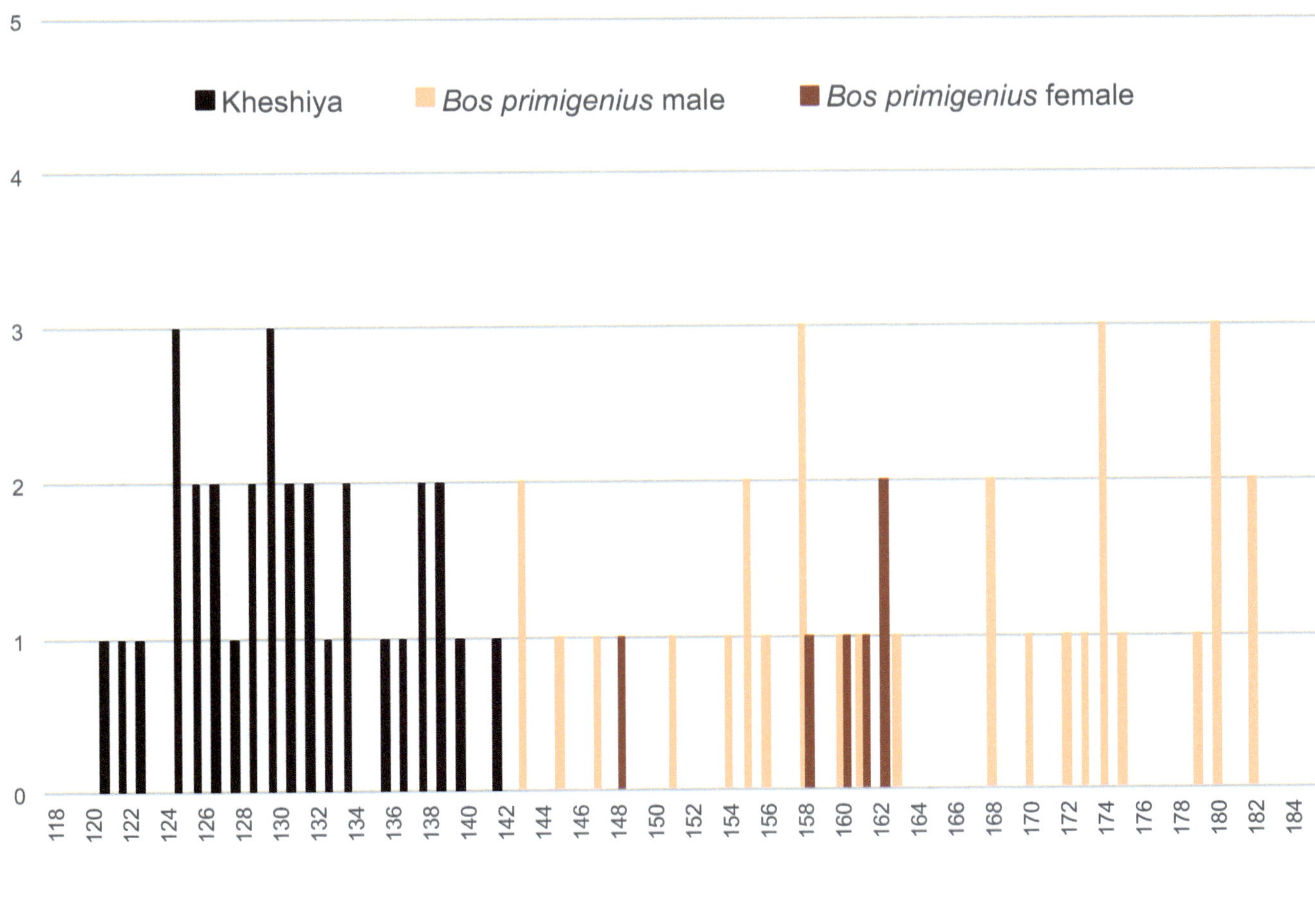

of the same species tending to show larger body size [e.g., Davis 1981; Wright and Viner-Daniels 2015; Zeder and Hesse 2000]), it is not at all surprising that Anatolian wild aurochs are much larger than Southern Arabian domestic cattle. In addition, the central Anatolian examples are from the Konya Plain, which is considered prime wild cattle territory and where one would expect maximum body size (Russell et al. 2005).

Size Comparison with Prehistoric Cattle in Europe
The European record offers some comparatives. The valuable biometrical database of wild and domestic cattle (Wright et al. 2016) includes mostly postcranial measurements, and those that are cranial consist mainly of mandibular and loose teeth. The most useful collection for comparison of cattle skulls is that described in detail by Degerbøl and Fredskild (1970) from Denmark, which includes large samples of both prehistoric *Bos primigenius* and *Bos taurus*, with the advantage that they derive from a restricted geographical area, although over relatively long time spans. Many skulls are part of whole skeletons (from bogs), and most have horns attached, meaning that they can be identified as male or female. It would certainly be expected that prehistoric Danish cattle (both wild and early domesticates) were larger than domesticates from distant Southern Arabia. Wright and Viner-Daniels (2015) have demonstrated that aurochs display morphological variation even across Europe during the Pleistocene and Holocene, with a south–north cline (increase) in body size evident. In brief, more southerly habitats display animals with smaller body sizes than northern areas, with tentative evidence also for a west–east cline. Whether this size cline is temperature-related alone (following Bergmann's rule) or regulated by indirect factors such as variations in seasonality and forage availability, Wright and Viner-Daniels (2015) cannot yet determine. Thus, while we clearly expect size differences between the Danish *Bos* skulls and the Southern Arabian Kheshiya skulls, it is nevertheless informative to view the Kheshiya sample alongside this larger sample of known status and known sex *Bos* skulls.

Figure 11.19. Maxillary tooth row length (von den Driesch Measurement 20) expressed in millimeters; Kheshiya cattle measurements (from appendix 11.4) compared to those of Danish *Bos taurus* males and females (Degerbøl and Fredskild 1970:85, table 9). *Illustration by Louise Martin.*

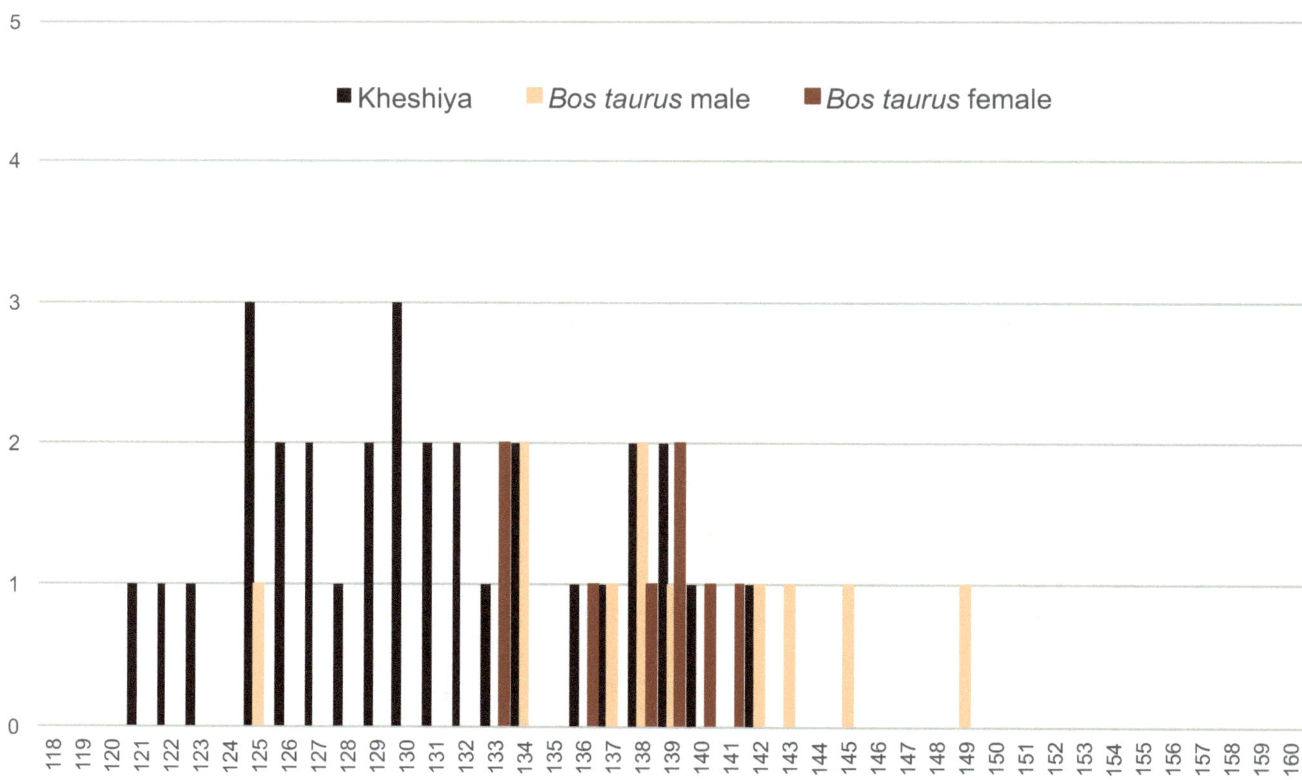

Figure 11.20. Least frontal breadth metric (von den Driesch Code 32) expressed in millimeters; Kheshiya cattle measurements (from appendix 11.3) compared to Danish *Bos primigenius* males and females (Degerbøl and Fredskild 1970:85, table 9). *Illustration by Louise Martin.*

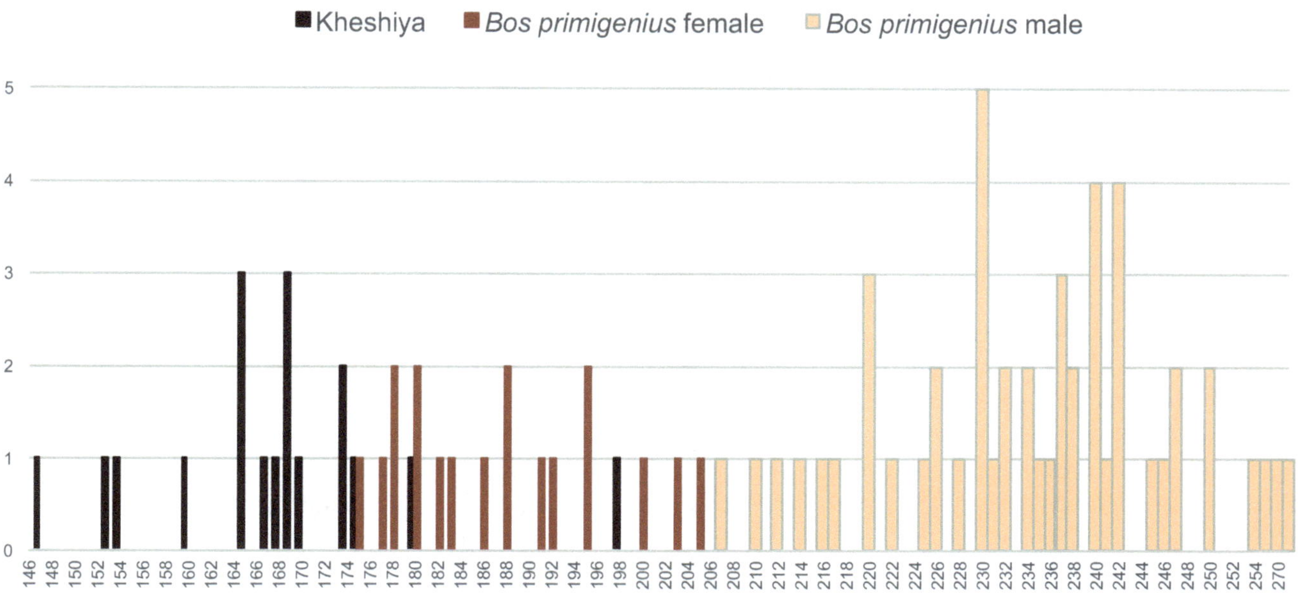

Figure 11.21. Least frontal breadth measurements (von den Driesch Code 32) expressed in millimeters; Kheshiya cattle measurements (from appendix 11.4) compared to Danish *Bos taurus* (Degerbøl and Fredskild 1970:68–69, table 2). *Illustration by Louise Martin.*

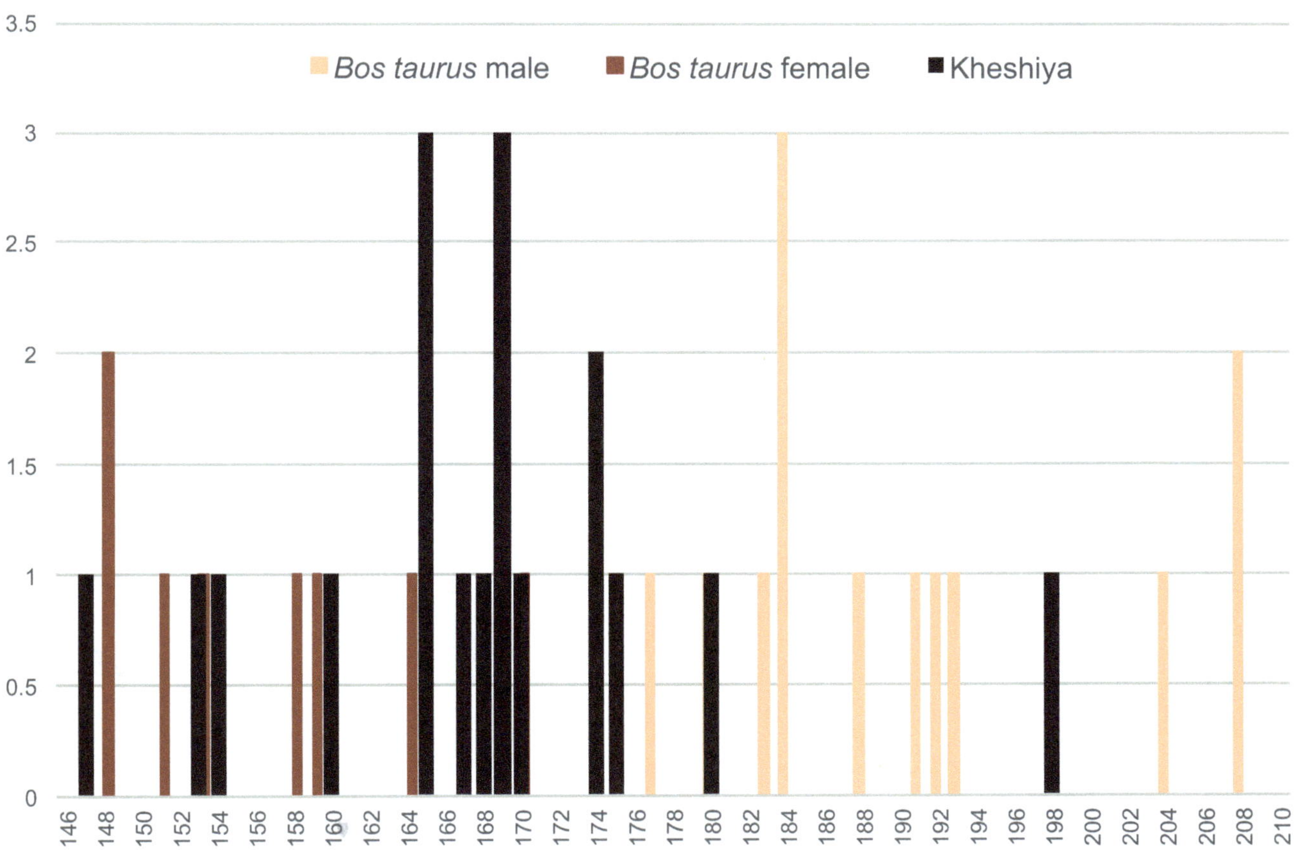

The following discussion draws on data comparisons with metrics published in Degerbøl and Fredskild (1970), unless otherwise stated.

Figure 11.18 shows a histogram that plots the length of maxillary tooth row measurement (Measurement 20) for the Kheshiya skulls alongside Danish *Bos primigenius* males and females. The *Bos primigenius* samples are, of course, much larger and do not overlap with Kheshiya sizes at all. The Danish samples display a wider range, probably because they derive from mixed locations and multiple time scales (and thus represent multiple breeding populations), with females at the lower end of the range. Grigson (1982) finds low sexual dimorphism in tooth or tooth row size, which is borne out in figure 11.18 by the complete overlap of the Danish male and female aurochs.

When Kheshiya tooth lengths are plotted against Danish *Bos taurus* data (figure 11.19), there is more size overlap between the two sample sets, although the Danish domesticates are still larger. Within the Danish sample, females again sit in the lower end of the range, although they completely overlap with males.

There are 18 Kheshiya data points for the least frontal breadth metric (von den Driesch Code 32) from the total of 35 skulls assessed. Figure 11.20 shows this measurement plotted for the Kheshiya specimens alongside Danish *Bos primigenius* males and females. Again, predictably, the Danish wild aurochs are larger than the Kheshiya cattle, although there is some overlap; the *Bos primigenius* males and females, however, completely separate using this measurement, indicating that the least frontal breadth is highly sexually dimorphic.

Although sample sizes for Danish domesticates are much smaller, the same dimorphic pattern holds when the least frontal breadth measurements are compared between the Kheshiya cattle and the Danish *Bos taurus*, with known males and females plotted separately (figure 11.21). It is notable that the overall size range is not dissimilar between the Danish and Southern Arabian cattle and that most Kheshiya measurements fall into the smaller (female) part of the range, with just one skull measurement firmly falling in the larger (male) part of the range.

Thus, while tooth row lengths are useful for highlighting overall skull size variation, they do not exhibit much sexual dimorphism in the cattle skulls plotted, but the least frontal breadth measurements exhibit sexual dimorphism in both wild and domestic cattle samples. Figure 11.22

Figure 11.22. Scatterplot showing least frontal breadth plotted against maxillary tooth row length (von den Driesch Codes 32 and 20) in millimeters; Kheshiya cattle measurements compared to Danish *Bos primigenius* sample (metrical data provided by Caroline Grigson, taken on same sample as Degerbøl and Fredskild 1970, but Grigson data allow least frontal breadth and length of tooth row data to be linked within the same skull). *Illustration by Louise Martin.*

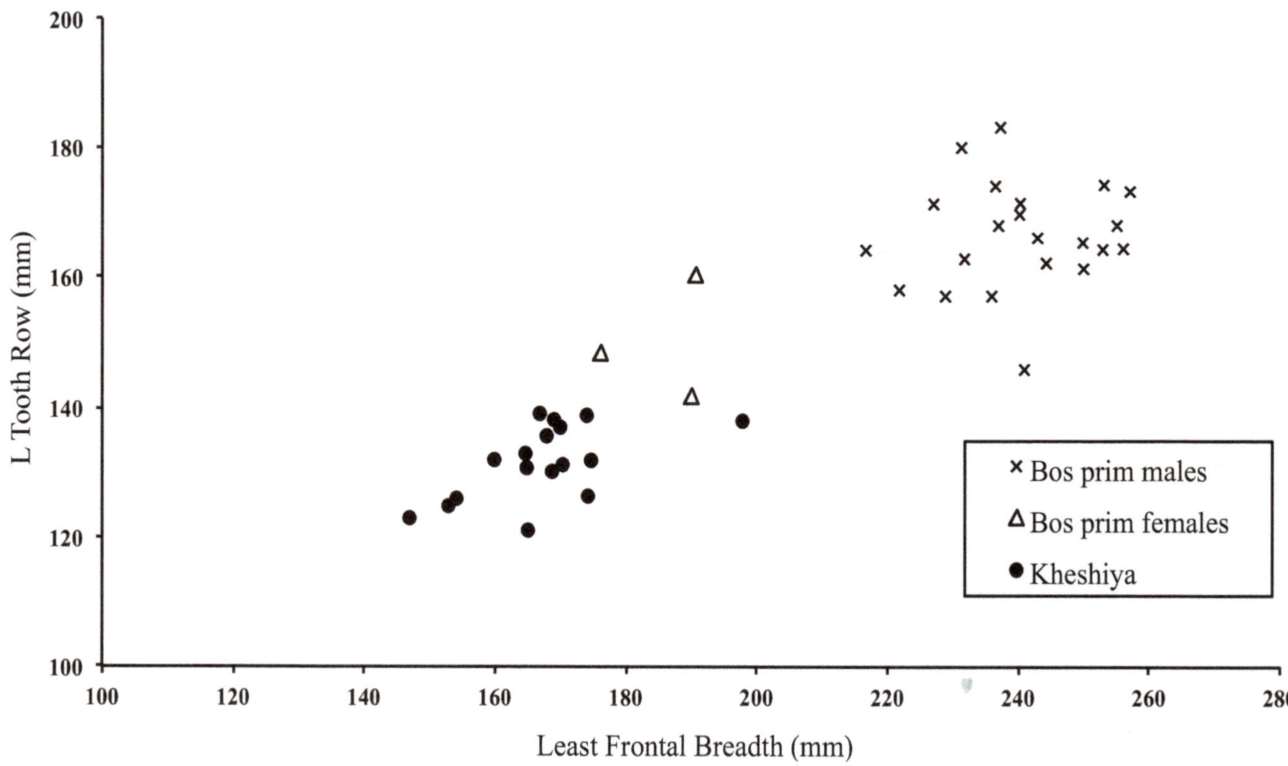

is a scatter plot showing the two measurements plotted together for the Kheshiya and *Bos primigenius* samples. Note that the interrelationship of these two metrics for individual cattle skulls is not easy to see in Degerbøl and Fredskild's published data (1970). The *Bos primigenius* data in figure 11.22, therefore, is from Caroline Grigson (2007), who independently measured the same Danish collections, allowing least frontal breadth and length of tooth row data to be linked within the same skull.

Interpretation of figure 11.22 requires some caution, since only 18 of the 35 assessable Kheshiya skulls provided both measurements and therefore could be plotted; the skull ring originally consisted of more than 40 skulls, so those shown in figure 11.22 represent less than half of those originally buried. Nevertheless, an interesting pattern emerges. The scatter plot shows the Kheshiya skulls, as expected, to be far smaller in both dimensions than wild *Bos primigenius* from Denmark, with hardly any overlap. As demonstrated above, the least frontal breadth dimension clearly displays sexual dimorphism in both wild aurochs and domestic cattle, and we can see how the *Bos primigenius* metrics in figure 11.22 separate clearly into males and females. The same separation exists in the Kheshiya skulls, with the cluster of smaller skulls most likely representing females and the single larger skull probably representing a male. It is intriguing to note that the one large skull interpreted as a male, with the frontal breadth of 198 mm, is the one located centrally in the cattle skull ring (Skull 39).

Statistical Size Comparison with Cattle in East Africa

To further explore the sexual dimorphic element of the Kheshiya skulls, a metrical and statistical comparison was also made with cattle skulls from Kerma in Sudan, where thousands of *Bos taurus* bucrania derived from grave contexts have been studied in detail osteometrically (Chaix 2007). This provides a large sample of measured domestic cattle skulls that has a closer proximity geographically and temporally than the European comparisons described above. One of the largest Kerma graves has been selected here, Grave 253, which dates to the Middle Kerma period (4050 to 3750 cal BP) and contains 1,217 measured cattle skulls, including males, females, and probably also castrates (Chaix 2007:175, table 2).

Figure 11.23. Histogram showing Kerma Grave 253 cattle horncore basal circumference measurements (data from Chaix 2007:175, table 2), with solid line showing the kernel density estimate (KDE) of the distribution. *Illustration by Joe Roe.*

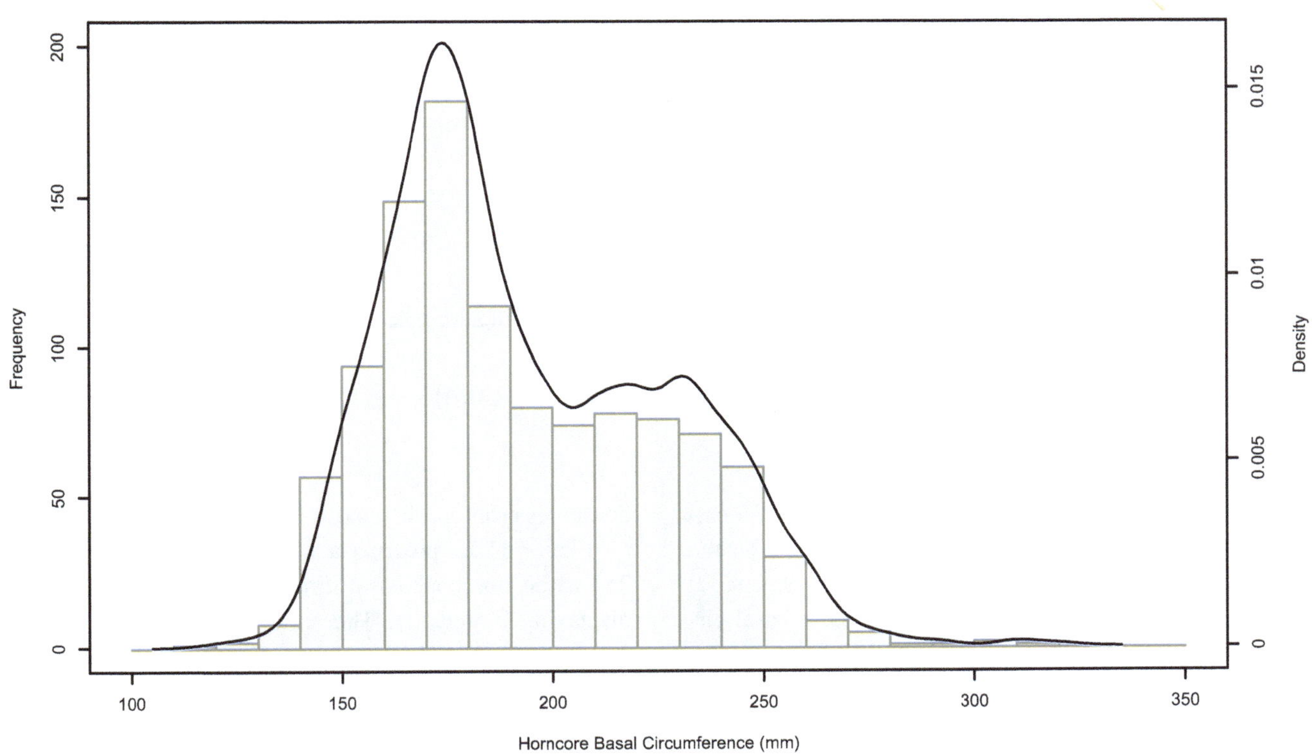

Figure 11.24. Biplot and 2D kernel density estimate of the horncore basal circumference and least frontal breadth measurements of the Kerma Grave 253 cattle skulls (data from Chaix 2007:175, table 2). *Illustration by Joe Roe.*

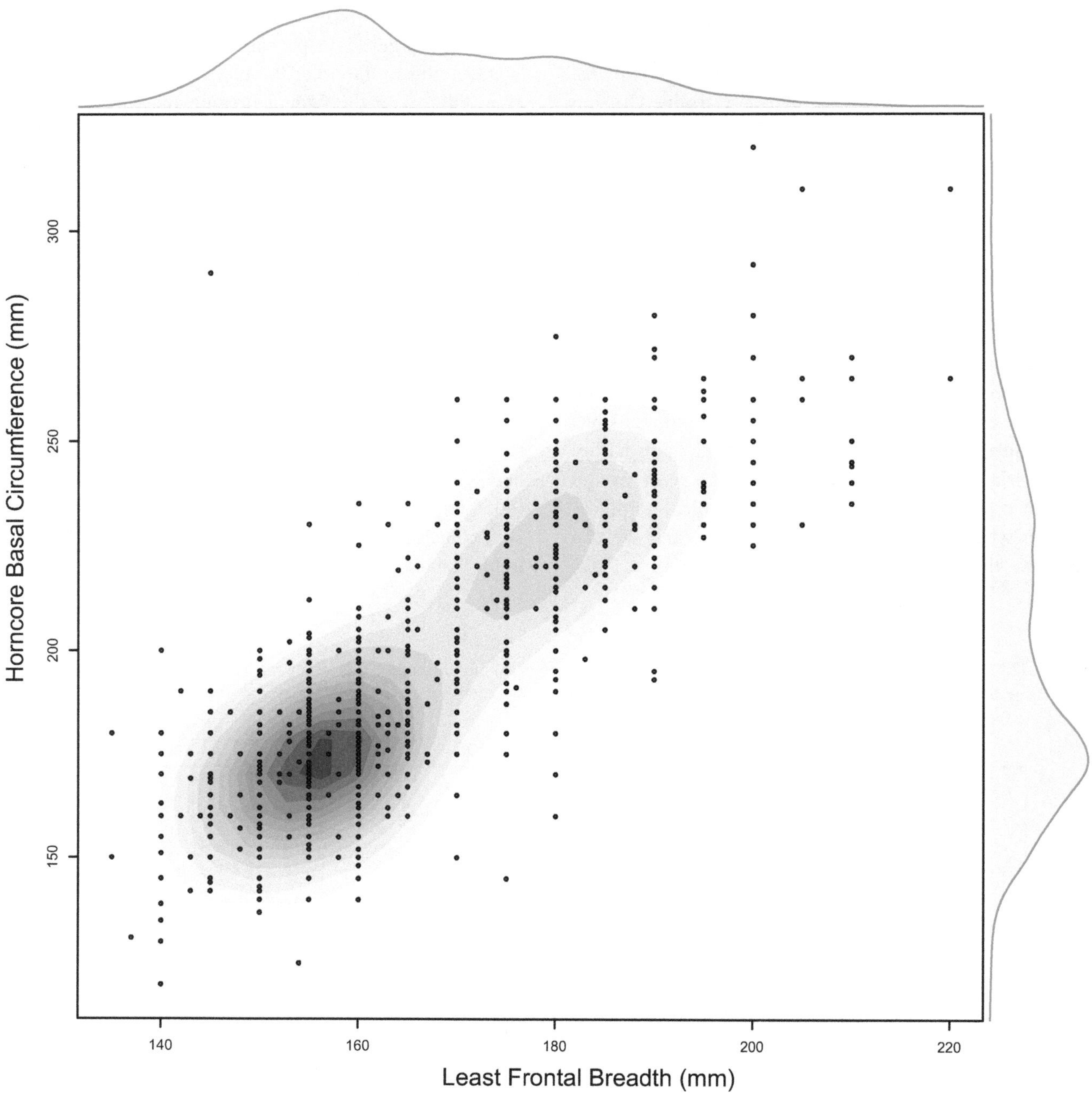

Two sets of osteometric data from the Grave 253 cattle were used (from Chaix 2007:216–28, table 13): the least frontal breadth (von den Driesch's Measurement 32; Chaix 2007:Measurement 5) and the horncore basal circumference (von den Driesch 1973:28, Measurement 44; Chaix 2007:Measurement 2). The first of these measurements can be compared with the Kheshiya sample; the latter provides confirmation of sexual dimorphism, because horncores sizes are distinctive between the sexes in cattle.

Figure 11.23 presents a histogram of Kerma Grave 253 cattle horncore basal circumference measurements, displaying bimodality. The solid line shows the kernel density estimate (KDE) of the distribution, which is essentially a smoothing of the histogram (following Beardah and Baxter 1996). Figure 11.24 is a biplot and 2D kernel density estimate of the horncore basal circumference and least frontal breadth measurements of the Kerma Grave 253 cattle skulls. Hartigan's dip statistic (Hartigan and Hartigan 1985) was used to verify that the distribution of measurements was not unimodal (as would be expected of a random variable from a single population), with the assumption that bi- or multimodality in a large sample of biometric data is likely to be a manifestation of sexual dimorphism. The distribution of horncore basal circumference, which is known to be sexually dimorphic in cattle (Grigson 1982), is significantly unlikely to be unimodal ($D = 0.030594, p < 0.001$). The distribution of least frontal breadth measurements is also unlikely to be unimodal at the same confidence level ($D = 0.06214$, $p < 0.001$), and it is positively correlated with horncore basal circumference (Pearson's $r = 0.842443$, $r^2 = 0.71$, $p < 0.001$). This confirms the findings of the metrical analyses based on European *Bos* data (described above) that the least frontal breadth measurement exhibits sexual dimorphism.

Figure 11.25 shows kernel density estimates for the least frontal breadth measurements from both Kheshiya and Kerma Grave 253 cattle skulls. Both samples show bimodality, with the Kerma sample showing more females and fewer males but still a fair proportion of the latter. The overall size range is larger at Kerma than Kheshiya, perhaps reflecting cattle that derived from different breeding populations (representing cattle tribute from across a wide landscape). The bimodality seen for the Kheshiya cattle metrics confirms a picture of mostly females and a single male skull, with the narrower range perhaps reflecting a tighter breeding group.

Dental Aging

Following approaches described in the "Methods" section above, table 11.5 shows the Grant (1982) cattle dental eruption and wear data for both left and right sides of the Kheshiya skull maxillary cheek teeth—P4, M1, M2, M3—where teeth were present. In 32 of the 35 total skulls, we could record early to middle aging dental data. While Grant's tooth wear system was intended for mandibular dentitions, there were no difficulties adapting the wear stages to maxillary teeth for this study; all observed wear could be matched with stages, despite obvious differences in tooth proportions between upper and lower teeth (maxillary being wider buccal-lingually), because underlying tooth structures are similar. Some teeth were missing (either fallen out and lost or removed as samples), but in most cases, except Skulls 15, 24 and 40, it was possible to create a maxillary wear score (MWS) based on Grant's

Figure 11.25. Kernel density estimates of the least frontal breadth measurements from Kheshiya (from appendix 11.4) and Kerma Grave 253 cattle skulls (data from Chaix 2007:175, table 2). *Illustration by Joe Roe.*

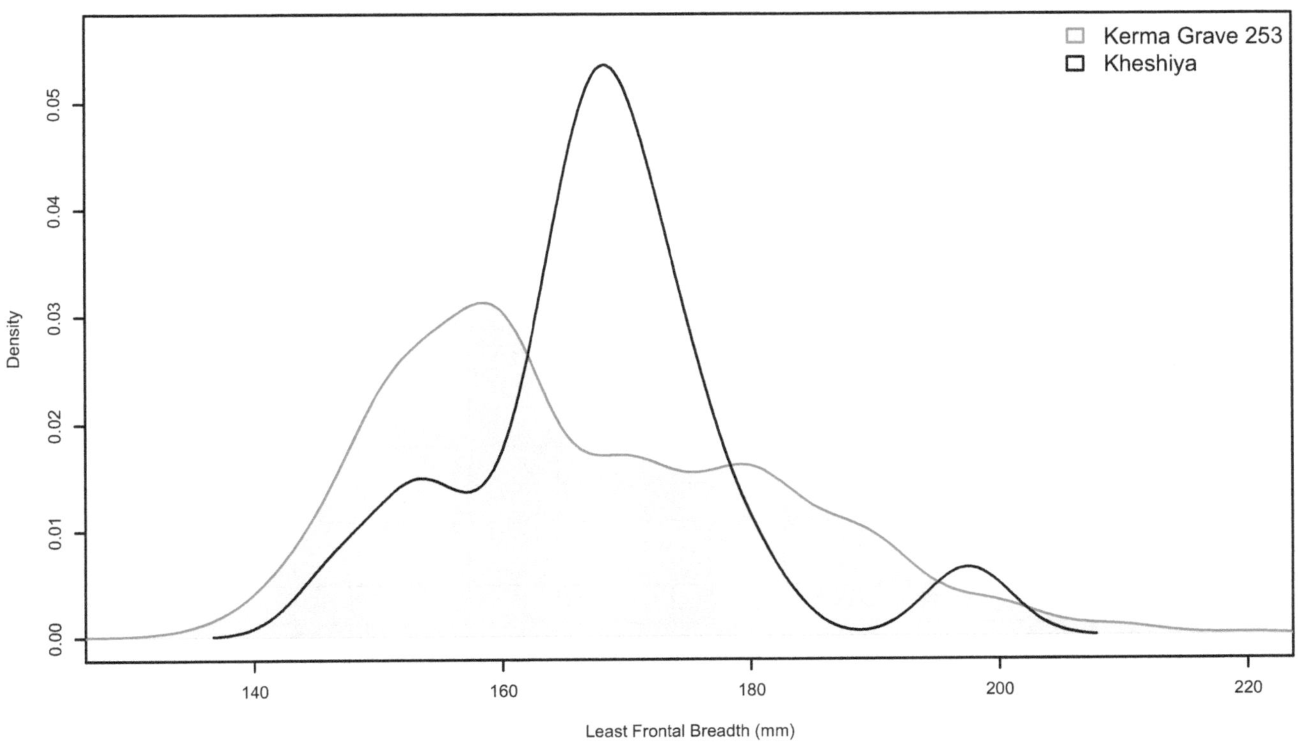

Table 11.5. The Kheshiya cattle skull maxillary tooth wear scores, following Grant (1982); Mandible Wear Scores (MWS) calculated by adding converted letter/numerical scores for M1, M2 and M3.

	Left Maxilla						Right Maxilla						
Skull/ Lot Number	**P4**	**M1**	**M2**	**M3**	**MWS**		**P4**	**M1**	**M2**	**M3**	**MWS**	**P4 Distal Wear into M1***	**Comments**
2	g	l	k	k	46							N	
3	g	k	j	g	41								
4	g	m	l	k	48							Y	Very heavy wear, enamel smooth, roots visible**
5	g	k	j	g	41							N	
11	g	l		k			g	l	k	k	46	N	
12		l	k	l	47		g	l	k	l			M3 heavier wear than M2
13	g	k	k	k	45								
14	g	l	k	k	46							Y	
15													No teeth present
16	g	l	k	k	46								
17	g	l	k	k	46							Y	
18	g	k	g	g	39								
19		k	g				d	k	g	c	35		
20	g	k	j	g	41								
21	g	l	k				g	l	k	k	46		
22	g	k	k				g	k	k	k	45	Y	
23	g	l	k				g	l	k	g	43		
24	g	l					g	l				Y	
25	g	k	k				g	k	k	j	44	Y	
26							g	k	k	k	45	Y	
27	g	l	l	l	48							Y	All teeth: lots of cementum on outer surfaces
28	g	l	k				g	l	k	k	46		
29	g	l	k	k	46							Y	
30	g	k	k	k	45								
31	g	l	k	k	46							Y	
32			k					k	k	k	45		
33	g	l	k	k	46							Y	
34	g	m	l	l	49							Y	M1 sides plus base of pillar worn; all roots visible
35													Mandible, no teeth
36	g	l	k				g	l	k	k	46		
37	g	k	k	k	45							Y	
38	h	k	h	f	39								
39	g	k	j	f	40								
40													No teeth, alveoli show adult dentition
41	g	k	k	j	44								
42	g	m	k	k	47							Y	

**Right maxilla tooth wear is only shown if left maxilla dentition is missing, incomplete, or if wear scores differ between the two sides. MWS shaded fields are those used for relative aging analysis. *This field not consistently noted. **Crown heights for Skull 4: P4, 17.3 mm, M1 13.5 mm.*

Figure 11.26. Dorsal view of Skull 31, showing left-side P4 distally worn into the anterior cusp of M1. *Photograph by Louise Martin and Lisa Usman, digitally enhanced by Stuart Laidlaw.*

Figure 11.27. The number of cattle skulls at each maxillary wear stage (MWS) (following Grant 1982), showing a narrow range between 35 and 49. *Illustration by Louise Martin.*

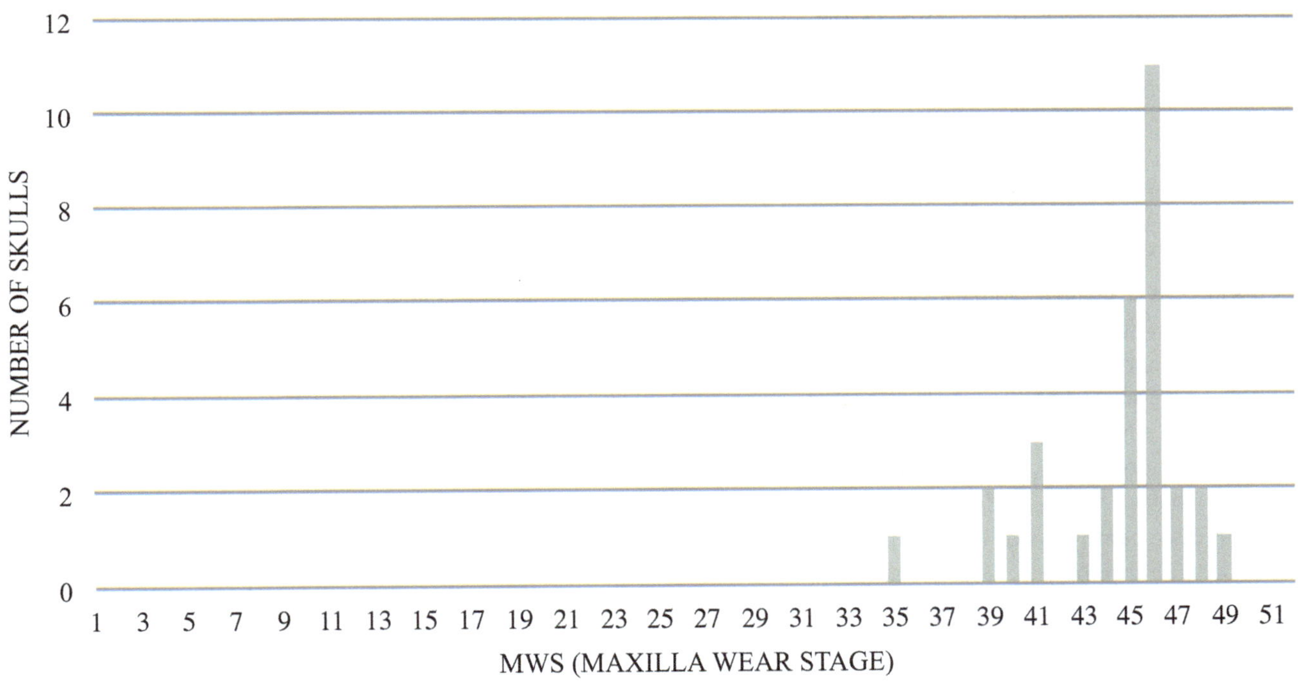

(1982) system, which sums the wear for the three molar teeth. Table 11.5 also shows instances where P4 was worn distally into the anterior cusp of M1. (Figure 11.26 shows an example of this in the left maxillary dentition of Skull 31.)

All skulls contain a full component of adult teeth, fully erupted and in wear. Some teeth have high individual wear stages (for example, Stage l or Stage m, in a range from a to p, where p is heavily worn). Maxillary wear stages range between 35 and 49 (figure 11.27), which is relatively narrow and reflects a cull that targeted adults (Grant 1982:98–99, table 2, finds adult cattle to have a range of 31 to 54). More than half of the skulls have a wear score of 45 or 46. On first appearances, the cattle seem tightly clustered in their ages at death.

Interpreting approximate ages of death is challenging in the unusual Kheshiya case, in large part because almost all studies that attempt to correlate recorded wear stages with actual age are based on mandibular teeth, which normally survive intact better than maxillae. The single study that draws on maxillary dentition (Andrews 1982) focuses only on young cattle. It is therefore necessary to use information on the timing of cattle mandibular wear stages for this study, knowing that although maxillary tooth stages follow the same patterns of increasing wear, the timing of those wear stages may not be mirrored between the two jaws. Indeed, Andrews and Wedderburn (1973) find that cattle maxilla teeth erupt later than mandibular counterparts.

In a study that tests various methods of scoring cattle tooth eruption/wear against known aged individuals, Jones and Sadler (2012a) found that Grant's wear stages correlate well with age in animals under three years of age, but in older animals wear stages cannot be linked to narrow age brackets. This means that the same MWS can be assigned to a wide spread of known aged animals (2012a:25, figure 13). For example, they find that MWS 46 has a particularly wide spread: It is assigned to animals killed anywhere between 6.2 and 13.3 years of age—a seven-year span—which makes MWS 46 difficult to interpret. Jones and Sadler, in contrast, find that MWS 47/48 predictably describe animals between 8 and 12 years old, and wear scores above 50 are fairly consistently recorded only in animals over 13 years of age.

What inferences can thus be made about the ages of death of the Kheshiya cattle? Following Jones and Sadler's findings (2012a:25, figure 13), one animal seems to have been killed as a younger adult, about 3 to 3.5 years old (MWS 35); a few skulls with MWS 39–41 could have been between 3.5 and 6.5 years old at death; the majority of skulls ($n = 20$) have an MWS between 42 and 46 and could have been anywhere between 6 and 13 years old at the time of death; five other skulls had an MWS between 47 and 49 and fall in the 8–12 bracket of the older age range.

To further narrow down the group of 20 skulls with the wide age range of 6 to 13 years, another scheme was applied. We examined individual tooth wear rather than the whole molar row (Jones and Sadler 2012a)—a scheme that can refine patterns. We see that 24 of the 32 Kheshiya skulls with dental early to middle aging data have M3s at Stage g or above (table 11.5), and of these, the majority of their corresponding M2s are at Stage k, which puts the animals in the oldest adult class, bordering the elderly stage (Jones and Sadler 2012a:15, table 2).

The terms *old adult* and *elderly* are relative to an animal's longevity, which itself can vary depending on breed and individual/herd life histories. Jones and Sadler (2012b) find no consistency in the records of cattle life expectancy, but their review of information for modern/recent *Bos taurus* breeds in Europe suggests that while some cattle can live 20 to 25 years, with occasional/rare females still breeding up to and above 15 years of age, domestic cattle aged above 20 are rare. In terms of age of last breeding, a study of early-twentieth-century dairy cattle showed that the last calving occurs generally around 12 years of age, with some females continuing until 13 to 15 years of age (Jones and Sadler 2012b:8).

The Kheshiya cattle are obviously distant in both time and space from well-studied modern European cattle populations, but we can use this information to build a picture of the relative ages of death within the Kheshiya assemblage, which are more realistic than actual ages. It is clear that none of the cattle were culled as juveniles or subadults, and only one is a young adult. The majority of cattle seem to have been older adults when culled—not elderly but in the upper range of their mature adult stages, perhaps at the ends of their reproductive lives, at least in the case of females. Bearing in mind that maxillary tooth wear stages are unstudied but thought to lag behind mandibular equivalents (and thus reflect older individuals), it seems wise not to attempt any more exact age assessments.

Cranial Suture Fusion

As with mammal long bone epiphyseal fusion, cranial suture fusion timings tend to have broad age ranges, and variation is expected within species due to animal breed, health, and nutrition (see Popkin et al. 2012). Data for the Kheshiya cattle cranial suture fusion is given in table 11.6, with approximate age estimates in the right-hand column (following Grigson 1982:20, appendix 1).

A fairly consistent picture of suture closure is evident: All but one of the skulls (Skull 24) have their frontal halves fused, which occurs in cattle over about seven years old (following Grigson's 1982 data); most skulls also have their medial palatine bones fused, which would place them in an older range, older than 10 years. That most facial sutures (maxillae, lacrimals, zygomatic) are unfused or fusing, however, indicates that animals were killed younger than about 15 years of age, which supports the dental wear results that show the cattle to have been culled generally as older adults, but not at extreme "elderly" age.

There are inconsistencies between the suture closure and dental wear data, however. Dental wear places Skull 19 as a young adult, but its skull shows no difference in cranial suture inferences from other individuals (estimated age at death 10–15 years?); similarly, the group identified through dental wear as "younger adults" (Skulls 3, 5, 20, 38, and 39) also shows no differences in suture closure from the overall trend. In zooarchaeological analyses, dental eruption and wear stages are considered more refined tools for estimating age at death than fusion analyses (e.g., Davis 1987), which may explain the variation seen here.

Table 11.6. Cranial suture closure data for the Kheshiya cattle skulls. Only selected sutures were recorded (those visible), on whichever side of the skull was preserved. *F = fused/closed; UF = unfused/open; JF = just fusing. Blank fields indicate no data available.

Skull/Lot Number	Basioccipital Area and Sphenoids	Parietals/ Temporals	Frontal Halves	Frontals/ Lacrimals (orbit)	Maxillae/ Lacrimals/ Zygomatics	Medial Palatine Suture	Frontal/ Lacrimals (on face)	Nasals, Fused Together	Age Estimate
Suture Closure Ranges	**2–3 Years**	**5–7 Years**	**7–10 Years**	**7–10 Years**	**10–15 Years**	**10–14 Years or > 15?**	**Extreme Age**	**Extreme Age**	**Years (very approximate)**
2					UF	F			10–15?
3			F		UF	F			10–15?
4						F			> 10?
5			F		UF	F			10–15?
11	F		F		UF/F	F			10–15?
12			F	F	JF	F			10–15?
13			F		JF	F			10–15?
14	F		F		UF	F			10–15?
15			F						> 7?
16	F				UF	F	UF	?	10–15?
17			F		UF	F			10–15?
18	F		F		UF	F		F?	10–15?
19					UF	F			10–15?
20					UF				< 15?
21			F		UF?	F?			10–15?
22			F		UF	F			10–15?
23	F			UF		F			about 10?
24	F		UF		UF	F			10–15?
25	F		F		UF	F			10–15?
26						F			> 10?
27					UF	F			10–15?
28	F		F		UF	F			10–15?
29	F		F		UF	F			10–15?
30			F		UF	F			10–15?
31			F		UF	JF			10–15 plus?
32	F				UF	F			10–15?
33			F		UF	F			10–15?
34			F		JF	F		UF	10–15?
mandible 35									
36	F	F	F		UF				10–15?
37	F		F		UF	F		UF	10–15?
38						JF			10–15?
39			F		UF				10–15?
40					UF			?	< 15?
41			F		UF	F			10–15?
42					UF	F			10–15?

Suture closure age ranges in right-hand column are from Grigson 1982:20, appendix 1, which summarizes data from other authors.

Table 11.7. Dental anomalies recorded in the Kheshiya cattle maxillary teeth, describing cases of premolar tooth rotation, malocclusion, and the single example of a tooth missing during the life of the animal. The right column shows the MWS for each skull, as in figure 11.27, indicating the relative age stages of the cattle.

Skull/ Lot Number	P4/P3 Tooth Rotation: Side and Degrees of Rotation	Malocclusion	Teeth Missing in Life	Comments	MWS
2				LHS P2 double ring of enamel	46
3	RHS P4 rotated 50–60°			(1) L+RHS P4 appear large in proportion to molars; rotation seems to be result of lack of space for tooth eruption? (2) Dentine stub between RHS P4 and M1 may be remnant of dp4, showing as very worn dentine pillar with tiny area of enamel. Interesting to note rotation only one side. (LHS is visible.)	41
4					48
5					41
11		RHS M3 worn into central peak.			46
12		?; see comment.		L+RHS M3 more worn than M2	47
13					45
14					46
15					
16					46
17					46
18					39
19					35
20					41
21					46
22					45
23					43
24					
25					44
26					45
27					48
28		RHS P4: steep anterior–posterior wear			46
29	LHS P4 rotated about 20°			LHS P4 pushing/wearing into M1.	46
30				RHS not visible.	45
31				LHS P4 pushing/wearing into M1.	46
32					45
33					46
34	LHS P3 rotated about 70°			Can't see if RHS also rotated. Premolars missing.	49
35					mandible
36					46
37					45
38	L+RHS P4 rotated 25–30°			Both P4s rotated, unlike Skull 3.	39

Table 11.7. Dental anomalies recorded in the Kheshiya cattle maxillary teeth, describing cases of premolar tooth rotation, malocclusion, and the single example of a tooth missing during the life of the animal. The right column shows the MWS for each skull, as in figure 11.27, indicating the relative age stages of the cattle. *(continued)*

Skull/ Lot Number	P4/P3 Tooth Rotation: Side and Degrees of Rotation	Malocclusion	Teeth Missing in Life	Comments	MWS
39					40
40					
41			LHS P3	LHS P3 seems missing in life. P2 erupted into its space. P4 alveolus present but not enough space for P3.	44
42		RHS P4 higher in jaw than adjacent M1; malocclusion and suggesting not enough space for P4 eruption.			47
Total: 35	4	4	1		

Figure 11.28. Skull 3, close-up of right-side maxilla, showing P4 with a high degree of rotation and a worn dentine/enamel stub between the P4 and M1, probably a remnant of dp4, indicating P4 maleruption. *Photograph by Louise Martin and Lisa Usman, digitally enhanced by Stuart Laidlaw.*

Figure 11.29. Skull 34, showing maxillary left-side P3 with a high degree of rotation. *Photograph by Louise Martin and Lisa Usman, digitally enhanced by Stuart Laidlaw.*

Figure 11.30. Skull 34, close-up of left-side P3 with a high degree of rotation, showing lack of space for tooth to erupt normally. *Photograph by Louise Martin and Lisa Usman, digitally enhanced by Stuart Laidlaw.*

Figure 11.31. Skull 38, showing maxillary left-side P4 with a low degree of rotation. (Right-side P4 was similarly rotated in this skull; not shown in the photograph.) *Photograph by Louise Martin and Lisa Usman, digitally enhanced by Stuart Laidlaw.*

Dental Anomalies

Dental abnormalities noted in the Kheshiya maxillary teeth are shown in table 11.7. Four instances of tooth rotation were observed (Skulls 3, 29, 34, and 38), each in premolars, with three P4s affected and one P3. Two cases show high degrees of rotation (figures 11.28, 11.29, and 11.30), and in one skull (38) P4 is rotated to a lesser degree on both the left and right sides of the jaw (figure 11.31).

Other dental abnormalities include malocclusion of P4 in Skull 28, an absence of P3 in Skull 41 (where there seems to be no space for it to erupt), and a dentine "stub" in Skull 3 (figure 11.28), which appears to be a remnant deciduous tooth (dp4). Each case seems to signify maleruption, and together with the tooth rotation examples, they suggest that some of the cattle experienced tooth crowding. Skulls 3 and 41, for example, show a lack of space for P4 to erupt into (noting that M1 is in place long before the premolars erupt).

Colyer's *Variations and Diseases of the Teeth of Animals* (1936; revised by Miles and Grigson in 1990) still serves as a useful reference for irregularities in mammalian dentition. Colyer found that major anomalies in ruminant jaws were fairly common (estimated at about 9 percent of reference jaws studied), and he describes how positional anomaly most often affects premolars, which can have extreme rotation because they erupt later than molars and sometimes find no space for eruption. There is also the suggestion that high proportions of tooth positional anomaly (about 30 percent), such as tooth rotation, can result from population isolation, likely due to founder effect (Miles and Grigson 1990) and breeding bottlenecks.

The Kheshiya sample size is small, but the overall proportion of dental irregularities they exhibit is 22 percent, with tooth rotations seen in 12 percent of assessable skulls. Following Colyer's study, the number of tooth anomalies in the Kheshiya cattle seems slightly higher than expected, which might hint at a degree of isolation in the Neolithic Southern Arabian cattle populations. Could this relate to a founder effect, or are the irregularities within the range of normal variation?

In either case, the Kheshiya skulls show clear evidence of tooth crowding, which itself results from jaw foreshortening, where teeth—not correspondingly reduced in size—are seen to touch, overlap, malerupt, or rotate. It has long been assumed that tooth crowding is one of a suite of markers of early domestication, especially in dogs and pigs but in other mammals too (Clutton-Brock 1999; Zeder 2012), wherein bone and tooth size reductions are out of sync. A recent study of wolves and dogs, however, where both the wild and domestic counterparts revealed tooth overcrowding (Ameen et al. 2017), shows that this idea needs reevaluation, and tooth crowding alone cannot identify domesticates. The same study found a high correlation between tooth crowding and tooth rotation, both traits that characterize the Kheshiya assemblage, whatever the underlying cause of the tooth crowding.

Condition of Skulls, Breakage, and Treatment

This section uses information on skull part presence, fragmentation, and treatment—recorded according to details described in the "Methods" section above—to examine how the skulls were originally deposited, subsequent site formation processes, what cranial elements survived, and what can be gleaned about any preburial treatment of the skulls.

Appendix 11.6 shows full data for each skull relating to condition, parts of the skull present, and treatment. Each skull was recorded for the presence of different skull areas, with the frontal eminence and occipital condyles representing the posterior, and nasals representing the anterior. The presence of maxillary dentitions was also recorded. The table in appendix 11.6 also notes the frequent cases in which information could not be assessed. Table 11.8 summarizes selected skull part presence from data in appendix 11.6.

As noted in the "Methods" section, initial examination of the skulls revealed a strong likelihood that they had all been initially buried intact, with no evidence that any cranial parts (apart from mandibles) had been removed prior to burial, even though skulls experienced much post-depositional fragmentation. Table 11.8 shows the presence of even the most fragile skull parts, the nasals, in 16 of the 35 skulls assessed (figure 11.14), while in the remainder they appeared broken off and probably fragmented beyond identification. This is not surprising given that the nasals, the most deeply buried skull parts, were pushed down into a deposit that hardened around them (chapter 10).

Counts of other cranial elements (table 11.8) show that rear skull areas survived more poorly than even the thin nasals, despite being more robust. The thick ridge of the frontal eminence (or parts of it) survived in only six skulls, and part of occipital condyles survived in only five skulls. As already noted in the discussion of skull morphology, horncores did not survive at all, but horncore bases were visible in four skulls and hinted at in another three. The whole rear skull area seems to have suffered from long-term exposure to the elements, or repeated burial/exposure to the point of complete degradation in most cases. This perhaps is not surprising given how close to the present ground surface they were found.

Table 11.8. Summary of the presence of selected cranial parts surviving for each of the Kheshiya skulls. The left column shows skulls that retain any evidence of the direction in which horncores leave the skull, based on fragments of horncore bases. Other columns indicate the survival of other skull extremities. The right column shows where maxillary tooth rows survived intact, on either one or both sides or partially. Data summarized from appendix 11.6.

Skull/Lot Number	Horncore Direction Visible? Y = Yes; (Y) = Partial	Frontal Eminence Present? Y = Yes; (Y) = Partial	Occipital Condyles Present? Y = Yes; (Y) = partial	Nasals Present? Y = Yes/Both; (Y) = Partial	Dentition Present? Y = L + R; L = Left; R = Right; (Y) = Partial
2					L
3					Y
4				(Y)	L
5					Y
11				Y	Y
12			(Y)		Y
13					Y
14					Y
15					
16				Y	Y
17	Y	Y	(Y)	Y	Y
18		Y		Y	Y
19					(Y)
20					Y
21	(Y)		(Y)		(Y)
22	Y	Y			(Y)
23	Y	(Y)		Y	Y
24					Y
25	(Y)	(Y)	(Y)	Y	Y
26					R
27				Y	Y
28					Y
29	Y				Y
30				(Y)	Y
31					(Y)
32					(Y)
33				Y	Y
34				Y	(Y)
36		Y	Y	(Y)	Y
37				Y	Y
38					Y
39				(Y)	Y
40				Y	
41	(Y)				Y
42				Y	Y
Total: 35	**4 (3)**	**4 (2)**	**1 (4)**	**12 (4)**	**27 (6)**

Figure 11.32. Skull 39, superior view, showing typical preservation of cranium and maxillae, with all skull extremities not surviving. *Photograph by Louise Martin and Lisa Usman, digitally enhanced by Stuart Laidlaw.*

Figure 11.33. Skull 14, showing a thermally-altered rock wedged into a break in the palatine wing area of the skull. *Photograph by Louise Martin and Lisa Usman, digitally enhanced by Stuart Laidlaw.*

Cheek teeth preserved better. Of 35 skulls, 33 had at least one side of maxillary cheek teeth present, even if some teeth were missing, and often both tooth row sides survived. In two skulls (26, 40) all teeth had fallen out, although the loose teeth of Skull 26 could be refitted. One skull (15) was too poorly preserved to show any tooth root sockets. That tooth rows so often survived intact is notable since the maxillary bone supporting them is not strong; survival probably results from quick burial of the skull to the level of the teeth and points to the relative stability of the burial environment thereafter. Figure 11.32 shows a typically preserved skull (39) with all extremities missing but with cranium and maxillae intact.

Weathering

All 35 skulls provided data on bone surface weathering, summarized in table 11.9, which shows that the majority of skulls have fairly consistent weathering stages (Stage 3) on their frontal bones, which is the most commonly surviving skull part for assessment. Adapting Behrensmeyer's 1978 stages for the Kheshiya skulls (table 11.10 and "Methods," above), Stage 3 indicates that bone surfaces are rough, with pitting and some round-edged cracking, but whether this resulted from surface exposure (not lengthy—the bone does not show deep cracking/splintering) or continuous wetting/drying after burial is hard to tell. The degree of uniformity between skulls and the occasional higher-weathering Stage 4 noted around the rear areas of skulls—which were more likely to be exposed (see table 11.9: intercornual ridge, around horncores, orbits)—supports the idea of differential weathering. The anterior parts of skulls remained buried in a relatively stable fashion after installation (albeit in shallow deposits), while posterior areas from approximately the orbits backward protruded aboveground for some time, with exposure eroding away the backs of the skulls, before deposits later covered and stabilized the remaining skull parts, preserving them in situ.

Burning

The absence of any sign of burning on the skulls is notable only because the surface deposit inside the skull ring feature was ashy and the skulls appeared to have been pushed into it. That none of the skulls showed charring, or even the characteristic "browning" suggestive of contact with heat, indicates either (1) that the skulls were installed after the internal deposits were burned—that is, they were not in place at the time of any fire in the circle—or (2) that the fires that created the ashy deposit did not affect the skulls, which probably were protected by a skin/hide covering.

Table 11.9. Bone surface weathering stages recorded for the Kheshiya cattle skulls, following the descriptors adapted for this assemblage (from Behrensmeyer 1978) shown in table 11.10. Weathering stages were assessed on all frontals and noted for other cranial elements only if they differed from frontals.

Skull/Lot Number	Bone Surface Weathering Stage: Frontals	Bone Surface Weathering Stage: Other Cranial Elements
2	3	
3	3	
4	3	
5	3	
11	3	
12	3	4 around orbits
13	3	4 around zygomatics
14	3	
15	3	
16	3	
17	3	
18	3	
19	3	
20	3	4 around intercorneal ridge and orbit
21	3	4 in patches
22	3	4 in patches
23	3	4 around horncore bases
24	3	
25	4	3 elsewhere
26	3	
27	3	4 on nasals and anterior maxilla
28	3	4 on basioccipitals
29	3	
30	3	
31	3	4 in patches
32	3	
33	3	
34	3	
35 mandible	3	
36	3	
37	3	
38	3	
39	3	
40	3	
41	3	
42	3	

Table 11.10. Behrensmeyer's (1978) weathering stages and descriptors for bone surfaces, alongside descriptor adaptations made for the recording of the Kheshiya cattle skulls.

Stage	Behrensmeyer's (1978) Weathering Stage Descriptors	Behrensmeyer's (1978) Estimated Years since Death	Descriptor Adaptations for Kheshiya Cattle Skulls
0	No cracking or flaking; greasy; soft tissue present.	0–1	Greasy, fresh bone
1	Longitudinal cracking in long bones	0–3/4	Very slight rough frontal surface
2	Surface flaking, cracks(?), exfoliation started	2–6 or 7	Slight pitting on frontal surface
3	Bone surface rough, fibrous, round-edged cracks	4–15 plus	Surface rough, pitting, some round-edged cracking
4	Bone surface course, rough, fibrous; splintering, deep cracks opening	6–15 plus	Surfaces course, rough, fibrous, deep cracks opening
5	Bone falling apart, very fragile	6–15 plus	Bone falling apart

Figure 11.34. Skull 29, showing breakage between the rear part of the skull and the anterior (maxillae) part, with breakage across the wings of the palate. There is also a clear shift in angle between the two parts of the skull. *Photograph by Louise Martin and Lisa Usman, digitally enhanced by Stuart Laidlaw.*

Fire-cracked rocks were found clearly lodged behind the skulls during excavation (see chapter 10, figure 10.5), with some stones finding their way into breaks in the skulls. In some cases, the stones appear as if they served as rough butchery tools wedged into chopped bone (figure 11.33, white arrow), although there is good evidence (presented below) that finer chipped stone tools were used in the butchery and preparation of the skulls. It is more likely that burned stones accidentally became incorporated into skull breaks through bioturbation of the earlier-laid ashy deposits.

Our interpretation is that the burned stones and ashy deposit relate to cooking/preparation activities of the cattle carcasses, and only later were the prepared skulls inserted into the burned surface to create the monument. (See chapter 10.) Of note here is that skulls must have been broken at the palatine wings (as in Skull 14, figure 11.33) before or during the process of skull installation for thermally-altered rocks to become wedged into this break. This is discussed further below.

Skull Breakage and Anterior/Posterior Slumping

During the process of cleaning the skulls for study, it was observed that in many, the rear parts of skulls were overhanging the anterior, as if pushed forward, which is observable as frontals that overlap lacrimals, nasals, and occasionally even maxillae bones. This contrasts with

Table 11.11. Occurrence of forward slumping of the rear parts of skulls and frontals over the anterior (nasals, lachrymals). The right column shows where the wings of the palate are broken and also the angle of slumping where evident. (Anterior angle of shift was examined when looking at maxillary teeth occlusally from anterior to posterior. Therefore, if a skull in the ground is described as "angled right," this means the cranium collapsed to its left.)

Skull/Lot Number	Forward Slumping of Posterior Skull, Frontals Shifted over Nasals/ Lacrimals; Y = Yes; (Y) = Likely	Posterior–Anterior Skull Breakage at Wings of Palatine, Anterior Angled Left (L), Right (R), or Indeterminate (INDET)
2		
3	Y	L
4		
5	(Y)	R
11	Y	R
12	no	
13	Y	
14	Y	R
15		
16	Y	L
17	Y	
18	Y	R
19	can't assess	
20	Y	L
21	Y	
22	Y	INDET
23	Y	
24	Y	R
25		
26	can't assess	
27	Y	R
28	Y	R
29	Y	L
30	Y	INDET
31	(Y)	
32	Y	R
33	Y	
34	Y	
36	Y	R
37	Y	INDET
38	can't assess	
39	Y	
40	can't assess	
41	Y	INDET
42	can't assess	
Total	**25/35**	**17/35**

Figure 11.35. Skull 37, anterior view, showing the frontals slumped forward over the nasals and lacrimals on both sides. *Photograph by Louise Martin and Lisa Usman, digitally enhanced by Stuart Laidlaw.*

finding all cranial elements flush at the sutures, which they should be anatomically. Table 11.11 shows that 25 of the assessable 35 skulls have this form of breakage and skull compression. Some display only slight shifting forward or "slumping" of rear skull parts (for example, 1 cm, Skull 28; see appendix 11.6); others have shifted more (4 cm, Skull 29). The wings of the palate on the underside of the skulls, which act as "bridges" between the heavier rear part of a skull and the lighter anterior portion, were also often broken (figure 11.34 and table 11.11), leading the skull to collapse, sometimes to one side or the other, or sometimes forward. The question posed here is whether the skull slumping resulted from natural processes or whether human butchery activities contributed?

We note first that in all skulls the lacrimal/zygomatic/maxilla sutures were not yet fused (or a few were just fusing; table 11.6), so facial bones were still separate in the skull and were not yet joined by advanced age. Slumping, therefore, occurred at areas of existing weakness in the skull. There is no suggestion that younger skulls collapsed more than older ones. (Some younger skulls are intact while some older specimens exhibit slumping.) Factors other than unfused sutures must have contributed.

The burial environment certainly played a part in skull breakage. Repeated wetting/drying of silts that built up internally in the skull cavities would have led to expansion and contraction of deposits, probably aiding the explosion of unfused facial bones and causing the frontals to shift over the nasals, as seen in Skull 37 (figure 11.35). Whether this shifting occurred soon after skull burial or over a longer term is not known. The heavier weight of the rear/upper part of the skulls, with horns attached, must also have added to forward/downward slumping after soft tissues had degraded, a process estimated to take anywhere from two to nine months in arid environments (Galloway 1997; Janaway 1996). As argued above in the discussion of weathering, if the rear/posterior parts of skulls (from the orbits backward) were unburied and exposed, gravity would exacerbate skull collapse.

Cut Marks, Butchery, Skull Preparation

In addition to natural processes, there is some evidence that human butchery practices impacted the skulls. Signs of skull processing and preparation are very few, but they probably reflect common wider practices. Table 11.12 shows three skulls (3, 32, 37) with evidence of cut marks on the wings of the palate, on one lateral side in each case. These multiple small cuts and notches appear to have been made from one side of the skull. They are too light and superficial to have intended to cut through bone itself; rather they suggest the removal of soft tissue (figure 11.36). A likely explanation for their placement is that they result from attempts to free and remove the tongue—a prized nutritious organ—from the skull, if the carcass/skull was lying on one side. If the palatine wings were not fragmented in so many of the Kheshiya skulls, perhaps more cut marks in this location would be seen.

The kind of cuts seen in figure 11.36 would not inflict any great damage to a skull, but if mandibles were separated from skulls before or after the tongue, as we know they were at some stage prior to burial, this could have caused greater damage. Mandible removal from cattle skulls often is achieved by chopping through the jaw's vertical ramus (see Rixon 1989:56) to smash the heel area of the mandible, thus freeing the mandible condyle from the skull. The single piece of identifiable bone from Kheshiya other than skull was a fragment of cattle mandible (Lot 35, figure 11.37) that

Table 11.12. Description of cut marks/butchery marks on the Kheshiya cattle skulls, alongside possible interpretations.

Skull/Lot Number	Cut Marks on Wings of Palate, Description	Cut Marks on Frontals/Nasals	Figures	Interpretation
3	On RHS lateral side of wing of palate, about eight small cuts and indentations seemingly made by a sharp cutting/ scraping implement, each less than 1 cm in length; clearly old since they have similar patination to rest of the bone surface, despite being close to a modern break (but wing of palate has an original break on this side too).			Too light for mandible removal and not in right location; more likely for tongue removal; cuts made from right side of skull.
32	On RHS lateral side of wing of palate, small notches posterior to the old break; not very clear/sharp.			Too light for mandible removal; more likely for tongue removal; cuts made from right side of skull.
33		Cut marks on the frontals, near their meeting point with the nasals and lacrimals. LHS: about five very fine parallel cut marks, 1–1.5 cm in length with other light traces of similar cuts adjacent; RHS: three deeper also parallel cuts, about 1 cm in length; both sets of cuts are distinct and separate from each other (although we can't see if nasals also had cuts, since they are pushed beneath frontals); cuts are angled on anterior–dorsal direction; characteristic of chipped stone tool cuts.	11.38, 11.39, 11.40	Skinning marks, to obtain hide including skull shape? Cutting facial arteries for bleeding?
37	On RHS wing of palate, on lateral and ventral surfaces, a series of five small cuts, 3–4 mm long, sharp as if made with a chipped stone tool and in parallel lines.		11.36	Likely for tongue removal? Perhaps for separation of mandible from skull but seem too light. Decapitation would not leave marks in this location. These cuts are "notches" as if something cut on them, supporting the idea of tongue removal.

Figure 11.36. Skull 37, close-up of palatine wings, showing multiple light notch-like cut marks on lateral side. *Photograph by Louise Martin and Lisa Usman, digitally enhanced by Stuart Laidlaw.*

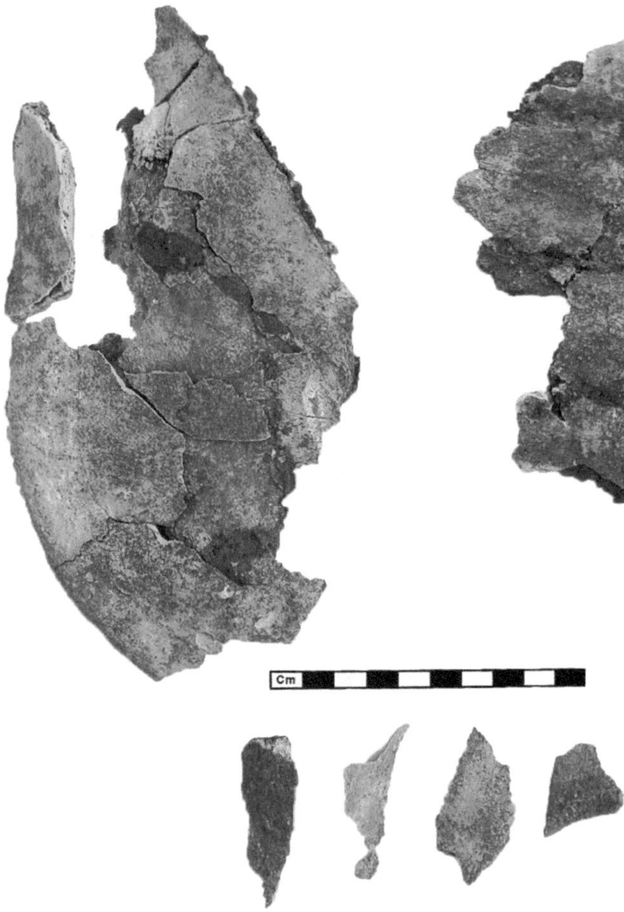

Figure 11.37. Lot 35 from Kheshiya, the only fragments of bone that are not cattle skulls, consists of fragments of (cf.) a cattle mandible, including a shattered mandibular heel fragment, an ascending ramus, and part of the coronoid process. *Photograph by Louise Martin and Lisa Usman, digitally enhanced by Stuart Laidlaw.*

Figure 11.38. Skull 33, superior view prior to surface cleaning, which revealed the cut marks. *Photograph by Louise Martin and Lisa Usman, digitally enhanced by Stuart Laidlaw.*

includes a shattered mandibular heel area, ascending ramus, and part of the coronoid process—matching expectations of breakage incurred during mandible removal. There is no evidence as to whether mandibles were subsequently processed for cheek meat removal or marrow, but that might be likely if nutrients were intensively extracted from the cattle carcasses. It is also likely that the process of mandible removal dealt heavy chops and blows to the sides of skulls, potentially causing additional damage to skull structure and integrity. Most skulls slumped in one direction or another—to their left or right (table 11.11), in no particular pattern—and one wonders whether the breakage of palatine wings, and the collapse of the skull to one side, was in part caused by structural weakness resulting from heavy blows to free the mandible—blows that need be applied to one side of the jaw only (often the mandibular hinge area).

The skulls show no evidence of horn removal, which often is visible as cuts around the base of the horncores. In the Kheshiya assemblage, this area often survives. We can therefore assume that horns were left on skulls for visual effect—these were, after all, the main features of the installation that protruded above ground level.

The only other cut marks observed were on Skull 33 (table 11.12, figures 11.38, 11.39, and 11.40). Figure 11.38 shows the skull before cleaning revealed the cut marks; figure 11.39 shows the multiple small incisions on both the left and right sides of the anterior frontals, close to the point where they meet the nasals and lacrimals. The short, light cuts appear as "hatch" marks, close together and parallel, that seem to have been intended to disconnect or cut specific soft tissues. They superficially mark the bone but do not break it, and the sharp edges and multiple cuts are characteristic of chipped stone tools cutting into fresh rather than dry bone (figure 11.40) (Greenfield 1999; Olsen 1988). Figure 11.39 shows how the frontals of this skull had shifted a few centimeters over the nasal bones, making it impossible to gauge whether the cuts continued across the nasals.

The most obvious interpretation of these cut marks is that they relate to careful skinning—for example, producing a hide complete with cattle skull shape. The removal of a bovid hide often results in a continuous piece that includes two strips of cheek hide, and sometimes a thin strip of face/frontal hide too, where careful skinning has circumvented the horns and peeled off these face pieces. The cuts on Skull 33 may relate to face hide removal,

Figure 11.39. Skull 33, close-up of anterior part of frontals, showing multiple small cut marks on both the left and right sides (described in table 11.12). *Photograph by Louise Martin and Lisa Usman, digitally enhanced by Stuart Laidlaw.*

Figure 11.40. Skull 33, close-up of cut marks on anterior part of frontals, showing multiple fine parallel cuts, characteristic of chipped stone cuts into fresh bone. *Photograph by Louise Martin and Lisa Usman, digitally enhanced by Stuart Laidlaw.*

although placement seems slightly too high for skinning the cheeks and perhaps too low for frontal hide removal. The skull is too badly damaged to see whether any characteristic skinning marks existed around the horn bases (e.g., Binford 1981:105–41).

A close look at cattle soft tissue anatomy might suggest an alternative interpretation for consideration. These "nicks" (seen in figure 11.39) are exactly at the location of the main facial artery on either side of a bovid skull. In brief, the common carotid artery (which supplies blood to the head) splits into several smaller arteries, with the facial artery winding above and beneath muscles and other soft tissues, over the maxillae, to run over the surface of the lacrimals and frontals, ending up in the orbit to provide the front of the face with blood. The cut marks on Skull 33 would be well located to target the main blood supply to the front of the animal's face, but for what reason? Cutting the facial arteries—both left and right—is certainly not an effective way to kill an animal, which normally is done by slitting the major common carotid artery in the neck/throat area. Neither does bleeding an animal to obtain blood to drink (e.g., Århem 1989, for accounts of Maasai practice) use this facial artery. Instead, it is done by nicking the jugular vein that returns deoxygenated blood to the heart—a procedure that is not life threatening to the animal. But slitting these smaller facial arteries on a live, stunned, or recently killed animal would produce strong spurts of blood just below the orbits. This, of course, remains completely speculative, but it is just possible that rather than signifying skinning marks, the frontal cut marks result from another practice that created bleeding from the face in a highly dramatic effect. Such a practice has not been found in any ethnographic literature on cattle ritual, but it is a reasonable suggestion to consider given cattle facial anatomy.

No other cut marks or butchery marks were found on the Kheshiya skulls, which does not mean that more were not originally present. Rather, poor bone condition has preserved only the marks reported here.

Discussion and Conclusions

The Kheshiya Cattle

The study of morphological traits of the Kheshiya skulls strongly suggests that they belong to taurine cattle, with no firm evidence for zebu (*Bos indicus*), although there does seem to be morphological variation within the assemblage. The identification of taurine cattle is not surprising given that zebu are not known to have dispersed westward into the Middle East area until the late sixth millennium cal BP, and they are not seen in Africa until a couple of millennia later.

Morphometric analysis shows that the Kheshiya skulls are far smaller than wild *Bos primigenius* equivalents, whether comparatives are larger samples from Pleistocene/Holocene Europe, smaller comparatives from Early Holocene Anatolia, or single comparisons made with Pleistocene East African aurochs. We can comfortably assume that the skulls come from domestic *Bos taurus*. In terms of skull size—using the limited preserved dimensions of the least frontal breadth of the skulls and maxillary tooth row lengths—the Kheshiya assemblage surprisingly overlapped with Holocene European cattle (from Denmark), although the latter have a larger range. Given an expected northwest-to-southeast size cline geographically (Wright and Viner-Daniels 2015), Arabian domestic cattle might have been expected to be much smaller than European counterparts. Kheshiya cattle are more similar in skull size to the Middle Kerma comparatives, which date to the early fourth millennium cal BP. Skull size is not a good indicator of overall body size in mammals (e.g. Dayan et al. 1991), and we have no postcranial elements from Kheshiya to allow cattle body size reconstruction. It is worth noting, however, that Chaix (2007:208) found that the Kerma cattle "possessed a strong build reaching an average stature of ca. 1.40 [m]," which might give an impression of the Kheshiya cattle height, if their morphologies broadly corresponded.

The identification of domestic *Bos taurus* at Kheshiya allows us to assume that the stock from which they derived originated in the Levant/Fertile Crescent area, or possible East Africa. The earliest domestic cattle finds from Southern Arabia—from eighth-millennium cal BP Manayzah (chapter 8 this volume; Martin et al. 2009)—do not allow species assignation. Based on Kheshiya evidence, it now can be assumed that the earlier Manayzah specimens, too, are taurines. The Manayzah specimens were present in the same region (Wādī Sanā) a millennium earlier. Likewise, it is tempting to think of the seventh-millennium cal BP cattle finds from Wādī ath-Thayyilah 3 in the highlands of Northern Yemen (Fedele 2008) as being of similar type stock.

In the Persian Gulf area, cattle remains from seventh-millennium cal BP Jebel al-Buhais 18 in Sharjah are too few and fragmentary to assess whether they belong to taurine or indicine cattle, although they too appear to be domestic (Uerpmann and Uerpmann 2008). By the fifth millennium cal BP, most sites in the Gulf region report the presence of European cattle, *Bos taurus*, while from Umm an-Nar on the Emirates coast, cattle are assigned to *Bos indicus*, although primarily on the basis of habitat expectations of dryness rather than cattle morphology.

Kheshiya, therefore, aids our understanding of domestic cattle stock origins in the south of Arabia, but to further document cattle introductions and dispersals, analysts need to develop more zooarchaeological and genetic research that supports archaeological evidence of trade and exchange networks.

The Herding System and Cattle Cull

Of the 35 cattle skulls that formed part of the zooarchaeological analysis, just over half provided metrical data that allowed assessment of sexual dimorphism. Results suggest that all the skulls in the outer part of the cattle ring that could be assessed were of narrower morphology and likely to be females; the skull in the center of the ring (Skull 39) was broader, is completely metrically separated from the others, and is very likely to be male. This is an intriguing finding, raising questions about the rationale and meaning behind the installation.

If we add to this picture the results of the dental aging analysis of the cattle, we find the cull to be highly focused on mature adult animals. There are no juveniles or subadults in the skull ring, and there was only one younger adult; the majority of cattle are older adults, not yet elderly, but as female cows they were probably beyond their useful reproductive lives. Dahl and Hjort (1976) report the life spans of recent African cattle to be 9 to 15 years (although an occasional animal living up to twenty years is known), and the dental aging evidence places the Kheshiya cattle in this age bracket. The Kheshiya cull can then be interpreted as consisting of mainly older adult cows, who were beyond calving—a "take-off" of expendable stock.

We have elsewhere extensively modeled the herd management systems lying behind the Kheshiya cattle ring cull or sacrifice (McCorriston et al. 2012), producing detailed ecological predictions of the herd sizes that could sustain a take-off of 40 head of cattle. We also modeled meat yields of such a sacrifice and worked through estimates of consumers and the social and subsistence implications for the Southern Jol herding landscape. The cattle age and sex data produced in this chapter contribute to our discussion; the evidence for the cull of older adult cows supports the idea of milk being a key component of the pastoralist strategy in the Southern Jol Neolithic. To ensure reproduction and safeguard milk production, pastoral nomads who rely on cattle milk keep as many females as possible (Dahl and Hjort 1976:35). In milking herds, fewer individuals survive into the older adult age classes (1976:48, table 2.5) because the predominantly female herd is slaughtered as individuals reach the end of their reproductive lives—an outcome seen in the Kheshiya females.

Returning to the single male skull in the center of the ring (adult, but at the younger end of the range), are we seeing here the sacrifice and special placement of a bull that probably served many cows in the herding landscape? While the majority of males are culled young for beef in most cattle pastoralist systems, a few would be selected for breeding purposes. Dahl and Hjort (1976:28) find that one bull in modern African herding systems regularly serves 50 or 60 cows, providing interesting thought for the Kheshiya context of a bull surrounded by post-reproductive cows.

We know from Henton's work on the oxygen isotopes in the cattle dental enamel (Henton et al. 2014) that groups of animals were herded in at least four distinct locations beyond the Wādī Sanā but within the wider Southern Jol landscape, which shows that the cattle cull was drawn from different herds with varied pasturing and mobility patterns. Henton also demonstrated through dental microwear analysis that the culled cattle all grazed on a diet of soft, clean forage in the weeks immediately prior to death, interpreted as being just after the monsoon season in late summer. The strong similarity in preslaughter diets among all culled animals is consistent with pasturing close to the site of Kheshiya (Henton et al. 2014:128) and is in marked contrast with the variety of herding regimes evident in the dental enamel data. The combined isotope and microwear evidence gives a picture that the Kheshiya cattle derived from different herds that converged in one vegetation zone—likely close to the site itself—in the post-flood/monsoon season, prior to being culled (2014:129). We speculate that the seasonal aggregation and ritual cattle slaughter seen at Kheshiya not only was an occasion for consolidating social networks through feasting, and negotiating access to grazing and other resources (Henton et al. 2014; McCorriston 2011) but may also have provided a context for exchanging cattle and organizing cattle breeding regimes. The installation of the bull skull surrounded by females might commemorate these activities. The dental abnormalities in the Kheshiya assemblage also argue for cattle exchange and interbreeding within the wider landscape, since these abnormalities tend to be characteristic of genetic bottlenecks.

Construction of the Monument

Finally, how does the bone surface modification and taphonomic data presented in this chapter add to our understanding of carcass processing of the Kheshiya cattle, skull preparation prior to installation in the monument, and skull ring depositional history? A few details add to McCorriston's (2011) rich description of how the sacrifice and ring construction took place.

The intact nature of the skulls when they were buried in the ring, with delicate nasals and teeth unbroken or chipped, strongly suggests that the cattle were culled nearby, not at other disparate locations and assembled here (cf. Davis and Payne 1993). Whether whole large herds accumulated at the Kheshiya location in the late-summer, post-monsoonal season (Henton et al. 2014), or whether just those animals selected for slaughter did so, we will never know, but the selection of these similarly aged animals implies intimate knowledge of the life stages of individual animals and careful herd management decisions (e.g., Galaty 1989).

Whether the frontal cut marks on a single skull (Skull 33) reflect skinning activities or the more intriguing suggestion of slitting the facial arteries to stimulate spurts of blood from the face (of a live or recently dead animal) is difficult to tell. But other cattle skinning marks seen in the zooarchaeological literature (e.g., Lisowski 2014) tend to show cuts farther back on the frontal, which might encourage a rethinking of the skinning interpretation.

Decapitation of carcasses left no visible signs—the occipital condyles were too poorly preserved, and the atlas/axis is absent. Only one skull shows evidence of careful face hide skinning (Skull 33). While all carcasses obviously would have been skinned, it is not clear if only this skull had its face hide removed or whether we cannot see it in other skulls because the preservation of their parts and surfaces is too poor for us to be certain.

Mandibles seem to have been removed by a heavy blow or chop across the ramus to release condyles, maybe serving to also weaken the side of the skull at this point. Mandible removal made the skull narrower for burial but also allowed easier access to the tongue—a delicacy—for which two skulls show evidence of extraction. Presumably, to keep the skulls intact, the nutritious brain was not extracted; crania were installed complete and show no signs of breakage.

McCorriston (2011) fleshes out discussion of the large quantities of fresh meat, blood, and other products that the Kheshiya cattle cull would have produced and considers in detail the alternatives of immediate feasting, preserving meat, or redistribution of joints. Because other cattle skeletal elements are absent at Kheshiya, interpreting consumption activities requires broader social and ecological approaches, as our synthesis shows (McCorriston et al. 2012).

As expected, skulls showed no evidence of horn removal; the smooth keratin horn sheaths presumably gave the desired effect to the whole installation. The horn form of these Neolithic Arabian cattle is not known, and taurine horns can be as variable as those of any cattle (Grigson 1978). The adult females in the outer part of the ring would have had long slender horns, certainly interlocking with those of their neighbors, while we can assume the central male skull carried maximum-size horns that protruded prominently into the space.

Apart from mandible removal, none of the skulls shows any modification. The point was not to fashion them into bucrania for household display, as seen, for example, in Neolithic domestic installations (Baird et al. 2016; Mellaart 1967), or just to remove the horn-carrying part of the skull, as seen in the Kerma examples (Chaix 2007), which allowed them to be laid flat with horns extending upward. The Kheshiya skulls fitted their purpose, with no further tailoring or shaping, perhaps indicating quicker manufacture and a shorter-term impact for mobile people.

Bone surface weathering patterns indicate both that skulls were buried relatively rapidly after they were prepared and that rear skull parts—from the orbits backward—suffered from surface exposure. Over time—whether the short or long term—occipitals, the intercornual ridge, and horns disintegrated completely.

Whatever activities took place in the center of the skull ring left no traces on the skulls; none show signs of burning, which might suggest that open fires were not nearby. It is estimated that within a year of burial (Galloway 1997), soft tissues would have degraded, cranial cavities filled with silts, and skulls collapsed downward, exacerbated by wetting and drying of the shallow deposits and the pull of gravity. After these routine processes of taphonomic decay, the skulls thankfully were stabilized by deposition, and they survived the 6,000 years until excavation.

References

Ameen, Carly, Ardern Hulme-Beaman, Allowen Evin, Mietje Germonpré, Kate Britton, Thomas Cucchi, Greger Larson, and Keith Dobney

2017 A Landmark-Based Approach for Assessing the Reliability of Mandibular Tooth Crowding as a Marker of Dog Domestication. *Journal of Archaeological Science* 85:41–50.

Andrews, A. H.

1982 The Use of Dentition to Age Young Cattle. In *Ageing and Sexing Animal Bones from Archaeological Sites*, edited by Caroline Grigson, Bob Wilson, and Sebastian Payne, pp. 141–54. British Archaeological Reports, British Series 109. Archaeopress, Oxford.

Andrews, A. H., and R. W. M. Wedderburn

1973 The Relationship of Age to First Molar Tooth Development in a Group of Cross-Bred Steers. *British Veterinary Journal* 129:512–17.

Arbuckle, Benjamin S., Sarah Whitcher Kansa, Eric Kansa, David Orton, Canan Çakırlar, Lionel Gourichon, Levent Atici, Alfred Galik, Arkadiusz Marciniak, Jacqui Mulville, Hijlke Buitenhuis, Denise Carruthers, Bea De Cupere, Arzu Demirergi, Sheelagh Frame, Daniel Helmer, Louise Martin, Joris Peters, Nadja Pöllath, Kamilla Pawłowska, Nerissa Russell, Katheryn Twiss, and Doris Würtenberger

2014 Data Sharing Reveals Complexity in the Westward Spread of Domestic Animals across Neolithic Turkey. *PLoS ONE* 9(6):e99845.

Århem, Kaj

1989 Maasai Food Symbolism: The Cultural Connotations of Milk, Meat, and Blood in the Pastoral Maasai Diet. *Anthropos* 84(1/3):1–23.

Armitage, Philip

1982 A System for Ageing and Sexing the Horn Cores of Cattle from British Post-Medieval Sites (17th to Early 18th Century) with Special Reference to Unimproved British Longhorn Cattle. In *Ageing and Sexing Animal Bones from Archaeological Sites*, edited by Bob Wilson, Caroline Grigson, and Sebastian Payne, pp. 5–61. British Archaeological Reports, British Series 109. Archaeopress, Oxford.

Baird, Douglas, Andrew Fairbairn, and Louise Martin

2016 The Animate House, the Institutionalization of the Household in Neolithic Central Anatolia. *World Archaeology* 9(6):e99845.

Baker, Polydura, and Achilles Gautier

1997 Site E71K12 and the Late Paleolithic Annual Round along the Nile in Upper Egypt. In *A Late Paleolithic Kill-Butchery-Camp in Upper Egypt*, edited by Fred Wendorf, Romuald Schild, Polydora Baker, Achilles Gautier, Laura Longo, and Amal Mohamed, pp. 43–56. Institute for the Study of Earth and Man, Department of Anthropology, Southern Methodist University, Dallas.

Beardah, C., and M. Baxter

1996 The Archaeological Use of Kernel Density Estimates. *Internet Archaeology.* http://intarch.ac.uk/journal/issue1/beardah/kdeia1.html, accessed August 25, 2017.

Beech, Mark, and Emily Glover

2005 The Environment and Economy of an Ubaid-Related Settlement on Dalma Island, United Arab Emirates *Paléorient* 31(1):97–107.

Behrensmeyer, Anna K.

1978 Taphonomic and Ecologic Information from Bone Weathering. *Paleobiology* 4(2):150–62.

Binford. Louis R.

1981 *Bones: Ancient Men and Modern Myths*. Academic Press, New York.

Boivin, Nicole, and Dorian Q. Fuller

2009 Shell Middens, Ships and Seeds: Exploring Coastal Subsistence, Maritime Trade and the Dispersal of Domesticates in and around the Ancient Arabian Peninsula. *Journal of World Prehistory* 22(2):113–80.

Carter, Robert, and Harriet Crawford

2003 The Kuwait-British Archaeological Expedition to as-Sabiyah: Report on the Fourth Season's Work. *Iraq* 65:77–90.

Chaix, Louis

2007 Contribution to the Knowledge of Domestic Cattle in Africa: The Osteometry of Fossil *Bos taurus* L. from Kerma, Sudan (2050–1750 BC). In *Skeletal Series and Their Socio-economic Context*, edited by Giselsa Grupe and Joris Peters, pp. 120–249. Documenta Archaeobiologiae 5. Verlag Marie Leidorf GmbH, Rahden, Germany.

Chen, Shanyuan, Bang-Zhong Lin, Mumtaz Baig, Bikash Mitra, Ricardo J. Lopes, António M. Santos, David A. Magee, Marisa Azevedo, Pedro Tarroso, Shinji Sasazaki, Stephane Ostrowski, Osman Mahgoub, Tapas K. ChaudhurI, Ya-ping Zhang, Vânia Costa, Luis J. Royo, Félix Goyache, Gordon Luikart, Nicole Boivin, Dorian Q. Fuller, Hideyuki Mannen, Daniel G. Bradley, and Albano Beja-Pereira

2009 Zebu Cattle Are an Exclusive Legacy of the South Asia Neolithic. *Molecular Biology and Evolution 27*(1):1–6.

Cleuziou, Serge, and Maurizio Tosi

1998 Hommes, climats et environnements dans la péninsule arabique à l'Holocène. *Paléorient* 23(2):121–35.

Clutton-Brock, Juliet

1999 *A Natural History of Domesticated Animals.* 2nd ed. Natural History Museum, London.

Dahl, Gudrun, and Anders Hjort

1976 *Having Herds: Pastoral Herd Growth and Household Economy*. Department of Social Anthropology, University of Stockholm.

Davis, Simon J. M.

1981 The Effects of Temperature Change and Domestication on the Body Size of Late Pleistocene to Holocene Mammals of Israel. *Paleobiology* 7(1):101–14.

1987 *The Archaeology of Animals.* Batsford, London.

Davis, Simon J., and Sebastian Payne

1993 A Barrow Full of Cattle Skulls. *Antiquity* 67(254):12–22.

Dayan, Tamar, Daniel Simberloff, Eitan Tchernov, and Yoram Yom-Tov

1991 Calibrating the Paleothermometer: Climate, Communities, and the Evolution of Size. *Paleobiology 17*(2):189–99.

Decker, Jared E., Stephanie D. McKay, Megan M. Rolf, JaeWoo Kim, Antonio Molina Alcalá, Tad S. Sonstegard, Olivier Hanotte, Anders Götherström, Christopher M. Seabury, Lisa Praharani, Masroor Ellahi Babar, Luciana Correia de Almeida Regitano, Mehmet Ali Yildiz, Michael P. Heaton, Wan-Sheng Liu, Chu-Zhao Lei, James M. Reecy, Muhammad Saif-Ur-Rehman, Robert D. Schnabel, and Jeremy F. Taylor

2014 Worldwide Patterns of Ancestry, Divergence, and Admixture in Domesticated Cattle. *PLoS Genetics* 10(3):e1004254.

Degerbøl, Magnus, and Bent Fredskild

1970 *The Urus (Bos primigenius Bojanus) and Neolithic Domesticated Cattle (Bos taurus domesticus Linné) in Denmark: with a Revision of Bos Remains from the Kitchen Middens: Zoological and Palynological Investigations*. Det Kongelige Danske Videnskabernes Selskab. Munksgaard, Copenhagen.

Drechsler, Philipp

2007 The Neolithic Dispersal into Arabia. *Proceedings of the Seminar for Arabian Studies* 37:93–109.

Fedele, Francesco G.

2008 Wādī aṯ-Ṯayyilah 3, a Neolithic and Pre-Neolithic Occupation on the Eastern Yemen Plateau, and Its Archaeofaunal Information. *Proceedings of the Seminar for Arabian Studies* 38:153–71.

Galaty, John G.

1989 Cattle and Cognition: Aspects of Maasai Practical Reasoning. In *The Walking Larder: Patterns of Domestication, Pastoralism, and Predation*, vol. 15, edited by Juliet Clutton-Brock, pp. 215–31. Routledge, London.

Galloway Alison

1997 The Process of Decomposition: A Model from the Arizona-Sonoran Desert. In *Forensic Taphonomy: The Postmortem Fate of Human Remains,* edited by Marcella H. Sorg and William D. Haglund, pp. 139–50. CRC Press, Boca Raton.

Grant, Annie

1982 The Use of Tooth Wear as a Guide to the Age of Domestic Ungulates. In *Ageing and Sexing Animal Bones from Archaeological Sites,* edited by Bob Wilson, Caroline Grigson, and Sebastian Payne, pp. 91–108. British Archaeological Reports, British Series 109. Archaeopress, Oxford.

Greenfield, Haskel J.

1999 The Origins of Metallurgy: Distinguishing Stone from Metal Cut-Marks on Bones from Archaeological Sites. *Journal of Archaeological Science* 26(7):797–808.

Grigson, Caroline

1974 The Craniology and Relationships of Four Species of Bos. 1. Basic Craniology: *Bos taurus* L. and Its Absolute Size. *Journal of Archaeological Science* 1(4):353–79.

1975 The Craniology and Relationships of Four Species of Bos. 2. Basic Craniology: *Bos taurus* L. Proportions and Angles. *Journal of Archaeological Science* 2(2):109–28.

1976 The Craniology and Relationships of Four Species of Bos. 3. Basic Craniology: *Bos taurus* L. Sagittal Profiles and Other Non-Measurable Characters. *Journal of Archaeological Science* 3(2):115–36.

1978 The Craniology and Relationships of Four Species of Bos. 4. The Relationship between *Bos primigenius* Boj. and *B. taurus* L. and Its Implications for the Phylogeny of the Domestic Breeds. *Journal of Archaeological Science* 5(2):123–52.

1980 The Craniology and Relationships of Four Species of Bos. 5. *Bos indicus* L. *Journal of Archaeological Science* 7(1):3–32.

1982 Sex and Age Determination of Some Bones and Teeth of Domestic Cattle: A Review of the Literature. In *Ageing and Sexing Animal Bones from Archaeological Sites*, edited by Bob Wilson, Caroline Grigson, and Sebastian Payne, pp. 7–24. British Archaeological Reports, British Series 109. Archaeopress, Oxford.

2000 *Bos africanus* (Brehm)? Notes on the Archaeozoology of the Native Cattle of Africa. In *The Origins and Development of African Livestock: Archaeology, Genetics, Linguistics and Ethnography*, edited by R. Blench and K. MacDonald, pp. 38–60. Routledge, London.

2007 *Database of Bos primigenius Measurements from Late Pleistocene and Early Holocene Denmark.* Manuscript on file, Institute of Archaeology, University College London.

Guagnin, Maria, Ceri Shipton, Louise Martin, and Michael Petraglia

2017 The Neolithic Site of Jebel Oraf 2, Northern Saudi Arabia: First Report of a Directly Dated Site with Faunal Remains. *Archaeological Research in Asia* 9:63–67.

Halstead, P.

1985 A Study of Mandibular Teeth from Romano-British Contexts at Maxey. In *Archaeology and Environment in the Lower Welland Valley,* vol. 1, edited by Francis Pryor and Charles French, pp. 219–24. East Anglian Archaeology, Norwich, UK.

Hanotte, Olivier, Daniel G. Bradley, Joel W. Ochieng, Yasmin Verjee, Emmeline W. Hill, and J. Edward O. Rege

2002 African Pastoralism: Genetic Imprints of Origins and Migrations. *Science* 296(5566):336–39.

Hansen, Henrik B., Peter B. Damgaard, Ashot Margaryan, Jesper Stenderup, Niels Lynnerup, Eske Willerslev, and Morten E. Allentoft

2017 Comparing Ancient DNA Preservation in Petrous Bone and Tooth Cementum. *PLoS ONE* 12(1):e0170940, https://doi.org/10.1371/journal.pone.0170940, accessed August 24, 2017.

Hartigan, J. A., and P. M. Hartigan

1985 The Dip Test of Unimodality. *Annals of Statistics* 13(1):70–84.

Henton, Elizabeth, Joy McCorriston, Louise Martin, and Eric A. Oches

2014 Seasonal Aggregation and Ritual Slaughter: Isotopic and Dental Microwear Evidence for Cattle Herder Mobility in the Arabian Neolithic. *Journal of Anthropological Archaeology* 33:119–31.

Horwitz, Liora Kolska, and Pierre Ducos

2006 Counting Cattle: Trends in Neolithic Bos Frequencies from the Southern Levant. *Revue de paléobiologie* 10:209–24.

Horwitz, Liora Kolska, and Nigel Goring-Morris

2004 Animals and Ritual during the Levantine PPNB: A Case Study from the Site of Kfar Hahoresh, Israel. *Anthropozoologica* 39(1):165–78.

Janaway, R. C.
1996 The Decay of Buried Human Remains and Their Associated Materials. In *Studies in Crime: An Introduction to Forensic Archeology*, edited by John Hunter, Charlotte Roberts, and Carol Heron, pp. 58–85. B. T. Batsford, London.

Jones, Gillian G., and Peta Sadler
2012a Age at Death in Cattle: Methods, Older Cattle and Known-Age Reference Material. *Environmental Archaeology* 17(1):11–28.
2012b A Review of Published Sources for Age at Death in Cattle. *Environmental Archaeology* 17(1):1–10.

Linseele, Veerle
2004 Size and Size Change of the African Aurochs during the Pleistocene and Holocene. *Journal of African Archaeology* 2(2):65–185.

Lisowski, Mikolaj
2014 Hides and Horn Sheaths: A Case Study of Processed Skulls and Horn Cores from the Early-Middle Neolithic Site of Kopydłowo 6, Poland. *Assemblage PZAF* (2014):32–41.

Lyman, R. Lee, and Gregory L. Fox
1989 A Critical Evaluation of Bone Weathering as an Indication of Bone Assemblage Formation. *Journal of Archaeological Science* 16(3):293–317.

MacHugh, David E., Mark D. Shriver, Ronan T. Loftus, Patrick Cunningham, and Daniel G. Bradley
1997 Microsatellite DNA Variation and the Evolution, Domestication and Phylogeography of Taurine and Zebu Cattle (*Bos taurus* and *Bos indicus*). *Genetics* 146(3):1071–86.

Magee, David A., David E. MacHugh, and Ceiridwen J. Edwards
2014 Interrogation of Modern and Ancient Genomes Reveals the Complex Domestic History of Cattle. *Animal Frontiers* 4(3):7–22.

Marshall, Fiona
2000 The Origins and Spread of Domestic Animals in East Africa. In *The Origins and Development of African Livestock: Archaeology, Genetics, Linguistics and Ethnography*, edited by Roger Blench and Kevin MacDonald, pp. 191–221. Routledge, London.

Marshall, Fiona, and Elisabeth Hildebrand
2002 Cattle before Crops: The Beginnings of Food Production in Africa. *Journal of World Prehistory* 16(2):99–143.

Martin, Louise, Joy McCorriston, and Rémy Crassard
2009 Early Arabian Pastoralism at Manayzah in Wādī Ṣanā, Ḥaḍramawt. *Proceedings of the Seminar for Arabian Studies* 39:271–82.

Matthews, Roger
2002 Zebu: Harbingers of Doom in Bronze Age Western Asia? *Antiquity* 76(292):438–46.

McCorriston, Joy
2011 *Pilgrimage and Household in the Ancient Near East.* Cambridge University Press, Cambridge.

McCorriston, Joy, Michael Harrower, Louise Martin, and Eric Oches
2012 Cattle Cults of the Arabian Neolithic and Early Territorial Societies. *American Anthropologist* 114(1):45–63.

McCorriston, Joy, and Louise Martin
2009 Southern Arabia's Early Pastoral Population History: Some Recent Evidence. In *The Evolution of Human Populations in Arabia: Paleoenvironments, Prehistory and Genetics,* edited by Michael Petraglia and Jeffrey Rose, pp. 237–50. Springer, Dordrecht.

Mellaart, James
1967 *Çatal Hüyük: A Neolithic Town in Anatolia.* McGraw-Hill, New York.

Miles, A. E. W., and Caroline Grigson
1990 *Colyer's Variations and Diseases of the Teeth of Animals.* Revised ed. Cambridge University Press, Cambridge.

Olivieri, Anna, Francesca Gandini, Alessandro Achilli, Alessandro Fichera, Ermanno Rizzi, Silvia Bonfiglio, Vincenza Battaglia, Stefania Brandini, Anna De Gaetano, Ahmed El-Beltagi, Hovirag Lancioni, Saif Agha, Ornella Semino, Luca Ferretti, and Antonio Torroni
2015 Mitogenomes from Egyptian Cattle Breeds: New Clues on the Origin of Haplogroup Q and the Early Spread of Bos taurus from the Near East. *PloS ONE* 10(10):e0141170.

Olsen, Sandra L.
1988 The Identification of Stone and Metal Tool Marks on Bone Artifacts. In *Scanning Electron Microscopy in Archaeology*, edited by Sandra L. Olsen, pp. 337–60. British Archaeological Reports, International Series 452. BAR, Oxford.

Park, Stephen D. E., David A. Magee, Paul A. McGettigan, Matthew D. Teasdale, Ceiridwen J. Edwards, Amanda J. Lohan, Alison Murphy, Martin Braud, Mark T. Donoghue, Yuan Liu, Andrew T. Chamberlain, Kévin Rue-Albrecht, Steven Schroeder, Charles Spillane, Shuaishuai Tai, Daniel G. Bradley, Tad S. Sonstegard, Brendan J. Loftus, and David E. MacHugh
2015 Genome Sequencing of the Extinct Eurasian Wild Aurochs, *Bos primigenius*, Illuminates the Phylogeography and Evolution of Cattle. *Genome Biology* 16(1):234.

Patel, Ajita K., and Richard H. Meadow
2017 South Asian Contributions to Animal Domestication and Pastoralism. In *The Oxford Handbook of Zooarchaeology*, edited by Umberto Albarella and Mauro Rizzetto, pp. 280–303. Oxford University Press, Oxford.

Popkin, Peter, R. W. Polydora Baker, Fay Worley, Sebastian Payne, and Andy Hammon,
2012 The Sheep Project (1): Determining Skeletal Growth, Timing of Epiphyseal Fusion and Morphometric Variation in Unimproved Shetland Sheep of Known Age, Sex, Castration Status and Nutrition. *Journal of Archaeological Science* 39(6):1775–92.

Peters, Joris, Angela von den Dreisch, and Daniel Helmer
2005 The Upper Euphrates-Tigris Basin: Cradle of Agro-Pastoralism? In *The First Steps of Animal Domestication*, edited by Jean-Denis Vigne, Joris Peters, and Daniel Helmer, pp. 96–124. Oxbow Books, Oxford.

Rachad, Madiha
2007 Chronologies et styles de l'art rupestre. In *Art Rupestre et Peuplements Préhistoriques au Yémen*, edited by Marie-Louise Inizan and Madiha Rachad, pp. 73–82. Centre Français d'Archeologie et de Sciences Sociales de Sanaa, Sana'a.

Rixon, D.
1989 Butchery Evidence on Animal Bones. *Circaea* 6(1):49–62.

Ross, Ann H., and Sarah L. Cunningham
2011 Time-since-Death and Bone Weathering in a Tropical Environment. *Forensic Science International* 204(1):26–133.

Russell, Nerissa, and Louise Martin
2005 The Çatalhöyük Mammal Remains. In *Inhabiting Çatalhöyük: Reports from the 1995–99 Seasons*, edited by Ian Hodder, pp. 33–97. British Institute of Archaeology at Ankara, London.

Russell, Nerissa, Louise Martin, and Hijlke Buitenhuis
2005 Cattle Domestication at Çatalhöyük Revisited. *Current Anthropology* 46(S5):S101–S108.

Russell, Nerissa, Katheryn C. Twiss, and David C. Orton
2013 More on the Çatalhöyük Mammal Remains. In *Humans and Landscapes of Çatalhöyük: Reports from the 2000–2008 Seasons*, edited by Ian Hodder, pp. 213–58. British Institute of Archaeology at Ankara, London.

Seguí, Maria Saña
1999 *Arqueología de la domesticación animal: la gestión de los recursos animales en Tell Halula (Valle del Éufrates, Siria): del 8.800 al 7.000 BP*. Departament d'Antropologia Social i Prehistòria, Universitat Autònoma de Barcelona Barcelona.

Stock, Frauke, and Diane Gifford-Gonzalez
2013 Genetics and African Cattle Domestication. *African Archaeological Review* 30(1):51–72.

Twiss, Katheryn C., and Nerissa Russell
2009 Taking the Bull by the Horns: Ideology, Masculinity, and Cattle Horns at Çatalhöyük (Turkey). *Paléorient* 35(2):19–32.

Uerpmann, Hans-Peter, and Margarethe Uerpmann
2003 The Capital Area of Northern Oman: Stone Age Sites and Their Natural Environment. In *Beihefte zum Tübinger Atlas des Vorderen Orient (TAVO)*, edited by Christian Hannss and Hans-Peter Uerpmann, pp. 182–84. Reihe A (Naturwissenschaften), Nr. 31/3. Dr. Ludwig Reichert, Wiesbaden.
2008 Neolithic Faunal Remains from al-Buhais 18 (Sharjah, UAE). In *The Natural Environment of Jebel al-Buhais: Past and Present. The Archaeology of Jebel al-Buhais*, vol. 2, edited by Hans-Peter Uerpmann, Margarethe Uerpmann, and Sabah Abboud Jasim, pp. 97–132. Kerns Verlag, Tübingen.

Uerpmann, Margarethe, Hans-Peter Uerpmann, and Sabah A. Jasim
2000 Stone Age Nomadism in SE-Arabia: Palaeo-Economic Considerations on the Neolithic Site of Al-Buhais 18 in the Emirate of Sharjah, UAE. *Proceedings of the Seminar for Arabian Studies* 30:229–34.

von den Driesch, Angela
1976 *A Guide to the Measurement of Animal Bones from Archaeological Sites*. Peabody Museum Bulletins 1. Harvard University, Cambridge, MA.

Wright, Elizabeth, and Sarah Viner-Daniels
2015 Geographical Variation in the Size and Shape of the European Aurochs (*Bos primigenius*). *Journal of Archaeological Science* 54:8–22.

Wright, Elizabeth, Sarah Viner-Daniels, Umberto Albarella, Martin Street, Daniel Makowiecki, Karlheinz Steppan, and Jean-Phillipe Brugal
2016 Biometrical Database of European Aurochs and Domestic Cattle. *Open Context*. http://opencontext.org/projects/1816A043-92E2-471D-A23D-AAC58695D0D3, accessed July 5, 2017.

Zeder, Melinda A.
2012 The Domestication of Animals. *Journal of Anthropological Research* 68(2):161–90.

Zeder, Melinda A., and Brian Hesse
2000 The Initial Domestication of Goats (Capra hircus) in the Zagros Mountains 10,000 Years Ago. *Science* 287(5461):2254–57.

Appendix 1

The Conservation and Treatment of the Kheshiya Cattle Skulls: Report of Procedures Undertaken in the Mukhalla Museum to Aid the Study and Stabilization of the Assemblage

Lisa Usman and Louise Martin

We are extremely grateful to ʿAbdalʿazīz Bin ʿAqīl, director of culture for Mukalla at the time of the study, for his many forms of assistance with work in the Mukalla Museum and for his enthusiastic and generous support of the project. We are also indebted to the staff of the museum for their kind assistance, providing space and equipment for the work.

The conservation aspect of the project aimed to prepare the 35 Kheshiya cattle skulls for zooarchaeological analysis, to provide elementary conservation and stabilization under time-limited conditions, and to pack the material for long-term storage. To this end, one room in the museum was converted into a temporary laboratory for the period of study, December 29, 2005–January 12, 2006.

Review of Methods of Packaging Skulls upon Excavation, On-Site, Spring 2005 (from Observation)

Upon excavation in spring 2005, each skull was lifted and packed for transport to the museum in Mukalla. The first layer of packing material was newspaper, followed by strips of sheet taped together with masking tape and finally a plaster of paris bandage to provide rigid support. The newspaper and sheet were wrapped around the skull as a barrier layer between the skull and the outer bandage, which was wet when applied. Each skull was labeled and packed in a metal box (normally five to a box) and supported with foam to prevent damage during transport. Given the constraints of time and materials in the field, the method proved extremely successful and the skulls were safely transported to the museum.

Note: Prior to packing, a single tooth was extracted from each skull to serve as a sample for scientific analyses (for example, DNA, ^{14}C, isotope, and dental microwear analyses). These samples were exported to the University College London Institute of Archaeology, where they underwent further analyses.

Opening Skull Packages

On unwrapping skulls for study, we removed the rigid plaster bandages using a scalpel, angled to avoid risk of the blade touching the bone. The bandages were prized apart and the skulls gently lifted out of the support onto plastic trays. The remaining packing material was cut open when the skulls were on the trays, leaving a layer of paper and sheet beneath them. Lifting and turning of skulls was avoided as much as possible due to their fragile state; trays allowed for movement and study to be carried out without the need to overhandle the objects.

Skull packages were opened with occlusal surfaces of maxillary teeth facing upward. Our assessment was that frontal bones stood a better chance of surviving with the weight of the skull resting on them than dentition.

Observation of Skull Condition

Most of the soil around the skulls was removed during excavation, but much deposit remained on the surfaces and inside the crania. In most cases, the internal deposit appeared to provide key internal support for the cranium, and therefore it was not removed. The deposit consisted of fine silty particles, compact and hard when dry. Teeth had survived in good condition, while the bones of the palatine and maxillae were mostly highly fragmented; nasals and pre-maxillae often were missing altogether. The posterior area of the palatine was mostly encased in deposit, which made assessment of this area difficult. There was no evidence that horncores had survived; indeed, posterior areas of the skulls (which would have been uppermost in the ground and may have been exposed) had suffered badly.

Cleaning

Due to time constraints, only selective cleaning of the skulls was undertaken. Focus was on areas of the skulls required for zooarchaeological recording, and cleaning attempted to maximize information collection. Deposit that acted as the internal "glue" and held skulls intact was not removed. If time permits, a comprehensive cleaning (involving the complete removal of deposit) could be carried out on the better-preserved skulls, with consolidation of fragments and adhering of joins as each fragment is separated from the soil. This is recommended if any skulls are to be displayed in the future.

For the most part, the soil was harder than the bone it adhered to. Removal was carried out using acetone and a pipette, and gentle scraping away of deposit using a wooden or plastic tool to prevent scratching of the bone's surface. For the teeth, for the most part, it was possible to brush soil away using a soft brush.

Bone fragments that fell away were consolidated using a dilute solution of Butvar. When they dried, they were wrapped in acid-free tissue paper and packed with the skull. It was decided not to consolidate the whole skull, as this would have made the soil even harder to remove, and it is hoped that further work may be undertaken on the skulls in the future. When breaks occurred on morphologically diagnostic features during cleaning, they were repaired using Paraloid B-72 in acetone.

Basal sides of skulls (or dorsal views) were cleaned, studied, and documented first. Then skulls were turned over onto a foam support covered in layers of acid-free tissue. Skull cleaning and study was then undertaken on ventral sides.

Packing

After completion of cleaning, study, recording, and documentation, the skulls were prepared for packing. (Specialist packing materials were limited.) At this stage, each skull was sitting on a foam support and on sheets of acid-free tissue on its own tray. Further layers of acid-free tissue were placed on the top, and the sheets from underneath were brought up over the sides and fastened to the top sheet using masking tape. Once securely sealed, the skull and tray were wrapped in cling film. This was chosen as it held the skull firmly in place, preventing movement if the tray was tipped. Because cling film is transparent, skull numbers and labels were wrapped into it, making labels readable without the need to open the wrapping and thus preventing the labels being separated from the skulls. Cling film also can be easily removed (and reapplied) without damaging the tissue paper, should further study be required. The plastic trays provided excellent support and allowed easy movement of the cattle skulls without causing any damage. The skulls were left in the Mukalla Museum to await placement in custom-made metal boxes for longer-term storage.

Appendix 2

Cattle Skull Recording Form 1

Basic

Site:
Skull number:

Recorded by:
Date(s) recorded:

Photographs:

Overall condition:

Surface weathering—highest:
Surface weathering—lowest:

Burning—degree and location?

Other surface modification? Describe:

Presence of skull parts				
	Left	Left	Right	Right
	> 50%	< 50%	> 50%	50%
Frontal				
Parietal				
Temporal				
Occipital				
Perioticum				
Interparietal				
Palatine				
Sphenoid				
Zygomatic				
Lacrimal				
Nasal				
Maxilla				
Premaxilla				

Horncore

Base				
Corpus				
Tip				

Presence of parts on diagram?

Yes		No	

Appendix 2

Cattle Skull Recording Form 2
Morphology

Site:
Skull number:

Recorded by:
Date(s) recorded:

Nonmetrical traits:

Sagittal Profile		**Complete**		**Estimated**		**Not Done**	

Orbital rim		**Flat/*indicus*:**		
		Sharp/*taurus:*		

Nasal–frontal suture		***taurus* shape:**		
		***indicus* shape:**		

Frontal–lacrimal suture			
	Frontal–lacrimal suture:	**straight?**	**bowed?**
	Lacrimal–jugal suture:	**straight?**	**bowed?**

Frontal profile from above:	**1**	**2**	**3**	**4**	**5**

Intercornual ridge:	**1**	**2**	**3**	**4**	**5**	**6**

Horncore shape:

Horncore direction:

Shape of horncore base:

Shape of posterior end of palate:

Comment:

Appendix 2

Cattle Skull Recording Form 3
Aging Data

Site:
Skull number:

Recorded by:
Date(s) recorded:

Dental Eruption and Wear (after Grant 1982, adapted for maxilla)

Left Side	Grant	Comment
	TWS	
dp2		
P2		
dp3		
P3		
dp4		
P4		
M1		
M2		
M3		

Right Side	Grant	Comment
	TWS	
dp2		
P2		
dp3		
P3		
dp4		
P4		
M1		
M2		
M3		

MWS	

MWS	

Horncores (after Armitage 1982, surface aging method)

Left core
Right core

Horncore rings?
Left
Right

Suture closures (after Grigson 1982)

Comment:

Appendix 2

Measurements on the Cranium of *Bos*

(after von den Driesch 1976)

Measurement	mm	Modifier Code
1		
2		
3		
4		
5		
6		
7		
8		
9		
10		
11		
12		
13		
14		
15		
16		
17		
18		
19		
20		
21		
22		
23		
24		
25		
26		
27		
28		
29		
30		
31		
32		
33		
34		
35		

Site:

Skull Number:

Code:	0-1	Standard measurement
	2	Estimated
	3	Influenced by pathology
	4	See comment
	5	Unfused/young
	6	Burned
	7	As preserved (for artifacts)

Tooth measurements, adapted from mandible

Measurement*	mm	Modifier
L of dp4		
B of dp4		
L of P4		
B of P4		
L of M1		
B of M1		
L of M2		
B of M2		
L of M3		
B of M3		

*all taken near biting surface

Crown heights—add here:

Appendix 3: Kheshiya Cattle Teeth Ancient DNA Pilot Study

Cecilia Anderung and Anders Götherström

Three cattle teeth from Kheshiya (SU151-1) in Yemen were selected from the samples imported to UCL, based on visual good preservation:

Kheshiya DNA id. RA1
Locus 009, Lot 37. Maxilla P3, P4, both worn

Kheshiya DNA id. RA2
Locus 009, Lot 32. Maxilla M1/M2, worn

Kheshiya DNA id. RA3
Locus 009, Lot 33. Maxilla M3, worn

Methods

The specimens were sampled in a dedicated ancient DNA facility at Uppsala University in Sweden. Bone powder was removed from the specimens using a dental drill, producing small holes with a diameter of 2–3 mm. The work surface was sterilized between each sampling procedure and a new drill bit was used for each sample.

About 70 mg of bone powder was incubated at 55°C with 100μg Proteinase K in 1 ml of 0.5M EDTA buffer. Thereafter the DNA was extracted using previously published methods (Bouwman and Brown 2002; Svensson et al. 2007; Yang et al. 1988).

The mtDNA control region was amplified in three overlapping fragments: 157, 176, and 139 bp, respectively. PCR was carried out using 2μl of extracted DNA, 2.5 units of HotStarTaq DNA polymerase (Qiagen), 1X Qiagen PCR buffer, 2.5mM MgCl, 200μM of each dNTPs, and 0.2μM of each primer in a total volume of 25μl.

Results

None of the three samples generated a readable sequence. Considering the geographic origin of the sample (Smith et al. 2003), future work could involve the designing of primers that will amplify shorter DNA fragments; this would probably increase the amplification success rate. As some samples produced a smell of collagen during sampling and collagen survival is correlated with DNA survival (Anderung et al. 2005), it is suggested that further DNA analyses of specimens from this region should involve preservation analyses.

References

Anderung, Cecilia, Abigail Bouwman, Per Persson, José Miguel Carretero, Ana Isabel Ortega, Rengert Elburg, Colin Smith, Juan Luis Arsuaga, Hans Ellegren, and Anders Götherström
2005 Prehistoric Contacts over the Straits of Gibraltar Indicated by Genetic Analysis of Iberian Bronze Age Cattle. *Proceedings of the National Academy of Sciences of the United States of America* 102:8431–35.

Bouwman, Abigail S., and Terence A. Brown
2002 Comparison between Silica-Based Methods for the Extraction of DNA from Human Bones from 18th to Mid-19th Century London. *Ancient Biomolecules* 4:59–63.

Smith C. I., A. T. Chamberlain, M. S. Riley, C. Stringer, and M. J. Collins
2003 The Thermal History of Human Fossils and the Likelihood of Successful DNA Amplification. *Journal of Human Evolution* 45:203–17.

Svensson, E. M., C. Anderung, J. Baubliene, P. Persson, H. Malmström, C. Smith, M. Vretemark, L. Daugnora, and A. Götherström
2007 Tracing Genetic Change over Time Using Nuclear SNPs in Ancient and Modern Cattle. *Animal Genetics* 38:378–83.

Yang, Dongya Y., Barry Eng, John S. Waye, J. Christopher Dudar, and Shelley R. Saunder
1998 Technical Note: Improved DNA Extraction from Ancient Bones Using Silica-Based Spin Columns. *American Journal of Physical Anthropology* 105:539–43.

Appendix 4: Kheshiya 151-1 Cattle Skull Measurements

Cattle Skull Metrics (following von den Driesch 1976)		Skull Numbers 2	3	4	5	11	12	13	14	15	16	17	18
Code	Description												
3	basal length, basion to prosthion												
4	basion to premolar		240*				250*	240*	235*		250*	255*	
5	premolar to prosthion												
12	greatest length of the nasal: nasion to rhinion												
17	dental length: postdentale to prosthion												
18	oral palatal length: palatinooral to prosthion												
19	lateral length of the premaxilla												
20	length of cheek tooth row (measured along the alveoli)	130	127.3	134	129.3	140*	133.2	128.8	124.5*		125.5	130.8	139*
21	length of molar row (measured along the alveoli on the buccal side)	75.8	74.6	83.3	76.9	73.2*	82.2	80	75.1		76.2	77.8	85
22	length of premolar row (measured along the alveoli on the buccal side)	51.3	52.7	51.9	55.2	55*	52.1	52.3	51.9*		46.7	54	53*
23	greatest inner length of the orbit												
26	greatest breadth of the occipital condyles												
24	greatest inner height of the orbit												
27	greatest breadth at bases of paraoccipital processes						156.8						
28	greatest breadth of the foramen magnum												
30	least occipital breadth						122.4						
31	least breadth between the bases of the horncores											140*	
32	least frontal breadth						164.6					169.6	167*
33	greatest breadth across the orbits = greatest frontal breadth						193*					196.5*	195-7*
34	least breadth between the orbits												
35	facial breadth across the tuberosities		145*		144*	152.7	147*	153.3*			139.7	148	149.3*
36	greatest breadth across the nasals					50**							
37	breadth across the premaxillae on oral protuberances											128.1	
38	greatest palatal breadth, measured across the outer borders of the alveoli		116.2		117.5		130*	130.9			126.6	130.8	116.6
46	least (dorso-basal) diameter of the horncore base												
LM1	P2-P2 internal least breadth		67.4*		64.6*			70.5*			63*	76.2*	
LM2	M1-M1 internal least breadth		77.6		75.2			83.5*			75.5	86.8	74
LM3	M3-M3 internal least breadth		75.3		75.1	79.1	79.1	83.4			77.6	77.3*	75*
Tooth Measurements (adapted from von den Driesch 1976 for mandible)		**2**	**3**	**4**	**5**	**11**	**12**	**13**	**14**	**15**	**16**	**17**	**18**
	measurements taken on left or right side?	L	L	L	L	R	R	R+L	L		R	R	L
L P4	length of P4 measured near the biting surface	17.6	17.3	18.3	18.4	18.6	20.6	16.8	17		16.4	18.2	19.2
B P4	breadth of P4 measured near the biting surface	17.4	18.4	20	16.5	18.1	22	16.3	17.7		17.1	16.6	14.1
L M1	length of M1 measured near the biting surface	21.6	21.6	19.6	24.3	23.6	20.1	24	18.6		22.5	21.8	27.5
B M1	breadth of M1 measured near the biting surface	23.1	20	22.4	18.9	21	24.4	19.6	20.6		21	18.6	20.2
L M2	length of M2 measured near the biting surface	26.3	24.5	27.2	25.6	26.4	27.1	27.8	25.4		26.3	26.3	28.7
B M2	breadth of M2 measured near the biting surface	22	19.2	23.5	17.9	20.3	23	19.4	20.8		21.4	19.6	19.3
L M3	length of M3 measured near the biting surface	30	27.7	29.8	28	26.6	30.7	27.6	26.1		28.5	29.4	29
B M3	breadth of M3 measured near the biting surface	22.5	18.8	23.4	17.9	20.7	22.1	19	20		21.2	19.6	18.1
	(lengths of teeth are often less than breadths because of heavy mesio-distal wear)												
Horncore Measurements Taken in the Field, in Situ (by M. Harrower)													
44	horncore basal circumference												
45	greatest diameter of the horncore base											5.7*	
46	least diameter of the horncore base											5.0*	

19	20	21	22	23	24	25	26	27	28	29	30	31	32	33	34	36	37	38	39	40	41	42	Total
																			430*				1
	245*	260*		260*		240*			250		240*	230*			230*	250*	240**	265*	280		290*		19
																			135*				1
																				2 -7*			1
																			270*				1
																			225*				1
																							0
	132	138*	125.5*	125.4	131.1*	136		124.7	120.5	132	129.5*	121.6	130*	134	122.6	127.5	138.5	141.8	138		137	127.2	31
	77.8	87.6	74.4	75.7	83.9	78.5	77.3	76.8	74.2	75.5	79.7	79	80.2	81.8	77.5	78.4	84.5	83.9	86		82	76.6	32
	53	56*	46.8*	51.3	52*	55.4		47	46.5	52.1	46.2	45.9	50**	53	46.8**	50.8	54.3	58.3	54.5		52.5	53.5	31
	47*																						1
																88							1
	56*																						
																							1
									49*							35.8							2
									117*							121*							3
		140*	110*	150*							127*					145-60							6
	160	169.1	154*	152.9	165**	168*			165	174.7	169.3		180**		147*	174*	174*		197.5*		169.4		18
	190.9*		230*	184*						205*	250**				197**	250*	197*		225*		197*		13
																			154*				1
	146.9			137	139*	125*			142*	148.8*	161.3*	146.3		146.8	135			160*	142.4		154*	145.5	22
		55*							47*	50**				45.8*	50*		51*		42*	52	41*		10
																							1
	121.3			120.5	124	114			124.5	125.4	131.6	130.4		131.3	118.4	141.7		133.2	128.4		121		21
			60*	42*																			2
				58.9		58.7				66*				70.1	65.1*	72.5*			69.2		57.7*		13
	76.6			75.1		72.5				79.9		83.6		84.9	75.7	94.7*			82.3		74.9*		16
	73.4*			71.5*		64.3*				76.1				74.6	67.8*	96*			82.5		74*		17
19	**20**	**21**	**22**	**23**	**24**	**25**	**26**	**27**	**28**	**29**	**30**	**31**	**32**	**33**	**34**	**36**	**37**	**38**	**39**	**40**	**41**	**42**	**Total**
R	R	L+R	R	R	L	R	R	L	R	R	L	L	R	L	L	L+R	L	L	L		R+L	R	
17.2	16.9	18.2	16.9	18	17.2	17.9	16.6	15.5	18.7	16.6	18.4	15.8		15.6	16.9	19.1	18.2	18.2	19		18	18	
12.7	14.2	20.8	16.7	17.9	18.2	16.4	19.2	17.5	19.2	20	15.9	18.6		19.5	19.5	20.7	18.4	15.5	16.4		16.6	19.1	
26	25.8	23.9	18.3	22	23.2	23.1	20.8	22.1	19.2	20.4	23.5	22.1	24.3	22.3	20	21.1	23.6	27.2	25.5		23.6	20.1	
18.6	18.3	23.6	19.1	19	21.2	20.6	21.4	21.6	20.7	20.6	19.1	21.7	20.1	21.7	22.2	22.8	20.3	20	19		20.2	20.9	
26.9	27	28.9	25.7	25.2	27.3	26.8	25.5	24.5	25.6	26.3	28.4	25.7	27.6	28.4	25.7	26.6	28.2	29	29.2		28.9	25.6	
16.3	18.2	22.8	20.2	21.2	21.4	21.9	20.7	21	21.4	21	20.4	21.9	22.1	22.3	22.6	22.9	20.6	20.2	19.3		23	22	
23.2	26.2	32.4	27.2	26.8	31.5	29	29.2	28.7	28.5	28.8	31.1	31.1	28.9	33.2	29.1	31.3	31.9	30	30.7		30.9	28.8	
13.8	16.5	21.9	20.2	18.5	21.1	20.1	21.1	21.5	20.3	20.1	20.4	22.6	22.3	22.6	21.8	22.4	20.2	19.6	19.6		18.9	22.5	

																17.4*						
				4.9*												5.6*	5.3*					
				4.3*												4.7*	5.1*					

*Estimated measurement

**Highly estimated measurement

Appendix 5

Skull/Lot Number	Sagittal Profile	Orbital Rim	Nasal/Frontal Suture	Frontal/ Lacrimal Suture	Lacrimal/Jugal Suture	Frontal Profile
2	can't assess	can't assess	can't assess	can't assess	can't assess	can't assess
3	can't assess	can't assess	can't assess	can't assess	can't assess	Type 2, relatively flat on anterior of frontals
4	can't assess	can't assess	can't assess	can't assess	can't assess	can't assess
5	can't assess	can't assess	can't assess	can't assess	can't assess	Type 2, flat across frontals
11	can't assess	can't assess	*taurus* shape	can't assess	straight	can't assess
12	can't assess	can't assess	can't assess	can't assess	straight	Type 2, flat across frontals
13	can't assess	can't assess	can't assess	can't assess	straight	Type 2, flat across frontals
14	can't assess	can't assess	can't assess	can't assess	straight	can't assess
15	can't assess	can't assess	can't assess	can't assess	can't assess	can't assess
16	can't assess	can't assess	can't assess	can't assess	straight	Type 2? Very flat between orbits; can't see posterior to that.
17	can't assess	can't assess	(*indicus* shape?)	can't assess	can't assess	can't assess
18	taurine	can't assess	(*indicus* shape?)	can't assess	bowed/straight intermediate	Type 3: slight boss
19	can't assess	can't assess	can't assess	can't assess	can't assess	can't assess
20	can't assess	can't assess	can't assess	can't assess	can't assess	can't assess
21	taurine?	can't assess	can't assess	can't assess	can't assess	Type 2 or 3
22	can't assess	can't assess	can't assess	can't assess	can't assess	Type 2
23	taurine?	can't assess	can't assess	can't assess	can't assess	Type 2 or 3
24	can't assess	can't assess	can't assess	can't assess	can't assess	Type 2
25	taurine	can't assess	can't assess	can't assess	straight	Type 3: slight boss
26	can't assess	can't assess	can't assess	can't assess	can't assess	can't assess
27	can't assess	can't assess	can't assess	can't assess	can't assess	can't assess
28	can't assess	can't assess	*taurus* shape?	can't assess	straight	Type 2
29	can't assess	can't assess	*taurus* shape?	can't assess	can't assess	Type 2?

Intercornual Ridge	Horncore Shape	Horncore Direction	Shape of Posterior End of Palate	Comments
can't assess	can't assess	can't assess	broken but broad and flat	
can't assess	can't assess	can't assess	medium width; U shaped	
can't assess	can't assess	can't assess	can't assess	
can't assess	can't assess	can't assess	narrow and V shaped, but U shaped where meets with wings of palate	
can't assess	can't assess	can't assess	broad and flat	
can't assess	can't assess	can't assess	broad and U shaped	
can't assess	can't assess	can't assess	narrow and V shaped; spines thin and pinched	
can't assess	can't assess	can't assess	broken but has one-half has narrow wing	
can't assess	can't assess	can't assess	can't assess	
can't assess	can't assess	can't assess	narrow and intermediate between V and U shaped	
Type 8	can't assess	angled backward?	V shaped	
can't assess	can't assess	can't assess	V shaped	
can't assess	can't assess	can't assess	narrow and V shaped	Check if narrow, V-shaped end of palate is age-related.
Type 4	can't assess	can't assess	broad and U shaped	
can't assess	can't assess	taurine	broad, straight, and U shaped	
Type 4 or 8	taurine	can't assess	can't assess	Horncores must be very small (about 4 cm) at base—female?
Type 8	taurine	angled backward?	narrow and V shaped	Horncores seem very small.
can't assess	can't assess	can't assess	very narrow and V shaped	
can't assess	taurine	can't assess	narrow, intermediate between V and U shaped	Seems to be a long narrow skull?
can't assess	can't assess	can't assess	medium width; U shaped	
can't assess	can't assess	can't assess	V shaped, thin walled, not similar to either of Grigson's forms	
can't assess	can't assess	can't assess	broad and U or W shaped	Left side of posterior end of palate flattened.
can't assess	taurine	can't assess	broad and U shaped/flat	

Appendix 5 *(continued)*

Skull/Lot Number	Sagittal Profile	Orbital Rim	Nasal/Frontal Suture	Frontal/ Lacrimal Suture	Lacrimal/Jugal Suture	Frontal Profile
30	taurine?	can't assess	can't assess	can't assess	straight	Type 3: slight boss
31	can't assess	can't assess	can't assess	can't assess	can't assess	Type 2
32	can't assess	can't assess	can't assess	can't assess	straight?	can't assess
33	can't assess	can't assess	can't assess	can't assess	straight?	can't assess
34	can't assess	can't assess	can't assess	can't assess	bowed?	Type 2?
35	mandible					
36	taurine?	can't assess	*indicus* type?	straight?	straight?	Type 2?
37	can't assess	can't assess	*indicus* type?	can't assess	can't assess	Type 2/3? Relatively flat with slight boss in center.
38	can't assess	can't assess	can't assess	can't assess	can't assess	can't assess
39	can't assess	can't assess	can't assess	can't assess	can't assess	can't assess
40	can't assess	can't assess	*indicus* type??	can't assess	can't assess	can't assess
41	can't assess	flat (*indicus* type?)	can't assess	can't assess	straight?	can't assess
42	can't assess	can't assess	can't assess	can't assess	straight?	can't assess

Intercornual Ridge	Horncore Shape	Horncore Direction	Shape of Posterior End of Palate	Comments
Type 2/4?	can't assess	can't assess	broad and U shaped	Horncore bases appear small, about 51 mm anterior/posterior on left side.
can't assess	can't assess	can't assess	narrow and U shaped (unlike either of Grigson's shapes)	
can't assess	can't assess	can't assess	broad and V shaped	
can't assess	can't assess	can't assess	narrow and V shaped	
can't assess	can't assess	can't assess	narrow and U shaped	
can't assess	can't assess	can't assess	can't assess	LHS horncore base seems very small, about 4 cm anterior/posterior.
can't assess	can't assess	can't assess	broad and V shaped	
can't assess	can't assess	can't assess	asymmetrical: left side more V shaped; right side more U shaped; unlike either of Grigson's forms	
Most like Type 1? But assessed anterior or ridge.	can't assess	can't assess	narrow and convex, re Grigson's characteristic shape for *indicus* (Grigson 1976:126, b)	
can't assess	can't assess	can't assess	can't assess	
Most like Type 1? But assessed anterior or ridge.	can't assess	Taurine? Poor preservation but must leave skull outward.	narrow and convex, re Grigson's characteristic shape for indicus (Grigson 1976:126, b)	
can't assess	can't assess	can't assess	broad and flat/U shaped, as Grigson 1976:126, a.	

Appendix 6

Skull/ Lot Number	Surface Weathering	Burning	Frontal Eminence Present	Occipital Condyle Present	Nasals Present	Dentition Present (Y = L+ R; L = Left; R = Right)	Breakage between Basal Tubercles and Palatine; Direction of Shift from Frontal Aspect	Thermally-Altered RocksEmbedded near Palatine
2	3	none	no	no	no	LHS	can't assess	can't assess
3	3	none	no	no	no	LHS + RHS	yes; anterior skull shifted left	no
4	3	none	no	no	yes, RHS	LHS	can't assess	can't assess
5	3	none	no	no	no	LHS/RHS	yes; anterior skull shifted right	can't assess
11	3	none	no	no	yes	LHS/RHS	yes; anterior skull shifted right	yes
12	3 and 4 around orbits	none	no	part	no	LHS/RHS	no	can't assess
13	3 and 4 around zygomatics	none	no	no	no	LHS/RHS	can't assess	can't assess
14	3	none	no	no	no	LHS/RHS	yes; anterior skull shifted right	yes
15	3	none	no	no	no	no	can't assess	can't assess
16	3	none	no	no	yes	LHS/RHS	yes; anterior skull shifted left	no
17	3	none	yes	yes (part)	yes	LHS/RHS	yes	no

Cut Marks on Wings of Palate	Cut Marks on Frontal/Nasals	Comments on Condition	Which Tooth Sampled for UCL
can't assess	can't assess	Posterior part of skull, behind palate, fallen apart. No palate, no premaxilla, no nasals; only maxillae present, but rest in highly fragmented state.	RHS M3
yes	can't assess	Breakage of post-palate wing midway between basioccipital and palatines, leading to slight shifting between anterior and posterior part of skull and some forward movement of frontals over lacrimals.	none
can't assess	can't assess	Just an LHS maxillary tooth row surviving intact (minus P2) in part of maxilla, with rest in fragments. Bone surface shows leaching.	LHS P2?
no	can't assess	Palates, maxillae, and tooth rows very well preserved; all teeth present; only RHS P2 missing (sampled?). Frontals very crushed, but there appears to be some forward movement of frontals over lacrimals (which are crushed and missing), indicative of breakage between anterior and posterior parts of skull.	RHS P2?
no	no	Both sides of dentition present (except LHS M2, probably sampled), with RHS premaxilla present and basioccipital but not occipitals; can see into cranial cavity. Anterior skull broken from posterior skull across post-palatine wings, across the zygomatics and orbits, leading frontals to shift forward about 1 cm. Two stone pieces lodged beneath LHS orbit.	LHS M2?
no	can't assess	One of best-preserved posterior skulls. Has parts of basioccipital present but still no frontal eminence, so can't assess profile or intercorneal ridge. Premaxilla and nasals not present. Dentition: RHS complete; LHS has only M1, M2, and M3. (Others may be sampled?) Notably no slumping or squashing. Is this an older skull, hence more fused?	LHS P2, P3, P4?
can't assess	can't assess	Frontals present but no intercorneal eminence, and orbits are missing, but some basioccipital present. Both LHS and RHS dentitions present. Some hint of asymmetry between anterior and posterior of skull but difficult to see. Not much slumping apparent but nasals (LHS) slightly pushed below frontal, indicating some slumping.	RHS M1
no	can't assess	Basioccipital to premaxilla is present, but frontal eminence missing, as are nasals. Both LHS and RHS dentitions present, with P2s missing both sides. It's notable that there is no burning, even though fire-cracked stone is wedged between RHS basioccipital and wing of palate. The stone is wedged deeply here, seemingly intentionally, and just where skulls are normally broken. There is slumping of this skull: the LHS orbit overhangs lacrimals and zygomatic.	LHS or RHS P2?
can't assess	can't assess	Very poorly preserved, just a 15 cm lump of soil matrix with some frontal fragments adhering and some parts of internal skull. No dentition and not much else visible.	assume none
no	no	Frontals present but no intercorneal eminence and no basioccipitals/occipitals. Nasals present both LHS and RHS; dentition present both LHS and RHS (but RHS premolars fallen). Appears to have had much post-depositional movements: nasals sunk between maxillae; frontals shifted over nasals; frontals overhanging lacrimals, which then overhang maxillae (on both sides). Seems related to breakage between anterior and posterior skull and state of (un)fusion.	RHS M2?
can't assess	no	Preservation relatively good, with basioccipitals, part of frontal eminence present, and posterior parts of nasals. Horncore direction can be assessed from RHS horncore base. Both LHS and RHS dentitions present. Frontals have shifted forward over nasals, and both are separated from the maxillae, with a wide gap between all sutures.	LHS M3

Appendix 6 *(continued)*

Skull/ Lot Number	Surface Weathering	Burning	Frontal Eminence Present	Occipital Condyle Present	Nasals Present	Dentition Present (Y = L+ R; L = Left; R = Right)	Breakage between Basal Tubercles and Palatine; Direction of Shift from Frontal Aspect	Thermally-Altered RocksEmbedded near Palatine
18	3	none	yes	no	yes	LHS/RHS	yes; anterior shifted right	no
19	3	none	no	no	no	yes, but many teeth missing	can't assess	can't assess
20	4 around posterior frontal and orbit; 3 anterior frontal and maxilla	none	no	no	no	LHS/RHS	yes; suggests anterior shifted left	yes
21	3 and 4	none	no	yes (part)	no	LHS/RHS, but many teeth missing	can't assess	no
22	4 highest and 3 lowest	none	yes	no	no	LHS/RHS, but many teeth missing	yes	no
23	4 around horncore bases; 3 elsewhere	none	partially	no	yes	LHS/RHS	none apparent	no
24	3	none	no	no	no	LHS/RHS	yes; anterior shifted right	no
25	4 on frontals; 3 elsewhere	none	yes (small part)	yes (part)	yes	LHS/RHS	yes; anterior shifted right	no
26	3	none	no	no	no	RHS but teeth missing	can't assess	can't assess
27	4 on parts of nasals and anterior of maxillae; 3 elsewhere	none	no	no	yes	LHS/RHS (some fallen)	yes; anterior shifted right; inferred, but see comment	can't assess

Cut Marks on Wings of Palate	Cut Marks on Frontal/Nasals	Comments on Condition	Which Tooth Sampled for UCL
no	no	Relatively good preservation, with some basioccipitals present and both nasals; can assess frontal and sagittal profiles; both LHS and RHS dentitions present but with some missing premolars. Breakage between anterior/ posterior of skull, across palatine wings, with shifting forward of frontals over lacrimals and both buckling under and over nasals. Slightly more overhang on RHS than LHS because of angle of break/slump.	RHS M3?
can't assess	can't assess	Very poor survival; just frontals and some anterior skull intact, including maxillae, but many teeth on both sides have been lost, and no skull parts anterior of maxillae are present. Maybe poorly surviving because appears younger?	?
can't assess	can't assess	RHS of skull better preserved; orbit present but rim broken off; no occipitals. Whole seems leached and highly fragmented. Posterior skull pushed forward over anterior; orbit broken and pushed over lachymals. Seems that posterior skull collapsed forward. There is fire-cracked rock stuck into LHS frontal, above orbit.	LHS M3?
can't assess	can't assess	Highly fragmented; RHS tooth row better preserved but LHS all fallen; maxillae fragmented. Some parts of frontal visible. Much leaching of bone. Frontals shift forward, more on RHS than LHS, and overhang maxilla on RHS by about 2 cm. Parts of occipitals preserved.	LHS M3?
no	can't assess	Highly fragmented, especially LHS and anterior of maxillae. The posterior part of palate hasn't survived, but this is the only skull where the intercorneal ridge survives. Teeth on both sides have fallen. Nasals are missing. Bone surface is leached; weathering high in places. Frontals have shifted forward over lacrimals and maxillae a few centimeters.	LHS M3?
no	no	Fair condition; most of skull length present but premaxillae absent and maxilla broken at anterior end; occipitals missing; can see traced shape of base of RHS horncore in soil. Maxillae and nasals mostly survive, as do lacrimals. Tooth rows complete and in good condition (observation made that if skulls were heated/burned, we may expect teeth to show cracking, which they don't). Front of skull seems relatively in place, with not much anterior shifting, although frontals have shifted over nasals slightly, in symmetrical fashion.	LHS M3
no	can't assess	Bone surface quite leached; preservation patchy, with fragments lost, but there are hints of the base of RHS horncore. Breakage between anterior and posterior of skull, with palatine wings broken and shift of 2-3cm between anterior and posterior parts. Frontals have separated by 2 cm along their fusion line but stay parallel. Did this occur though wetting and drying, with expansion and contraction pulling them apart? Frontals were unfused anyway. Frontals (especially LHS, because of slumping being more on this side) also overhang maxillae, so overlapping lacrimals too.	RHS M3?
no	no	This skull preserved well enough to take intercorneal ridge morphology and sagittal profile. Premaxilla doesn't survive but occipital is still present (but without complete condyles); tooth rows complete. Note dental anomalies on both LHS and RHS M1.	LHS M3
can't assess	can't assess	Very little surviving intact except RHS molar row, with premolars missing; part of palatine is in place, but otherwise soil holding tooth row in place.	?
can't assess	no	Very fragmentary. Nothing survives posterior of palate and forward of maxilla. LHS has full tooth row, but P2 and P3 are fallen (modern breaks); RHS all teeth fallen. Only nasals and maxillae present, with nasals sunk under maxillae, left more so than right, where there's a 1 cm gap, indicating shifting forward of frontals (not present) over anterior part of skull and with that some asymmetry, probably with anterior skull broken from posterior and shifting right (from inference).	RHS M3

Appendix 6 *(continued)*

Skull/ Lot Number	Surface Weathering	Burning	Frontal Eminence Present	Occipital Condyle Present	Nasals Present	Dentition Present (Y = L+ R; L = Left; R = Right)	Breakage between Basal Tubercles and Palatine; Direction of Shift from Frontal Aspect	Thermally-Altered RocksEmbedded near Palatine
28	4 on basioccipitals; 3 elsewhere	none	no	no	no	LHS/RHS	yes; anterior shifted right	no
29	3	none	no	no	no	LHS/RHS	yes; anterior shifted left	no
30	3	none	yes	no	RHS present	LHS/RHS	not visible but probable (see comment)	no
31	3 and 4 in places	none	no	no	no	LHS/RHS (some missing)	yes; anterior shifted right	no
32	3	none	no	no	no	LHS (M2 missing); RHS (only M1, M2, M3 present)	yes, anterior shifted right	no.
33	3	none	no	no	yes, LHS/ RHS	LHS/RHS	not visible but likely (see comment)	no

Cut Marks on Wings of Palate	Cut Marks on Frontal/Nasals	Comments on Condition	Which Tooth Sampled for UCL
no	can't assess	This skull has more of the basioccipital region present than most and is in good condition, but back of frontals and top of occipitals are gone, so can't assess frontal eminence of sagittal profile. Occipitals are clear in this specimen, particularly ventral (underside). Posterior of skull broken from anterior on a slight angle, with a break showing on RHS palatine wing. Posterior skull has shifted forward. For example, zygomatics overhang maxillae and lacrimals by about 1 cm on each side.	LHS M3
no	can't assess	Full set of teeth on RHS. LHS has only P2 missing (maybe sampled?). Premaxilla is present but covered in soil. Basioccipital present but occipital condyles are not; can see foramen mangnum even though flattened off. Whole of occipitals seem shaved off vertically, so we see a cross section of the back of the skull, including shape of horncore bases as they leave the skull. Breakage between anterior and posterior of skull has led to shift forward of posterior skull at an angle; can see RHS maxilla and zygomatics, but they are buried deeply in deposit on LHS. This results in RHS frontal and orbit overhanging maxilla by about 4 cm (hiding lacrimals). The angle shift of the front of the skull seems to have "twisted off" the nasals.	LHS P2?
can't assess	no	Basioccipital present but only fragments of occipital; can see intercorneal ridge, and although sagittal profile is not complete, it can be estimated. Frontals have shifted over maxillae, leaving an overhang. Zygomatics are broken, probably due to this shifting. RHS nasal is present but ruckered at the suture with the frontals. RHS of skull more squashed than LHS. Can assess small part of horncore shape at base.	RHS M3
no	can't assess	Leaching on surface of bone; teeth have gritty deposit on occlusal surface. Parts of basioccipitals present but seemingly little else of the occipital area. Palatines and maxillae survive, but everything anterior of the maxillae (teeth ara) is broken. LHS tooth row is complete but P2 fallen; RHS: M3 presumably taken as sample. P4, M1, and M2 present. (P2, P3, P4 fallen but would have been there originally.)	RHS M3
Yes; small notches on wings of palate, not very clear/sharp.	can't assess	From occipital view, very little visible surviving. Posterior of skull shifted forward over anterior and large overhang must have been present because much soil fills the "overhang" (estimated shift of 2-3 cm). Zygomatics are present, shifted over RHS and LHS maxillae. Posterior to zygomatics, frontals have fragmented a lot, showing mainly the soil within the cranium; bone seems to have broken off.	not clear
can't assess	Yes; there are cut marks on the frontals, near their meeting point with the nasals and lacrimals. LHS: c5 very fine parallel cut marks, 1–1.5 cm in length, with other light traces of similar cuts adjacent; RHS: 3 deeper also parallel cuts, about 1 cm in length. Both sets of cuts are distinct, separate from each other (although can't see if nasals also had cuts, since they are pushed beneath frontals). See figures 11.39, 11.40	Fair condition. Skull is present until midway along diasterma. RHS premaxilla is present but fragmented (modern breaks), indicating that it would have all been present originally. Posterior skull does not survive—nothing of occipitals/basioccipitals. Teeth all present (although LHS P2 fallen and RHS M3 sampled). Frontals (fused) have moved forward/anterior over nasals, lacrimals, and maxillae by about 3 cms, but don't appear to be "twisted" between anterior and posterior of the skull as some are. Nasals appear pushed together.	RHS M3

Appendix 6 *(continued)*

Skull/ Lot Number	Surface Weathering	Burning	Frontal Eminence Present	Occipital Condyle Present	Nasals Present	Dentition Present (Y = L+ R; L = Left; R = Right)	Breakage between Basal Tubercles and Palatine; Direction of Shift from Frontal Aspect	Thermally-Altered RocksEmbedded near Palatine
34	3	none	no	no	yes, LHS/ RHS	LHS/RHS (some fallen)	no	no
35	3	none	mandible; not skull					
36	3	none	yes	yes	yes, part of LHS	LHS/RHS	yes; anterior skull shifted right	no
37	3	none	no	no	yes, LHS and part of RHS	LHS (M2/M3 fallen); rest missing; RHS present	Yes; not clear which direction, since there isn't angle difference between anterior and posterior skull. Note in comments that palatine wing area is not broken. Therefore forward slumping still possible if palatine wings are intact.	no

Cut Marks on Wings of Palate	Cut Marks on Frontal/Nasals	Comments on Condition	Which Tooth Sampled for UCL
no	no	This skull has less leaching than some. There is no premaxilla; no occipitals present; all modern breaks. Tooth row LHS is complete; RHS M3 taken for sample(?), with only M1 and M2 remaining. Has a dental anomaly: LHS P3 rotated. No evidence of skull being at different angles between anterior and posterior, and little collapse forward is evident, except that zygomatics are pushed forward slightly (about 1 cm), especially on RHS but not on lacrimals or nasals.	RHS M3
		This is a mandible fragment, not a skull.	
no	no	Skull has complete occipital condyles (showing that even spongy bone preserves—maybe this skull was buried more deeply than most?) and enough of the occipitals that we can see the base of the horncores (but no horncore circumference) and to the sagittal profile. RHS teeth complete but whole tooth row fallen. Occipitals very fragmented. Anterior part of skull gone; even maxillae fragmented. Anterior and posterior parts of skull broken apart at different angles, with anterior shifted right and more buckled up. There is some shift of frontals forward, but they don't overhang much. Frontals are generally well preserved.	LHS M3
Yes; series of five small cuts, 3-4 mm long, sharp, as if made with a stone tool, and in parallel lines, on RHS basioccipital area, lateral side of palatine wings. Function? Are these for tongue removal? Separation of mandible from skull? Decapitation would not leave marks here. These cuts are noted as "notches" as if something cut on them—supporting the idea of tongue removal?	no	Teeth: LHS has complete row, P2–M3; RHS M2, M3 fallen and rest missing (maybe one was sampled?). Good length of skull but very little survives around occipitals. Can see into cranial cavity. RHS maxilla badly broken. Part of frontal may have adhered to Skull 36 adjacent, since that had extra frontal fragments stuck to occipital condyle area. This skull is unusual in that palatine wings are not broken, but there is still forward movement of frontals over nasals and lacrimals by 3–4 cm.	?

Appendix 6 *(continued)*

Skull/ Lot Number	Surface Weathering	Burning	Frontal Eminence Present	Occipital Condyle Present	Nasals Present	Dentition Present (Y = L+ R; L = Left; R = Right)	Breakage between Basal Tubercles and Palatine; Direction of Shift from Frontal Aspect	Thermally-Altered RocksEmbedded near Palatine
38	3	none	no	no	no	yes, LHS/RHS	can't assess	no
39	3	none	no	no	yes, LHS	yes, LHS/RHS	no	no
40	3	none	no	no	yes, LHS/ RHS	RHS alveoli only	can't assess	can't assess
41	3	none	no	no	no	yes, LHS/RHS	yes; can't tell angle of shift	no
42	3	none	no	no	yes	LHS/RHS (LHS has P2/ P3 missing)	can't assess	can't assess

Cut Marks on Wings of Palate	Cut Marks on Frontal/Nasals	Comments on Condition	Which Tooth Sampled for UCL
no	can't assess	Present from basioccipital to palate/maxilla, just at point of diastema. All teeth present except RHS P2 (maybe taken for sample?); P3 fallen but present. Otherwise, dentition in good condition, although encrusted with grit. Bone has leaching on surface. Note: strange post-palate-area asymmetry (morphological, not relating to breakage) and rotation of P4 on both LHS and RHS.	P2 RHS?
no	no	Whole length of skull present (roughly), but premaxillae are broken off and whole posterior skull area very fragmentary. So basioccipital present but occipitals crumbled away. No horncores. Teeth in excellent condition—RHS M3 removed for sample. That posterior skull is much more fragmentary indicates that it was exposed, whereas the anterior is not. Bone surface pitted; doesn't seem to be root etching but there is leaching. Frontals overlap nasals through slippage, and nasals are pushed back into frontals, but frontals, zygomatic, and lacrimals are "flush," not collapsed on an angle. Field/lab observation: this skull seems narrow; tooth rows appear closer together than on some skulls; teeth themselves more gracile. Curiously, metrical analysis doesn't match this observation. It is interesting that this is noted as the "longest" skull and is in the center of the circle.	RHS M3
can't assess	not on nasals; can't assess frontals	Highly fragmentary. Only maxilla parts survive; both LHS and RHS and parts of palatines. RHS maxilla has some alveoli, seemingly of P4, M1, and M2, but no dentition. There is nothing surviving posterior of the maxillae. Surface shows erosion that may be root etching or normal exposure weathering.	none?
no	no	Condition fairly good (has evidence of field conservation). From occipital to diastema present, although very fragmented around posterior end of skull. No horncores but there are hints of horncore direction; can't take skull profile. Has full sets of teeth, except RHS M3 was taken for sample, and LHS P2 seems to have been absent in life. Frontals are in place but have shifted anterior over the lacrimals/maxillae, and probably this movement broke the zygomatics. Hence the orbits seem too far forward and there is an "overreach" between the frontals and maxilla. This movement probably pushed off the nasals, which are missing.	M3 RHS
can't assess	no	Mainly tooth rows and maxillae held together. Nasals are fallen and fragmented but present. Nothing survives anterior of the palate/maxilla and some fragments of premaxilla. There are also some fragments of the posterior part of the skull but no clear bone surfaces. Tooth rows in good condition (complete, except for LHS P2 and P3 are missing—maybe one taken as sample?). Lots of small fragments collected; probably represent broken skull.	LHS P2/ P3?

Chapter 12

Neolithic Stone Platforms

Survey and Excavations

Joy McCorriston

After the day's notes were done, penned under a raw bulb shedding singed insects onto the pages, we could watch the night sky. In Wādī Sanā, I learned to tell time by the stars, judging just where Al-Thurayyā (the Pleiades) and its neighbors had shifted over a midwinter evening. The generator had ceased throbbing and our desert camp was quiet, except for the crackle of voices at Saʿid Al-ʿAlīy's stone circle. As our camp guard, he stayed close, and his family brought him food and settled into his windbreak, a ring of small boulders on the gravel terrace beyond our cook tent, from which he provisioned a steady stream of visitors. On moonlit nights we had no stars, but I could hear the footfalls of bedouin children and watch their shadows cross my tent wall. They came and went all night long, ferrying milk, bowls, and dead sticks to the men seated around the fire in Saʿid's stone circle.

Survey of Neolithic Platforms

Survey revealed a concentration of Neolithic platforms in the Khuzma as-Shumlya vicinity (figure 12.1). We excavated a number of stone circles and subcircular stone structures. They finished as platforms but contain remnants of hearths and occupational debris in their initial stages. Chapters 10 and 11 report on excavations at the Kheshiya cattle ring site and subsequent analysis of cattle bone, which revealed an exceptionally well-preserved association of a Neolithic stone monument with the remains of sacrificed animals. As table 12.1 demonstrates, this excavated example is but one of the many stone platforms documented through comprehensive survey and landform taphonomic analysis in Wādī Sanā. Survey results show that Neolithic platforms all lie at relatively low elevations near to the modern wadi channel and are situated either on (or eroded from) Early Holocene silt terraces or on nearby lower-elevation gravel or bedrock terraces. Excavations show that they ended as platforms but began as stone rings, perhaps much like Saʿid's home space in our camp.

The association of animal bone with Neolithic platforms is reasonably strong (McCorriston et al. 2012). On silt terraces near the telltale limestone blocks and concentrations of boulders accumulated by human hands, one finds chips of animal bone, charcoal flecks, much thermally altered rock, chert flakes, and sometimes obvious accumulations of animal bone embedded in and eroding from silt (table 12.2). While the bone collected by RASA survey reflects a nonsystematic strategy and extremely small numbers of identifiable specimens, the identifications nonetheless indicate the presence and utilization of a range of other animals besides the cattle so evident at Kheshiya.

Excavations at Khuzma Platform SU37-3

Beyond the results from survey, at the southwest corner of Khuzma as-Shumlya, Neolithic monument SU37-3 is one of a constellation of stone structures embedded in the upper layer of the silt terrace and near to several stone platforms on the lower bedrock and gravel terraces.

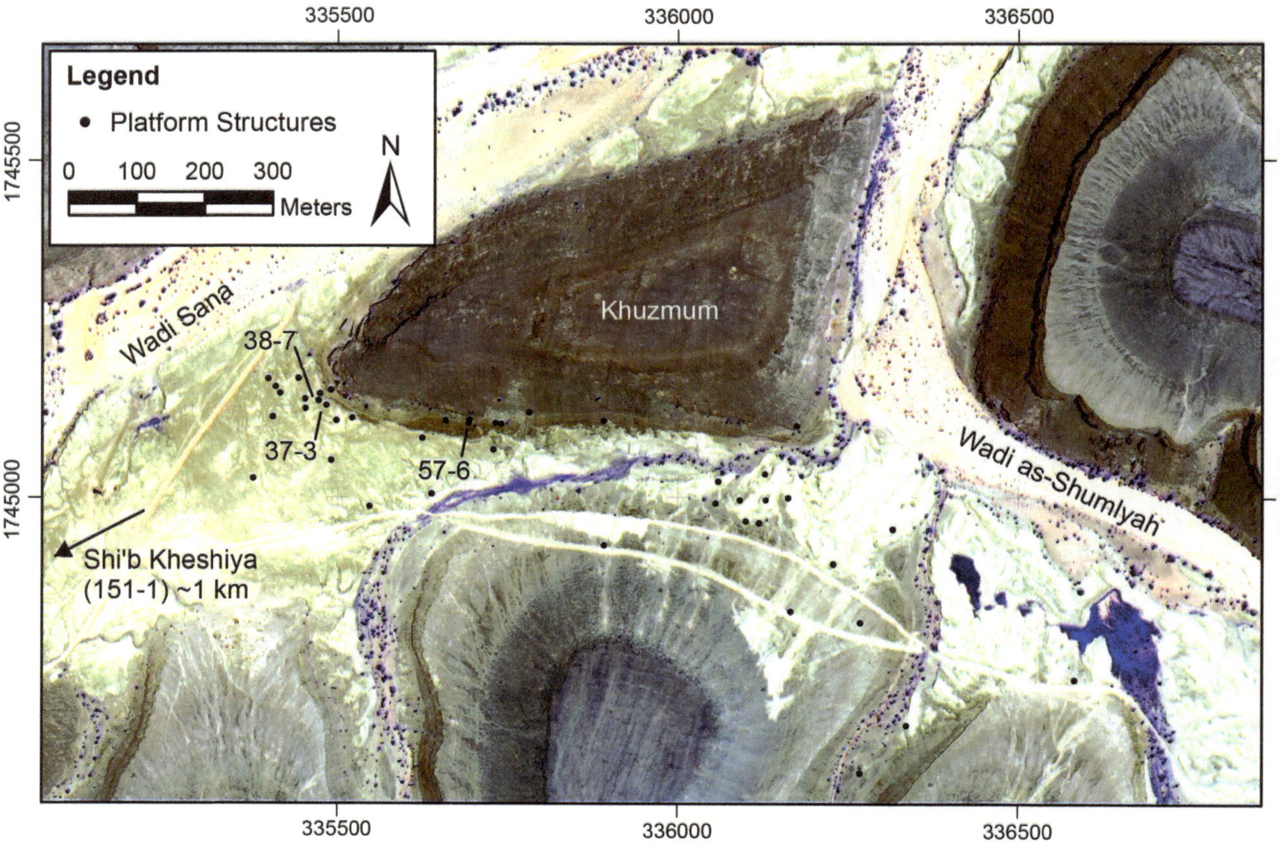

Figure 12.1. Map of Neolithic platforms surrounding Khuzma as-Shumlya. For coordinates, see table 12.1. *Illustration by Michael Harrower.*

Discovery

On an exploratory survey in 1996, Burkhard Vogt of the German Archaeological Institute (DAI) Survey guided Joy McCorriston to Wādī as-Shumlyah; the region had been explored in the (unpublished) pipeline archaeological survey conducted by DAI for CANOXY-Yemen. Through mapping and intensive survey (chapters 4 and 6), the RASA team examined in greater detail the stone structures Vogt had included in a generalized description of a "Hadramawt Megalithic Complex." In the course of RASA survey, we generated the sketches and scale drawings of structural detail and surface preservation that ultimately led to a broader understanding of taphonomy and preservation on different landforms (chapter 6). Intensive survey also confirmed a casual observation that the spatial concentration of structures is particularly dense at this corner of the Khuzma as-Shumlya.

SU37-3 is well defined by a surface perimeter of upright limestone blocks. The structure was almost entirely exposed by sediment erosion at its northeast corner, while in the southeast of the structure, a layer of light yellow–brown silty sediment covered the perimeter and a small area of interior fill. Overlying the entire structure as a top layer of the rubble fill and well contained within the protrusive limestone slab perimeter was a concentration of thermally altered clastic limestone pebbles (figure 12.2). A standing stone with a worked top still protrudes from the unexcavated sediments 1.5 m to the east of SU37-3.

Methods

In the 2000 season, a small excavation tested the possibility that these were occupational structures rather than tombs, as had been previously suggested. We chose SU37-3 for testing because its plan showed on the surface, it stood in relative isolation from other structures promising a simpler chronological sequence, and the clear surface plan of intact uprights suggested good preservation of underlying deposits (figure 12.3).

In 2000 the RASA team opened a test excavation of 1 x 3 m using the stratigraphic excavation methods, sieving, and sampling for charred plant remains and

Figure 12.2. Neolithic platform SU37-3 before excavation in 2000. *Photograph by Joy McCorriston.*

organics for radiocarbon dating as described in chapter 10. In 2005 Nisha Patel, Ghufrān Aḥmad, and Margaret Wilson expanded this excavation to remove most of the interior deposits and to test exterior deposits abutting the structure's upright slabs. Finally, in 2008, a Nexen contractor, Arrow Geophysics, had a brief opportunity to conduct a limited ground-penetrating radar study of a 10 x 10 m exterior surface immediately southwest of the perimeter. This study explored the possibility of another buried bone deposit (without conclusive results tested by any further excavation).

For excavations in 2000, a brass screw referred to as Main 37-3 Khuzma Datum was cemented to the bedrock approximately 25 m northeast of 37-3 at E 335498.83, N 1745159.77 (UTM Zone 39 North WGS84), elevation 689.17 m MSL (EGM96) as measured on March 8, 2000, by a Trimble Pathfinder Pro XRS GPS using Omnistar real-time correction (elevation 670.03 m HAE). For excavations in 2005, the same brass screw was found still cemented to the bedrock and measured as E 335497.41, N 1745161.87 (UTM Zone 39 North WGS84) elevation 695.72m MSL (EGM96) on 4 February 2005 by the same Trimble XRS Pro GPS using Omnistar real-time correction (elevation 676.58 m HAE). Given the expected submeter accuracy of our GPS configuration, we cannot explain the 6.55 m vertical difference between elevation measurements in 2000 and 2005, but it may relate to U.S. selective availability scrambling of GPS signals, which was deactivated on May 1, 2000. Importantly, elevation readings reported in this chapter are based on the MSL measurement taken on February 4, 2005.

Results

This monument was initially published as the stratigraphy inside a habitation (McCorriston et al. 2002), and it was only after further excavation in 2005 that its complex life history of was apparent, including its habitation, multiple abandonments, deliberate filling, and possible revisits. What follows is a description of each excavation quadrat, any extensions, and the stratigraphy of the stone monument.

Table 12.1. Neolithic monument locations in Wādī Sanā (coordinates in UTM Zone 39 North WGS1984; elevation in MSL EGM96).

Survey Unit #	Site Number	Preservation	Northing	Easting	Elevation	Description	In GIS
003	x	medium	1744186	337293	701	Circular-oval low platform of irregular limestone slabs, about 7 m diameter, possible platform structure on scree slope; many of the large limestone slabs have been reused in probably more recent camp circles nearby.	x
C00	1	medium	1745156	335474	699	Drop-shaped platform of upright limestone slabs filled with cobbles; recorded in the 2005 cairn survey.	x
C10	12	good	1755541	376271	536	Teardrop-shaped platform of upright limestone slabs; recorded in the 2005 cairn survey.	x
C10	x	medium	1755542	376279	533	Ring of upright slabs without residual fills; recorded in the 2005 cairn survey.	x
C11	x	medium	1752757	342988	656	Stone ring filled to make a platform; recorded in the 2005 cairn survey	x
C11	x	medium	1752348	342895	662	Platform; recorded in the 2005 cairn survey.	x
015	x	poor, deflated	1744733	336583	698	Circular mounded feature with four flat-lying limestone slabs and one upright. (*Mounded* in this instance means that stone has resisted erosion of nearby silts, leaving the feature higher than the surrounding terrace surface.) No GPS point was taken on this structure, so this coordinate uses a corner of the survey unit.	x
016	1	poor, deflated	1744664	336337	686	Possible platform structure, heavily eroded, but exhibits characteristics of better-preserved nearby structures: circular, about 4 m diameter, 0.4 m in height, slabs covering nondeflated silt mound. May have been source of stone reused in nearby rock piles/campsites.	x
017	1	poor, deflated	1744903	336228	695	Distribution 8.2 x 3.6 m of limestone slabs, medium and small angular boulders atop a prominent mound of uneroded silt. Stones have slid down especially to the north. Animal bone eroding out of the surface from sediments protected under the slab/structural cap.	x
017	8	poor, eroded	1745000	336163	689	Circular distribution; 12 x 10 m of large, medium, and small boulders on and around a 1–1.5 m high mound of uneroded silt. This feature resembles better-preserved structures to the west and suggests that as the elevation of uneroded silt increases through deflation of surrounding silts, the limestone slabs and platform fill slid downslope to ring a mound of silt.	x
017	11	poor, protrusive	1744954	336316	692	Curvilinear partial structure with a tangent of 1.6 m preserved. Built of exterior limestone slabs, with most fallen away. Medium and small boulders and cobbles also used for fill, and the interior included pebble fill, perhaps as an original surface.	x
022	3	poor	1744594	336268	697	Large circular rock mound, about 2 m tall, 3 m north–south, 5 m east–west, probable reuse of all small and medium boulders to construct other stone rings nearby, but three large boulders are present; 1 boulder is 1.5 m in length.	
024	3	poor, deflated	1744966	336100	701	Circular 3.8 m scatter of limestone slabs, medium and small boulders, cobbles, and pebbles sliding down the slopes of a mound of silt. Limestone slabs include two uprights. Mound is relict from the protective effect of a stone structure as the silt terrace around it erodes/deflates.	x

Survey Unit #	Site Number	Preservation	Northing	Easting	Elevation	Description	In GIS
024	4	poor, eroded	1744997	336091	698	Circular distribution; 5.80 x 8.40 m of medium and small boulders on slopes of a mound of uneroded silt, probably formed by retention under a stone structure that eroded once the surrounding terrace surface had eroded. Spatial association with lithic scatter of thermally altered rock and oolitic chert flakes and chips.	x
024	6	poor, highly eroded surface	1744997	336129	689	One of many mounds in the immediate area. This 20.6 m x 12.0 m mound 2 m high has large angular pieces of boulder and only a few limestone slabs. Medium and small boulders and cobbles also present. Stones are concentrated on the northwest and west lower slopes (downstream) of a mound of uneroded silt, probably formed by retention under a stone structure that eroded once the surrounding terrace surface had eroded.	x
024	2	poor, eroded	1744964	336119	699	Mound of silt, heavily eroded on south side by an adjacent gully. Thought to have been once circular, the mound is presently elongate: 3.5 m east–west x 1.5 m north–south. Medium and small boulders are present along with cobbles, of which about 30 percent are limestone slabs eroding into the gully. The mound was probably formed by retention of the natural silt terrace under a stone structure that eroded once the surrounding terrace surface had eroded.	x
024	7	poor, eroded	1745035	336128	693	Equals SU025-1. Mound of silt 16 m in diameter at edge of gully in silt terrace. Large limestone slab boulders, medium and small boulders, and cobbles lie at the base of the mound on the gully side, where they have probably accumulated through erosion. The mound was probably formed by retention of the natural silt terrace under a stone structure that eroded once the surrounding terrace surface had eroded.	x
025	2	poor, highly eroded	1744992	336056	690	Ovoid; 14.0 m north–south x 10 m east–west; mound of silt heavily eroded on east side by an adjacent gully. The mound is bare on top but has a ring of eroded stones around its base. On the east (lower) side are flat limestone slab boulders, while smaller boulders and cobbles lie to the south, west, and north. The mound was probably formed by retention of the natural silt terrace under a stone structure that eroded once the surrounding terrace surface had eroded. Lithics of oolitic chert are spatially associated with the rubble.	x
026	1	poor, eroded	1745024	336059	689	Mound of silt; 16 m in diameter, about 5 m high, with large limestone slabs, medium and small boulders (may be fragmented from large slabs, flat-lying on sides of mound); top of mound is clear. Heavy concentrations of thermally altered rock (80–90 percent, except limestone slabs) also eroding downslope. Top of mound is clear of stones but is a burned surface capped by 0.2 m of silt in 3.6 m diameter ring. Large animal bone (not human) is eroding out of burned surface. The mound was probably formed by retention of the natural silt terrace under a stone structure that eroded once the surrounding terrace surface had eroded.	x

Table 12.1. Neolithic monument locations in Wādī Sanā (coordinates in UTM Zone 39 North WGS1984; elevation in MSL EGM96). (*continued*)

Survey Unit #	Site Number	Preservation	Northing	Easting	Elevation	Description	In GIS
033	10	poor, eroded	1745089	335624	697	Irregular 3.2 m diameter pile of medium and small boulders in spatial proximity to a stone ring and other stone concentrations. All boulders on the silt terrace were transported by human agents, and in this case, reuse and robbing have likely damaged the original structure(s).	x
033	6	good, 70%	1745071	335728	693	Platform structure 2.3 m in diameter, 0.44 m high. North side eroded into the drainage, buried in silt terrace. Constructed of large boulder-size standing stones (limestone slabs), creating an outer border filled with cobble and small boulder-size stones, silt. Limestone slab lies flat where eroded. Up to 15 chipped stone fragments scattered around on the present-day surface; also spatially associated with a scraper (15 m distant).	x
033	18	good, 60%	1745007	335638	694	Semicircular platform abruptly truncated by gully cut on its south side; 2.5 m east–west and once probably about 3 m north–south; upright limestone slabs with a cobble fill, 0.74 m high. On the east side, a second, outer margin of limestone slabs forms a double arc of uprights. Few poorly preserved bone fragments observed eroding from the cobble fill.	x
034	2	poor	1744930	335891	697	Platform of semicircular shape, with possible disturbance and deflation of what was the north side of a once complete subcircular structure. Built of limestone slab, upright small boulders with small boulder, cobble, and pebble fill. Existing structure 2.7 m east–west, preserved to a height of 0.3 m. The location on a bulldozed bedrock terrace offers poor association with other nearby monuments, such as a trilith (034-001) and pecked Old South Arabian letters or animal imagery on a limestone slab, possibly once part of the platform structure.	x
035	1	good	1744833	336166	697	Circular to subcircular platform structure constructed of limestone upright, medium boulder slabs with cobble and silt fill. Preserved to a height of 1.2 m. More than 50 percent appears intact, with likely removal or erosion of upright perimeter slabs. Existing structure is 2.9 m southwest–northeast; the largest upright is 0.8 m in height.	x
035	2	poor	1744817	336268	695	Large oval-shaped structure with boulder-size limestone slab uprights as exterior defining a 7.07 m roughly west–east perimeter, 3.30 m north–south and 0.8 m in height. Structure possibly once contained more that the upright slabs, some of which are flat-lying. Angular clastic cobbles piled against and upslope of the exterior of uprights may be due to slope erosion rather than relict of original construction. Oolitic chert lithics also occur upslope.	x
037	1	poor, heavily eroded	1744988	335547	693	Circular silt mound, 4.5 m north–south x 4.2 m east–west and 0.9 m high, capped by angular boulder and cobble fill residue. Flat-lying slabs eroding down the side of the mound suggest that a platform was constructed by medium to large boulder-size limestone slab uprights as the outer border filled with cobble. This construction is mounded, probably due to heavy erosion of the surrounding silt terrace while the stone structure retained underlying silts.	x

Survey Unit #	Site Number	Preservation	Northing	Easting	Elevation	Description	In GIS
037	2	poor, very eroded	1745119	335521	693	Sub-rectangular structure, 3 m west–east x 2.7 m north–south and 0.85 m high, constructed of upright large and medium boulder-size limestone slabs filled with small boulders. Cobble fill (platform) may be buried within silt terrace. Within the perimeter is some animal bone and a burned surface, relict from a widespread and now eroded surface of the terrace. Cobbles and bone have eroded with silt into the drainage immediately adjacent to the northeast. One obsidian flake was on the terrace outside the structure; a core and flake of brown chert were inside the slab upright perimeter.	x
037	3	good, 75%, but surrounding area is highly eroded	1745138	335483	696	Sub-rectangular, D-shaped platform; 4.2 m north–south x 4.1 m east–west and 0.82 m high; constructed with upright limestone, boulder-size slabs as a perimeter filled with medium and small boulders, cobbles, and pebbles. A standing limestone slab to the northeast has a pecked/worked upper edge (most of the slab is buried in silt). Animal bone detected on the surface outside the structure at the lowest (eroded) north corner, at the gully edge of the silt terrace.	x
037	4	poor, very heavily eroded	1745115	335497	699	Stone piles about 1.9 and 1.3 m in diameter, composed of medium and small boulders and cobbles on a heavily eroded silt terrace. At one time there were probably about three stone platforms. Some mounding of the silt terrace is evident where structures may have stood. Erosion (and possibly robbing) has resulted in cobbles and small boulders now scattered in lower areas between the silt mounds.	x
037	6	poor, eroded	1745056	335490	699	Circular cluster, 3.2 m in diameter, of medium and small boulders with a few limestone slabs. The stone must have been transported by humans to the middle of a silt terrace. The original structure has suffered from erosion and stone-robbing.	x
038	7	poor, eroded	1745145	335471	699	Circular/sub-rectangular 5.8 m diameter platform ringed by medium and small boulders, filled with cobbles, pebbles, and small boulders. Structure abuts other eroded features, including small to large boulder-size alignments of upright limestone slabs to the northwest and a possible terrace or platform on the north side. Located 20 m from the base of Khuzma as-Shumlya, this complex may be related temporally to other nearby platform structures similarly eroding from early Holocene silts.	x
038	8	intact	1745179	335442	689	Originally recorded as a cairn, this is a circular platform that may have some reuse (hearth, *madhbaḥ*?) or accretive construction. The original structure is a 2.10 x 2 m platform of upright slabs, bifurcated by an alignment east–west of four upright slabs. A smaller platform on top is 2.10 m north–south x 1.1 m east–west.	x

Table 12.1. Neolithic monument locations in Wādī Sanā (coordinates in UTM Zone 39 North WGS1984; elevation in MSL EGM96). (*continued*)

Survey Unit #	Site Number	Preservation	Northing	Easting	Elevation	Description	In GIS
038	10	poor, eroded	1745148	335452	689	A 27 m north–south complex of multiple structures, possibly all platforms standing 0.8 m high. Rockfall and erosion from the silt terrace makes the original plans somewhat unclear. There appear to be two subcircular limestone slab, upright, ringed platforms to the north, 2.9 m diameter and 1.4 m diameter. Aligned farther south are two likely trapezoidal/sub-rectangular apsoidal platforms, 5.0 m and 3.2 m length, each with a small 2 m circle outlined with angular small–medium boulders and filled with pebbles and small boulders. Adjacent to each at the west is a rock pile of closely placed flat slabs, which could be interpreted as remnants of iconoclasm (destroyed standing stones from the smaller stone circles before the platforms). This complex probably represents three, not one, platforms, possibly with standing stones once associated with each one. A nearby hearth may be associated.	x
038	11	poor, eroded	1745134	335452	690	A circular 7.5 m diameter mound of silt with flat-lying slabs at its basal perimeter. A few small and medium-size angular boulders are on top. A hearth at the northwest perimeter may be a later feature. The mound was probably formed by retention of the natural silt terrace under a stone structure that eroded once the surrounding terrace surface had eroded.	x
038	12	good	1745121	335404	692	A 9.2 m north–south x 5.8 m east–west, 0.70 m high stone platform constructed of upright limestone slabs around the perimeter with medium and small boulder-size slabs, cobbles, and pebbles in the silted fill.	x
038	13	good	1745029	335375	692	A subcircular 1.2 m north–south x 0.8 m east–west pile of angular small boulder-size stones, one course high, with a single rectangular standing stone 0.40 m high embedded in silt at 0.53 m east of the structure. A very patinated chert flake found on surface 0.6 m east of the standing stone; much thermally altered rock on surface.	x
039	1	good	1745156	335415	687	A roughly 8 m deflated/eroded distribution of small and medium boulder-size stones, best preserved as a 3.2 m diameter section. Much of the top central portion is "hollow," although there is a 0.8 m height to the stone pile overall. Flat limestone slabs of small boulder size lie on the surface to the east; to the west there were likely also such slabs, which now lie in the nearby wash bottom. This stone pile resembles other deflated structures represented as mounds of silt.	x
039	2	poor	1745165	335408	687	A linear emplacement of two parallel lines of upright limestone slabs, 1 m apart, aligned roughly north–south with one standing stone, worked by pecking, to the northeast. The east alignment is three slabs. The west alignment is two slabs. An interpretation is that these alignments are the remnants of a platformed structure with standing stone (always worked) in front.	x

Survey Unit #	Site Number	Preservation	Northing	Easting	Elevation	Description	In GIS
039	3	poor, very heavy erosion from goat traffic	1745177	335398	689	Platform of ovoid shape, likely constructed with multiple additions. The overall dimensions are about 7 m north–northwest/south–southeast x 2.7 m wide and 0.8 m high, constructed with upright limestone, boulder-size slabs as a perimeter to the east. An original platform or construction filled with medium and small boulders, cobbles, and pebbles may have been added (to the east), with additional alignments of limestone uprights filled with stone: there are four parallel alignments of slabs within the current structure. A single limestone slab to the east may be a standing stone. Bone was observed on the structure's surface, and there are three hearths to the east.	x
052	1	poor, deflated	1743943	338172	703	Mound of silt, 4.8 m north–south by 4.5 m east–west, 1.6 m height, capped by few medium and small boulders, some as flat-lying slabs, and pebbles. This mounded feature was likely formed by the retentive effect of a stone platform that prevented erosion of underlying silts. The original structure has been very badly deflated and its condition is poor. Within the silt mound, traces of two burned layers and/or hearths were noted.	x
052	2	poor, deflated	1743936	338164	701	Very small mound of silt, less than 4 m diameter, capped by several medium-size boulders. Probably formed by the retentive effect of a stone platform that prevented erosion of underlying silts. The original structure was not preserved, and its reconstruction as a platform is conjecture based on its current condition, proximity very near to 052-001, and better-preserved examples. In this location, the remnant silt mound sits directly on bedrock terrace and is the last vestige of the Holocene silt terraces.	x
057	2	good	1745161	335490	691	Trapezoidal/sub-rectangular plan platform constructed like 038-7, 038-10, and 037-3 nearby. Outer perimeter defined by upright limestone, medium and small boulder-size slabs filled with two trisecting east–west regular lines of upright limestone slabs. (It is possible that this platform was constructed in stages, with the central one-third constructed first). East–west dimension at south end 2.1 m; at north end 1.6 m; 3 m in length north–south. Cobble limestone clasts are in fill, with pebbles and silt. Compares well to trapezoidal monuments elsewhere, such as D038-3 (Hanun, Dhufar) (McCorriston et al. 2014) .	x
057	5	poor, eroded	1745115	335658	694	Upright limestone slab as stela standing 0.97 m in height downslope of a semicircular ring of upright limestone, medium boulder-size slabs, with a total estimated tangent of 2.3 m (probably was 2.5 m in diameter). This is a remnant of what was probably three limestone slab–faced stone platforms 20 m apart in an east–west line along the south slope of Khuzma as-Shumlya (057-5, 057-8; 057-6, 057-7). The stela uprights were south of the platform construction in each case. The limestone slab source appears to be a ridge of tabular limestone about 5 m away.	x
057	6	good	1745111	335692	694	Stela before a platform. See SU057-8.	x

Table 12.1. Neolithic monument locations in Wādī Sanā (coordinates in UTM Zone 39 North WGS1984; elevation in MSL EGM96). (*continued*)

Survey Unit #	Site Number	Preservation	Northing	Easting	Elevation	Description	In GIS
057	7	good	1745112	335732	694	Stela before a platform. See SU057-8.	x
057	8	good	1745116	335693	695	D-shaped platform facing south on the south slope of Khuzma as-Shumlya. Perimeter defined by upright large and medium boulder-size slabs of limestone; these contain flat-lying slabs stacked three high above the uprights and boulder and cobble fill. The back perimeter (north side) is covered by scree so that dimensions cannot be taken; the platform appears to be 5 m north–south at its widest and about 1 m high. About 1 m to the south is a pair of upright limestone stelae, 1 m high, buttressing each other and as the northernmost of a small 1.1 m diameter ring of limestone cobbles and small boulders (057-6). This platform and associated stelae are one of three in an east–west line.	x
057	9	poor	1745127	335781	697	Originally recorded as a rock pile with medium boulders. Multiple revisits and the presence of large, displaced limestone slabs bring consensus that this structure is a very poorly preserved former platform, now collapsed downslope, retaining a 2.4 x 2.4 m concentration of unworked cobbles.	x
057	10	poor	1745115	335891	694	Collapsed platform now consisting of a pile of cobbles and small boulders ringed by outward-leaning limestone slabs collapsing downslope from a bedrock ridge. North to south. Current dimension is about 4 m. A noticeable density of thermally altered rock and hearths is adjacent.	x
057	11	poor	1745108	336175	692	A roughly north–south alignment, possibly double alignment of limestone slab uprights, 1 m in height, approximately 4 m long. This remnant of an unknown plan may be part of a construction that included a pile of cobbles, small boulders, and downslope-tilted limestone slabs situated upslope on a natural bedrock outcrop, possibly forming a natural prominence for a platformed structure that has since eroded.	x
057	12	very poor	1745110	335740	695	Three in situ medium boulder-size limestone uprights that may have been a perimeter of a ring or platform formed of limestone slabs. Flat-lying slabs lie to the east and are eroding or sliding downslope. These slabs roughly form the east side of a semicircle of uprights, with a single boulder in front (south).	x
077	1	good	1744452	336679	702	Subcircular structure of collapsed medium boulder-size limestone slabs; exhibits same ring of flat-lying slabs as other poorly preserved platform structures. Cobbles noted, but not specified as fill. Structure is 4.60 m east–west and 3.0 m north–south; slabs collapsed to 0.5 m height.	x

Survey Unit #	Site Number	Preservation	Northing	Easting	Elevation	Description	In GIS
C7	1	intact	1744545	334133	705	Large, highly visible D-shaped platform, 1.1 m high, 6.2 m north–northeast/south–southwest by 5.2 m; straight side faces east. Constructed of an outer perimeter of a few limestone, upright large boulder slabs and mostly medium–small boulder blocks. Top of platform is beginning to collapse. In plan, it appears that the final plan form may be the result of accretive building phases, with an inner orthostat-defined horseshoe/semicircular monument added across the front to lengthen the north–south and, finally, another east face of limestone blocks. The terrace surface east of the platform shows signs of burning (widespread thermally altered rock may come from ancient hearths). A north–south single course of stone, alignment 2.75 m east of the east face, may be later reuse of stone for a water management *shrūj*. This platform monument is exceptional in its visibility, preservation, and size (= 134-004).	x
C7	x	poor	1744235	333892	707	Rectangular structure, registered as a platform by the 2005 cairn survey.	x
C7	x	poor	1744027	333789	708	Linear structure, probably remnants of a platform with upright slabs; a standing limestone unworked stela associated.	x
150	1	good?	1737771	328772	729	One of three adjacent limestone slab perimeter structures or platforms. Slabs protrude about 20–30 cm from gravel terrace and define the perimeter of what may have been platforms with eroded cobble fill. Set adjacent to each other, these three structures may have belonged to a single trapezoidal platform (cf. SU038-10) with transverse inner divisions. Overall, the three-platform complex measures 5.5 m across.	x
C12	2	good?	1737775	328773	730	Second of three adjacent limestone slab perimeter structures or platform. Slabs protrude about 20–30 cm from gravel terrace and define the perimeter of what may have been platforms with eroded cobble fill. Set adjacent to each other, these three structures may have belonged to a single trapezoidal platform. The construction on a gravel terrace that was likely covered in Early Holocene silt beds (judging from the adjacent sediment islands) suggests that this was either very much earlier than or postdates the majority of Middle Holocene platforms situated on silt terrace surfaces. (= 150-1 second structure and C12-2 in the 2005 cairn survey.)	x
C12	3	good?	1737778	328774	729	Third of three adjacent limestone slab perimeter structures or platforms. Slabs protrude about 20–30 cm from gravel terrace and define the perimeter of what may have been platforms with eroded cobble fill. Set adjacent to each other, these three structures may have belonged to a single trapezoidal platform. All three platforms recorded as one site, SU150-1, but counted as three. (= 150-1 third structure and C12-3 in the 2005 cairn survey.)	x
C12	1		1737771	328770	729	Trapezoidal structure of upright limestone slab perimeter with cobble fill.	x

Table 12.1. Neolithic monument locations in Wādī Sanā (coordinates in UTM Zone 39 North WGS1984; elevation in MSL EGM96). (*continued*)

Survey Unit #	Site Number	Preservation	Northing	Easting	Elevation	Description	In GIS
151	1	intact	1744647	334319	697	Platform partially buried in silts. Excavated in 2005. Appeared on surface as silt-filled tear-shaped ring of limestone slab uprights. Upon excavation it proved to be D-shape in plan, filled with flat-lying limestone boulders, slabs, and cobbles, with stela(e?) outside to the southeast and a ring of cattle skulls embedded in sediments to the east. See excavation reports for details.	x
151	2	poor	1744625	334438	695	Three poorly preserved stone piles of cobbles and pebbles, probably remnant of an eroded, quarried, and reused stone platform built on silts. Cobbles around the perimeter recorded as a "foundation" were possibly placed as chock supports inside limestone uprights, but any trace of the latter has disappeared. Cobble piles have prevented deflation of underlying silts, resulting in a raised relict of the original height of the silt terrace.	x
151	3	poor	1744599	334408	693	A cobble and pebble placement that may have been the core of a stone platform. The stones have prevented erosion of underlying silts, resulting in a raised relict of the original silt terrace, 1.2 m across. Cobbles retain distinct concentric arrangement not typical of hearth construction. (It's possible that limestone slabs were removed from some structures to build others: later platforms, hearths. This sequence was evident in excavation of SU151-1.) Some of the stones appear thermally altered.	x
152	1	good	1744815	334194	688	Stone platform 5.5 m diameter preserved 0.95 m height, constructed of limestone small-boulder size slabs slumped outward or eroding off a raised relict of the former silt terrace height. The interior is packed with smaller cobbles and pebbles.	x
152	2	poor	1744784	334199	685	An elongated oval of medium boulder-size rectangular limestone slabs protruding 0.39 m from the surface. Pebbles and thermally altered rock are scattered in and around the structure, which measures 3.25 m east–west x 1.4 m north–south.	x
152	3	poor	1744786	334106	687	A roughly circular 5.80 x 5.60 m formation of large boulder-size rectangular limestone slabs, with smaller boulder slabs. Many are flat-lying, suggesting possible collapse of a platform structure through the erosion of underlying silts. Thermally altered rock is scattered throughout and at the perimeter of the slabs. While this is possibly a campsite reuse of original platform materials, the alignment of perimeter slabs conforms to platform construction techniques.	x
W14	x	poor	1744860	334170	685	Circular remains of large limestone upright blocks, some still placed in upright formation.	x
C34	1	poor	1699084	280612	961	Low structure on bedrock terrace, 3.50 diameter, 0.30 height.	x
C39	2	poor	1709172	295484	958	Structure on plateau, 5.20 diameter, 0.45 height.	x
C56	15	poor	1725681	245569	998	A platform of stone; rectangular, filled with pebbles, 2.40 diameter, 0.50 height.	x
C66	8	poor	1670576	328724	1308	Many more small boulders and slabs present, but only those clearly in situ are drawn. Structure maybe a platform, 2.20 diameter, 0.30 height.	x

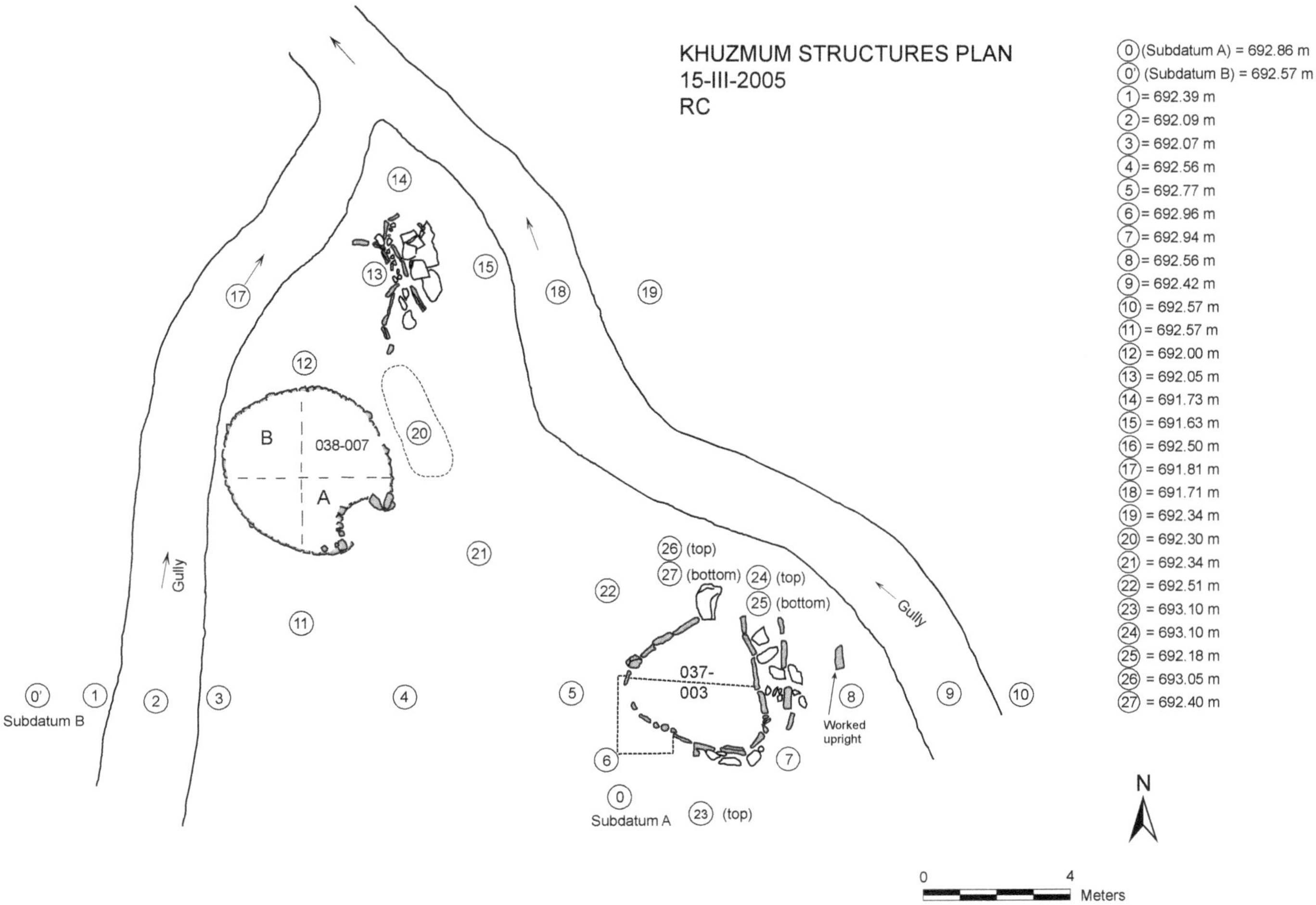

Figure 12.3. Silt terraces with platform structures 38-7 and 37-3 at the southeast corner of the Khuzma as-Shumlya inselberg. The dotted outline (half-buried stones) around elevation 20 was designated in the field as a potentially modern grave but is now considered to be an unknown component of the Neolithic structures in this area. *Drawing by Rémy Crassard and Clara Hickman.*

Quad A: Test Excavation

In 2000 the team opened a 1 m east–west x 3.5 m north–south trench (Quad A, figure 12.4), which uncovered a rubble layer (figure 12.5, Test Pit A, west section, ①, ②,③) overlying occupational surfaces with thermally altered rock, charcoal, and a discrete, central concentration of ash (figure 12.6). A radiocarbon date on charcoal from this ashy concentration (AA38547, 5616 ± 84) established an occupation date of 6683–6331 cal BP. Underlying this surface was alluvial sandy silt (④). The earliest datable charcoal fragments come from a hearth in A-009; these established a terminus post quem of 6797–6500 cal BP (AA38544, 5806 ± 64) for the structure's occupation but do not date its construction. A small test undertaken in 2000 in the northeast corner established that in alluvial sediments underneath one of these slabs was an earlier hearth, Quad A-010, from which emerged a winged point crafted prior to the monument's construction (figure 12.7). This hearth yielded no datable charcoal.

There are three possible explanations for the distinct layer of thermally altered pebbles that covered the surface of SU37-3, confined by limestone uprights. None of these explanations can be eliminated with current evidence. First, these pebbles may be relicts of fires deliberately set on top of this monument, either by visitors deliberately referencing the events that the platform commemorated or by passersby who found it a convenient surface for a campfire. Second, these burned pebbles may have accrued across a broader area of the surface of the silt terrace, perhaps through the erosion of multiple adjacent hearths, marking the passage of people after the platform was well buried by alluvial sedimentation. In this case, the pebbles atop the platform have been marooned there by the constraining protrusive limestone slabs. Other thermally altered rock

Table 12.2. Faunal remains recovered from silt terraces during survey. Analysis by Louise Martin.

Bag Number	Element	Part	Taxa	Side	Additional Information	Context	Estimated Cal BP
2000-017-1	mandible	posterior portion of the articular process	*Bos*	left	About 3 cm in size. Breaks are post-depositional.	occupation under Neolithic (?) platform monument	6500
2000-017-1	mandible	coronoid process	*Bos*	right	Breaks are post-depositional.	occupation under Neolithic (?) platform monument	6500
2000-017-1	tibia	anterior portion of proximal articulation	*Bos*	right	Proximal end unfused. Breaks are post-depositional.	occupation under Neolithic (?) platform monument	6500
2000-013-1-1	mandible	Two fragments; portion containing P2, P3, and P4 and articular process.	caprine	left	Annie Grant TWS:h. Breaks are both fresh and post-depositional.	Recovered during survey; embedded in surface of middle Holocene terrace; near spatial association with water management canal.	5300
2000-026-1-1	maxillary M3	-	*Bos*	right	Annie Grant TWS:g. Roots of tooth broken off post-depositionally.	occupation under Neolithic (?) platform monument	6500
2000-026-1-1	metacarpal	distal end and about one-fifth of shaft	camelid	-	Possible cut marks on anterior. Distal end unfused. Mixture of fresh and post-depositional breaks.	occupation under Neolithic (?) platform monument	6500

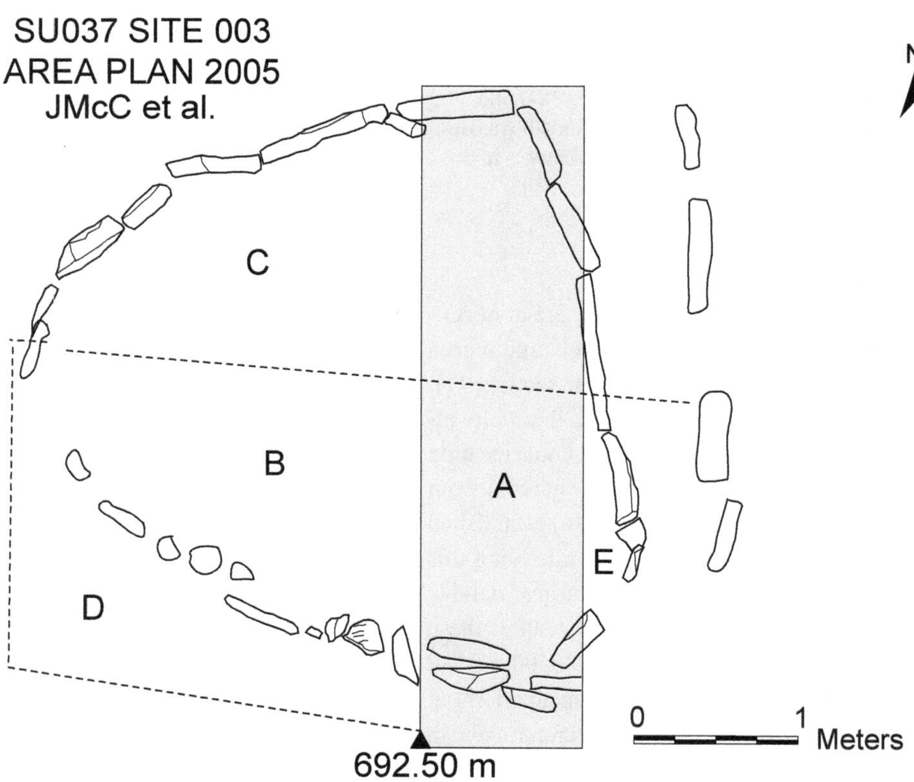

Figure 12.4. Excavation quadrats of 37-3. *Drawing by Joy McCorriston, Michael Harrower, and Clara Hickman.*

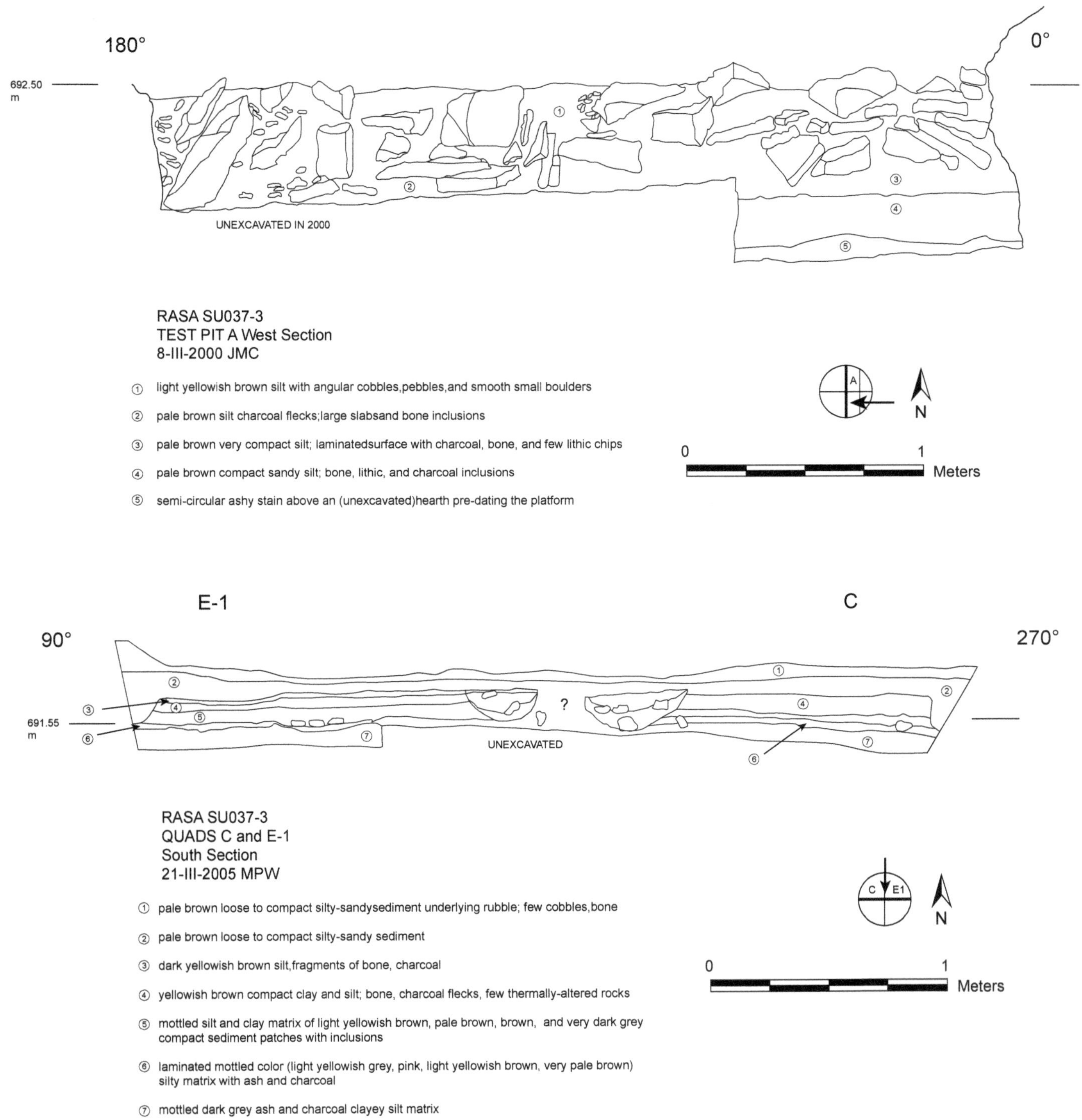

Figure 12.5. Platform 37-3, Quad A, west section (2000) and Quads B and C, south section (2005). *Drawing by Joy McCorriston, Margaret Wilson, and Clara Hickman.*

Figure 12.6. Final view of SU37-3, Quad A. *Photograph by Joy McCorriston.*

Figure 12.7. Hearth in SU37-3, Locus 010. It underlies the installation of the perimeter limestone uprights of the structure. *Photograph by Joy McCorriston.*

from hearths would have mostly washed into the nearby gully that separates the current silt terrace from the rock terrace and scree slopes of Khuzma as-Shumlya. Finally, and related to the second explanation, numerous hearths built on the rock and scree slopes may have eroded downward prior to the formation of this gully, forming a broad fan across the alluvial terrace, including the area of the buried platform. The gully bottom contains an abundance of grey, thermally altered limestone that may have come from hearths in the second or third scenario. Whatever the case, it is apparent that many hearths now long eroded beyond individual recognition attest to the repeated passage of people near 37-3 and adjacent structures.

The stone platform at 37-3 was built of perimeter limestone uprights sunk into the shallow sides of a slight pit or hollow. As with Kheshiya 151-1, the limestone uprights of the structure perimeter were placed with corners or angles pointing down, indicating that slabs were sunk or pushed into supporting sediments. From surface view, there appeared to be a break in the slab perimeter, possibly a door, at the southwest angle of the platform, but subsequent excavations revealed a full perimeter of limestone slabs (see "Quad D," below). Sandy silt sediments mixed with occupational debris—bone, charcoal, thermally altered rock—filled the lower structure. After occupation ceased, the structure was filled with cobbles and pebbles, into which were also cast small amounts of unidentifiable fragments of bone, none of which appeared to be human.

Quad B and Quad C: Interior Rubble and Occupation

Excavations in 2005 sectioned 37-3 into Quad B (southwest) and Quad C (northwest) as extensions of the interior excavations in Quad A. In 2005 the team expanded the excavations of the platform interior with Quad B (2 m north–south at its eastern, greatest extent x 2 m east–west) in the southwest area (figures 12.8 and 12.5). The southwestern limits of Quad B appeared almost immediately upon removal of an upper layer, 2–10 cm deep, of light yellow–brown sediment that had accumulated largely through alluvial flooding over the top layers of cobble and pebble fill. In the case of SU37-3, this fill consisted of large cobbles and small boulders rather than flat-lying slabs.

Three small deposits of unidentified animal bone fragments lay among the rubble fill in Quad B. Excavators also noted that rubble fill in this southwest area of SU37-3 contained larger angular cobbles than the eastern section of the platform fill (Quad E), where fill included more flat-lying, tabular limestone cobbles and boulders (table 12.3). A silty sediment had accumulated throughout the cobble fill, and within this matrix, excavators recovered small amounts of fragmented animal bone and fewer than 30 nondiagnostic chips and flakes of chert. Tiny flecks of charcoal also appeared infrequently throughout the matrix. To preserve an east–west section across the interior of SU37-3, excavation in Quad B stopped after the removal of rubble fill, about 0.40 m depth, noting the presence under fill of a hearth (Locus 006). This hearth included in situ burning of the underlying sediments, many thermally altered limestone clast pebbles, ash, charcoal fragments, and a darkened matrix. The hearth yielded three charcoal fragments (*Acacia* sp.), dated by radiocarbon assay to 6683–6374 cal BP (AA66862, 5682 ± 47). Note that this date, also establishing the structure's occupation, is contemporaneous with the dated hearth under rubble fill in Quad A (above).

In Quad C, a similar sequence of rubble fill overlay occupation layers, which included multiple hearths. Directly underlying the rubble fill, excavators encountered brown silts several centimeters thick. Under these lay a more orange layer (Loci 008, 009, and 010), 5–6 cm deep, including fragmented bone, charcoal flecks, and thermally altered pebbles. A less compact brown silty layer (Locus 011), about 3 cm deep and also including occupational debris, underlay the orange layer. Cut into this brown silty surface were several hearths (including C-012), shallow-sided pits about 50 cm in diameter, with a compacted burned lining, rich charcoal and ash contents, and dense thermally altered cobbles and pebbles over the charcoal. Throughout the matrix of the brown silty layer occurred dense thermally altered rock.

Below the brownish silts, excavators noted more grey sediment (Loci 013, 014, 015, and 016), still containing bone, charcoal, and thermally altered rock, with a hearth (Quad E, Locus 014) cut from these sediments. Grey ashy sediments with occupational debris were about 20–30 cm deep.

A lower layer (Locus 017) of orange–yellow/brown sediment, still containing ash, charcoal flecks, and thermally altered rock in small quantities, extended across the entire excavated interior of SU37-3. Excavators noted another hearth within this sediment, but the surface from which the hearth was cut was not differentiated as a distinct occupational surface. Hearth 019 was partially truncated by the cut of an upper, later hearth (C-014), indicating reuse of a structural feature and suggesting a source for the ashy stains and localized grey deposits throughout this otherwise orange-yellow silty alluvium.

Below were more ashy layers, notably grey ashy sediment with bone fragments and thermally altered rock pebbles sandwiched between two thin grey layers with higher clay content. These overlay another yellow-brown layer,

Figure 12.8. Platform 37-3 after excavations in 2005. *Photograph by Michael Harrower.*

also containing ash and thermally altered rock and notable for a concentration of unidentified animal bone.

These excavations largely confirmed the stratigraphic probe of Quad A with greater detail and clearer reconstruction of the structure history of 37-3 (table 12.3). No door was present after all; here, as in Kheshiya SU151-1, the structure was entirely ringed by a perimeter of limestone slab uprights, into which people deposited a deliberate fill. This deliberate fill created a platform in a structure that had housed multiple episodes of occupation, interlayered with alluvial sedimentation, presumably seasonal episodes in this seasonally flooded wadi system. As animal bone from this excavation remains in the Mukalla Museum, the opportunity to analyze any identifiable fragments is, for the moment, limited, but excavators noted no human bone.

Quad E: Extension of Interior Excavations

The stratigraphic sequence of Quad E, the interior of the structure east of the 2000 test excavation (Quad A), mirrors that of Quad C. As noted in 2000, sediments with traces of anthropogenic activity (charcoal, ash, animal bone fragments, thermally altered rock) underlie the bases of limestone slab uprights, suggesting that the construction of the structure took place in an area repeatedly revisited.

Quad D: An Exterior Probe

Under the surface of yellow-brown silt, excavators revealed an ashy deposit (Locus 002) that partially covered several buried limestone uprights of the platform's perimeter. Ash had been distributed throughout the dark ashy sediment that buried this portion of 37-3. The exterior part of this deposit was flecked with charcoal and reddened clay. No charcoal from this deposit was available for dating, and no extensive excavation of the exterior of the structure took place.

Ground-penetrating radar survey across a 10 x 10 m area in 2008 suggested a circular configuration of dense objects about 1 m below surface to the southwest of the platform structure. No further excavations were possible to test this result.

The southwestern part of Quad B and its extension, Quad D, straddled the location where surface remains led us to expect a doorway. With excavation, it was clear that the upright slabs actually formed a continuous perimeter without an entry gap. The distribution of ash outside the structure postdated the construction of the 37-3 platform. Yet the actual burning and use of an ill-defined pit may have been earlier, in sediments adjacent to and cut by erecting the uprights. This deposit—probably an earlier hearth—with its concentrated evidence of burning, probably was a source of ashy debris distributed through flooding and redeposition within alluvial sedimentation. Flooding and slackwater sedimentation could have buried a skull ring like that excavated at 151-1, but any testing of GPR results must await new fieldwork.

Faunal Remains

While there were many fragments of animal bone recovered from the rubble fill and underlying occupation levels of the pithouse-platform SU37-3, most were unidentifiable. None of the bone was human. Identified bones ($n = 13$) are too few for robust quantitative analyses (table 12.4). A single identifiable bone from SU38-7 (below) has been included in this limited inventory. (Remaining bone from SU38-7 was unidentifiable.) The range of taxa includes domesticates—certainly sheep, possibly goats, and domesticated cattle. Equids were probably wild, for native onagers still roam in remote Arabian oases today. Cut marks on the SU37-3 bones attest to human butchery. Perhaps one of the most significant implications from this limited inventory is the range of domesticate and wild taxa represented. The preservation of food remains in domestic contexts and domestic refuse—the usual source of economic information—is scant. From the excavations and interpretation of the dramatic cattle skull ring at SU151-1 (Kheshiya), it would be all too easy to assume that domesticated cattle at the center of people's social and ideological practices were also at the center of their economic lives. This could have been the case. Yet the wider range of taxa (beyond cattle) in the domestic refuse and fill of other, contemporary Neolithic platforms underscores the ritual and ideological role of cattle in sacrifice and the broader economic practices of hunting and mixed-herd pastoralism.

Objects

There was minimal lithic debris from the platform fill and underlying surfaces. None of the flakes or debitage were diagnostic to period or industry (table 12.5).

Discussion

People constructed the slab ring 37-3 in an area already frequented, at least intermittently. The erection of the limestone uprights sunk them into an earlier hearth (A 009/010) and probably disturbed another (D 002). Although untested stratigraphically, it is also possible that the upright limestone slabs used to construct the perimeter and perhaps some of the fill were robbed from earlier structures, for there are many nearby vestiges of other platforms, some poorly preserved.

To summarize the stratigraphic sequence at the well-preserved platform SU37-3, a D-shaped perimeter of limestone boulder slabs set upright on their angular tips was either pressed into soft mud or, more likely, pressed into the edges of a shallow scoop or pit. This technique required the labor of a collective—at least five or six people working simultaneously, based on our experience with stone removal—and would support the uprights during the multiple seasons that the interior was occupied. Multiple reoccupations of the interior occurred, with well-constructed hearths as shallow pits filled with limestone pebbles and cobbles to aid in retaining heat. Hearths were cut from different strata within the interior. There were at least four discrete phases of occupation, possibly annual seasons, between which orange-brown aeolian and alluvial sediments accumulated between distinctly grey and grey-brown habitation layers mixed with bone and minor lithic debris. No evidence suggests any roofing construction or materials. As with bedouin shelters today, a low stone barrier provided a windbreak and retained some warmth after sundown. If the occupants were herders, the shelter served also to exclude their animals and may have provided, with a fire, some protection from predators (such as hyenas, canids, or leopards).

Like SU151-1, the upright slabs were deliberately filled to create a platform after occupations ceased. Fill for platform construction might have required a minimum of a few people over several days but could have also been a group task in a shorter time. There was no interior human burial, and all the recognizable bone was animal bone. Rubble inside the slabs did not accrue outside, as would occur through roof or wall collapse, and the form of the well-preserved platform conforms to that of other well-preserved examples nearby, some intact on bedrock terraces, others protruding from the Early Holocene silt terrace. The exterior standing stone, or betyl, strongly suggests a commemoration and sanctity in the construction of the SU37-3 platform.

Table 12.3. Stratigraphic sequences for SU37-3, concordance with figure 12.4.

Stratigraphic Descriptions, SU037-003 Khuzma Platform Excavated in 2000					
Quad	**Locus**	**Under**	**Over**	**Equals**	**Description**
A	000	-	001	surface	Silt matrix over and between limestone gravel and small boulders; unsorted. This locus number was also used for cleaning the profile after removing backfill in 2005.
	001	000	002		Very pale brown silt matrix with few boulders and many thermally altered gravel pebbles; bone, lithic, charcoal, and land snail shell inclusions.
	002	001	003	E003	Very pale brown to light yellowish–brown fine silts with small boulder and cobble-size limestone inclusions, some as flat-lying slabs.
	003	002	004		Light yellowish–brown silty matrix with small angular cobble and pebble inclusions and more weathered limestone small boulders as flat-lying slabs; land snail inclusions.
	004	003	005		Light yellowish–brown silty matrix with pebble and cobble inclusions among tumbled limestone slabs, some greater than 10 cm thick and flat-lying. Other inclusions are land snails and bone fragments.
	005	004	006		Pale brown silty matrix with charcoal flecks among large limestone slabs; bone inclusions.
	006	005	007		Pale brown very compact silt with large charcoal inclusions, bone fragment, and a few lithic chips; few boulders and no smaller stones; a surface on which larger rubble fill is resting.
	007	006	008		Pale brown compact silt in laminations; a surface without stone inclusions except concentrated thermally altered rock as probable remains of a hearth.
	008	007	009		Arbitrary removal of lower layers of pale brown compact sandy silt with ashy lenses about 15 cm in diameter; bone and one flake inclusions.
	009	008	unexcavated		Pale brown compact sandy silt within the interior of upright limestone slabs at the northern end of Quad A; inclusions are bone, a bifacially worked lithic point (lost), and charcoal; a semicircular stain appeared at the base of (below) limestone uprights and probably is the top of a hearth predating the platform.
Excavated in 2005					
A	100	backfill	E009	008, E008	Very pale brown compact clayey silt below rubble fill of platform, excavated in 2005 before Quad A became Quad E.
E	001	surface	002	backdirt	Loose pale brown silt with lithic and bone inclusions; backdirt from bedouin digging.
	002	001	003	backdirt	Loose to compact pale brown silt with lithic and bone inclusions; backdirt from bedouin digging.
	003	002	004	A002	Pale brown loose silt matrix among clastic cobble and pebble fill of platform; inclusions of bone and lithic chips.
	004	003	005	A003	Pale brown loose to compact silt matrix among clastic small boulder, cobble, and pebble fill of platform; inclusions of lithic chips and bone.
	005	004	006	A004	Pale brown loose to compact silt matrix among clastic limestone boulders, cobbles, and pebble fill of platform; inclusions of bone.
	006	005	007	A005 and A006	Compact silt matrix among clastic limestone boulders; bone inclusions.
	007	006	008	A007	Light yellowish–brown clayey silt layer with charcoal and bone inclusions and an ashy stain, 40 cm diameter.
	008	007	unexcavated		Brownish-yellow to light yellowish–brown compact clayey silt concentration, about 30 x 70 cm diameter; inclusions of ash and charcoal and lithics; no cobbles except multiple pieces of thermally altered rock; a hearth.

Quad	Locus	Under	Over	Equals	Description
B	001	-	002	backdirt	Loose light yellowish–brown sandy backdirt tossed onto Quad B by bedouin digging in backfilled Quad A. Few thermally altered rocks came from surface of Platform 037-3 under backfill.
	002	001	003	A001	Light yellowish–brown compact loam matrix around limestone rubble; inclusions of bone, lithic chips, and land snails; some thermally altered rock present.
	003	002	004	A002, E003	Loose to compact light yellowish–brown silt and sand matrix among clastic cobble and pebble fill of platform; inclusions of bone and lithic chips.
	004	003	005	A003, E004	Light yellowish-brown loose to compact loamy matrix among medium dense clastic cobbles over flat-lying limestone slabs; inclusions of bone.
	005	004	006	A005 and A006, E006	Pale brown loose to compact silty matrix among flat-lying limestone slabs overlying surface and medium density of clastic cobbles; increased inclusions of bone and frequent charcoal fragments.
	006	005	unexcavated		Hearth with pockets of very dark grey ash in a matrix of light grey and light yellowish–brown sediment with abundant charcoal inclusions and thermally altered clastic pebbles.
C	001	surface	002	backdirt, B001	Pale brown loose silty sediment with few disturbed lithic chips and bone; backdirt from bedouin digging.
	002	001	003	B002, A001	Pale brown loose clayey silty matrix among fairly clastic dense cobbles and pebble fill; several boulder size limestone blocks; lithic and bone inclusions; land snails.
	003	002	004	B003, A002, E003	Pale brown loose to compact silty matrix among limestone rubble of fairly dense, unsorted cobbles, pebbles, and small boulders; lithic and bone inclusions; land snails.
	004	003	005	B004, A003, E004	Pale brown loose to compact silty matrix among a (deeper layer of) limestone rubble fill composed of unsorted clastic cobbles, pebbles, and boulders; boulders slightly larger than most of the overlying fill; inclusions of charcoal, lithic chips, and bone; land snails.
	005	004	006	B005, A005, E006	Pale brown loose to compact silty matrix with greater sand inclusions; among limestone rubble of unsorted clastic, a few cobbles, and mostly small boulders; inclusions of lithic chips and bone; one land snail.
	006	005	007	A006	Pale brown loose to compact silty sandy sediment underlying rubble fill with moderate cobbles; bone inclusions; one land snail.
	007	006	008	A007, E007	Dark yellowish–brown layer, 1–3 cm thick, of silt sediment directly underlying rubble fill; no cobble inclusions; small fragments of bone inclusions and charcoal flecks.
	008	007	008	A007	Yellowish brown compact clay and silt layer, 1–7 cm thick; inclusions of bone fragments, charcoal flecks, few thermally altered rocks (small cobble size).
	009	008	010, 012	A007	Very pale brown matrix of silt and clay with patches of pale brown and brown sediment concentrated around thermally altered rock pebbles with small charcoal flecks; bone inclusions; animal burrows about 5 cm in diameter.
	010	009	011	A008?	Mottled silt and clay matrix of light yellowish–brown, pale brown, brown, and very dark grey compact sediment patches with ash, charcoal, lithic chips, and bone inclusions; some thermally altered rock; root holes, insect burrows, and larger animal burrows.
	011	010	012	A008?	Compact silt layer 1–2 cm thick, slightly less compact than overlying layer; inclusions of ash, charcoal fragments, lithic, bone, and thermally altered rock (cobbles); animal or plant root passage horizontal across layer.

Table 12.3. Stratigraphic sequences for SU37-3, concordance with figure 12.4. (*continued*)

Quad	Locus	Under	Over	Equals	Description
C	012	008	018	A009, E-1-012	Pit 30 cm deep filled with compact brown and light brown sediment and inclusions of many rounded, thermally altered rock pebbles and charcoal fragments. Pit had a light yellowish–brown very compact clayey silt (baked) rind; a hearth.
	013	011	015	A009, E-1-013	Laminated mottled (light yellowish–grey, pink, light yellowish–brown, very pale brown) silty matrix, 3 cm deep, with ash and charcoal inclusions; few lithic chips and bone fragments.
	014	013	015		Pit 9 cm deep; thin oxidized brown rind with fill of black charcoal and ash mixed with light grey and pale brown clayey silt; inclusions of thermally altered rock cobbles; a hearth.
	015	013	016	A009, E-1-015	Mottled dark grey ash and charcoal clayey silt matrix mixed among very pale brown hard laminated clay with patches of light brown–backed silt around Hearth 012, containing light brownish–grey ashy charcoal fill.
	016	015	017	E1-016	Mottled dark grey ash and charcoal clayey silt matrix mixed among pale brown compact ashy matrix with charcoal and bone inclusions; rodent hole or posthole, 10 cm diameter.
	017	016	018	E1-017, 018	Pale brown layer (interpreted as occupation level) with dark greyish–brown patches, bone inclusions, ash and charcoal evident; together with 018, 5 cm deep.
	018	017	020	017,	Light brownish ("orange") grey compact to hard laminated clayey silt with ash and charcoal inclusions, also bone and lithic inclusions
	019	017	020		Pit 5 cm deep; 30 cm diameter; dark greyish-brown fill with ash, charcoal, and thermally altered rock, clastic pebble inclusions; a hearth.
	020	018	021		Pale brown compact laminated clayey layer, 2–5 cm deep, greyish-brown in patches, inclusions of ash, charcoal, rootlets, insect burrows, thermally altered rock cobbles.
	021	020	unexcavated 022		Very pale brown (?) clayey layer (interpreted as occupation level) with scattered bone inclusions and pebble-size, clastic, thermally altered rock; possible hearth area (unexcavated) in northwest corner.
	022	021	unexcavated		Unexcavated; very pale brown clayey silt layer.
E-1	009	E008, A100	010	A009, C009	Very pale brown compact silt with charcoal, bone, and thermally altered rock inclusions; several lenses of darker grey ash.
	010	009	011	A009, C010	Very dark grey to grey to yellow mottled compact silt matrix with thermally altered rock and charcoal inclusions and ash; several root or postholes detected.
	011	010	013	A009, C011	Pale brown clay matrix with ash and charcoal inclusions; part of the mottled occupation consistent with A Locus 009.
	013	011	015	A009, C013	Yellow to grey compact laminated clayey silt with ash and charcoal inclusions; insect burrow disturbance.
	015	013	C016 = E 1-016; E-1 excavated with C	A009, C015	Grey compact laminated clayey silt
D	001	surface	002		Light yellowish–brown to brownish-yellow loose, surface loam matrix layer, 1–7 cm deep outside and overlying limestone uprights of structure 037-003; land snail shell and bone inclusions; some ash.
	002	001	004	003	Yellowish-brown silty layer, about 1 cm deep, with ash, charcoal, and thermally altered rock inclusions.
	003	001	004	002	Pale brown layer with ash, extends across top of limestone uprights.
	004	001, 002, 003	unexcavated		Very pale brown to yellow (?) compact silty layer with charcoal fleck inclusions, adjacent to 002 and 003.

Table 12.4. Faunal remains from SU37-3. Analysis by Louise Martin.

Bag Number	Element	Part	Taxa	Side	Additional Information	Context	Estimated Cal BP
2005-037-3-B-002-lot3-bag2	metapodial	distal end	*Bos*	-	Distal end unfused	Rubble deliberately filling platform	6500
2005-037-3-B-002-lot2-bag3	seismoid	-	*Bos*	-	-	Rubble deliberately filling platform	6500
2000-037-3-A-002	femoral head	head and part of neck	caprine	-	About 1 x 1 cm in size	Rubble deliberately filling platform; near surface.	6500
2005-037-3-B-004-bag1	distal phalanx	half of distal articulation	*Equus*	-	2 cm in width. Fresh break.	Rubble deliberately filling platform	6500
2005-037-3-B-005-bag1	astragalus	-	caprine	left	Chopped through the medial side at an oblique angle from the caudal to cranial surfaces.	Rubble deliberately filling platform	6500
2005-037-3-E-005-bag1	distal phalanx	complete	caprine	left	-	Rubble deliberately filling platform	6500
2005-037-3-E-004-bag1	tibia	distal articulation	caprine	right	Distal end unfused	Rubble deliberately filling platform	6500
2000-037-3-A-007-bag20	astragalus	complete	caprine	right	Note on label: single piece bottom 006/top 007.	Rubble deliberately filling platform	6500
2000-037-3-A-008-bag22	mandibular dp4	-	*Bos*	right	Extremely worn. Annie Grant TWA:past n++ with mesiodistal wear.	Neolithic occupation levels prior to deliberate rubble fill of platform	6500
2000-037-3-A-008-bag22	metacarpal	half of distal articulation	*Ovis*	-	Distal end fused. Post-depositional break down medial.	Neolithic occupation levels prior to deliberate rubble fill of platform	6500
2005-037-3-C-015-bag3	carpal trapezoid	-	*Bos*	left	Medial side cut through on cranial	Neolithic occupation levels prior to deliberate rubble fill of platform	6500
2005-037-3-C(+E)-021-bag5	femur	shaft diaphysis fragment	*Bos/ Equus*	-	Large fragment, about 10 cm. Fresh breaks along edges; dry break at proximal end.	Neolithic occupation levels prior to deliberate rubble fill of platform	6500
2000-037-3-A-010-bag28	vertebrae	broken	*Bos*	-	Broken longitudinally. Unfused.	Neolithic occupation levels prior to deliberate rubble fill of platform	6500
2005-038-7-A-007-lot12-bag3	maxillary M1, M2, and M3	-	caprine	left	Annie Grant TWS:g; m3 is fragmentary.	Neolithic occupation levels prior to deliberate rubble fill of Neolithic platform	6500

Excavations at Khuzma Platform SU38-7

About 8 m to the northwest of SU37-3, a neighboring complex of structures includes: (1) a 5 m diameter perimeter of angular boulders filled with more boulders and cobbles; (2) a 2 m north–south x 1 m east–west tumulus of cobbles (tentatively identified as an Islamic grave with upright markers, Elevation 20 in figure 12.3); and (3) eroded remnants of one or several platform structures built, like SU37-3 and SU151-1, of limestone slab uprights with cobble and slab fill. The Early/Middle Holocene silts of the terrace are very shallow (less than 0.40 m depth) in this location, underlain by sandy deposits.

Discovery

The RASA intensive survey of 2000 documented this structure as one of many on the silt terraces at the southwest corner of Khuzma as-Shumlya (chapter 6). Surface indications provide no proof that this structure was contemporary with others nearby, but the boulder ring and platform remnants are embedded in the same Early/Middle Holocene silts (figure 12.9).

Methods

Excavations used the procedures, recording, and sampling described in chapter 10 and used at SU37-3. We selected

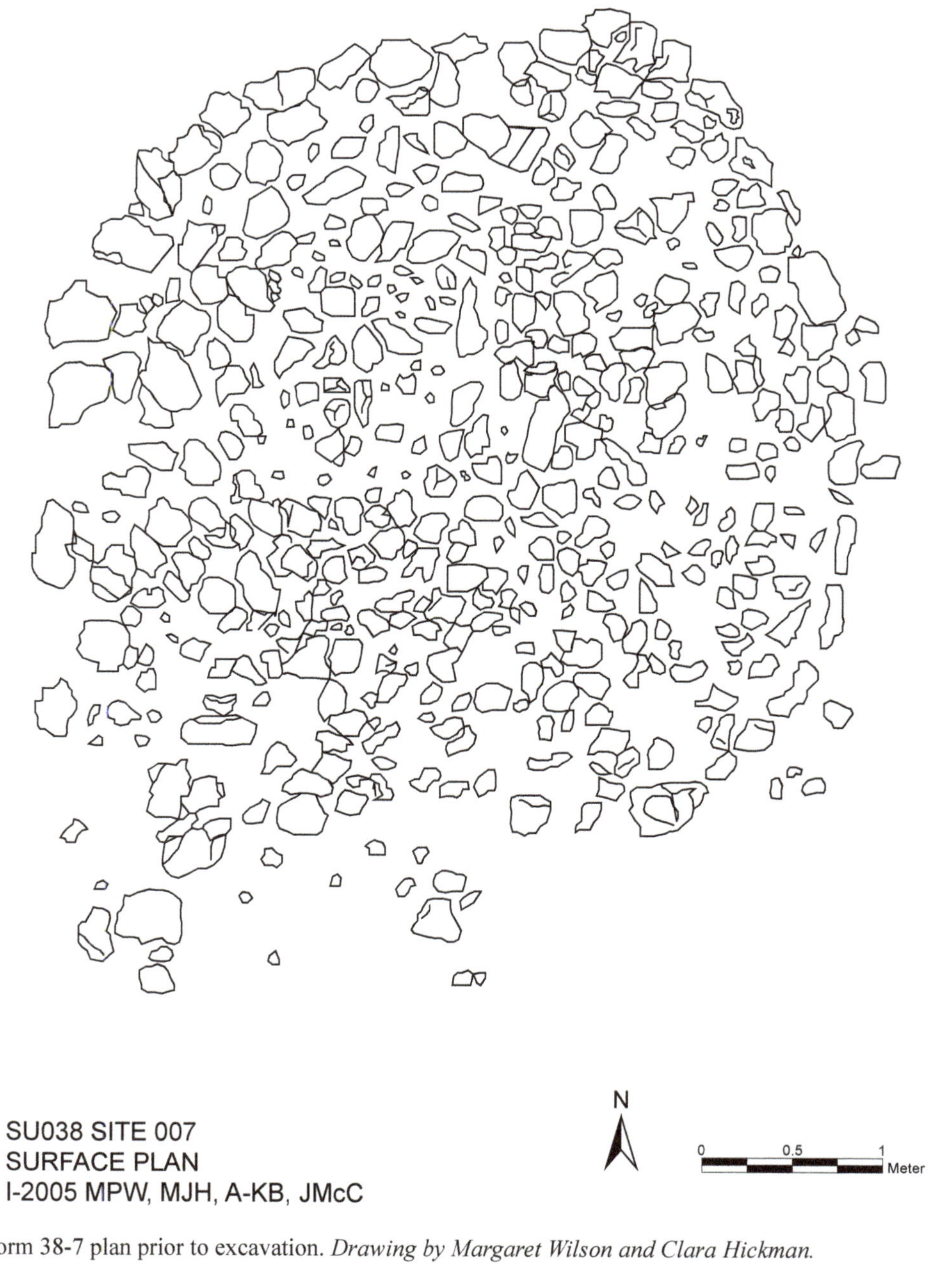

Figure 12.9. Platform 38-7 plan prior to excavation. *Drawing by Margaret Wilson and Clara Hickman.*

the boulder ring with its cobble fill for its readily identifiable surface plan and preservation. Excavators sectioned the 5 m boulder ring into four pie-shaped quadrats, digging into opposing northwest (Quad B) and southeast (Quad A) quadrats and leaving a continuous section north–south and another east–west across the interior of the structure (figures 12.10 and 12.11).

From surface view, the nearby tumulus of cobbles cited above appeared quite similar to Islamic graves found throughout the region. Although a lone Islamic grave

Figure 12.10. Platform 38-7, Quad B, excavations of Locus 002, showing one-quarter of the boulder-ringed platform with cobble fill exposed before its removal. *Photograph by Joy McCorriston.*

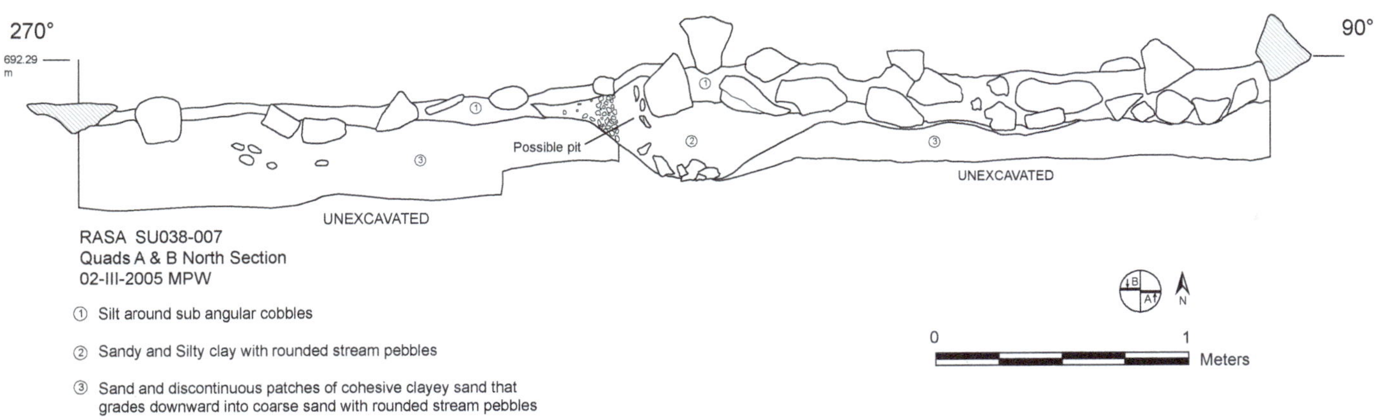

Figure 12.11. Platform 38-7 composite section across boulder-filled ring, showing north profile of Quad A and south profile (mirrored here) of Quad B. *Drawing by Margaret Wilson and Clara Hickman.*

Table 12.5. Lithic artifacts from excavations at SU37-3. Analysis by Joy McCorriston.

						Products			Tools			
Date	Quad	Site	Locus	Lot	Bag #	Flakes	Primary Flakes	Obsidian Flakes	Other	Utilized Flakes	Arrowheads	Misc. Tools
3/6/05	A	037-3	101		2				1			
3/6/05	A	037-3	102		2							1
2/6/05	B	037-3	002	1	3				2			
2/7/05	B	037-3	002	2	2				11			
2/8/05	B	037-3	002	3	1				2			
2/9/05	B	037-3	003	3	2	2			1			
2/10/05	B	037-3	003	5	2				1			
2/17/05	C	037-3	001		2				2			
2/17/05	C	037-3	002		1				8			
2/21/05	C	037-3	003		2	1			3			
2/22/05	C	037-3	004		3	1			5			
2/28/05	C	037-3	005		2	1						
2/8/05	C	037-3	010		5				1			
3/9/05	C	037-3	011						1			
3/10/05	C	037-3	011		8				1			
3/9/05	C	037-3	011		4				1			
3/13/05	C	037-3	013		6				1			
2/18/05	E	037-3	001		2	1			1			
2/19/05	E	037-3	002		1	1	1					
2/20/05	E	037-3	003		2	1						
2/27/05	E	037-3	004		2				1			
3/6/05	E	037-3	008		1	1			2			
3/17/05	E1	037-3			1	1			1			
2/23/05	F	037-3			1	3						
3/18/05		037-3	018		6	1						

associated with pre-Islamic ruins seems highly unlikely, to avoid any chance of disturbing a Muslim burial, no excavations took place.

Results

The rubble fill of SU38-7 proved to be no deeper than 40 cm, and in some areas it was only one cobble's thickness. An outer boulder perimeter clearly visible from the surface was set slightly deeper than the fill, indicating that the structure had filled several centimeters deep with sediment—clayey silt overlying the sandy base—before an inner ring of slightly smaller boulders and a cobble fill were added (table 12.6; figures 12.12 and 12.13). Although several 10 cm deep concentrations of ashy discoloration appeared within the rubble fill (Quad A, Loci 002 and 003), nothing was recovered that would yield a radiometric date associated with the fill of SU38-7 or its construction. At least one pit, measuring 0.7 m east–west, had been excavated and refilled in the central portion of the rubble fill, but the pit contained no cultural material or bone. The structure's proximity to Neolithic platforms and entrenchment within the Early Holocene silts provides a best clue to its antiquity.

As was the case with SU37-3, the rubble fill was deliberate: excavators found no evidence of roofing materials or superstructure, and the fill was entirely confined to the structure's interior (as would not be the case had perimeter walls collapsed). The fill included limited animal bone and chert flakes, and atop the sandy bed on which the structure was built lay more flakes, indicating human activity before construction (table 12.7).

Cores					
Cores for Flakes	Core Trim Fragments	Lithics Total	Weight (all bag contents)	Weight (tools only)	Observations
		1	0.3 g		
		1	5.2 g		Retouched, notched truncated flake
		2	30.7 g		Unknapped chunks and chips
		11	12.5 g		Unknapped chunks and chips
		2	3.2 g		Unknapped chunks and chips
1		4	23.8 g		
		1	0.8 g		
	1	3	27.7 g		
		8	4.7 g		Unknapped chunks and chips
		4	9.5 g		
		6	18.1 g		
		1	0.3 g		
		1	0.6 g		
		1	0.0 g		Unknapped
		1	0.3 g		Unknapped
		1	0.0 g		Heavy fraction; thin chip unknapped(?)/scatter sample from "sterile."
		1	0.9 g		
		2	2.1 g		
		2	9.9 g		
		1	0.6 g		
		1	0.6 g		Very battered flake?
1		4	6.2 g		
		2	6.8 g		
2		5	121 g		Disturbed surface; one nonworked chunk.
		1	9.2 g		

Objects

Apart from a single maxillary caprine molar included in finds from 37-3, above (table 12.4), there were no finds even of charcoal fragments suitable for radiocarbon dating. The only objects recovered were lithics (table 12.7). These were few, and none were diagnostic points. The occurrence of one broken foliate or biface is consistent with the Neolithic industry studied from stratified sites in the Wādī Sanā (chapters 8 and 9).

Discussion

The original construction and activities that produced SU38-7 remain enigmatic. Perhaps a cobble-filled ring of boulders represents the lowest, uneroded layer of a once more substantial platform. In this case, the boulder ring served not as a perimeter but as inner chock stones for a face of upright limestone slabs. If so, these slabs may since have been robbed to build any of the many Neolithic platforms nearby, including SU37-3. Or the low, boulder-ringed, cobble-filled platform evident today may be faithful to an original construction for purposes unknown. A central pit contains no human bone, and the multiple small stake holes around this pit may have preceded or postdated this activity, all completed before the boulder ring was filled. One guess—interpolated from the Kheshiya cattle skull ring—is that participants in sacrifice and feasting set up an array of skulls on this low cobble platform, designed to chock upright the skulls and provide a more permanent base for their display.

Table 12.6. Stratigraphic sequences for SU38-7, concordance with Figure 12.10.

Quad	Locus	Under	Over	Equals	Description
A	001		surface	-	Light yellowish–brown loose well-sorted surface silt matrix with thermally altered rock and large, smooth cobbles. Matrix unexcavated; no artifacts on surface.
	001	surface	002, 003	009	Light yellowish–brown compact well-sorted silt matrix, 1–2 cm deep, with rounded cobbles; basalt core and chert flakes, animal bone, and tooth fragment inclusions; land snails.
	002	001	004	-	Greyish-brown loose silt in a distinct thin subcircular patch, < 1 cm deep and 40 cm radius within perimeter of cobble platform and overlying cobbles; chert flake, ash, land snail inclusions.
	003	001	004	-	Greyish-brown loose silt in a distinct patch, about 5–7.5 cm deep and 20 cm radius within the perimeter of the cobble platform and overlying cobbles; chert flake, ash, land snail, and insect burrows. This and Locus 002 are remains of possible hearth areas on the platform?
	004	001	005	010; B002	Light yellow–brown compact well-sorted silt matrix, about 20–25 cm deep over and around angular cobbles, fill of platform; inclusions of animal bone and chert flakes, a few stray wood charcoal fragments, and a small concentration of thermally altered rock.
	005	004, 001	006	B003	Light yellowish–brown to grey clayey well-sorted layer with many rounded pebbles; a noted increase in quantities of small chert flakes and small animal bone fragment inclusions, charcoal fragments, and land snails; many small 4–6 cm diameter stake or postholes (very shallow) near center of platform structure.
	006	005	007	B004	Light yellowish–brown compact clayey layer, 0.5–5 cm thick, with a definable compaction as interface between this layer and overlying matrix; preserved as about 1 x 0.8 m surface at the center of the platform as liner of a pit; inclusions of chert flakes and charcoal.
	007	005, 006	unexcavated	011	Light yellowish–brown mixed compact and loose sand and silt sediment with few rounded cobbles and rounded pebbles, about 15 cm depth; inclusions of charcoal, flakes, a steatite and an obsidian chip, animal bone and teeth, land snails, and a few uncommon terrestrial shells. Interpreted as a layer in which a compact occupational pit lining was developed.
	008	006	unexcavated	B004	Light yellowish–brown compact sandy matrix without artifacts except thermally altered rock in one concentration, about 50 cm northeast–southwest along base of central pit.
	009	surface	010	001	Light yellowish–brown silt matrix, about 15 cm deep over and around angular cobbles, fill of platform at the inner perimeter of the boundary boulders.
	010	009	011	004	Light yellowish–brown compact clayey silty matrix and angular cobbles, 6–15 cm depth along inner perimeter of boundary boulders; also fill of platform; inclusions of ash, land snails, and insect burrows, with one chert flake.
	011	010	unexcavated	007; B010	Light yellowish–brown mixed compact and loose loamy sediment, about 1–15 cm depth with few flakes of chert and one small animal bone, possibly a burrowing rodent; charcoal flecks. This is the underlying layer in which pit and platform were constructed.
B	001	surface	002	008; A005	Light yellowish–brown loose well-sorted silt matrix, about 30 cm depth over and around angular cobbles inside boulder perimeter; charcoal fragment, two chert flakes, and fragments of animal bone and teeth inclusions. Platform fill.
	002	001	006	A004	Light yellowish–brown well-sorted silt matrix around angular cobbles; about 25 cm depth of platform fill.
	003	002	007	A005	Light yellowish–brown compact clayey silty matrix with pockets of sandy sediment and rounded pebbles; inclusions of charcoal fragments, roots, land snails, insect and small animal burrows, and chert flakes. Many shallow postholes or insect burrows, about 6 cm width, evident at base of this 7–10 cm thick layer; also evident in plan and section is the pebble-rich fill of a central pit.

Quad	Locus	Under	Over	Equals	Description
B	004	001		A006, A008	Light yellowish–brown clayey silty layer with a pile of small stones, about 15–35 cm thick at the center of the platform. This sediment, relatively rich in chert flakes, animal bone fragments, and charcoal, likely represents the fill of a pit also apparent in Quad A, center.
	005	002	unexcavated	A006	Light yellowish–brown clayey silty layer in a 30 cm deep sub-excavation at the inner perimeter of the boulder ring. Platform fill includes charcoal, chert flake, and animal bone fragments.
	006	003	unexcavated	A007	Light yellowish–brown loose and compact mix of clay and sandy patches with very few rounded cobbles and angular pebbles. This is the sedimentary matrix over which the platform was constructed and includes a chert flake and land snails, including several burned examples; postholes/insect burrows observed at the bottom of 003 extend deeper into this layer.
	007	003	unexcavated	A006	Light yellow–brown to very pale brown to light grey compact clayey layer, about 15 cm thick in center of platform. This probably is a lining in a central pit.
	008	surface	009	001; A009	Sedimentary matrix among angular cobbles and boulders at inner perimeter of platform, about 10 cm in depth; inclusions of few chert flakes and small animal bone fragments.
	009	008	010	A010	Light yellow–brown loose clayey silt sedimentary matrix around dense angular cobbles at inner perimeter of platform against boulders; fill of platform and possibly also dense as chocking support or constructed inner ring of boulder perimeter. Inclusions of land snails and one animal bone fragment.
	010	009	unexcavated	A011	Compact clayey silt matrix below platform fill abutting inner perimeter of boulder enclosure; inclusions of angular pebbles and insect burrows.

Figure 12.12. Platform 38-7, end of excavations. *Photograph by Michael Harrower.*

Table 12.7. Lithic artifacts from excavations at SU38-7. Analysis by Joy McCorriston.

						Products				Tools		
Date	**Quad**	**Site**	**Locus**	**Lot**	**Bag #**	**Flakes**	**Primary Flakes**	**Obsidian Flakes**	**Other**	**Utilized Flakes**	**Arrowheads**	**Misc. Tools**
2/7/05	A	038-7	001	2	1	12			13			
2/8/05	A	038-7	001	3	3				1			
2/9/05	A	038-7	002	4	1	1			1			
2/9/05	A	038-7	003	5	1				1			
2/12/05–2/13/05	A	038-7	004	7	1	1						1
2/13/05	A	038-7	004	8	1	6			1			1
2/15/05	A	038-7	005	10	9	5	1		11			
2/22/05	A	038-7	006	11	5	3			5			
2/22/05	A	038-7	007	12	4	12		1	41			
2/28/05	A	038-7	010	15	1	2			4			1
3/1/05	A	038-7	011	16	2	5			8			
2/10/05	B	038-7	001	1	2	7	1		6			
2/12/05	B	038-7	001	2	1	1						
2/12/05	B	038-7	002	3	1	2			3	1		
2/14/05	B	038-7	002	4	1	3			1			
2/6/05	B	038-7	002	1	3	1			37			
2/16/05	B	038-7	003	5	3	3			6			
2/22/05	B	038-7	004	7	4	2			2			
2/20/05	B	038-7	006	8	1	4			14			
3/1/05	B	038-7	007	9	1	2			3			
2/26/05	B	038-7	008	10	1	3			4			
2/28/05	B	038-7	010		1	1			4			
2/17/05	B1	038-7	005	6	1	3			3	1		

Neolithic Platforms beyond Wādī Sanā

Intense survey and excavations in middle Wādī Sanā reveal important details and highlight for the first time the significance and function of platform structures across Arabia. Platform structures occur widely in Arabia, and perhaps because they are enigmatic, old, lie on low terraces, and often resemble inchoates heap of stones, they may be substantially underreported in the archaeological literature, or at least unrecognized as a linked, widespread, cultural manifestation. After all, it can be difficult just to discern their original form and construction in survey, especially as their preservation is highly contingent on the original depositional environment and the passage of later peoples. Prominent, upright stones also make an attractive target, not only as a quarry for construction material for later structures but also because their ancient and widespread association with pre-Islamic religious beliefs and practices makes them a target for deliberate iconoclasm. In Wādī Sanā, this iconoclasm has been apparent as both an ancient practice (McCorriston et al. 2011) and a modern one observed in our visits.

An increasing body of evidence points to autochthonic Neolithic culture in Arabia, despite the adoption of domesticated animals most probably introduced from the Levant or Africa. Not only did some domesticates—certainly among them sheep—arrive from other cultural regions (Martin et al. 2009), but there is material evidence of a wide network of contacts in the Neolithic period. Obsidian moved across Arabia from East African and highland western Yemen sources (Khalidi et al. 2013), and marine shell occurs far inland. Long-distance movement of materials and animals occurred within a

Cores					
Cores for Flakes	Core Trim Fragments	Lithics Total	Weight (all bag contents)	Weight (tools only)	Observations
		25	61 g		Includes four pieces limestone
		1	0.8 g		
		2	0.8 g		
1		2	7.2 g		
		2	10.4 g	8.6g	Biface (?) knife resharpened
		8	20.4 g	3.9	(Broken) end scraper
		17	28.6 g		
		8	2.1 g		
		54	33.3 g		
		7	10.7 g	5.7	One broken foliate/biface
		13	4.9 g		Five thin pressure flakes
		14	30.7 g		
		1	2.0 g		
		6	7.0 g		
		4	6.2 g		
		38	33.6 g		Mostly other elements of non-knapped shatter
		9	4.9 g		
		4	0.7 g		Two thin resharpening (pressure?) flakes
		18	6.8 g		Three thin resharpening (pressure?) flakes
		5	1.2 g		One thin resharpening (pressure?) flake
		7	9.7 g		One thin pressure (?) trimming flake
		5	0.9 g		One thin pressure (?) trimming flake
		7	5.5 g		Three thin resharpening (pressure?) flakes

broad cultural context, with technological and stylistic continuity across much of Southern and Eastern Arabia from 8000 cal BP onward. The emergent and developed Neolithic projectile point traditions of Southern Arabia show no links with a Levantine Neolithic (Charpentier and Crassard 2013; Crassard 2008), even as they do manifest cultural connections among Arabian groups over much of the peninsula. And the modern genetic profile of Arabian populations suggests a significant continuity and ancestral affinity with distinct haplotype groups in Dhofar today (Al-Abri et al. 2012; Černý et al. 2011; Rose et al. 2013). Arabian populations may have lingered through the hyperarid phases of the Pleistocene to repopulate the interior during the wetter era of the Early Holocene (Rose et al. 2013), and they adopted domesticates into an established Arabian cultural framework of mobile hunter-gatherer groups with significant mobility (Charpentier 2008; Cleuziou and Tosi 1997).

In this context, it is important to note the presence of Neolithic platform monuments across a wide geographic area. There are platform monuments documented outside the Wādī Sanā by the HDOR Project (Braemer et al. 2003), and the archaeological literature provides undated examples with structural similarity from as far as the Negev (Avner 2002), eastern Jordan (Gebel and Mahasneh 2013), and eastern Saudi Arabia (Zarins et al. 1979). Excavated examples in Dhofar, Oman, were firmly Neolithic in date. One stone platform on a bedrock terrace overlooking access to one of the oasis springs at Al-Mudhai measured 8 m across, and excavation showed that it was constructed of upright limestone slabs chocked by interior boulders and filled with boulder and cobble rubble. The rubble contained large animal bone

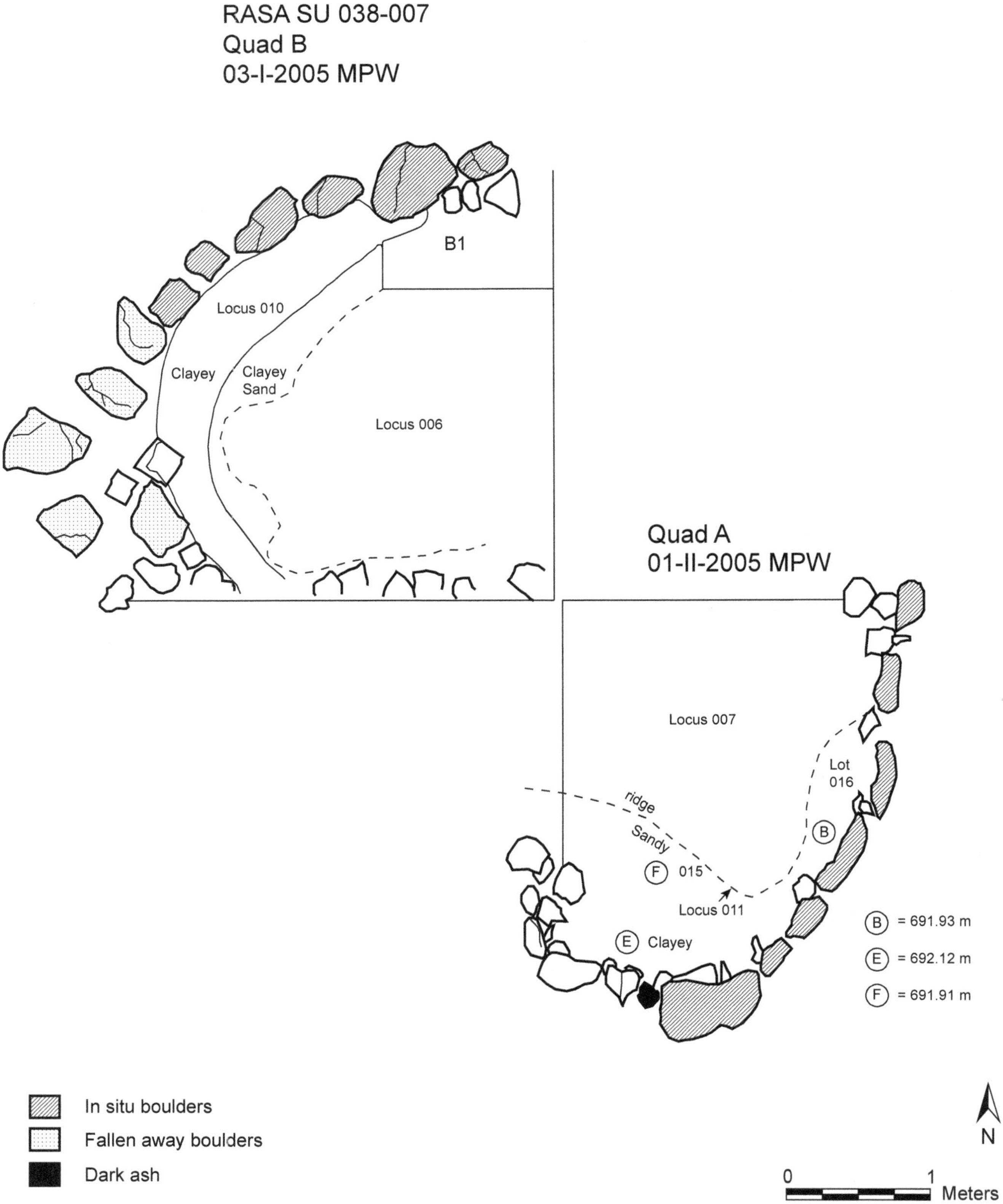

Figure 12.13. Platform 38-7; final plans of Quads A and B after excavation. *Drawing by Margaret Wilson and Clara Hickman.*

yielding a date of 6569–6536 cal BP (AA95064). A platform at Hānūn, less easy to date, also suggested a Neolithic period construction and conforms closely to the structural forms found in the Negev and Saudi Nejd (McCorriston et al. 2014:122–24). There are more platforms, unexcavated, across the Dhofari Nejd. Even where stone was unavailable, as at the site of Aqab in the United Arab Emirates, people constructed a commemorative platform. Like the skull ring at Kheshiya, a dugong bone monument at Aqab is entirely constructed of animal bone, placed with organized care, mounting the skulls on top, and representing the carcasses and meat of many animals (Méry et al 2009).

Conclusions

Neolithic platforms probably attest to a widespread practice of social constitution through intermittent gatherings, performance of sacrifice, and feasting, a conclusion explored in greater depth elsewhere (McCorriston 2011). This constellation of activities could be supported at times of grazing plenty but could not sustain a large population of pastoralists in one place in permanent residence (Henton et al. 2014; McCorriston et al. 2012). The number of platform structures in middle Wādī Sanā indicates that the area was revisited numerous times in antiquity. In turn, revisitation suggests that the sacrifices at Shi'b Khishiyah were economically sustainable and not a freak event devastating to a pastoral economy (figure 12.14). Not only does archaeological survey provide the traces of many periods in Wādī Sanā, but the construction and reconstruction of more than 50 platforms, some within a few meters of one another and many possibly robbed for later building, attest to the many times and many generations of return to middle Wādī Sanā to fulfill religious and ideological imperatives while renewing and inscribing social ties and engaging in the economic exchanges that moved cattle, obsidian, marriage partners, finished tools, knapping knowhow, and exotic materials along social networks.

There existed many Neolithic platforms in middle Wādī Sanā as part of a broader Neolithic tradition of social gatherings with sacrifice and feasting, commemorated by permanent monuments. Not all platform structures in Wādī Sanā can be dated by absolute or relative methods, but many can. Of those excavated, two (151-1 and 37-3) yielded radiocarbon dates in the middle seventh millennium cal BP, within 500 years of a platform structure in Dhofar dated to the beginning of the seventh millennium cal BP. Other platform structures in Wādī Sanā are embedded in Middle Holocene sedimentary terraces that buried them through alluvial processes that ended about 5,000 years ago (chapters 3 and 18). Thus, at a minimum, these platform structures predate the Bronze Age (circa 5000–3300 cal BP). For platforms buried in Middle Holocene alluvium, a difference in patina appears where the stones have been long buried and the upper parts exposed. This upper patina—apparent, for example, at SU151-1, at SU37-3, and within the structures buried in the Gravel Bar Site (SU0, Quads A, B, and C)—provides another, albeit uncalibrated, indication of the antiquity of platform structures. Undated and probably undatable platform structures on bedrock and gravel terraces share with dated examples their construction details and a relatively low elevation in their local landscapes. Such platforms are also plausibly Neolithic in date.

In a number of examples, the association of platform structures with adjacent standing stones is apparent. In Wādī Sanā, SU57-6, SU57-8, SU35-1, SU35-2, SU151-1, and SU37-3 have a standing stone placed before the platform, generally to the northeast, southeast, or north. Some of these examples are on rock terraces that permit no stratigraphic and therefore no indisputable contemporaneity of construction—a standing stone could have been erected well before or well after the platform. In the excavation at Kheshiya, a standing stone with a worked top probably predated the platform and cattle skull ring, possibly belonging to an earlier structure and set of activities at a site revisited and reconsecrated multiple times.

Survey and excavation have also revealed an association between platform structures and large mammal bone, in some cases identifiable as cattle. Adjacent to SU151-1 was an entire ring of cattle skulls (chapter 11), the excavation of which suggests that monuments commemorated animal sacrifices and feasts. The evidence from Kheshiya prompted a closer analysis of the survey results. At Kheshiya, survey work prior to excavation documented several other buried cattle skulls eroding out of the gully to the west of SU151-1. At unexcavated survey sites SU26-1, SU-34-2, and SU33-18, the surrounding sediments or the structure itself contained fragments of large mammal bone. In one case, large, very badly preserved bone had been ejected from a recent fox burrow into the fill of a platform structure. Animal bone occurred in the excavated fill of SU37-3, and SU38-7, and bone was evident in tiny fragments eroding from the Early/Middle Holocene sedimentary deposit surface between them.

Even as these structures ended as buried platforms, they were structures with life histories. Platform structures appear in locations where prior remains abound, preserved in Early/Middle Holocene sediment terraces. Predating platforms are standing stones, buried hearths, redeposited ash layers, layers of in situ burning, and the debris from retooling projectile points and skinning game.

Figure 12.14. Neolithic platforms SU57-6 and SU57-8 (C54-6 and C54-8) and standing stones still intact on a scree slope on the southeast side of Khuzma as-Shumlya. *Drawing by Tara Steimer and Clara Hickman.*

Nearby rockshelters (for example, Khuzmum SU45-1 and SU45-10) provided ideal camping spots for short-term, repetitive visits, whether in pursuit of game or in an annual round to pasture domesticated herds. Lithic assemblages and radiocarbon dates show that these shelters were used long before Late Neolithic herders converged to construct platforms (Crassard 2008; McCorriston et al. 2002).

Once constructed as shelters in frequented regions, the rings of uprights were occupied, sometimes briefly, as at Kheshiya, sometimes repeatedly over many years, as at Khuzma as-Shumlya (SU37-3). Interior stratigraphy suggests episodic and short-term uses and reuses. They accumulated little interior debris from occupation besides hearths and their contents, a circumstance compatible with semipermanent occupation, where residents keep living interiors clean and dispose of trash outside (Hayden and Cannon 1983). No evidence of roofing exists, suggesting that these were unroofed campsites, perhaps constructed well with the intent to create a monument therein. However brief or repeated a residency, when it ended, these structures were deliberately filled with slabs, cobbles, and boulders, or a combination of fill, to create a protrusive platform, monumentalizing the house and commemorating activities that had taken place there.

There were no human burials under or within platform structures. Likewise absent are tools and broken tools. A small number of chert flakes and chips signals a minimal amount of stone tool use and the use of multiple raw materials, including obsidian, but it is unsurprising that these sharp articles should be absent from interior surfaces on which people were, however briefly, living. Nothing diagnostic in lithic type or technology was found from strata clearly dated to the occupation and construction of monuments.

In tandem with the broader evidence from archaeological survey in Wādī Sanā and elsewhere, the cattle skull ring and associated monument at Kheshiya and excavated platforms at Khuzma as-Shumlya provide strong evidence to suggest that the transformation of ringed structures into platforms was commemorative of significant feasting events. Save one example on a gravel terrace where downslope scree suggested the erosion of fill, no abandoned rings of stone uprights were found unfilled. The archaeological evidence suggests that all occupied structures were subsequently transformed into monumental stone platforms (or completely dismantled). They appear to commemorate a ritual sacrifice of cattle and the constitution of social identity that a grand sacrifice facilitated, and stone rings were constructed at the outset with such commemoration as the culminating purpose. Standing stones may have provided guideposts for the locales and the repetition of these events (McCorriston 2011:figures 29–31; McCorriston et al. 2012), but the monuments themselves attest to the significance of these gatherings, sacrifices, and feasts in the social lives of pastoralists who built no lasting houses.

References

Al-Abri, Abdulrahim, Eliška Podgorná, Jeffrey I. Rose, Luísa Pereira, Connie J. Mulligan, Nuno M. Silva, Riad Bayoumi, Pedro Soares, and Viktor Cerný

2012 Pleistocene-Holocene Boundary in Southern Arabia from the Perspective of Human mtDNA Variation. *American Journal of Physical Anthropology* 149(2):291–98.

Avner, Uzi

2002 Studies in Material and Spiritual Culture of the Negev and Sinai Populations, during the 6th–3rd Millennia BC. PhD dissertation, Institute of Archaeology, Hebrew University, Jerusalem.

Braemer, Frank, Serge Cleuziou, and Tara Steimer

2003 Dolmen-like Structures: Some Unusual Funerary Monuments in Yemen. *Proceedings of the Seminar for Arabian Studies* 33:169–82.

Černý, Viktor, Connie J. Mulligan, Verónica Fernandes, Nuno M. Silva, Farida Alshamali, Amy Non, Nourdin Harich, Lotfi Cherni, Amel El Gaaied Ben Ammar, Ali Al-Meeri, and Luísa Pereira

2011 Internal Diversification of Mitochondrial Haplogroup R0a Reveals Post-Last Glacial Maximum Demographic Expansions in South Arabia. *Molecular Biology and Evolution* 28(1):71–78.

Charpentier, Vincent

2008 Hunter-Gatherers of the "Empty Quarter of the Early Holocene" to the Last Neolithic Societies: Chronology of the Late Prehistory of South-Eastern Arabia (8000–3100 BC). *Proceedings of the Seminar for Arabian Studies* 38:93–115.

Charpentier, Vincent, and Rémy Crassard

2013 Back to Fasad . . . and the PPNB Controversy. Questioning a Levantine Origin for Arabian Early Holocene Projectile Points Technology. *Arabian Archaeology and Epigraphy* 24(1):28–36.

Cleuziou, Serge, and Maurizio Tosi

1997 Hommes, climats et environnements de la Péninsule arabique à l'Holocène. *Paléorient* 23(2):121–35.

Crassard, Rémy

2008 *La préhistoire du Yémen. Diffusions et diversités locales, á travers l'étude d'industries lithiques du Hadramawt.* British Archaeological Reports, International Series 1842. Archaeopress, Oxford.

Gebel, Hans G. K., and Hamzeh M. Mahasneh

2013 Disappeared by Climate Change: The Shepherd Cultures of Qulban Beni Murra (2nd Half of the 5th Millennium BC) and Their Aftermath. *Syria* 90:127–58.

Hayden, Brian, and Aubrey Cannon

1983 Where the Garbage Goes: Refuse Disposal in the Maya Highlands *Journal of Anthropological Archaeology* 2(2):117–63.

Henton, Elizabeth, Joy McCorriston, Louise Martin, and Eric A. Oches

2014 Seasonal Aggregation and Ritual Slaughter: Isotopic and Dental Microwear Evidence for Cattle Herder Mobility in the Arabian Neolithic. *Journal of Anthropological Archaeology* 33:119–31.

Khalidi, Lamya, Bernard Gratuze, Marie-Louise Inizan, and Rémy Crassard

2013 Considering the Arabian Neolithic through a Reconstitution of Interregional Obsidian Distribution Patterns in the Region. *Arabian Archaeology and Epigraphy* 24:59–67.

Martin, Louise, Joy McCorriston, and Rémy Crassard

2009 Early Arabian Pastoralism at Manayzah in Wādī Ṣanā, Ḥaḍramawt. *Proceedings of the Seminar for Arabian Studies* 39:271–82.

McCorriston, Joy

2011 *Pilgrimage and Household in the Ancient Near East.* Cambridge University Press, Cambridge.

McCorriston, Joy, Michael J. Harrower, Louise A. Martin, and Eric A. Oches

2012 Cattle Cults of the Arabian Neolithic and Early Territorial Societies. *American Anthropologist* 114:45–63.

McCorriston, Joy, Michael Harrower, Tara Steimer, Kimberly D. Williams, Matthew Senn, Masʿūd Al Hādhari, Masʿūd Al Kathīrī, ʿAli Aḥmad Al Kathīrī, Jean-François Saliège, and Jennifer Everhart

2014 Monuments and Landscape of Mobile Pastoralists in Dhufar: The Arabian Human Social Dynamics (AHSD) Project 2009–2010. *Journal of Oman Studies* 19:117–43.

McCorriston, Joy, Eric A. Oches, Dawn Walter, and Kenneth Cole

2002 Holocene Paleoecology and Prehistory in Highland Southern Arabia. *Paléorient* 28(1):61–88.

McCorriston, Joy, Tara Steimer-Herbet, Michael J. Harrower, Kimberly A. Williams, Jean-François Saliège, and ʿAbdalaziz Bin ʿAqil

2011 Gazetteer of Small Scale Monuments in Prehistoric Hadramawt, Yemen: A Radiocarbon Chronology from RASA Project Research 1996–2008. *Arabian Archaeology and Epigraphy* (22):1–22.

Méry, Sophie, V. Charpentier, G. Auxiette, and E. Pelle

2009 A Dugong Bone Mound: The Neolithic Ritual Site on Akab in Umm al-Quwain, United Arab Emirates. *Antiquity* 83(321):696–708.

Rose, Jeffrey I., Viktor Černý, and Riad Bayoumi

2013 Tabula Rasa or Refugia? Using Genetic Data to Assess the Peopling of Arabia. *Arabian Archaeology and Epigraphy* 24(1):95–101.

Zarins, Juris, Moawiyah Ibrahim, Daniel Potts, and Christopher Edens

1979 Saudi Arabian Archaeological Reconnaissance 1978: The Preliminary Report on the Third Phase of the Comprehensive Archaeological Survey Program, the Central Province. *Atlal* 3:9–26.

Chapter 13

Water Management and Irrigation along Wādī Sanā

Michael J. Harrower

It started raining at around 5:30pm, just after we had returned to our Wādī Sanā tent camp from fieldwork. It had rained a little last week but nothing like this. There were very light showers at 4:30, and then we saw dark clouds and lightning approaching. An hour later we fought tent-ripping wind as we scrambled to get everything under cover and repair the tent stays. Water started to fill our camp. As darkness loomed, we worried that if rains continued through the night, we faced a new danger. We were camped on a safe terrace—"Safe 'til the thousand-year flood," Rick said—yet we were only a little more than 100 meters away and a few meters above the main Wādī Sanā channel. The rain slowed. Walīd the cook had salvaged dinner, and a sodden crew grateful for hot rice and sauce tumbled from cars and the scant shelter of a rocky overhang. Just afterwards, I collapsed exhausted in a damp bed, only to waken hours later to the rumbling sound of the sayl flashflood rushing down the wadi with uprooted trees and boulders. It was another full day before the flow and floodwaters eventually receded, and we were able to continue fieldwork.

—Adapted from Michael Harrower field notes, March 6, 2005

In archaeologists' long efforts to find the world's earliest agriculture, studies in the Near East have played a central role. Although domesticated plants and animals appear in Yemen far later than in the Levant or Mesopotamia, evidence of ancient agriculture in Yemen nevertheless offers a range of important insights, not only regarding the unique histories of Yemen itself but more broadly pertinent to ancient agriculture worldwide.

For more than 50 years, researchers presuming that agriculture originated in a few core areas and then spread to surrounding, supposedly peripheral, regions have studied ancient agriculture as a problem of origins and spread, (Sauer 1952; Vavilov 1951). Although this simple model has served as a useful introduction to the problem, it has come to inhibit more sophisticated understanding of factors that were involved in, and were responsible for, transitions to agriculture. Most notably, the unique environmental and social contexts of different regions necessitated that agriculture was readapted, reformulated, and reinvented to fit new, atypical circumstances. Further research in less well-known regions, including Yemen, is thus crucial to understanding agriculture's diverse beginnings.

RASA's search for early agriculture has not yet yielded direct evidence of ancient domesticated crops, but the focal area of the RASA Project—Wādī Sanā—has generated some of the earliest domesticated animals in Arabia (chapter 8) as well as evidence of some of the earliest irrigation in the region. This chapter recounts results of investigations of irrigation, which convey substantial insights about early agriculture. Irrigation began in Wādī Sanā at least as early as the sixth millennium BP, at about

Figure 13.1. Wādī Sanā showing the receding *sayl* flood on March 7, 2005. The only significant rain event during six seasons of RASA project fieldwork happened the previous evening. *Photograph by Michael Harrower.*

the same time as terrace agriculture (Wilkinson 1999) and domesticated crops (Ekstrom and Edens 2003) first appeared in the highlands of western Yemen. Recently funded phytolith research on sediment samples collected by RASA holds promise for clarifying the botanical history of the area, but currently we do not know what was first irrigated or if early water management structures were used to capture water for domestic purposes or for animals. Nevertheless, we do know that Wādī Sanā was one of the areas in which small-scale water management technologies appeared very early in Arabia's history. Over the next few thousand years, people in the region developed a wide range of irrigation design, construction, operation, and management expertise, and this laid the agricultural foundations for the rise of Yemen's ancient desert kingdoms. Wādī Sanā yields significant evidence of the very early, small-scale beginnings of Southwest Arabian water control technologies.

Irrigation and Early Agriculture in Southwest Arabia

Archaeologists have long recognized the significance of water in ancient Near Eastern agriculture (e.g., Childe 1952; Flannery 1969), but relatively few studies over the past half century have focused on ancient water use technologies and geographies. Although irrigation is commonly viewed as an innovation that developed after rain-fed farming, this supposition is not warranted (Sherratt 1980). Cultivation in many arid regions first targeted water-rich areas, such as near springs, periodically flooded drainages or riverbanks, or high-water-table bottomlands. And as Robert McC. Adams (2006:17) cogently noted, irrigation (that is, supplemental watering) in the Near East may be as old as the earliest cultivation. Indeed, Australia (Tindale 1977) and the American Southwest (Bean and Lawton 1993; Steward 1930) have even yielded evidence of irrigation before cultivation of domesticates; and Wādī Sanā contributes important evidence to the issue of water control at the cusp of agriculture.

Irrigation technologies suitable in Yemen are helpfully described along a continuum of increasing technological complexity and labor intensity (Hunt 2007:105–28; Varisco 1996). The simplest forms of supplemental watering involve cultivation in water-rich areas, such as near springs or periodically inundated drainages. Digging wells or cisterns, building small barrages or channels to divert water, and terracing require slightly more complex design expertise and somewhat more labor, yet such systems can still readily be devised, constructed, operated, and maintained at the household level. Larger-scale flash flood (*sayl*) irrigation systems and underground infiltration galleries (*qanats*) require design skills and construction expertise that supersedes the household scale and often necessitates coordination at the community (village) level or beyond. These various strategies appear in a variety of contexts and combinations across Yemen, and different technologies appear to have originated in different places during the sixth millennium BP (Wilkinson 2006). Interestingly, these appearances occur at a time of dramatic sociopolitical change and state formation across the Near East and also during the interval of earliest crop agriculture in Southeast Arabia (United Arab Emirates and Oman). The temporal coincidence suggests that interregional factors may be partly responsible for changes in Yemen.

Remnants of Water Management and Irrigation in Wādī Sanā

As one of the contexts in which irrigation first began in Southwest Arabia, Wādī Sanā yields considerable insights regarding social and environmental contexts of early agriculture. In recent times a variety of attempts to capture water for agriculture are evident in bulldozed embankments along the margins of middle Wādī Sanā, but these constructions reportedly never successfully sustained crops. A tube well equipped with a diesel pump was installed near Khuzma as-Shumlya by Canadian Occidental Petroleum in the late 1990s. Water from this well supplied herded sheep, goats, and camels for a short interval, but the pump quickly fell into disrepair and was abandoned. These forsaken efforts attest to scarcity of water in the area and the profound challenges of contemporary pastoralism and agriculture. Larger-scale floodwater irrigation agriculture operates along other drainages of the Hadramawt, such as Wadi Do'an, but it relies on more technologically advanced and labor-demanding technologies that are ill-suited to Wādī Sanā. Irrigation today along Wādī Sanā is restricted to headwaters at Ghayl Bin Yumain and Wādī Sanā's outlet at the village of Sanā, where diesel pumps tap groundwater. Thus the present population density is extremely low, with perhaps only 50 people living from Ghayl Bin Yumain to Sanā. Fortunately for archaeologists, this low population density has contributed to excellent preservation of archaeological remains, including ancient waterworks.

The deeper past of Wādī Sanā exhibits evidence of Early Holocene pluvial conditions and the ingenuity of early peoples in adapting to water-scarce environments. While many of the small-scale water management structures evident along Wādī Sanā are difficult to date, geoarchaeological field examination, radiocarbon dating, and mapping help reveal their likely uses and approximate ages.

In 1998 and 2000 the RASA Project investigated a number of waterworks along Wādī Sanā and its tributary Wādī as-Shumlyah. Interest focused on two check dams (designated 1998-000-B and 2000-009-1). Optically stimulated luminescence (OSL) dates on underlying sediments (10,400 ± 4,500 years ago for 1998-000-B on the Gravel Bar Site and 7,300 ± 1,500 years ago for 2000-009-1) suggested an early Holocene age (McCorriston and Oches 2001). However, the wide error ranges of these dates (due to the general lack of quartz and feldspar grains in Wādī Sanā alluvium) only very approximately situated them in time. (See more on subsequent dating of Channel 9-1 below.)

After these preliminary investigations, many questions about irrigation remained unanswered. How many similar structures were along Wādī Sanā? How old were they and how were they used? Was a chronological development of agricultural technologies evident in the record? This chapter abbreviates the lengthier documentation of survey, excavation, results, and analysis found in Michael Harrower's (2006) dissertation and subsequent publications (e.g., Harrower 2008a, 2008b, 2016).

Methods

In 2004, with support of a National Science Foundation Dissertation Improvement Grant for research on ancient irrigation, more concerted efforts were made to discover, map, and date irrigation along Wādī Sanā. In such an arid area, one would expect that irrigation strategies would have been highly structured by opportunities to harness scarce water resources. Therefore, in terms of both type and location, irrigation structures would be highly correlated with hydrological variables, including spatial patterns of water availability. But irrigation technologies were also subject to critical social contingencies, including the ability of communities to cooperatively or coercively mobilize labor, coordinate construction and maintenance activities, and mitigate disputes. This work thus tested the hypothesis that ancient irrigation systems along Wādī Sanā were highly correlated with quantifiable hydrological variables reflecting close behavioral ties to environmental conditions. Spatial analyses gauged the strength of associations between water, arable land, and irrigation, and they revealed hidden details of ancient water management (Harrower 2006, 2008a).

Along Wādī Sanā, a total of 174 irrigation structures were mapped with a Trimble Pathfinder Pro XRS GPS backpack system using real-time differential correction. These structures were also photographed, and details of their construction, stratigraphic associations, and preservation were recorded.

In conjunction with archaeological survey, samples were collected from irrigation structures and their immediate vicinity to gather information about construction techniques and to collect samples for radiocarbon and OSL analysis. In comparison with most contexts archaeologists excavate, irrigation structures present extraordinary dating challenges (Wilkinson 2003:71–100). Our efforts to determine the age of irrigation along Wādī Sanā sought contexts where structures were exposed on the ground surface so they could be identified but were also partially buried or associated with nearby buried materials. Such contexts are rare, and datable materials in sediments deposited in, around, and over irrigation structures are not necessarily the same age as a structure's date of construction and use. Three main strategies were used to establish the age of irrigation structures: (1) excavating the structures themselves (W5, W19); (2) excavating sediments surrounding irrigation structures (9-1, 13-1); (3) sampling hearths nearby and in association with irrigation structures (W1, W5, W6, W13, W23). In the absence of additional buried irrigation structures that could be excavated for terminus ante quem (must be older than) ages, our strategy of sampling of hearths adjacent to irrigation structures was used to retrieve terminus post quem (must be younger than) age estimates that would clarify the maximum possible age of irrigation (see Harrower 2008b).

Results

Types of Water Management Structures

A wide variety of remnant water management structures were identified along Wādī Sanā. In addition to the 174 diversion channels and check dams mapped, we observed a range of less common structures that are often difficult to situate in time but nevertheless yield important insights about experimentation with water control.

Diversion Channels (*Shrūj*) and Check Dams

Most water management structures identified can be divided into two basic types: (1) diversion channels that redirect runoff from rocky hillsides (figure 13.2) and (2) check dams that slow water flows so that moisture and nutrient-rich sediments accumulate (figure 13.3). These small-scale systems are no longer used in Wādī Sanā but are known and are locally called *shrūj* irrigation. Local Ḥumūm bedouin recognize them but insist that such technologies have never worked successfully in living memory because of insufficient rainfall.

Storage Reservoirs (*Krif*)

No ancient wells or cisterns were identified along Wādī Sanā, but large earthen *krif* storage reservoirs are found in

Figure 13.2. Water diversion channel on rocky plateau above Wādī Sanā. *Photograph by Michael Harrower.*

Figure 13.3. Check dam in small tributary drainage along Wādī Sanā. *Photograph by Michael Harrower.*

a number of locations (figure 13.4). For example, a large earthen reservoir 2 km west of Khuzmah as-Shumlyah captures runoff from a small, upland 22 ha area. Although this reservoir and others like it show bulldozer scars, indicating they were recently worked, it seems likely that some have been rebuilt from earlier constructions. Although all reservoirs along Wādī Sanā were dry during RASA fieldwork, imagery on Google Earth dated September 5, 2009, showed the aforementioned reservoir and one other nearby filled with water. This suggests that they do occasionally fill and probably are used to supply animals.

Abatement Walls

Rough dry-laid stone walls built parallel to hillslopes are found in a number of places, including just south of the Khuzma in an area designated WATER 5 (figure 13.5). Given the complete lack of sediment on upland areas, these walls are definitely not agricultural terraces like those found in northern Yemen. Instead, most probably were built to slow runoff and prevent erosion of arable sediments below. Their function is illustrated by comparable modern walls built on the plateau above Wadi Do'an. These walls similarly slow runoff and are equipped with

Figure 13.4. Earthen *krif* water reservoir. *Photograph by Michael Harrower.*

Figure 13.5. Stone wall built perpendicular to the hillslope along Wādī Sanā (W5-2). *Photograph by Michael Harrower.*

Figure 13.6. Stone wall built to slow runoff and prevent erosion above Wadi Do'an. *Photograph by Michael Harrower.*

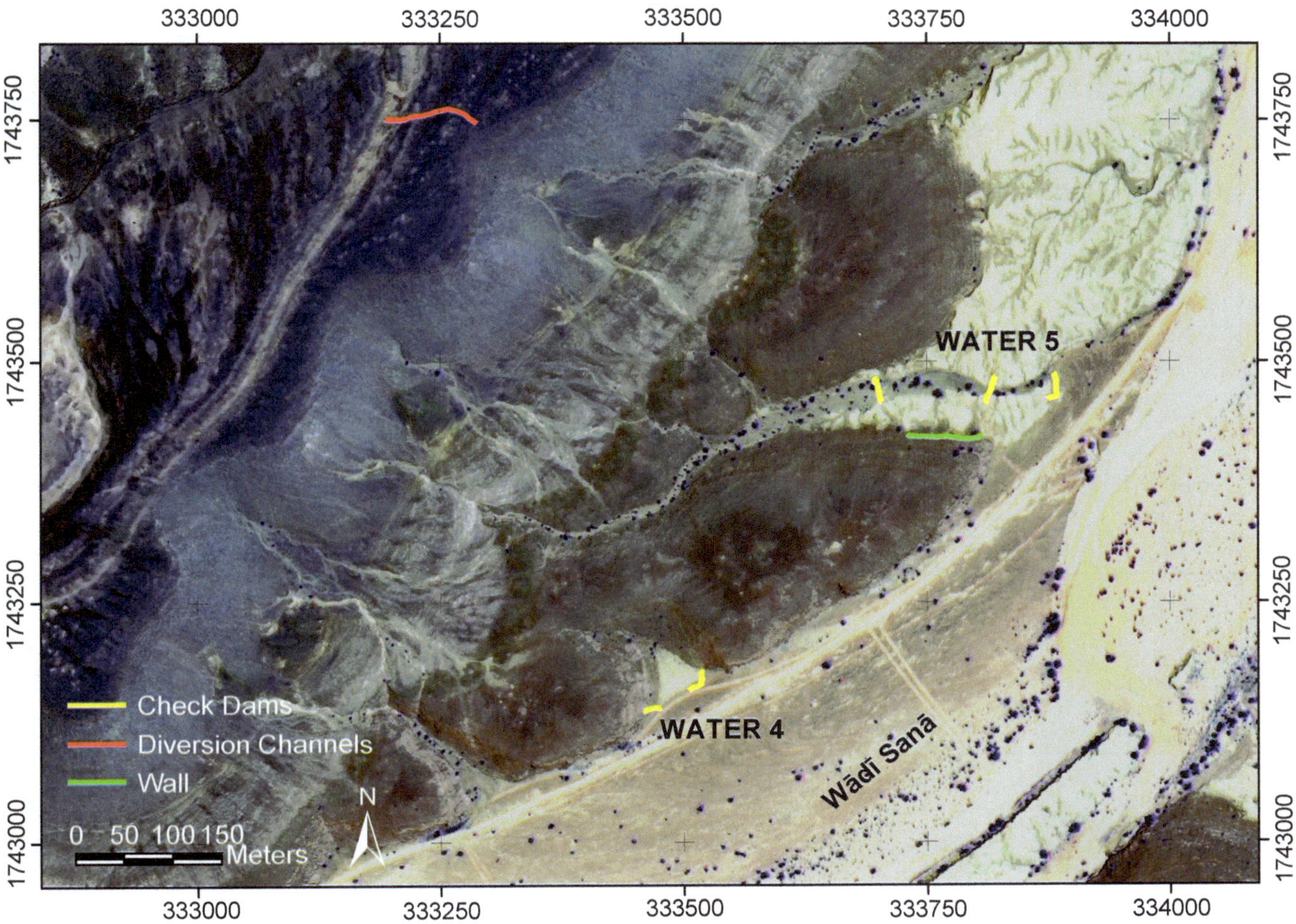

Figure 13.7. Map of W5 (pan-sharpened QuickBird). *Illustration by Michael Harrower.*

periodic holes to allow water to slowly percolate through to fields (figure 13.6). Comparable walls found near Bronze Age occupations in the Ad-Dhahīrah region of northern Oman suggest that this strategy is widespread and of considerable, yet indeterminate, antiquity.

Chronology of Water Management Structures

Many of the contexts with potentially helpful chronological information offered terminus post quem (the irrigation structure must be younger than the dated material) rather than terminus ante quem (the irrigation structure must be older than the dated material) ages. Moreover, given their wide error ranges and divergence from radiocarbon assays, the OSL dates, which ranged from 8.06 kya to 0.97 kya, offered relatively little to clarify the chronology of irrigation in Wādī Sanā (tables 13.1 and 13.2) and are excluded from Bayesian modeling presented in chapter 18.

Excavations of Water Management Structures

Earthen check dams W5-1 and W19-1 were breached, creating natural cross-section profiles that were easily cleaned and examined (figures 13.7, 13.8, and 13.9). At W5-1, radiocarbon samples were recovered from a thin layer of charcoal covering the upstream face of the dam, and OSL samples were taken above and below this layer. Collectively these samples suggest that the dam may have been constructed after 4,600 years ago (OSL-5, lab no. 653) and was in use, or at least was surmounted by a layer of burned material, during the last half century (OSL-4, lab no. 652, and AA60247, 293 ± 39). At W19-1, two OSL assays suggest that the check dam may have been built roughly 2,500 years ago. Yet the stratigraphically lower sample yielded a younger date, suggesting inaccuracy of OSL dating at this location.

Table 13.1. Radiocarbon ages from Wādī Sanā irrigation structures.

#	Sample #	Lab #	Sample	Context	UTM Northing	UTM Easting	Age For	^{14}C Year BP	Calibrated BP	Posterior Mean (cal BP)
RC-1	2004-W5-1-T1-6	AA60247	charred plant debris	check dam burned layer	1743471	333882	terminus ante quem for W5-1	293 +/- 39	471–158	405
RC-2	2004-009-1-T3-7	AA60245	wood charcoal	sediment above rock-bordered canal	1744589	337163	terminus ante quem for 9-1	4471 +/- 42	5297–4970	5185
RC-3	2004-009-1-T1-2	AA59569	wood charcoal	sediment above rock-bordered canal	1744567	337175	terminus ante quem for 9-1	4475 +/- 36	5292–4976	5175
RC-4	2004-W23-1-H1-1	AA60251	wood charcoal	hearth in silt below water diversion channel	1747736	337933	terminus post quem for W23	5637 +/- 44	6499–6308	6453
RC-5	2004-W13-1-H1-1	AA60250	wood charcoal	hearth in silt section near check dam	1743925	333798	terminus post quem for W13-1	5783 +/- 44	6677–6467	6618
RC-6	2004-W6-1-H2-1	AA60249	wood charcoal	hearth in silt section near check dam	1740674	329536	terminus post quem for W6-1	5923 +/- 44	6880–6657	6788
RC-7	2004-W1-4-H1-1	AA60246	wood charcoal	hearth on silts near check dam	1742559	333515	terminus ante quem for W1-1?	6168 +/- 51	7241–6936	7111
RC-8	2004-W5-3A-3	AA60248	wood charcoal	hearth in silt section near check dam	1743486	333749	terminus post quem for W5-3A	6232 +/- 45	7259–7006	7186

Note: Posterior distribution means are from Bayesian modeling. See chapter 18, table 1.

Table 13.2. Optically stimulated luminescence ages from Wādī Sanā irrigation structures.

Field #	Survey Unit/Site #	Lab #	Description	UTM Northing	UTM Easting	Elevation (m)	Depth below Surface	Age For	IRSL Age (kya)	Radiocarbon Date Posterior Mean (cal BP)
OSL-1	009-001-T2	649	Fluvial deposits burying irrigation structure	1744581	337168	697	0.36 m	terminus ante quem for 009-1	3.11 ± 1.30	N/A
OSL-2	W1-1-T1	650	Fluvial deposits beneath irrigation structure	1742559	333515	693	0.45 m	terminus ante quem for W1-1?	8.06 ± 1.91	7111 (AA60346)
OSL-3	W1-1-T2	651	Fluvial deposits beneath irrigation structure	1742541	333522	693	0.35 m	terminus ante quem for W1-1?	2.27 ± 1.73	N/A
OSL-4	W5-1-T1	652	Construction fill of irrigation structure (fluvial)	1743471	333882	690	0.25 m	terminus ante quem for W5-1	0.97 ± 0.84	405 (AA60247)
OSL-5	W5-1-T1	653	Construction fill of irrigation structure (fluvial)	1743471	333882	690	0.80 m	terminus post quem for W5-1	4.67 ± 1.48	N/A
OSL-6	009-001-T1	654	Fluvial deposits burying irrigation structure	1744567	337175	693	0.24 m	terminus ante quem for 009-1	1.65 ± 1.02	5175 (AA59569)
OSL-7	009-001-T3	655	Fluvial deposits burying irrigation structure	1744589	337163	695	0.21 m	terminus ante quem for 009-1	1.37 ± 0.89	5185 (AA60245)
OSL-8	W19-1-T1	656	Construction fill of irrigation structure (fluvial)	1739933	328756	720	0.77 m	terminus post quem for W19-1	2.30 ± 0.86	N/A
OSL-9	W19-1-T1	657	Construction fill of irrigation structure (fluvial)	1739933	328756	720	0.60 m	terminus ante quem for W19-1	2.44 ± 1.13	N/A

Note: Radiocarbon age is posterior distribution mean from Bayesian modeling. See chapter 18, table 1.

Figure 13.8. Cross-section excavation of check dam W5-1. *Photograph by Michael Harrower.*

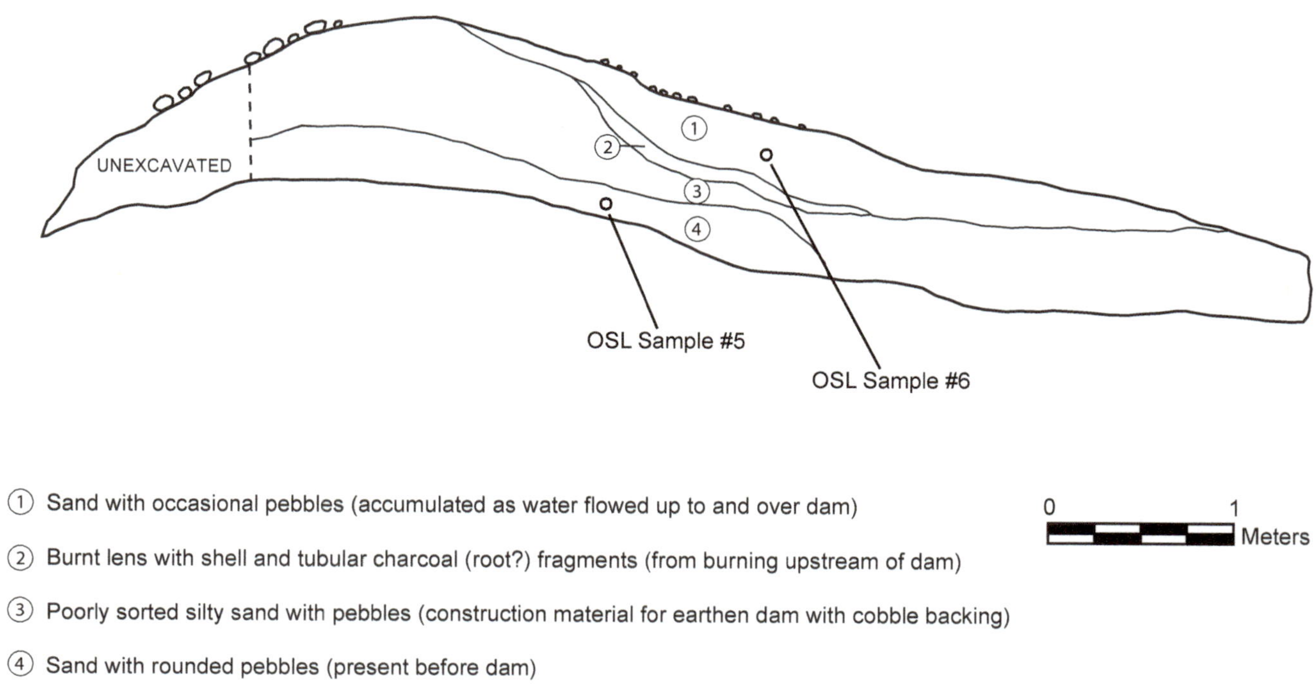

Figure 13.9. Profile drawing of W5-1 excavation showing burned layer and OSL sampling locations. *Illustration by Michael Harrower and Clara Hickman.*

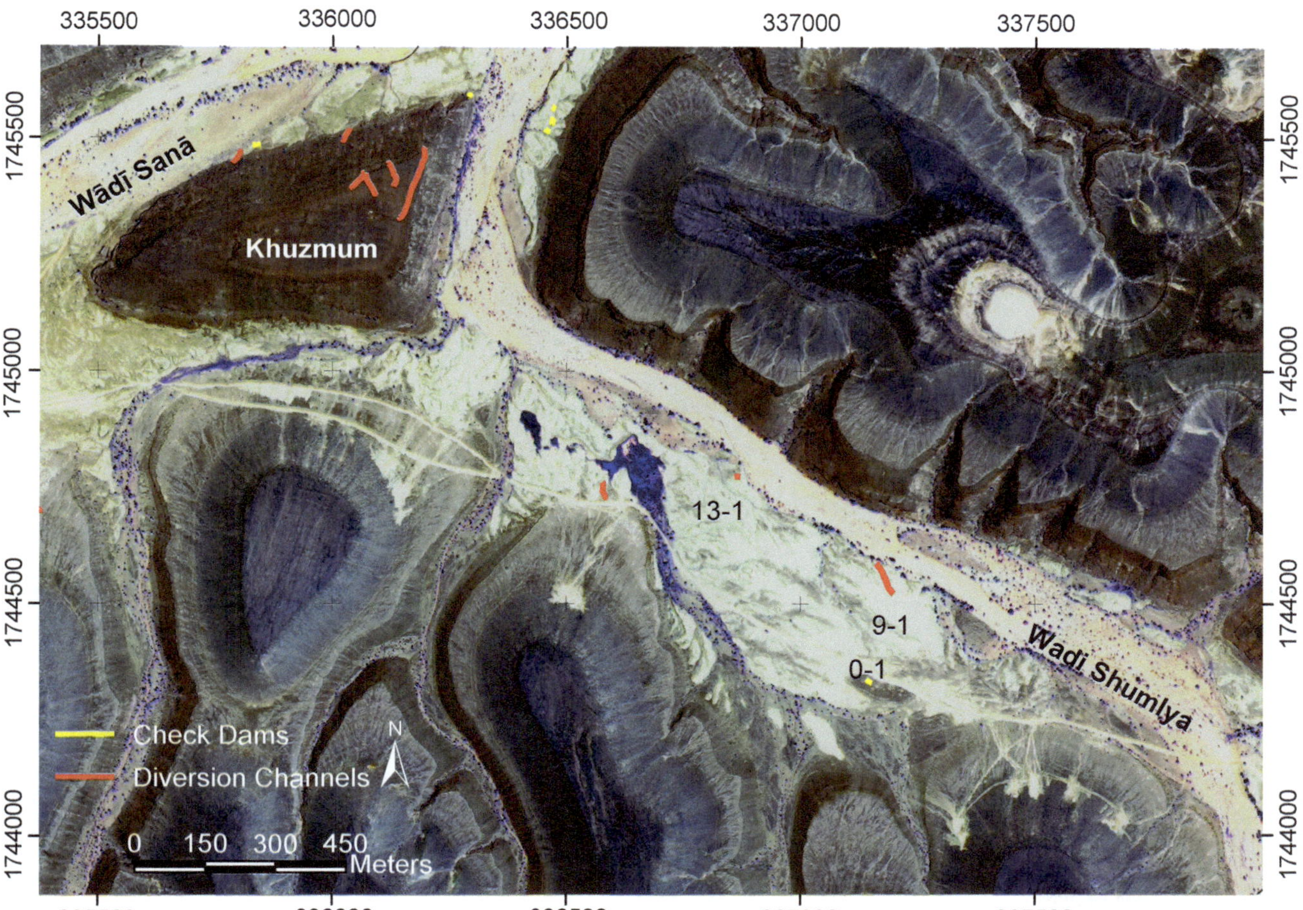

Figure 13.10. Map of the Khuzma area showing 13-1 and 9-1 (pan-sharpened QuickBird). *Illustration by Michael Harrower.*

Excavations of Sediments Surrounding Irrigation Structures

This second strategy was used for two rock-bordered canals found along Wādī as-Shumlyah. The first of these, 13-1, is comprised of single exposed line of small boulders, 12 m long (figures 13.10 and 13.11). A small 50 cm wide and 45 cm deep test pit was dug beneath one of the stones. A compact layer of silty sand was found, and then a lower layer of very loose sand filled with *Melanoides tuberculata* shells. This species, a freshwater snail that feeds on algae and can tolerate brackish water, attests to marshy conditions (Al-Safadi 1991). A sample of these shells was collected for possible radiocarbon dating, but because of the complexities of carbon uptake, samples of snail shell from this and other areas were not dated.

Another irrigation structure we excavated, a rock-bordered canal designated 9-1, proved by far the most informative site we examined (Harrower 2008b). This structure was originally studied in 1998 and identified as a check dam (McCorriston and Oches 2001); more thorough examination and modeling of water flow in the area, along with test excavations, revealed that it was in fact a small rock-bordered canal traceable for 75 m. Two lines of small boulders running in a northwesterly direction (332 degrees) are visible on the surface for 41 m, after which the stones continue underneath the ground surface another 34 m (figures 13.12 and 13.13). These two lines of stones, about 1 m apart, demarcate the outer edges of a small earthen canal that would have carried water redirected from Wādī as-Shumlyah—a major tributary of Wādī Sanā. Even though many of the stones exposed on the ground surface are loose and dislodged from their original locations, the area surrounding 9-1 is devoid of large stones, making the structure's original alignment recognizable. Moreover, stones that are embedded in sediment and thus are still in situ were documented in six locations, including

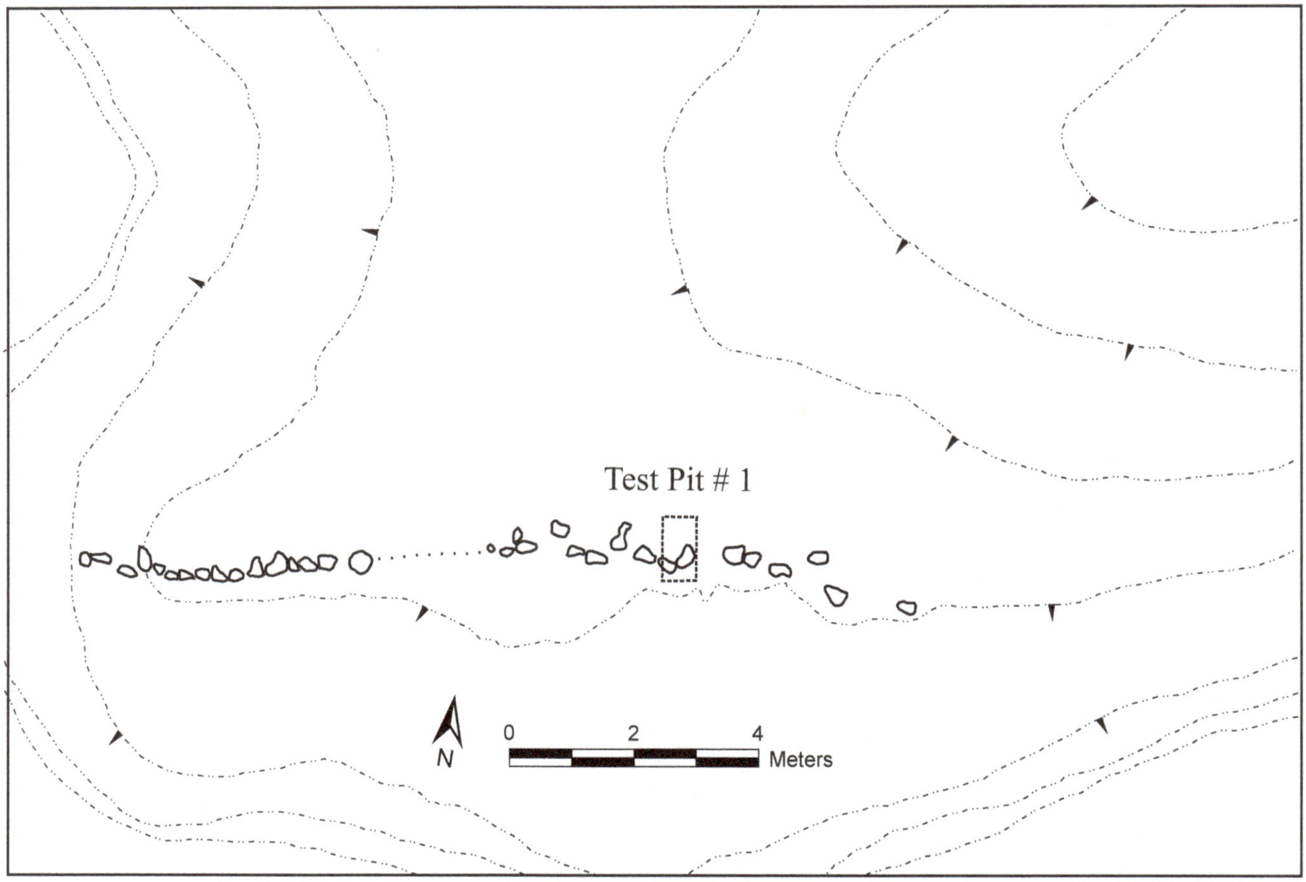

Figure 13.11. Plan of diversion channel 13-1. *Drawing by Michael Harrower and Clara Hickman.*

Figure 13.12. Rock-bordered canal 9-1 looking northwest from the southeast end, with test excavations in the background. *Photograph by Michael Harrower.*

Figure 13.13. Rock-bordered canal 9-1 looking northwest at embedded stones (foreground), the section sampled in 1998 (mid-ground), and the 2004 test excavations (background). *Photograph by Michael Harrower.*

Figure 13.14. Rock-bordered canal 9-1 looking southeast from the northwest end, showing test excavation T3 in foreground with excavations T2 and T1 in background. *Photograph by Michael Harrower.*

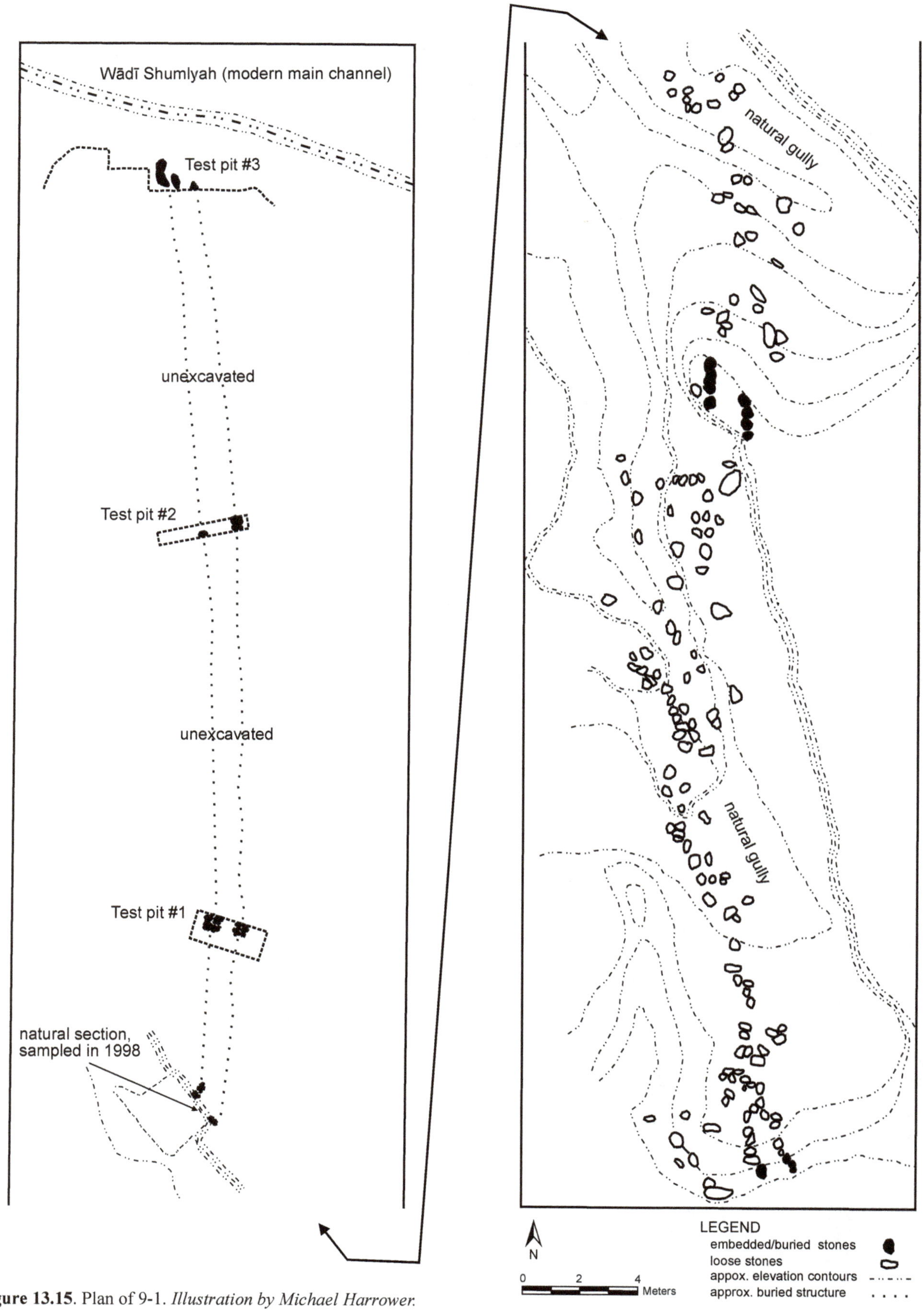

Figure 13.15. Plan of 9-1. *Illustration by Michael Harrower.*

Figure 13.16. Rock-bordered canal 117-1, comparable to 9-1. *Photograph by Michael Harrower.*

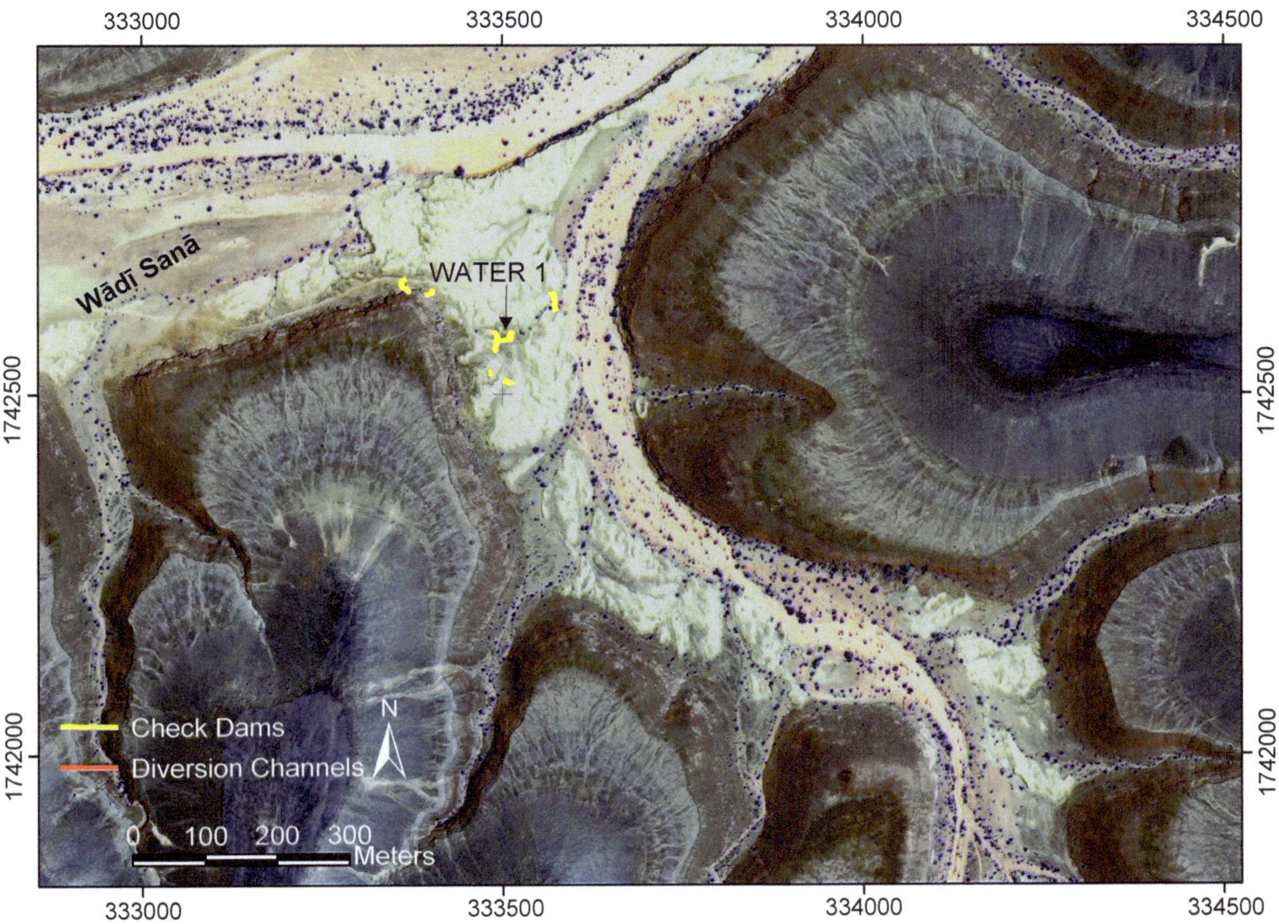

Figure 13.17. Map of W1 (pan-sharpened QuickBird). *Illustration by Michael Harrower.*

Figure 13.18. Dam at W1 with hearths in foreground. *Photograph by Michael Harrower.*

Figure 13.19. Hearth H1 at W1. *Photograph by Michael Harrower.*

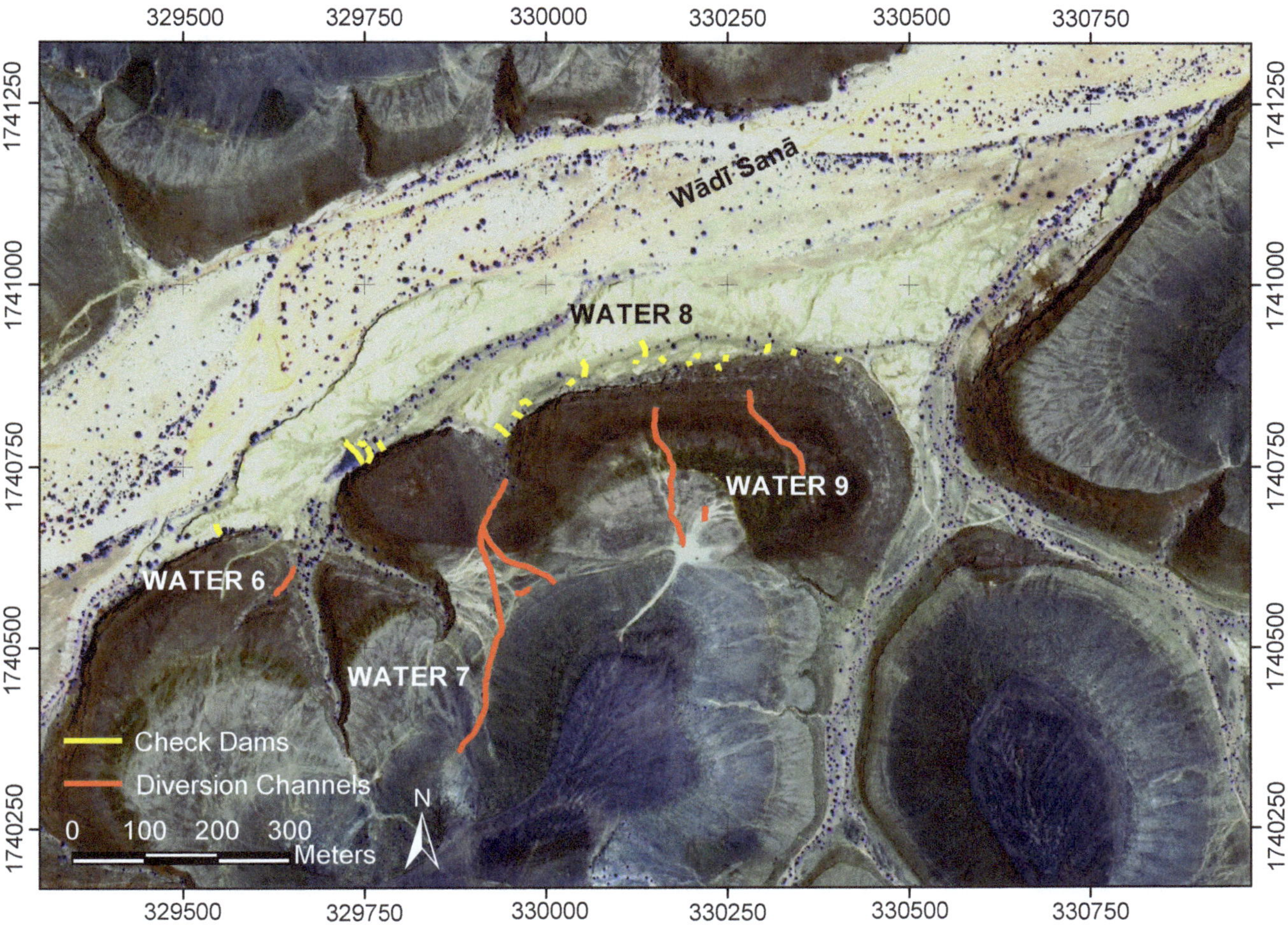

Figure 13.20. Map of W6 (pan-sharpened QuickBird). *Illustration by Michael Harrower.*

three test pits (figures 13.14 and 13.15). Two radiocarbon samples (AA55969, 4475 ± 36; AA60245, 4471 ± 42) collected from Test Pits 1 and 3 (T1 and T3) yielded dates that when calibrated fall near 5200 cal BP. These dates are among the latest of more than 30 radiocarbon dates from wadi silts along Wādī Sanā and provide a strong terminus ante quem for 9-1. After fieldwork in 2004, one OSL assay (lab nos. 646, 654, and 655) was run from each of the three test pits along 9-1. These yielded age estimates between 1,370 and 3,110 years ago, with wide error ranges (± roughly 1,000 years). These are not only inconsistent with the aforementioned radiocarbon assays, but they also contradict the OSL date of 7,300 ± 1,500 collected in 1998. Therefore we consider the radiocarbon ages the more reliable age estimate. (See also chapter 18.) This important structure thus offers some of the earliest evidence of irrigation in Southern Arabia, as early, or perhaps somewhat earlier, than terrace agriculture in western Yemen (Wilkinson 1999, 2006). A canal structure of very comparable design was found farther north along a smaller tributary of Wādī Sanā (117-1), but it lacked opportunities for dating (figure 13.16).

Strategic Sampling of Hearths in Association

We used strategic sampling of hearths in association with irrigation structures to date irrigation along Wādī Sanā in five areas (W1, W5, W6, W13, W23). At W1, two hearths just upstream of a dam in the channel of a tributary of Wādī Sanā were sampled for radiocarbon, and OSL samples were also collected immediately beneath the two hearths (figures 13.17, 13.18, and 13.19). It was originally hypothesized that these samples would provide terminus ante quem age estimates for the dam, as we presumed that these hearths could not have existed immediately upstream of the dam when it was operable. The radiocarbon sample from W1 (AA60246, 6168 ± 51) submitted

Figure 13.21. Hearth at W6 (pointed to by Nasser Al-ʿAlīy) sampled for radiocarbon analysis. *Photograph by Michael Harrower.*

Figure 13.22. Hearth at W5 (foreground) sampled for radiocarbon dating. *Photograph by Michael Harrower*

for dating yielded an age of 7070 cal BP, and the two OSL samples (lab nos. 650 and 651) yielded ages of 8,060 ± 1,910 years ago and 2,270 +/- 1,730 years ago . This again raises questions about the accuracy of OSL dating. Even though two of these three age estimates are exceptionally early, we do not believe they conclusively demonstrate irrigation before the mid-Holocene.

Sampling of hearths at W5 and W6 illustrates our terminus post quem approach. As discussed in chapter 3, alluvial sediments infilled along Wādī Sanā during the Early Holocene and a prominent shift to incision and erosion began during late sixth millennium BP. Sporadic islands of alluvial silt were created along middle Wādī Sanā, with channels eroded into silt along cliff lines at the outer margins of Wādī Sanā. These secondary backwater channels offered attractive opportunities for small-scale irrigation; water from adjacent hillslopes could be diverted and slowed to distribute moisture and nutrients (figure 13.20). Hearths embedded (in but not on top of) silts along these channels must have formed when silt was still infilling during the Early Holocene and, therefore, must be older than irrigation structures built within the backwater channels (figures 13.21 and 13.22). Radiocarbon assays from such hearths yielded ages of 6750 cal BP at W6 (AA60249, 5923 ± 44) and 7141 cal BP at W5 (AA60248, 6232 ± 45), indicating that irrigation in these areas dates to the Middle to Late Holocene at the earliest.

Strategic radiocarbon dating similarly yielded terminus post quem ages for irrigation at W13 and W23. Hearths in the vicinity of irrigation structures produced radiocarbon dates of 6415 cal BP (AA60251, 5637 ± 44) at W23 and 6584 cal BP (AA60250, 5783 ± 44) at W13, further supporting a Middle/Late Holocene age for the irrigation along Wādī Sanā.

Conclusions

Our investigations along Wādī Sanā substantially clarify the beginnings of water management and irrigation in Yemen and offer important insights not only for Southern Arabia but for the origins of agriculture generally. Although dating water control systems presents significant challenges, our efforts reveal irrigation during the mid-sixth millennium BP. These findings complement and extend results in western Yemen, where research, independent of investigations in Wādī Sanā, similarly shows terrace agriculture (Wilkinson 1999) and crops (Ekstrom and Edens 2003) as early as the sixth millennium BP. Additionally, Bayesian modeling of radiocarbon dates presented in chapter 18 of this volume conclusively pushes the age of irrigation in Wādī Sanā even earlier, to 5837 cal BP, making it arguably the earliest dated irrigation in Southern Arabia (cf. Desruelles et al. 2016).

The earliest dates for irrigation in ancient Yemen are much later than the earliest crop agriculture elsewhere in the Near East (for example, the Levant, Anatolia, and Mesopotamia), but Yemen nevertheless reveals key characteristics of early farming. Irrigation has long been viewed as an innovation added long after the earliest cultivation, yet Yemen shows water control from the very beginnings of agriculture and calls attention to the possibility that the earliest farming in other regions may have similarly involved more than merely rainfall. Irrigation was not a requirement of crop agriculture in Yemen; many areas of western Yemen receive more than enough rainfall for dry farming. Early farmers turned to water control not because it was an unavoidable requirement but because it increased productivity and reliability. In essence, it made sense to employ techniques that were efficient, dependable, and sustainable—and water control techniques tailored to unique social and environmental circumstances helped ensure the long-term viability of food production.

References

Adams, Robert McC.

2006 Intensified Large-Scale Irrigation as an Aspect of Imperial Policy: Strategies of Statecraft on the Late Sassanian Mesopotamian Plain. In *Agricultural Strategies*, edited by Joyce Marcus and Charles Stanish, pp. 17–37. Cotsen Institute of Archaeology Press, Los Angeles.

al-Safadi, M. M.

1991 Freshwater Macrofauna of Stagnant Waters in Yemen Arab Republic. *Hydrobiologia* 210:203–08.

Bean, Lowell J., and Harry W. Lawton

1993 Some Explanations of the Rise of Cultural Complexity in Native California with Comments on Proto-Agriculture and Agriculture. In *Before the Wilderness: Environmental Management by Native Californians*, edited by Thomas C Blackburn and Kat Anderson, pp. 27–54. Ballena Press, Menlo Park, CA.

Childe, V. Gordon

1952 *New Light on the Most Ancient East: The Oriental Prelude to European Prehistory*. 4th ed. W. W. Norton, New York.

Desruelles, Stéphane, Eric Fouache, Wassel Eddargach, Cecilia Cammas, Julia Wattez, Tara Beuzen-Waller, Chloé Martin, Margareta Tengberg, Charlotte Cable, Christopher Thornton, and Andrew Murray

2016 Evidence for Early Irrigation at Bat (Wadi Sharsah, Northwestern Oman) before the Advent of Farming Villages. *Quaternary Science Reviews* 150:42–54.

Ekstrom, Heidi, and Christopher Edens

2003 Prehistoric Agriculture in Highland Yemen: New Results from Dhamar. *Yemen Update: Bulletin of the American Institute for Yemeni Studies* 45:23–35.

Flannery, Kent

1969 Origins and Ecological Effects of Early Domestication in Iran and the Near East. In *The Domestication and Exploitation of Plants and Animals*, edited by Peter Ucko and G. W. Dimbleby, pp. 73–100. Aldine, Chicago.

Harrower, Michael J.

2006 Environmental versus Social Parameters, Landscape, and the Origins of Irrigation in Southwest Arabia (Yemen). PhD Dissertation, Department of Anthropology, The Ohio State University, Columbus.

2008a Hydrology, Ideology, and the Origins of Irrigation in Ancient Southwest Arabia (Yemen). *Current Anthropology* 49(3):497–510.

2008b Mapping and Dating Incipient Irrigation in Wādī Sanā , Hadramawt (Yemen). *Proceedings of the Seminar for Arabian Studies* 38:187–202.

2016 Water Histories and Spatial Archaeology: Ancient Yemen and the American West. Cambridge University Press, Cambridge.

Hunt, Robert C.

2007 *Beyond Relativism: Comparability in Cultural Anthropology*. Altamira Press, Lanham, MD.

McCorriston, Joy, and Eric Oches

2001 Two Early Holocene Check Dams from Southern Arabia. *Antiquity* 75:675–76.

Sauer, Carl O.

1952 *Agricultural Origins and Dispersals*. American Geographical Society, New York.

Sherratt, Andrew

1980 Water, Soil and Seasonality in Early Cereal Cultivation. *World Archaeology* 11(3):313–30.

Steward, Julian H.

1930 Irrigation without Agriculture. *Michigan Academy of Sciences, Arts, and Letters Papers* 12:149–56.

Tindale, N. B.

1977 Adaptive Significance of the Panara Grass Seed Culture of Australia. In *Stone Tools as Cultural Markers*, edited by R. V. S. Wright, pp. 345–350. Australian Institute of Aboriginal Studies, Canberra.

Varisco, D. M.

1996 Water Sources and Traditional Irrigation in Yemen. *New Arabian Studies* 3:238–57.

Vavilov, N. I.

1951 *The Origin, Variation, Immunity, and Breeding of Cultivated Plants: Selected Writings*. Chronica Botanica, Waltham, MA.

Wilkinson, Tony J.

1999 Settlement, Soil Erosion and Terraced Agriculture in Highland Yemen: A Preliminary Statement. *Proceedings of the Seminar for Arabian Studies* 29:183–91.

2003 *Archaeological Landscapes of the Near East*. University of Arizona Press, Tucson.

2006 Highland to Desert Yemen. In *Agricultural Strategies*, edited by Joyce Marcus and Charles Stanish, pp. 38–70. Cotsen Institute of Archaeology, Los Angeles.

Part V

Middle to Late Holocene: The Social Life of Pastoralists

Chapter 14

Survey and Excavation of Small-Scale Monuments

Joy McCorriston, Michael J. Harrower, Tara Steimer-Herbet, Kimberly Williams, and Jennifer Everhart

From the earliest RASA survey in 1996, wherever we looked for houses and settlement as hallmarks of agricultural practice and village life, we found tombs and monuments. All too often there were a good many monuments with no existing trace of permanent settlement, recent or ancient. Wādī Sanā is a good example: for 80 km, from the narrowed gorge at Qārah Ḥabshiyah to the modern village of Sanā at the mouth, one finds no villages or hamlets. Yet there are high circular tombs (HCTs) along its high terraces; wall tombs, triliths, and platforms at lower elevations; rockshelters with pecked images and text; hearths; and irrigation structures that attest to the construction of environment and society in ancient times. The ephemeral nature of campsites, the indistinct nature of stone rings (which could have been windbreaks, sleeping spaces capped with brush or skins, tent bases, corral bases, or goat pens), and the lack of durable modifications to most rockshelters are sharp contrasts to the durability of small-scale stone monuments that make up the most enduring and visible remnants of ancient life. This chapter outlines results of field research on small-scale monuments, including archaeological survey and excavation.

History of Research

In 2005 the Roots of Agriculture in Southern Arabia (RASA) Project completed a pilot study on small-scale monuments along Wādī Sanā. In 2008 a new endeavor—the Arabian Human Social Dynamics (AHSD) Project—focused more specifically on small-scale monuments as the archaeological residue of landscapes shaped by human communities from 7000 cal BP. The AHSD project completed one field season in Hadramawt, Yemen (2008), and two subsequent field seasons in Dhofar, Oman (2009 and 2010). For the purposes of this volume, we report the results of our team's field survey and excavations in Yemen as one regional culture history and long-term socioecological system dynamic integrated into the RASA Project. Even if logistically funded by multiple proposals and heuristically named for different projects at the time of fieldwork, the results contribute to an integrated view of Hadramawt prehistory, and it matters little what we call the project in retrospect—whether RASA, AHSD, or both. With the AHSD Project our research questions shifted somewhat, but our results contribute to the interpretations we developed out of the RASA Project analyses. Therefore and without further elaboration of the AHSD objectives and design, we provide a summary of the field results of small-scale monument survey and excavation in Yemen (see Bin ʿAqil and McCorriston 2009; Harrower et al. 2013; McCorriston et al. 2011; Schuetter et al. 2013). The Oman data is reported elsewhere (McCorriston et al. 2014), and the full scope of the AHSD Project and results will be published anon.

In Hadramawt, the built environment is comprised of a distinctive range of small-scale stone monuments. As further described below, these stone constructions include water management structures, tombs, platforms, triliths, dolmens, and *madhābiḥ*. To impose a rigid division of economic and ideological purpose on their construction and functions would invoke a cumbersome theoretical distinction, probably unrecognizable to their builders. The former inhabitants probably experienced Wādī Sanā and their place within it in terms of practical habits that reinforced their social world and perpetuated their existence. For example, the RASA survey team distinguished between water management structures (chapter 13) with an important role in incipient food production (Harrower 2016) and platform monuments (chapter 12) marking social events and ideological import (McCorriston 2011; McCorriston et al. 2012). Although attempting to disentangle the economic from the social purpose of such constructions is probably a futile exercise, the following discussion embraces all remaining durable stone markers (with their wide range of functions and meanings) within the small-scale stone monuments category.

Methods

The Cairn Survey Pilot Study

High circular tombs make up a sizable proportion of the identifiable monuments, but they were not a target of the RASA survey from 1996 to 2004. As a known category of mortuary structures, seemingly unlikely to contain full burials in poor-preservation environments on rock surfaces, they initially seemed to offer little promise for research on the introduction and adoption of food production. Survey teams recorded HCTs as encountered but made no effort to place targeted survey units around or near them, even though the tombs are highly visible monuments to ancient human presence. Although monuments certainly were not our initial focus, the results of canonical correspondence analysis (chapter 6) suggest that random strategies have strong value for helping explain HCT distributions. Additionally, early systematic survey and intensive coverage made it clearly evident that wider-scale survey focused on highly protrusive monuments and employing less intensive strategies would document a much greater sample from which one might draw meaningful qualitative and quantitative observations about the distribution of ancient human activity.

In 2005, Tara Steimer, Michael Harrower, and Nisha Patel designed a pilot study (the RASA Cairn Survey) to identify and categorize the types of small-scale monuments present in Wādī Sanā and evaluate the feasibility of satellite imagery–based mapping. This effort was prompted by the realization that tombs and other monuments represented a critically important and somewhat neglected source of information. It was also inspired by acquisition in August 2004 of QuickBird (60 cm spatial resolution) satellite imagery, in which HCTs were visible. We recognized that mapping the many hundreds of tombs along Wādī Sanā, and many thousands more across Southern Arabia, might not require physically hiking to them and recording by hand. Rather (if appropriate methods could be developed), tombs could be mapped with satellite imagery.

Using field-collected ground control points (GCPs), Harrower and Patel geo-rectified QuickBird imagery of Wādī Sanā and overlaid coordinates of HCTs and other small-scale stone monuments already visited in the field. Harrower and Patel noted that the dark circular shape of HCTs was frequently emphasized by a crescentic shadow cast by tombs with height. Color differences between tombs and trees visible in infrared imagery made HCTs easy to distinguish from vegetation. Our team found it almost impossible to see platforms in QuickBird imagery, yet the hearths in front of trilith monuments were frequently recognizable as a distinctive line of dots.

With QuickBird image maps and the locales targeted by Patel over the course of six months of visual image study, the team carried out the cairn survey pilot study of ground-truthing in winter 2005. We visited both targeted locales (where we could see HCTs on the ground) and randomly selected quadrats to assess the success of visual image searching. The team found considerable success in identifying HCTs in images, but there were many false positives, and many HCTs had also been missed in the visual study (false negatives). With a focus on likely areas, such as rocky spurs and the confluence of several drainages, other small-scale monuments also were recorded. The survey identified 13 different formal monument types, the most numerous of which were HCTs ($n = 31$), triliths ($n = 16$), and structures differentiated by plan shape ($n = 20$), many of which are probably Neolithic platforms (which vary in plan). There were many unknown stone piles ($n = 111$) and stone rings ($n = 13$), most of which we subsequently identified as camp complexes. Most importantly, the qualitative results of the pilot study, including the preliminary monument typology and challenges in exact overlay of GPS coordinates for monuments on satellite imagery, provided critical feedback for the design of further monument survey.

Arabian Human Social Dynamics (AHSD) Project

Following the 2005 pilot study, funding from the U.S. National Science Foundation Human Social Dynamics Program (NSF-HSD) supported the AHSD Project, a longer-term effort to examine small-scale monuments as indicators of tribal social identities and territorial behaviors. This included support to: (1) survey a much larger sample and generate more detailed field records about HCTs and other monuments; (2) excavate monuments to collect materials for analysis and radiocarbon dating; and (3) acquire new scenes of QuickBird imagery, which at the time offered the highest spatial resolution available (60 cm), to develop imagery-based methods for tomb detection. A wide range of publications draw on data generated by AHSD work in Yemen (Bin 'Aqil and McCorriston 2009; Harrower 2016; Harrower et al. 2013; McCorriston 2011; McCorriston et al. 2011; Schuetter et al. 2013). Although the larger, long-term goal of examining tribal social identities and territorial behaviors is still ongoing, the following pages summarize field data generated and describe key findings.

Excavation and Dating of Small-Scale Monuments

Small-scale monuments are among the most visibly prominent remains left by ancient hinterland peoples of Arabia. In the Hadramawt, even the casual traveler notices the striking lines of circular tombs—variably referred to over the years as beehives, turret tombs, and pillbox cairns—that bear resemblance to and perhaps share cultural heritage with monuments in Saudi Arabia and Oman, and even those as far away as the Sinai and the Levant. Archaeological survey certainly provides valuable information about monuments, but without excavating and dating, observations made during survey about types, ages, and other characteristics of monuments are difficult to substantiate. In addition to the Neolithic platform monuments (chapter 12) and water management structures (chapter 13), the AHSD team excavated and obtained radiocarbon ages on material from a range of monuments (table 14.1). Tara Steimer-Herbet guided the excavation strategies, Kimberly Williams excavated and analyzed human remains, and Jennifer Everhart analyzed faunal bone under the guidance of Katheryn Twiss. These studies helped define a monument typology used in the processing of survey data and for distributional analysis.

In the reporting of results below, a reader will discern the use of terms *level* and *layer* to refer to stratigraphy. These terms follow the use by excavators in their notebooks and reflect differences in the strategies of recording by different excavators. (All excavations followed natural stratigraphic

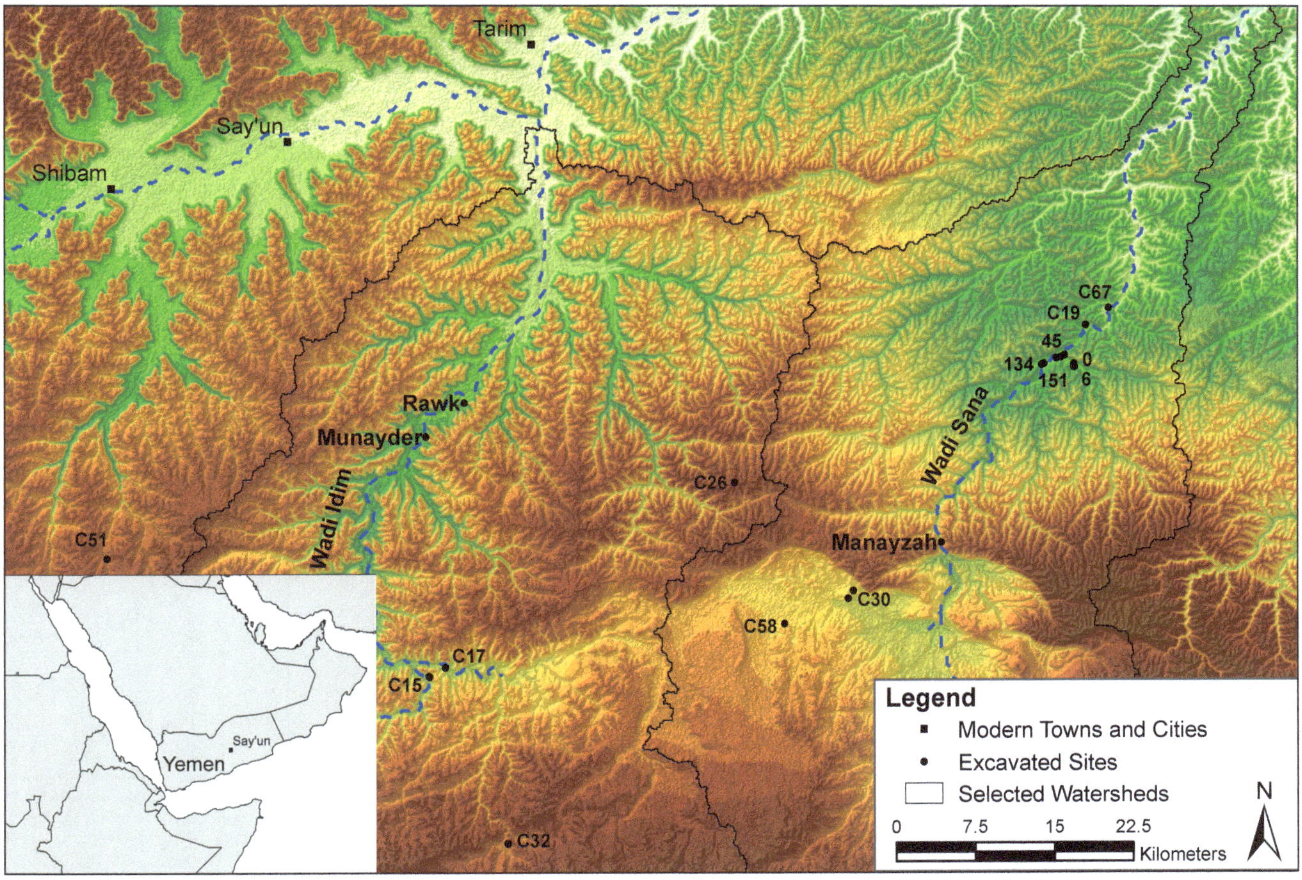

Figure 14.1. Map of RASA–AHSD small-scale monuments excavated. Topography and and hydrology derived from Shuttle Radar Topography Mission v4. *Illustration by Michael Harrower.*

Table 14.1. Hadramawt small-scale stone monument types (after McCorriston et al. 2011). See chapter 18, table 18.1 for further details on individual radiocarbon age samples.

Category	Description	Distribution	Excavated Examples	Dates (use, not construction)	Age in Cal BP	Reuse in Cal BP
Platforms (chapter 7)	Initially constructed as subcircular (drop-shaped, trapezoidal, or D-shaped) rings of upright slabs, platforms were subsequently filled. They may have had a structural life history first as houses and later as monuments. Platforms may have inner, accretive, or concentric alignments of upright stones, and they may also have (or have had) standing stones outside. They do not include graves or interments but do include bone fragments and charcoal flecks with their deliberate fill. Dated examples have radiocarbon ages in the seventh millennium cal BP.	Platforms occur on lower gravel, bedrock, or silt terraces in locations that have not been substantially inundated since the Middle Holocene. They are preserved today in the drainages of upper Wādī Sanā (south of Ghayl Bin Yumain) and throughout the middle Wādī Sanā . Comparable structures occur in the Northern Jol and Dhofar Nejd (McCorriston et al. 2014), but RASA members encountered none to the west in Wādī ʿIdim and its upstream tributaries.	SU151-001, SU037-003, SU038-007	5514 ± 48 (AA66861). 5616 ± 84 (AA38547), 5682 ± 47 (AA66862), 6010 ± 69 (AA69755)	6406–6213, 6631–6624	none dated
Dolmens	Constructed of upright limestone slabs supporting a limestone slab roof, these monuments enclosed a hollow chamber, which sometimes was compartmentalized. Roof support was provided by strong corners and internal subdividing slabs. A low platform of slabs around the exterior may also have buttressed the structure, generally rectilinear in plan. Dolmens frequently carry pecked ornamentation as geometric shapes and wavy lines on their exposed surfaces. A survey team from the Deutsches Archäeologisches Institut in Sana'a verbally reported an interior burial surviving in one case.	Geometric-decorated slabs and dolmens are rare; two examples are in Wādī Sanā (one intact, 1998-21= SU001-001 = C9-001, next to the Gravel Bar Site; one shattered presumed dolmen, C067-2). They sit on lower bedrock and silt terraces, respectively. More widely, there are dolmens known from Jibal Jidrān in Shabwa Province and the Northern Jol (Wadi Sarr HDOR-431), and these may have cultural links with widespread megalithic construction across the Near East at this time.	C067-2	5702 ± 51 (AA81814), 5603 ± 67 (AA81816), 5709 ± 45 (AA81815)	6633–6355	none dated

Category	Description	Distribution	Excavated Examples	Dates (use, not construction)	Age in Cal BP	Reuse in Cal BP
Tower tombs (high circular tombs; HCTs)	There is little doubt that these dry-walled structures were intended as tombs. Built with an inner, subcircular, 1–2 m wide chamber of uprights or corbelled blocks and an outer face of undressed, regular blocks or horizontal slabs, the circular wall had a rubble core, a corbelled roof built of flat slabs to a height of about 2 m, and a capstone. There was no side entry in the Hadramawt examples. These tombs may have "tails," alignments of small cairns or supported uprights evenly spaced and normally visible along a cliff edge or ridgeline. Their interiors apparently filled quickly with aeolian silt, and there are no cases of deliberate interior fill; bodies were placed directly on the existing ground surface, and the chamber was sealed. Evidence of reuse appeared in all excavated examples, which had later interments, charcoal, or disturbance of first interments, leaving behind fragmentary residues or only a few beads.	HCTs occur in high locations, demonstrated in the results of systematic survey (chapter 3) and qualitatively evident throughout Hadramawt. They occur along the edges of plateaus so that they are highly visible from the wadi bottoms and also accessible; along the middle principal drainages of Wādī ʿIdim and Wādī Sanā , HCTs line the upper bedrock terraces but not the uppermost, generally inaccessible plateau. In broader geographic distribution, HCTs conform to a widespread phenomenon of dry-walled circular tombs observed through the arid margins of the Levant, Yemen, the Dhofari Nejd, and in parts of northern Oman and the UAE.	C15-2, C15-3, C30-3, C30-4, C32-1, C32-2	158 ± 38 (AA83495), 1559 ± 38 (AA83496), 1733 ± 39 (AA83500), 1868 ± 35 (AA79762), 2407 ± 40 (AA86370), 2474 ± 38 (AA81817), 2489 ± 39 (AA81818), 2776 ± 54 (AA83498), 3067 ± 39 (AA90838), 3912 ± 39 (AA83492), 4288 ± 47 (AA83494), 4525 ± 66 (AA90336-on marine shell)	4499–4185, 5030–4659	287 . . . , 1538–1369, 1730–1549, 2737–2381, 3000–2761, 3367–3175
Wall tombs	These structures consist of a low wall up to 1 m high and about 1 m wide, constructed with undressed slabs used as exterior facing for a rubble core. In plan, the sides are parallel, 7–10 m long, with ends either squared or rounded. A central feature is a small box chamber about 0.6 m long, built of slab uprights and presumably once sealed with capstones or corbelling. All RASA-excavated examples were robbed, and the chambers filled with aeolian silt. One disturbed chamber contained worked marine shell artifacts and fragments of human bone, suggesting that human burial was indeed one purpose of these monuments.	Wall tombs appear on the lower gravel and bedrock terraces of the upper and middle reaches of Wādī Sanā and Wādī ʿIdim. While there may be HCTs nearby, HCTs usually are situated higher and have greater visibility than wall tombs, which are fewer. Wall tombs do not appear on the escarpment and coastal plain, but they do fit within a broader geographic distribution, including the arid southern Levant, interior Yemen, western and northern Hadramawt Jol, Mahra, and western Dhofar.	C17-2, C19-1, C26-2	3686 ± 41 (AA83497)	4148–3902	none observed

Table 14.1. Hadramawt small-scale stone monument types (after McCorriston et al. 2011). See chapter 18, table 18.1 for further details on individual radiocarbon age samples. *(continued)*

Category	Description	Distribution	Excavated Examples	Dates (use, not construction)	Age in Cal BP	Reuse in Cal BP
Tumuli (with stelae)	All tumuli reported as such were low heaps of stone cobbles or slabs, sometimes ringed with limestone stelae or boulders. They offer no surface indication of a central chamber. Generally more than 5 m in diameter, these are difficult to differentiate as a distinct category unless accompanied by upright stelae, some shaped and pecked to indicate a bearded human figure with long hair or arms, a bared torso, a belt, and a crescent-handled dagger. Many stelae have been reported out of context, but of the few in-situ examples, only two have been excavated. The RASA tumulus example had a central chamber 0.8 m across, defined by the inner faces of large cobbles. This chamber contained a burial, and the tumulus was formerly ringed with limestone uprights.	Because the sample of in situ stelae is small, the distribution of tumulus monuments associated with stelae is unknown. Known examples occur in remote passes of the escarpment and highest plateau. Reported examples of stelae out of context include Thamud (desert interior) and the Shabwa highlands. Anthropomorphic stelae appear widely in the Bronze Age from 2500 BCE in highland East Africa, the Mediterranean, and Northern Arabia at a time when metal and other precious materials circulated along widespread trade routes.	C51-1	2216 ± 55 (AA79767), 3663 ± 41 (AA83499)	4142–3877	2347–2069
Triliths	Triliths are an unmistakable formations of multiple long, low platforms constructed on the surface to support groups of uprights (in alignments of upright formations of one, three [triliths], or four slabs). Multiple platforms, each with multiple triliths, form an alignment. In parallel alignment with the trilith platforms are square arrangements of boulders and alignments of raised hearths carefully constructed with a ring of large cobbles filled with smaller, usually thermally altered stones. Despite many suggestions, no widely accepted interpretation of these monuments exists. All published radiocarbon ages on triliths come from hearths and date the latest hearth uses around 2,000 years ago.	Triliths occur on low terraces of gravel, bedrock, or silt alongside a track or watercourse. The long axis often follows a landscape feature (track, watercourse, or terrace edge), and the stone blocks and hearths often lie between the landscape feature and the trilith platforms. From their placement, it seems that the long axis was intended to be visible or accessible from important routes and that the working side—the hearths—faced these routes. Triliths occur in middle and upper Wādī Sanā and more widely through the arid coastal and mountainous terrain of eastern Hadramawt, Mahra, Dhofar, and eastern Oman.	SU134-3, C58-1, SU006-1	1749 ± 35 (AA79768), 2026 ± 35 (AA79769)	1774–1559, 2106–1894	none observed

Category	Description	Distribution	Excavated Examples	Dates (use, not construction)	Age in Cal BP	Reuse in Cal BP
Islamic graves	Recognizable in clusters of ovate cobble rings filled with mounded smaller cobbles, Islamic graves also typically include a "witness" stone or several such uprights indicating male or female interment, according to local custom.	Islamic graves occur far from other monuments and in clusters. They are on low elevations near the modern wadi channel.	not excavated		after 1300	
Madhābiḥ	An oval or round platform built of unworked dry-stone outer facing and a rubble core. In elaborate examples, this platform may be a meter in height and support a smaller oval or circular hearth made of a ring of boulders with a fill of thermally altered pebbles and cobbles, often smooth stones carried up from the streambed. *Madhābiḥ* are differentiated from other hearths by their larger scale, up to 5 m in length or diameter. This scale and that of the stones involved imply communal labor in their construction. Some examples include internal divisions or uprights.	These monuments sit on low terraces or slopes, never on the high plateau. They also occur near rockshelters and other signs of camping or temporary shelter.	C30-27, C30-near27a	74 ± 33, (AA79766), 158 ± 33, (AA79765), 178 ± 33 (AA79764)	24–262, 298–286	not known

layers.) Where *level* appears in C15-3, it reflects post hoc assessment of depositional events, numbered from bedrock or sterile upward. (Level 1 is lowest and earliest.) Elsewhere, the term *level* refers to field notation, and levels are stratigraphically numbered from top to bottom. Where *layer* appears, the term reflects field labels and the numbering is from top down. (Layer 1 is highest and latest.) The decision to follow notebook terminology here is not an elegant solution. It requires readers to forgive excavators' (sometimes) interchangeable uses (*layers* for one tomb, *levels* for another; see below). While this is an inelegant format for presentation, this decision maintains concordance with the primary data (such as notebooks, bag labels, and radiocarbon assay submission forms) made available as online digital format.

Results

Dolmen Monuments

The term *dolmen* has long been used to describe structures built of large slabs of stone, often presumed to be tombs, found in many regions, including the Near East (Braemer et al. 2003; Swauger 1966). Survey by the RASA and AHSD teams encountered three dolmen monuments:

- One recorded in three seasons as 98-1/SU1-1/C9-1 (E 337304.00, N 1744243.43)
- One recorded as C67-2 (E 343073.08, N 1758777.80)
- One recorded in two seasons as C11-11 and C71-1 (E 340511.51, N 1749754.96)

All three of these dolmen monuments were constructed in low-lying areas using large limestone slabs (originally upright) in conjunction with an array of smaller limestone blocks; and all three dolmens include pecked artwork (pictographs or geometric designs).

The first dolmen (C9-1) is located on a low bedrock terrace along Wādī as-Shumlyah, east of the Gravel Bar Site, within view of the Khuzmum inselberg. Today open to the elements, this dolmen exemplifies the construction of other dolmens less well preserved, but it provides no

Figure 14.2. Dolmen in Wādī as-Shumlyah (C9-1). *Photograph by Tara Steimer-Herbet.*

clear indication of its date of construction (figures 14.2 and 14.3).

Another highly noteworthy remnant (C67-2) is located along Wādī Sanā approximately 8 km downstream (north) from the Khuzmum. In this instance, the team documented iconoclasm: a presumed dolmen was deliberately destroyed in antiquity. It is impossible to be certain about the exact original form of this structure, as the large limestone slabs from which it was built were deliberately shattered in antiquity (figure 14.4). However, the slabs bore geometric motifs consisting of double lines of triangles (chevrons) pecked into a limestone slab surface. The geometric design was patinated, but the shatter breaks were far fresher than the patina on the design. The RASA team refitted multiple fragments from a single shattered slab, including one fragment recovered from a hearth (figure 14.5). Several of these fragments were surfaces finds, but others were partly buried in a Holocene silt terrace. There were several in situ limestone slabs protruding from silts, which presumably represent the remnant uprights of the dolmen. From a stratified hearth (C67-2, Hearth 1) containing a fire-reddened fragment of the geometric decoration, Jean-François Saliège obtained a radiocarbon age on charred bone (AA81814) at 6640–6355 cal BP, a terminus ante quem dating the destruction of the original structure to which the geometric slab belonged. Other hearths revealed by Late Holocene gullying through the terrace were at the same level as the in situ uprights, and these hearths yielded charcoal ages between 6633 and 6283 cal BP (AA81815, AA81816). Radiocarbon samples and observations were obtained in the course of survey, with a limited excavation to reveal details of the hearths. No excavation probed the upright slabs—their architectural and stratigraphic associations. The stratified context permits a very clear date of destruction and less certain chronological association between the in situ hearths and the buried monument. The patination suggests that the geometric designs were exposed for some time before the slab was burned and deliberately shattered, with rapid subsequent burial in alluvial sediments. The hearths also contained a cattle bone and unidentified large caprine or small bovid bones (table 14.2).

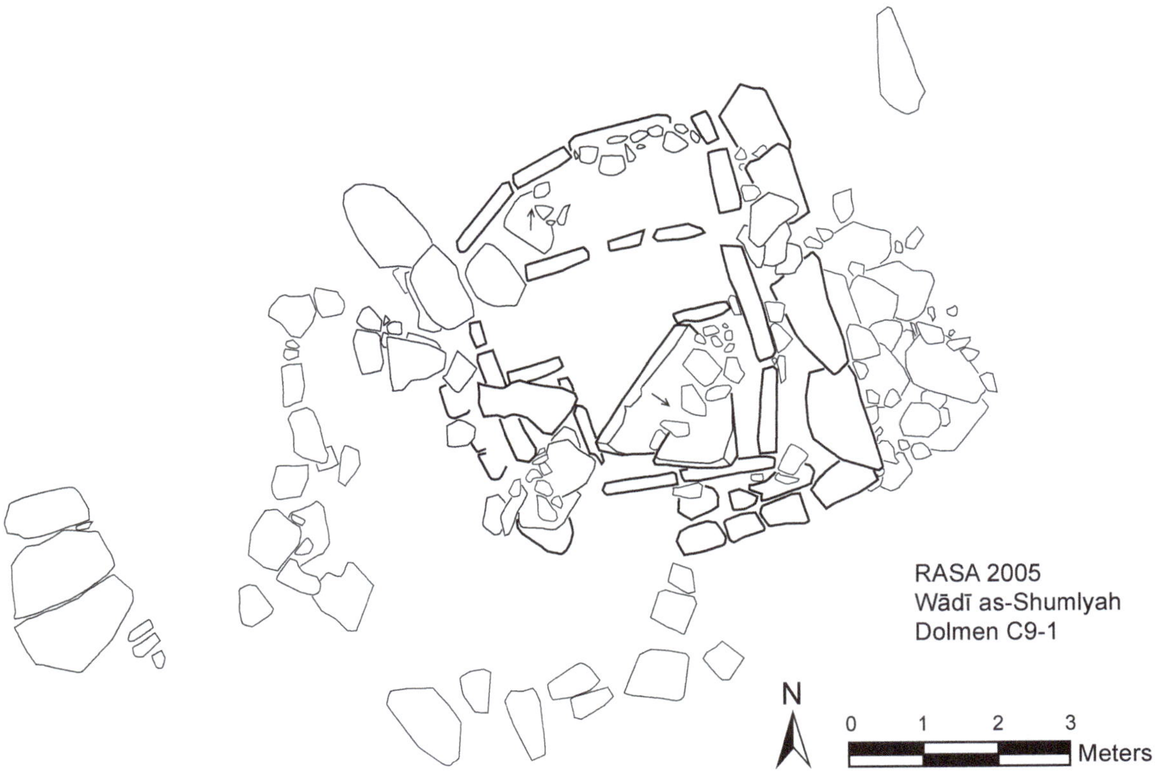

Figure 14.3. Plan drawing of Dolmen C9-1.
llustration by Tara Steimer-Herbet.

Figure 14.4. Broken slabs of Structure C67-2.
Photograph by Michael Harrower.

Figure 14.5. Geometric designs on Structure 67-2.
Photograph by Michael Harrower.

Table 14.2. Faunal remains from monuments, not including an entire camel buried in HCT C30-4. Identifications by Jennifer Everhart.

Survey Unit	Site	Locus or Layer	Element	Taxon	Monument Type
15	3	2	indeterminate	sheep size (medium dog to medium sheep)	HCT
19	1	1	indeterminate	indeterminate	wall tomb
26	2	3	long bone (shaft fragment)	sheep size (medium dog to medium sheep)	wall tomb
26	2	3	indeterminate	medium dog to wild boar	wall tomb
26	2	3	long bone (shaft fragment)	sheep size (medium dog to medium sheep)	wall tomb
30	3	1	tooth fragment	medium sheep to medium cattle	HCT
30	3	1	tooth fragment	medium sheep to medium cattle	HCT
30	3	1	tooth fragment	cow size	HCT
30	3	1	tooth fragment	medium sheep to medium cattle	HCT
30	3	1	tooth fragment	medium sheep to medium cattle	HCT
30	3	1	tooth fragment	medium sheep to medium cattle	HCT
30	3	1	tooth fragment	medium sheep to medium cattle	HCT
30	3	1	tooth fragment	medium sheep to medium cattle	HCT
30	3	1	tooth fragment	medium sheep to medium cattle	HCT
30	3	1	tooth fragment	medium sheep to medium cattle	HCT
30	3	1	tooth fragment	*Ovis/Capra*	HCT
30	3	1	tooth fragment	medium sheep to medium cattle	HCT
30	3	1	tooth fragment	medium sheep to medium cattle	HCT
30	3	1	tooth fragment	medium sheep to medium cattle	HCT
30	3	1	tooth fragment	sheep size (medium dog to medium sheep)	HCT
30	3	1	tooth fragment	medium sheep to medium cattle	HCT
30	3	1	tooth fragment	medium sheep to medium cattle	HCT
30	3	1	tooth fragment	indeterminate	HCT
30	27	4	rib	medium artiodactyl	*madhbaḥ*
30	27	4	skull	medium artiodactyl	*madhbaḥ*
30	27	4	indeterminate	indeterminate	*madhbaḥ*
30	27	4	indeterminate	indeterminate	*madhbaḥ*
30	27	6	calcaneus	*Ovis/Capra*	*madhbaḥ*
30	27	6	astragalus	*Ovis/Capra*	*madhbaḥ*
32	2	1	femur	*Capra*	HCT
32	2	1	femur	*Capra*	HCT
32	2	3	indeterminate	sheep size (medium dog to medium sheep)	HCT
32	2	3	first phalanx	*Ovis/Capra*	HCT
32	2	4	indeterminate	indeterminate	HCT
32	2	4	indeterminate	indeterminate	HCT
32	2	4	calcaneus	medium artiodactyl	HCT
32	2	5	indeterminate	indeterminate	HCT
32	2	6	long bone (shaft fragment)	indeterminate	HCT
32	2	7	long bone (shaft fragment)	indeterminate	HCT
32	2	7	indeterminate	indeterminate	HCT
32	2	8	indeterminate	indeterminate	HCT

Survey Unit	Site	Locus or Layer	Element	Taxon	Monument Type
32	2	8	indeterminate	indeterminate	HCT
32	2	8	central + fourth tarsal (naviculo-cuboid)	*Ovis/Capra*	HCT
32	2	8	indeterminate	indeterminate	HCT
32	2	8	indeterminate	indeterminate	HCT
32	2	8	indeterminate metapodial	*Ovis/Capra*	HCT
32	2	9	phalanx	*Ovis/Capra*	HCT
32	2	9	indeterminate	indeterminate	HCT
32	2	9	long bone (shaft fragment)	indeterminate	HCT
32	2	10	radius	*Ovis/Capra*	HCT
32	2	10	indeterminate	indeterminate	HCT
32	2	10	indeterminate	indeterminate	HCT
32	2	11	indeterminate metapodial	*Ovis/Capra*	HCT
32	2	11	indeterminate metapodial	*Ovis/Capra*	HCT
32	2	11	indeterminate	indeterminate	HCT
32	2	11	phalanx	indeterminate	HCT
32	2	11	long bone (shaft fragment)	indeterminate	HCT
32	2	11	phalanx	sheep size (medium dog to medium sheep)	HCT
32	2	12	first phalanx	*Ovis/Capra*	HCT
32	2	12	third phalanx	*Ovis/Capra*	HCT
32	2	12	intermediate carpal (semilunar)	*Ovis/Capra*	HCT
32	2	12	second phalanx	*Ovis/Capra*	HCT
32	2	12	radius	*Ovis/Capra*	HCT
32	2	17	second phalanx	*Ovis/Capra*	HCT
32	2	17	rib	sheep size (medium dog to medium sheep)	HCT
32	2	17	calcaneus	*Ovis/Capra*	HCT
32	2	17	phalanx	*Ovis/Capra*	HCT
32	2	17	second phalanx	*Ovis/Capra*	HCT
32	2	17	loose lower tooth	small carnivore	HCT
32	2	17	second phalanx	*Ovis/Capra*	HCT
32	2	17	indeterminate	indeterminate	HCT
32	2	17	tooth fragment	sheep size (medium dog to medium sheep)	HCT
32	2	18	thoracic vertebra	medium artiodactyl	HCT
32	2	18	second phalanx	*Capra*	HCT
32	2	19	astragalus	*Ovis/Capra*	HCT
32	2	19	indeterminate	indeterminate	HCT
32	2	19	tibia	*Ovis/Capra*	HCT
32	2	19	indeterminate	indeterminate	HCT
32	2	19	astragalus	*Ovis/Capra*	HCT
51	1	3	long bone (shaft fragment)	sheep size (medium dog to medium sheep)	tumulus
51	1	3	long bone (shaft fragment)	sheep size (medium dog to medium sheep)	tumulus

Table 14.2. Faunal remains from monuments, not including an entire camel buried in HCT C30-4. Identifications by Jennifer Everhart. *(continued)*

Survey Unit	Site	Locus or Layer	Element	Taxon	Monument Type
51	1	3	indeterminate	sheep size (medium dog to medium sheep)	tumulus
51	1	5	rib	medium artiodactyl	tumulus
51	1	8	indeterminate	indeterminate	tumulus
51	1	8	indeterminate	indeterminate	tumulus
51	1	8	indeterminate	indeterminate	tumulus
51	1	10	tooth fragment	*Ovis/Capra*	tumulus
51	1	10	indeterminate	indeterminate	tumulus
51	1	10	indeterminate	indeterminate	tumulus
51	1	10	indeterminate	indeterminate	tumulus
51	1	10	indeterminate	indeterminate	tumulus
51	1	10	indeterminate	indeterminate	tumulus
51	1	10	indeterminate	indeterminate	tumulus
51	1	10	indeterminate	medium sheep to medium cattle	tumulus
51	1	10	indeterminate	indeterminate	tumulus
51	1	10	indeterminate	indeterminate	tumulus
51	1	10	tooth fragment	indeterminate	tumulus
51	1	10	indeterminate	indeterminate	tumulus
51	1	10	indeterminate	indeterminate	tumulus
51	1	10	indeterminate	indeterminate	tumulus
51	1	16	long bone (shaft fragment)	indeterminate	tumulus
51	1	16	long bone (shaft fragment)	sheep size (medium dog to medium sheep)	tumulus
51	1	16	long bone (shaft fragment)	sheep size (medium dog to medium sheep)	tumulus
51	1	16	long bone (shaft fragment)	sheep size (medium dog to medium sheep)	tumulus
51	1	16	long bone (shaft fragment)	sheep size (medium dog to medium sheep)	tumulus
51	1	18	indeterminate	indeterminate	tumulus
51	1	22	tooth fragment	*Ovis/Capra*	tumulus
51	1	22	humerus	*Capra*	tumulus
51	1	22	indeterminate	indeterminate	tumulus
51	1	23	long bone (shaft fragment)	sheep size (medium dog to medium sheep)	tumulus
51	1	23	indeterminate	sheep size (medium dog to medium sheep)	tumulus
51	1	23	indeterminate	sheep size (medium dog to medium sheep)	tumulus
51	1	23	indeterminate	sheep size (medium dog to medium sheep)	tumulus
51	1	23	indeterminate	sheep size (medium dog to medium sheep)	tumulus
67	2	1	indeterminate	indeterminate	unknown/dolmen
67	2	1	tooth fragment	medium sheep to medium cattle	unknown/dolmen
67	2	1	skull	medium sheep to medium cattle	unknown/dolmen
67	2	3	tooth fragment	medium sheep to medium cattle	unknown/dolmen
67	2	3	indeterminate	indeterminate	unknown/dolmen
67	2	4	skull	large bovid	unknown/dolmen
67	2	7	indeterminate	indeterminate	unknown/dolmen

Figure 14.6. Overview of Dolmen C71-1. *Photograph by Michael Harrower.*

The third dolmen (C71-1) is located on a low inselberg in Wādī Sanā, approximately 19 km downstream (north) of the Khuzmum (figure 14.6). It was visited and recorded in both 2005 and 2008. It consists of large upright slabs with pecked designs that are difficult to interpret but clearly represent intentional messaging, with meaning long obscured by wind and sand. The dolmen is surrounded by *madhābiḥ* hearths (see below) and stone rings of enigmatic purpose.

Discussion of Dolmens

A geometric design on a dolmen in the middle sixth millennium BP is unsurprising. There are a number of parallel examples for geometric decoration in Arabia sometime after 5000 cal BP. A dolmen in Jibal Jidrān shows pecked circles on one of the uprights; another bears a lattice of lozenges. In the drainages of the Northern Jol are limestone uprights, also possibly once belonging to dolmens, one with crenelated lines and another with a lattice of rectangular panes. Although unexcavated in Hadramawt, these are widely assumed, like other dolmens, to have once been tombs. On the bodies of rockart images and anthropomorphic figures also appear geometric designs (Braemer et al. 2003; McCorriston et al. 2011:8, figure 5), whose significance may include domesticated or piebald animals or tribal brands (Khan 2000; Nayeem 2000:343; Zarins 1992:27) and may correspond to a broader Bronze Age sociological and symbolic system (Newton and Zarins 2000; chapter 15 this volume).

High Circular Tombs

Small circular cairn tombs roughly dated to the third millennium BC are known in a wide variety of different forms across a broad swath of the Near East, including the Arabian Peninsula, the Sinai, and Syria. Following on the work of Tara Steimer-Herbet (2004) in classifying them, we refer to examples documented in Yemen as HCTs. We excavated six HCTs in 2008, revealing in all cases a complex set of uses and reuses of these prominent funerary monuments.

Tomb Construction (C15-2 and C15-3)

These two HCTs sit high on the plateau edge, commanding a view of the juncture of the Wādī ʿAtuf tributary with Wādī ʿIdim. An intact tower tomb more than 5 m in diameter, with straight sides and a flat top, C15-3 (E 275483.51, N 1715980) has several tails—alignments of smaller stone cairns evenly placed in a line emanating from the tomb. One tail has 6 elements; the other 25. In the 1960s, Brian Doe, the Aden Protectorate antiquities inspector, noted this site (1971:236–37, 1983:59), which has been a prominent landmark surely for many ages. Survey of the surrounding plateau shows a concentration of tower tombs along the upper plateau edges of Wādī ʿAtuf. Many of these tombs appear to be in remarkably good condition, and some have prominent tails. On the basis of good structural preservation and the hope that a small opening in the capstones of the chamber might signify an unrobbed burial, Tara Steimer-Herbet directed an excavation of C15-3. The team also excavated adjacent, badly damaged HCT C15-2 (E 275444.57, N 1716033.42), presumed to have been robbed of stone to construct its better-preserved neighbor.

HCT C15-2 was a low, rubble-filled circular wall with an outer diameter of 3.25 m north–south and 3.4 m east–west in its modern, poorly preserved form (figure 14.7). Situated about 65 m south–southeast of its more impressive neighbor C15-3, C15-2 was preserved to a height of only about 0.5 m. To reveal the form of its circular wall, constructed of unshaped cobbles, excavators removed an overlying rubble of collapsed, unsorted pebbles and cobbles mixed with loose silt, sieving the upper 0.40 m of deposit without recovering any artifacts, bone, or charcoal. The outer face of this tower tomb, if it had once had one, had been robbed, and an inner chamber roughly 1 m in diameter was defined by large, unshaped blocks set upright and in rough courses (figures 14.8 and 14.9). Entry was presumably by the roof. No trace of side entry at ground level was found. A limestone slab along the north face of the inner chamber remained upright, with intact human bone at its base (figure 14.10). The chamber fill was of unsorted pebbles, cobbles, horizontal slabs, and silt to a depth of 0.7 m. Near the base of the inner chamber, excavators resumed sieving and recovered a small amount of unidentifiable human bone fragments on an old land surface 0.10 m above bedrock. Jean-François Saliège pioneered a bone apatite dating method (Saliège et al. 1995) to yield a radiocarbon age of 5030–4659 cal BP (4288 ± 47, AA83494). This sample dates the latest human burial in the tomb but not the subsequent disturbance nor the original construction. This tomb is nevertheless earlier both in construction and final use than its neighbor, C15-3.

HCT C15-3 showed much better preservation, with perpendicular outer faces of unshaped stone in rough courses about 1.65 m high (figure 14.11). Although the outer face had collapsed, with a rubble core spilling down on the southeast side, the inner chamber retained most of its corbelled roof, with only a capstone missing on the top (figure 14.12). Given the narrow breadth of roof entry from a missing capstone, excavators hoped to find intact contents of this HCT undisturbed since initial sealing. The team excavated only the inner chamber after removing several layers of corbelling to allow safe access.

Figure 14.7. HCT C15-2 partially excavated in foreground, facing northwest. *Photograph by Tara Steimer-Herbet.*

Figure 14.8. HCT C15-2 chamber fully excavated to original ground surface, showing upright blocks of inner chamber. East is at top of image. *Photograph by Tara Steimer-Herbet.*

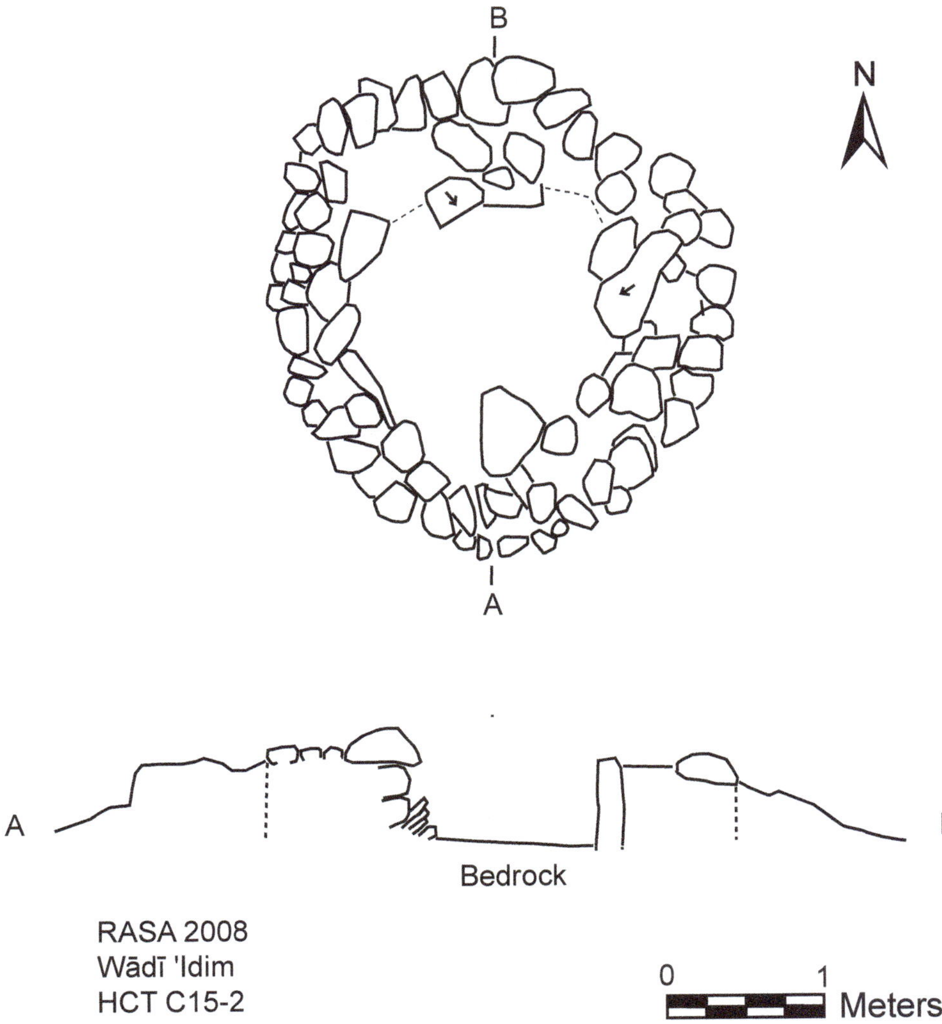

Figure 14.9. Plan and elevation of HCT15-2 (after McCorriston et al. 2011:9, figure 6). *Illustration by Tara Steimer-Herbet.*

Figure 14.10. HCT C15-2; human bone in burial chamber. *Photograph by Catherine Heyne.*

Figure 14.11. HCT C15-3, viewed from the southeast. *Photograph by Tara Steimer-Herbet.*

Figure 14.12. HCT C15-3; entry at top of chamber before excavation. *Photograph by Tara Steimer-Herbet.*

Burials in HCT C15-3

Inside the chamber, the upper level, Level 3, about 20 cm deep, contained a mix of aeolian silt with very fragmentary human bone, some unidentifiable; fragmentary adult teeth; tarsal fragments; and very fragmented vertebrae. Because they were close to the surface and fragmentary and there were several underlying strata with human bone, these were not submitted for bioapatite assays. Below Level 3, unidentifiable bone fragments also appeared in the silt matrix of Level 2, along with one badly eroded sacral element and several fragmentary adult and subadult teeth. A few charcoal fragments also occurred in Level 2. In the north half of C15-3, excavators encountered an increased number of charcoal fragments, thermally altered rock, and clastic limestone cobbles, presumably from partial roof collapse with the removal of one capstone. This deposit, registered as "intermediate between Level 1 and Level 2," also contained two fragments of animal bone, one of which was sheep/goat size. Under this deposit were multiple human burials in Level 1, with a minimum of three adults and at least one subadult. The remains were commingled, possibly through later disturbance, and the remains include elements from limbs, crania, and torso. There were nonetheless within Level 1 two distinct layers of burial separated by a sterile deposit of aeolian silt. A radiocarbon age on human bone taken from the lower burial—likely one inhumation event on ground surface—yielded a 4499–4185 cal BP (AA83493, 3912 ± 39) terminus ante quem for tomb construction.

Kimberly Williams's analysis reveals that there were at least four interments in this tomb. Three adults are represented by very fragmentary and commingled remains, but a MNI of three was established by the identification of three right femora. Because of the very fragmentary nature of the bone, no age or sex estimation for these individuals beyond assessment as adults was possible. At least one of the adult interments experienced antemortem posterior tooth loss and minor degenerative changes to the vertebrae during life. The subadult was represented by an unfused proximal femur head (age less than 15.5–19.5 years) and unfused proximal and distal tibia epiphyses (age less than 16–18 years).

In the upper burials of Level 1, stratigraphically later than this first interment, another bioapatite sample was taken from subadult remains to ensure that interments of different individuals were being dated. This bone yielded a radiocarbon age of 3001–2761 cal BP (AA83498, 2776 ± 54), at least a thousand and perhaps several thousand years after the earliest burial. Reuse of the tomb in the Iron Age probably accounts for its excellent state of preservation and also probably explains the discrepancy in dates from charcoal in Level 2 and intermediate Levels 1 and 2. Charcoal ages between 2737 and 2379 cal BP (AA81818, 2489 ± 39; AA81817, 2474 ± 38) are about 300 years later than the later burials. The radiocarbon ages on deposited bone and charcoal are important in constraining the construction date of C15-3. Construction of C15-3 used limestone upright slabs to build an inner chamber, capped with a corbelled vault built of flat-lying stones and cobbles. Collapse at the outer southeast corner of the HCT reveals a dry-stone rubble core of an outer facing of unshaped cobbles not laid in regular courses (figure 14.13). Excavation did not section the tomb, and one may only surmise that the outer cobbles were a rebuilt facing to an earlier tomb. The original construction probably occurred after C15-2, by robbing stone from C15-2 to build a second Bronze Age tomb (C15-3) and therefore dating construction to 5,000 to 4,500 years ago, between the last interment in C15-2 and the earliest in C15-3.

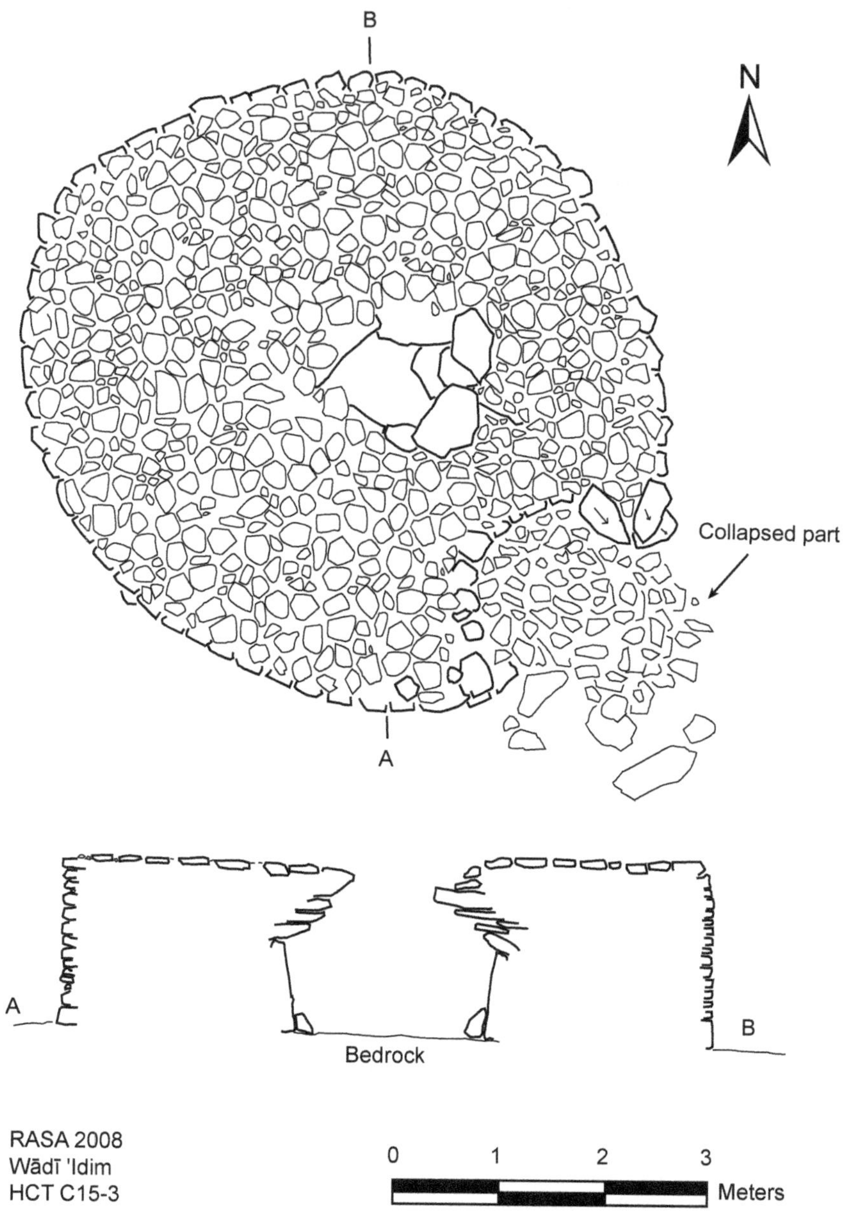

Figure 14.13. Plan and elevation of HCT 15-3 (after McCorriston et al. 2011:9, figure 6). *Illustration by Tara Steimer-Herbet.*

Objects in HCT C15-3

The silty matrix in C15-3 also included many beads of various materials, including especially shell, steatite, and an unknown grey stone. Most objects came from Layer 1, and most of the smaller beads were recovered through sieving, probably having escaped the predation of looters. Most of the shell and steatite beads were small, between 3 and 5.5 mm in diameter and only a few millimeters thick. The matrix also included a 29.5 mm long awl of copper or bronze; five shell ornaments and pendants, each longer than 20 mm; and an unmodified oyster shell. While not strictly diagnostic, tiny steatite beads are typical of Bronze Age interments and may date to the earliest interment. Obsidian may have been traded or imported from a great distance, possibly from the west, and appears to be concentrated with the later disturbance in the upper levels. Including several nondiagnostic chert flakes, obsidian, beads, pendants, and metal, all objects from C15-3 have been deposited in the Mukalla Museum, along with all human bone.

Table 14.3. Artifacts from HCT C15-3.

Level	Shell Bead	Shell Object	Unmodified Oyster Shell	Steatite/Stone Bead	Copper Awl	Obsidian Flakes	Chert Flakes
1/S	15	3	0	1	0	0	0
1	28	2	1	15	1	1	0
Interment 1 and 2	2	0	0	1	0	5	2
Interment 1 and 2/N	43	0	0	0	0	0	0
2	1	0	0	0	0	1	1

Figure 14.14. Selected artifacts from HCT C15-3, Level 1. Left: six examples of shell beads; right top: copper awl; right middle: six examples of steatite stone beads; right bottom: three shell objects. *Photographs by Catherine Heyne.*

HCT C30-3

Several other HCTs chosen for excavation had a similar appearance of good preservation, standing with intact outer faces to a good height, with a narrow roof entry open to the inner chamber, sometimes with a few capstones in place. In the Wādī Ḥarū west of Ghayl Bin Yumain, a high gravel terrace isolated by two channels of more recent down-cutting forms a ridge within a good view of the silt terraces and main traffic to the south. Along this ridge are many HCTs, some preserved several meters in height, as well as other monuments and high-quality chert outcrops. Two HCTs held promise for undisturbed contents, since they were all but closed and in one case closed at the top (figure 14.15).

HCT C30-3 (E 315890.30, N 1723462.39) was a small tomb, about 1.8 m in diameter and nearly closed at the top, with a narrow opening about 0.40 m at its widest point (figure 14.16). The roof was corbelled using angular blocks and limestone slabs, and through the narrow opening from which a capstone was missing, one could see a fill of silt and large cobbles about 1 m below the capstones of the roof. HCT 30-3 was well preserved to a height of 1.5 m and was constructed of dry-laid blocks of undressed limestone. Unlike the inner construction of many other HCTs, C30-3 did not use upright limestone slabs to define an inner chamber. Instead the wall was constructed of an outer facing of undressed limestone blocks in rough courses, with an inner facing also of undressed limestone blocks, each about 0.20–0.40 m wide, in courses. A rubble core of smaller cobbles and pebbles had been packed between these faces. The excavation did not section a wall. Therefore the cross section of the wall could not be examined for construction details. Instead, excavators removed two top courses of the corbelled chamber roof, which began at about 0.70 m height inside the chamber. By widening the roof entry, excavators could access and excavate the fill

Figure 14.15. HCT C30-3 (right) and C30-4 (left) during excavation. *Photograph by Jennifer Everhart.*

Figure 14.16. HCT C30-3 before excavation, facing north; Matthew Senn operating GPS equipment. *Photograph by Jennifer Everhart.*

of the 0.48 m diameter inner chamber, the construction of which was otherwise left intact. All contents were sieved through a fine screen (about 2 mm mesh).

In the upper level of chamber fill (Layer 1, as termed by the excavators), with its surface some 0.30 m below the corbelled vault, excavators removed a layer of large angular cobbles within a silty matrix, presumed to be the result of partial collapse of the capstone(s) and upper vault. Aeolian silt had accumulated around this collapse.

Finds and Objects in HCT C30-3

Within Layer 1, excavators recovered fragmentary animal teeth of large caprine or small bovid size and of cattle, fragmentary but otherwise well-preserved subadult human molars, and a number of scattered beads in a 0.04–0.05 m deep matrix of large rock, small pebbles, and loose silt. Underlying Layer 1, Layer 2 was a 0.16 m depth of aeolian silt, containing more beads, also of a great variety (figure 14.17). Sedimentation through aeolian activity covered very small, unidentifiable human bone fragments, the discovery of which arbitrarily defined a bottom layer, Layer 3. This layer was about 0.28 m deep and included several more beads. The base of the chamber was paved within the chamber walls with 16 flat, close-fitting stones (figures 14.18 and 14.19).

In addition to the inventory of beads, probably all from one necklace and shifted through later disturbance, there was a bronze or copper clasp in eight fragments and one chert flake also in Layer 1 (table 14.4). The tomb showed signs of reuse, probably contributing to the disturbance. A radiocarbon age between 3367 and 3175 cal BP (AA90838, 3067 ± 39), obtained on the subadult human teeth of upper Layer 1, is more than 2,000 years later than the date obtained on beads from Layer 2. Beads in Layer 2 were found in concentration above human bone elements, suggesting that they belonged to jewelry partially in place

Figure 14.17. Six tubular *Turbinella pyrum* (Indian chank shell)–type beads from a lower burial in HCT C30-3. *Photographs by Jennifer Everhart.*

with the earliest burial. The radiocarbon age of 5447–4963 cal BP (AA90336, 4525 ± 66) on marine shell beads (probably *Turbinella pyrum,* or Indian chank) is probably 200 years too old if one allows for a marine carbon correction factor (Saliège et al. 2005), but there is still a substantial gap between the first interment and the subadult human teeth at the top of the tomb.

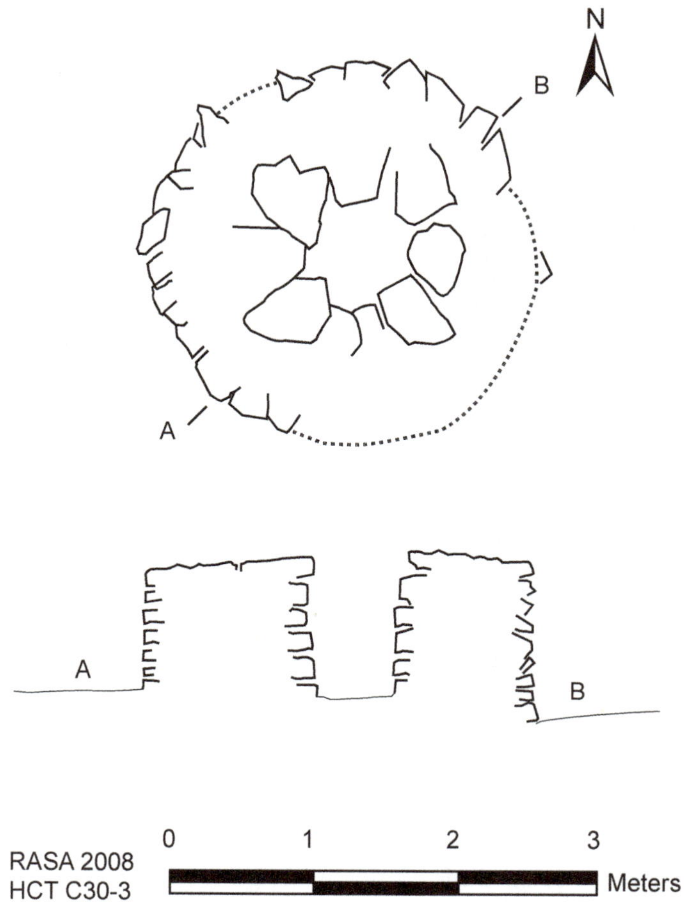

Figure 14.18. Plan and elevation of HCT C30-3. *Illustration by Kimberly Williams and Tara Steimer-Herbet.*

Table 14.4. Artifacts from HCT C30-3.

	Large Beads, 10–15 mm					Small Beads, 3–10 mm				
Layer	Disk *Turbinella pyrum* Type	Tubular *Turbinella pyrum* Type	Disk *Conus* Type	Disk *Spondylus* Type	Disk Mother-of-Pearl	Annular Steatite/ Stone	Tubular Gastropod Type	Disk Frit Type	Disk Amber/ Agate Type	Copper/ Bronze Clasp Pieces
1	2	0	0	10	0	37	1	2	2	8
2	0	7	5	24	1	0	0	1	0	0
3	0	0	0	0	0	6	0	0	2	0

Figure 14.19. HCT C30-3 fully excavated, showing paver floor. *Photograph by Kimberly Williams.*

Camel Burial in HCT Tomb C30-4

Only 10 m to the west of C30-3 lies another well-preserved HCT, also constructed with an outer dry-wall facing of undressed limestone blocks in rough courses. Neither blocks nor slabs were naturally available in the immediate vicinity of the tombs on the gravel terrace but must have been brought through the wadi from bedrock faces 500 m to the north. The top of C30-4 (E 315880.19, N 1723466.21) had a capstone in place, and its removal left only a narrow opening about 0.40 m wide. A secondary feature, a very rough heaping of loose stones about 0.20 m high, encircled the top of C30-4. To create a space safe for further excavation, the excavators removed loose stones, the capstone, and three layers of corbelled limestone slabs that roofed an interior chamber. The interior chamber was constructed as a circle of large limestone slabs, set upright on their flat ends for stability on the ground surface. These inner chamber slabs were used as support for corbelling slabs stacked flat on top of them. Although excavators did not section the tomb wall, it is clear that construction of this HCT mirrored other examples with an inner chamber of upright slabs, an outer face of undressed blocks, and a core of smaller cobbles, pebbles, and sediment to make a wall about 0.60 m wide (figure 14.20).

Inside the chamber, the surface of the surviving deposit was about 0.50 m below the corbelled roof. Chamber fill was a matrix of aeolian silt with few pebbles and cobbles, probably collapsed from the walls and packing over the corbelling. Uppermost Layer 1, about 0.20 m in depth, consisted of fine silt and cobbles with a few adult human tooth fragments, bone fragments, and a single human rib fragment. A radiocarbon age from collagen in the Layer 1 rib (AA83495, 158 ± 38) extends out of calibration range but is clearly recent. The rib also looked recent, with good preservation and a greenish tinge. The underlying Layer 2 was about 0.10 m in depth of fine silt only, suggesting that the tomb remained intact long enough to trap a substantial silt deposit before the top level of roof fall. Also about 0.06–0.10 m deep, Layer 3 also had an aeolian silt matrix, and the deposit included pockets of sterile silt, pebbles of thermally altered rock, and several pieces of animal bone. Layer 4 was an underlying deposit with a concentration of thermally altered rock and animal bone in the west half of the chamber, where a few nondiagnostic chert flakes were also recovered. There were also several large slabs, possibly either rock fall or brought into an open chamber in connection with a firing event. The concentration of thermally altered rock suggests an actual fire on or in the tomb. An alternate explanation is that pebbles from ancient hearths on the gravel terrace were gathered up as part of the rubble core in wall construction, only later collapsing into the chamber as it filled with aeolian silt. This is unlikely, for thermally altered rock was concentrated within one area of Layer 4. Layer 4, still with a matrix of aeolian silt, was about 0.23 m deep and overlay Layer 5, which was defined by the presence of poorly preserved animal bone still within a silty matrix. Excavators defined as

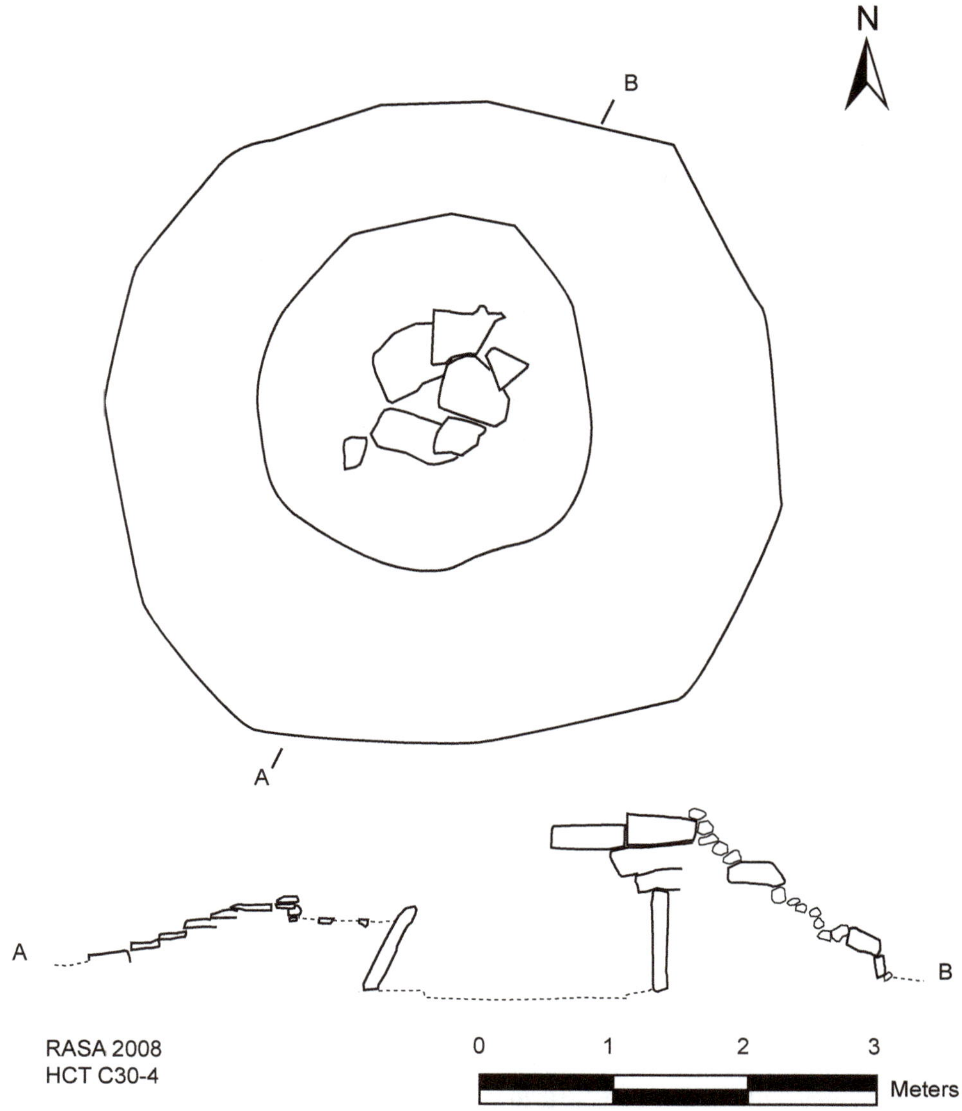

Figure 14.20. HCT C30-4, plan and elevation. *Illustration by Kimberly Williams and Tara Steimer-Herbet.*

Layer 6 the silt matrix that included intact and articulated bone from an entire camel, of which only the cranial and neck bones (in Layer 5) lacked good preservation.

The entire camel had been couched in the tomb facing northwest, its legs folded up beneath it and its head and neck drawn back over its left shoulder (figure 14.21). Beside it at the north wall of the inner chamber was a smooth flat disk of limestone—perhaps a cosmetic palette—placed over a cockleshell. Below the area of the camel cranium and against the south wall was a single cowrie shell button; it could have been attached to a bridle or trappings. There was a deposit of desiccated animal dung rich in plant fibers. This deposit lay beneath the abdominal area close to the hind limbs. Finally, a single tiny bone bead appeared in the final sieving of aeolian sediment from the bottom fill of the tomb, which had been constructed on the gravel terrace surface (table 14.5).

With the collaboration of Jean-François Saliège, RASA submitted three radiocarbon samples from the same camel, which returned a wide range of age estimates (McCorriston et al. 2011). The most reliable age, between 1538 and 1369 cal BP (AA83496, 1559 ± 38), comes from tooth enamel, the least likely of the samples to have been undergone diagenesis. The dung, 1881–1717 cal BP (AA 79762, 1868 ± 35), was likely evacuated at the death of the animal and is most susceptible to contamination by environmental organic materials such as micro-fauna and decomposers. An age estimate on long bone collagen at

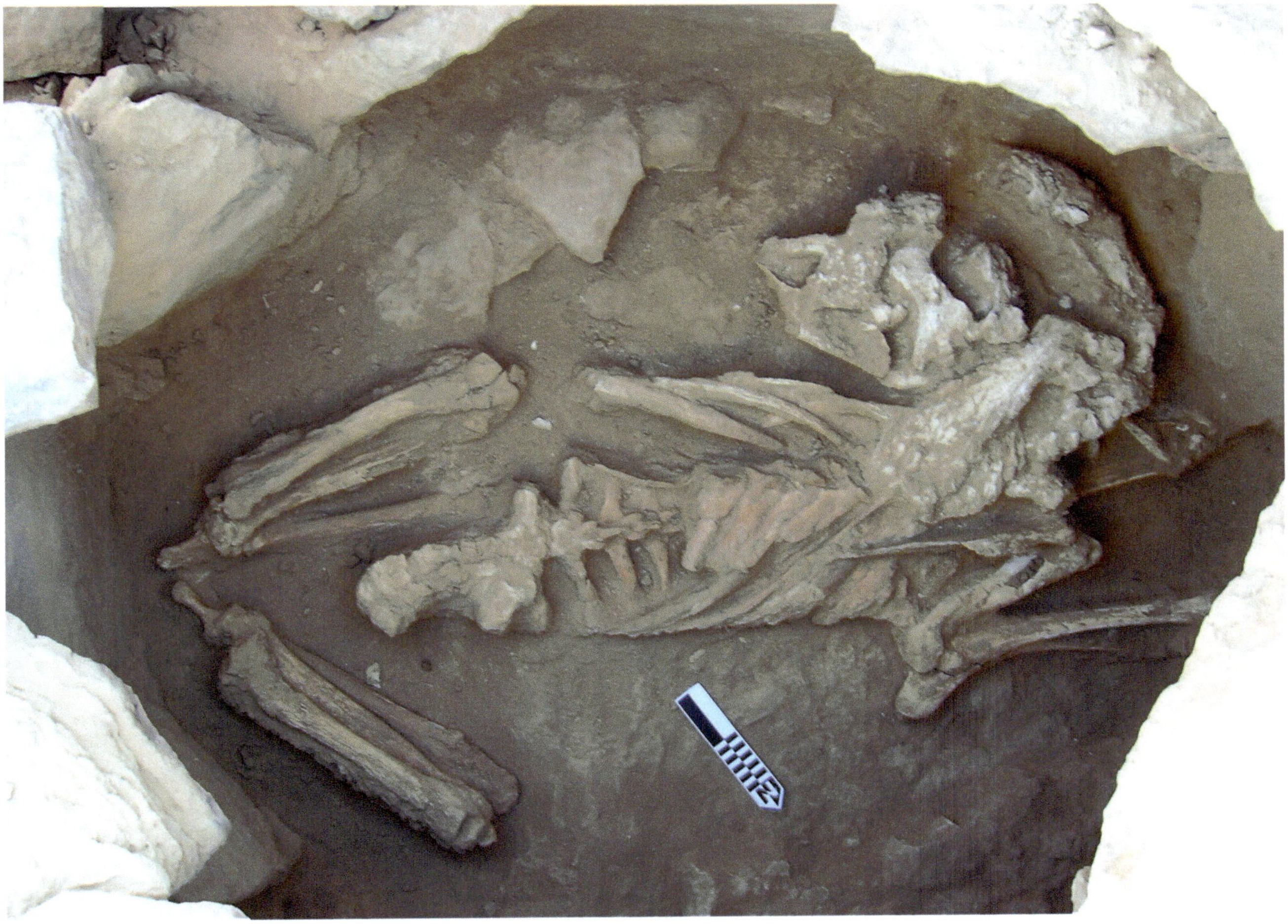

Figure 14.21. HCT C30-4; camel burial as reuse of an earlier tomb. The animal was deliberately bound, crouched, with its head drawn back up over its shoulder. *Photograph by ʿAbdalKarīm Al-Burkānī.*

Table 14.5. Artifacts from HCT C30-4.

Layer	Cockleshell	Cowrie Bead	Marine Gastropod	Bone Bead	Chert Flake	Palette
1	0	0	0	0	0	0
2	0	0	0	0	0	0
3	0	0	0	0	0	0
4	1	0	1	0	5	0
5	0	0	0	0	0	0
6	1	1	0	1	2	1

2699–2346 cal BP (AA86370, 2407 ± 40) is perhaps 1,200 years too old. While diagenesis may explain some of the range (Zazzo and Saliège 2011), some human error likely plays a role.

The camel burial and bone in Layer 1 date two uses of the tomb, but there is no evidence to date the construction of C30-4. A single, tiny bone bead, typical of beads in C30-3, hints at an earlier possible interment in C30-4.

HCT C32-1

In Wādī Ḥarād, an upstream tributary to Wādī ʿIdim, two HCTs about 42 m apart along a bedrock terrace had a medium level of preservation, standing about 0.80 m high, with chambers filled with sediment and rubble. Both C32-1 (E 282734.60, N 1700124.61) and C32-2 (E 282692.81, N 1700131.51) were excavated by removing the fill of the chambers, which like other HCTs regionally were entered only from the roof. The team cleared rubble from the outer faces in a small section only to determine construction and preservation. It appears that undressed limestone outer facing blocks had been removed at upper levels, exposing and enhancing the collapse of a rubble core about 3.4 m in diameter. They were intact in the bottom outer courses.

The inner face of the chamber of C32-1 was of upright limestone slabs, set in a roughly oval shape 2.10 m across, with pointed ends formed by abutting edges of uprights (figure 14.22). The upper layers of fill were a sequence of sediment and small cobbles overlying large slabs and small boulders. Several slabs sloped down inward and showed signs of cracking, as if a heavy force had been placed on top of them while flat. The matrix was a fine aeolian silt around limestone rubble, presumably from the collapse of a roof. This fill (Levels 1 and 2) was about 0.60 m deep.

Below the collapse, the chamber contained an aeolian silt layer, Level 3, about 0.08 m deep. Within Level 3 excavators identified a central deposit of several dozen smooth cobbles and pebbles, characteristic of a hearth but without signs of thermal alteration. An adjacent deposit of charcoal and a few thermally altered rocks marked the differentiation of Level 4, which proved to be only 0.05 m deep and overlying the bedrock on which the dry-walled tomb was constructed.

Apart from charcoal fragments, sieving in Levels 3 and 4 recovered only nine small bone beads typical of the 5,200-year-old beads recovered from C30-3. A single, unidentifiable animal bone fragment, thought possibly to come from an animal burrow, occurred in Level 2, and charcoal samples disintegrated in radiocarbon assay pretreatment.

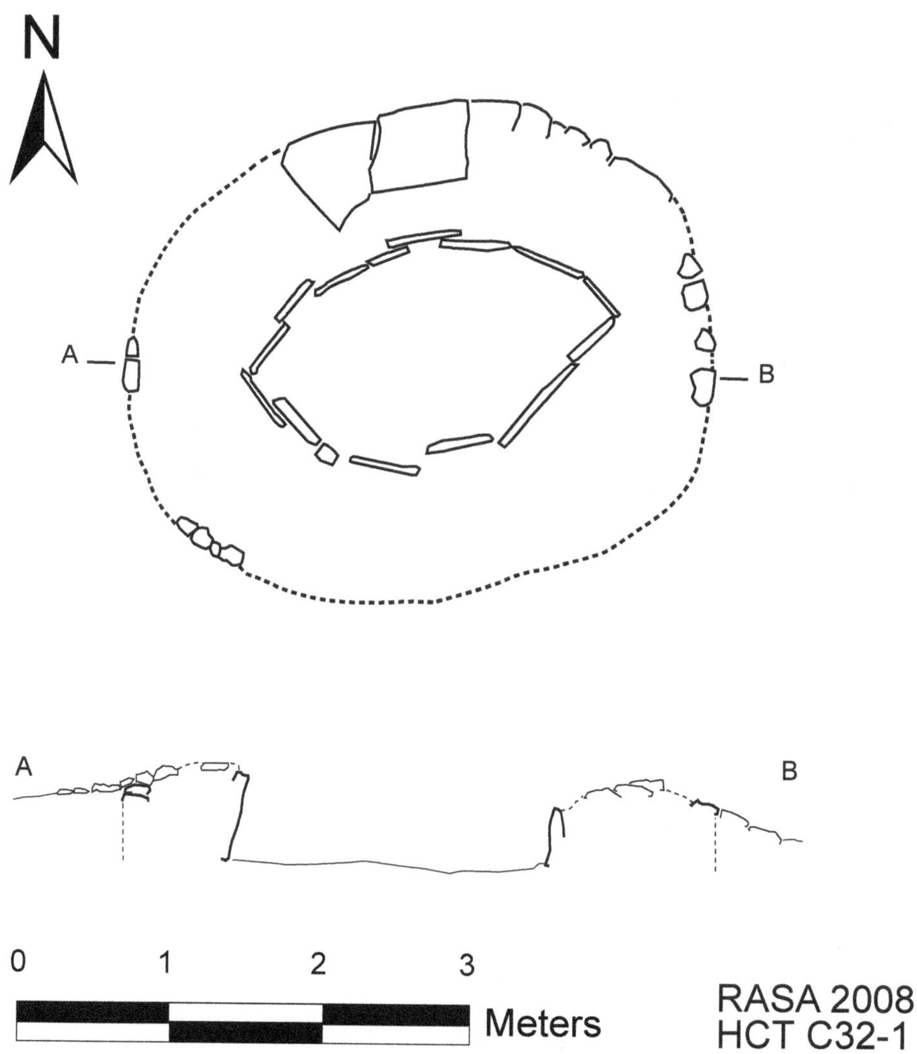

Figure 14.22. HCT C32-1, plan and elevation. *Illustration by Joy McCorriston and Tara Steimer-Herbet.*

HCT C32-2

This HCT had a tail with five tail elements of dry-stone piles, each about 0.90 m across and standing to a height of about 0.20 m (figure 14.23). C32-2 showed a construction similar to C32-1, with an outer facing of undressed limestone slabs in rough courses, an inner chamber constructed of limestone slabs set upright, and a rubble wall core of pebbles and cobbles with a silty fill. Limestone slabs were available naturally from a tabular limestone bedrock outcrop nearby. A detail of construction in C32-2 was the use of large blocks set as supports against the inner basal faces of the limestone slab uprights defining the chamber (figure 14.24). The outer diameter of C32-2 was 3.55 m, preserved to a height of 1.30 m (figure 14.25).

Like C32-1, the upper Levels 1 and 2 in C32-2 consisted of broken limestone slabs, cobbles, and silty fill. After removal of about 0.40 m of debris, excavators encountered the remains of animal bone pressed against the side wall of

Figure 14.23. HCT C32-2 after first clearing of rubble, showing collapsed limestone slabs. View toward south, showing tail in background. *Photograph by Tara Steimer-Herbet.*

Figure 14.24. HCT C32-2; view toward east during excavation of Level 3, with inner support blocks and limestone uprights in shadow. *Photograph by Joy McCorriston.*

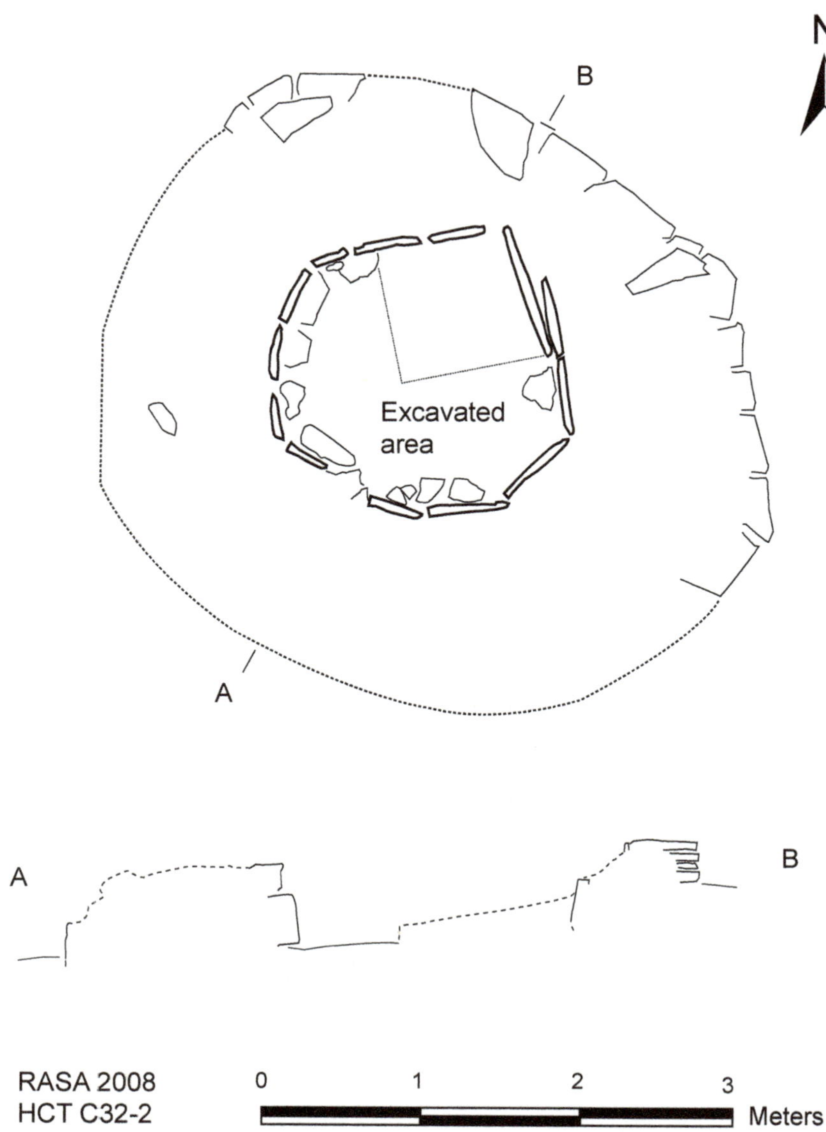

Figure 14.25. HCT C32-2, plan and elevation. *Illustration by Tara Steimer-Herbet.*

the chamber. A human subadult with an unfused femur head lay on the articulated animal bones. Bones were in very poor condition and were not lifted intact, fragmenting into many hundreds of pieces. Animal bone included articulated, right hindquarters of a caprine and large caprine- or small bovid-size bone (figure 14.26). By mistake, the intended sample for radiocarbon analysis (subadult human bone) remains in the Mukalla Museum. The radiocarbon age of 1730–1549 cal BP (AA83500, 1733 ± 39) that was obtained came from a sample of unidentifiable bone, possibly from two human individuals and possibly also including animal bone fragments. Given the stratigraphic position of these remains in the sediment of Level 1, between two layers dense in large limestone slabs, the human and animal bone belonged to the same (contemporary) deposit.

These burials and animal bone overlay Level 2, about 0.14 m in depth and again characterized by heavy limestone slabs and cobbles within a silty matrix. Under Level 2, Level 3 was a moist, silty brown fill with fewer cobbles. Level 3 filled the bottom of the chamber to the level of the tops of the boulder blocks supporting the inner upright slabs, and Level 3 contained tiny, uncollectable, and unidentifiable fragments of fragile bone. These lay at the base of Level 3.

Figure 14.26. HCT C32-2; right hindquarters of a caprine. *Photograph by Kimberly Williams.*

Level 4 was a very thin, silty sediment about 0.01 m deep overlying bedrock, and it underlay and predated the construction of the tomb. One may therefore conclude that the original inhumation, of which only uncollectable bone traces remained, was placed on an unprepared ground surface over bedrock, the same ground on which its builders constructed C32-2. Three tiny shell beads also occurred with the bone traces in Level 3.

There are therefore apparently two incidents of interment in C32-2; the first was earliest and accompanied by beaded goods on the original ground surface. The remains of this adult burial were extremely poorly preserved, such that nothing else can be learned from the bones. After a substantial period during which aeolian silt and collapse built up a chamber floor height about 0.40 m, there was another burial of at least one subadult human, sufficiently preserved that Kimberly Williams could identify a fragmentary but unfused proximal femoral head (less than 15.5–19.5 years old). When this subadult was buried, a meaty part of sheep or goat was also deposited in the tomb. Covering rocks may have been deliberately laid over this second burial—in this case perhaps robbed from an outer facing originally constructed to full height—but from the angle of the slabs, it appears that a hollow chamber roof collapsed inward.

High Circular Tomb Reuse

Through the excavation of six HCTs, the RASA Project not only substantially expanded the inventory of excavated Bronze Age tombs in Hadramawt but also noted some important patterns useful in interpreting survey records.

First, HCTs that present a well-preserved appearance show signs of reuse and reconstruction; some of the best-preserved examples, like C15-3 and C30-4, have Iron Age or later burials and may have been reconstructed during this later time. The tails of C15-3 and C32-2 might date to the time of reuse, bringing into question the contemporaneity of Bronze Age tombs and the tails associated with them. We still cannot discount the simplest explanation (null hypothesis): that the tombs and their tails date to the time of the tombs' original use in the Bronze Age.

Second, the internal sedimentation suggests a significant lapse of time since interment. The fill of tombs includes neither fine gravels nor graded sizes and lacks the small organic or anthropogenic inclusions that might be expected if these fills were carried in for deliberate burial. Instead, the fills consist of clastic limestone pebbles and cobbles unsorted in fine silt matrix. The evidence suggests that bodies were deposited within an empty chamber constructed on unprepared surfaces, the chamber was then corbelled and capped, and sedimentation occurred through aeolian transport and the natural trap created by a still space. When capstones collapsed or were unsealed, clastic rocks and more silt were introduced from the packing of walls or roofs.

Regardless, all RASA-excavated HCTs contain some evidence—even a few tiny beads—pointing to initial burials, and presumably tomb construction, in the Early Bronze Age, 5,200 to 4,200 years ago. The initial burials tend to be very poorly preserved, with bone preservation better after 3,000 years ago, except where bone has been exposed in upper or very shallow sediment. Subsequent burials and reuse of these tombs in the Iron Age and through the South Arabian period may have had important implications for social constitution and commemoration in mobile societies, a topic explored elsewhere (McCorriston 2013). There is clear evidence for new burials in HCTs C15-3, C30-3, and C32-2 (and in Tumulus C51-1, below). These were accompanied by animal bone, lithics, and fire in C15-3; by caprine hindquarters in C32-2; by fire in C32-1; and by a camel burial, utilized flakes, and fire in C30-4. These were likely the remains of offerings, perhaps sacrifices or commemorative meals with revisits long after interment. Animal bone and charcoal in C15-3 is at least 20 years younger than the latest burials, and the gap may have been much greater. It is highly plausible that the quarrying of C15-2 stone took place at the time of reinterment and revisiting C15-3.

The case of the entire camel in C30-4 deserves special comment. Evacuation of dung and the awkward placement of the animal's head suggest interment of a live bound animal, as described in early Islamic texts (Bin 'Aqil and McCorriston 2009; King 2009). Written sources indicate that this sacrifice was intended as a mount for a dead person. The

Figure 14.27. Wall tomb C4-1 overlooking Wādī Sanā, with Tara Steimer-Herbet in background. This example was not excavated but provided a particularly informative overview image. *Photograph by Michael Harrower.*

practice of late pre-Islamic *al-balīyah*, or camel sacrifice, has been well documented elsewhere through archaeological discoveries in Eastern Arabia (Uerpmann 1999) and Hadramawt (Vogt and Sedov 1994). In the case of C30-4, future excavators might look for a contemporary human burial in C30-5, a nearby (unexcavated), well-preserved HCT.

While there are some hints of fairly modern inclusions, there is no systematic evidence of recent bedouin reuse of these tombs, and bedouin in Wādī Sanā vehemently report that they eschew ancient tombs as burial places. Except for the (unidentified, possibly animal) bone fragments from upper C30-4, the latest burials are pre-Islamic. But the evidence for robbing, rebuilding, reuse, and revisiting considerably blurs any calculus of human occupation and landscape in the Bronze Age based only on survey records of these characteristically Bronze Age tombs.

Wall Tombs

An unusual type of tomb often found near HCTs, a wall tomb consists of a wall roughly 4 to 12 m long and about 1 m wide with a central burial chamber (figure 14.27). The RASA Project excavated three wall tombs, in each case removing overlying rubble to ascertain the exterior tomb dimensions, reveal the inner dimensions of a central chamber, and excavate the central chamber's fill. There are other reported excavations of wall tombs in the Sinai (Eddy and Wendorf 1998), Negev (Rothenberg 1979), Dhofar (McCorriston et al. 2014), and Hadramawt (Crassard and Hitgen 2007; Steimer-Herbet et al. 2006; Vogt 1997), none of which recovered entire inhumations.

Wall Tomb C17-2

This wall tomb is located on a bedrock terrace overlooking Wādī ʿAtuf (E 277004.33, N 1716845.02). It is aligned at 236 degrees, with an overall length of 10.75 m and a width of 1.2 m, preserved to a height approximately 0.60 m. Two areas of the outer facing have collapsed on its southern side, and a secondary feature, a semicircular wall, was later built against the center of the northern side. By observing the collapsed areas of the monument, the excavators could determine that the wall was built at least eight rough courses high of an outer facing of undressed limestone boulders and cobbles and filled with a rubble core of pebbles, cobbles, and sediment. The inner chamber, measuring about 1 m north to south and 0.5 m east to west, was defined by an inner facing of small boulders and cobbles. This facing appeared to have been repaired on the northern side, probably predating construction of the semicircular secondary feature. The team excavated the entire chamber, removing its homogeneous, unsorted fill of aeolian silt and limestone cobbles and pebbles, about 0.40 m depth, without intact roofing or capstones (figures 14.28 and 14.29). In the bottom 0.09 m of silty fill over the bedrock surface on which the tomb was constructed, excavation revealed a human humerus and rib with eight pieces of cut and smoothed Indian chank shell (*Turbinella pyrum*) nearby (figures 14.30 and 14.31). A radiocarbon age on the human bone of 4148–3902 cal BP (AA83497, 3683 ± 41) provided a terminus ante quem for the wall tomb construction and its latest burial, of which these finds were probably only a remnant. Secondary features such as a single cobble thickness (patch) wall on the northern side of the inner chamber and a lack of roofing suggested to excavators that the tomb had been opened, perhaps in antiquity. Rock collapse had refilled the chamber, trapping aeolian sediment thereafter.

Wall Tombs C19-1 and C26-2

Two other wall tombs were excavated; both were built with similar construction and to approximately the same dimensions. In both cases, excavation emptied central chambers, which contained only animal bone. Wall Tomb C19-1 (E 338330.57, N 1748189.36) was found in the vicinity of multiple HCTs on the first plateau (an intermediate bedrock terrace) overlooking Wādī Sanā, approximately 3.7 km downstream (north) of the Khuzmum. The tomb was oriented 245 degrees and was roughly 12.0 m long, 1.0 m wide, and 1.5 m high but was not very well preserved in part due to the rounded cobbles with which it was constructed (figure 14.32). The chamber contained a single unidentifiable animal bone. Wall Tomb C26-2 (E 304734.16, N 1733853.98) was found on a low plateau overlooking a broad valley. It was 7.95 m long, 0.65 m wide, 0.85 m tall, and oriented at 12 degrees (figure 14.33). The central chamber (33.5 x 42.5 cm) was filled about 35 cm deep with sediments and a fragment of the broken capstone. Two large slabs held the interior walls of the chamber, which contained two sheep/goat-size bones and one large caprine/small bovid–size bone (figure 14.34).

Tumuli

Discovery of C51-1

Apart from a tumulus in the Wādī ʿArf photographed and briefly described in museum exhibition catalogs (Vogt 1997), there exist no published excavations of tumuli with stelae. The tumulus C51-1 (E 244992.61, N 1727582.44) from Wādī Sissib is the first excavated example. The excavations described below are relevant for surface finds of anthropomorphic stelae throughout Hadramawt (chapter 15). C51-1 has been greatly damaged by a secondary

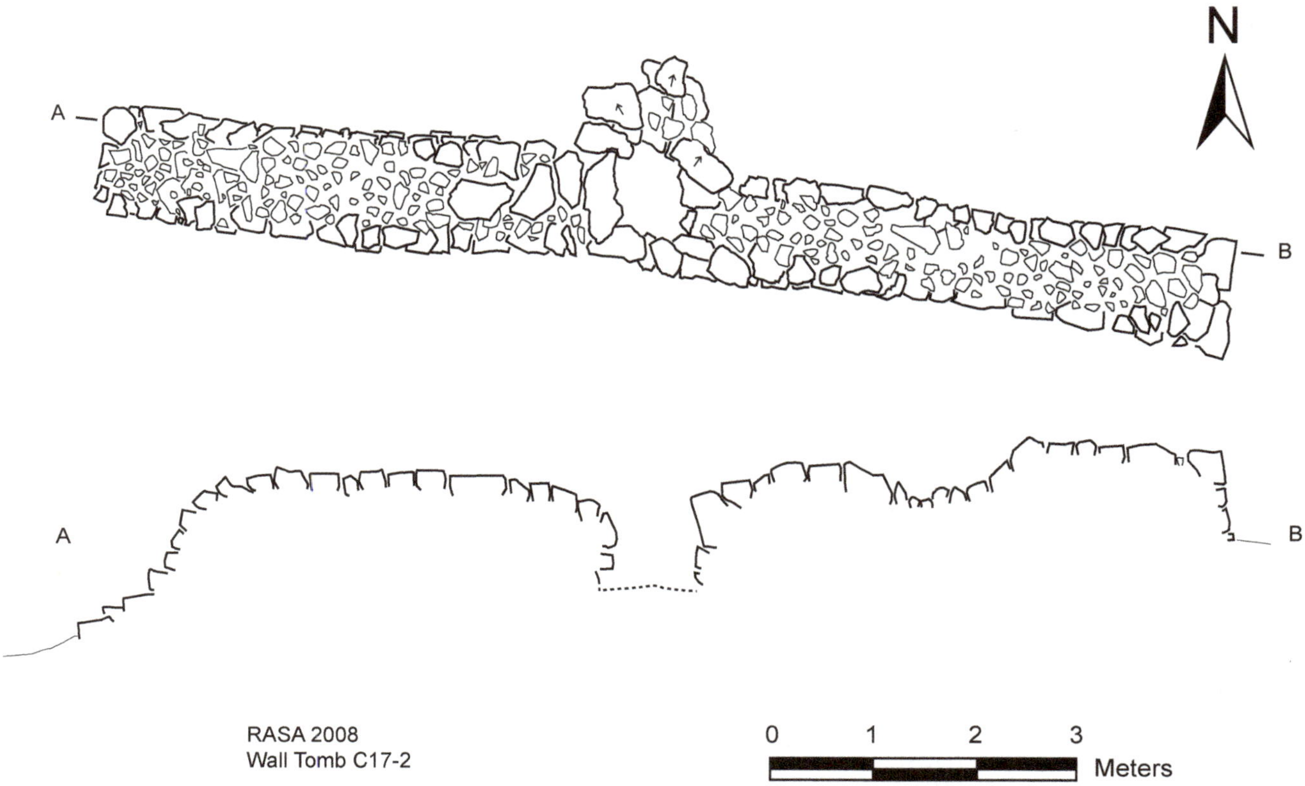

Figure 14.28. Wall tomb C17-2, plan and elevation. *Illustration by Tara Steimer-Herbet.*

Figure 14.29. Wall tomb C17-1 under excavation. *Photograph by Catherine Heyne.*

Figure 14.30. Wall tomb C17-2; bone and shell artifacts in situ during excavation. Photograph by Tara Steimer-Herbet.

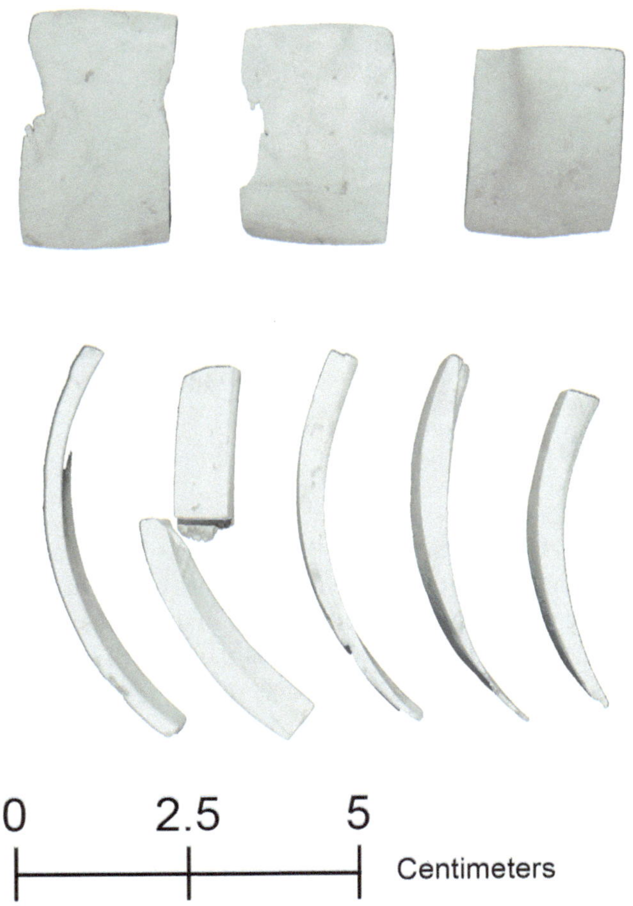

Figure 14.31. Indian chank shell artifacts from wall tomb C17-2.

construction that all but effaced surface indications of a tumulus situated at the southern end of a steep inselberg. This location is a remnant of an ancient bedrock terrace capped by colluvium and now isolated by erosion within the drainage channel of a broad, upland tributary to the Wādī Bin ʿAlī. Today there is a single *krif* upstream, and downstream lies a sandy canyon bottom filled with *Nannorrhops* palm. The wadi is a braided dry streambed studded with acacia trees and intersected by the dry bunds of opportunistic farmers. None live permanently in this area, but the signs, ancient and modern, of temporary visitors were clear.

The most recent activity at C51-1 was the construction of a fairly modern, double-face, rubble-core, C-shaped, dry-stone wall standing about 1 m high (figure 14.35). Probably a large hunting blind, it had clearly been constructed in haste and was unlikely to last a century in its current form. On the surface within its 5 m perimeter were tin cans, animal bone, a riveted iron knife, glazed ceramics, and two oblong limestone slabs. The survey crew flipped over the slabs to find that one is a simple—or perhaps unfinished—anthropomorphic stela. Nose and hair are in relief where the stone has been pecked away, and like many similar stelae, the lower third was unworked where it would have been buried. The sculptor had allowed a natural ferrous nodule in the limestone to serve as one of the eyes and another as a navel. On the outside of the hunting blind was the other 1 m long limestone slab,

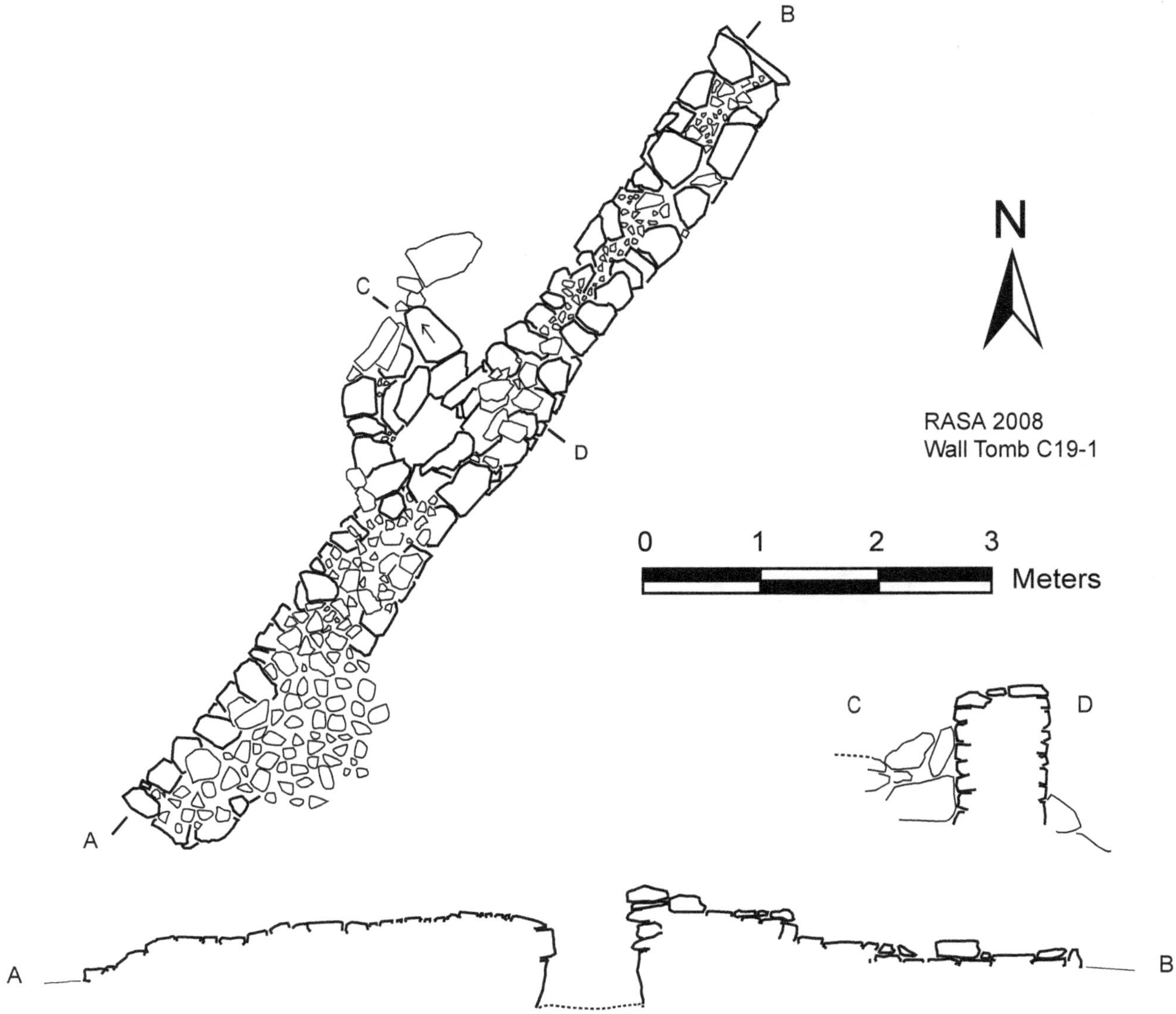

Figure 14.32. Wall tomb C19-1, plan and elevation. *Illustration by Tara Steimer-Herbet.*

also lying on the surface; this example was unworked and, like the worked stela, was far from its geological source.

No natural process could account for the presence of exogenous limestone slabs on this knoll; they were brought here by humans. A closer look showed that the surface contained stone from mixed sources—small and large limestone clasts, stream-rolled cobbles of chert and limestone, fragments of tabular limestone slabs—that could only have come together by human portage. Where tabular limestone slabs appeared stacked in several courses, the RASA team opened an excavation, hoping to recover evidence of a tumulus where the stelae had stood. Although the stelae were not in primary context, the tumulus from which they surely derived was only 4 m away, still ringed with the fragments and stumps of broken limestone uprights, an association demonstrated in the tumuli visited in 1996 (chapter 4) and in the Wādī ʿArf tumulus. The C51-1 tumulus was badly robbed, presumably to build the hunting blind.

Methods at C51-1

At the outer margin of the tumulus, at least as preserved, there were no more than two courses of stone atop a natural accumulation of ancient colluvium on the terrace. An initial excavation trench measuring 1.78 m north–south and 2.75 m east–west was quickly extended to the east as

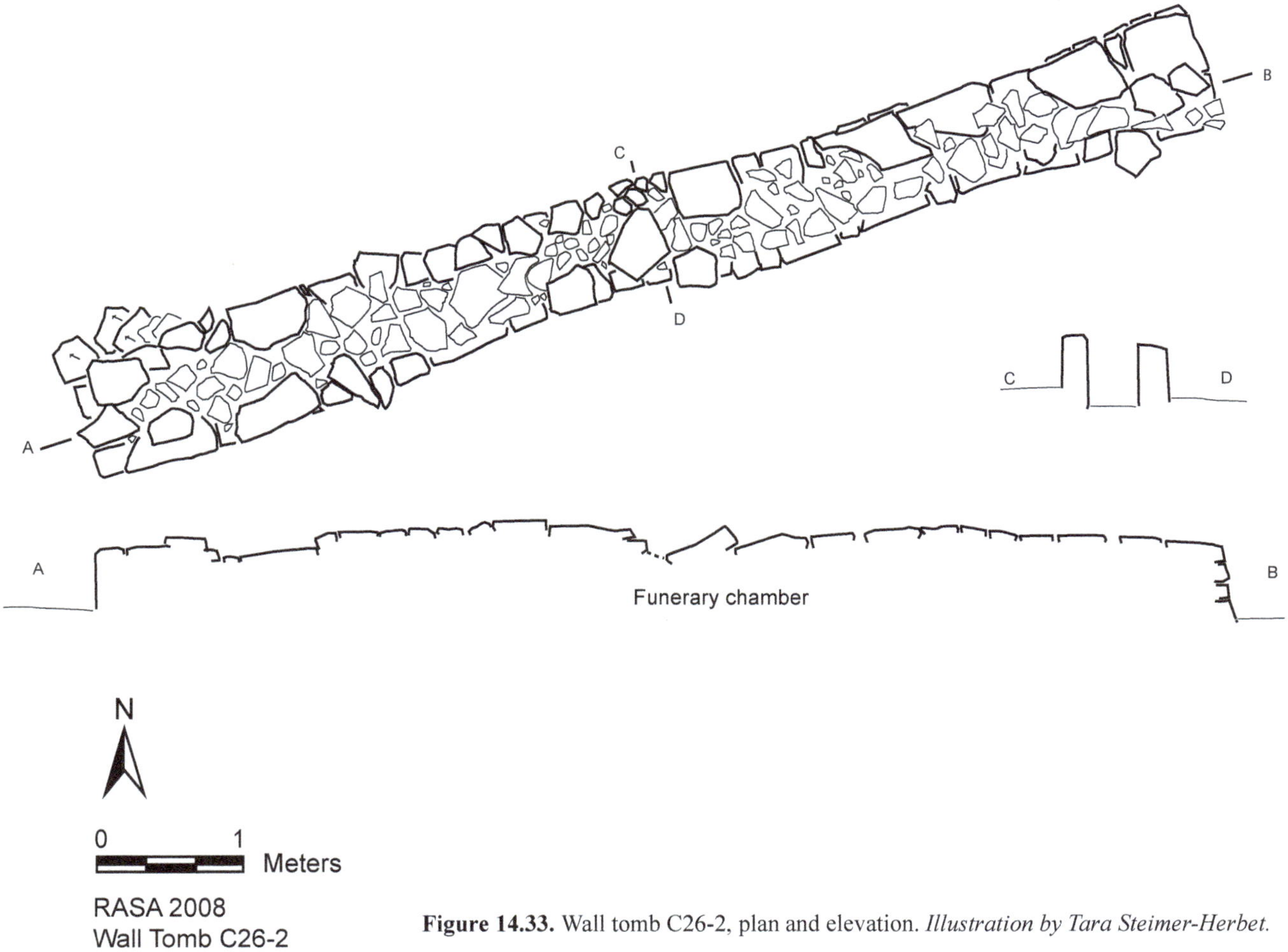

Figure 14.33. Wall tomb C26-2, plan and elevation. *Illustration by Tara Steimer-Herbet.*

Figure 14.34. Wall tomb C19-1; inner chamber after excavation. *Photograph by Tara Steimer-Herbet.*

Figure 14.35. Tumulus C51-1; view toward southwest. RASA team member J. Everhart stands on the center of an ancient tumulus. Several of its broken uprights left stumps, which are embedded in the terrace in front of her. To her right is a modern hunting blind with a stela (not in situ), which is turned to show its worked face. *Photograph by Matthew Senn.*

excavators realized that the trench had not intersected an outer perimeter of the ancient tumulus. Surface clearing included the removal of stone from the overlying modern hunting blind and the removal of bone, chert flakes, and modern debris (figure 14.36).

Finds in Context (C51-1)

Below surface, Level 1 was a 0.15 m deep mix of loose stone—some badly degraded—animal bone, and nondiagnostic lithics in dark brown matrix. Underneath, Level 2 included unsorted cobbles and pebbles within which pockets of bone, charcoal, and undiagnostic lithic debris indicated human deposition. A radiocarbon age on one charcoal fragment from within Level 2 (and sealed by the overlying construction of the hunting blind) is 2347–2069 cal BP (AA79767). This date either (1) is a terminus post quem on tumulus construction, related to fill scooped from adjacent camping on the terrace to create the tumulus, or (2) stems from a revisitation event (terminus ante quem) in which new activities accrued on top of an existing tumulus. Bones of sheep/goat, goat, and large caprine or small bovid size were within or atop the ancient tumulus construction (Level 2).

Removal of Level 2 exposed another large limestone slab, lying flat on the terrace surface. At the base of Level 2, about 0.30 m below surface, excavators encountered the sterile ancient terrace, an unsorted conglomerate of pebbles consolidated with reddish, presumably ferric, sandy sediment.

The difficulty in differentiating human tumulus construction from the natural terrace prompted excavators to section the tumulus across an east–west profile, which showed the original terrace surface to be highly uneven and which revealed a clear chamber about 0.60 x 0.60 m across, filled with reddish silty sediment (also excavated as Level 2), unsorted pebbles, and thermally altered rock and containing at its base, about 0.30 m depth, extremely fragile bone (figure 14.37). Cobbles and small boulders in courses formed the chamber walls. The tumulus was so poorly preserved that its perimeter today in the southern,

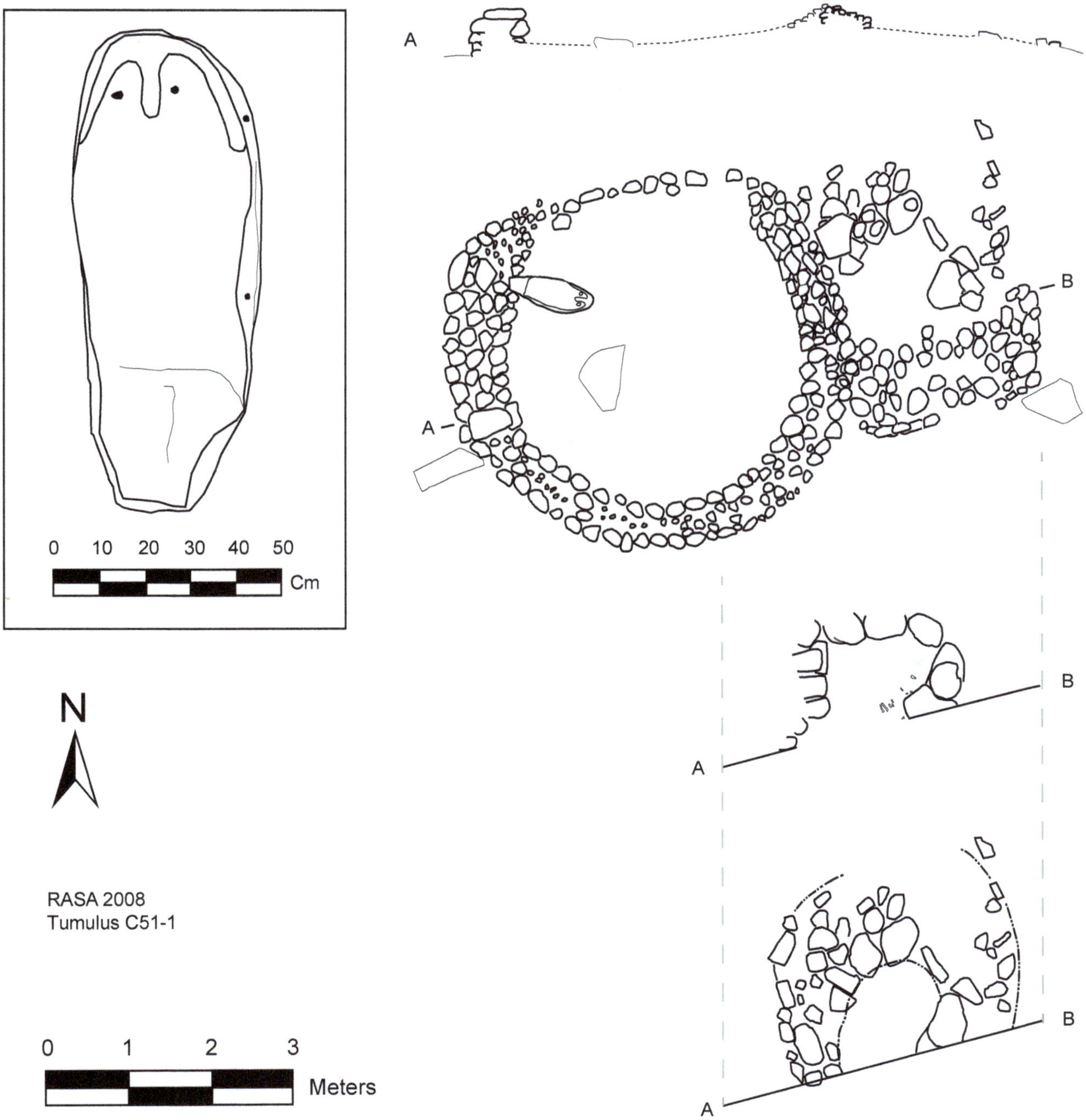

Figure 14.36. Tumulus C51-1, plan and elevations with enlarged view of nearby stela. *Illustration by Tara Steimer-Herbet, Joy McCorriston, Catherine Heyne, Jennifer Everhart, and Clara Hickman.*

excavated area may be only an unconsolidated slump. In its northeastern area, the broken stumps of limestone slab uprights suggest a clear perimeter about 2 m radius from the central chamber.

A tiny rodent bone indicates the potential for bioturbation, but the chamber was sealed with Level 1 overlying cobbles and a lighter brown matrix between them. The chamber fill was clearly defined as a reddish-brown matrix, and within the upper 0.10 m were poorly preserved, unidentifiable animal bone fragments associated with thermally altered rock and a few stray flecks of charcoal. While bone disintegrated immediately upon exposure and could not be lifted, skull, jaw, rib, and long bone fragments were observed.

Figure 14.37. Tumulus C51-1. View toward north of excavated central chamber. *Photograph by Joy McCorriston.*

This deposit overlay about 0.15 m of reddish-brown sediment, which in turn overlay an earlier deposit of thermally altered limestone pebbles and unidentifiable animal bone. This deposit, possibly a dump from burning elsewhere because it included no charcoal, overlay a partly articulated human skeleton, too fragile to lift. Like the animal bone, human bone within the chamber was exceptionally poorly preserved, seemingly in worse condition than animal bone and perhaps older. Therefore, without a bioarchaeology expert on hand, excavators decided to rebury exposed skeletal material in situ for future study. Subsequent events in Yemen have prevented return for such study.

Discussion of C51-1

The chamber clearly had a human burial placed on the original terrace surface, covered by a deposit of thermally altered rock and animal bone, with a radiocarbon age of 4142–3877 cal BP (AA83499) on bioapatite from animal bone within the chamber. This predates 2,000-year-old charcoal (above) found higher in Level 2, outside the chamber, and stratigraphically higher in the tumulus. Together, the two dates suggest that uncharred animal bone and hearth debris was deposited within the chamber about 4,000 years ago, perhaps but not certainly a second event in which the chamber was opened and new offerings placed within it. The age of the earliest human burial remains uncertain, as does the age of the tumulus construction itself. The upper deposit of bone and charcoal sealing the chamber is later, and considered with the chamber evidence, is most likely from a revisit event that involved fire and animal bone.

Triliths

In Wādī Sanā and more widely in Hadramawt, Mahra, Dhofar, and Eastern Arabia one finds a distinctive monument built of very low, rectangular or ovoid platforms aligned end to end, each supporting multiple groups of uprights, sometimes associated as a triad, sometimes monolithic. Generally one also encounters a parallel line of hearths, each constructed as a ring of boulders or cobbles filled with smaller, thermally altered stones, often from the riverbed. The best-preserved examples have four small boulders laid as corners of a square between each platform and its associated hearths. Triliths—the epithet describes the entire monumental assemblage, including the hearths and boulders—present an enigmatic ensemble with various ascribed functions and meanings, most of which can never be tested archaeologically (Bin ʿAqil and McCorriston 2009). Survey in Wādī Sanā and Wādī ʿIdim drainages found triliths only on lower silt, gravel, or bedrock terraces, never on the high plateau, nor with HCTs. Examples from Eastern Arabia (De Cardi et al. 1977) and Dhofar (Al-Shahri 1991; Cremaschi and Negrino 2002; Zarins 2001) have yielded radiocarbon ages around 2300–1700 cal BP, always from samples of charcoal, which date the latest use of the hearths rather than construction of the monument.

The RASA Project excavated hearths from three triliths in 2008 to recover datable material:

Trilith SU6-1

Also recorded as 98/23, this trilith (E 337219.42, N 1744349.66) was constructed on the Holocene silt terrace and was very visible from the Wādī as-Shumlyah east–west track toward Mahra (figure 14.38). About 16 m in length along a 120-degree orientation, the monument is composed of three oval-ended low platforms, each about 0.30 m high, as a ring of large cobbles retaining gravel fill. This fill supported alignments of limestone slab uprights, patterned the same way on each platform (east–west): each has two triad sets of uprights, followed by two single uprights, then two more triad sets of uprights. Within the fill composed of clasts and smooth wadi-bed pebbles, a team member found a trilateral Neolithic projectile point, suggesting that some material for construction came from the nearby Gravel Bar Site (chapter 9), the nearest source of gravel. Each of the platforms has a hearth to the northeast, about 3.5 m from its associated platform and measuring 1.7 to 1.9 m in diameter.

The construction of the monument is very evident from surface view, and so there is little reason to excavate such a monument. A terminus post quem for its construction is the silt terrace on which it sits. Silt terraces stopped aggrading after 5000 cal BP (chapters 3 and 18). To provide a terminus ante quem for construction, Hearth 3, the westernmost of three associated hearths, was excavated with a 0.5 x 0.5 m square in its center. Excavation revealed a 0.05 m deep upper layer of loose gravel. Below was about a 0.30 m depth of gravel with a fine, light yellow/brown silty matrix, probably deposited through aeolian activity after the hearth went out of use. There was

Figure 14.38. Trilith SU006-001, view from south, with ʿAbdalKarīm Al-Burkānī. Hearth 2 is in center background. *Photograph by Joy McCorriston*

no charcoal recovered, nor did all stones appear heavily altered by heating, but there were fragments of charred, unidentified bone.

Because Hearth 3 yielded poor material for radiocarbon dating, excavators turned to Hearth 2, central in the alignment. A slightly larger trench, 1.55 m northwest–southeast x about 0.50 m southwest–northeast extended across most of the northern interior area of the hearth. Within gravel at about 0.20 m depth below surface, excavators recovered a single dark clayey fragment, which proved upon close examination not to be charcoal. The hearth was about 0.45 m deep and sat on a sandy layer of natural sedimentation of the terrace before the hearth was built. Combined charcoal fragments from 0.50 m depth within this stratigraphically earlier sandy layer yielded a radiocarbon age of 4967–4825 cal BP (AA79763), consistent with the stray deposition of charcoal within Holocene sediment terraces. Continued excavation into the silt terrace itself yielded further charcoal at 0.70 m depth, but there was an associated uncharred *dom* seed (*Ziziphus leucoderma*) and evidence of rodent burrowing, rendering the sample unreliable.

Trilith SU134-3

A second trilith (E 334090.64, N 1744580.08) on a low bedrock terrace overlooks the tiny Kheshiya tributary to Wādī Sanā. Its original formal layout has been subsequently altered by robbing uprights and laying them against the full length of the platforms to shape a water diversion channel (*shrūj*), and the trilith itself offers a terminus post quem for this water management feature, of a type that is elsewhere usually undatable. The best preservation of the entire 45 m trilith construction is the eastern end, where four distinct platforms are evident in a 340-degree alignment. These platforms contain six or seven uprights each, but it is impossible to know their original pattern as triads and monoliths. The easternmost hearths are also the best preserved. An effort to extract charcoal by excavating inside Hearth 2 in 2004 failed to recover any charcoal, and excavators met with similar disappointment in 2008. Although well preserved, the hearth contains silty sediment likely trapped by aeolian forces. Rootlets and casts suggest that overlying gravel trapped enough moisture to contribute to decay of any charcoal.

Counting westward, Hearth 11 also appeared to contain a substantial depth of thermally altered rock that could trap and preserve charcoal. An excavation with a radius of 1.4 m emptied the southern half of the hearth within its perimeter cobbles. At a depth of about 0.05 m below surface loose thermally altered rock, there was a dark, ashy patch without intact charcoal. Below this, a fine yellow-brown sediment, probably aeolian in origin, had formed a matrix around the gravel. Finally, at 0.30 m depth, the matrix changed to a mottled, light-and-dark ash with intact charcoal pieces. One of these yielded a radiocarbon age of 2106–1894 cal BP (AA79769) for the last use of this hearth. A similar stratigraphic sequence characterized a comparable excavation of the southern section of Hearth 14. No definitive evidence links the age of the hearth's last use to the construction of the full trilith platforms, so it is only a premise that these events were contemporary.

Trilith C58-1

A third trilith (E 309298.36, N 1720468.01), along a drainage leading into Wādī Ḥarū, yielded a charcoal fragment from a hearth with a radiocarbon age of 1774–1559 cal BP (AA79768). The monument itself is mostly gone, recently robbed for stone by the neighboring hamlet in Wādī Ḥarū and effaced by the marks of a modern bulldozer over the gravel terrace on which it sat (figure 14.39). Its overall length was about 116 m, composed of a series of platforms, presumably also supporting uprights. From east to west remain 19 hearths spaced approximately 5–6 m apart, each with a diameter about 1 m. Hearths consisted of a ring of smooth, large cobbles containing a heaped fill of thermally altered gravel derived from the bed of the wadi.

Because it appeared best preserved, Hearth 10 (counted 1-19 from east) was sectioned north–south and the western half was excavated. Only a single piece of charcoal appeared at shallow depth, along with modern plastic and pervasive evidence of bioturbation (thick spiderwebs). Hearth 5 provided a more promising context. A similar excavation in its western half encountered ashy sediment about 0.30 m below the level of gravel fill. This sediment yielded a large intact charcoal fragment (AA79768, above) among many smaller pieces, dating the last use of the hearth.

Discussion of Triliths

Thus the last use of trilith hearths in Wādī Sanā appears to have occurred around 1,500 years ago. Despite the integral appearance of complete triliths with platforms, uprights, boulders, and hearths, these combinations could have accrued over time or been in use for longer than the 600-year span, a period consistent with radiocarbon ages obtained in Dhofar and Eastern Arabia.

Islamic Graves

There were no excavations of Islamic graves, but al-ʿAlīy (Ḥumūm) bedouin offered some comments about their

Figure 14.39. Trilith C58-1; view toward northwest. Stone piles in the background are modern, probably robbed from the trilith. *Photograph by Joy McCorriston.*

general location and form. First, modern bedouin never reuse ancient monuments. Even when someone must be buried on the plateau, a burial must be sited away from ancient ruins. Second, bedouin seek to bury their dead at locations where other Muslim graves are already found, in the sediments of the lower terraces, where they cover graves with rocks and set up a *shahīd* (witness stone)—one for a man, two for a woman. Ethnographer ʿAbdalʿazīz Bin ʿAqīl documented different customs among Mehri speakers to the east, where modern graves have two uprights for men and three for women. Finally, rural graves are dug into sediment as simple pits without the *lahd* (niche at the bottom) found in urban burials.

Madhābiḥ

Madhābiḥ are circular or subcircular rings of boulders or slabs filled with thermally altered cobbles and pebbles selected from the smooth, rounded gravel of the wadi bed. It is clear that these structures supported fire—there is discoloration of stone, including the inner faces of the perimeter stones, heat-shattered clasts, and ash and charcoal among the gravel. Modern grilling hearths are still constructed like *madhābiḥ*, a term the bedouin use to denote structures where communities reaffirm their social allegiances with a gathering for a communal feast of grilled fresh meat (Bin ʿAqil and McCorriston 2009) (The singular term, *madhbaḥ* derives from the Arabic verb "to slaughter.") Survey alone can neither date these structures nor entirely differentiate them from an ordinary hearth (*mūwqaḍ*) used for any kind of cooking. Size and form may differ—larger structures and those with prominent uprights and internal divisions probably served as *madhābiḥ*.

Some *madhābiḥ* are monumental, several meters in diameter and built with a platform and upper tier. One example, C30-27 (E 315401.00, N 1722741.24), measured 4.5 m northeast–southwest x 3 m southeast–northwest and stood a prominent 2 m above the surface of the bedrock terrace on which it was constructed (Figure 14.40). It provided a panoramic view of the Wādī Ḥarū camping ground, with nearby rockshelters, camp complexes, and very good flint sources.

Madhbaḥ C30-27

Excavations proceeded using the structural divisions of the upper tier, which was constructed of limestone slab uprights secured in oval formation by a gravel fill. Within this fill, two transverse sets of uprights divided the upper tier into three chambers, excavated as Layers 1a, 1b, and 1c. Each of these was about 10–30 cm deep and contained thermally altered rock clasts, yellow ashy sediment, and abundant charcoal, and in Layer 1b, two charred bones of sheep or goat. A charcoal fragment from Layer 1a has a radiocarbon age of 178 ± 33 (AA79764), which extends out-

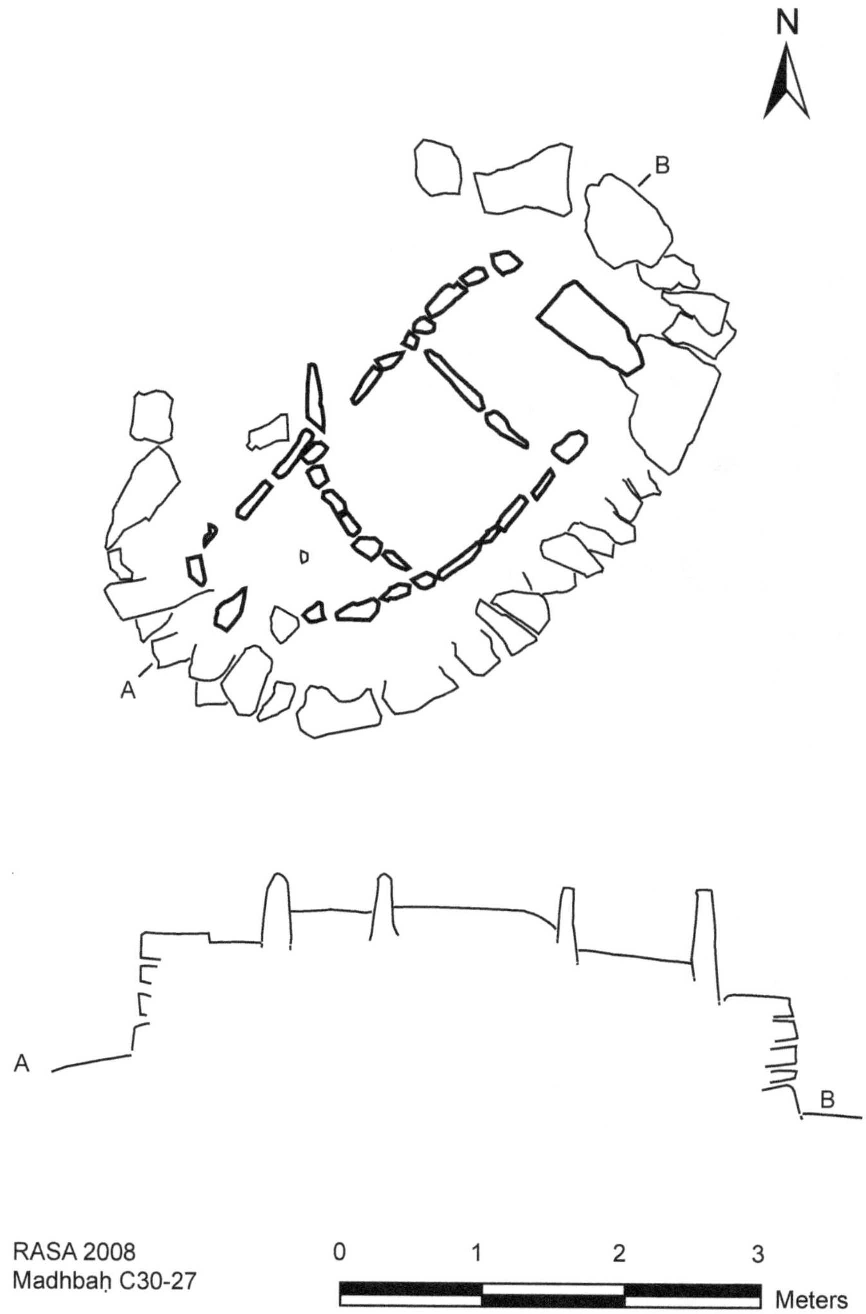

Figure 14.40. *Madhbaḥ* C30-27, plan and elevation. After McCorriston et al. 2011:16, figure 14. *Illustration by Joy McCorriston and Tara Steimer-Herbet.*

side the range of calibration but is very recent. Below the *madhbaḥ* tier, the fill of the platform itself contained a few more sheep/goat-size bones in its upper level (Layer 2). The matrix of Layer 2 was even richer in ash and charcoal and dark grey in color. A charcoal fragment recovered 0.60 m below surface from the western base of Layer 2 returned a radiocarbon age of 158 ± 33 uncal BP (AA79765), also outside the range of calibration.

The *madhbaḥ* upper tier was constructed on top of an oval platform built of large undressed boulders raised as rough courses against a scree slope. This retaining wall was filled with a rubble core of clastic limestone cobbles and pebbles, with a matrix of charcoal and ash. This matrix probably settled into place from the use of the upper tier as a *madhbaḥ*. The radiocarbon ages from the top of the upper tier and bottom fill of the lower platform are contemporaneous, suggesting single or short-term use of the structure.

There were multiple *madhābiḥ* in the area, all with a ring of upright angular limestone slabs and several built on small platforms. To recover charcoal, a second *madhbaḥ*—C30-27a, near C30-27—was also excavated in its eastern hemisphere, leaving intact the perimeter slabs. A single charred branch was collected from 0.17 m below the surface of thermally altered limestone cobble clasts, and this sample returned a radiocarbon age of 74 ± 33, also outside the calibration range. Like its neighbor, C30-27a is also a fairly recent *madhbaḥ*.

Spatial Analysis of Small-Scale Monuments Survey Data

Excavation and radiocarbon dating of the monuments described above provide a strong foundation for understanding the results of archaeological survey. While archaeological survey is sometimes considered an endeavor that precedes excavation, in our case, both efforts took place in tandem and contributed new insights and understanding in parallel. As our familiarity with sites and monuments accumulated over time, our ability to interpret, refine, and (re) classify survey records generated over 10 years of survey, from 1998 to 2008, correspondingly improved. Since our last fieldwork in 2008, hundreds of hours of consultation, discussion, and debate among team members and colleagues generated a standardized classification scheme and lexicon of terminology used for Microsoft Excel, Microsoft Access, and ArcGIS records about sites and monuments.

The explicit focus on monuments during fieldwork in 2008 was preceded by lengthy conversations with AHSD project statistician Prem Goel about different sampling strategies and their impact on potential future spatial analyses. From the outset, a number of key characteristics of small-scale monument patterning in Hadramawt were clear: (1) monuments are very sporadically distributed and most often appear in clusters, particularly along drainages and (2) the Southern Jol study area is rugged and offers few roads, making many areas very difficult to access. These two characteristics inhibit (if not prohibit) many traditional survey strategies, including random (or stratified random) sampling and regional full-coverage survey (Kowalewski 2008), which both would have re-

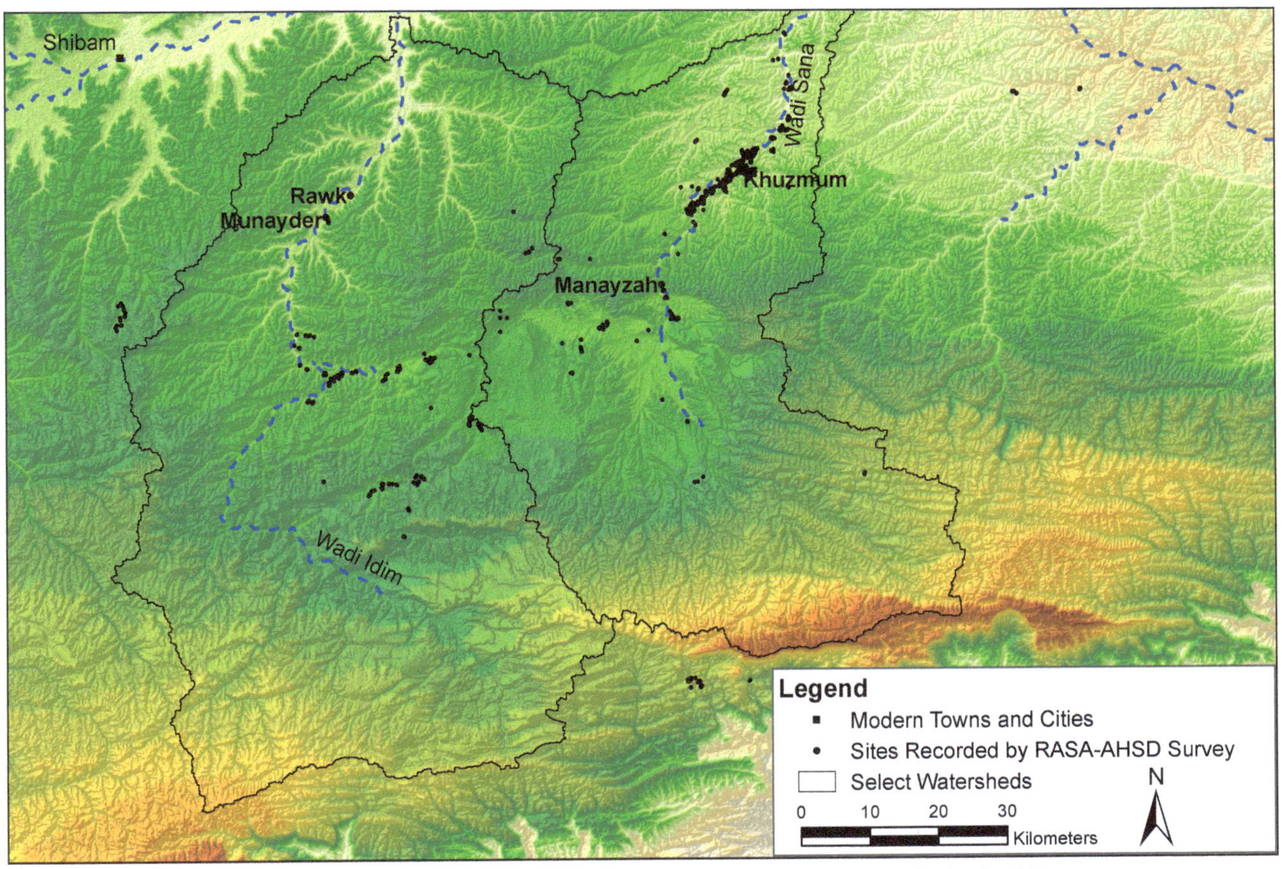

Figure 14.41. Map of monuments recorded by RASA–ASHD from 1998 to 2008. *Illustration by Michael Harrower.*

Table 14.6. Sites and monuments recorded by the RASA–AHSD Project from 1998 to 2008.

Type		Quantity
Monuments		
	high circular tomb (HCT)	466
	wall tomb	69
	trilith	29
	madhbaḥ	26
	tumuli	15
	Islamic grave	5
	dolmen	3
		613
Other Sites		
	water management structure	185
	lithic cluster	73
	platform structure	67
	camp complex	45
	bedouin shelter	29
	rockshelter	23
	building	21
	petroglyph	17
	structure	3
	other	792
Subtotal		1255
Total		**1868**

quired expending the majority of the team's time traversing difficult terrain where relatively little was to be learned about monuments. Instead, we opted for a satellite imagery–based sampling strategy in which strategically located QuickBird satellite images would serve as a sample of a region (the Southern Jol) that spans more than 20,000 km^2. Polygons of QuickBird imagery were ordered for areas that: (1) had some form of backcountry road access so that survey would not require many kilometers of hiking into remote areas before work could commence and (2) had some indication that tombs or other monuments might be present, because the area had been previously visited and/or because potential tombs were visible on Google Earth in at least some part of the prospective polygon. This strategy was meant to ensure that we would not end up with expensive imagery of remote areas that would be extremely difficult to access and might yield little or no information about monuments. Both because they tend to contain roads and because tombs and other archaeological remains noticeably concentrate along drainages, a lesser degree of preference was also made for purchasing polygons of imagery that included drainages. Collectively, this strategy was not meant to generate a complete or random sample but rather a collection of widely distributed polygons of areas that promised considerable new information about monuments. We anticipated nearly full coverage of high-visibility monuments along Wādī Sanā and that polygons of QuickBird imagery would serve as wider-area samples that would be surveyed in as close to entirety as possible. Due to the deterioration of security conditions in Yemen, particularly from 2008 onward, we were not able to complete our planned program, and our work shifted from eastern Yemen to western Oman in 2009 and 2010 (figure 14.41).

Collectively, the RASA–AHSD dataset is a substantial compendium of information about ancient sites across the Southern Jol that offers opportunities for further analysis. Here we report some basic descriptive statistics and general observations. From 1998 to 2008 we recorded 1,868 archaeological sites, including 613 monument sites (table 14.6). Survey and recording strategies for field seasons in 1998, 2000, 2004, and 2005 are reported in chapters 4 and 5. In 2008 we employed a more sophisticated approached focused on monuments. This included recording each monument on a specially designed monuments form along with photographs and measurements. We also adopted a more advanced system for field mapping that involved running our own GPS base station (Trimble 5700 model with zephyr antenna) at Canadian Nexen's central processing facility (CPF), along with two GPS rovers (both Trimble 5700 models with zephyr antennas). This kinematic GPS system provided better than 30 cm accuracy in a 100 km radius from Nexen's CPF. This level of precision was required to ensure that monument locations would overlay correctly on high-resolution satellite imagery and serve as training data for the development of tomb-detection algorithms (Harrower et al. 2013). Concordantly, our data holdings also include 17 polygons of QuickBird imagery covering areas from 50 to 124 km^2 for total coverage of 1,350 km^2. Unfortunately, 4 of these 17 polygons could not be surveyed due to security concerns, but survey records within the remaining 13 polygons provided baseline data for the development of tomb detection algorithms reported elsewhere (Schuetter 2010; Schuetter et al. 2013).

HCTs are by far the most common monument type (n = 466) and are far more frequent in our dataset than the

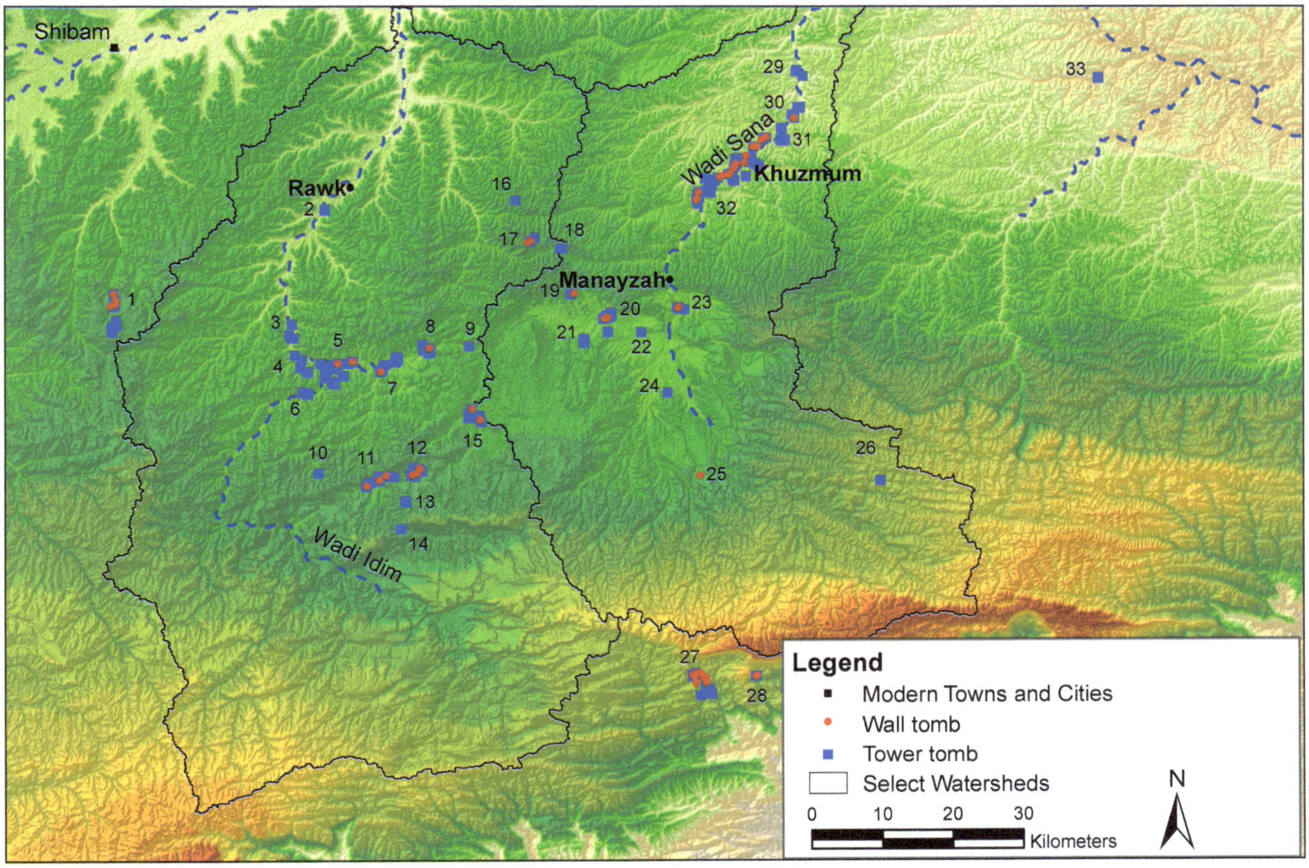

Figure 14.42. Map of major high circular tomb and wall tomb clusters recorded by RASA–AHSD. *Illustration by Michael Harrower.*

second most common type: wall tombs ($n = 69$). Interestingly, these two types often occur together, even though WTs date substantially later than the earliest HCTs. Given the patchy and sporadic distribution of our sample, we divided these two types into clusters for analysis. Clusters were defined by groups in which there was no more than a 1 km gap between monuments (figure 14.42). In other words, any gap greater than 1 km would define a new cluster. In our dataset of 535 HCTs and WTs, this procedure defined 33 clusters that consist of between 1 and 86 monuments (table 14.7). There are, of course, many more HCTs, and only 16 of the 33 clusters contain WTs. Notably, Wādī Sanā forms four clusters (29, 30, 31, and 32), with most WTs concentrated south of the Khuzmum inselberg, where they constitute, for example, 29 percent of the 80 monuments in Cluster 32. Noting that HCTs and WTs often seem to appear together, we calculated (in GIS) the distance from each WT to the nearest WT and nearest HCT. The average distance from a WT to the nearest WT (1058 m) is far greater than the average distance from a WT to the nearest HCT (285 m). This indicates that rather than building WTs side by side or in groups, ancient residents of the Hadramawt purposefully built WTs near HCTs.

Triliths—the third most common monument recorded by RASA–AHSD survey—are far less plentiful than the first two categories ($n = 29$). Triliths invariably occur along wadis or comparable low-lying areas, where they are highly visible. Analyses of data for 116 triliths surveyed in Dhofar, Oman, showed that their orientations are nonrandom and they are often oriented parallel to drainages, perhaps to increase their visibility (Harrower et al. 2014). Although our sample size is presently too limited to conduct comparable quantitative analysis for RASA–AHSD triliths, these observations also seem to hold true for the Hadramawt. Moreover, as in Oman, triliths frequently appear in clusters, with notable clusters of six triliths near the Khuzmum inselberg and 14 triliths approximately 32 km east of Wādī Sanā. However, triliths do not appear to cluster in the vicinity of HCTs or WTs, which may be related to chronology. (Triliths date much later in time.)

Table 14.7. Clusters of high circular tombs and wall tombs.

Cluster #	Wall Tomb	High Circular Tomb	Total	% Wall Tomb
1	6	23	29	21%
2	0	8	8	0%
3	0	13	13	0%
4	0	13	13	0%
5	2	84	86	2%
6	0	21	21	0%
7	1	29	30	3%
8	1	14	15	7%
9	0	1	1	0%
10	0	3	3	0%
11	4	24	28	14%
12	5	19	24	21%
13	0	4	4	0%
14	0	4	4	0%
15	2	24	26	8%
16	0	1	1	0%
17	2	6	8	25%
18	0	3	3	0%
19	1	2	3	33%
20	2	39	41	5%
21	0	8	8	0%
22	0	2	2	0%
23	3	7	10	30%
24	0	2	2	0%
25	1	0	1	100%
26	0	2	2	0%
27	13	21	34	38%
28	2	2	4	50%
29	0	6	6	0%
30	1	8	9	11%
31	0	12	12	0%
32	23	57	80	29%
33	0	4	4	0%
Total	**69**	**466**	**535**	

Conclusions

Extensive and intensive surveys coupled with excavations have provided new information on the major structural remains of prehistoric or historic peoples in Hadramawt. Most people left no physical record of themselves besides the construction of monuments. Our work studying, mapping, excavating, and dating tombs and monuments over a substantial region greatly benefited from high-accuracy GPS field data and high-resolution satellite imagery. Our findings provide humanistic insights, and we anticipate that spatial data will prove crucial for the development of semi-automated object-detection algorithms that will expand our understanding of geographies across the Arabian Peninsula and beyond.

Spatial data needs temporal calibration. Not all monuments, even of the same type, reflect contemporaneous communities. Excavations have helped refine local and regional chronology (chapter 18) and provide important evidence about the use and reuse of monuments. Tombs and other small-scale monuments are an important material record of ancient social relations and affiliations in ancient Hadramawt. Death, burial, ritual, and ancestors—often physically marked by tombs, monuments, and rockart—provide mnemonic devices for oral tradition, which is in turn an important tool in constituting and reproducing social frameworks. Ancestors and the memory of ancestors recalled through spatial landmarks are particularly important to tribal sociopolitical communities because tribespeople often use narratives of descent to constitute kinship relations that create social links.

One of the surprising and significant results of our excavations of HTCs is the repeated reuse of these monuments. A Bayesian analysis (chapter 18 has further details) of radiocarbon ages established on events that date the use of HCTs shows a near-continuous reuse of these monuments (figure 14.43). By revisiting tombs, reopening and removing earlier burials, or placing later burials and offerings within, people reshaped the narratives—including kinship and ancestry—associated with these monuments. Such inscription into an existing place is part of the continuous formation and socio-spatial maintenance of landscape. Markers, memories, and narratives of existing place—landscape history—shape and structure people's actions. In the clustering of wall tombs and HCTs revealed in survey and in the reuse of HCTs evident in excavations, the RASA Project documented a spatial and temporal landscape shaped by monuments in Hadramawt, even as the specific forms and distributions of monuments changed through time.

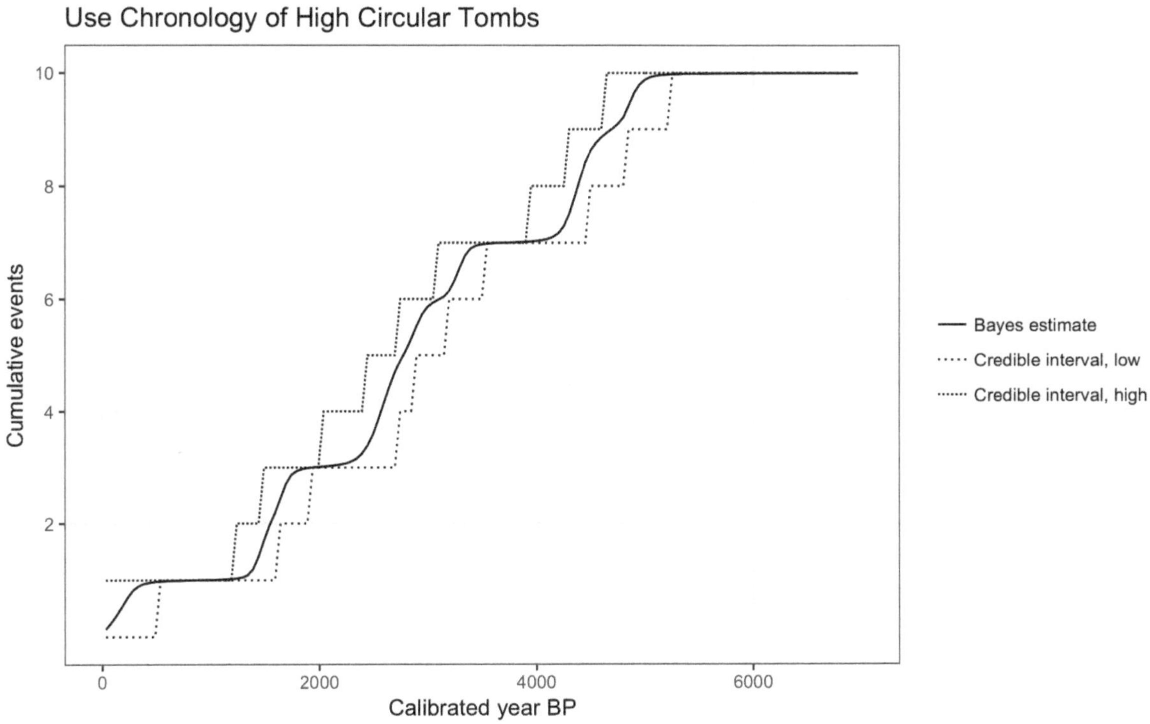

Figure 14.43. Posteriors on events associated with use and reuse of high circular tombs (excavated by RASA in Hadramawt). For extensive discussion of the Bayesian analysis of RASA Project dated events, refer to chapter 14 of this volume. *Illustration by Thomas Dye.*

References

Al-Shaḥri, ʿAli Aḥmad Mahash
1991 Recent Epigraphic Discoveries in Dhofar. *Proceedings of the Seminar for Arabian Studies* 21:173–91.

Bin ʻAqil, Abdalaziz, and Joy McCorriston
2009 Convergences in the Ethnography, Semantics, and Archaeology of Prehistoric Small Scale Monument Types in Hadramawt (Southern Arabia). *Antiquity* 83:602–18.

Braemer, Frank, Serge Cleuziou, and Tara Steimer-Herbet
2003 Dolmen-like Structures: Some Unusual Funerary Monuments in Yemen. *Proceedings of the Seminar for Arabian Studies* 33:169–82.

Crassard, Rémy, and Holger Hitgen
2007 From Safer to Balhaf: Rescue Excavations along the Yemen LNG Pipeline Route. *Proceedings of the Seminar for Arabian Studies* 37:1–17.

Cremaschi, Mauro, and Fabio Negrino
2002 The Frankincense Road of Sumhuram: Paleoenvironmental and Prehistorical Background. In *Khor Rori Report I*, edited by Alessandra Avanzini, pp. 325–63. Edizioni Plus, Pisa.

De Cardi, Beatrice, B. Doe, and S. P. Roskams
1977 Excavation and Survey in the Sharqiyah, Oman 1976. *Journal of Oman Studies* 3:17–33.

Doe, Brian
1971 *Southern Arabia*. Thames and Hudson, London.
1983 *Monuments of South Arabia*. Falcon-Oleander, Naples.

Eddy, Frank W., and Fred Wendorf
1998 Prehistoric Pastoral Nomads in the Sinai. *Sahara* 10:7–20.

Harrower, Michael J.
2016 *Water Histories and Spatial Archaeology: Ancient Yemen and the American West*. Cambridge University Press, Cambridge.

Harrower, Michael J., Jared Schuetter, Joy McCorriston, Prem K. Goel, and Matthew J. Senn
2013 Survey, Automated Detection, and Spatial Distribution Analysis of Cairn Tombs in Ancient Southern Arabia. In *Mapping Archaeological Landscapes from Space*, edited by Douglas C. Comer and Michael J. Harrower, pp. 259–68. Springer, New York.

Harrower, Michael J., Matthew J. Senn, and Joy McCorriston
2014 Tombs, Triliths and Oases: Spatial Analysis of the Arabian Human Social Dynamics Project (AHSD) Archaeological Survey 2009–2010. *Journal of Oman Studies* 18:145–52.

Khan, Majeed

2000 *Wusūm: Tribal Symbols of Arabia*. Ministry of Education of Saudi Arabia, Riyadh.

King, Geoffrey

2009 Camels and Arabian Balīya and Other Forms of Sacrifice: A Review of Archaeological and Literary Evidence. *Arabian Archaeology and Epigraphy* 20(1):81–93.

Kowalewski, S.

2008 Regional Settlement Pattern Studies. *Journal of Archaeological Research* 16:225–85.

McCorriston, Joy

2011 *Pilgrimage and Household in the Ancient Near East*. Cambridge University Press, Cambridge.

2013 Pastoralism and Pilgrimage: Ibn Khaldūn's Bayt-State Model and the Rise of Arabian Kingdoms. *Current Anthropology* 54:607–40.

McCorriston, Joy, Michael Harrower, Louise Martin, and Eric Oches

2012 Cattle Cults of the Arabian Neolithic and Early Territorial Societies. *American Anthropologist* 114(1):45–63.

McCorriston, Joy, Michael J. Harrower, Tara Steimer, Kimberly D. Williams, Matthew Senn, Masʿūd Al Hādhari, Masʿūd Al Kathīrī, ʿAli Aḥmad Al Kathīrī, Jean-François Saliège, and Jennifer Everhart

2014 Monuments and Landscape of Mobile Pastoralists in Dhofar: Arabian Human Social Dynamics (AHSD) Project, 2009–2011. *Journal of Oman Studies* 12:117–44.

McCorriston, Joy, Tara Steimer-Herbet, Michael Harrower, Kimberly Williams, Jean-François Saliège, and 'Abdalaziz Bin 'Aqil

2011 Gazetteer of Small Scale Monuments in Prehistoric Hadramawt, Yemen: A Radiocarbon Chronology from RASA Project Research 1998–2008. *Arabian Archaeology and Epigraphy* 22:1–22.

Nayeem, M. A.

2000 *The Rock Art of Arabia: Saudi Arabia, Oman, Qatar, the Emirates and Yemen*. Hyderabad Publishers, Hyderabad.

Newton, Lynne S., and Juris Zarins

2000 Aspects of Bronze Age Art of Southern Arabia: The Pictorial Landscape and Its Relation to Economic and Socio-Political Status. *Arabian Archaeology and Epigraphy* 11:154–79.

Rothenberg, Beno

1979 *Sinai*. Binns, Washington, DC.

Saliège, Jean-François, Anne-Marie Lézine, and Serge Cleuziou

2005 Estimation de l'effet réservoir ^{14}C marin en mer d'Arabie. *Paleorient* 31(1):35–43.

Saliège, Jean-François, A. Person, and F. Paris

1995 Preservation of 13C/12C Original Ratio and ^{14}C Dating of the Mineral Fraction of Human Bones from Saharan Tombs. *Journal of Archaeological Science* 22:301–12.

Schuetter, Jared M.

2010 Cairn Detection in Southern Arabia Using a Supervised Automatic Detection Algorithm and Multiple Data Spectroscopic Clustering. PhD Dissertation, Department of Statistics, The Ohio State University.

Schuetter, Jared M., Prem K. Goel, Joy McCorriston, Jihye Park, Matthew J. Senn, and Michael Harrower

2013 Autodetection of Ancient Arabian Tombs in High-Resolution Satellite Imagery. *International Journal of Remote Sensing* 34(19):6611–35.

Steimer-Herbet, Tara

2004 *Classification des sépultures à superstructure lithique dans le Levant et l'Arabie occidentale: (IVe et IIIe millénaires avant J.-C.)*. BAR International Series 1246. Archaeopress, Oxford.

Steimer-Herbet, Tara, Frank Braemer, and Gourguen Davtian

2006 Pastoralists' Tombs and Settlement Patterns in Wādī Wash'ah during the Bronze Age (Ḥaḍramawt, Yemen). *Proceedings of the Seminar for Arabian Studies* 36:257–65.

Swauger, James L.

1966 Dolmen Studies in Palestine. *Biblical Archaeologist* 29(4):105–14.

Uerpmann, H.-P.

1999 Camel and Horse Skeletons from Protohistoric Graves at Mleiha in the Emirate of Sharjah (U.A.E.). *Arabian Archaeology and Epigraphy* 10:102–18.

Vogt, Burkhard

1997 La fin de la préhistoire au Hadramawt. In *Yémen au pays de la reine de Saba*, edited by Christian J. Robin and Burkhard Vogt, pp. 30–33. Institut du monde arabe—Flammarion, Paris.

Vogt, Burkhard, and Alexander V. Sedov

1994 *Surveys and Rescue Excavations in the Hadhramawt Governate, Republic of Yemen*. Manuscript on file, Deutches Archäologisches Institut, Sana'a.

Zarins, Juris

1992 Pastoral Nomadism in Arabia: Ethnoarchaeology and the Archaeological Record, a Case Study. In *Pastoralism in the Levant, Archaeological Materials in Anthropological Perspectives*, edited by Ofer Bar-Yosef and Anatoly Khazanov, pp. 219–40. Prehistory Press, Madison, WI.

2001 *The Land of Incense: Archaeological Work in the Governorate of Dhofar, Sultanate of Oman, 1990–1995*. Sultan Qaboos University Publications.

Zazzo, A., and J.-F. Saliège

2011 Radiocarbon Dating of Biological Apatites: A Review. *Palaeogeography, Palaeoclimatology, Palaeoecology* 310:52–61.

Chapter 15

Rawk: Statue-Menhirs and Anthropomorphic Statues of Ancient Wādī ʿIdim

Tara Steimer-Herbet

Translated from the French by Joy McCorriston

Like Wādī Sanā, Wādī ʿIdim flows toward the north and feeds into Wādī Ḥaḍramawt. The modern village of Rāwik sits on the slopes of the first terrace, about halfway along the main channel's eastern bank, some 50 km to the west of Wādī Sanā. In November 2005, salvage excavation of a group of upright stones situated in the heart of the village (E 279159, N 1741745 UTM Zone 39 North) brought to light a group of inhumations with particularly rich goods, including an anthropomorphic statuette (figure 15.1). A survey of the surrounding zones inventoried three statue-menhirs and a dozen tower tombs. Even though tower tombs are widely distributed across the Arabian landscape (Steimer-Herbet 2004), the locations where one finds anthropomorphic statuettes and statue-menhirs are confined to the margins of the Ramlah as-Ṣabʿatayn desert. Contrary to previous understanding, the anthropomorphic statuettes discovered at Rawk show that more than 5,500 years ago, human representations were frequent and varied. The rural engravings in Saada (Garcia et al. 1991) and at Jarf an-Nabīrah near ad-Ḍāliʿ (Crassard 2013) already show a broad array of prehistoric templates. The Rawk anthropomorphic statuettes and statue-menhirs innovate upon and enrich this repertoire with sculpting and relief. The study of archaeological *art mobilier*, with morphological and comparative analysis of the Wādī ʿIdim examples, provides unfiltered information on the changes that affected social structures just before 5,500 years ago.

History of Research and Discovery

The scientific program of archaeological projects in Yemen has always been subject to local political conditions, without a long and uninterrupted flow of research. The Jawf region was only visited briefly by Serge Cleuziou in 1988–1992 before being definitely avoided, with unpublished documentation and several abbreviated studies (Cleuziou et al. 1988, 1992a, 1992b). In the Khawlān, still accessible in the 1980s, with excavations directed by Alessandro de Maigret (2002), the requirements for investment in site and artifact protections became overwhelming for research budgets. For this and other reasons, the French archaeological program shifted in 2002 toward the eastern Hadramawt. While the American (RASA) team focused on Wādī Sanā and Wādī ʿIdim, the French team (HDOR) studied the Wādī Washaʿah to the north (Steimer-Herbet et al. 2006).

As this publication shows, Hadramawt is a true archive of pre- and protohistorical remains. Funerary and stone monuments are visible on the edges of high and low terraces overlooking wadis or on the Holocene silt terraces, where also appear the traces of occupation. With its high plateau cut by a sequence of deep wadis, eastern Hadramawt is difficult to access. The key to archaeological discoveries in Yemen has often been erosion, looting, or lucky finds that result in further survey and scientific excavation. These were key to the discoveries of 16 of the 17 known anthropomorphic statuettes and the 35 statue-menhirs documented in the current study (figures

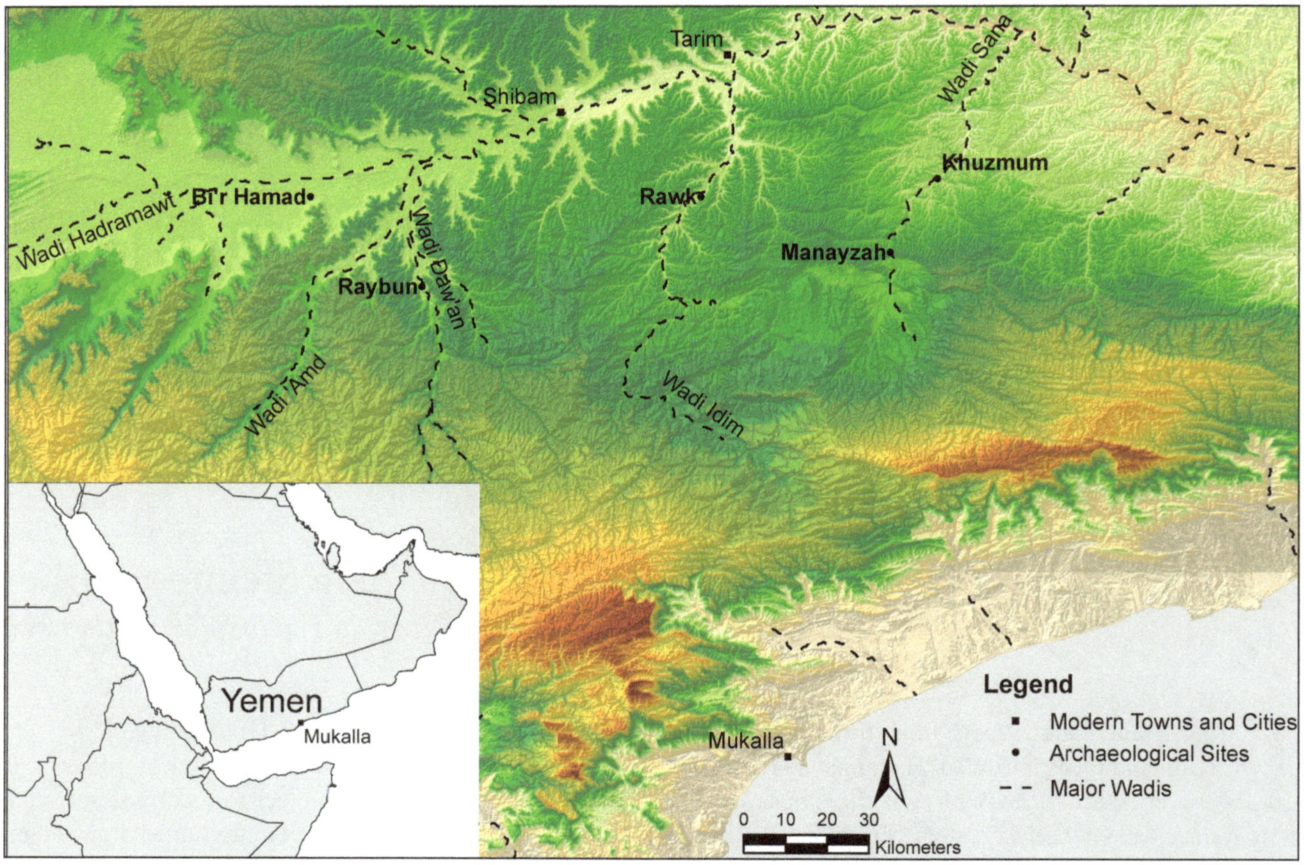

Figure 15.1. Map of Wādī ʿIdim with Rawk. *Illustration by Michael Harrower.*

15.2 and 15.3). These figures summarize an inventory inadvertently multiplied elsewhere by poor reproduction and inversions of illustrations; it has not been an easy task to arrive at an accurate count. For example, some uncertaintly lingers about one of the statuettes described by de Maigret (2002:figure 77). Worse, there are mentions of unillustrated, unphotographed anthropomorphic statuettes that reappear in the collections of museums or private hands—there are three such cases in the literature.

Statuettes

The first two anthropomorphic statuettes discovered came from Bronze Age sites in Wādī Ṣulāyb (Khawlān). The first was delivered by villagers to A. de Maigret and can be found at the Sana'a Museum; the other was resold to a private collector in Kuwait (de Maigret 1990). There appears to be no photo or drawing of this statuette. In 1989 excessive rain and the collapse of an earthen bank in the Rāwik village (Wādī ʿIdim) brought to light three statuettes that were consigned by village inhabitants to ʿAbdalraḥman as-Saqqāf and his colleague Alexander Sedov, who published them (Sedov and as-Saqqaf 1992). In 2004 villagers indicated to as-Saqqaf, now director of the Sayʾūn Musuem, that a fourth statuette had been discovered. A salvage excavation in 2005 brought to light a fifth statuette in situ, this time in the area of the other discoveries. Nine of the statuettes shown in figure 15.3 were brought to the Antiquities Services by individuals. Information concerning their place of discovery is either highly summarized or entirely absent (compare Groups 1, 3, and 5). The quality of illustrations, moreover, is not equal for all statuettes. An illustrator specializing in stonework drew those of Rawk, bringing out numerous details, but most other illustrations are but schematic sketches of the most obvious traits, mostly produced from photographs. Two of the nine appeared in the international antiquities market: one was acquired by the Metropolitan Museum of Art in New York, and the other figured in the catalog of Pierre Bergé and Associates in 2009 (Audouin 2005).

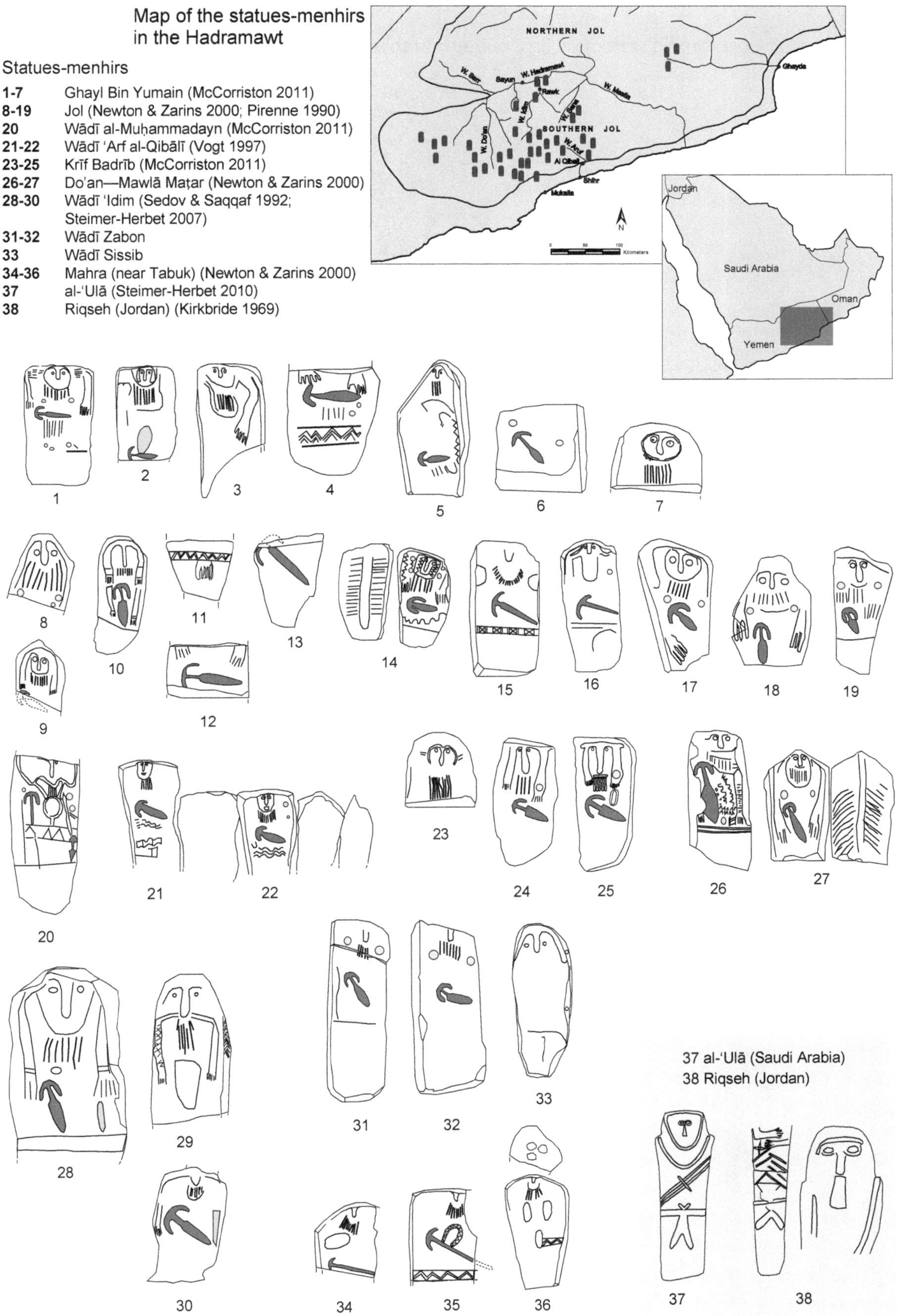

Figure 15.2. Statue-menhir find spots in Southern Arabia. *Illustration by Tara Steimer, E. Guscher, and Clara Hickman.*

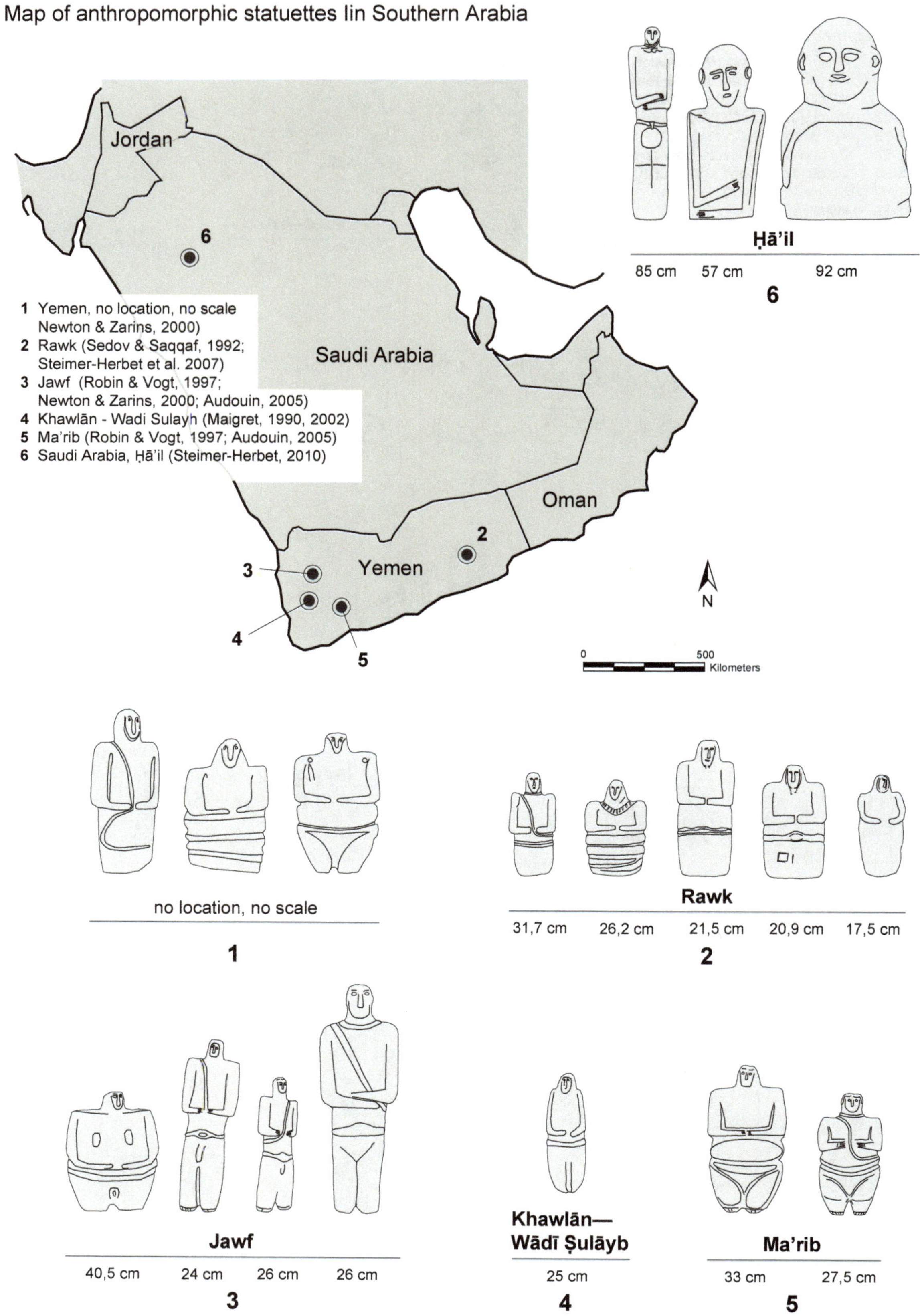

Figure 15.3. Anthropomorphic statuette find spots in Southern Arabia. *Drawing by Tara Steimer-Herbet.*

Statue-Menhirs

To the work of Jacqueline Pirenne (1990:28, plate xxxi) we owe the recovery and preservation of a full dozen statue-menhirs from near Shurruj Bakeli in Wādī Jirdān (Shabwa Province, west of Hadramawt). These discoveries were interspersed between the Mukalla Museum and the Aden Museum. Still in highland territory at al-Qibālī (Wādī ʿArf, north of Shiḥr), Christian Robin and Burkhard Vogt (1997) photographed and published an oval monument whose perimeter was formed of upright statue-menhirs. Lynne Newton and Juris Zarins (2000:165, figure 8) compiled 24 drawings of statue-menhirs discovered in the Southern Jol, including most particularly examples from Hadramawt—Mawlā Maṭar, Wādī al-Muḥammadayn—and near Thamūd in Mahra Province. ʿAbdalʿazīz Bin ʿAqīl and Joy McCorriston completed this inventory with new examples discovered north of Al-Mudhainab at Krīf Badrīb (McCorriston 2011:160, figure 2) and inventoried in the southern watershed of Wādī Sanā. Chapter 14 describes an excavated example, C051-1, in the southern watershed of Wādī Bin ʿAlī.

Stone Monuments as Ritual Structures

The Southern Jol survey (McCorriston et al. 2011) and survey in Wādī Washaʿah (Steimer-Herbet et al. 2006) documented numerous Bronze Age sites constructed along watercourses—sites that were trapped under colluvium from upper slopes. The Rawk monument fortuitously escaped being buried under sediments and also escaped several architectural intrusions, like a stable in the 1970s built against a line of upright stones that forms the eastern wall of the monument. Still visible at the beginning of the 2005 salvage excavations, these were five upright stones (designated Pierre [P1 to P5]) and four limestone slabs, superimposed in pairs (designated Dalle [D1 to D4]) (figure 15.4). The slabs form a bench and are located to the west of P4 and P5. The western extent has been effaced by the seasonal erosion of the watercourse.

Methods

The excavations at Rawk have already been published (Steimer-Herbet 2007a, 2007b) and here receive only brief mention. Within an extremely brief season of fieldwork, excavators tested inside and outside the residual surface remains (stone uprights), focusing on documentation or architectural plans, stratigraphy, and stratigraphic context of associated remains.

Results

The characteristics of the architectural remains and the associated finds clearly differentiate this monument from an ordinary house lacking the traces of ritual activities (Steimer-Herbet 2007a, 2007b). The upright stones and the slabs were embedded in a layer of very compact soil consisting of a mix of rounded pebbles and ashy inclusions. This sand-and-gravel matrix derived from slope erosion. Within the fill adjacent to the base of P1 (10 cm above the base of P1) were discovered several fragments of an adult's skull (age 16 to 30 years) and the almost complete skeleton of a child (between six months and one year old). These bones were associated with a small bone pendant, an obsidian blade, a knapped chert tool, a small chert nucleus, and an anthropomorphic statuette (ST 05), set vertically with its face looking toward the southwest. A few centimeters from the base of P2 was found the lower molar of a bovid. To the west of P4, on Slab D1, lay a chert flake and two pottery sherds, which later conjoined. Under D3 was found a copper fragment of unidentifiable shape (11.6 x 7.58 mm). On the sandy interface to the west of the bench had been laid a skull and some bone fragments (part of a clavicle) belonging to a newborn (age zero to four years). The position of the lower part of the mandible indicates a primary inhumation in a sealed context. An unstable reddish deposit was visible on some parts of the skull.

Rawk's Anthropomorphic Statuettes

Statuettes Found in 1989 and 2004 from Unspecified Contexts

Tara Steimer-Herbet and Olivier Lavigne (a master stone carver whose observations are used in the descriptions below) have been able to reexamine two of the four small statues, ST 02 and ST 04, both discovered by the inhabitants of Rāwik and kept in the Sayʾūn Museum (figure 15.5). ST 01 and ST 03 were unavailable due to their inclusion and storage within the traveling exhibit of Yemen's antiquities.

Statuette 02 is made of hard and fine-grained limestone (317 mm height, 132 mm width, 55 mm thickness) (figure 15.6). Its general shape is trapezoid, and the top of the head is flattened. The long face is marked by fine striations, all slightly parallel. The nose is straightened and quite large. The eyes appear as depressions and the mouth as a line visible only when slightly illuminated from the sides. Under the chin, a slight depression is visible, representing an almost pointed goatee. The shoulders are round and symmetrical. In the middle part, the arms form a right angle and end in stumps. A belt with two bands was carved in the lower part. The base is massive and rectangular in shape. This statuette was not thoroughly cleaned, and we were able to note the undeniable presence of strong red pigments spread over the artifact's entire surface (except

Figure 15.4. Rawk monument and view from the south. *Drawing and photograph by Tara Steimer-Herbet.*

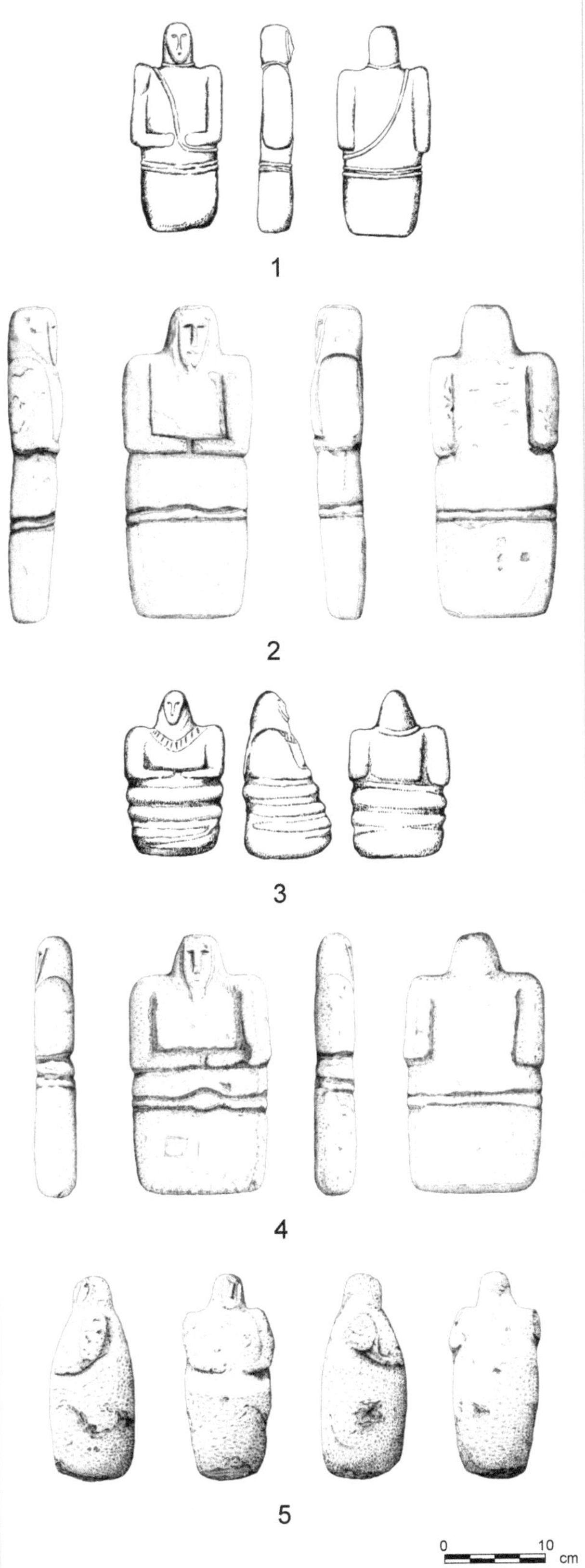

Figure 15.5. Five anthropomorphic statuettes from Rawk. *Drawing by Olivier Lavigne.*

the head). This coloration seems to have been intentional. Dark residual deposits were detected only on the frontal part of the object, near the elbows and arms, and at the level of the belt buckle. The artifact was made without any preliminary preparation work. The statuette's general appearance is that of a polished object, its manufacture is quite elaborate, and its fineness of grain makes it an item worthy of analysis. Three types of tools were used: first an abrasive (a piece of sandstone, or wood with sand) was applied to create hollows; then a sharpened object, doubtless a stone tool; and finally a small metal implement.

Statuette 04 was carved in greyish-white granite veined with black (262 mm height, 144 mm width, 43 mm thickness) (figure 15.7). This stone is not local. The object was made from a tabular stone without side faces (that is, it was almost totally polished and flattened). The head is in the shape of a trapeze, and its top part is slightly flattened. The eyes are shown as hollows and the nose is straight and wide. The mouth appears as a fine line. The elongated face ends in a little pointed goatee. The chin is not emphasized because of the goatee. The arms form a straight angle and end in stumps. The statuette's lower part was covered by a belt, with only a single band and an ovoid buckle. In the basal part of the sculpture, a series of incisions was recorded: a square (21.3 mm wide) and a vertical line of identical length. The statuette's base is rectangular and well proportioned, just like the other parts of the object. In the lowest part, minute traces of red pigment were found. One can observe traces of blackened residue in all the statuette's cracks and hollows; this is less obvious on the back. Facial features were either drawn or outlined with the use of a very fine abrasive, which could imply that a pointed metal tool was used. These traces are clearly visible on the oval of the face. In the middle part of the statue, the inside of the elbows was badly executed. While attempting to carve a sharp angle, the implement met an obstacle and penetrated inside the stone. Under the right arm there were characteristic traces of a metal tool used with an abrasive material, such as sand. This tool was used to carve the part of the statuette between the elbow and the stump; it had a working surface of 0.8 mm. The buckle and the incisions were created with the same type of tool, and the buckle itself was completed using a very fine abrasive.

The following description comes from a publication (Sedov and as-Saqqaf 1992) and from observations of the pictures of the catalog of the exhibition (Robin and Vogt 1997). ST 01 is a statuette in grey-brown granite. The statuette is 213 mm in height, 95 mm in width, and 31 mm thick, following the dimensions given by Sedov and as-Saqqaf (1992). Worked in the round, it has many pecked

Figure 15.6. Statuette ST 02. *Photograph by Thomas Sagory.*

Figure 15.7. Statuette ST 04. *Photograph by Thomas Sagory.*

and engraved details. Viewed from the front, the head is slightly trapezoidal, without a defined neck. The oval face shows slight depressions for eyes, a line for the nose, and one for the mouth. The statuette possesses a defined necklace. The shoulders are asymetrical and rounded. On one elevated shoulder rests a sash or cordon, which cuts across the chest from the upper right to lower left, passing between two hands represented by stumps. The foreams, depicted level in an equestrian posture, are disproportional and too short. The lower part of the statuette shows a double-banded belt, and below that is the rectangular base.

ST 03 (Sedov and as-Saqqaf 1992) is a statuette of light-grey granite, and it measures 170 mm high, 103 mm wide, and 45–78 mm thick. Worked in the round, the details are also pecked and engraved. The trapedoidal head rests on rounded shoulders, with clavicles marked by depressions. The elongate face ends in a pointed chin. The eyes are cavities, and the straight nose is prominent. The statuette possesses a large pectoral collar marked with vertical stripes. The foreams, depicted level in an equestrian posture, end in only slightly detailed hands. The base is flat, larger than other statuettes from the group. The indication of a horizontally folded cloth leaves one to suppose that the statuette is depicted in a seated position or with legs crossed.

The fifth statuette, ST 05, is made of yellow granite, without any traces of polishing or painted decoration (209 mm height, 95 mm width, 90 mm thickness) (figure 15.8). The head is rounded, and the face, whose features are very faint, is asymmetrical and gazes slightly toward the left. The nose is straight and thick. The mouth and eyes are not shown. The shoulders are not symmetrical, the left one being slightly higher and more rounded. The right shoulder is flattened and seems to have been badly carved. In the middle part of the body, the arms form a right angle but are nonidentical in length. The hands end in stumps and do not join. In the lower part of the body, no motif was carved; the base is rounded.

Wādī ʿIdim's Statue-Menhirs

The Rawk statue-menhir was discovered only 20 m from the excavated monument (figure 15.9). Thus isolated, it served out of context as construction material for a modern dwelling. The worked face was exposed on the exterior facade of the house, thus allowing its retrieval during the French project of 2005 in the village. The statue-menhir was carved in a soft, fine-grained limestone, a material particularly appropriate for the precise rendering of human features and costume detail, especially the dagger and its sheath. The dagger is the only element in relief that was smoothed with an abrasive; the rest was pecked. The stone is rounded in its upper part. Partly obscured under modern mortar, the upper part has iconography that is difficult to describe and interpret. One can nevertheless distinguish the rounded shape of the shoulders, which join in the

Figure 15.8. Statuette ST 05. *Photograph by Thomas Sagory.*

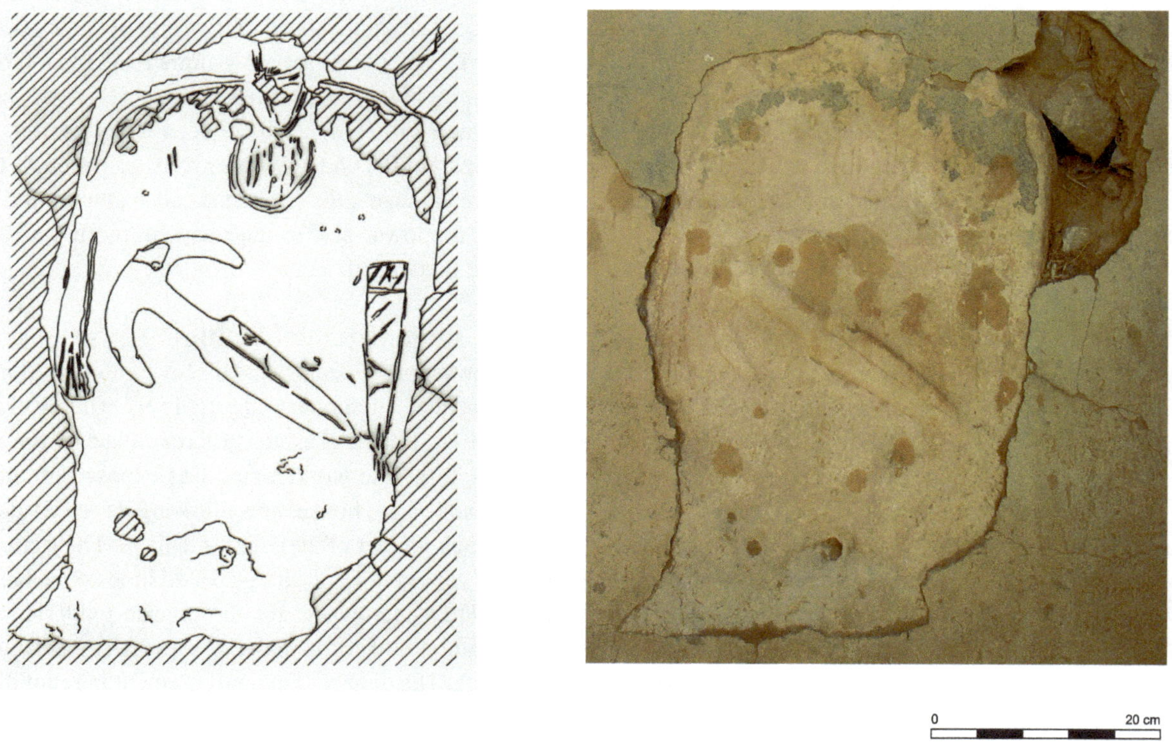

Figure 15.9. The Rawk statue-menhir. *Drawing by Olivier Lavigne. Photograph by Tara Steimer-Herbet.*

middle part and touch a slightly off-axis motif (perhaps representing a pointed beard). A pendant is visible underneath. It consists of seven beads, all circumscribed within an oval. The right arm only is preserved but is not adorned with a bracelet. It ends in a hand, where five fingers were presumably incised. The sheath to the left is decorated with oblique incisions, and the weapon's hilt is visible. The dagger with hilt and its triangular blade is quite evident and fills the entire width of the stone. Its details were depicted with much care. It is the only part carved in the round and smoothed with abrasive; all the other parts of the statue were made by striking the stone with a sharp tool. Traces of a red pigment have also been found.

At the end of the Rawk monument excavations, the inhabitants of Ṣunah, a village located several kilometers to the north of Rawk, near Rudūd, brought to our attention another statue-menhir (figure 15.10). The material is grey limestone. Its general form is more elongate, with a similar rounding of the upper margin. It is in an overall poor condition, with a broken base from which cracks extend throughout the entire block (detectable through the ring of the stone when gently tapped). Nonetheless, it was possible to do a scale drawing.

The ventral face was entirely worked to create a flat surface to facilitate the representation of traits; the other faces of the block are rounded. There is no distinction between the head and shoulders. Over the eyes, eyebrows join to form the nose, which descends to the level of the shoulders. Below the nose are seven vertical incised lines, probably indicating a beard. On the abdomen one can detect a vaguely triangular object with no apparent function. Covered by crosshatching, the right arm ends in a four-fingered hand. The left arm lacks ornamentation and possesses a complete hand. On the left side, the sculptor incised chevron patterns. The lower part is too degraded to detect other traits. Two techniques were used in creating the statue-menhir: the more coarse traits were shaped by pecking, and the finer and deeper incisions were the product of a metal tool.

The statue-menhir of Khushum as-Sinām lay about 8 km to the northwest of Rawk in the Wādī Buyūt, a tributary of Wādī ʿIdim (figure 15.11). The statue-menhir was displayed in the Say'ūn Museum, and we benefited from our studies of statuettes ST 03 and ST 04, stored in the museum, to make a scale drawing of the statue-menhir (1.6 m height, 0.75 m width). The stone is a grey-tinted

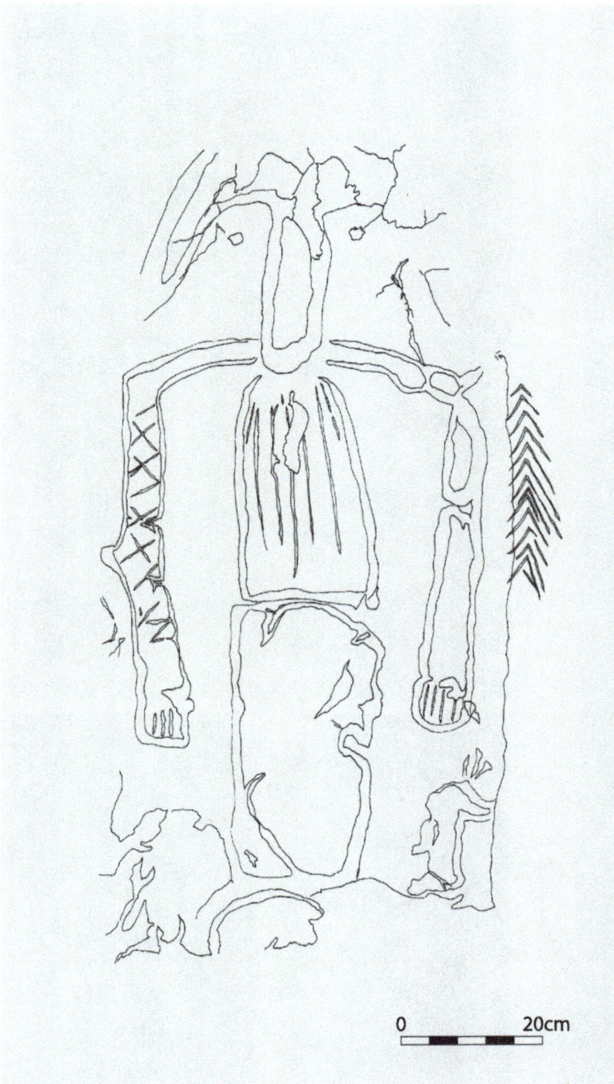

Figure 15.10. The Ṣunah statue-menhir. *Drawing by Olivier Lavigne. Photograph by Tara Steimer-Herbet.*

limestone taken from the cliffs that tower over the wadi. Its upper part is rounded. The face has been degraded, but it was possible to reconstruct the face of a man with a strongly defined nose. The bottom of his face is underlined by a deep line, underscored by another line that can be interpreted in one of two ways: either it marks a collar with multiple pendants or it is a schematic beard. The arms along the body end in hands; the left one is large with all the fingers represented. Two bands at each of the wrists might portray bracelets or marks like tattoos or scarifications. On the abdomen is located a dagger with a lunate pommel like that of the statue-menhir at Rawk. Under the left hand is located an object that could be a sheath, but the dimensions are much smaller than those of the blade. A belt band covers the base of the stone, leaving scant place to represent the bottom of the body. The sculpted elements were achieved by pecking, with the dagger as a focus of workmanship, where the sculptor used abrasive. Black traces are visible on the rough stone, and these seem to be the same kind as traces noted on ST 04.

Tower Tombs (High Circular Tombs) and Stone Circles Overlooking Rawk

A quick survey of the lower bedrock terraces immediately overlooking the village of Rāwik enabled us to record 29 high circular tombs, or tower tombs. The entire

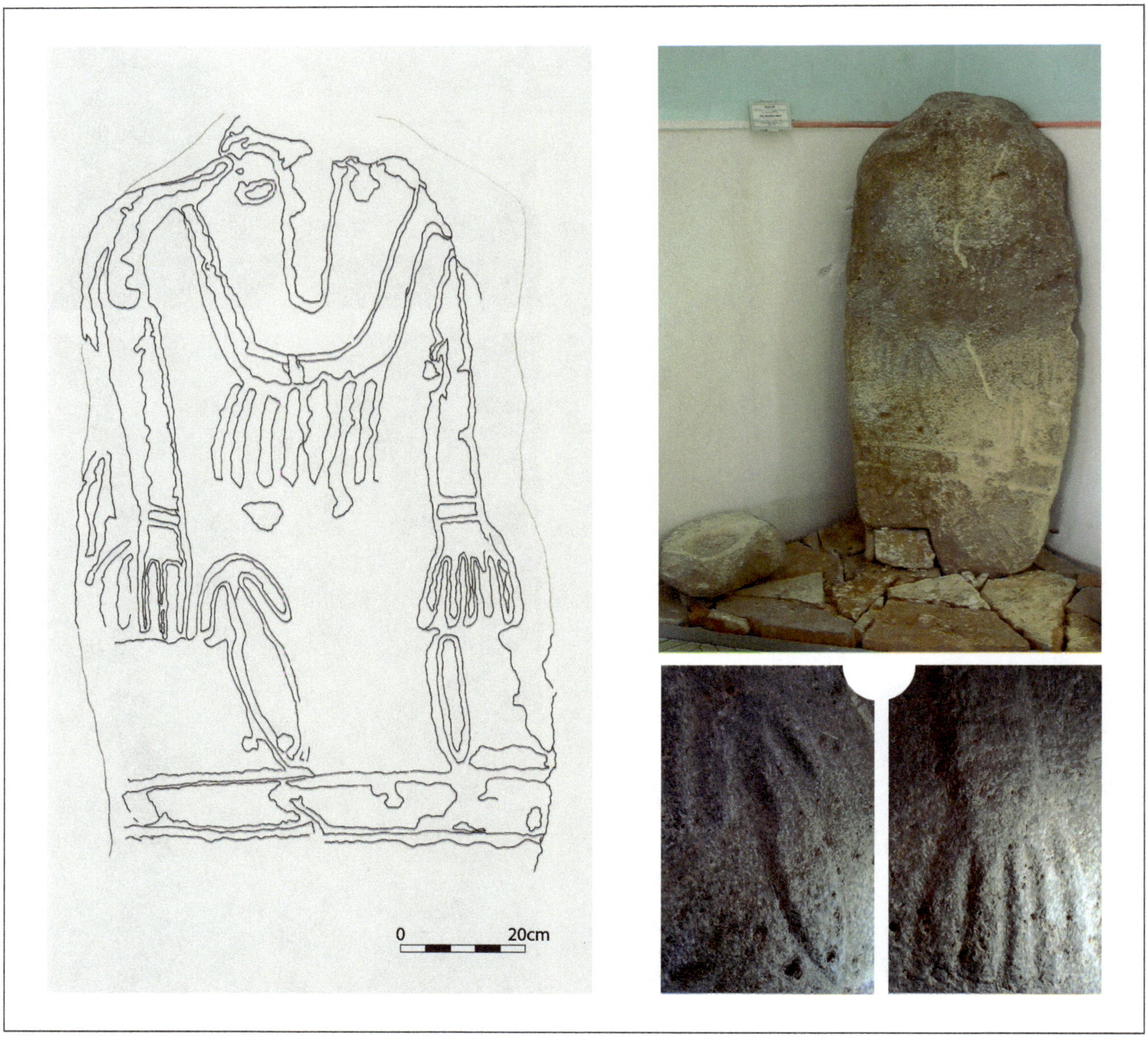

Figure 15.11. Khushum as-Sinām statue-menhir (Museum of Say'ūn). *Drawing by Olivier Lavigne. Photograph by Thomas Sagory.*

group was distributed within a radius of approximately 1 km (figure 15.12). Two types of what most likely are graves appear quite clearly: those of an average size (2.5 m in diameter; Type T1) and of modest height (maximum 1 m), and the large tombs (3.5–5 m in diameter and 1.5–2 m in height; Type T2). The smaller type had only one course of stones enclosing the circular chamber and the external wall, while the larger tombs had walls made of several (six or seven) courses of coarse fieldstones and a corbelled roof over the chamber. None of the structures revealed the presence of a door. One of the high circular tombs (T2) is associated with two ancillary piles of stones that form small round structures (1 m in diameter and 0.5 m in height); this HCT belongs to a type that has peripheral extensions.

The tombs overlooking Rawk have never been excavated. The rough local materials used in their construction offer no possibility of fine architectural contours like HCT C15-3, excavated about 30 km to the south (figure 15.13, chapter 14). Instead, their morphology resembles those detailed in the Shabwa region (Inizan et al. 1998) and at Munayder (chapter 16; McCorriston 2000), about 5 km to the south. Two undated stone circles measuring 3.4 m and 3.8 m in diameter have been recorded near the

Figure 15.12. Map of the Rawk monuments: C) high circular tombs, B) statue-menhir, and A) sanctuary.
Illustration and photographs by Tara Steimer-Herbet.

HCT. These structures are in poor condition and were likely looted to build the HCT.

Absolute, Relative, and Inferential Dating at Rawk

Radiocarbon Ages

Radiocarbon ages on charcoal samples taken from numerous hearths at Rawk reveal that on this small promontory overlooking the wadi there were at least two phases of occupation—the first more than 7000 cal BP and the second more than 3000 cal BP, with a gap of about 2,300 years between them (Steimer-Herbet 2007a:10) (table 15.1).

During the earlier period, humans occupied Rawk multiple times, as attested by the numerous hearths scattered across a gentle slope downward east to west. This terrace is made up of a yellow sedimentary layer of calcareous matrix. Two ashy pockets were sampled, prepared by Jean-François Saliège (Université Pierre et Marie Curie, Paris Sorbonne 4), and yielded radiocarbon ages of 5015 ± 50 ^{14}C BP (Pa 2384) and 6795 ± 90 ^{14}C BP (Pa 2389). These sampled hearths lay on either side of the wall of the structure and were documented in previous publications (Steimer-Herbet 2007a, 2007b). They indicate that the earlier occupation is contemporary with the Neolithic hunter-herders described by the RASA Project in Wādī Sanā (chapter 8; Crassard 2008; Crassard and Hitgen 2007; Crassard et al. 2006; Martin et al. 2009). The placement of a structure with its upright stones took place only 2,300 years later, between 5500 and 5200 cal BP. The stones were held fast in a hardened mixure composed of sand, clastic gravel, and ashy material. This sandy gravel matrix was a result of water runoff from nearby slopes. Its compact nature made it impossible to observe (or sample) the original ditches associated with setting the uprights in place.

Inferential Chronology

There are several sites to which Rawk may be compared, and these sites span a long chronological range. The rare published examples of skulls associated with upright stones in funerary contexts are found in the Negev (Eilat, Muyat Galla, Wadi Zalaqa, and Kfar Hahoresh) (Avner 1984, 2001, 2002) and at Riqseh in Jordan (Kirkbride 1969a, 1969b). At Eilat, the circular tomb possesses two standing stones in its center, one of which is anthropomorphic. At the foot of the latter, a deposit of six skulls was discovered. The construction was dated to 8000–7000 cal BP. At Riqseh, Diana Kirkbride discovered a likely sanctuary containing menhir-type statues associated with circular or rectangular graves. The only radiocarbon date (K1467, 6010 ± 120) (Kirkbride 1969b:195) was obtained from a charcoal sample from a large circle located at the center of the site.

Figure 15.13. HCT C15-3 in Wādī ʿIdim. *Photograph by RASA Archive.*

Table 15.1. Radiocarbon dates from Rawk. These samples were prepared and submitted by Jean-François Saliège, and their laboratory numbers are no longer available. These analyses were supported by the French Archaeological Mission to Jawf-Hadramawt.

Archaeological Sounding Number	Sample Matter	Sample Number	1σ Range Uncal BP	Cal BP (OxCal 4.3) 2σ
16	charcoal piece	Pa2384	5015 ± 50	5901–5651
18	charcoal piece	Pa2388	6820 ± 80	7839–7518
13	ashy spread of charcoal fragments	Pa2389	6795 ± 90	7833–7496
"lump"	human bone (skull fragment, female?)	Pa2392	4605 ± 45	5480–5066
12	ashy spread of charcoal fragments	Pa2394	4435 ± 100	5317–4842

Closer to Rawk, in Yemen, the construction most reminiscent of our site are those of the "public space, ritual space" type described by Alessandro de Maigret (2002:145, plate 10) in the al-Aʾrush region of the Khawlān. These structures are rectangular in shape, with walls made of monoliths set upright and slightly rounded in their upper parts. Benches made of flat stones were found inside along three walls. For structure L14, the archaeologist describes four monoliths, or betyls, without any iconography, set in an upright position (de Maigret 1990:66, 214, figure 30). Excavation of this building uncovered a substantial mass of animal bone (de Maigret 2002:155). Structures of this type have been attributed to the second half of the fourth millennium BCE. According to the radiocarbon ages, the structure at Rawk falls within the chronological range established by these two other sanctuaries.

The Rawk ritual structure has yielded important information on burial traditions and modes of funerary deposition (primary and secondary), ancient technology (metals, sculpture, and pottery), burial offerings placed jointly with the deceased (a bovid tooth, a statuette, an obsidian blade, a pendant, and possibly pigments), and imports into Hadramawt from the Yemeni highlands, both of objects (statuettes) and of raw materials (granite and obsidian). The artifacts discovered at Rawk show numerous common features with those recovered from sites in the Khawlān, including anthropomorphic statuettes, grinding slabs, bronze objects, and faunal remains. The singularity of Rawk lies in the presence of an adult cranium and two newborns. The adult's and child's bones in the immediate proximity of ST 05 have yielded a radiocarbon age on bioapatite with the range of about 5500–5200 cal BP (Pa 2392).

Inferential Date of the Rawk Statue-Menhir

Although found out of context, the statue-menhir was only several meters from the sanctuary. The use of an abrasive and a metal tool for sculpting the decorative elements, as well as the partial coating with red pigment, are signicant reasons to associate the statue-menhir with the sanctuary structure. There is no unequivocal dating evidence for this find; the excavator associates it by proximity and technology with the fourth millennium BCE sanctuary. The statue-menhir perhaps faced outward, as was the case with statue-menhirs in situ at al-Qibālī (Robin and Vogt 1997) and at Krīf Badrīb (chapter 4; McCorriston 2011).

Morphological Features and Comparatives Studies of Yemen's Anthropomorphic Statuettes and Statue-Menhirs

Morphological Features of Anthropomorphic Statuettes in Yemen

At least 17 anthropomorphic statuettes have been discovered in Yemen: five in Hadramawt (Wādī ʿIdim), two in the Khawlān region, four in the Jawf, two in Mar'ib, and three unprovenienced. All the statuettes were sculpted in the round using tabular limestone or granite cobbles. The material may sometimes be locally available or may be exogenous. Incisions, pecking, and polishing were in turn used to form the arms, noses, eyes, eyebrows, beards, other anatomical characteristics, and elements of dress (necklaces, collars, belts, and pendants). The reduction of volume focused on details of the belt, the arms, and the shaping of the head. The working of the basal half of the statuettes is more sparing, with the sculptor's work limited to shaping the overall form. In general, the statuettes are smoothed and completed. Thanks to the fine grain of granites and limestones, three types of tools can be identified: an abrasive (a piece of sandstone or a piece of wood with sand abrasive can easily hollow into these materials), a stone tool, and a small tool in metal.

Two statuettes from the Wādī ʿIdim group have traces of red pigment and black residues. In the case of ST 02, the red pigment covers the entire body except the head, and the black residue is concentrated across the front, above the elbows to the belt.

The height of the Yemen statuettes varies between 17 and 40 cm, and they are 8 to 18 cm in diameter. The thickness of the statuettes is linked to the choice of material, whether tabular stone or a rounded cobble. Three distinct morphologies appear:

1. Rectangular, flat, and more or less elongated
2. Elongated oval, with a rounded end
3. Rectangular, with a rounded end and a flat base

All the statuettes have the same position, with the arms squarely set. The body is schematically rendered, whereas the head and its features are more realistic. The form of the head is oval, trapezoidal, or squared. Hair is unrepresented except in two cases that have incised lines on the dorsal face, face, and sides. The edge of the face is sometimes marked by a line or a deep furrow. The facial style in the form of a T representing nose, eyes, and eyebrows is fairly frequent. Ears are absent. The eyes and the nose are formed with hollows. The head so overscales the body that it leaves the sculptor sufficient space to shape out the nose. This last feature seems to be a major characteristic of the face, placed large and straight; it may also be slightly potato-shaped. The goatee was also worn during this period. Generally the head and neck are undifferentiated, but in three cases, the statuettes have a well-articulated neck. Around the neck are necklaces or collars shown as simple rounded cords or pendants.

The shoulders are squared and as wide as the hips. Shoulders are rounded off or pointed, with a hollow marking the clavicle. Usually this hollow is fashioned by a band. It crosses the torso from the right to the left or vice versa, passing between the arms, resting on the buttocks, continuing around to the front again. In one case the band ends between the hands; in another the band passes under the arms. Torsos are flat. To keep the elbows close to the body, the sculptors have foreshortened the forearms. These end either in stumps or in fingers represented very schematically with only two, three, or even four incised lines.

The belt may be a simple band, double band, or a belt with buckle (shown as a medallion). The statuettes from Ma'rib have an abdominal protruberence that approaches obesity. Below the belt, the lower body is left intact, polished, or marked by folds or large furrows that frame the thighs. When represented, the legs are joined and usually massive. Two cases have toes. The base of the statuette may be rectangular, circular, or convex, but only two statuettes can stand upright without support.

In one corpus, these 17 statuettes present a relatively homogeneous facies across several regions. Only three statuettes from the Jawf are clearly identifiable by sex—one feminine and two masculine—through their genital appendages; the 14 others have flat buttocks, no nipples, bands, belts, and collars, all criteria that help to define them as masculine. (Statuette ST 16 is problematic: we were unable to obtain a photograph, and the drawing is of poor quality. It does record more or less two small circles near the shoulders that might be interpreted either as a youthful chest or as a decorative element.)

Regional Characteristics

The first group is that of Rawk with five statuettes. Three of them are of general rectangular form worked from a fairly thin slab. The other two were produced on cobbles and have a more rounded appearance. The statuette found in situ at Rawk was unfinished by its sculptor and therefore possesses no details of dress. In contrast, ST 02 has a folded cloth that covers the entire base of its body. Statuettes ST 01–ST 03 and ST04 wear a flat, belted cloth that hides the lower part of the statuette.

While Wādī ʿIdim lies in eastern Hadramawt, the other three regions rich in anthropomorphic statuettes are located 500 km to the west. These are Jawf, Khawlān, and Ma'rib. The statuettes from these three regions form a coherent ensemble, to which one can add the statuette of unknown provenience. (Looting has been greater on this side of Hadramawt.) Wearing a cloth pareo held by a belt is a common trait in these three western regions. The lower part of the statuettes is marked by a vertical line indicating the legs. Nudity is represented in the Jawf by a feminine statuette (with vulva and breasts) and two masculine ones (with penises). Even when sexual organs are visible, the statuettes wear a belt. Certain ones have a true tendency toward fat, especially around the abdomen. Such details highlighted in clothing, treatments of the body, and faces in eastern and western regions—especially the specific manner in the Jawf of showing nudity, or obesity in Ma'rib—reveal a true desire for realism.

Other Parallels in Yemen

Three very schematic statuettes were discovered by Diana Pickworth (2005) and Vittoria Buffa (Buffa and Vogt 1999) (figure 15.14). One from Wādī Tuban delta at Miqaḥala near Sabr (Tihāmah) was a stylized statuette with a polished surface, sculpted from a white quartz cobble. Even though the statuette is largely acephalous, it is possible to detect the nose, however much flattened. The arms end in stumps and the bust has been fashioned through pecking. The belt is marked with an indentation. Traces of red pigment are visible. Buffa argues that the statuette dates to 5000–4000 cal BP (Buffa and Vogt 1999).

At Bint al-Methul in the Ramlah as-Sabʿatayn, Diana Pickworth (2005:420, figures 1–2) recovered a statuette fashioned in granite. Without much contextual explanation, she associates the statuette with nearby high tower tombs presumed to date between 5300 and 4000 BCE.

Statuettes from Northern Arabia

The Arabian exhibit at the Louvre in 2010 offered an opportunity to document two other anthropomorphic statuettes, which confirm their presence from eastern Hadramawt (Yemen) to Ḥā'il in Saudi Arabia. Statuette ST 18 of Ḥā'il is a standing man, 92 cm high (Steimer-Herbet 2010). The head and the neck are well articulated. The facial details are very finely worked; the eyes bulge, with overarching eyebrows that lengthen into a straight and strong nose. Only a necklace decorates the trunk, which seems to be otherwise nude, as the sculptor has represented the clavicles. The position is solemn and regal, the arms folded over the stomach. Fingers are represented. The man wears a belt, to which is attached a circular object. Legs and knees are marked. The overall bearing, the form of the head, the necklace at the throat, and the belt are elements well recognized from the known Yemeni examples. The position of the arms seems to be a characteristic more properly of the north. At Rawk, it was possible to observe different skills and stages of the sculpting technique; with a remarkable symmetry, the Ḥā'il statuette was without doubt executed by a highly skilled artist. The circular object shown has no parallel in the corpus of statuettes. There are several examples of statue-menhirs where daggers and circles are associated. Nevertheless, the representation most closely parallel is from the rock engravings in the Sa'ada region, where the muzzle of a bovine is restrained by a circular form (Nayeem 2000). Humans of 6000 cal BP were sometimes settling down and widely practiced herding.

Also from Ḥā'il, statuettes ST 19 and ST 20 are distinct from the Southern Arabian group through their representation of only the bust and head of a man (Steimer-Herbet

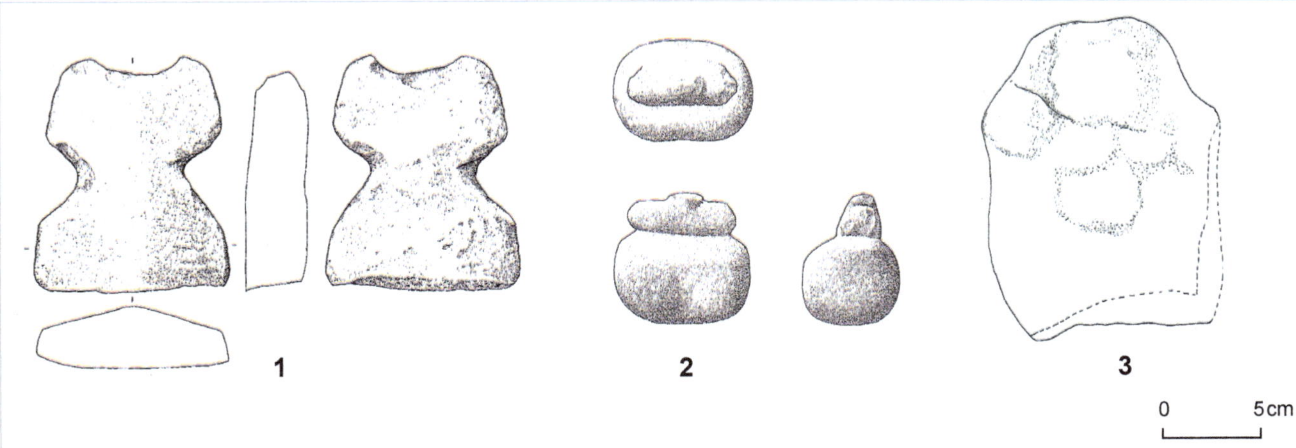

Figure 15.14. Schematic statuettes of Sabr (1 and 2) and Bint al-Methul (3) (after Buffa and Vogt 1999; Pickworth 2005). *Illustration by Tara Steimer-Herbet.*

2010). The two were of notable height, with a measurement of 57 cm for ST 19. Its rounded head is proportionally greater than heads of other statuettes, with a squat neck. It seems to have a notably regional representation of ears and mouth, both well depicted. The bust is truly different; for the upper part, the relief is achieved with curved lines, while the torso and the arms are straight. The closest parallels remain in Riqseh (Kirkbride 1969a, 1969b), where during the same period one finds representations of standing humans, busts, or simply heads.

The dimensions of the Yemeni anthropomorphic statuettes are relatively modest compared to that of their Saudi cousins (Ḥā'il). The Yemeni statuettes could be held in a hand and easily transported.

Statue-Menhirs

Statue-menihrs seem to be concentrated in the Jol and somewhat to the north of Mahra near Tabuk. To date, no statue-menhir has been described to the west in the Jawf, Ma'rib, or the highlands of Yemen. The function of these objects, isolated or found in groups, has not been fully determined. At Rawk, the statue-menhir probably sat in the facade of a ritual structure, perhaps defining a circulation path, as at Krīf Badrīb (McCorriston 2011).

The raw material of statue-menhirs was a limestone slab, less than 1 m in height (with the exception of the example in the Say'ūn Museum). Decorative elements were incised or worked by pecking. All statue-menhirs are masculine. In contrast with anthropomorphic statuettes, only the upper part of the body is represented, with the basal part unworked, probably because it was buried in the ground.

One finds great stylistic similiarities with the rock-art documented in the region of Sa'ada and in the far north of Saudi Arabia (Inizan and Rachad 2007; Nayeem 2000). Shared anatomical motifs include the nose, eyebrow arch, and eyes (all in the form of a T). The mouth is missing, and the bottom of the face is underscored by a curved line and a beard, if not a necklace or pectoral ornament. Hair is shown on several statue-menhirs. The arms, with or without hands, are along the body. An important object crosses the abdomen—a dagger with a crescent-shaped pommel. Sometimes a sheath is associated, often with smaller dimensions than the blade it should cover. The belt is also important. Represented most often by a simple line, the belt may also have a decoration in fine chevrons. Two statue-menhirs show markings on their dorsal sides; these markings represent the vertebral column and ribs.

North Arabian Statue-Menhirs

Not until one comes to Northern Arabia, in the region of al-'Ulā, and southern Jordan, at the site of Riqseh, does one find another concentration of statue-menhirs. The 1 m tall statue-menhir of al-'Ulā (Steimer-Herbet 2010) represents a standing man. Both flat sides have been sculpted, but the emphasis is on the frontal aspect, notably the face, the chest, and the upper legs. The trapezoidal-shaped head rests on its shoulders, indicated by an indentation. Its base is narrow and irregular. The outline of the face is marked by a fine relief, framing the close-set eyes and a long, inset nose. A collar and two chest bands, to which is fixed a sort of awl, occupy the torso. A double-bladed dagger is held by a large belt. The chest bands and belt are carried out on the back side. At Riqseh, Diana Kirkbride (1969b:195) discovered a broken statue-menhir on which one can see a double-bladed dagger and an awl. On a rock face at Tabūk in Wādī Damm, two human silhouettes of the Late Neolithic with similar attributes have been identified (Nayeem 2000). There are also Syrian examples, one exhibited in the Museum of Ma'rrah an-Nu'mān and a statuette from Tell Brak, which appears in the Louvre Museum

Conclusions

The human groups populating Hadramawt 6000–5000 cal BP left us two views of themselves. One, a martial view, is of male statue-menhirs armed with daggers—attributes of authority. The other is more peaceful, of statuettes generally presumed to be males, with their folded arms in a respectful pose.

The site structure at Rawk allows one to perceive some preoccupations of its builders; the site was for burial of the dead and also for receiving the living, who used the bench. In this shared space, what symbolic role should one attribute to the anthropomorphic statuettes? In light of their morphology and the deliberate burial of ST 05, they were destined for interment more than for exposure, in contrast to the statue-menhirs. Furthermore, on the skull of the newborn, on statuettes, and on statue-menhirs, the presence of pigments shows that people of this era treated their dead and their representations—large or small—with the same regard. The facial details and depicted clothing on the statuettes convey realism.

The presence of the dagger on statue-menhirs raises a number of socially linked questions. For one, was the dagger, which represents coercive power, a symbol of authority? Newton and Zarins (2000) have argued that the dagger symbolized access to rare but widely circulating metal exotics and that its association with an individual demonstrated access to the wealth and status associated with consensual authority in chiefdoms. Control of a material resource often accompanies authority (Earle 1987; Giddens 1984). World systems theory, long invoked to explain long-distance trade from 5100 to 4200 cal BP (Edens 1993, 1994; Frank 1999), suggests that the highly finished products of developed societies—like bronze weaponry—were exchanged for bundled raw products in the periphery, engendering the differentiation of wealth there that promoted social authority. (This differentiation in technical acumen applies poorly to most of the ancient Near East [Stein 1999].)

Another issue is the dagger as a manifestation of long-distance trade and contact. Were these daggers actually possessed, or do they depict desired objects? Present on virtually all the statue-menhirs or accompanying numerous painted or engraved human images on rock faces, daggers have seldom been brought to light in excavations of sites attributed to this specific period. Nevertheless, one knows that people of Hadramawt in the second half of the fourth and third millenniums BCE (Early Bronze Age) knew metal, for fragments were discovered beneath the bench of the Rawk structure and also in a wall tomb in Wādī 'Idim (chapter 14) and an HCT in the Jibal Jidrān necropolis (Steimer-Herbet 2001). What became of the daggers depicted on statue-menhirs? Although no forges, crucibles, molds, or other traces of metalworking have ever been found in Yemen, so precious a resource as metal was probably melted down for new objects.

The Early Bronze Age seems to have been an important phase in social development. The Rawk sanctuary is the outcome of a small work group, and its influence need not have reached a large radius beyond the hamlet it presumably occupied (of which there is no remaining trace), probably effaced by modern construction. Such a sanctuary surely must imply a notion of stability and a link to territory that corresponds to a sedentary population. The haphazard discoveries (Ṣunah, Khushum as-Sinām) at regularly spaced intervals in Wādī 'Idim show that sites like Rawk could have dotted the wadi and its tributaries, although hard evidence linking statue-menhirs to sanctuaries remains elusive. Many such sites have disappeared through the effects of erosion or under a thick layer of sediment (Wādī Washa'ah). At a regional scale, these locations participated in a coherent system of motifs and symbols used from Mahra to al-'Ulā, a minimum range of 2,000 km. The skill and expressiveness of the sculptors transmit a rather clear image of the appearance of ancient folk of the era and their regional characteristics. If a code for physical attitude has been widely observed, dress traditions nevertheless varied from one region to another, just as they do for modern Yemeni tribes; one can see today that Hadramis prefer the *futa*, a cloth wrap skirt, over the *galabiya*, the long tunic liked by men of Ma'rib, Jawf, Khawlān, and the highlands of Yemen. Even as men of northern Yemen all wear a traditional *jambiya* dagger, southerners carry none but wear a belt that may vary in price according to the status of its owner.

Even though the archaeological data are fragmentary, it seems that toward 5500–5200 cal BP, people had ceased to represent themselves, or, as seen in the statuettes recovered by Vittoria Buffa and Diana Pickworth, did so only in schematic fashion.

References

Audouin, Rémy

2005 Une "Venus" steatopyge au Metropolitan Museum de New York. In *Sabaean Studies: Archaeological, Epigraphical and Historical Studies in Honour of Yusuf M. 'Abdallah, Alessandro de Maigret and Christian J. Robin on the Occasion of Their 60th Birthdays*, edited by Amida Sholan, Sabina Antonini, and Mounir Arbach, pp. 359–75. De Boccard, Paris.

Avner, Uzi

1984 Ancient Cult Sites in the Negev and Sinai Deserts. *Tel Aviv* 11(2):115–31.

2001 Sacred Stones in the Desert. *Biblical Archaeology Review* 27(3):30–41.

2002 Studies in the Material and Spiritual Culture of the Negev and Sinai Populations, during the 6th–3rd Millennia B.C. PhD dissertation, Hebrew University, Jerusalem.

Buffa, Victoria, and Burkhard Vogt

1999 Sabir: Cultural Identity between Saba and Africa. In *Migration und Kulturtransfer*, edited by Ricardo Eichmann and Hermann Parzinger, pp. 437–50. Akten des Internationalen Kolloquiums, Bonn.

Cleuziou, Serge, Marie-Louise Inizan, and Rémy Audouin

1988 *Rapport préliminaire de la mission de prospection sur le peuplement pré- et protohistorique du Yémen dans le Wâdî Markha*. Archives équipe Du village à l'état, Maison de l'archéologie et de l'ethnologie, UMR ARSCAN, Nanterre, France.

Cleuziou, Serge, Marie-Louise Inizan, Rémy Audouin, Michel Mouton, and Gilles Tosello

1992a *Rapport préliminaire de la mission de prospection sur le peuplement pré- et protohistorique du Yémen dans le Wâdî Markha* (16 septembre–6 octobre 1992). Archives équipe Du village à l'état, Maison de l'archéologie et de l'ethnologie, UMR ARSCAN, Nanterre, France.

Cleuziou, Serge, Marie-Louise Inizan, and Bruno Marcolongo

1992b Le peuplement pré- et protohistorique du système fluviatile fossile du Jawf-Hadramawt au Yémen (d'après l'interprétation d'images satellite, de photographies aériennes et de prospections). *Paléorient* 18(2):5–29.

Crassard Rémy

2008 The "Wa'shah Method": An Original Laminar Debitage from Ḥaḍramawt, Yemen. *Proceedings of the Seminar for Arabian Studies* 38:3–14.

2013 Prehistoric Rock Art in Hadramawt: The Site of Ali-1 in Wâdî Bin 'Alî (Yémen). *Raydan* 8:67–72.

Crassard R., and H. Hitgen

2007 From Ṣāfer to Bālḥāf: Rescue Excavations along the Yemen LNG Pipeline Route. *Proceedings of the Seminar for Arabian Studies* 37:43–59.

Crassard, Rémy, Joy McCorriston, Eric Oches, 'Abd Al-Aziz Bin 'Aqil, Julien Espagne, and Mohammad Sinnah

2006 Manayzah, Early to Mid-Holocene Occupations in Wādī Ṣanā (Ḥaḍramawt, Yemen). *Proceedings of the Seminar for Arabian Studies* 36:151–73.

de Maigret, Alessandro

1990 *The Bronze Age Culture of Khawlān at-Ṭiyyāl and Al-Ḥadā (Republic of Yemen)*. ISMEO, Rome

2002 *Arabia Felix: An Exploration of the Archaeological History of Yemen*. Stacey International, London.

Earle, T.

1985 Prehistoric Economics and the Evolution of Social Complexity. In *Prehistoric Production and Exchange: The Aegean and the Eastern Mediterranean*, edited by A. B. Knapp and T. Stech, pp. 106–11. Monograph 25. UCLA Institute of Archaeology, Los Angeles.

1987 Chiefdoms in Archaeological and Ethnohistorical Perspective. *Annual Review of Anthropology* 16:279–308.

Edens, Christopher

1993 Indus-Arabian Interaction during the Bronze Age: a Review of Evidence. In *Harappan Civilization: A Recent Perspective*, edited by Gregory Possehl, pp. 335–63. Oxford and IBH South Asia, New Delhi.

1994 On the Complexity of Complex Societies: Structure, Poser, and Legitimation in Kassite Babylonia. In *Chiefdoms and Early States in the Near East: The Organizational Dynamics of Complexity*, edited by Gil Stein and Mitchell Rothman, pp. 209–23. Prehistory Press, Madison.

Frank, Andre Gunder

1999 Abuses and Uses of World Systems Theory in Archaeology. In *World-Systems Theory* in Practice, edited by P. N. Kardulias, pp. 275–96. Rowman and Littlefield, Lanham, MD.

Garcia, Michel, Madiha Rachad, Dijillali Hadjous, Marie-Louise Inizan, and Michel Fontugnes

1991 Découvertes préhistoriques de l'art rupestre de la région de Saada. *Comptes Rendus de l'Académie des Sciences de Paris* 313(2):1201–06.

Giddens, Anthony

1984 *The Constitution of Society: Outline of the Theory of Structuration*. Polity Press, Cambridge.

Inizan, Marie-Louise, Frank Braemer, A.-M. Lézine, and Tara Steimer

1998 *Rapport préliminaire de la prospection dans les régions de Saada, Shabwa et du Ramlat as-Sabaatayn*. Archives équipe Du village à l'état, Maison de l'archéologie et de l'ethnologie, UMR ARSCAN, Nanterre, France.

Inizan, Marie-Louise, and Madiha Rachad
2007 *Art rupestre et peuplement préhistoriques au Yémen.* Centre français d'archéologie et de sciences sociales, Sana'a.
Kirkbride, Diana
1969a Ancient Arabian Ancestor Idols. Part 1, The Discovery of the Sanctuary at Risqeh. *Archaeology* 2:116–21.
1969b Ancient Arabian Ancestor Idols. Part 2, The Interpretation of the Sanctuary at Risqeh. *Archaeology* 3:188–95.
Martin, L., Joy McCorriston, and Rémy Crassard
2009 Early Arabian Pastoralism at Manayzah in Wādī Ṣanā, Ḥaḍramawt. *Proceedings of the Seminar for Arabian Studies* 39:271–82.
McCorriston, Joy
2000 Early Settlement in Hadramawt: Preliminary Report on Prehistoric Occupation at Shi'b Munayder. *Arabian Archaeology and Epigraphy* 11:129–53.
2011 Anthropomorphic Stelae in Ancient Hadramawt (Yemen). In *Standing Stones: Anthropomorphic Stelae and Dolmens*, edited Tara Steimer, pp. 159–68. British Archaeological Reports, International Series 2317. Archaeopress, Oxford.
McCorriston, Joy, Tara Steimer-Herbet, Michael Harrower, K. Williams, and ʿAbdālʿaziz Bin ʿAqil
2011 Gazetteer of Small-Scale Monuments in Prehistoric Hadramawt, Yemen: A Radiocarbon Chronology from the RASA-AHSD Project Research 1996–2008. *Arabian Archaeology and Epigraphy* 22:1–22.
Nayeem, Muhammed Abdul
2000 *The Rock Art of Arabia.* Hyderabad Publishers, Hyderabad.
Newton, Lynn S., and Juris Zarins
2000 Aspects of Bronze Age Art of Southern Arabia: The Pictorial Landscape and Its Relation to Economic and Socio-Political Status. *Arabian Archaeology and Epigraphy* 11:54–79.
Pickworth, D.
2005 A Very Old Lady: A Female Figurine from Gawat Bintayn Metūl. In *Sabaean Studies: Archaeological, Epigraphical and Historical Studies in Honour of Yusuf M. ʿAbdallah, Alessandro de Maigret and Christian J. Robin on the Occasion of Their 60th Birthdays*, edited by Amida Sholan, Sabina Antonini, and Mounir Arbach, pp. 417–21. De Boccard, Paris.
Pirenne, Jacqueline
1990 *Les témoins écrits de la région de Shabwa et l'histoire.* Institut Français du Proche-Orient, Bibliothèque archéologique et historique 134. Librairie orientaliste Paul Geuthner, Paris.
Robin, Christian J., and Burkhard Vogt
1997 *Yémen, au pays de la reine de Saba: Exposition présentée à l'Institut du Monde Arabe du 25 oct. 1997au 28 fév. 1998.* Flammarion, Paris.
Sedov, Alexander, and A. as-Saqqaf
1992 Stone Idols from Wādī 'Idim (Inner Hadramawt). *Yemen studi archeologici, storici e filologici sull'Arabia meridionale* 1:125–29.
Steimer-Herbet, Tara
2001 Results of the Excavation in Jabal Jidran (February 99). *Proceedings of the Seminar for Arabian Studies* 31:221–26.
2004 *Classification des sépultures à superstructure lithique dans le Levant et l'Arabie occidentale: (IVe et IIIe millénaires avant J.-C.).* BAR International Series 1246. Archaeopress, Oxford.
2007a Open Sanctuary and Anthropomorphic Statuettes at Rawk during the 4th Millennium BC (Wādī 'Idim, Yemen). *Antiquity* 81:312.
2007b Sanctuaire ouvert du IVe millénaire avant l'ère chrétienne (Wâdî 'Idim). *Chroniques yéménites* 14:1–8.
2010 Stèles d'Arabie. In *Routes d'Arabie, Trésors archéologiques du royaume d'Arabie Saoudite*, edited by André-Salvini, Sophie Makariou, and Françoise Demange, pp. 166–69. Somogy-Éd. d'art, Paris.
Steimer-Herbet, Tara, Frank Braemer, and Gourguen Davtian
2006 Pastoralists' Tombs and Settlement Patterns in Wādī Wa'shah during the Bronze Age (Ḥaḍramawt, Yemen). *Proceedings of the Seminar for Arabian Studies* 36:257–65.
Stein Gil
1999 Rethinking World Systems: Power, Distance, and Diasporas in the Dynamics of Inter-regional Interaction. In *World Systems Theory in Practice: Leadership, Production, and Exchange,* edited by P. Nicholas Kardulias, pp 153–77. Rowman and Littlefield, Lanham, MD.
Vogt, Burkhard
1997 La fin de la préhistoire au Hadramawt. In *Yémen, au pays de la reine de Saba'* edited by Christian J. Robin and Burkhard Vogt, pp. 30–33. Institut du monde arabe—Flammarion, Paris.

Chapter 16

Testing at Munayder (Wādī ʿIdim)

Joy McCorriston

In 1998 the RASA team systematically surveyed seven minor tributaries to Wādī ʿIdim and Wādī ʿAtuf and conducted broad reconnaissance survey throughout Wādī ʿIdim's upstream channels—Wādī Risib and Wādī ʿAtuf—and the upstream slopes of the main Wādī ʿIdim channel. Although the survey encountered many monuments, hearths, and other remains that subsequently informed research, the focus of 1998 survey was settlement, of which the team found best evidence at Shiʿb Munayder in the form of independent circular structures spaced along natural rock and gravel terraces along the main Wādī ʿIdim channel, south of Ghayl ʿUmar. This chapter summarizes the previous publication (McCorriston 2000) of the survey and test excavations and provides an updated synthesis of these results with subsequent RASA research and a broader archaeological record of the Southern Jol and Southern Arabia.

Munayder Résumé (from McCorriston 2000)

Discovery of Settlement Architecture and Surface Description

Named for a shallow tributary feeding the main Wādī ʿIdim-channel, Shi'b Munayder lies about 2 km northeast of major traditional passes leading up to the east and west plateaus and about 5 km north of the shrinking date palm groves at Ghayl ʿUmar. Modern date palm cultivation at Ghayl ʿUmar and the density of human occupation at Sikdān and Sāh farther south have effaced whatever traces might have persisted of former occupation. This taphonomic problem exists at the rare permanent sources of water in the Southern Jol and has been observed also by RASA survey at Ghayl ʿUmar, Ghayl Bin Yumain, Risib, and Ghayl Bā Wazīr, as well as at Wādī Ḥajar and Aīn Bā Maʿbad (Shabwa Province).

As the RASA extensive survey team moved northward in Wādī ʿIdim, we encountered a high density of small-scale irrigation structures—check dams in the gullies cut through silt terraces, diversion walls along the plateau and slopes—within the main drainage of Wādī ʿIdim. Tributaries Wādī Sikdān (RASA 1998, Waypoints 013 and 014; E 277789, N 1731017 and E 278663, N 1730233) and Wādī Kuwwa (RASA 1998, Waypoint 015; E 273445, N 1738445) have surface traces of such activities close to the main Wādī ʿIdim channel, but upstream, there is scant evidence of human activities other than hearths and graffiti. By the late afternoon, lengthening shadows along the eastern talus slopes and bedrock terraces of Wādī ʿIdim showed clear indications of structures, even at a distance of 0.5 km. These caught the eye of ʿAbdalʿazīz Bin ʿAqīl, who directed the drivers to the gravel terrace at the foot of the looming incline. Strung along the edge of a bedrock terrace, the survey team discovered a relatively dense suite of archaeological remains, including what appeared to be houses.

The Munayder site (RASA 1998, Waypoint 020; E 274980, N 1738716) includes a surface density of lithic knapping debris, sometimes very obviously spatially associated with circular structures built of boulders and upright slabs on rock terraces. Other remains at the site include terraced retaining walls, high circular tombs, small and hollow conical cairns (possibly goat pens or storage facilities), a 12 m diameter enclosure, and linear walls running downslope. There are no clear site boundaries to the north and south, where the density of remains thins near the eastern pass to the plateau and date palm groves. The archaeological remains of structures lie on the lower bedrock and gravel terraces of Wādī ʿIdim (figure 16.2). Adjacent silt terraces of the Wādī ʿIdim are even lower in elevation and have many Neolithic formal tools (bifaces, points) and the remains of hearths on surfaces and in erosional gullies around the dried sediments of ancient marshlands. Neolithic tools are extremely rare on the surfaces of elevated gravel and bedrock terraces of the Shiʿb Munayder site, perhaps because the scree slopes are high-energy environments. Tufa deposits along the edge of the modern Wādī ʿIdim channel (chapter 2) attest to the former presence of now dried springs, probably a northward extension of the spring system at modern Ghayl ʿUmar (figure 16.1).

Another widely recognized cultural type is the high circular tomb (HCT) present at Shiʿb Munayder, generally 10–50 m upslope of house structures (figure 16.3). Elsewhere in Hadramawt and across Arabia, these tombs date to the Early Bronze Age (Giraud and Cleuziou 2009; Harrower et al. 2013; McCorriston et al. 2011, 2014; Steimer-Herbet 2004; Williams et al. 2014). At Munayder, they appear mostly collapsed or robbed, without intact capstones, and as has been demonstrated upstream in Wādī ʿIdim, they may have been reused (McCorriston et al. 2011). From survey examination, there is no indication that tombs were robbed of stone to build houses, nor that structures were destroyed to build tombs (figure 16.4). Sometimes these remains occur in good condition in close proximity.

Figure 16.1. Tufa deposits at the base of the Munayder slope accumulated during a period when springs were active as far north as Munayder. The figure at middle foreground is in front of "fossil springs," with banded (paleosol) sediments of Wādī ʿIdim in the background. *Photograph by Joy McCorriston.*

Figure 16.2. View southward along the Munayder terraces. On the upper terrace are tombs and isolated houses; on the lower, gravel terrace are larger enclosures and houses. *Photograph by Joy McCorriston.*

Figure 16.3. A small cairn (foreground) and Bronze Age tomb (silhouette) at the Sh'ib Munayder site. *Photograph by Joy McCorriston.*

Figure 16.4. A structure, probably a house, on the lower, gravel terrace at Munayder. *Photograph by Joy McCorriston.*

Some house structures are standing to 1 m in height and have clear entrances flanked by orthostat uprights, which were placed perpendicular to the wall line. Construction methods vary: some structures have a single thickness of upright slabs to form a perimeter; some have boulders rather than slabs; others consist of doubled uprights as faces packed with a rubble core. All are subcircular and freestanding in plan, without connecting compounds or apparent subdivisions.

Methods

A surface survey of Munayder recorded more than 80 structures and architectural features. These were tallied and formed the basis for the types of structures listed above, but no formal map of the site was drawn. Surface lithics were most densely concentrated on the interior surfaces of structures and could be found in less concentrated accumulations outside and downslope of structures. Structure doorways mostly faced downslope (facing the

Figure 16.5. Excavated interior of Munayder Structure 1 (house) with the central hearth (between scales) and interior wall in the background. *Photograph by Joy McCorriston.*

Figure 16.6. Munayder 1, Quad B; exterior of house wall and upslope retaining wall (under signage). *Photograph by Joy McCorriston.*

main channel of Wādī ʿIdim). Instead of mapping this site of unknown date, the team dedicated its limited efforts in 1998 to test excavations, with the objective of recovering datable material in stratigraphic context.

Test excavations opened only a small sounding, 2 x 1 m, to bedrock in Munayder 1 (1998 Munayder-1-A) (figure 16.5). A second test trench (Munayder-1-B) upslope and outside the house revealed the 0.8 m depth of colluvium and construction of an upslope terrace wall (figure 16.6), and a third trench (Munayder-1-C) probed the depth and fill of the terrace (figure 16.7).

Results

The most important and conclusive results came from Test Pit A. Lithics were found on the surface. Underneath, a sterile, aeolian, sandy silt fills the structure interior and overlies a secondary interior division or rockfall, which in turn overlies interior fill with cultural deposits. The earliest cultural fill (Level III) included a central hearth (Level IV) on the primary surface (Level VI), knapped flakes, bone fragments, and flecks of charcoal. This lower cultural fill is significantly and clearly separated from later knapping debris on the interior surface of the structure, and there are clearly at least two, possibly three, phases of use (McCorriston 2000:140–41).

The structure wall of Munayder 1 is built with a base of single upright slabs capped by three courses of undressed boulders and cobbles (figure 16.5). This structure was built on a prepared terrace (Test Pit C), and upslope was the retaining wall of another terrace (Test Pit B), buried, like the structure wall, under colluvium. There was no evidence of roofing material or construction.

Figure 16.7. Munayder 1, Quad C; terrace fill excavated to bedrock, looking downslope from house site. *Photograph by Joy McCorriston.*

A radiocarbon age on the hearth (OS16724) of 3463–3361 cal BP suggests a use in the Late Bronze Age era, when there are few other remains in Hadramawt and, indeed, across much of Southern and Eastern Arabia (de Maigret 2002; Magee 2014).

If roofed at all, houses may have been covered in antiquity by timber and branches, as was the case in Dhofar, seen ethnographically (Janzen 1986) and archaeologically in the (unpublished) results of excavations at Shakeel and Halqoot in the Dhofar mountains.

Surface artifacts provided weak chronological indication of when the structure was reused. Nor does the last use of a central hearth date the house construction. Reuse of structures, particularly by mobile people, is common in Arabia. Despite problems in dating these structures, the radiocarbon age from one of a concentration of houses offers one of the few indications of coresident communities of households prior to the emergence of urban centers in Wādī Ḥaḍramawt.

Objects

Knapped stone is mostly debitage and includes a few nonformal tools like utilized flakes. A wide array of raw sources is represented. One fragment of soapstone bowl, mended with hand-smithed metal, attests to fairly recent use, as these artifacts and repairs are common in Yemeni *salta* restaurants today. Other pieces, such as microliths, suggest Iron Age knapping. These microliths were surface finds.

From excavations it was clear that almost all lithics are surface accumulations. Knapped stone on the surface, clearly a later accumulation after sterile fill covered base occupation, may represent one or many reoccupations. Taphonomic factors may account for our observation of denser concentrations of surface lithics inside houses: the walls of structures may retain artifacts that do not survive on the outside surfaces because of high-energy slopes.

Munayder in Context: Aceramic Settlements of Southern Arabia's Bronze and Iron Ages

Unlike the Bronze Age ceramic sites of highland northern Yemen (de Maigret 1990; Edens 1999), Munayder is aceramic. Likewise, if microliths and signs of reoccupation signal an Iron Age reuse of Munayder's houses, and possibly also its high circular tombs, the reoccupants also left no ceramic refuse. Bronze Age cultures of northern Yemen clustered in building complexes that show contiguous perimeter walls and enclosures very different in plan from the isolated houses at Munayder.

The communities at Munayder left behind no ceramics but do seem to belong to a broader pattern of late prehistoric settlement near permanent water sources. In plan and in associated material culture, the remains at Munayder most closely resemble the aceramic concentration of (undated) houses and HCTs in Wādī Kharshīt, Dhofar (Sultanate of Oman) (McCorriston 2000; Zarins 2001). Recent excavations led by McCorriston at Shakeel and at Halqoot, Dhofar, also revealed a settlement of similar double-faced, rubble-core, circular house walls constructed of limestone uprights and boulders and also lacking ceramic debris. These Shakeel houses have multiple associated radiocarbon ages around 2000 cal BP and appear to be associated with seasonally mobile pastoralists on the Dhofar escarpment.

Unlike Munayder and Wādī Kharshīt, there are no adjacent tombs at Shakeel, but there are major stone monuments, probably Bronze Age and pre-Islamic, interspersed

Figure 16.8. Large enclosure (C12) of standing stones near Ghayl Bin Yumain (Wādī Sanā), viewed from the south. *Photograph by Tara Steimer-Herbet.*

Figure 16.9. Structure on the gravel terraces (C11) north of the modern extent of irrigated palms near Ghayl Bin Yumain (Wādī Sanā). *Photograph by Tara Steimer-Herbet.*

with houses and corrals only 2 km to the north at Ḥalqoot. The association in Dhofar (for example, Ḥalqoot, Wādī Kharshīt) of houses near major monumental sites is also found in Wādī ʿIdim. Shiʿb Munayder and similar house remains line the slopes of Wādī ʿIdim only 4–5 km southeast of Rawk's ancient sanctuary (chapter 15). One must remember that most of these sites are poorly if at all dated and that occupants may have shared little or no temporal or cultural affinities.

Much closer in location to Munayder are poorly preserved structures in Wādī Sanā. Noted in the RASA survey around the dried marshes several kilometers north of Ghayl Bin Yumain's spring-fed date palm groves, these remains are comparable to the standing stone circle at Munayder and house remains there and in Dhofar. The Wādī Sanā location, on the lower terraces of Wādī Wasā's main drainage near formerly permanent water sources, is today heavily disturbed, not least by dirt tracks and small vehicles removing natural and archaeological stone for village expansion at Ghayl Bin Yumain. Archaeological remains observed in 1996 today have vanished, but the RASA survey (1996) and the CANOXY pipeline survey noted two very large circles of standing stones (like one at Munayder) and the remains of smaller circular structures built of limestone slab uprights and boulders, remnants perhaps of houses beside spring-fed wetlands. Several were redocumented in the RASA 2005 pilot cairn survey (C12 "grand cercle" and C11 "habitats," figures 16.8 and 16.9). Like Munayder, the surface of these sites is aceramic—indeed, in the entire Wādī Sanā survey, only a handful of stray ceramic sherds was found, and several of these came from a single pot (not associated with the C12 or C11 remains).

Ceramics in the Southern Arabian Highlands

Ceramics are one of the classic material markers of different cultural groups, but their distributions across Southern Arabia's chronology and geography remains poorly understood. Some of the earliest levels excavated under Hadramawt's antique South Arabian urban centers include distinctive ceramics and mudbrick architecture of the second millennium BCE, suggesting perhaps that cultural groups implicated in the founding of these regional centers were more closely related to communities along the littoral or in north Yemen's highlands (Badre 1991; Breton 1996; Schiettecatte 2006; Sedov 1996). As many archaeologists have noted, the presence, form, technologies, and functions of ceramics may differ according to economic, social, ideational, chronological, contextual, and taphonomic variables at play in their use and discard. It is difficult and potentially misleading at this stage of knowledge about Hadramawt's prehistory to ascribe to the lack of ceramics at Munayder a marginal role of its inhabitants in wider social interactions (McCorriston 2013).

Conclusions

In chapter 15 of this volume, Tara Steimer-Herbet signals the importance of an emergent and widespread Early Bronze Age iconographic repertoire of anthropogenic representations across the Arabian Peninsula. By the middle second millennium, when people settled at (or reused) houses at Munayder, the downstream sanctuary at Rawk had been extant for at least 1,000 years (and possibly already abandoned). Statuary and symbolic representation recovered at Rawk suggest iconographic conventions shared with semisedentary farming communities in northern Yemen whose ceramic types never appeared in Hadramawt. There are (undated) ceramic sites in the Southern Jol (chapter 4), near the escarpment, with multicellular enclosures and clusters, ovate plans, and dry-stone walling reminiscent of Khawlān-area Bronze Age hamlets in northern Yemen (de Maigret 1990) and littoral western Dhofar. These sites bear little resemblance in location, architecture, and layout to the house sites clustered around rare sources of permanent water in the upper main channel of Wādī ʿIdim and Wādī Sanā.

References

Badre, Leila

1991 Le Sondage Stratigraphique de Shabwa, 1976–1981. *Syria* 68:229–314.

Breton, Jean-François

1996 Quelques dates pour l'archeologie sud-arabique. In *Arabia Antiqua: Early Origins of South Arabian States*, edited by Christian Robin and Iwona Gajda, pp. 87–110. IsMEO, Rome.

de Maigret, Alessandro

2002 *Arabia Felix: An Exploration of the Archaeological History of Yemen*. Stacey International, London.

de Maigret, Alessandro (editor)

1990 *The Bronze Age Cultures of Hawlān at Ṭiyāl and al-Ḥada (Republic of Yemen): A First General Report*. ISMEO, Rome.

Edens, Christopher

1999 The Bronze Age of Highland Yemen: Chronological and Spatial Variability of Pottery and Settlement. *Paléorient* 25(2):105–28.

Giraud, Jessica, and Serge Cleuziou

2009 Funerary Landscape as Part of the Social Landscape and Its Perceptions: 3,000 Early Bronze Age Burials in the Eastern Jaʿlān (Sultanate of Oman). *Proceedings of the Seminar for Arabian Studies* 39:163–80.

Harrower, Michael J., Jared Schuetter, Joy McCorriston, Prem Goel, and Matthew J. Senn

2013 Survey, Automated Detection, and Spatial Distribution Analysis of Cairn Tombs in Ancient Southern Arabia. In *Mapping Archaeological Landscapes from Space*, edited by Douglas C. Comer and Michael J. Harrower, pp. 259–68. Springer, New York.

Janzen, Jörg
1986 *Nomads in the Sultanate of Oman: Tradition and Development in Dhofar*. Westview Press, Boulder.

Magee, Peter
2014 *The Archaeology of Prehistoric Arabia: Adaptation and Social Formation from the Neolithic to the Iron Age*. Cambridge University Press, Cambridge.

McCorriston, Joy
2000 Early Settlement in Hadramawt: Preliminary Report on Prehistoric Occupation at Shi'b Munayder. *Arabian Archaeology and Epigraphy* 11:129–53.
2013 Pastoralism and Pilgrimage: Ibn Khaldun's Bayt-State Model and the Rise of Arabian Kingdoms. *Current Anthropology* 54(5):607–41.

McCorriston, Joy, Tara Steimer-Herbet, Michael J. Harrower, Kimberly A. Williams, Jean-François Saliège, and 'Abdalaziz Bin 'Aqil
2011 Gazetteer of Small Scale Monuments in Prehistoric Hadramawt, Yemen: A Radiocarbon Chronology from RASA Project Research 1996–2008. *Arabian Archaeology and Epigraphy* (22):1–22.

McCorriston, Joy, Michael Harrower, Tara Steimer, Kimberly D. Williams, Matthew Senn, Mas῾ūd Al Hādhari, Mas῾ūd Al Kathīrī, ῾Ali Aḥmad Al Kathīrī, Jean-François Saliège, and Jennifer Everhart
2014 Monuments and Landscape of Mobile Pastoralists in Dhofar: The Arabian Human Social Dynamics (AHSD) Project 2009–2011. *Journal of Oman Studies* 19:117–43.

Schiettecatte, Jérémie
2006 Villes et urbanisation de l'Arabie du Sud à l'époque préislamique. PhD dissertation, Université Panthéon-Sorbonne 1, Paris.

Sedov, Alexander V.
1996 On the Origin of the Agricultural Settlements in Hadramawt. In *Arabia Antiqua: Early Origins of South Arabian States*, edited by Christian Robin and Iwona Gajda, pp. 67–87. ISMEO, Rome.

Steimer-Herbet, Tara
2004 *Classification des sepultures a superstructures lithiques dans le Levant et l'Arabie meridionale (IVe et IIIe millenaires avant J.-C.)*. BAR International Series 1246. Archaeopress, Oxford.

Williams, Kimberly D., Tara Steimer-Herbet, Lesley A. Gregoricka, Jean-François Saliège, and Joy McCorriston
2014 Bioarchaeological Analyses of 3rd Millennium BC High Circular Tower Tombs from the Arabian Human Social Dynamics (AHSD) Project in Dhofar, Oman. *Journal of Oman Studies* 18:153–73.

Zarins, Juris
2001 *The Land of Incense: Archaeological Work in the Governate of Dhofar, Sultanate of Oman 1990–1995*. Project of the National Committee for the Supervision of Archaeological Survey in the Sultanate, Ministry of Information, Sultanate of Oman.

Chapter 17
Graffiti and Pictographs

Joy McCorriston, ʿAbdalʿazīz BinʿAqīl, and Alessia Prioletta

Classic Arabian archaeology in the centers of the Kingdom of Hadramawt, at sites like Raybūn, Makaynūn, and Shabwa, has established that writing in alphabetic scripts had appeared in the Hadramawt region by 2,700 years ago. While there may once have been an extensive corpus of written documents on palm leaf ribs and other perishable materials, with few exceptions, only the stone and bronze inscriptions have been preserved. Rockart—that is, images created through the modification of rock surfaces—is widespread across Arabia, even as its study has nowhere received comprehensive treatment. In the course of survey, the RASA team found a number of examples of alphabetic graffiti and inscriptions on stone, sometimes accompanying images pecked onto the surface of bedrock faces and limestone blocks. This chapter adds documented examples to the small corpus from Southern Arabia but attempts no comprehensive treatment of regional rockart.

Methods

Our small team of prehistorians and paleoecologists never included a dedicated expert in epigraphic studies and the Arabian languages one might plausibly encounter—Arabic, Sabaic, Hadramitic, and the yet-undeciphered Dhofari script. Nonetheless, the team did basic documentation, including oblique photography (not photogrammetric) from vantages easily reached by the survey team, sketches and not-to-scale interpretive renderings, context descriptions, and field assessment of techniques and relative patination. In the renderings presented here, approximate scales have been included as available; most photographs include a scale. Consistent with our survey designs (chapter 4), these rockart and graffiti examples have GPS points within 1 m accuracy.

Wādī Sanā

In the course of survey, the RASA team documented four sets of rockart and graffiti on the bedrock faces of Khuzma as-Shumlya and other rock images, lettering, and inscriptions in other locations along Wādī Sanā and Wādī ʿIdim. At several other locations in the middle Wādī Sanā, pecked designs and letters were also left on portable blocks of limestone lying on the surfaces of silt and gravel terraces, close to the current incision of the wadi bed (figures 17.1 and 17.2)

One of these is the dolmen site (C67-1) discussed in chapter 14 and not further mentioned here. Because the taphonomic circumstances clearly indicate that these images were part of the rock surface when it was deliberately smashed (more than 5,000 years ago), the description of the rockart belongs with the documentation of the structure it once decorated. Other rockart of uncertain age, and especially alphabetic lettering, is described below.

In middle Wādī Sanā, limestone blocks in SU003-1 and in SU033-17 had graffiti pecked onto a well-patinated surface, and these examples of graffiti in turn show less patination than the surrounding surface of limestone (figure 17.3). The limestone slabs were otherwise unworked.

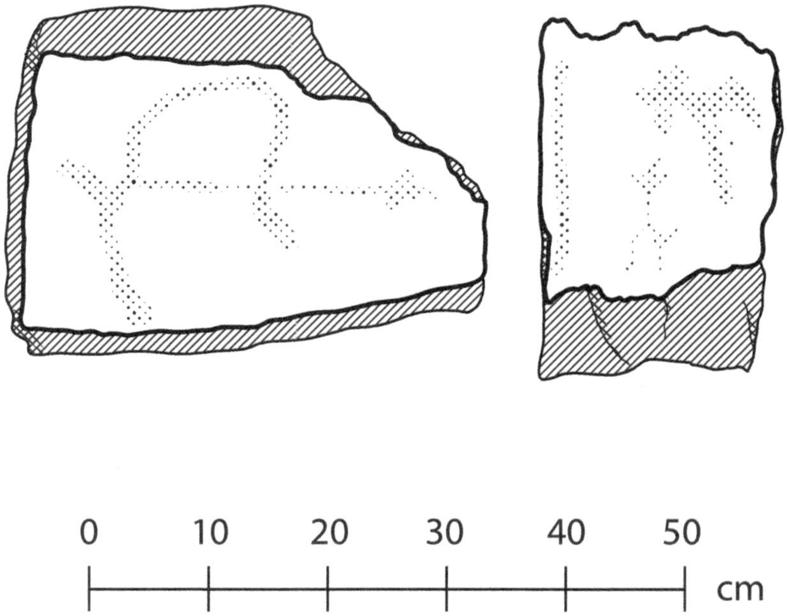

Figure 17.1. Limestone blocks from a surface arrangement in SU000-001. Field drawing, not to scale (scale here inserted for overall dimensions) by ʿAbdālazīz Bin ʿAqīl; identified by Alessia Prioletta as probably a *wusūm*, or a tribal/animal brand. *Drawing by Michael Harrower. Illustration by Clara Hickman.*

Figure 17.2. Limestone block with letters from W15. *Photograph by Michael Harrower.*

On the block from SU003-1, graffito probably are to be read from right to left. The script is of an undefined type. The first glyph resembles the form for *ḥ* in some Ancient North Arabian (ANA) scripts such as Hismaic and Thamudic C. The second glyph has been found in the Dhofari script, although its phonemic value has not been established; Jamme has suggested that it should be read as *d* in one graffito from Hammah, in Oman (JaT 97). The last letter resembles the Ancient South Arabia (ASA) *r* (table 17.1).

Another block on the silt surface in SU033 bears a two-line graffito, possibly in ASA script. The first line is probably read from left to right, with the *n* being back to

Figure 17.3. Limestone block in a stone ring in SU003-001. Reading by Alessia Prioletta. *Photograph by Joy McCorriston.*

front: *tn[...]r*. The second line should be read from right to left, with *n* being back to front: *t[.]nr* (figure 17.4).

Khuzmum Rockshelters

The Khuzma as-Shumlya inselberg is the sole location in which the RASA survey noted ancient graffiti and rockart images on bedrock faces in Wādī Sanā. These appear to be associated with rockshelters, including the eastward-facing Khuzmum rockshelters tested and described in chapter 9.

Panel SU045-7

Between rockshelters SU045-10 and SU045-9 is a bedrock panel, SU045-7, with an array of pecked images and lettering, all bearing lighter patination than the natural bedrock surface but none apparently fresh and none overlapping other images or letters. ʿAbdalʿazīz Bin ʿAqīl documented the writing and sketched the images (figure 17.5), which are not easily legible in overview photographs of the entire panel. The panel contains a group of five uniden-

Table 17.1 Possible non-ASA characters in Khuzmum rockshelters. *Table by Jason Weimar and Alessia Prioletta.*

Surface	Letter	Possible Scripts and Letters	
SU003-1		Thamudic C and D *ḥ*	
SU003-1		Dhofari *d*? (JaT 97)	
SU045-7		Ancient South Arabian s^2 or *r*	Ancient North Arabian *f*

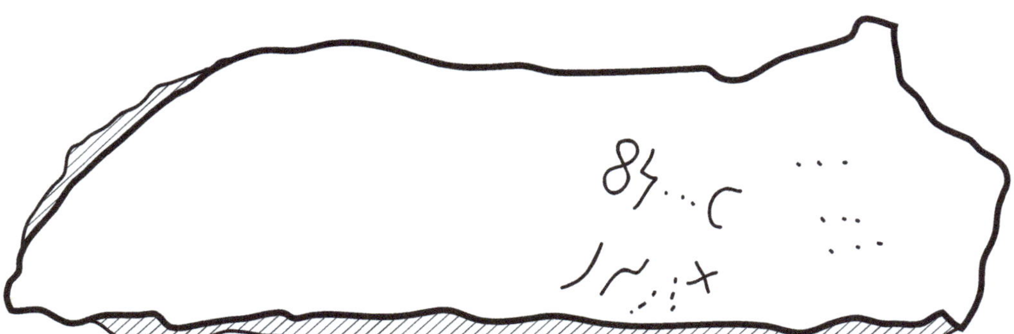

Figure 17.4. Limestone block on silt surface in SU033-017 with Ancient South Arabian graffiti. Field drawing not to scale by ʿAbdalʿazīz Bin ʿAqīl. Overall block length is approximately 0.75 m. *Illustration by Clara Hickman.*

Figure 17.5. Khuzmum Rockshelter SU045-007, Ancient South Arabian graffiti, and pecked images. Field drawing, not to scale, by ʿAbdalʿazīz Bin ʿAqīl, numbered for reference to text and images. The distance between ibex (far left) and camel (no. 9, center) is 3 m. Five quadrupeds, possibly dogs, at far right of panel. *Illustration by Clara Hickman.*

Figure 17.6. Close-up view of SU045-007, no. 9. Camel and graffiti of different patination pecked into the limestone face. *Photograph by Jennifer Everhart.*

tified quadrupeds, possibly dogs, one perhaps straddling a curvilinear feature (far right). There is a separate pair of ibex, identifiable by their backward-curving horns (far left), and there are camels (no. 9 and also one omitted from figure 17.5 under no. 8). The graffito adjacent to the camel (no. 9) is in undefined script (figure 17.6). It is read from the top to the bottom, with the two last letters having a horizontal stance. If we assume that the text is in ASA script, the reading is: s^2ms^1, and s^2 would have an unusual shape. Another possibility is to read the first glyph as an *r* with reversed ends, which would produce the reading rms^1. Besides, the glyph also resembles the form of the letter *f* in various ANA scripts (such as Thamudic, Hismaic, and Safaitic); a similar glyph has also been found in the Dhofari script.

Khuzmum or Ḫušmum?

To graffito no. 8 in Figure 17.5, ʿAbdalʿazīz Bin ʿAqīl gave an initial reading of the name Khuzmum, with the *–mūm* ending providing a designation of place in ASA languages. The RASA Project therefore adopted this name for the sites around Khuzma as-Shumlya. A revision by Alessia Prioletta reads this graffito in ASA script (read from right to left) as $ḫs^3mm$ (figure 17.7). This graffito is carved in a long-abandoned alphabet on the face of the inselberg in a place that is today still referred to by local bedouin as the Place of Khuzma. And when asked by ethnographer ʿAbdalʿazīz Bin ʿAqīl, local bedouin identified the place-name Khuzma as meaning the nose ring that passes through the septum of a camel, consistent with contemporary Arabic.

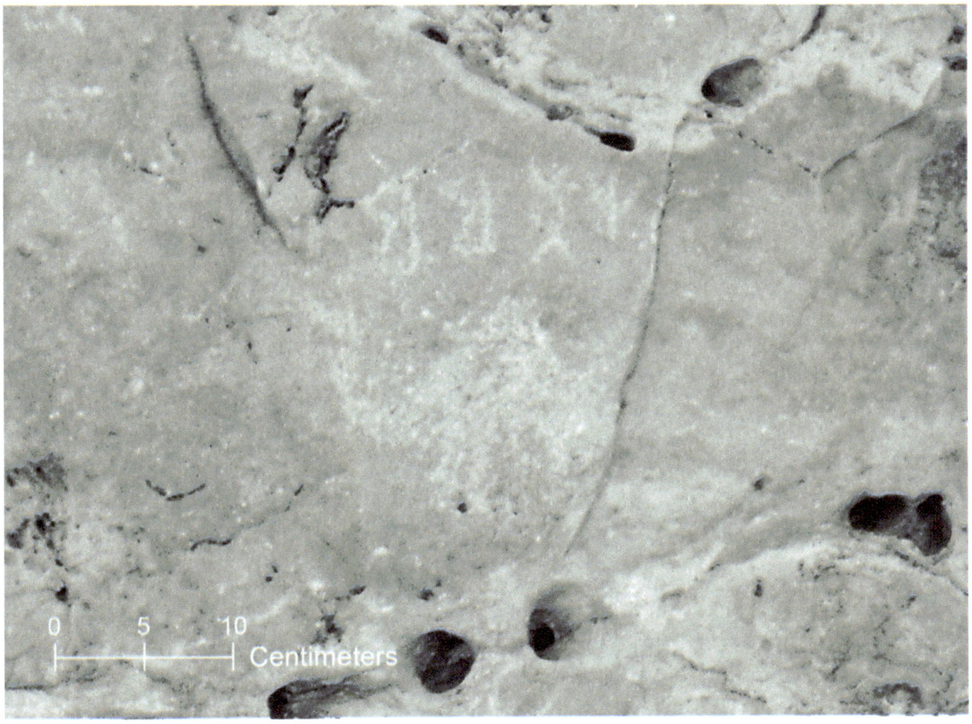

Figure 17.7. ASA text ḥs³mm (close-up of Figure 17.5, no. 8) pecked into the limestone face at SU045-007. *Photograph by Jennifer Everhart.*

Perhaps the meaning was not always this, if there have been phonetic substitutions over time. A transformation of the hard *Ḥ* to a *Kh* can happen among modern Arabic speakers. For example, archaeologists hearing locals pronounce the name of the Ḥabūr River in northern Mesopotamia as *Khabūr* have long substituted *Kh* in its spelling. Even with phonetic and semantic substitutions, "Place of Ḥušma" and "Place of Khuzma" attest to an impressive continuity of geographic and territorial tradition among the inhabitants of Wādī Sanā. Whoever first wrote the name Khuzmum or *ḥs³mm*, its persistence as place-name to this day conveys the persistence of a cultural sense of commemoration as people literally inscribe themselves into landscape.

Panel SU045-5

The bedrock panel SU045-5 is nearby, between the rockshelters SU046-6 and SU045-8. This panel has two evident levels of patination over the pecked graffiti, with the upper left set more heavily patinated. All the lettering appears to be the same style, and there are no images to accompany the names or short texts that appear (figure 17.8). ʿAbdalʿazīz Bin ʿAqīl, who recorded these texts, identified a composition of 12 words and word groups, some of which may plausibly be names. These inscriptions certainly postdate the Neolithic occupations sealed in Early Holocene terraces below rockfall (chapter 9), and they attest to a long use of the Khuzmum rockshelters. They also hint at literacy among rural pastoralists—if these were the persons who inscribed the texts here. Caravanners may also have passed along and camped on this route, but their presence need not exclude the literacy of rural pastoralists, for caravanners may themselves have been of the desert peoples through whose territories they passed. While it is impossible to discern by whom and for whom these graffiti were inscribed, their presence strongly suggests wider literacy and its use in rural contexts beyond the limited purview of an urban, scribal elite. Further discussion of literacy among desert peoples can be found in Macdonald (2005, 2015), who notes the many thousands of alphabetic inscriptions across Arabian deserts.

Panel SU045-5 is a group of graffiti in undefined script. Most of them are read vertically from top to bottom, with the letters having vertical stance. Figure 17.8a, nos. 11 and 12, are read horizontally from right to left. Most of the letters resemble those of the ASA script, except for the glyph also found in the text of figure 17.3 (second letter), which is found in the Dhofari script.

The preliminary readings (right to left) are based on the sketch made by ʿAbdalʿazīz Bin ʿAqīl and must be considered as highly uncertain (See readings top p. 489).

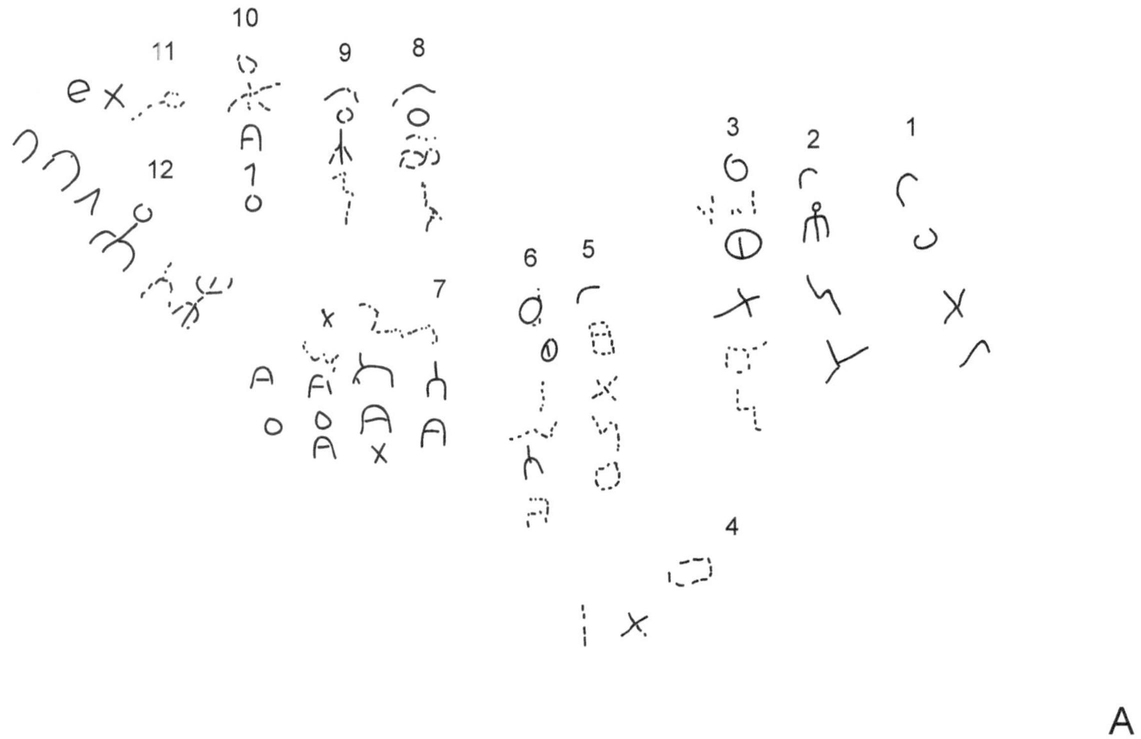

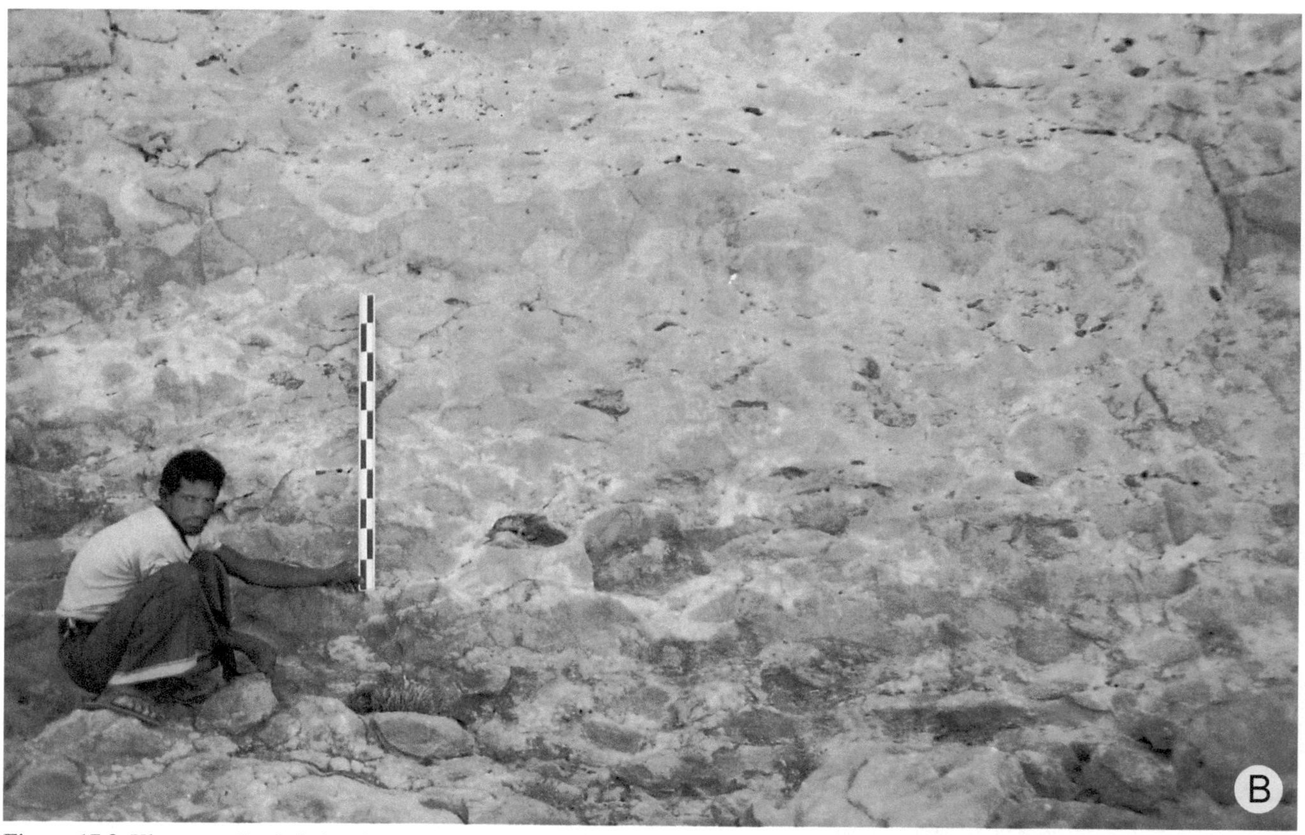

Figure 17.8. Khuzmum Rockshelter SU045-005 ASA texts pecked into the limestone bedrock face: field drawing not to scale by ʿAbdalʿazīz Bin ʿAqīl (a); overview of panel with Nasser Al-ʿAlīy (b). Graffiti blends with natural rock but can be seen in the lower right and over Nasser's head at the top of the meter bar (text number 12 from A above; texts numbers 1, 2, and 3 visible at the upper right). *Illustration by Clara Hickman. Photograph by Joy McCorriston.*

*r*ʿtl

rṣnh

ʿwt(*yn)*

ʿt[.] or *wt[.]*

*rḍtn*ʿ

ʿw[..]s^1.

rs^1*d* (7a)

[.]kdt (7b)

ts^2*d*ʿ*d* (7c)

*d*ʿ (7d)

*r*ʿṯh

rṣn

ʿ[.]dlʿ

htw

.*[..]ṣlbb*

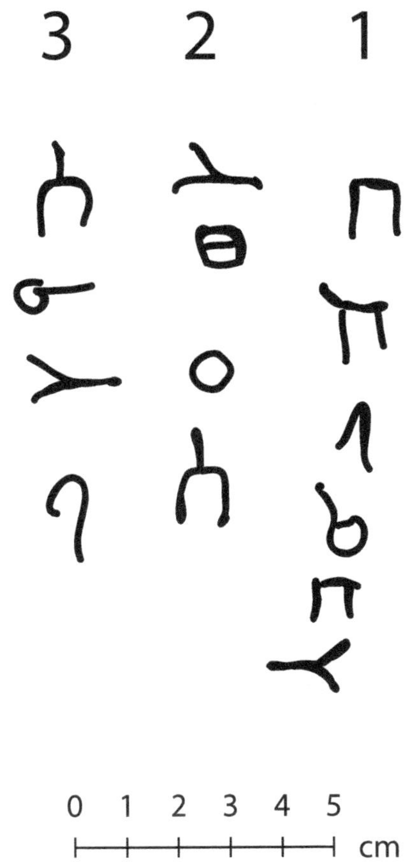

Figure 17.9. Graffiti panel SU056-012 from Khuzmum Rockshelter SU056-013. Field drawing not to scale (a scale inserted here for overall dimensions), by ʿAbdalʿazīz Bin ʿAqīl. *Illustration by Clara Hickman.*

Other Panels

Other panels also occur on bedrock at the Khuzma as-Shumlya. In the northeast corner of the inselberg is a sheltered lookout, about 8 m above the lowest terrace and many times revisited. About 11 m away, on a rock ledge along the high-energy bedrock slope, the team recovered an intact trihedral Neolithic point. The rockshelter (SU056-13) has several small curtain dry-stone walls, constructed to protect a lookout from view. These could also shelter supplies or kid goats, but there are no datable materials aside from the graffiti and the Neolithic point. A bedrock panel (SU056-12), 0.08 x 0.08 m, pecked into the bedrock face under the *abri*, is clearly associated with this rockshelter (figure 17.9). The panel shows three names, perhaps individual and tribal, according to ʿAbdalʿazīz Bin ʿAqīl, who recorded this text. Alessia Prioletta's reading adds that three graffiti are possibly in ASA script. They are read from top to bottom, with some letters having horizontal stance (for example, the *y*'s and the *h*'s), being upside down (*y* of the first text), or having the normal vertical stance. Her preliminary reading according to the ASA script is: *bklybh* (1); *hw*ʿ*s*^{1} (2); *s*^{1}yhl (3).

A second panel, SU056-14, is on the face of a rocky ledge over the top of the *abri*. There are three camels pecked in crude outline into the face. None of these

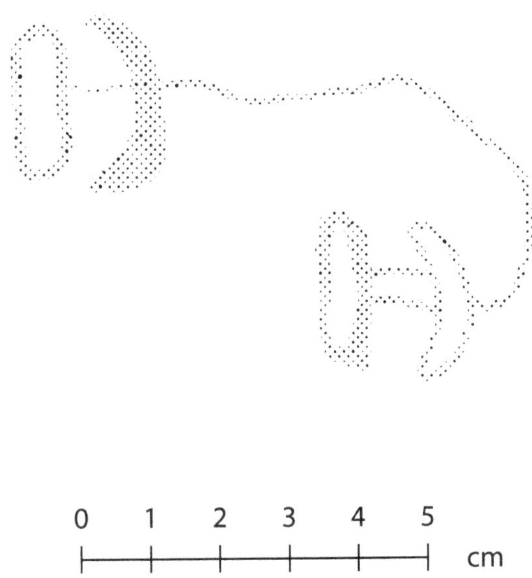

Figure 17.10. Enigmatic images pecked into limestone bedrock near Rockshelter SU056-002. Field drawing not to scale (a scale inserted here for overall dimensions), by Michael Harrower. *Illustration by Clara Hickman.*

camels measures more than 0.10 m long, and they share the ledge with three oblong bedrock mortars. The camel images are patinated and likely date sometime after the introduction of these pack animals around 3,000 years ago. This substantially postdates a Neolithic projectile point, and the mortars remain undated and surely were many times reused.

Finally, there are pecked images in a small 0.08 x 0.10 m panel (rockart site SU056-2), located on a smooth limestone face of a rockshelter at the western corner of the Khuzma as-Shumlya inselberg (figure 17.10). Found low near silt terraces, the images are protected under an overhang. The pecked work shows a pair of unidentified objects linked with or along a curving line. These may be *wusūm*, distinctive markings used as animal brands and found on rock faces that are tribal or group identifiers.

Comparison with Rockart outside Wādī Sanā

Wādī Sāh is a tributary to Wādī ʿIdim where there are also sites with rock art and *wusūm*. *Wusūm* appear in various forms on rocks across Arabia and most likely signal social identity, proprietorship, and territory (Khan 2000). Most *wusūm* decorate immense boulders that themselves form shelter or are near shelters, or the glyphic work appears on rock faces over or beside rockshelters. Wavy lines may represent snakes, water, or abstract *wusūm* (for example, C84-2). Other *wusūm* show concentric circles; variants on a figure eight, including one with three tails; a circle with a tail; a boomerang with a dot; and other oval, triangular, or cross-based designs (for example, C75-2, C75-3, and C75-4). The sites in Wādī Sāh have attracted modern Arabic graffiti in charcoal or pecking and appear very different in content from the Khuzmum sites.

Wādī Sāh also has pecked motifs that include many examples of ibex, identifiable from their backward-curving horns and often in pairs. There are other quadrupeds, sometimes shown upside down, and one image shows a man on a camel, perhaps on a saddle.

Downstream in another, more northern tributary to Wādī ʿIdim is an ancient inscription in Hadramitic dialect of the ASA languages, brought to the attention of RASA archaeologists by Canadian Nexen Petroleum Yemen engineers. This inscription, published by Sergey Frantsouzoff (FR-ʿIdim 1; see Frantsouzoff 2014), records the sale of two lots of dates by one Yaṣduq'il dhū Ilīqašam during a time of warfare in Hadramawt and attests to a broader corpus of literate records across the Southern Jol. This theme is similarly evident in the inscription left by tribesmen in Wādī Sanā at Qārah Ḥabshiyah (Beeston 1962; see chapter 14).

Conclusions

The graffiti at Khuzmum attest, as do such graffiti elsewhere in Arabia, to literacy among mobile people, whether caravanners, bedouin herders, or both. Graffiti also mark a socio-ideological landscape, with irretrievable experience and meaning associated with place. In Wādī Sanā, parietal rockart and graffiti appear at Khuzmum and nowhere else. Canids and ibex are widely employed across Arabia, and while often interpreted as hunting scenes in materialist terms (e.g., Anati 1972, 1974; Khan 1993), they may have other symbolic, perhaps cosmological, significance, widely read and understood (Avner et al. 2017). *Wusūm* engraved on rock were likely ephemeral brands on stock and possibly tattooed on people, and they may inscribe people deeply into landscape places and territories imbued with social and historical meaning (Khan 2000; McCorriston 2011:124–27; van Gennep 1902). Such practices persist ethnographically (like the place-name of Khuzma and the use of camel brands today), and parallel material expressions may suggest their occurrence in the past. As behavior and thought, social and historical meaning can never be conclusively reconstructed from the archaeological record, but the concentrations of graffiti and rockart at the Khuzmum rockshelters and their immediate surroundings suggest that this was a meaningful place indeed, long after specific details of cattle sacrifices and gatherings of the Neolithic era had faded from people's memories.

References

Anati, Emmanuel

1972 *Rock-Art in Central Arabia: Corpus of the Rock Engravings I and II*, vol. 3. Universite Catholique de Louvain, Louvain-la-Neuve, Belgium.

1974 *Rock-Art in Central Arabia: Corpus of the Rock Engravings III and IV*, vol. 4. Universite Catholique de Louvain, Louvain-la-Neuve, Belgium.

Avner, Uzi, Liora Horowitz, and Wayne Horowitz

2017 Symbolism of the Ibex Motif in Negev Rock Art. *Journal of Arid Environments* 143:35–43.

Beeston, Alfred F. L.

1962 Epigraphic and Archaeological Gleanings from South Arabia. *Oriens Antiquus* 1:41–42.

Frantsouzoff, Sergey

2014 Deux inscriptions rupestres du Wādī ʿIdim (Ḥaḍramawt oriental). In *Arabian and Islamic Studies: A Collection of Papers in Honour of Mikhail Borishovich Piotrovskij on the Occasion of His 70th birthday*, edited by Alexander V. Sedov, pp. 92–98. Gosudarstvennyi muzei Vostaka, Moscow.

Khan, Majeed
1993 *The Prehistoric Rock Art of Northern Saudi Arabia: A Synthetic Approach to the Study of the Rock Art from Wadi Damm, Northwest of Tabuk.* Ministry of Education, Kingdom of Saudi Arabia, Riyadh.
2000 *Wusūm: Tribal Symbols of Arabia.* Ministry of Education of Saudi Arabia, Riyadh.

Macdonald, Michael C. A.
2005 Literacy in an Oral Environment. In *Writing and Ancient Near Eastern Society: Papers in Honour of Alan R. Millard*, edited by Piotr Bienkowski, Christopher Mee, and Elizabeth A Slater, pp. 49–118. T&T Clark International, New York.
2015 On the Uses of Writing in Ancient Arabia and the Role of Palaeography in Studying Them. *Arabian Epigraphic Notes* 1:1–50.

McCorriston, Joy
2011 *Pilgrimage and Household in the Ancient Near East.* Cambridge University Press, Cambridge.

Van Gennep, M. A.
1902 Les "Wasm," ou marques de propriété des arabes. *Internationales Archiv für Ethnographie* 15:85–98.

Part VI

Synthesis and Conclusions

Chapter 18

A Bayesian Approach to Chronology of the Southern Jol

Joy McCorriston and Thomas S. Dye

RASA fieldwork gently reminded us that we archaeologists have far to go in educating even a well-educated public about archaeological methods and epistemologies. In 1998 the team struggled in archaeological terra incognita; a pit found in section could be thousands of years old or contain last year's refuse. When we returned to quarters at Canadian Nexen's camp, employees would ask us whether we'd found anything interesting. Our answer was too often, "I'm not sure." Canadian Nexen's generosity was huge, and its staff often asked if we needed anything, to which I once replied, "An accelerator mass spectrometer would be useful." I spoke in jest, but one of the camp administrators reflexively suggested that I give him some diagrams and he'd see if the resident engineers could build us one! Ultimately we have the NSF-Arizona AMS Laboratory to acknowledge for most of our radiocarbon ages and Manfred Frechen for OSL ages on sediments. The University of Minnesota Limnological Research Center prepared some of the targets for AMS radiocarbon age estimates generated by Woods Hole, and Kenneth Cole submitted hyrax midden samples for conventional (gas scintillation) dating by Beta Analytic. Together these methods have yielded a total of 116 age estimates, providing the basis for a chronological model of Wādī Sanā's environmental and cultural history.

On the subject of educating about radiocarbon ages, this chapter embraces a Bayesian approach to chronology, analytically still unfamiliar to some archaeologists and a steep learning curve for the original RASA team. A Bayesian approach builds upon the implications of individual ages by constraining their a priori probabilities within a chronological model; the outcome often narrows temporal parameters of an event or phase of archaeological interest. Using the data entry and model construction framework of Chronomodel 1.5.0 (available at www.chronomodel.fr; see Buck et al. 1999; Lanos et al. 2015), it was possible to use extrinsic information such as a stratigraphic sequence and geomorphological context to constrain the range of age probabilities. Within phases, there are several known sequences, like the stratigraphic ages at Manayzah and Kheshiya. This approach allowed us to refine prior distributions (Bronk-Ramsey 1995). Many of the radiometric assay results and OSL ages discussed here appear in earlier chapters, which provide context, descriptions, and evaluations of the dated materials as they clarify sites and sequences. Beyond their immediate contextual and individual applications, presented together these geochronological results offer a new and exceptionally large dataset, useful in a regional perspective beyond Wādī Sanā and potentially beyond Yemen. In this interest, the RASA ages presented here deserve a dedicated chapter analysis (table 18.1).

Table 18.1a. Radiocarbon samples from the RASA Project. Because analyzing laboratory numbers are unknown and because analysis focuses on Wādī Sanā, this table excludes samples from Rawk in Wādī ʿIdim. Rows in light grey are nonanthropogenic samples taken to date paleoenvironmental processes. Asterisks denote charcoal analysis done on samples from the same archaeological context; some such samples yielded few identifiable fragments and have no further reporting. All ages calibrated using Oxcal 4 online program (with the Libby radiocarbon half-life).

Map ID	Lab #	Material	Charcoal Analysis*	AMS Dated	Off-site ID	Site (Grid)	Quad	Locus	Lot	Bag
1	AA79766	charcoal		x	Wādī Ḥarū	C30-near27a				3
2	AA83495	bone, possibly human, collagen		x	Wādī Ḥarū	C30-4		Level 1		2
3	AA79765	charcoal		x	Wādī Ḥarū	C30-27		Layer 2		8
4	AA79764	charcoal		x	Wādī Ḥarū	C30-27		Layer 1		1
5	AA60247	monocotyledonous charcoal fragment		x	Wādī Sanā water structures	W5-1	T1	006		
6	A11778	*Procavia* pellets			Wādī Sanā 6A	hyrax midden				
7	OS16947	charcoal; *Ziziphus* sp.	x	x	Wādī as-Shumlyah–Wādī Sanā confluence	profile		H13		
8	A11779	*Procavia* pellets			Wādī Sanā 7A	hyrax midden				
9	AA83496	enamel, bioapatite		x	Wādī Ḥarū	C30-4		Layer 6		9
10	AA83500	human bone, bioapatite		x	Wādī Ḥarāḍ, near BAK 51	C32-2		Layer 1		16
11	AA79768	charcoal		x	Wādī Ḥarū	C58-1		Hearth 5		1
12	AA79762	desiccated camel dung		x	Wādī Ḥarū	C30-4		Layer 5		10

Northing	Easting	Comments on Context (quotes ["..."] refer to excavation notes)	^{14}C Years BP	Cal BP (OxCal 4)	Phase	Theta	Posterior Distribution Mean (cal BP)	Credibility Interval Start (cal BP)	Credibility Interval End (cal BP)
1722741	315401	Smaller *madhbaḥ* adjacent to large excavated C30-27. Warning: Date may extend out of range.	74 +/- 33	262–24	MDH	92	175	394	-65
1723466	315880	HCT, bone fragments, possibly human, from upper level of tomb. Warning: Date may extend out of range.	158 +/- 38	287– . . .	HCT	87	216	449	-63
1722741	315401	Lower gravel fill of *madhbaḥ* structure, lower terrace of Wadi Harou. Warning: Date may extend out of range.	158 +/- 33	286– . . .	MDH	90	298	613	36
1722741	315401	Upper gravel fill of *madhbaḥ* structure, lower terrace of Wadi Harou. Warning: Date may extend out of range.	178 +/- 33	298– . . .	MDH	91	142	303	-56
1743471	333882	Layer C, a burned layer above the constructed water dam; postdates structure. Warning: Date may extend out of range.	293 +/- 39	471–158	OTH	96	405	657	145
1745497	336974	No anthropogenic remains.	415 +/- 40		HYX	67	487	725	233
1745745	336705	Wādī as-Shumlyah Hearth 1998 13; on rock terrace, hearth probably intruded and was in a gully?; Charcoal in vial split for ^{14}C. Date seems incompatible with stratigraphic position. A resample is desirable.	680 +/- 35	684–559	OTH	93	647	816	464
1745497	336974	No anthropogenic remains.	690 +/- 45		HYX	66	662	843	476
1723466	315880	HCT, camel skull fragments from camel burial in presumed reuse of tomb. Same animal, same date as AA79762, AA 86370. (Note 300-year difference, explained by J.-F. Saliège as probable diagenesis of dung.)	1559 +/- 38	1538–1369	HCT	86	1486	1683	1306
1700132	282693	HCT, human remains (adult, poorly preserved). Bag 16 was accidentally substituted for Bag 15 intended for dating. Not certain if this dates earliest inhumation in HCT because the deposit was disturbed. May include animal bone scraps (possibly not contemporary with burial?).	1733 +/- 39	1730–1549	HCT	85	1682	1902	1470
1720468	309298	Hearth from in front of trilith/ asafy.	1749 +/- 35	1774–1559	TRI	77	1694	1909	1489
1723466	315880	Dung from pelvic/belly area of camel sacrifice. Should be same date (same animal!) as AA83496, AA86370 (note 300-year difference, explained by J.F. Saliège as probable diagenesis of dung.)	1868 +/- 35	1881–1717	excluded				

Table 18.1. *(continued)*

Map ID	Lab #	Material	Charcoal Analysis*	AMS Dated	Off-site ID	Site (Grid)	Quad	Locus	Lot	Bag
13	A11776	*Procavia* pellets			Wādī Sanā 8	hyrax midden				
14	AA79769	charcoal		x	Wādī Sanā	SU134-3		Hearth 11		2
15	AA39070	linen fabric		x	Wādī Sanā 8	hyrax midden				
16	AA79767	charcoal		x	Wādī Sissib (south of Wādī Bin ʿAlī)	C51-1		Layer 2		27
17	AA86370	camel bone		x	Wādī Ḥarū	C30-4		Layer 6		9
18	AA81817	mixed fragments charcoal		x	Wādī ʿAtuf/ Wādī ʿIdim	C15-3		Levels 1–2		5
19	AA81818	multiple fragments charcoal		x	Wādī ʿAtuf/ Wādī ʿIdim	C15-3		Level 2		6
20	AA83498	bone, bioapatite		x	Wādī ʿAtuf/ Wādī ʿIdim	C15-3		Layer 1		16
21	AA90838	human teeth		x	Wādī Ḥarū	C30-3		Layer 1		2 &3
22	OS16724	wood charcoal; *Ziziphus* sp.		x	Wādī ʿIdim	Munayder I	A	005		
23	AA83499	faunal bone, bioapatite		x	Wādī Sissib (south of Wādī Bin ʿAlī)	C51-1		Layer 2		25

Northing	Easting	Comments on Context (quotes ["..."] refer to excavation notes)	^{14}C Years BP	Cal BP (OxCal 4)	Phase	Theta	Posterior Distribution Mean (cal BP)	Credibility Interval Start (cal BP)	Credibility Interval End (cal BP)
1745620	336888	No anthropogenic remains.	1975 +/- 50		HYX	65	1991	2246	1741
1744580	334091	Hearth from in front of trilith/ athafy.	2026 +/- 35	2106–1894	TRI	76	2031	2235	1811
1745620	336885	Found with three amber beads; hyrax midden incorporates burial or cached garments.	2159 +/- 45	2310–2009	OTH	97	2205	2544	1812
1727583	244993	From upper deposit sealed in chamber of "tumulus--stela cairn" excavated in Wādī Sissib. Should be about the same date as AA83499 (on faunal bone in same level). (Note gap in dates.)	2216 +/- 55	2347–2069	TUM	89	2151	2382	1907
1723466	315880	HCT, camel bone from camel burial. Same animal as AA79762 and AA 83496 (but 300- to 1,200-year range in dates due to diagenesis on dung and collagen, according to J.F. Saliège).	2407 +/- 40	2699–2346	excluded				
1715980	275484	HCT, charcoal from north part of chamber; from an "intermediate level" between burials and upper fill. Probably related to activity prior to or during fallen rock episode and later than earliest burials. Source of charcoal is unknown but sufficiently abundant to be attributed to a local fire in or on the tomb.	2474 +/- 38	2720–2379	HCT	82	2573	2889	2260
1715980	275484	HCT, charcoal from under AA81817 and under a layer of "fallen rock containing many pieces of charcoal." Probably related to activity postdating earliest burials. Source of charcoal unknown but sufficiently abundant to be attributed to a local fire in or on the tomb.	2489 +/- 39	2737–2381	HCT	81	2585	2881	2280
1715980	275484	HCT, subadult remains separated from adult remains. Stratigraphically this date on burial should be earlier than AA81817, AA81818 if the charcoal was generated in situ. (Note 300-year difference.)	2776 +/- 54	3001–2761	HCT	80	2945	3165	2712
1723462	315890	HCT, upper find of well-preserved infant teeth in upper aeolian fill.	3067 +/- 39	3367–3175	HCT	84	3303	3530	3078
1738582	275500	Munayder I; sample taken from central hearth in structure.	3190 +/- 30	3463–3361	OTH	94	461	3566	581
1727583	244993	Tumulus associated with unfinished stela-*menhir.* Age is on animal bone placed in tomb after original burial and before an event that deposited TAR and closed the chamber. Should be same age or earlier that AA79767. (Note gap in ages, but AA83499 is the more sure context, as bone was placed in the tomb.)	3663 +/- 41	4142–3877	TUM	88	301	3784	3419

Table 18.1. *(continued)*

Map ID	Lab #	Material	Charcoal Analysis*	AMS Dated	Off-site ID	Site (Grid)	Quad	Locus	Lot	Bag
24	AA83497	bone, bioapatite		x	Wādī ʿAtuf/ Wādī ʿIdim	C17-2				1
25	AA83493	human bone, bioapatite		x	Wādī ʿAtuf/ Wādī ʿIdim	C15-3		Level 1		17
26	A11052	*Procavia* pellets			Wādī Sanā 2000-24A	hyrax midden				
27	AA83494	human bone fragments, bioapatite		x	Wādī ʿAtuf/ Wādī ʿIdim	C15-2				1
28	AA79763	charcoal fragments		x	Wādī Sanā	SU006-1		Hearth 2		2
29	Beta-208493	charcoal		x	05-WIY-03D	profile				
30	A11780	*Procavia* pellets			Wādī Sanā 14	hyrax midden				
31	AA60245	charcoal		x	Wādī Sanā water structures	009-1	T3	1		7
32	AA59569	dicotyledonous hardwood charcoal fragment		x	Wādī Sanā water structures	009-1	T1	2		
34	AA39071	*Procavia* pellets		x	Wādī Sanā 17A	hyrax midden 2000-17A				
35	Beta-208492	charcoal		x	05-WIY-01-915	profile				
36	AA90336	marine shell beads		x	Wādī Ḥarū	C30-3		Layer 2		5-x
37	AA61078	charcoal		x	04-WS-4(1)	profile				
38	A11777	*Procavia* pellets			Wādī Sanā 4A	hyrax midden				
39	GX24614	*Procavia* pellets			Wādī Sanā 1	hyrax midden				
40	AA38421	*Procavia* pellet		x	Wādī Sanā-23A2	hyrax midden				

Northing	Easting	Comments on Context (quotes ["..."] refer to excavation notes)	^{14}C Years BP	Cal BP (OxCal 4)	Phase	Theta	Posterior Distribution Mean (cal BP)	Credibility Interval Start (cal BP)	Credibility Interval End (cal BP)
1716845	277004	Wall tomb. Contained disturbed bone fragments and turbo shell jewelry.	3686 +/- 41	4148–3902	OTH	95	4062	4333	3788
1715980	275484	HCT; human remains most likely from single inhumation event at time of construction of tomb. Should be earlier than AA81817, AA81818.	3912 +/- 39	4499–4185	HCT	79	4370	4625	4129
1755583	335147	No anthropogenic remains.	4230 +/- 50		HYX	64	4801	5136	4453
1716033	275445	HCT; assumed earlier and partially destroyed by building C15-3 (nearby). Extremely small amount of bone recovered.	4288 +/- 47	5030–4659	HCT	78	4937	5196	4673
1744350	337219	Hearth from in front of trilith/athafy (excavated by ʿAbdalKarīm Al-Burkānī; sample was removed at depth too deep for hearth and must be underlying charcoal in silts).	4294 +/- 38	4967–4825	4	50	4925	5117	4730
1735536	273982	Charcoal found in sand lens, 2.5 m below surface of tufa exposure; southernmost extend of tufa mounds in Wādī ʻIdim.	4400 +/-70	5285–4848	excluded				
1745801	337422	No anthropogenic remains.	4425 +/- 115		HYX	63	5166	5851	4448
1744589	337163	Rock-lined canal buried in silt terrace, Wādī as-Shumlyah; sample from sediments overlying feature.	4471 +/- 42	5297–4970	4	46	5185	5603	4740
1744567	337175	Rock-lined canal buried in silt terrace, Wādī as-Shumlyah; sample is isolated piece from sediments, 12 cm overlying top stone of feature.	4475 +/- 36	5292–4976	4	45	5175	5569	4771
1745545	335209	No anthropogenic remains	4490 +/- 44		HYX	61	-1301	5587	4789
1739386	275484	Charcoal in hearth in wadi sand/silt with interbedded marl, 9.15 m above wadi cobbles, Wādī ʻIdim.	4520 +/- 40	5310–5046	excluded				
1723462	315890	HCT; marine shell beads from in situ jewelry on human skeletal elements too fragile to collect. This sample dates the earliest interment in the tomb.	4525 +/- 66	5447–4963	HCT	83	4504	4900	4091
1735715	328876	Wādī Sanā; uppermost wadi silt in PSI infilling (Wādī Sanā Cave III).	4545 +/- 45	5435–5046	5	53	5229	5609	4888
1746137	337722	No anthropogenic remains.	4555 +/- 60		HYX	60	5264	5795	4759
1742017	331966	No anthropogenic remains.	4590 +/- 90		HYX	59	5335	5970	4625
1745583	335147	No anthropogenic remains.	4602 +/- 45		HYX	58	5369	5877	4879

Table 18.1. *(continued)*

Map ID	Lab #	Material	Charcoal Analysis*	AMS Dated	Off-site ID	Site (Grid)	Quad	Locus	Lot	Bag
41	OS16958	charcoal: *Balanites* or *Tamarix*	x	x	Cave Site-1, 0.30–0.40	cave sediment				
42	AA59756	charcoal		x	04WS-17(1)	profile				
43	AA59757	charcoal		x	04WS-17(4)	profile				
44	OS18691	*Acacia* charcoal	x	x	Cave Site-2	cave sediment				
45	AA60239	charcoal		x	Wādī Sanā survey	SU 80-1		H1		1
46	AA60240	charcoal	x	x	Wādī Sanā	110-6			1	
47	AA60243	charcoal		x	Wādī Sanā	110-4	A	005	1	
48	AA38420	*Procavia* pellets		x	Wādī Sanā 11	hyrax midden				
49	AA38384	wood charcoal fragments	x	x	00-WI-10	Wādī ʿIdim tufa				
50	AA38385	wood charcoal fragments	x	x	00-WI-12	Wādī ʿIdim tufa				
51	Beta-208494	charcoal		x	05-WSX-16A2	profile				
52	AA59763	charcoal		x	04WS-7(0.7)	profile				
53	AA59761	charcoal		x	04-WS-6	profile				
54	AA38380	wood charcoal	x	x	P2000-8A-0.25	profile				
55	AA66861	four cf. *Acacia* sp. charcoal fragments	x	x	Wādī Sanā	151-1	E	003	7	3
56	AA60241	charcoal		x	Wādī Sanā survey	SU 80-1		H2		1
57	AA81816	multiple fragments charcoal		x	Wādī Sanā	C67-2		Hearth 4		8

Northing	Easting	Comments on Context (quotes ["..."] refer to excavation notes)	^{14}C Years BP	Cal BP (OxCal 4)	Phase	Theta	Posterior Distribution Mean (cal BP)	Credibility Interval Start (cal BP)	Credibility Interval End (cal BP)
1742313	331838	Wādī Sanā Cave I; 0.30–0.40 below top of silt; LRC 1090.	4610 +/- 45	5568–5067	EG5	51	5364	5707	4977
1740346	328512	Wādī Sanā; PSI wadi silts, 20 cm below surface; youngest silt deposition.	4633 +/- 40	5569–5296	EG5	55	5372	5574	5134
1740346	328512	Wādī Sanā; PSI wadi silts, 90 cm below surface; youngest silt deposition.	4721 +/- 56	5585–5321	EG5	54	5532	5736	5354
1742607	331746	Wādī Sanā Cave II; layer 0.25 m below top of silt; LRC 1093.	4800 +/- 60	5651–5327	EG5	52	5550	5803	5295
1743766	333795	One of three hearths dating terrace silt accumulations and burned surface in SU 80.	5081 +/- 70	5985–5656	EG3	28	5902	6154	5634
1739231	328531	From hearth in silt terrace section abutting SU110 rockshelter.	5161 +/- 43	5997–5754	EG3	32	5969	6206	5724
1739219	328531	From 0.99 m below ground surface (ground surface at datum A); 2.37m E of datum A; less than 1mm thick layer	5182 +/- 58	6177–5751	EG3	31	6020	6337	5704
1745896	336977	No anthropogenic remains.	5236 +/- 55		HYX	57	6064	6420	5696
1736705	274128	Wādī ʿIdim; embedded in tufas from southern extent of fossil spring deposits (upstream).	5280 +/- 52	6191–5928	excluded				
1735785	274020	Wādī ʿIdim; embedded in tufas from southern extent of fossil spring deposits (upstream); = 05-WI-05.	5288 +/- 52	6202–5932	excluded				
1744569	334347	Charcoal in channel fill clay (paleochannel) 50 cm below surface; Wādī Sanā near cattle skulls.	5320 +/- 50	6271–5948	EG4	47	6131	6457	5818
1744605	337149	Wādī as-Shumlyah; uppermost burned horizon in 98-WS3 profile; 0.7 m below ground surface.	5329 +/- 42	6269–5992	EG3	20	6138	6413	5879
1744471	334264	Top edge of paleochannel filling.	5402 +/- 42	6294–6020	EG4	48	6237	6514	5945
1745663	336469	Khuzma as-Shumlyah, in wadi silts close to top of profile, 0.25 m below ground surface; dates the end of sedimentation/level to which modern surface has eroded.	5485 +/- 64	6413–6121	EG3	11	6327	6613	6030
1744647	334319	Upper ashy layer formed after (postdates) cattle skull ring and predating a pit cut to set uprights for platformed structure	5514 +/- 48	6406–6213	EG3	24	6346	6539	6143
1743766	333795	One of three hearths dating terrace silt accumulations and burned surface in SU 80.	5545 +/- 43	6410–6280	EG3	29	6371	6531	6209
1749752	340514	Hearth stratigraphically same level and assumed contemporary with dolmen-like structure, earlier than AA81814.	5603 +/- 67	6534–6283	EG3	17	6541	6805	6305

Table 18.1. *(continued)*

Map ID	Lab #	Material	Charcoal Analysis*	AMS Dated	Off-site ID	Site (Grid)	Quad	Locus	Lot	Bag
58	AA38547	charcoal		x	Wādī Sanā	037-3	A	006		018
59	AA60251	charcoal		x	Wādī Sanā	W23	1	H1		1
60	AA66862	charcoal	x	x	Wādī Sanā	037-3	B	006	2	
61	AA81814	bone and charred bone		x	Wādī Sanā	C67-2		Hearth 1		2
62	AA81815	multiple fragments charcoal		x	Wādī Sanā	C67-2		Hearth 3		5
63	OS16934	wood charcoal		x	Wadi Section 4	profile		H10		
65	AA61077	charcoal		x	04-WS-18	profile				
66	AA60250	charcoal		x	Wādī Sanā water structures	W13	1	H1		1
67	Beta-208495	charcoal		x	05-WSX-19W	profile				
68	AA38544	wood charcoal fragments		x	Wādī Sanā	037-3	A	009		
69	AA59760	charcoal		x	04WS-7(3.6)	profile				
70	OS16933	wood charcoal (*Calotrpis procera*)	x	x	98WS-1, Wadi Section 1	profile				
71	OS16689	wood charcoal	x	x	98WS-2, Wadi Section 2	profile				
72	AA60249	charcoal		x	Wādī Sanā water structures	W6-1		H2		1
73	AA60244	charcoal		X	Wādī Sanā survey	SU 80-1		H3		1
74	AA59762	charcoal		x	04-WS-8	profile				

Northing	Easting	Comments on Context (quotes ["..."] refer to excavation notes)	^{14}C Years BP	Cal BP (OxCal 4)	Phase	Theta	Posterior Distribution Mean (cal BP)	Credibility Interval Start (cal BP)	Credibility Interval End (cal BP)
1745143	335483	Hearth area on uppermost (latest use) occupational surface inside platformed structure	5616 +/- 84	6631–6280	EG3	26	6444	6691	6167
1747736	337933	Hearth near *shrūj*; assumed to predate structure.	5637 +/- 44	6499–6308	EG3	33	6453	6683	6229
1745138	335483	In FCR/ash layer in silt terrace beside Khuzmum platform structure.	5682 +/- 47	6631–6324	EG3	27	6489	6685	6295
1749755	340512	Hearth stratigraphically postdating destruction of a dolmen-like structure; fragments of structure smashed and burned in hearth.	5702 +/- 51	6640–6355	EG3	18	6396	6602	6094
1749753	340514	Hearth stratigraphically same level and assumed contemporary with dolmen-like structure, earlier than AA81814.	5709 +/- 45	6633–6407	EG3	16	6581	6797	6390
1744738	336889	Wādī as-Shumlyah Hearth 1998-10; LRC 1063; no further charcoal available; middle of wadi silt (2 m above gravel terrace). Associated with bones, including animal jaw, and angular stone blocks.	5750 +/- 45	6659–6443	EG3	34	6587	6848	6343
1735240	333387	Wādī Sanā tributary; charcoal in hearth in middle of wadi silt deposition.	5765 +/- 45	6665–6452	EG3	40	6602	6829	6338
1743925	333798	Hearth near check dam; its assumed this ^{14}C sample predates dam; context uncertain due to loose soil.	5783 +/- 44	6677–6467	EG3	41	6618	6867	6389
1744629	334393	Charcoal from hearth adjacent to paleochannel of Wādī Sanā; hearth near cattle skulls, about 0.20 m below ground surface.	5800 +/- 50	6731–6488	EG3	42	6644	6900	6367
1745143	335483	Khuzmum platform structure; lowest occupation debris in surfaces within platform.	5806 +/- 64	6747–6449	EG3	25	6709	6955	6498
1744605	337149	Wādī as-Shumlyah; lower exposed silt in 98-WS3 profile; 3.6 m below ground surface.	5842 +/- 43	6772–6508	EG3	19	6694	6921	6446
1744597	337220	Wādī as-Shumlyah Hearth 1998-2, LRC#1062; a bell-shaped pit with hearth in bottom, 3.9 m below ground surface, 1.1 m above gravel terrace.	5870 +/- 45	6792–6561	EG3	43	6730	6976	6488
1744579	337174	Wādī as-Shumlyah burned horizon, about 2 m below ground surface, 1998 WS-2; +45 cm over flat stone 0.0.	5880 +/- 55	6851–6547	EG3	21	6746	7029	6436
1740674	329536	Hearth buried in sediments; assumed to predate nearby check dam.	5923 +/- 44	6880–6657	EG3	44	6788	7019	6547
1743786	333811	One of three hearths dating terrace silt accumulations and burned surface in SU 80.	5953 +/- 45	6892–6672	EG3	30	6826	7083	6572
1744554	334330	Basal infill of channel.	5970 +/- 72	6995–6645	EG4	49	6878	7268	6486

Table 18.1. *(continued)*

Map ID	Lab #	Material	Charcoal Analysis*	AMS Dated	Off-site ID	Site (Grid)	Quad	Locus	Lot	Bag
75	Beta-208491	charcoal		x	05-WIY-01-180	profile				
76	AA69755	charcoal		x	Wādī Sanā	151-1	B/E	010	41	
77	AA69754	charcoal		x	Wādī Sanā	151-1		H2		3
78	OS16950	wood charcoal; *Acacia* sp.	x	x	Wādī as-Shumlyah-Wādī Sanā confluence	profile		H14		
79	OS16935	wood charcoal, *Zygophyllum* sp.	x	x	98WS-6 Pieter Vlag's Wadi Section 6	profile		H16		
80	AA59571	sediment/charred		x	Wādī Sanā	151-1	H11	001		
81	AA60246	charcoal		x	Wādī Sanā water structures	W1-4		H1		1
82	AA60248	charcoal		x	Wādī Sanā water structures	W5-3	A	005		
83	AA38381	wood charcoal fragments	x	x	P2000-8A-1.95	profile				
84	AA38546	charcoal		x	Wādī Sanā	044-25		001		
85	AA59764	charcoal		x	04WS-3(a)	profile				
86	AA38383	wood charcoal fragments	x	x	00-WI-9	Wādī ʿIdim tufa				
87	AA38382	wood charcoal fragments	x	x	00-WI-5	Wādī ʿIdim tufa				

Northing	Easting	Comments on Context (quotes ["..."] refer to excavation notes)	^{14}C Years BP	Cal BP (OxCal 4)	Phase	Theta	Posterior Distribution Mean (cal BP)	Credibility Interval Start (cal BP)	Credibility Interval End (cal BP)
1739386	275484	Charcoal in cross-bedded sand and gravel, 1.8 m above wadi cobbles, Wādī ʿIdim.	5970 +/- 50	6930–6676	excluded				
1744647	334319	Kheshiya cattle skull ring; lower ashy layer formed before (predates) cattle skull ring and platformed structure.	6010 +/- 69	7149–6671	EG3	23	6859	7094	6586
1744647	334319	Kheshiya Hearth 2, collected by Catherine Heyne in 2004; lower and stratigraphically earlier than cattle skull ring and platformed structure.	6069 +/- 48	7156–6787	EG3	22	7055	7340	6808
1745745	336705	Wādī as-Shumlyah Hearth 14; 1998-14 in silt section northeast of Khuzma as-Shumlya, about 0.70 m below Hearth 1998-13, apparently in situ.	6070 +/- 40	7151–6793	EG3	35	6971	7265	6670
1744604	336383	Wādī as-Shumlyah Hearth 1998-16; LRC#1082; no waypoint, 1.1 m above gravel terrace, about 4 m below ground surface.	6080 +/- 55	7156–6795	EG3	36	7006	7416	6633
1744647	334319	Hearth high in silt terrace on opposite face of gully at Kheshiya platform.	6097 +/- 39	7157–6859	EG3	37	7026	7351	6690
1742559	333515	Hearth in area in front of dam; relationship unclear; assumed hearth postdates dam, as it would have been washed away earlier; or this may be an earlier hearth sedimented before the dam was built; "dam" may also be the eroded remains of a canal built into silts that contained an earlier hearth	6168 +/- 51	7241–6936	EG3	38	7111	7427	6798
1743486	333749	Sample from Hearth 1 in sediments; assumed to predate water dam.	6232 +/- 45	7259–7006	EG3	39	7186	7510	6868
1745663	336469	Khuzma-as-Shumlya, in wadi silts; 1.95 m below top of section (modern ground surface).	6246 +/- 58	7287–6991	EG3	10	7185	7499	6831
1745281	336251	Khuzmum rockshelters; hearth, in upper ashy layer SU 44-29, about 1 m below ground surface. No Neolithic tools or clearly Neolithic debitage from ashy layer or above; stratigraphically later than the Neolithic bifacial industry in lower 044-028 ashy layer.	6352 +/- 57	7418–7173	EG3	15	7320	7610	7037
1740964	329773	Wādī Sanā; charcoal from hearth near base of wadi silts (onset of silt deposit).	6387 +/- 61	7428–7177	EG3	13	7349	7598	7066
1737632	274694	Wādī ʿIdim, embedded in tufas from northern extent of fossil spring deposits (downstream).	6586 +/- 56	7573–7424	excluded				
1739630	275497	Wādī ʿIdim, embedded in tufas from northern extent of fossil spring deposits (downstream).	6859 +/- 57	7825–7590	excluded				

Table 18.1. *(continued)*

Map ID	Lab #	Material	Charcoal Analysis*	AMS Dated	Off-site ID	Site (Grid)	Quad	Locus	Lot	Bag
88	AA 59570	charcoal fragments, all *Ziziphus leucoderma*	x	x	Wādī Sanā	155-2; K9 north half, Hearth 1		Layer 9		
89	AA66684	cf. *Tamarix* sp. charcoal piece	x	x	Wādī Sanā	155-2; I14	C	009	10	
90	AA66683	*Acacia hamulosa/ mellifera* charcoal piece	x	x	Wādī Sanā	155-2; L9	A	010	15	
91	AA66685	10 charcoal fragments, at least five *Ziziphus* sp.	x	x	Wādī Sanā	155-2; K9		Layer 17 upper		
92	AA38543	charcoal		x	Wādī Sanā	045-1	A	004		
93	AA38545	wood charcoal fragments		x	Wādī Sanā	044-20		002		
94	AA38548	wood charcoal fragments		x	Wādī Sanā	045-1	A	009		30
95	AA66686	three *Acacia* sp. charcoal fragments	x	x	Wādī Sanā	155-2; K9		Layer 20		
96	AA59765	charcoal		x	04WS-10(4b)	profile				
97	Beta-208490	shell		x	05-GBY02-1.0	profile				
98	AA59768	shell		x	04WS-3(b)	profile				

Northing	Easting	Comments on Context (quotes ["..."] refer to excavation notes)	^{14}C Years BP	Cal BP (OxCal 4)	Phase	Theta	Posterior Distribution Mean (cal BP)	Credibility Interval Start (cal BP)	Credibility Interval End (cal BP)
1727877	324343	Manayzah; hearth in stratigraphic context in Test Pit K; first indirect relative date on lithics; stratigraphically younger than AA66685, AA66686.	6902 +/- 41	7834–7665	MZH	70	7856	7991	7727
1727877	324343	Manayzah; arbitrary layer 20 cm deep in compact ashy silts with charcoal, bone, and lithic inclusions. Younger than AA57570.	6981 +/- 51	7932–7679	MZH	72	7689	7901	7370
1727877	324343	Manayzah; grey, compact sediment overlying good surface. Contains ash, charcoal, and burned stone. Younger than AA59570. Provides indirect date on use of fluting technique.	6987 +/- 57	7935–7695	MZH	71	7801	7940	7669
1727877	324343	Manayzah; brown-grey ashy sediment layer with dense charcoal, containing significant new lithic technology—arrowhead "concorde plane"; stratigraphically younger than AA6686; stratigraphically older than AA59570.	7133 +/- 51	8035–7845	MZH	69	8012	8182	7855
1745417	336258	Khuzmum rockshelters; ashy layer with Neolithic bifacial points and tools; stratigraphically over AA38548.	7403 +/- 70	8370–8047	KZM	74	8241	8522	7943
1745259	336246	Khuzmum rockshelters; hearth in lower ashy layer, about 2–3 m below ground surface; includes Neolithic bifacial points; stratigraphically under and earlier than AA38546.	7432 +/- 60	8384–8066	KZM	75	8290	8571	8014
1745417	336258	Khuzmum rockshelters excavation; lower ashy layer with multiple hearths; stratigraphically under AA38543.	7723 +/- 87	8723–8365	KZM	73	8624	8971	8329
1727877	324343	Manayzah; lowest layer (yet excavated) in which occurs datable charcoal; lithics below this layer differ in patination; this sample stratigraphically under AA66685, AA57570	8072 +/- 79	9254–8653	MZH	68	9057	9593	8466
1749987	341256	Wādī Sanā; single large charcoal fragment from lower third of wadi silts (1.4 m above base of section; 4 m below top).	9252+/-52	10565–10268	EG3	14	10356	10817	9774
1722793	324905	Aquatic gastropod shell, 1.0 m below surface in tufa-capped marl, about 500 m north of Ghayl bin Yumain.	10220 +/- 40		excluded				
1740964	329773	Wādī Sanā; shells from base of wadi silts (onset of silt deposit).	10254 +/- 55		excluded				

Table 18.18b. OSL samples from the RASA Project

Map ID	Field #	Material	Lab #	Northing	Easting	Environment (Wādī Sanā)
99	04WS-3c	sandy silt	658	1740964	329773	sand in channel gravel within lower wadi silt; ~ 20cm below 04-WS3 hearth (^{14}C)
100	04WS-11	sand lens	659	1745675	336175	sand lens in gravel terrace, 1m below contact with overlying wadi silt
101	04WS-13	sand lens	660	1750174	341411	sand lens in gravel terrace, 1.5m below contact w/ wadi silt; ~ 3m below 04WS10-4 (^{14}C)
102	04WS-14	sand lens	661	1753665	343185	sand lens in gravel terrace, 1.0m below contact with overlying wadi silt
103	04WS-15	sand lens	662	1762582	341802	sand lens in gravel terrace, 0.5m below contact with overlying wadi silt; mouth of Wadi Himeri
104	04WS-19	sand lens	663	1742299	333712	sand lens in gravel terrace, 1.0m below contact with overlying wadi silt; mouth of small tributary to W. Sana
105	04WS-22-1	sandy silt	664	1745675	336183	P2000-8 basal wadi silt
106	04WS-22-2	sand lens	665	1745675	336183	P2000-8 sand at transition between overlying wadi silt and underlying gravel terrace
107	04WS-22-3	sand lens	666	1745675	336783	P2000-8 sand lens in upper gravel terrace underlying wadi silt
108	04WS-23	sand lens	667	1740909	329736	sand lens ~ 0.5m below wadi silt/gravel terrace contact

IRSL Age	Phase	theta θ	Posterior Distrbution Mean (cal BP)	Credibility Interval Start (cal BP)	Credibility Interval End (cal BP)
7,040 ± 1,100	EG3	12	8454	10220	7261
10,600 ± 1,500	EG1	4	12594	17089	9913
22,400 ± 3,300	EG1	3	21294	27944	14047
14,000 ± 2,700	EG1	2	15371	22110	10314
11,700 ± 1,600	EG1	1	13135	17736	10153
6,500 ± 1,400	EG2	8	6622	11640	1492
7,140 ± 1,220	EG3	9	8434	10293	7094
9,230 ± 970	EG1	7	11147	12973	9890
11,200 ± 1,400	EG1	5	13230	17649	10359
9,160 ± 2,290	EG1	6	13275	19842	9903

What Do RASA Radiocarbon Ages Represent?

Environmental Processes and Events of Human Activity

Many RASA ages are on proxies of environmental processes. All the OSL ages (chapter 3) sought to establish the sedimentation chronology of Wādī Sanā and Wādī ʿIdim and were from deposition events unrelated to human occupation or indirectly related (for example, OSL 649, fluvial deposits burying an irrigation structure). Likewise, a number of the radiocarbon ages, including ages on terrestrial and freshwater mollusk shell, were unrelated to archaeological features (for example, Beta-208490, AA59768). *Procavia* pellets, the desiccated droppings of rock hyraxes, also yielded radiocarbon ages on environmental processes independent of human occupation. These 13 radiocarbon ages have been shaded in table 18.1 to differentiate them from ages on materials generated through human activity and directly associated with human occupation of Wādī Sanā and Wādī ʿIdim. Additionally, all OSL dates from water management contexts are excluded because of imprecision, inconsistency, and divergence from more reliable radiocarbon assays, including many from the same contexts (see chapter 13).

One result from the hyrax middens indisputably marks human activity. On a linen fragment embedded with three amber beads in a hyrax midden, AA39070 arguably dates a nearby burial or cache otherwise long since decayed. RASA also dated hearths, burned layers, human bone, and animal remains (such as bone, feces, and marine shell) in archaeological contexts. Some radiocarbon ages are from stray charcoal fragments, collected to contribute to chronological definition of sedimentation profiles. While such samples' accuracy in dating sedimentation can best be evaluated in light of other characteristics of the profile (including other ages), charcoal itself is indisputably the result of human activity in Wādī Sanā, where natural brushfires are not components of the vegetative evolutionary ecology, and plant adaptations do not include fire resistance and postfire regeneration. We argue that charcoal samples not found in primary burned context—that is, not found in a hearth or ashy layer containing hearths—still represent human activity.

A Sample of Human Occupations

Although we sought to apply principles of random sampling in our survey, like all archaeological research dictated by stratigraphic and site biases, RASA excavations were targeted. Twenty-nine charcoal samples, whether stray or from hearths, came from paleoecological sampling; other samples were generated through archaeological sampling and excavations. In most cases, we selected for dating a particularly large single piece of charcoal based on its size and preservation, even when presented with the option of numerous small fragments (potentially of different woods and dates) collected via flotation. Both in site selection and sample selection, archaeologists ignored random principles. Yet our radiocarbon sample selection does represent a range of biases framed through different research approaches and priorities over multiple seasons and funding cycles. For example, we dated 32 hearths chosen for different purposes. McCorriston targeted early human occupation; Oches targeted chronological anchors for sedimentation rates; Harrower targeted irrigation structure ages; Crassard dated stratigraphic association with lithic technologies; Steimer-Herbet dated funerary events and reuses of tombs and monuments. Although not collected with random principles, the accumulated sample of radiocarbon ages is nevertheless associated with a diverse range of human activities in Wādī Sanā.

Chronology Not Demography

RASA ages offer an important set for chronological modeling but are not numerous enough to quantify regional population dynamics. Following Williams (2012:581) a general recommended sample size for a regional chronology of human populations (e.g., Athens et al. 2014; Dye 2012; Waters and Stafford 2007; Williams 2012;) would be 200 to 500 ages, far short of what we—or any imminent research—will provide. We could generate more than 200 ages from samples held in reserve, but these would duplicate many assays already run and would not date 100-plus new human events to add to our current inventory. Nor is there a standard desired number; fewer or more radiocarbon ages may be appropriate to smaller or larger regions, shorter or longer temporal spans. Lower sample sizes may be valid if the mean standard error in laboratory radiocarbon ages is low (Michczÿska and Pazdur 2004; cf. Williams 2012:580). With future archaeological research in Yemen and new ages, there is a good probability that any demographics we propose would be modified with more radiocarbon samples. Another complicating problem in reconstructing a demographic trend from the RASA ages is the inherent bias introduced through taphonomic loss of samples.

Taphonomic Loss of Samples

The overall effect of taphonomic bias is difficult to assess. Paleoecological studies have established an inherent taphonomic bias in charcoal and other organics preservation in Wādī Sanā and other drainages of the Southern Jol (e.g., Berger et al. 2012; Harrower et al. 2012). During the first half of the Holocene, enhanced and more frequent rainfall resulted in sedimentation in the major drainages, whereas increasing aridity and intervals between storms resulted in later Holocene down-cutting and erosion. The opportunities for charcoal preservation therefore differed dramatically during the Early and Late Holocene. The best charcoal preservation is in sandy silt terraces of the Early Holocene. Earlier Levallois-type technologies found unstratified on the plateau surfaces represent a compelling example of human presence unrepresented in the radiocarbon record. RASA targeted Late Holocene contexts in which human activities were also recorded—tombs, which established local aeolian sediment traps around organics, caprine dung mats in rockshelters, hyrax middens, and monumental hearths. As many of these occur near the modern surface and are more clearly visible archaeologically, their inherent susceptibility to erosion is offset by a greater likelihood that they will be observed in archaeological survey. In many probability studies of radiocarbon datasets, taphonomic bias works against the likelihood of earlier ages being fully represented (e.g., Williams 2012:581), but in southern Yemen, this bias is balanced by good environmental conditions for preservation of organics.

Given these factors, does the distribution of RASA radiocarbon ages suggest a constant representation across the environmental and climatic changes that characterize the Middle/Late Holocene? This would be testable with a null hypothesis predicting that radiocarbon ages on anthropogenic events are randomly distributed through time. Using mean intercepts of 84 RASA anthropogenic ages, a chi-square test of the count distribution shows significant difference from what was expected. Even though the RASA Project did collect evidence of human activity throughout the Holocene (over a range of 9304–41 ^{14}C years BP), the numbers of ages appear significantly skewed toward the Early to Middle Holocene (before 4290 ^{14}C years BP), prior to aridification ($\chi 2$, $p = 0.0005$, $df = 1$) (table 18.2).

Table 18.2. Chi-square table of counts of radiocarbon ages obtained on samples from Early and Late Holocene contexts.

	Ages Expected	Ages Observed
Early Holocene	45.05	61
Late Holocene	38.95	23

Calibration Blurs Chronological Resolution

One of the most important periods during which "an occupation signal will be dampened" is 7000–5500 cal BP (Williams 2012:583), during which fall calibrated intercepts of nearly half (41 of 97) of the radiocarbon ages. Due to the effects of calibration with a flat curve, there would be poor chronological resolution of human occupation dynamics within this Early/Middle Holocene time frame. That the bulk of radiocarbon ages fall within this period also contributes to a poorly resolved chronological framework within the Early Holocene, which ended stratigraphically in Wādī Sanā around 5000 cal BP.

Bayesian Analysis: Building a Chronological Model

A Bayesian analysis employs prior chronological information—that is, "a probabilistic representation of what was known about the parameters before the current data was collected" (Buck and Messon 2015:5)—to constrain calibration results and ensure that they are archaeologically interpretable. Bayesian analysis implies that researchers have thoughtfully employed ancillary information in modeling parameters for the statistical range into which a radiocarbon age most probably falls. Whereas each radiocarbon age has a prior distribution inherent in counting decay events or remnant radioactive carbon molecules, a Bayesian analysis generates posterior distributions constrained by factors extrinsic to the sample's molecular chemistry. Buck and Meson (2015:6) succinctly argue that the analyst's role in good Bayesian analysis includes the following:

1. Include prior knowledge in the inference process. Prior knowledge may include information on stratigraphy, archaeological context, and chronological relationships to other samples.
2. Understand intuitively the nature of the prior distributions that the software programmer has coded.
3. For parameters for which one has informed prior knowledge, select suitable parameterization of the chosen distribution. For example, the primary context of a branch charcoal fragment lifted from a well-preserved hearth strongly suggests that the charcoal (closely associated with human activity—that is, latest hearth use) can be assigned temporal parameters constrained by the latest possible age from a hearth stratigraphically below and the earliest possible age from a hearth stratigraphically above.
4. When selecting parameterization, be clear whether one is expressing personal expertise or collective prior knowledge. In the RASA case, the former would be a sense that hearths containing trihedral section points are early in date; the latter would be the broad consensus

among Arabian archaeologists that high circular tombs date to the Bronze Age.

5. Provide a formal substantial description explaining the prior distributions using either clear qualitative statements or direct statistical ones (to avoid including information from the data in constructing the prior). For RASA data, this led to the construction of a formal chronological model (below).
6. Computation methods should combine probability distributions using multiple trials (Markov Chain Monte Carlo [MCMC] methods) to check model convergence. By model convergence, Bayesians mean that the multiple trials using randomly selected dates within the prior distributions of each radiocarbon age generate posterior distributions within model parameters a statistically significant percentage of the time.

With 97 radiocarbon ages taken on diverse materials ranging over 10,000 years and 10 OSL ages, RASA has generated a rich dataset with which to develop and refine chronological parameters for the stratigraphic and relative chronologies of human activities and environmental processes in Wādī Sanā. These activities and processes are already situated in a regional sedimentary and geochronological framework for Wādī Sanā and Wādī ʿIdim, outlined in chapter 3. This framework provides an important context in which to assess the preservation of archaeological sites and their temporal relationships. Sites as widely dispersed as Manayzah, Khuzmum, Kheshiya, Munayder, and the many small-scale stone monuments across the Southern Jol can be better understood within a broader taphonomic and stratigraphic context that, in turn, serves as the basis for a provisional chronological model (figure 18.1). Such a model constrains the probability ranges of individual radiocarbon ages, allowing them to refine the broader chronology with Bayesian statistics. Constructing a chronological model for Wādī Sanā involves three important steps:

1. Reconstruction of regional paleoecosystem dynamics (chapter 3; Harrower et al. 2012; McCorriston et al. 2002) and the taphonomic processes affecting organic preservation and the availability of radiocarbon samples
2. Removing from the analytical dataset any radiocarbon ages that do not inform us about human activities and environmental processes in Wādī Sanā itself. (That is, leave out AA38383, AA38382 ages on tufa accumulations in Wādī ʿIdim.)
3. Judicious scrutiny of context and stratigraphy in grouping ages, using prior knowledge (personal expertise and collective prior knowledge) to assign radiocarbon and OSL ages to phases.

By constructing a model for Wādī Sanā—one catchment in the highlands of the Southern Jol—we offer a chronology with broadly regional potential, one that can be tested and compared to other catchments in a regional system.

From the analysis and synthesis of sedimentary profiles, hyrax middens, charcoal assemblages, and burned surfaces, the Holocene paleoecology of Wādī Sanā has become relatively clear, as outlined in chapter 3 and synthesized briefly here. Increased moisture from the Early Holocene strengthening of the Southwest Asian monsoon resulted in greater precipitation in the Wādī Sanā catchment, and this precipitation was likely distributed more evenly over a longer annual window than occurs today (Harrower et al. 2012). Aquifer recharge was enhanced, resulting in a greater spatial extent of springs and formation of calcareous tufa deposits at Ghayl Bin Yumain and in nearby Wādī ʿIdim. Although there was certainly greater annual flow volume than today, the runoff events were less violent than modern spate discharge. Widespread pooling occurred, especially where the channel was constricted and where inflow from tributaries like Wādī as-Shumlyah met the main Wādī Sanā drainage. Slowed streams dropped the sediment load derived from the shales and surface sediments south of Ghayl Bin Yumain. Fluvial slackwater deposits, mixed with aeolian silt, began aggrading in the middle Wādī Sanā after 12,200 cal BP and show signs of multiple reworkings as flow waxed and waned. Stable land surfaces with vegetative cover appeared many times, alternating with periods during which sediments were removed through enhanced localized flow. Within the marshy confines of the Wādī Sanā margins, oxbow channels held water during the dry season, an attractive semipermanent source for a spectrum of winter wildfowl, wild animals, domesticated herds, and people. This landscape changed most dramatically in the Middle Holocene, after about 5000 cal BP, when substantial weakening of the monsoon reduced precipitation, now concentrated in a shorter period of the year, probably in a few great rains. The resultant spate flow instigated down-cutting, which impoverished the soil moisture across the Early Holocene sediment bed. Vegetation shriveled, possibly exacerbated by overgrazing, and landscape surfaces were no longer stable or enriched by organics and soil nutrients. Erosion and deflation began removing large sections of the sediment terraces in profile and across their surfaces. Oxbow ponds and wetlands filled in with sediment and dried. The reduced vegetation and lack of standing water meant a reduced animal biomass, and during critical parts of winter drought, people and animals may have concentrated around springs and natural rock pools of water.

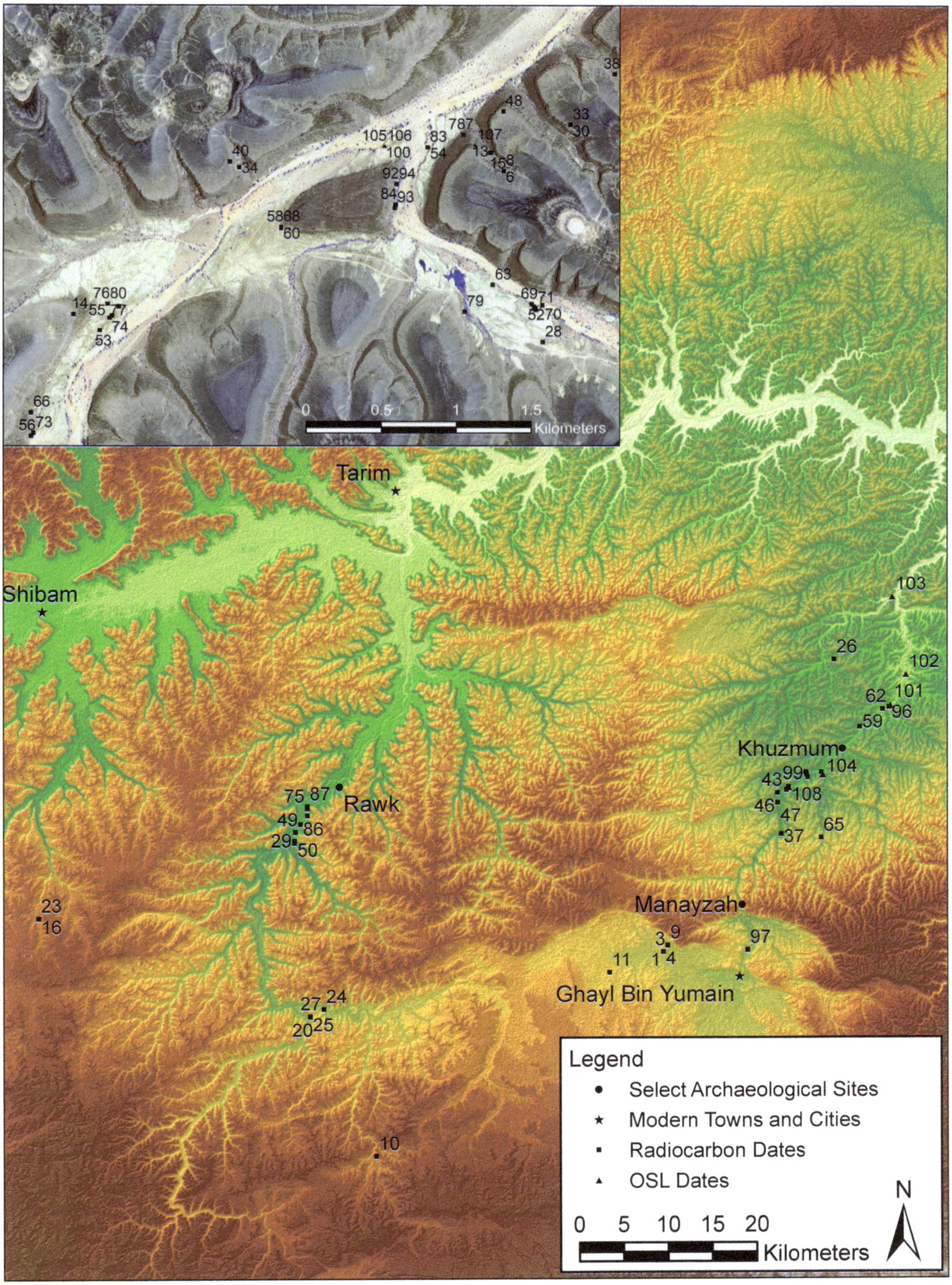

Figure 18.1. Map of radiocarbon and OSL date locations along Wādī Sanā and Wādī ʻIdim. (Numbers correspond to map ID numbers in Table 18.1; background image is topography from ASTER GDEM; inset image is QuickBird imagery of Khuzmum area). *Illustration by Michael Harrower.*

Most of the radiocarbon ages and OSL ages from the RASA Project track human activities and environmental processes in Wādī Sanā. To construct a chronological model, we excluded seven radiocarbon ages that show progressive drying of the springs and other sedimentary sequences in Wādī ʿIdim (for example, AA38385, AA38384, and Beta-208491), which may differ in timing and process from those of Wādī Sanā (Berger et al. 2012:159). We also ignored a single radiocarbon age (Beta 208490) from aquatic gastropod shell in the springs at Ghayl Bin Yumain and terrestrial shell (AA59768) for which there is no clear correction factor for old carbon. As Oches et al. write in chapter 3, the uptake of older ambient carbon during gastropod growth or shell reworking from older sediments likely explains a wide discrepancy between shell ages and charcoal ages from the same contexts. Another two widely disparate ages on the camel burial in HCT C30-4 were also dropped, following the expert opinion of Jean-Francois Saliège, who argued that the best sample with the least possibility of diagenesis is AA83496. This triage left a corpus of 86 ages (plus 10 OSL ages). Many of these—notably from the contents of monuments and tombs excavated widely across the Southern Jol (McCorriston et al. 2011; chapter 14)—did not actually come from Wādī Sanā. But they were included in this model because they represent chronological markers for activities that were observed in Wādī Sanā, such as the construction and reuse of high circular tombs, triliths, and *madhābiḥ*.

As further described below, radiocarbon and OSL ages were constrained within phases, constructed through reference to stratigraphic position, spatial location, archaeological context, and regional geomorphology (table 18.3). Most of the radiocarbon and OSL ages lie within a group of phases stratigraphically defined by geomorphological markers in the Wādī Sanā. These include environmental-geomorphological (EG) phases: streambed gravels (EG1), tributary gravels (EG2), silt terraces (EG3), silt terraces reworked (EG4), and terminal silts (EG5). One

Table 18.3. Bayesian model posterior distribution credibility intervals. Mean start and mean end numbers show estimated starts and ends of phases. Inferior interval start and end numbers show estimated beginnings of phases (alpha; α), while superior interval start and end numbers show estimated ends of phases (beta; β). All dates are in cal BP using the 95 percent confidence interval. Negative numbers reflect age estimates more recent than 1950.

	Phase	Mean Start (cal BP)	Mean End (cal BP)	Credibility Interval Inferior (start)	Credibility Interval Inferior (end)	Credibility Interval Superior (start)	Credibility Interval Superior (end)
Environmental-Geomorphological (EG) Phases							
streambed gravels	EG1	22018	10613	27949	16189	11938	9913
silt terraces	EG3	10385	5837	10906	9675	6003	5650
silt terraces reworked	EG4	6892	4885	7292	6479	5126	4612
terminal silts	EG5	5614	5154	5803	5448	5454	4794
Environmental (E) Phases							
hyrax middens	HYX	6631	473	6934	6260	682	269
Cultural (C) Phases							
Manayzah occupations	MZH	9057	7689	9593	8466	7901	7370
Khuzmum rockshelters	KZM	8631	8188	8964	8329	8415	7924
high circular tombs	HCT	4949	216	5212	4659	449	-63
tumulus monuments	TUM	3599	2151	3784	3419	2382	1907
trilith monuments	TRI	2037	1688	2238	1804	1909	1481
madhbaḥ	MDY	331	95	651	59	229	-65
other	OTH	4067	395	4349	3796	639	177

Note: Phase EG2 tributary gravels not shown, as this category included only one OSL date. See table 18.1.

environmental phase (E) designates accumulation of hyrax middens (HYX) independent of geomorphological or cultural processes. Finally, a series of seven cultural phases (C) designates occupations of the site of Manayzah (MZH), Khuzmum rockshelters (KZM), use of high circular tombs (HCT), tumulus monuments (TUM), trilith monuments (TRI), *madhābiḥ* (MDY), and an additional "other" (OTH) category. Ages in the headings below are the mean for the beginning and end of phases from Bayesian modeling (table 18.3).

Environmental-Geomorphological (EG) Phases

EG Phase 1: Pre-Holocene Streambed Gravels Deposited in Wādī Sanā "Streambed Gravels" (22,018 to 10,613 cal BP)

Gravel Terrace Formation Prior to the Holocene Silt Terraces

Throughout Wādī Sanā, the remnants of Holocene silt terraces are stratigraphically above the gravel terraces that mark an earlier bed of the Wādī Sanā. Extensive survey reveals no human remains within the gravel beds, which are remnants of high-energy environments that are unlikely to preserve sites and artifacts (chapter 3). OSL samples from sand lenses embedded within gravel were extracted to frame the gravel deposition (target event) ages. OSL measurements date the last time these sands were exposed to light. The evident unconformity in the gravel–silt transition in many locations suggests that silts may have been eroded and redeposited, plausibly repeatedly. Thus the surfaces of gravel terraces may have been exposed to light (dated event) long after their deposition (target event). The dated event may not be the age of the gravel terraces. OSL measurements from gravel show a weak correlation between downstream location and decreasing age (Pearson's $r = 0.46$, $r^2 = 0.21$), suggesting reworkings of the gravel bottom even before silt deposition and Holocene reworkings of the silts.

Where there exist OSL measurements or radiocarbon ages in the silts overlying OSL measurements from sandy lenses in the terraces, these upper measurements have been designated in the model as later than the underlying measurements from gravel terraces. The remaining four OSL measurements from Wadi sections 04WS-11, 04WS-14, 04WS-15, and 04WS-23 have no known stratigraphic relationship to one another. System-wide geomorphological evidence suggests that these all precede silt deposition.

Following standard notation for chronological models in archaeology, the start of Phase 1 is denoted α_1 and the end of Phase 1 is denoted β_1. The calendar ages of the target events assigned to Phase 1 are represented as theta 1 to theta n, as seen in table 18.1

$$\alpha_1 > \theta_1 .. \theta_4, \theta_6 > \beta_1$$

$$\alpha_1 > \theta_7 > \theta_5 > \beta_1$$

EG Phase 2: Persistent High-Energy Flow in Small-Basin, Steep-Gradient Tributaries to the Main Stream Channel "Tributary Gravels" (6622 cal BP)

Based on the geomorphology, collective prior knowledge predicts that higher-gradient, tributary wadis would be more energetic at any time than the trunk stream, or main wadi. Therefore, Rick Oches's expertise led him to expect that these tributary wadis continued depositing coarse gravel later than the onset of silt deposition in the main channel. Oches often observed lenses of coarse gravel in side channels much higher in the silts than what he observed in Wādī Sanā. Unfortunately, the team recovered a very small number of datable samples from the tributary sections, here designated as Phase 2 in the model. For Phase 2, there is only one OSL age (04WS-19) on a sand lens (θ_8) within a gravel terrace. Given the geomorphic setting, this tributary gravel probably continued to form after the beginning of silt deposition in the main wadi. Nonetheless, in cases such as this one, in which the chronological relationship is unknown, it is best to model the phase as independent to allow the radiocarbon age distributions to determine the relationship between phases.

$$\alpha_2 > \theta_8 > \beta_2$$

EG Phase 3: Silt Terraces (10,385 to 5837 cal BP)

A large group of radiocarbon samples from secure archaeological contexts date the earlier and middle-period deposition of silt beds across Wādī Sanā. These beds are now relict as silt terraces, the surfaces of which are manifestly eroded, indicating that even a short-life radiocarbon sample (for example, death of a branch as a dated event) from a secure context near the modern surface would not indicate the end of silt terrace formation (target event). Within the terraces are a number of local stratigraphic sequences, some of which constrain site occupations and anthropogenic constructions. Stray charcoal fragments attest to the reworking of materials with annual flooding, aeolian transport, and seasonal drying of the bottomlands. As components of a stratigraphic series, constrained by multiple ages on stray charcoal, OSL ages, and radiocarbon ages from materials in primary context, stray charcoal fragments offer support for a regional chronology of silt formation (chapter 3) and are included in Phase 3.

$\beta_1 > \alpha_3 > \theta_{14}, \theta_{21}, \theta_{28} .. \theta_{44} > \beta_3$

$\alpha_3 > \theta_{16} > \theta_{18} > \beta_3$

$\alpha_3 > \theta_{17} > \theta_{18} > \beta_3$

$\alpha_3 > \theta_{19} > \theta_{20} > \beta_3$

$\alpha_3 > \theta_{22} > \theta_{23} > \theta_{24} > \beta_3$

$\alpha_3 > \theta_{25} > \theta_{26} > \beta_3$

$\alpha_3 > \theta_{25} > \theta_{27} > \beta_3$

In addition to the charcoal ages produced by human burning, OSL measurements frame the onset of silt deposition at the beginning of the Holocene. The transition between the underlying cobbles and the silts is an unconformity that was probably deposited and eroded many times, as suggested also by OSL ages on gravels (that is, the sands deposited between gravel, the latter potentially laid down much earlier in a different fluvial environment, Phase 1) and overlying silts (chapter 3). There is also probably a taphonomic loss of early radiocarbon samples from human use of the Wādī Sanā bed and lower terraces because the basal silts, along with ages obtained from them, may not record the earliest habitable bottomlands in Wādī Sanā.

Silt Terrace Chronology at 2004's Wadi Section 3 (04WS-3)

This section yielded three superimposed samples, two of which were from in situ deposits. The third, a stray charcoal fragment recovered from the basal silts (θ_{14}), also underlay a substantial overburden of 4 m depth. A radiocarbon age on terrestrial shell (AA59768) from the base of the silts is omitted (see above and chapter 3). In this model, we ignore the terrestrial shell in favor of the OSL age indicating when silty infiltrations of the upper layer of gravel terraces were last at the ground surface (and hence exposed to light). Hearth sample 04WS-3a (θ_{13}) came from an in situ burning event near the base of the silt terraces and overlay the OSL measurement from this section. Together these samples establish the beginning of silt terrace formation at this location.

$\alpha_3 > \theta_{12} > \theta_{13} > \beta_3$

Silt Terrace Chronology at 2000's Profile 8 (P2000-8 = 04WS-22)

Collective prior knowledge of stratigraphic superposition places these samples in a chronological order. An OSL sample (Phase 1 θ_5) from a sand lens in the upper gravel terrace constrains the beginning of slackwater sedimentation at this location. The overlying OSL age (θ_7) will have a later posterior distribution as it was taken from sand deposited at the transition between the gravel terrace and overlying wadi silt. As one of the few locales where a sandy deposit attests medium flow, this profile may best capture an untruncated geomorphological record of the changing hydrology of Early Holocene Wādī Sanā (chapter 3). An OSL age (Phase 3, θ_9) from the base of the silt terrace marks the earliest extant deposit of silt at this location. Several stray charcoal fragments (θ_{10}) embedded within the overlying silt terrace section probably collectively relate to burning on a vegetated paleosurface, but these charcoal fragments cannot be decisively determined to lie exactly where they were burned. Instead, Oches's expertise suggests that they most likely rest close to their original conflagration site and have been somewhat disturbed by the slackwater ponding that deposited the silty matrix. They most likely provide an age estimate slightly prior to siltation at that depth, about 1.95 m below the modern terrace surface. Finally, a radiocarbon age on a wood charcoal fragment (θ_{11}) embedded at 0.25 m depth from the modern surface of the silt terrace offers a terminal date on the extant sedimentation at this location. As a stray charcoal, its sample age may not correspond to the age of the target event (sedimentation date at this level); the charcoal may be substantially older than the end of sedimentation.

$\alpha_3 > \theta_9 > \theta_{10} > \theta_{11} > \beta_3$

Silt Terrace Chronology at 2004's WS10/WS1

This chronological sequence also relies on stratigraphy to constrain posterior distributions. A Phase 1 OSL age (θ_3) from a sand lens in the gravel terrace lies 1.5 m below the interface with the silty terrace and about 3 m below a radiocarbon sample of a stray charcoal fragment from 04WS-10b (θ_{14}, AA59765). Collective prior knowledge shows that the high-energy environment in which gravel deposits appear may cause substantial depth of gravel in a short interval/one episode. The dated event is the last time the sand was exposed to light before a subsequent overlying deposit of gravel, and this sets a terminus post quem for the target event—deposition of silt over the gravel terrace. The dated event does not itself date either the last deposit of gravel or the beginning of silt deposition but constrains these target events. The earliest dated event after the gravel terrace ceased forming at this location is the burning that charred a single large charcoal fragment (AA59765). It was recovered from the lower third of the silt terrace. The charcoal was not in situ where it burned; embedded 1.0 m above the gravel–silt unconformity and 4 m below the top of the silt, it must be earlier than the silt deposition above it, but its precise relationship to the gravel deposition below it remains unknown. In this model, our expert opinion is that: (1) charcoal would not last long on the surface without being buried, and (2) it would not last or be deposited in a high-energy environment like the gravel streambed. Therefore we conclude that the charcoal burning (dated event) occurred after the transition to a low-energy slackwater regime in this location.

$$\theta_3 > \beta_1 > \alpha_3 > \theta_{14}$$

Finally, the construction of a trilith (Phase TRI, below) on top of the silt surface provides a terminus ante quem for sedimentation in Wādī Sanā (however long the gap between the end of sedimentation and construction).

Human Activities during Aggradation of the Holocene Silt Terraces

(Note: the Khuzmum rockshelter Middle Neolithic phase has been modeled separately as cultural phase KZM below. Other archaeological sites stratified in silt terraces have been included within Phase 3, with their local stratigraphic relationships embedded in the model).

Human Camp Sites

Hearths were sampled to develop a regional chronology of human activity during the wetter phase of the Holocene while silts were forming; some of these hearths provided a terminus post quem for nearby water channels and dams (Harrower 2008a). Charcoal from these hearths comes from primary context, so the dated event (life of tree) is known to be close to the age of the target event (human campsite). These radiocarbon ages also contribute to estimating the length of silt deposition, which cannot end sooner than the life of the trees charred in hearths that are buried within it.

Khuzmum Rockshelter Occupations Lacking Middle Neolithic Technology

As the stratigraphy at the Khuzmum rockshelters demonstrates (chapter 9, table 9.1), the upper hearths from the SU044 profile lack Middle Neolithic tools and blanks. A few hearths included several micro-chips, which alone are unreliable indications of the formal Middle Neolithic types. There is a single radiocarbon age (AA38546) from the stratified, upper hearths at Khuzmum. As the only stratified sequence of hearths with Middle Neolithic technology underlying hearths lacking this technology, the SU044 Khuzmum rockshelter profile is one key to a chronological model that posits an earlier Middle Neolithic phase (see KZM, below) in relation to many of the monuments, water channels, and campsites found in and on silt terraces throughout Wādī Sanā. There is inherent weakness in relying upon the lack of evidence as prior knowledge. In other words, a lack of Middle Neolithic technology associated with the hearth (θ_{15}) yielding AA38546 or other hearths in the same level is problematic as evidence to differentiate Phase KZM and Phase 3, but other criteria, such as depth of silt, are even less reliable.

Monuments, Hearths, and Constructions of the Early/Middle Holocene

Other radiocarbon ages on samples from in situ archaeological contexts embedded in silt terraces were placed in Phase 3. These include a wider variety of contexts that lack Arabian Middle Neolithic stone tools or their recognizable debitage, and the current analysis tests the hypothesis that they date human activities in the region after the cessation of formal stone tool types (McCorriston 2013). The samples include charcoal from hearths that consist of thermally altered pebbles and small cobbles in a shallow scoop or depression without a liner or retaining stones (Kimiaie and McCorriston 2014:35, figure 2) and charcoal from in situ burned surfaces. Both are primary contexts with dated events close in time to their target events. Several local posterior distributions of radiocarbon ages can be constrained by stratigraphic prior knowledge.

$$\theta_{75} > \theta_{15} > \beta_3$$

Construction of Monument C067-2

Several hearths stratigraphically constrain the construction and destruction of a monument embellished with pecked geometric motif (chapter 14). Hearths buried nearby and at the same level in the silt terraces as the monument's base were in use before or during the monument's use (θ_{16}, θ_{17}) (chapter 14). These hearths contained no formal lithic tools or debris from their manufacture. The dated events are the living wood used as charcoal; their relationship to construction (target event) is assumed to be close in time.

Monument C067-2 Destruction

One hearth (AA81814) contains smashed and burned fragments from in situ iconoclasm of geometric-decorated slabs (θ_{18}). This iconoclasm (target event) postdates the use of the monument and therefore its construction.

Burning Vegetation

Why humans burned dense vegetation in Wādī Sanā is still a matter of investigation, but the chronology of these targeted events (θ_{20}, θ_{21}) can be resolved with two radiocarbon ages (AA59763, OS16689) from charcoal in situ within burned layers (chapter 3) and from a bell-shaped pit with fire remains (θ43) at its base (OS16933) cut from a burned layer. Dated events are the lives of the trees, here thought to be very close in time to the (targeted) burning events. Burned layers are embedded in silt terraces, and in several cases there is (collective) prior knowledge from stratigraphy that constrains the posterior distributions in a sequence of samples.

Burning Vegetation at Wādī Sanā 2004's WS-7 Profile (= 98WS-3)

From silts underlying the four superimposed burned layers, a sample of loose charcoal (AA59760) constrains the beginning of the earliest incident of burning (θ_{19}). From a clearly burned layer, which marks the in situ remnants of vegetation burning (chapter 3), geomorphologists extracted charcoal fragments that provide a radiocarbon age (AA59763) for the vegetation (dated event) very close in time to the (target event) burning (θ_{20}). The remnants of this upper, fourth incident of burning at WS-7 were quickly covered in flooding episodes that deposited at least 0.7 m of silt. About 500 m east of 04WS-7, in the Wādī as-Shumlyah tributary to Wādī Sanā, geomorphologists documented another series of three superimposed burned layers in situ. The uppermost of the three superimposed burned layers (0.45 m above the flat-lying 0.0 m stone in the middle of the profile and 1.85 m below the modern silt surface) yielded charcoal fragments (OS16689) nearly in primary context, from vegetation close in age to the target event (burning of vegetation θ_{2}).

Neolithic Cattle Skull Ring and Platforms

Two dated events, dead wood (AA69754) from a basal hearth (θ_{22}) and dead wood (AA69755) from an ashy layer (θ_{23}) overlying that hearth, are stratigraphically earlier than the insertion of the cattle skulls at Kheshiya (chapter 10). An ashy layer (dead wood, AA66861) built up (target event θ_{24}) around the eye sockets of the inserted cattle skulls. This ashy buildup event is stratigraphically later than the construction event of the cattle skull ring. We assume this construction event predates construction of the slab platform monument at Kheshiya (chapter 10). Only at a different site, Khuzmum 037-3, do we have samples that constrain the occupation use of a Neolithic platform, which is a target event, dated by radiocarbon samples stratigraphically later than construction of the slab perimeter ring. We assume that the monuments commemorate and therefore are later than the sacrifices, and our use of Khuzmum 037-3 tests this hypothesis. Platform monuments are described in chapter 12 and in the archaeological literature (McCorriston et al. 2002, 2012) and can be readily differentiated from other monuments by their construction and use. Radiocarbon ages are from contexts associated within or with the actual uses of the stone structures; at Khuzmum 037-3, two charcoal samples (AA38547, AA66862) from different hearth uses (θ_{26}, θ_{27}) during late occupation are stratigraphically later than charcoal in situ in the hearth (AA38544) on an underlying occupational surface (θ_{25}). Platforms lie at or near the modern surface; they probably represent relatively late activities of the Middle Holocene. Moreover, these platforms were open-air sites, away from the canyon walls and rockshelters, another feature that distinguishes them from the occupations at the Khuzmum rockshelters and Manayzah. The differentiation of rockshelter and open-air sites is an early hypothesis from the Wādī Sanā survey (McCorriston et al. 2002).

EG Phase 4: Silt Terraces 2 "Reworked Sediments" (6892 to 4885 cal BP)

This group of radiocarbon ages comes from paleoecological sampling to provide a terminus ante quem of silt terrace formation. These are radiocarbon ages from stray charcoal (for example, AA59762), except those in stratigraphic sequences (for example, AA38380; see Phase 3). One must remember that these may come from reworked sediments; hence such samples have been modeled within their own Phase 4 in recognition that the dated event may be distant in time from the target event (*θ* reworking). Stray charcoal fragments have been blown or washed from primary context and cannot directly date the local sediment and the structures buried in them. Given the localized reworking of silt deposits and natural channel incisions and refillings, absolute and relative elevations in the silt sediments do not provide a regional-scale chrono-stratigraphic sequence. Furthermore, the flattened calibration curve between 7000 and 5500 cal BP also dampens resolution of radiocarbon ages during terrace formation, so there is a long calibrated range and greater possibility of overlap for each age, making the application of Bayesian methods truly informative.

One sample labeled from trilith SU006-001 actually derives from charcoal buried deeply below the trilith hearth. The excavation through a trilith hearth penetrated deep into the silt terrace on which it was constructed (chapter 14). Radiocarbon sample AA79763 came from an ashy layer buried in silt (θ_{50}) and has been therefore included in Phase 4.

Human-Made Water Channels

A sequence of stray charcoals from excavations of human-made water channels was also included in Phase 4 (below) because these samples offer support for a regional chronology of silt formation. The radiocarbon ages AA55969 and AA60245 have been used tentatively and conservatively to propose a 5000 cal BP terminus ante quem for water channel 9-1 (Harrower 2006, 2008a; chapter 13), but it is the silt deposited as overburden that provides the best terminus ante quem of water channel construction in antiquity. The use of human-made water channels, which appear at the top of the silt terraces, may have been relatively late, but their chronology can best be constrained by the stratigraphic associations of datable hearths and sediments. There are no direct ages on channels themselves. Two stray charcoal samples, which might derive from substantially older activities (θ_{45}, θ_{46}), lie embedded in silt deposited over and sealing water channels, whose use must have ceased before the end of widespread sedimentation in Wādī Sanā.

$$\beta_1 > \alpha_4 > \theta_{45} .. \theta_{50} > \beta_4$$

EG Phase 5: Terminal Silts (5614 to 5154 cal BP)

Terminal Silt Formation in Wādī Sanā

This phase ends at a date that can best be approximated from samples in the highest level, least-eroded remnants of silt terraces, arguably the latest manifestation of sedimentation. Two samples come from in situ burning events embedded in profiles under rockshelter overhangs (cave sediments [CS]). These are ages from charred dung mats under or in water-laid silts preserved by shallow caves. Other samples (CS-3, CS-4) are on loose fragments of charcoal that may have been reworked into sediment deposits. All dated events are on the death of the plant sampled, but in the case of in situ dung mats, charcoal formation is close in time to sedimentation (the target event), which buried and preserved fragile ash. The in situ burning events in CS-1 and CS-2 provide the most reliable ages, but Oches's expert opinion is a basis for including loose charcoal samples, especially those from CS-4 where stratigraphic superposition also serves as check on the relative ages of charcoal.

CS-1 (Cave Sediment 1, OS16958)

This context preserved a stranded block of silt high overhead, where the wadi channel had scoured away all other sediments. Similar conditions occurred at CS-2 and at the site of Manayzah. Geomorphological analysis at a regional scale suggests that these high-elevation remnants must postdate platforms, monuments, and water channels in Phase 3 silt terraces. The sampling itself was carried out to constrain the chronology of silt aggradation (θ_{51}).

CS-2 (Cave Sediment 2, OS18691)

With this rockshelter is another radiocarbon sample from a charred layer within the silt terrace remnant. Because there is no direct stratigraphic link among these caves (actually rockshelters), the chronological relationships between these samples is unknown. If one considers the regional geomorphology as collective prior knowledge, these cave sediment samples date the latest uneroded silt sedimentation (θ_{52}).

Post-sedimentation Infilling in CS-3

Oches's expert opinion argues that the post-sedimentation infilling in CS-3 represents the "best chance for a date on the end of the aggradation phase" (θ_{53}). This single radiocarbon age (AA61078) on charcoal was from the near top of a profile of water-laid wadi silts in a scour notch at the south end of a rockshelter. The sediment plug was near the high-water scour mark. The sample could not be directly stratigraphically integrated with similar samples from CS-1 and CS-2 (above).

CS-4

CS-4 came from a north-facing rockshelter in which a 1.5 m deep section of sediment was preserved high above the modern level of sediment terraces and the incised modern wadi bed. This was PSI 04WS-17, a section of alternating laminae of fine, sandy, porous silt and laminated silt, appearing to be windblown or water-laid. A radiocarbon sample (AA59757) at 0.90 m below surface was the lower of two stratigraphically related samples. Like the upper CS-4 sample, the dated event is the living tree burned for charcoal, and its relationship to the target event—the final deposition of silts (θdry)—is presumed close but actually unknown. From this same cave sediment in 04WS-17 came an upper radiocarbon sample on charcoal (AA59756), recovered 0.20 m below surface. It stratigraphically overlay the previous sample (AA59757) and is likely, but not a priori certain, to be younger. These charcoal fragments were deposited during the latest episodes (θ_{54}, θ_{55}) of sediment aggradation in Wādī Sanā. The frequency of charcoal (available in all layers) in CS-4 suggests that its original source and primary deposition were close, possibly human activity within the rockshelter. There is no direct stratigraphic link among among the sediments from different caves.

$$\beta_3 > \alpha_5 > \theta_{54} > \theta_{55} > \beta_5 > \theta\text{dry}$$

$$\alpha_5 > \theta_{51}, \theta_{52}, \theta_{53} > \beta_5$$

Environmental (E) Phases

E Phase HYX: Hyrax Midden Deposition in Wādī Sanā (6631 to 473 cal BP)

Hyraceum Preservation in Wādī Sanā

Hyraceum preservation was probably better in the arid later Holocene, so we anticipate that all rock hyrax midden radiocarbon ages postdate Phases 3, 4, and 5. There is no alternate to radiocarbon assays to date these rock hyrax middens in the Near East. Paleoecologist Kenneth Cole's long experience with the American Southwest and with the visual characteristics of packrat middens led him to an expert opinion that the Wādī Sanā middens were post-Pleistocene. The hyraceum neither smelled nor looked as early as Pleistocene packrat middens. Nor were the easily accessible locations, mostly at low elevations, typical of the Pleistocene contexts he knew. Preservation conditions were surely more favorable after aridification, but in the end, the radiocarbon ages themselves indicate that hyrax middens are middle/late deposits of macerated and urine-impregnated vegetation defecated in the caves of middle Wādī Sanā ($\theta_{56} .. \theta_{67}$).

Rock hyraxes are shy creatures that flee from humans today and are appreciated as tasty meat permissible to Muslims. They never inhabit caves while they are used by humans, but they do inhabit caves for many consecutive hyrax generations (hence the accumulation of hyraceum). Therefore, radiocarbon ages on hyraceum indicate when humans were either absent from particular caves or used them as burial or caching locations (for example, AA39070, included in Phase OTH). There is no stratigraphic relationship between Phase HYX and other phases, but the inclusion of a (presumed) burial fabric in a hyrax midden suggests that hyraxes continued to inhabit Wādī Sanā after burials in caves ceased. There are no cave burials in living memory, but the bedouin do remember hyrax populations.

$$\alpha_{HYX} > \theta_{56} .. \theta_{61} > \beta_{HYX}$$

$$\alpha_{HYX} > \theta_{63} .. \theta_{67} > \beta_{HYX}$$

Cultural (C) Phases

C Phase MZH: Manayzah Occupations (9057 to 7689 cal BP)

The occupations at Manayzah (chapter 8) represent the earliest documented human use of the Wādī Sanā marsh ecosystem. Although there was abundant lithic evidence on the high plateau surfaces of Paleolithic occupation (chapter 7; Crassard 2008), no diagnostic lithics were found in the high-energy gravels laid down under the lowest silt deposits. No instances of in situ charcoal burning provide an age associated with Paleolithic tools. Likewise, there are no OSL ages stratigraphically associated with lithic assemblages. The high-energy environment of the pre-Holocene channel may have eliminated most or all early sites within the active channel, and the gravel terraces were built of very rolled and smooth stones, suggesting a low likelihood of preservation for in situ knapped lithic assemblages or discard. The best taphonomic environment for radiocarbon samples to associate a dated event with a target event lies in the wadi silt terraces. Examination of many silt terrace profiles through archaeological survey and paleoecological studies revealed no sign of knapped lithics in basal silt layers, and the diagnostic bifacial technology of the Arabian Neolithic first appears higher in silt accumulations.

First Occupation at Manayzah (Manayzah 1)

A distinctive lithic assemblage stratified under the Neolithic occupation at Manayzah characterizes the earliest (known) site occupation (θ_{68}). One charcoal sample from the bottom of excavations at Manayzah (AA66686) was in association with unique lithic materials (chapter 8). Although excavations at this depth (1.8 m below surface) exposed only a 1.0 x 0.5 m area, analysts noted a clear difference from other lithic types in both raw material and patination. From limited exposure, insufficient lithics were recovered for a full description of technology and facies, but the contrast between these lowest lithics and the overlying bifacial Neolithic materials is striking (Crassard 2008). Basal Manayzah, which cannot be much higher in elevation than the gravel terrace under the site, seems to represent a cultural antecedent, to which belong the substantial overlying deposits at Manayzah.

Arabian Bifacial (Middle) Neolithics at Manayzah (Manayzah 2)

This subphase is defined at Manayzah by the association of knapped stone technology (broadly construed as the Arabian bifacial Neolithic) producing bifacially worked projectile points of a variety of typological forms on blanks and cores (Charpentier 2008). The rich archaeological contextual information from Manayzah (chapter 8), including the construction of hearths as pits with slab linings, the layout of camps, the absence of platforms and monuments, and associations of bifacially worked projectile points, allowed four of Manayzah's radiocarbon ages to be assigned to a Middle Neolithic cultural expression (Manayzah 2) later than its stratigraphic antecedent (Manayzah 1). Crassard's expert opinion finds a correlation in projectile point types found at the Neolithic Manayzah and Khuzmum rockshetlters, making it likely that the two sites overlap in time. Four radiocarbon ages from primary contexts, like a hearth (θ_{70}) and ashy surfaces near hearths (θ_{69}, θ_{71}, θ_{72}) at Manayzah, can be arranged using prior knowledge of lithic industry and stratigraphy to constrain posterior distributions.

$$\alpha_{MZH} > \theta_{68} .. \theta_{72} > \beta_{MZH}$$

C Phase KZM: Khuzmum Rockshelter Neolithic (8631 to 8188 cal BP)

Khuzmum rockshelter occupations contained bifacially worked tools in a techno-typological tradition identified by experts as Middle Neolithic (Charpentier 2008; Crassard 2008). Both Manayzah and Khuzmum rockshelter archaeological deposits underlay substantial silt deposits and contained short-term contexts (for example, hearths) with radiocarbon samples likely to yield ages close to the targeted event (site occupation). In the case of the Khuzmum rockshelters, most radiocarbon-dated hearths were several meters below present surface level or trapped below old rockfall. Wādī Sanā's active fluvial and aeolian environment has led to stratigraphically unpredictable exposures of the Middle Neolithic. For example, the Gravel Bar Site (chapter 9) is an in situ surface accumulation of unsorted and unrolled Middle Neolithic tools and debris that has experienced minimal spatial disturbance since deposition. Techno-typological analysis places these tools and their makers as contemporaries of the Manayzah and Khuzmum occupations (Crassard 2008:118–20). There are no radiocarbon ages associated with the Gravel Bar Site, with its surface accumulation. Given stratigraphic disparity and the unconformity at the base of wadi silts, depth of sample is not a reliable indicator of Middle Neolithic ages. Association with the Middle Neolithic knapping strategies differentiates hearths in the KZM phase and Manayzah 2 phase from other hearths embedded in silt terraces.

The Manayzah and Khuzmum rockshelters are modeled separately to take maximum advantage of the relative chronology yielded by stratigraphy. Phase KZM includes the stratigraphically lower sample 044-20 (AA38545) from the sequence of hearths excavated at one of two profiles in the Khuzmum rockshelters and two radiocarbon ages from excavations at 045-1A (AA 38548, AA38543). These three Khuzmum radiocarbon ages came from hearths (θ_{73}, θ_{74}, θ_{75}) that also contained broken Middle Neolithic points and therefore may be used to constrain age probabilities for KZM. Additionally, Hearth 044-20 (Phase 3, θ_{15}) stratigraphically underlies Hearth 044-25 (AA34546).

$$\beta 1 > \alpha_{KZM} > \theta_{73} > \theta_{74} > \beta_{KZM}$$

$$\alpha_{KZM} > \theta_{75} > \beta_{KZM}$$

$$\theta_{75} > \theta_{15}$$

C Phase TRI: Trilith Monuments (2037 to 1688 cal BP)

Some radiocarbon ages belonged to the last use of hearths (θ_{76}, θ_{77}) associated with a trilith monument. Triliths typically have hearths aligned on a terrace before them (chapter 14), and in every other reported instance in Arabia, charcoal from these hearths dates to about 2300–1700 cal BP (Al-Shahrī 1991:193; Cremaschi and Negrino 2002:342; De Cardi et al. 1977:28; Zarins 2001:134). In Dhofar, Oman, some trilith monuments are capped by stones with Ancient South Arabian writing. These graffiti also appear upside down and half hidden on stones robbed from the triliths to build adjacent boat-shaped graves. A radiocarbon age on other boat-shaped graves in Dhofar suggests they were built before 1100 cal BP (Zarins 2010:232), providing an implicit terminus ante quem for robbed triliths. Collective knowledge implies that these are younger than the sedimentation in the Wādī Sanā, even where direct stratigraphic link to silt terraces is absent.

$$\beta_3 > \alpha_{TRI} > \theta_{76}, \theta_{77} > \beta_{TRI}$$

$$\beta_5 > \alpha_{TRI} > \theta_{76}, \theta_{77} > \beta_{TRI}$$

C Phases Unrelated to Silt Terraces

Radiocarbon samples from contexts outside the silt terraces belong to other phases. Many samples were derived from tombs and monuments situated on rocky terraces and plateaus. From widespread investigations across Yemen and Arabia, archaeologists date the first appearance of monumental tombs, buildings, sedentary occupations, and trilith monuments to the millennia postdating widespread aridification in Arabia (Lézine et al. 2010). Tombs and other constructions on rocky surfaces in Wādī Sanā have been assigned to phases based on comparable monuments excavated and dated elsewhere. In most instances, there is no stratigraphic evidence to provide a relative chronological sequence with local radiocarbon samples embedded in silt terraces. An exception is the twinned construction of water management structures on the plateau with a check dam below, the latter sometimes relatively dated by organics (as terminus post quem) embedded in contiguous sediments and the accompanying *shrūj* on the rocky plateau above (chapter 13). Excavated tombs reveal a local depositional environment that often preserved organic and bioapatite samples for radiometric dating (chapter 14). After a dry-wall tomb was constructed on the old surface and a burial was placed within, the tomb served as a trap for aeolian sediment and filled, often fairly quickly. Radiocarbon ages (for example, AA83494, AA83493, AA81817) confirm that some tombs were reused, as was apparent also from their better state of preservation and the robbing of adjacent comparable structures. There is rarely any stratigraphic or secure independent basis for attributing tomb construction or construction of other stone monuments to a particular millennium. Therefore, discrete architectural types (chapter 14) have been assigned to different phases in an effort to build a regional chronology.

C Phase HCT: High Circular Tombs (4949 to 216 cal BP)

High Circular Tombs (HCT)

Situated on rocky terraces and on the plateau, HCTs seldom offer any evidence to establish a chronological posterity for their construction (target events) in relation to the aggradation and infilling of the wadis with silt (Phases 3, 4, and 5). Collective knowledge has regularly inferred that HCTs postdate silt aggradation (e.g., Cleuziou 2002; Giraud 2010; Lézine et al. 2010), but this has never been demonstrated in Hadramawt. These tombs have a central burial chamber of upright slabs or blocks in courses and were fitted with a corbelled vault and a capstone roof sealing a circular chamber wall with an outer facing and rubble core (chapter 14). The HCT construction chronology is also complicated by the strong possibility that later interments may be placed in tombs constructed and possibly used much earlier (e.g., de Maigret and Antonini 2005) and by the problematic use of charcoal fragments clearly out of primary context but sometimes used as the only possible dating evidence.

Most stratigraphic samples constrain tomb use rather than tomb construction. The 10 radiocarbon ages from five HCTs in the Southern Jol provide dates for the range of use of these tombs, with the construction of at least one—C15-3—within the phase.

Construction of HCT C15-2 and C15-3

For these target events (θconstr C15-2, θconstr C15-3), the dated event is the latest burial use of HCT C15-2. Although stratigraphically not directly linked to its near neighbor, a very badly looted and mostly destroyed HCT (C15-2) sits in a very isolated location perched high above the confluence of Wādī ʿAtuf and Wādī ʻIdim. A nearby HCT (C15-3) is preserved in much better condition and to full height. It is logical to assume that the walls of C15-2 were quarried to build or rebuild C15-3 and/or its tail elements, as there is no alternate destination for the stone from C15-2. The constraints of radiocarbon ages from materials recovered in both tombs test this hypothesis with the caveat that the dated events (bone, charcoal) may not closely match the target events (construction of the HCT). We infer as expert opinion that the last interment (bone, AA83494) in C15-2 ($\theta_{78)}$ predates the last interment (bone, AA83498) in C15-3 (θ_{80}).

Construction of HCT C15-3

This target event must predate the earliest burial in C15-3, θ_{79}, from which an apatite-based radiocarbon age on human adult bone is the dated event.

Use of HCT C15-3

The latest burials yielded radiocarbon ages that date the use of HCT C15-3. The dated event was the death of a subadult buried in a mixed deposit of multiple individuals, who may have died at different times. The radiocarbon age on subadult bone is almost certainly younger than the construction of the tomb itself (target event).

Revisit of HCT C15-3

The dated event for revisit/reuse (θ_{81}, θ_{82}) is wood burned in fires after interments at HCT C15-3. As described in chapter 14, there was in situ burning on top of tombs or on top of burials after sedimentation occurred. This suggests an elapse of time between burials and burning. Although evidence for in situ burning is reasonably strong, the dated event (death of tree) may differ from the target event (burning events after interment).

$$\alpha_{HCT} > \theta_{78} > _{79} > \theta_{80}, > \theta_{81}, \ \theta_{82} > \beta_{HCT}$$

Use of HCT C30-3

The lower human burial in C30-3 crumbled to the touch, but a radiocarbon age on marine shell beads (AA 90336, with correction factor ± 215) is a dated event (life of marine animal) presumed to be close in time to the target event (human interment in HCT C30-3, θ_{83}). With no sample underlying the tomb construction itself, one can only assert that tomb construction predated the earliest interment. Since a bead exotic to this region could have a long use life as an heirloom, its date could be earlier than that of the individual with which it was buried. Construction of C30-3 must have preceded the first interment.

Use of HCT C30-3

The later sample (AA90838) consists of a few well-preserved infant teeth deposited after an episode of aeolian sedimentation inside the tomb. This must be a dated event (infant death, θ_{84}) well after the target event (earliest use of the tomb).

$$\alpha_{HCT} > \theta_{83} > \theta_{84} > \beta_{HCT}$$

Camel Burial in HCT C30-4

Since the evidence suggests that this animal was interred alive, the dated event (death of the camel) and the target event (Interment θ_{86}) should be synchronous. It is unfortunate that three radiocarbon samples on the same animal yield three different ages. In the expert opinion of Jean-François Saliège, the fragments dated using bioapatite in tooth enamel are the least susceptible to diagenesis, and this sample (AA83496) should be regarded as the most accurate.

Reuse of HCT C30-4

Although the construction of the monument (θconstr C30-4) remains impossible to date, there is stratigraphic evidence for later reuse of C30-4 (θ_{87}), with human bone and tooth fragments in an upper fill over aeolian sedimentation.

$$\alpha_{HCT} > \theta_{86} > \theta_{87} > \beta_{HCT}$$

Reuse HCT C32-2

Although the ideal target event is the construction of HCT C32-2 (θconstr C32-2), there is no datable material recovered from the earliest interment phase. The second inhumations (θ_{85}, AA 83500) establish a terminus ante quem for tomb construction, but there was clearly a considerable lapse of time involved.

$$\alpha_{HCT} > \theta_{85} > \beta_{HCT}$$

C Phase TUM: Tumulus Monuments (3599 to 2151 cal BP)

Construction of Tumuli with Statue-Menhirs

Like many of the monument types clearly differentiated from surface remains, few tumuli yielding radiocarbon ages have been excavated. Some tumuli are associated with statue-menhirs (chapters 14 and 15). Most statue-menhirs now exist out of context and without any means of dating. The radiocarbon ages (AA83499, AA79767) in this chronology come from the Wādī Sissib example, C051-1, which has no stratigraphic relationship with any other monument types yet is the first published radiocarbon age on an excavated tumulus associated with statue-menhirs. There is discrepancy with radiocarbon ages from chapter 15 (Rawk), where Tara Steimer-Herbet argues that statue-menhir creation (target event) must be associated with the (dated event) radiocarbon ages from a nearby late-fourth/early-third-millennium BC sanctuary. The tumulus dates here cannot resolve this discrepancy about the shaping and erection dates of statue-menhirs.

The later radiocarbon ages from C051-1 may be due to samples taken on reuse (θ_{88}, θ_{89}) of the chamber (dated event) rather than an original interment close to the time of construction (θconstr C051-1, target event). One sample from the chamber yielded a radiocarbon age on animal bone (animal sacrifice is the dated event) that must postdate the construction (target event).

Reuse of Tumuli with Statue-Menhirs

From the stratigraphic evidence, there was a secondary deposit of bone and charcoal (dated event) over the tumulus chamber, providing a terminus ante quem for burial and for tumulus construction.

$$\alpha_{TUM} > \theta_{88} > \theta_{89} > \beta_{TUM}$$

Cultural Phase MDY: *Madhābiḥ* (331 to 95 cal BP)

Madhābiḥ

These are large, ceremonial grilling hearths that in recent times were built where tribesmen gathered and prepared a communal feast. Constructed mostly on rocky terraces, *madhābiḥ* have no stratigraphic relationship to other monuments. Their differentiation from ordinary hearths depends on their greater size and prominence (chapter 14). From the excavation of *Madhbaḥ* C30-27, a pair of stratigraphically related charcoal samples yielded radiocarbon ages with collective prior knowledge about their posterior distributions. Two fire events (θ_{90}, θ_{91}) atop the *madhbaḥ* must be later than the *madhbaḥ* construction (θconstr C30-27).

$$\alpha_{MDY} >> \theta_{90} > \theta_{91} > \beta_{MDY}$$

$$\alpha_{MDY} > \theta_{92} > \beta_{MDY}$$

C Phase OTH: Other Miscellaneous Radiocarbon Samples (4067 to 395 cal BP)

Other radiocarbon ages from a variety of other contexts offer important isolated chronological markers but do not belong to phases:

$$\alpha_{OTH} > \theta_{93} .. \theta_{97} > \beta_{OTH}$$

Construction of Wall Tombs

These are described in chapter 14. They consist of a small chamber at the midpoint of a long low wall faced with undressed limestone and filled with a rubble core. They are readily identifiable as distinct monuments, but as most chambers have been discovered empty, they are difficult to place in a chronology of monument types (McCorriston et al. 2011). A single radiocarbon age (AA83497) from a disturbed burial (θ_{95}) gives a use (dated event) that must postdate construction of C17-2 (target event).

The Occupation at Munayder (OS16724) (Chapter 16)

The houses at Munayder sat on bedrock terraces also occupied by high circular tombs, elsewhere known to have been constructed in the third millennium BC (5000–4000 cal BP), and there was no surface indication that either houses or tombs had been dismantled or quarried to build the other. Some poorly preserved and likely comparable structures exist at the northern margins of the springs at Ghayl bin Yumain (Wādī Sanā), and it seems to us a likely proposition that Wādī Sanā's inhabitants continued to cluster near remnant sources of water in the Late Holocene, as did the occupants at Munayder. Our differentiation of the radiocarbon age from Munayder, dating the latest use of a hearth (θ_{94}) and not the house construction of Munayder 1, reflects this conviction.

Burials and Caching in Caves

Bedouin do not today use caves for burial and do not remember doing so, although the older Jibālī herders of Dhofar do remember such a tradition among their herders. Bedouin do frequently cache items in caves. The linen fabric (AA39070) and accompanying amber beads from a hyrax midden suggest a burial (θ_{97}) with jewelry goods, later incorporated into a hyrax midden.

Miscellaneous Human Activities of the Later Holocene

There are many examples of hearths and organics recovered from the surface or within unstratified contexts on the rocky terraces. Because of the uncertainty of surface finds, the RASA Project dated very few such contexts, even where organics were preserved. Several exceptions include a hearth (θ_{93}, OS16947) in the modern gully bottom at the base of a silt terrace section and a deposition of a layer of burned debris (θ_{96}, AA60247) overlying and providing a terminus ante quem for a water dam.

Analytical Outcomes

As is common in Bayesian analysis, development of the formal model above, with its encoded prior knowledge, went through multiple iterations in which we reassessed prior knowledge before reaching model convergence with the Chronomodel program. To allow the adaption phase to finish successfully, we set the MCMC iterations to 2000. The posterior distributions generated by our model point to several important chronological refinements in Wādī Sanā with implications for the prehistoric sequence in the Southern Jol. These posterior distributions are reported as cal BP (age before 1950) in tables 18.1 and 18.3.

First Arabian Domesticates at Manayzah (8012 cal BP)

One of the significant outcomes of RASA Project research is the early Southern Arabia occurrence of domesticated herd animals, chronologically secured by stratigraphic association with AMS radiocarbon-dated charcoals (chapter 8; Martin et al. 2009). As the earliest confirmed domesticated cattle bone in Arabia, the domesticated *Bos* proximal radius bone, from a context fairly high in the site's stratigraphy (C-009), is of critical interest. Calibration of the prior distributions (OxCal 4) gave a range of 7932–7679 cal BP for charcoal associated with this bone, which is not the earliest *Bos* bone in the sequence but is the earliest bone clearly identifiable as domesticated cattle. Posterior distributions revise the age estimate on AA66684 from C-009, with a posterior distribution mean of 7689 cal BP and a confidence interval (95 percent) of 7901–7370 cal BP (table 18.1).

An incontestably domesticated sheep astragalus, which excavators recovered near the bottom (Level K9-16) of the dated portion of the Manayzah sequence, is even earlier. A radiocarbon age (AA66685) on charcoal (Level K9-17 upper) just underlying the sheep astragalus gave a prior distribution of 8035–7845 cal BP. The posterior distribution for this date has a mean of 8012 cal BP with a credibility interval of 8182–7855 cal BP (table 18.1). It is clear that even the most conservative interpretation of the range places domesticated sheep at the site before 7855 cal BP, very near the often-cited 6000 BCE introduction of herd domesticates into Arabia (Cleuziou and Tosi 1997; Magee 2014:49–50). And if we accept the mean of 8012 cal BP, domesticated sheep would fall just before 6000 BCE. Indeed, given the depth of the Manayzah sequence and the occurrence of domesticates even in the small faunal assemblages recovered, it is likely that domesticated fauna probably were at the site at least several centuries before 6000 BCE. In any case, the posterior distributions at Manayzah clearly confirm this site as the earliest occurrence of domesticated herd animals in Southern Arabia and underscore the puzzle of how they appeared there before they occurred in other regions.

Middle Holocene Shift to Aridity (5154 cal BP)

The Middle Holocene climatic changes of the monsoon system clearly produced regional environmental shifts (Harrower et al. 2012), dated in Wādī Sanā with posterior distributions on the Phase EG5 terminal silts. Our model output suggests that the Phase EG3 silt terraces began at 10,906–9675 cal BP, with a mean phase duration of 4,548 years (posterior distribution mean beginning to mean end), and that the Phase EG5 terminal silts ended around 5154 cal BP, with an end credibility interval of 5454–4794 cal BP. Because erosion may have removed the uppermost sediments even in cave shelters, Phase EG5 terminal silts end may not date the latest possible sediment aggradation. In other words, the critical environmental change in Wādī Sanā following upon the monsoon recession might have occurred some decades after 4794 cal BP.

The posterior distributions yield a maximum period during which silts were aggrading dating from 10,906 to 4794 cal BP, or 6,112 years. A minimum calculation from 9675 to 5454 cal BP is 4,221 years for the duration of Phase EG3 silt terraces through Phase EG5 terminal silts (table 18.3).

While we have been unable from the posterior distributions to pinpoint or substantially narrow an age estimate of the shift from aggradation to incision along Wādī Sanā (Harrower et al. 2012), the current analysis does shift our posterior distributions to a younger age range (figure 18.2). This shift highlights the lag between the onset of aridity in the climatological records (Fleitmann et al. 2003, 2007; Lézine and Cleuziou 2012; Lézine et al. 2007, 2010; Parker et al. 2006) and the expression of what may be a range of factors (including decreased rainfall and anthropogenic water management practices) in the record of subsequent environmental change at the end of Phase 5 terminal silts. This lag between climate and environment may have been as long as 700 years and is important in understanding human adaptations in ancient Wādī Sanā.

Figure 18.2. Graph of the Bayesian posteriors (2-sigma range) showing a 5,000-year period of silt bed formation (after the end of Phase 1 streambed gravels) and the onset of Middle Holocene environmental changes with the end of Phase 5 terminal silt formation. *Illustration by Thomas Dye.*

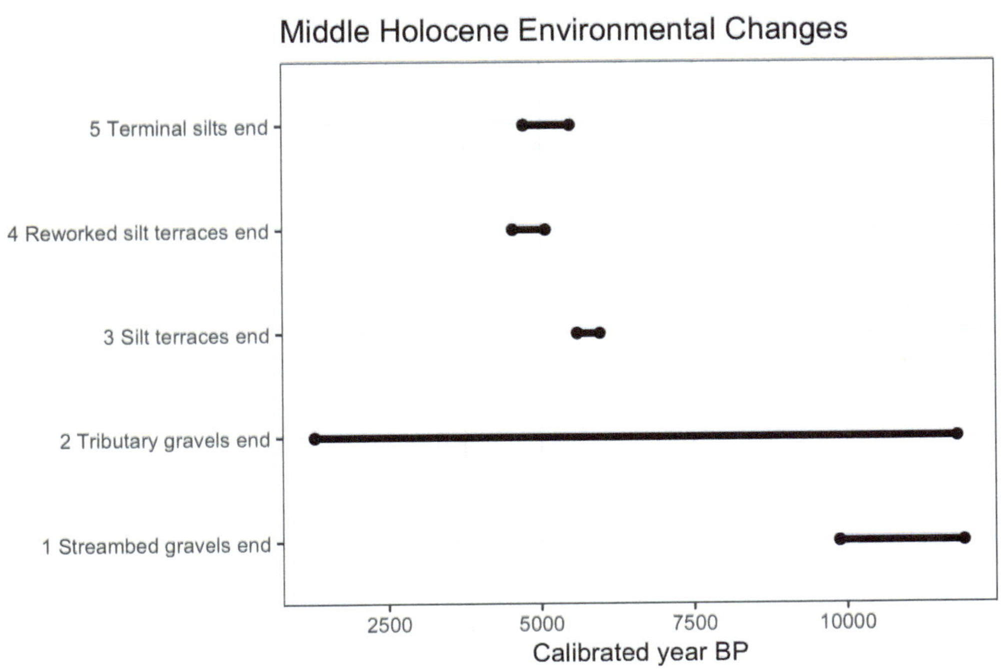

Water Management in Wādī Sanā (5837 cal BP)

The end of silt aggradation plays an important role in our understanding of the age of water management activities in Wādī Sanā. Irrigation structures, particularly Channel 9-1 (see chapter 13), are buried in silt. As there is no sample that can directly date the use of irrigation structures themselves, one must argue that irrigation structures:

a. are older than the latest levels of the silt terraces (Phase EG3) in which they are buried (that is, older than 5837 cal BP), or
b. contain potentially older charcoal eroded from higher terraces/elevations and carried by flooding into canals, aggrading into silts infilling canals. The canals then could be considerably younger than the charcoal; the two charcoal samples from sediments in which Channel 9-1 is buried have credibility intervals of 5603–4740 cal BP (AA60245) and 5569–4771 cal BP (AA55969).

Previous reporting used the date of the charcoals (b) as a terminus ante quem proxy for the date of the irrigation canals (Harrower 2008a). Here we argue the first approach (a), based on regional geomorphology, careful regional paleohydrology (Harrower et al. 2012), and Bayesian analysis that constrains the posterior distribution of phases and replaces our reliance on two stray charcoal dates.

Thus we now suggest that the canals are no later than the silt terraces (Phase EG3) that have completely covered them (Channel 9-1 lies up to 0.40 m deep in silt terraces) and thus are older than the end of Phase EG3. Using the posterior distribution mean for the end of Phase EG3 silt terraces places the irrigation canals prior to around 5837 cal BP (figure 18.3). This method reduces reliance on deposition of stray charcoals. At the same time, our analysis establishes a new early date of early–middle fourth millennium BCE (3887 BCE) for highland irrigation activities in Southern Arabia.

Social Collectives in Southern Arabia

In previous publications, we suggested that a widespread abandonment of highly skilled technical expertise in making tanged, fluted, trihedral, and finely retouched projectile points marks an important shift toward signaling participation in broadened social collectives (McCorriston 2013; McCorriston et al. 2012, 2014). Without reiteration of these arguments here, we highlight instead that such a shift is potentially captured within the chronological sequence at the Khuzmum rockshelters (chapter 9). In the

Figure 18.3. Graph of the Bayesian posteriors (2-sigma range) showing the age of water management structures—earlier than the end of Phase 3 silt terraces (mean 5837 cal BP), which cease aggrading before the beginning of Phase 5 terminal silts (mean 5614 cal BP). *Illustration by Thomas Dye.*

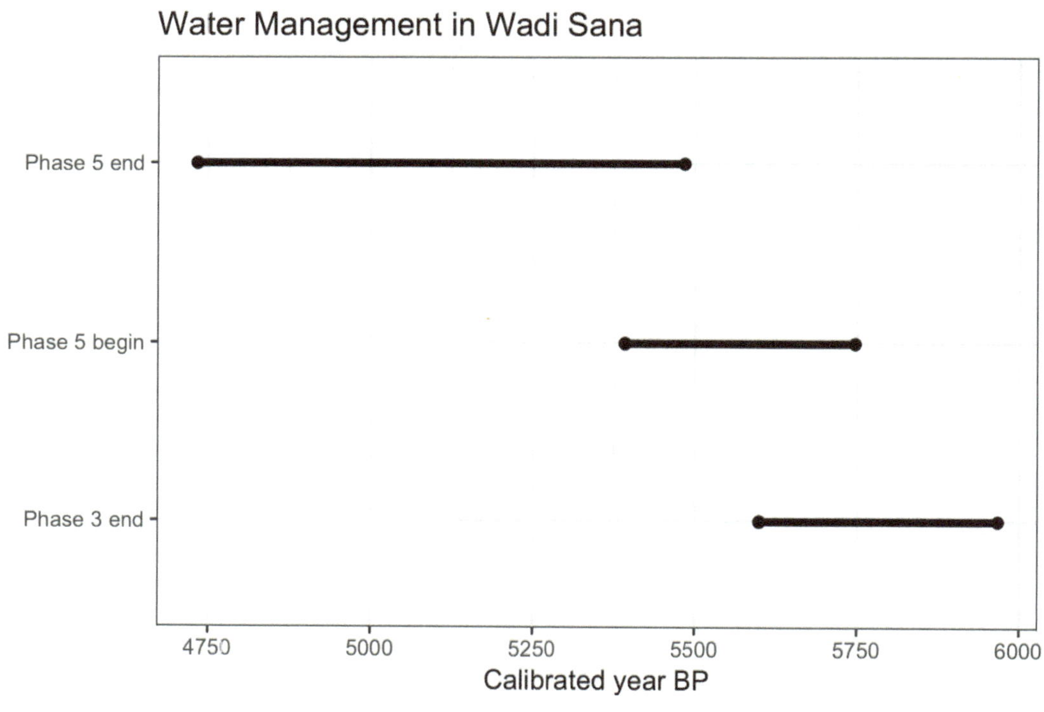

Figure 18.4. Graph of Bayesian posteriors (2-sigma range) showing the end of the Khuzmum phase with its skillfully knapped points and the earliest dated use of Neolithic monuments (θ23 and θ25). θ15 is the 2-sigma age range of the upper hearth at Khuzmum (lacking projectile points). While the gap between the cessation of skilled knapping and the use of monuments could be greater, the RASA evidence narrows the chronological minimum to only 300 years. *Illustration by Thomas Dye.*

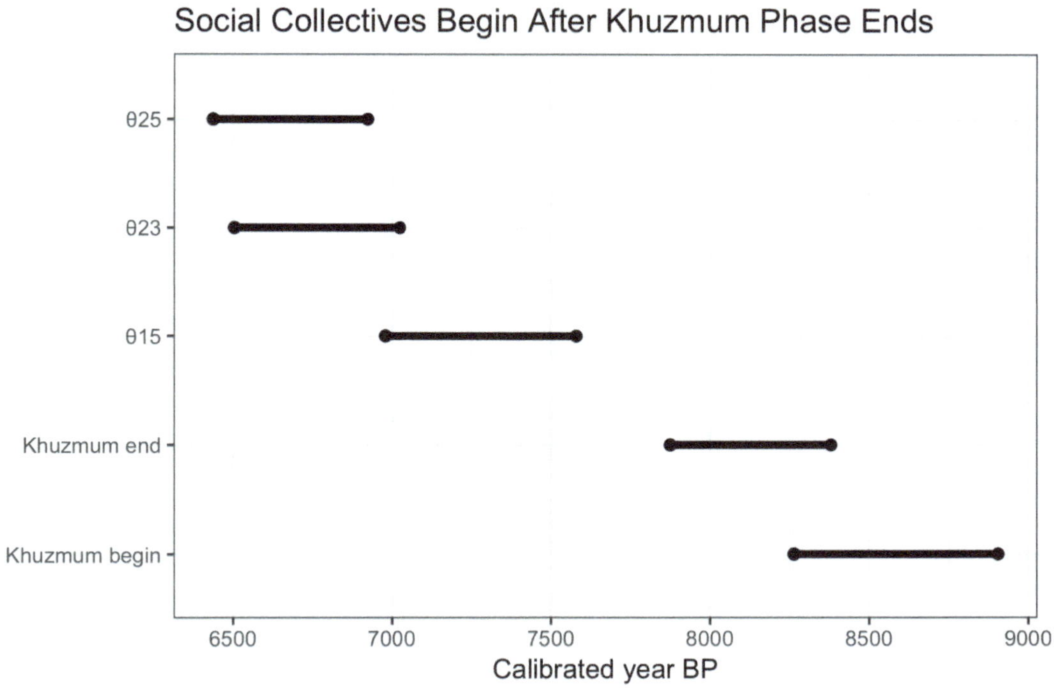

model, the latest manifestation of such knapping (end of Phase KZM) has a posterior distribution credibility interval of 8415–7924 cal BP, and the posterior distribution beginning age credibility interval on upper, knapping-debris-free hearths (AA38546) is 7610–7037 cal BP. The Khuzmum rockshelters sequence, then, may narrow the chronology of a putative shift to within a 300-year range (7924–7610 cal BP) (figure 18.4).

Even if one accepts a lengthier chronological period for this shift (8415–7037 cal BP), from the earliest possible date for the end of Phase KZM to the latest possible date for the beginning of AA38546, the shift still just precedes the earliest ages on the use of Neolithic platform monuments, hallmarks of a social collective (McCorriston 2011; McCorriston et al. 2012). Neolithic platforms have posterior distribution credibility intervals of 7094–6586 (AA69755) and 6955–6498 cal BP (AA38544).

Use Chronology of High Circular Tombs in Arabian Prehistory (4949 to 216 cal BP)

Researchers have previously observed radiocarbon ages in the distinctive high circular tombs clustering within Bronze Age (3200–1900 BCE) and Iron Age (1200–300 BCE) phases, often reflecting a later reuse of these structures (Crassard et al. 2010; McCorriston et al. 2011, 2014). This is evident if one considers the calibration intercepts before modeling HCT use. But a view of the posterior distributions shows near-continuous use of HCTs from 5,200 years ago (chapter 14, figure 14.43), with a slight break of about 400 years between about 3,900 and 3,500 years ago. Observed use for individual HCTs does indicate stratigraphically earlier and later burials, as in HCT C15-3, but as a collection of ages in a Phase HCT, the use period is nearly uninterrupted, with a duration of 4,210 to 5,275 years (credibility interval of posterior distributions). The outcome here challenges a pulse hypothesis for the use and reuse of tombs (McCorriston 2013). This outcome leaves unexplained the observed phenomenon of earlier and later burials with evidence of long disuse between them (chapter 14) and glosses distinct breaks in stratigraphy and chronology of individual tombs. It begs further research programs to refine the chronology of use and to date the chronology of HCT construction.

There is a long-standing suggestion that tomb builders signaled territorial use of oases and agricultural sediments by placing ancestors in view of these important resources

(Cleuziou 2002; Giraud 2010; Harrower 2008b). Phase HCT begins between 5212 and 4659 cal BP, while the Phase EG5 terminal silts end between 5454 and 4794 cal BP, according to the credibility intervals of posterior distributions. Recall that the ages in Phase HCT are all of use, not construction, of these monuments and that construction always preceded use. If we could date it, HCT construction probably pushes the beginning of Phase HCT slightly earlier. Equally, the Phase 5 terminal silts may not capture samples from the latest possible aggradation events. If Phase 5 terminal silts end later and Phase HCT begins earlier, there is an even stronger basis for the already evident contiguity between these phases, supporting a hypothesis that territorial signaling associated with the end of sediment accumulation spurred the construction of HCT along highly visible terraces and plateaus (Harrower 2008b; Harrower et al. 2012).

Conclusions

With posterior distributions, it is now possible to probe the chronological contiguity of activities and phases long conceptually linked. As the Khuzmum rockshelters sequence indicates, the emergence of supra-family social collectives occurred some two millennia before the environmental effects of Middle Holocene climate changes were felt. It is already clear that the climate changes of 5,500 years ago contributed to a later onset of alluvial erosion and degradation along Wādī Sanā, forcing archaeologists to consider a more complex and perhaps geographically variable response to changing climate than previously envisaged. If a lag was indeed as long as 800 years, it provided ample generations for even conservative and small societies to adjust and amplify their reliance on incipient water management systems. These water management systems could have been initiated up to 1,000 years before stream incision rendered many of them useless, and they certainly appeared before the first documented uses of high circular tombs, linked not to broader social collectives but perhaps to a reorganized collective access as the cumulative environmental impacts of climate, human landscape management, herding, and irrigation began to be felt.

References

Al-Shahrī, ʿAli Aḥmad Mahāsh
1991 Grave Types and "Triliths" in Dhofar. *Arabian Archaeology and Epigraphy* 2:182–95.

Athens, J. Stephen, Timothy M. Rieth, and Thomas S. Dye
2014 A Paleoenvironmental and Archaeological Model-Based Age Estimate for the Colonization of Hawaiʻi. *American Antiquity* 79(1):144–55.

Berger, Jean-François, Jean-Paul Bravard, Louise Purdue, Anne Benoist, Michel Mouton, and Frank Braemer
2012 Rivers of the Hadramawt Watershed (Yemen) during the Holocene: Clues of Late Functioning. *Quaternary International* 266:142–61.

Bronk-Ramsey, C.
1995 Radiocarbon Dating and the Analysis of Stratigraphy: The OxCal Program. *Radiocarbon* 37(2):425–30.

Buck, Caitlin E., J. Andrés Christen, and Gary N. James
1999 Bcal: An On-Line Bayesian Radiocarbon Calibration Tool. *Internet Archaeology* 7. http://intarch.ac.uk/journal/issue7/buck/, accessed June 10, 2017.

Buck, Catlin, and Bo Meson
2015 On Being a Good Bayesian. *World Archaeology* 47(4):567–84.

Charpentier, Vincent
2008 Hunter-Gatherers of the "Empty Quarter of the Early Holocene" to the Last Neolithic Societies: Chronology of the Late Prehistory of South-Eastern Arabia (8000–3100 BC). *Proceedings of the Seminar for Arabian Studies* 38:93–115.

Cleuziou, Serge
2002 Présence et mise en scène des morts à l'usage des vivants dans les communautés protohistoriques: L'exemple de la péninsule d'Oman à l'âge du Bronze ancien. In *Primi Popoli d'Europa*, edited by Manuel Molinos and Andrea Zifferero, pp.17–31. All insegna del Giglio, Florence.

Cleuziou, Serge, and Maurizio Tosi
1997 Hommes, climats et environnements de la Péninsule arabique à l'Holocène. *Paléorient* 23(2):121–35.

Crassard, Rémy
2008 *La préhistoire du Yémen. Diffusions et diversités locales, á travers l'étude d'industries lithiques du Hadramawt.* British Archaeological Reports, International Series 1842. Archaeopress, Oxford.

Crassard, Rémy, Hervé Guy, Jérémie Schiettecatte, and Holger Hitgen
2010 Reuse of Tombs or Cultural Continuity? The Case of Tower-Tombs in Shabwa Governorate (Yemen). In *Death and Burial in Arabia and Beyond*, edited by Lloyd Weeks, pp. 173–90. British Archaeological Reports, International Series 2107. Archaeopress, Oxford.

Cremaschi, Mauro, and Fabio Negrino
2002 The Frankincense Road of Sumhuram: Paleo-environmental and Prehistorical Background. In *Khor Rori Report*, edited by Alessandra Avanzini, pp. 325–63. Edizione Plus, University of Pisa, Pisa.

De Cardi, Beatrice, B. Doe, and S. P. Roskams
1977 Excavation and Survey in the Sharqiyah, Oman 1976. *Journal of Oman Studies* 3:17–33.

de Maigret, Alessandro, and Sabina Antonini (editors)
2005 *South Arabian Necropolises: Italian Excavations at Al-Makhdarah and Kharibāt al-Ahjur (Republic of Yemen).* Istituto Italiano per L'Africa e L'Oriente, Rome.

Dye, Thomas S.
2012 Hawaiian Temples and Bayesian Chronology. *Antiquity* 86(334):1202–06.

Fleitmann, Dominik, Stephen J. Burns, Augusto Mangini, Manfred Mudelsee, Jan Kramers, Igor Villa, Ulrich Neff, Abdulkarim A. Al-Subbary, Annett Buettner, Dorothea Hippler, and Albert Matter
2007 Holocene ITCZ and Indian Monsoon Dynamics Recorded in Stalagmites from Oman and Yemen (Socotra). *Quaternary Science Reviews* 26(1–2):170–88.

Fleitmann, Dominik, Stephen J. Burns, Manfred Mudelsee, Ulrich Neff, Jan Kramers, Augusto Mangini, and Albert Matter
2003 Holocene Forcing of the Indian Monsoon Recorded in a Stalagmite from Southern Oman. *Science* 300:1737–39.

Giraud, Jessica
2010 Early Bronze Age Graves and Graveyards in the Eastern Ja'alan (Sultanate of Oman): An Assessment of the Social Rules Working in the Evolution of a Funerary Landscape. In *Death and Burial in Arabia and Beyond*, edited by Lloyd Weeks. pp. 71–84. British Archaeological Reports, International Series 2107. Archaeopress, Oxford.

Harrower, Michael J.
2006 *Environmental versus Social Parameters, Landscape, and the Origins of Irrigation in Southwest Arabia (Yemen).* PhD dissertation, Department of Anthropology, The Ohio State University, Columbus.
2008a Mapping and Dating Incipient Irrigation in Wādī Sanā, Hadramawt (Yemen). *Proceedings of the Seminar for Arabian Studies* 38:187–202.
2008b Hydrology, Ideology, and the Origins of Irrigation in Ancient Southwest Arabia (Yemen). *Current Anthropology* 49(3):497–510.

Harrower, Michael J., Eric A. Oches, and Joy McCorriston
2012 Hydro-Geospatial Analysis of Ancient Pastoral/Agro-Pastoral Landscapes along Wādī Sanā (Yemen). *Journal of Arid Environments* 86:131–38.

Kimiaie, Masoumeh, and Joy McCorriston
2014 Climate, Human Palaeoecology and the Use of Fuel in Wādī Sanā , Southern Yemen. *Vegetation History and Archaeobotany* 23:33–40.

Lanos, P., A. Philippe, H. Lanos, and P. Dufresne
2015 ChronoModel: Chronological Modelling of Archaeological Data Using Bayesian Statistics. ChronoModel. http://www.chronomodel.fr, accessed June 17, 2017.

Lézine, Ann-Marie, and Serge Cleuziou
2012 Climate Change and Human Occupation in the Southern Arabian Lowlands during the Late Quaternary. In *Aux marges de l'archeologie: hommage à Serge Cleuziou*, edited by Jessica Giraud and Guillaume Gernez, pp. 329–42. Travaux de la Maison Rene Ginouves 16. Editions De Boccard, Paris.

Lézine, Anne-Marie, Christian Robert, Serge Cleuziou, Marie-Louise Inizan, Frank Braemer, Jean-François Saliège, Florence Sylvestre, Jean-Jacques Tiercelin, Rémy Crassard, Sophie Méry, Vincent Charpentier, and Tara Steimer-Herbet
2010 Climate Change and Human Occupation in the Southern Arabian Lowlands during the Last Deglaciation and the Holocene. *Global and Planetary Change* 72(4):412–28.

Lézine, Anne-Marie, J.-J. Tiercelin, Christian Robert, Jean-François Saliège, Serge Cleuziou, Marie-Louise Inizan, and Frank Braemer
2007 Centennial to Millennial-Scale Variability of the Indian Monsoon during the Early Holocene from a Sediment, Pollen and Isotope Record from the Desert of Yemen. *Palaeogeography, Palaeoclimatology, Palaeoecology* 243(3–4):235–49.

Magee, Peter
2014 *The Archaeology of Prehistoric Arabia: Adaptation and Social Formation from the Neolithic to the Iron Age.* Cambridge University Press, Cambridge.

Martin, Louise, Joy McCorriston, and Rémy Crassard
2009 Early Arabian Pastoralism at Manayzah in Wādī Ṣanā, Ḥaḍramawt. *Proceedings of the Seminar for Arabian Studies* 39:271–82.

McCorriston, Joy
2011 *Pilgrimage and Household in the Ancient Near East.* Cambridge University Press, Cambridge.
2013 The Neolithic in Arabia: A View from the South. *Arabian Archaeology and Epigraphy* 24(1):68–72.

McCorriston, Joy, Michael J. Harrower, Louise A. Martin, and Eric A. Oches
2012 Cattle Cults of the Arabian Neolithic and Early Territorial Societies. *American Anthropologist* 114:45–63.

McCorriston, Joy, Michael Harrower, Tara Steimer-Herbet, Kimberly Williams, Matthew Senn, Masʿūd Al-Hādhari, Masʿūd Al-Kathīrī, ʿAlī Aḥmad Al-Kathīrī, Jean-François Saliège, and Jennifer Everhart
2014 Monuments and Landscape of Mobile Pastoralists in Dhufar: the Arabian Human Social Dynamics (AHSD) Project 2009–2011. *Journal of Oman Studies* 19:117–43.

McCorriston, Joy, Eric A. Oches, Dawn Walter, and Kenneth Cole
2002 Holocene Paleoecology and Prehistory in Highland Southern Arabia. *Paléorient* 28(1):61–88.

McCorriston, Joy, Tara Steimer-Herbet, Michael Harrower, Kimberly Williams, Jean-François Saliège, and 'Abdalaziz Bin 'Aqil
2011 Gazetteer of Small Scale Monuments in Prehistoric Hadramawt, Yemen: A Radiocarbon Chronology from RASA Project Research 1996–2008. *Arabian Archaeology and Epigraphy* (22):1–22.

Michczÿnska, Danuta J., and Anna Pazdur
2004. Shape Analysis of Cumulative Probability Density Function of Radiocarbon Dates Set in the Study of Climate Change in Late Glacial and Holocene. *Radiocarbon* 46(2):733–44.

Parker, Adrian G., Andrew S. Goudie, Stephen Stokes, Kevin White, Martin J. Hodson, Michelle Manning, Derek Kennet
2006 A Record of Holocene Climate Change from Lake Geochemical Analyses in Southeastern Arabia. *Quaternary Research* 66:465–76.

Waters, Michael R., and Thomas W. Stafford
2007 Redefining the Age of Clovis: Implications for the Peopling of the Americas. *Science* 315:1122–26.

Williams, A. N.
2012 The Use of Summed Radiocarbon Probability Distributions in Archaeology: A Review of Methods. *Journal of Archaeological Science* 39(3):578–89.

Zarins, Juris
2001 *The Land of Incense. Archaeological Work in the Governate of Dhofar, Sultanate of Oman 1990–1995.* National Committee for the Supervision of Archaeological Survey in the Sultanate, Muscat.
2010 Funerary Monuments of Southern Arabia: The Iron Age-Early Islamic Traditions. In *Death and Burial in Arabia and Beyond*, edited by Lloyd Weeks. pp. 225–36. British Archaeological Reports, International Series 2107. Archaeopress, Oxford.

Chapter 19

Conclusions

A Landscape History of the Southern Jol

Joy McCorriston

Early in our field season of 2008, as Tara and I were seated in a front-end loader bucket drinking army tea from tin cans, Al-Qaeda operatives ambushed a nearby convoy of Belgian tourists, killing their Yemeni driver and Yemeni guide, killing two Belgian women, and injuring three others.

Tara and I were at a minor checkpoint being graciously looked after by soldiers while our driver, ʿAbdallah, struggled to fix a broken axle. The afternoon sun fingered through the dust-scored steel teeth overhead, and the gravel track crackled whenever something approached. Our hosts stopped every passing vehicle to commandeer one socket wrench after another, but late traffic is rare after Friday prayers. The attack took place only a few hours away.

As so often happens in fieldwork, something had changed our morning plans—either the maps were not ready or the GPS had acted up or we had to fix something in camp, which delayed our start. Our movements were never predictable like those on a regular tourist route. Now, after 12 calm years of fieldwork, our roaming was over. Geopolitical terrorism had come to Hadramawt.

Like our hosts at Canadian Nexen, we hunkered down for a few days, listened to the local sources for information and advice and decided to continue the season under much more constrained circumstances, cognizant not only of the danger to ourselves but also the danger that our presence brought to our Yemeni colleagues and escorts. It was to be our last season in Yemen, and as we said our good-byes, we knew they would be long-term. Good-bye to Nasser and ʿUbayd Al-ʿAlīy; to ʿAbdalʿazīz Bin ʿAqīl, Khālid Bā-Dhufāry, and Ietha Al-ʿmari; to ʿAbdalBaset Nʿoman and his brother Thābit Nʿoman; to ʿAbdalKarīm Al-Burkānī; to ʿAbdallah Ṣarām and ʿAly ʿAlwān; to Neil Bennet, Hassan Samawatī, Mohamad Lardy, Dave, and his back-to-back coworker Kevin; and to all the caterers, mechanics, operations and logistics staff, camp management, gatekeepers, and other many folks at Nexen who gave our team so much support and kindness.

Even in departure, we knew we had already acquired good data to address many of the questions that had launched the project and secured its funding. In that last, 2008 season, we focused on the small-scale stone monuments that shape a Hadramawt landscape populated by mobile people, for we had learned that we would find no prehistoric settlements, agricultural surpluses, emergent differentiation of wealth and power visible in the material record of Wādī Sanā. With an emergent record of South Arabian prehistory, our surveys and excavations in the Southern Jol definitively showed important new patterns linked to the introductions of domesticates, the construction of a socioecological niche, and the constitution of Arabian society.

Results of RASA Research

1. *Prehistoric peoples of the Southern Jol were predominantly mobile.* The earliest domesticates they acquired were cattle and sheep—the latter from a northern source such as the Zagros or the Levant. Their introduction—however it occurred—significantly influenced Southern

Arabia's subsequent historical ecology. The need for sufficient grazing lands kept people on the move, with herd animals assuming a significant material role in practices of social constitution and territoriality.

2. *Practices of niche construction shaped Hadramawt's socio-ecological landscape*. For example, in constructing a niche in which cattle herders could thrive, herders adapted the use of fire in marshlands and valley bottoms, managing a narrow vegetative zone for maximum growth of tender, easily digested fresh grasses and forbs while expelling and hunting competing wild fauna. In turn, this practice excluded other herds and herding households from scorched patches, marking resource territories through physical manipulation.
3. With the appearance of monument construction and collective sacrifice replacing individualized displays of skill, *enhanced social cooperation appears earlier in the archaeological record than settlement or agricultural technologies.*
4. *The first agriculture appeared long thereafter, with indigenous irrigation technologies appropriate to retention of summer rainfall.* Middle Holocene climate change and aridity, with reduced and concentrated rainfall, meant reduced grazing land for cattle, expanded distances between permanent water, and diminished quantities of standing water.
5. *There were virtually no prehistoric settlements in the Southern Jol.* Agricultural villages with staple crop storage, redistribution, and wealth-based power are not inevitably linked to the domesticated crops and animals as they spread from the Fertile Crescent, and such houses as do occur are widely dispersed and intermittently occupied.
6. *Where they occurred at all, settlements appeared late.* Like the fortress at Qārah Ḥabshiyah at the neck of upper Wādī Sanā, settlement is intrusive and reflects cooperation among different social and economic groups.
7. *Ceramics were not used.* The RASA Project collected only a handful—six surface sherds to be precise—and most of them fit together. This paucity of a plastic, durable material culture limits opportunities for relative dating and typological chronology.
8. *Lithics provide the basis for an Early Holocene typology.* The RASA Project supported the construction of Rémy Crassard's lithic typology with stratigraphic, relative dates and associated radiocarbon ages.
9. *A Bayesian model of Holocene chronology.* With more than 100 radiocarbon and OSL ages on archaeological remains, and regional paleoecological records, RASA has developed a regional chronological sequence through Bayesian analysis.

A synthesis of this record allows us to explain cultural continuities and change in Southern Arabia as a context for adoption and adaption of domesticates in the tropical and subtropical perimeter of the Indian Ocean.

A Landscape History of the Southern Jol

If one considers the main channel of Wādī Sanā a representative sample of Hadramawt's Southern Jol, then robust conclusions are possible from the survey and excavations described in this report. Without exception, the former inhabitants and wayfarers of Wādī Sanā lived lives poor in durable material culture. They constructed few or no permanent settlements, established no specialized industries to produce ceramics, metals, worked stone, or other crafted items. If they once worked in leather, wool, feathers, dyes, and wood, they left no specialized tools that would indicate such purposes—no vats, spindle whorls, burins, gravers, grinders; no middens or wasters. They built no storehouses; stashed no surplus grain or dates.

Wādī Sanā's ancient inhabitants did archive and retrieve information, a resource stored and selectively distributed. Where they lacked houses, they built tombs, monuments, memorials, shrines, and rocky epigraphs that encode memories, narratives, negotiations, convocations, appropriations and assertions, to name just a few of the social information contexts that these material remains reference. Technologies such as the highly skilled knapping to produce a fluted point (chapter 8) or the capture and manipulation of water to enhance a niche for plant production (chapter 13) were embedded in social action and agency. RASA research offers significant insight into the hitherto tabula rasa of South Arabian prehistory and into a broader understanding of the constitution and maintenance of Neolithic and post-Neolithic societies. This process was achieved as much through the flow of information as through the technologies that sustained it.

Mobile Lifestyles

Wādī Sanā is today and apparently has always been a domain of mobile people. On the upper plateau one finds Pleistocene lithics, and it is unsurprising that these locales retain little else of human passage that can be securely dated. Where they exist at all, the remains of hearth circles no longer retain charcoal, and the clearance of stones for camel crouches and other camping comfort could be relict of passage at many times. For the most part, the upper plateau retains a desolate character, and there has been no soil to support life here for 10,000 years or more. Nonetheless, surface lithic collections from the upper plateau have shown the passage of humans in Pleistocene times. Lithic data from surface sites gained in informative importance over the course of RASA research and now provide complementary support for data collected in datable contexts. In the absence of dating means, we considered an interest in surface sites "deferred," meaning that the RASA survey team deferred the collection of tools, cores, and debitage when not needed (especially Holocene assemblages) while nevertheless increasing the lithic knowledge from select surface contexts. Some surface sites, by their singularity, qualitative richness, or quantitative richness, also constitute bases for reflection that should not be ignored when there is no dated reference. Plateau surface scatters from the Paleolithic are an example. The informative value of surface sites is thus variable. The data that can be gathered are often meager, but when they are hitherto unseen or contribute to increased information about a new or poorly known operational scheme, they become primordial, as has been the case with Paleolithic open-air unstratified remains.

Accordingly, Middle Paleolithic Levallois industries, albeit impossible to accurately date in Wādī Sanā, reveal particular characteristics that facilitate their dating and interpretation when a stratified site is discovered. The *immediate* scientific impact of analysis of Middle Paleolithic lithics analyzed by the RASA Project is a preliminary step to the future determination of a chrono-cultural framework (chapter 7); we cannot yet say whether these earliest documented inhabitants in the Southern Jol were migrants from Africa, Levantine populations, or a local reservoir of innovators. Nor is it clear when and whither the Levallois informational *chaine opératoire* traveled along social networks across Southern Arabia. Mobility surely played a role, yet a final interpretation, and thus the ultimate significance of these lithic assemblages, is—here again—*deferred.*

Instead of constructed houses, for living quarters, mobile people and their flocks used rockshelters. Upper elevations and the rocky plateau are not used as camp sites today, and none of the rockshelters examined on upper slopes contained sedimentary deposits or evidence of contemporary camping. Rockshelters that are used are at lower elevations, close to the modern primary wadi channel, where vegetation for pasture grows sparsely along the lower terraces and slopes. Bedouin today favor rockshelters with adjacent terraces, where they build small circular pens and corbelled chambers for goats and their kids. Exploratory survey in 1998 suggested that rockshelters in Wādī Sanā and elsewhere in the Southern Jol seldom contain sedimentary deposits on the floor and talus. High-energy fluvial activity in the monsoon season scours the floors, and limestone overhangs and roofs periodically shed spalls and large fragments, seldom forming true caves.

In Wādī Sanā, rockshelter floors are bare rock or covered by a dung mat accumulated by goats kept inside for shade and at night. Today, bedouin occupants intentionally fire these dung deposits to clean the shelter for future use. (With bedouin still using these sites, it is difficult to differentiate the residues of recent camping from ancient camps, and too much prying in one's neighbor's home is hardly good survey etiquette.)

Neolithic Technologies for Mobile Societies

Manayzah and the Technologies of Early Holocene Hunter-Herders

Not all rockshelters are always reoccupied, offering the RASA team possible sites to explore what traces may remain from former inhabitants. At Manayzah, a low gravel terrace adjacent to the main stream channel provided a slight rise in the morning shade of a towering cliff, whose slight cleft offered only a tiny shelter. The channel of Wādī Sanā is at this point deeply scoured and constricted; it could have been lethal for herders to stay there in summer flooding. High on the cliff, where porous upper layers meet a harder limestone base, water seeps laterally and exits cliff springs, providing rare and probably intermittent surface water. In the case of the Manayzah rockshelter, the cliff face and a slight overhang provide morning shade. The slight terrace against the cliff face contained a stratified, well-preserved Early Holocene campsite dating at least as far back as 9000 cal BP. Clearly a desirable campsite near a natural rock pool, Manayzah was surely reoccupied multiple times over at least 600 years, and a stratigraphic and chronological break attests to at least one major gap in occupation (chapters 8 and 18). The record from Manayzah offers to date the earliest securely identified bones from domesticated animals in Southern Arabia. While the bone assemblage is small and few specimens are securely dated and identifiable, these are nonetheless significant data.

Domesticated sheep could only have arrived in Southern Arabia from a northern Levantine–Iranian domestication, but their discovery only raises new questions about routes and circumstances of adoption, the peoples who adopted domesticates, and their contacts with herders to the north. Likewise, the adoption of domesticated cattle, or a local cattle domestication (plausible but undocumented) from Arabian Pleistocene wild stock, provokes questions unanswered by the lonely data from Manayzah. Apart from the Khuzmum rockshelter (chapter 9), the Manayzah site is surrounded by vast regions without stratified, excavated archaeological sites, leaving it to future archaeological projects to examine the spatial networks, population dynamics, and regional variation in Early Holocene Hadramawt.

Manayzah accumulated deep layers of ash, presumably from charred animal dung. From these 8,000-year-old remains we infer that hunter-herders constrained their livestock in camps. Constraining animals may have kept them safe from predators, animal and human. While animal predators—wolves, hyenas, leopards—could be kept away by fire and thorny enclosures, a sharp spear or arrow would also deter human poachers. The technologies of stone knapping showcase the socio-ecological niche of Wādī Sanā, sparsely populated by hunter-herders. Manayzah was part-time home to master stone knappers, who traded and modified raw materials and honed the highly developed knapping skills that craftsmen exhibited in the fluted points and jasper dagger. Rémy Crassard's studies of Neolithic knapping schemes clearly show that these were indigenous technologies, not strategies inherited with domesticated animals. Knapping is a solo task. Such technological achievements may have communicated individual prowess in a social environment as effectively as Greek athletics or reading Nordic rune bones in other cultural contexts. Superb stone tools may have been as effective as technologies of communication—communicating experience, resolve, and one's broader social networks (in the access to rare raw materials)—as they were at killing animals or threatening poachers.

While we are still left with significant questions about the communities that adopted domesticates into a prior hunting and foraging mobile lifestyle, RASA research does highlight herding as an adaptation that structured a niche and perpetuated construction of Neolithic societies. Because mobility imposed constraints on the accumulation of material goods and on the archaeological preservation of discarded remains, our developing perception of a dynamic Neolithic landscape is necessarily viewed through sometimes subtle changes in long-standing technologies that did survive. Ethnoarchaeology of pastoral sites across the Near East has suggested that the technologies of pastoral nomads are conservative, and our local interviews and observations provided many insights about site formation and the scant material culture of traditional herders in Wādī Sanā. First detected at Manayzah, the confining of herd animals continued throughout prehistory. For example, several rockshelter sites, WS Cave 1 and WS Cave 2, suggest that firing dung mats dates back at least to the Middle Holocene. High relict deposits of wadi silts are preserved as terminal silt archives of the upper, presumably final layers of sedimentation before Middle Holocene down-cutting began (chapters 3 and 18). Embedded within these sediments are 5,500-year-old charred layers with visible seeds and the characteristic round pellets of charred goat dung. Firing dung mats offered herders a cleaner environment for living and rearing children by reducing disease, infection, and such pests as disease-bearing ticks and flies. They also may have signaled a presence—a social territory to which one might return.

Other Fire Technologies

Nor were rockshelters the only locus of niche construction through burning. A practice of deliberate firing of grass and brush may have been widespread in time and place. RASA survey results show only indications of Middle Holocene burns as burned surfaces. That there is a clear correlation between burned surfaces and wadi silts is unsurprising, for wadi silt beds are the only predictable local preservation environment for charcoal. Burned surfaces may have once extended elsewhere during periods when enough contiguous vegetation could sustain a fire, but such vegetation needed to be rooted in soil, long since gone from the high plateau and slopes. Where silts have completely disappeared, as in the upper reaches of Wādī Sanā, in narrow gorges, or the modern day channel, no burned surfaces appear. With a shift in precipitation and subsequent down-cutting from the Middle Holocene, the vegetative cover that would sustain burning shrank considerably.

Hearths are another feature that show conservative technologies with subtle changes. Hearths were constructed on terraces outside rockshelters and as components of open campsites on the lower terraces and slopes. Some areas of the plateau, notably the top of the Khuzma as-Shumlya inselberg, also have the remains of campfires, old and new. Survey and limited excavation revealed that three types of hearths were sequentially in use in Wādī Sanā, and many of these yielded charcoal sufficient for fuel analysis of the last use episode of a hearth. In the absence of permanent settlement, these hearths may be safely associated with mobile peoples. Possibly the distinct shift in construction techniques over time corresponds to changing cooking practices, with depressions filled with cooking stones used as ovens and the aboveground style seen today used for grilling meat. Oven types might have been preheated with a wood fire in situ. Then food was placed on a moist grass or leaf liner, and the whole was sealed with hot rocks and mud to prevent the heat escaping. In Arabia, early travelers describe cooking bread by burying dough in hearth ashes, a practice we observed Wādī Sanā's al-ʿAlīy bedouin doing today.

A unique feature is a 1 m wide, bell-shaped pit exposed in a natural erosional cut in the silt terraces in middle Wādī Sanā. Cut from a burned surface in the Phase 3 silt terraces (chapter 18), this pit antedates the earliest water management structures, a canal embedded in overlying sediments that probably formed over many hundreds of years. A bell shape, clear in section, suggests that this pit may have originally served as storage—of what would be utter speculation, since there are no indications of agricultural tools or domesticated plants in the local archaeological record of that period. An in situ oven or hearth was its last use, with thermally altered smooth stream gravel and charcoal well preserved at the bottom. One interpretation is that the pit was dug as a large-scale oven, perhaps to accommodate cooking a large feast. Charcoal recovered and analyzed from this feature was 7,000 years old and included *Cadaba sp.* and *Tamarix sp.*, both common in hearths throughout Wādī Sanā.

In Wādī Sanā, hearth construction changed with the appearance of raised grilling hearths. RASA survey documented no dated examples of the raised grilling hearths earlier than 2,400 years old (chapter 14). There are no raised grilling hearths preserved in the Early Holocene silt terraces. Grilling meat on hearths constructed with a ring of cobbles and fill of smooth pebbles is widespread in Hadramawt and continues today at rural restaurants. Poor preservation of charcoal in structures on bedrock and gravel terraces severely limits the possibility for dating hearths on these landforms, and the widespread scatter of thermally altered rock suggests that many such features have been destroyed. Whatever their mundane use, it is evident also that raised hearths—and surely by inference the meat shared from a large, fresh kill—featured as a prominent component of trilith monuments, whose highly symbolic structural layout conveyed and conferred archived information to transient people.

It is therefore important to track changing technologies of hearth construction, for hearths served a social as well as an economic-culinary purpose: as hearths changed, so also changed the social context of their use. In the shallow scooped pits filled with thermally altered stream cobbles, RASA excavations recovered burned animal bone, strongly suggesting that these Middle Holocene hearths were also used for grilling meat. Many such hearths were found in the vicinity of Shiʿb Kheshiya, where the monumentalized skulls of sacrificed cattle point to a socially constituting event of broader landscape significance than the routinely functional transformation from raw to cooked meat. *Madhābiḥ* are a special form of raised hearths—larger and often built upon an artificial platform—used for cooking sacrificed animals on ceremonial and social occasions (chapter 14). RASA research suggests that such historical tribal gatherings have a long prehistory as collective actions shaping and maintaining Hadramawt's Holocene socio-ecological landscape.

Shiʿb Kheshiya, the Social Collective, and Land Management Technologies of the Middle Holocene

Even as RASA research shows long-term continuities in pastoralists' manipulation of Wādī Sanā's resources (wood fuel selection, rockshelter use), research also documents the development of a Neolithic social landscape. While some economic activities—herding cattle and caprines, mobility, persistent hunting of gazelles and wild equids—clearly continued to structure the size and duration of campsites, there were important later changes in Neolithic social groups, their territories, and the ways these shaped each other.

RASA excavations and analysis in Wādī Sanā have shown the abandonment of highly skilled lithic technologies—the individual display of knapping skills—and the emergence of monument construction by social collectives practicing gatherings, sacrifices, and feasts. Neolithic social collectives are on display at sites like Shiʿb Kheshiya, with its dramatic cattle skull ring, and at the numerous other platform monuments in the vicinitiy of Khuzma as-Shumlya.

People occupied these rings of upright boulders, sometimes for multiple seasons, before transforming them into commemorative monuments (chapters 10 and 12). In the Middle Holocene, marshy grasslands with stagnant oxbows stretched across the middle Wādī Sanā as an inviting winter pastureland, rich in hummock grasses and tender browse, where man-made fires had reduced old brush to ash. Tender growth sprang up after the summer floods and slick mists cleared. Perhaps in the way of cattle herders in adjacent Dhofar, Hadramawt's ancient herdsmen roofed these shelters with the warped trunks of termite-resistant acacias and *Anogeissus*, completing a roof with lashed thatch or matting. Or perhaps they left such campsites open to the sky, trusting the warmth and warning of a central fire, as their bedouin successors do today.

The skull ring at Shiʿb Kheshiya testifies to a broader social collective than a herding family that could huddle in one stone circle. Based on the isotopic chemistry in cattle molars, at least three different cattle herds contributed to the sacrifice. The volume and off-take of meat suggest that many more than three herds gave up females for the feast (chapter 11). To recall such events (perhaps to memorialize the groups who contributed animals and journeyed in from adjacent drainages, or to commemorate ceremonies and negotiations concluded), work groups carted stone slabs, boulders, and wadi cobbles to fill the adjacent stone shelters. They erected east-and southeast-oriented worked-stone uprights outside. Were these stones among the earliest Arabian betyls, or *bayt-al*—the dwellings of divinities? Might they be named or recognized by any passerby, who could then reference himself to the mute witness of formation and attestation of a social community? Some of these platforms nestled among thickets, reeds, and hummock grass; all were low-lying in and near the tangle of the wadi bed. Retrieving and manipulating an associated narrative of historical landscape meant knowing and seeking a hidden place.

Technologies of History

Monuments of the Bronze Age

Not so hidden were subsequent stone monuments, some placed high on the ridges and cliff tops overlooking the main channel and visible for miles around (chapter 14). Michael Harrower has argued that the high circular tombs of the fifth millennium BP overlook the best agricultural land, improved with small-scale check dams and diversion channels that fed modest patches of ever-shrinking silt terraces. In Wādī Sanā, the presence of such tombs also reflects a selective narrative of landscape history. Not everyone who lived in the three millennia of HCT use and reuse was buried in such a tomb—the low numbers of tombs and their broad spatial distribution cannot account for an entire, viable, and self-reproducing mobile population over the span of time represented. Instead, these HCTs contain some individuals whose stories, along with one's relation to the interred, were authoritative resources to manipulate, retrieve, and negotiate through landscape places. Although the details of such narratives are lost, salient characteristics are that:

1. They were monuments revisited for subsequent burials and offerings.
2. They were highly visible from the most frequented major routes and vegetated lands.
3. They were archives of genealogies and kin relations through the dead, a characteristic manifestly different from the undifferentiated group commemorated as a collective in previous monuments.
4. They draw upon history and chronology as the basis for social participation, a new manipulation of monuments. and social constitution.
5. They belong to a wider tradition of fifth millennium BP tombs across Arabia, attesting a widespread tradition of kinship relations in the service of social negotiation and constitution.

Tara Steimer-Herbet's analysis of the Rawk sanctuary and its statuettes, consistent with a repertory of anthropoid iconography (chapter 15) across Arabia, suggests nodes in a widening network of shared narrative and oral traditions. The ambiguity of tombs in full view with hidden offerings provides ambiguous information that may be authoritatively retrieved and manipulated on a local level—for who but the informed can know who lies within? The sanctuary at Rawk represents a rare building in a format without local parallel (as far as is known). With its (hidden) anthropoid statuettes and unusual architectural form, Rawk references a broader cultural tradition, one surely widely known and also conferring broad social membership in a system of beliefs. No manifestation of this system is evident in Wādī Sanā itself, where no sanctuaries, anthropoid figures, or statuettes have been found; Wādī ʿIdim and the western lands of the Southern Jol seem to mark a cultural boundary in the sixth and fifth millennia BP. And Rawk itself appears to articulate as a node within a broad network of beliefs and practices.

Water Management

Where the RASA Project assiduously sought charred plant remains, plant-food processing technologies, and the traces of field management (axes, burning crop stubble, ploughing, terracing, fertilizing, water diversion) in the field surveys and excavations, we found no dated agricultural remains earlier than the sixth millennium BP (chapters 13 and 18). This mirrors archaeological reports elsewhere in Southern Arabia, where the earliest irrigation technologies appear some centuries later than in highland Wādī Sanā. For reasons that remain obscure—whether due to climatic and environmental changes or shifts in social constitution and territoriality—Wādī Sanā's inhabitants were among the first in Southern Arabia to practice irrigation. While still herders—now focusing on caprines—Wādī Sanā's inhabitants built small diversion channels and dams to concentrate surface water on small plots of silt terraces stranded by wadi down-cutting. It may be that these technologies were originally invented to enhance and perpetuate a niche for livestock, perhaps enriching pools and standing water for drinking. As yet there is no direct archaeobotanical evidence for the earliest agriculture, but the technologies of supplemental watering preadapted Wādī Sanā's residents to farming dates, cereals, indigo, sorghum, cotton, and other crops whose later introductions have been attested elsewhere. Wādī Sanā's irrigation technologies are indigenous and appear in concert with Middle Holocene monsoon recession, as surface water dwindled, as people shifted to caprine herding, and as people relied on highly visible monuments to genealogically ordered social communities to dominate their irrigated land.

Late Settlements

Whatever its purpose, irrigation did not usher in permanent settlements. Rockshelters were used through time, and at some rockshelters, people produced graffiti and modest rockart as the only record of their passage. There are virtually no native permanent settlements in the Wādī Sanā primary channel and tributaries upstream of Ghayl Bin Yumain, now or in times past. The one exception is a South Arabian fortress, locally known as Qārah Ḥabshiyah, which sits atop a strategic inselberg where the main drainage narrows. The team visited this site on several occasions but undertook no formal description of the ruins and no test excavations of the several room blocks on the plateau, which may be plainly seen in satellite imagery. Rough-hewn square-block masonry walls at intervals around the upper slopes complete the already formidable defenses of the sheer rock. Several rooms have abundant charcoal, suggesting destruction of roofs by conflagration. A scatter of prehistoric materials, including Neolithic projectile points, attests to other occupations of this high site, before which pass all southern routes into Wādī Sanā's main drainage. Visibility is extraordinarily good, and the site would have served well as a control and defense of overland traffic during the era of Old South Arabian kingdoms (from the third millennium BP). A partial inscription in Sabaean text (circa 1,500 years ago) describes the coalition of different clan members who presumably participated in building the fort. Several are known Sabaean clan names; others are presumed to be local collaborators (chapter 17). The inscription is no longer present at the site (Mukalla Museum no. 157), but it confirms the intrusive nature of this fortress-settlement into Wādī Sanā, which had no permanent settlements of its own.

It has proven challenging to interpret structural remains around the shrinking springs at Ghayl Bin Yumain, where millennia of date palm cultivation and a small village have accreted a sedimented overburden and reworking of cultural materials. Perhaps Ghayl Bin Yumain, with its precious permanent water, played a long-term and even nodal role in the human landscape, but traces of ancient permanent settlement are effaced or unreadable today. Stone is robbed and recycled, while the bunds and channels of irrigation in silty orchards leave no surface trace of former cultivation. (As the date palms die off, stricken by modern overuse of ground water, it may be someday possible for archaeology to probe underlying sediments in abandoned orchards.) At the other geographical extreme, at the mouth of Wādī Sanā lies the village of Sana, situated not far from ancient ruins once part of the Kingdom of Hadramawt and, like the modern village, part of the cultural territory of Wādī Hadramawt. These extremes—Ḥabshiyah in the south and Sanā in the north—monitor a major passage through the Southern Jol, but these few traces of settlement are relatively late in date, without known prehistoric antecedents.

Of course, like modern groups, the ancient inhabitants of Wādī Sanā also participated in cultural traditions that spanned an east-to-west expanse. Across the regional landscape of the Southern Jol, there are widely separated modern settlements where springs exist (for example, Ghayl ʿUmar, Ghayl Bā Wazīr, and Ghayl Bin Yumain). Modern settlements with gentle topography and broad sediment traps also occur closer to the coastal escarpment, so that annual rainfall provides reliable catchments for seasonal floodwater farming (for example, Raidah al-Maʿārrah and al-ʿUlayb, Risib). At none of these did the RASA Project detect traces of early settlement. The houses at Munayder overlooking the broad valley floor of Wādī ʿIdim and its defunct springs were the earliest indications of dedicated domestic constructions (chapter 16).

Cultural Continuities and Change in Southern Arabia

The not-so-remarkable encounter with ʿUbayd Al-ʿAlīy, the bedouin workman who awaited us at al-Faqqāsh as we arrived unannounced, was hardly unusual (chapter 1). Within two days of establishment of our Wādī Sanā tent camp in 2005, Saʿid Al-ʿAlīy appeared. Our elderly camp guard from previous seasons, Saʿid totes an antique flintlock that his father used to fire upon British RAF bombers in the 1930s. The knowledge that theft or harm to us would incur the retribution of a tribal elder made Saʿid's ancient gun extremely effective, even with its dubious firepower and accuracy. As our guard, Saʿid held a social gathering every night. Visitors to his fire also passed through our circle, hovering in the dim periphery of our single electric bulb to watch the team play at cards or write out notes. All that we had was in plain view. Never did Saʿid need to patrol, chase, or stand watch. Our belongings, even the small gadgets and useful tools left in the open, and the cash-stuffed chest in my tent, were never touched. Had I known of Saʿid's imminent return, I'd again have hired him. Crestfallen to find we'd hired another tribesman as guard, Saʿid explained that he'd walked from (Wādī) Hadramawt, some 60 km to the north, because he'd heard of our return. News travels that fast.

Information is an intangible resource, and yet it is integral to the negotiations and socially constituting actions of mobile pastoralists in Wādī Sanā. RASA archaeological research shows that there was a relatively poor material culture, at least in terms of durable, portable items. In a desert setting moderately poor in material culture yet rich in cultural continuities, the flow of information along social networks has shaped human niches and structured a landscape history of Hadramawt. Albeit challenging to detect archaeologically, social networks and information traffic have left material cues recognizable in Wādī Sanā's archaeological record. Neolithic herders in Wādī Sanā used technologies not only as functional adaptations to an arid environment but also as sophisticated signals to communicate, collaborate, and constitute social collectives critical to the long-term success of human populations and societies.

The RASA Project embarked on a research program to document and explain the transition to food production in Southern Arabia. The lack of preserved material culture contrasts markedly with the traces of Natufian and Early Neolithic settlements and their contents in the Fertile Crescent, and thus the Southern Arabian record led the RASA team toward a focus on information flow and networks as the basis for cultural continuity and change. Information is a precious resource, and it was transferred from person to person like the know-how to flute a Manayzah point; it was also passed through institutions like pilgrimage gatherings, sacrifices, and feasts that left platform monuments. Information was stored and retrieved through mnemonic devices like genealogical-territorial markers with buried kin; like trilith alignments; like rock images, *wusūm*, and alphabetic writing.

Albeit difficult to trace archaeologically, information could be manipulated as an authoritative resource significant in social negotiations. In this, information is as much a tool of social identity as manipulated allocative resources. Where the contribution and conspicuous consumption of allocative resources, like cattle to sacrifice or the interment of exotic bronze and shell in a tomb, leaves material trace, the manipulation of information may not. Nonetheless, like the acquisition of domesticates, the development of penning strategies, the enhancement of graze through burning and water management, and the marking of landscape with monuments, information technologies are adaptive tools in a changing world.

And in tandem with global processes of climate change, Wādī Sanā experienced regional manifestations of broad-scale environmental changes. There were major shifts in precipitation, both in quantity and intensity, and a system-wide, seasonal, slackwater regime emerged in the Early Holocene. This marshy backwater largely disappeared in the Middle Holocene as precipitation declined, vegetation cover decreased, and spate-water runoff intensified in short bursts of channel down-cutting. Wādī Sanā's environment was already anthropogenic. Gone, or severely reduced today, are the leopards, Arabian cats, hyenas, wolves, ibex, hyraxes, baboons, gazelles, and wild cattle that once inhabited these regions. In this anthropogenic landscape, it is important that we recognize that not all anthropogenic change is degradation and resource depression. Humans had constructed a niche of burned grasslands, diverted water, removed predators, improved rangeland populated with domestic stock, and socially proscribed territories, laying the foundations for palm oases and settlements at the base of Arabian civilization.

Glossary

ʿAbdalʿazīz Bin ʿAqīl: regional director for Hadramawt Province, General Organization for Antiquities and Museums, Yemen (retired), and a key contributor to RASA Project fieldwork and publication. Correct orthography given here.

Acheulian: tool type and term referring to the Lower Paleolithic period; generally hand axes

ʿĀd: See Land of ʿĀd.

Aden: Anglicized spelling of ʿAdin, port city at the southwest tip of Arabia

al-ʿAlīy Bedouin: A subgroup of the Hummūm bedouin in southern Yemen

al-balīyah: a rite described by early Islamic writers in which a riding animal (a camel) was sacrificed by binding and interring it live beside a grave

al–Faqqāsh: military checkpoint at the western pass down into middle Wādī Sanā

al-Hawa: Anglicized (published) orthography for al-Ḥawwā, a playa or dried lake sediments in the Ramlah as-Ṣabʿatayn region of Yemen

al-Manāhīl: a bedouin group in Southern Arabia

AMS: accelerator mass spectrometry, a technique for direct measurement of residual radioactive carbon (^{14}C) atoms in an organic sample

anthracology: the study of charred wood, generally from archaeological sites

Aqab: an archaeological site in United Arab Emirates; Anglicized orthography

ASTER (advanced spaceborne thermal emission and reflection radiometer): a satellite-mounted high-resolution (up to 15 m pixel) multispectral sensor

Baraqish: an archaeological site in northern Yemen; a ruined town of the Minaean Kingdom

Bayda: Anglicized spelling of al-Bayḍāʿ, a town in central Yemen in the governorate of the same name

bedouin: an Anglicized version of *badawī*, literally a dweller in the wilderness; widely used to describe mobile pastoralists in the Middle East

betyl: from the Arabic *bayt-al*, meaning "hous[ing] God"

Bint al-Methul: an Anglicized site name used by Diana Pickworth; no correct orthography from the Arabic available; exact location undocumented

Biʾr Barhūt (Well of Barhut): a legendary well from which jinn issue forth; mentioned in Qurʿān

CA: correspondence analysis, a multivariate statistical pattern-searching approach with results projected through ordination

CANOXY: Canadian Occidental, based in Calgary; original logistical sponsors of the RASA Project

CANOXY-Yemen: one of the former corporate identities of Canadian Nexen Petroleum Yemen (CPNY) and its corporate partner, Nexen Petroleum

CCA: canonical correspondence analysis, a multivariate statistical approach that assesses the fit between a dataset and a model (pattern) constrained by independent variables

***chaine opératoire*:** operational sequence; technical steps in manufacture or task sequence. Used widely to describe a tool-making sequence.

Deutsches Archäologisches Institut: DAI, the overseas research network of German archaeology

Dhofar: Dhufār, the westernmost province of southern Oman

DigitalGlobe: the company that formerly owned the QuickBird satellite produced its high-resolution digital imagery (up to 0.6 m pixels).

dolmen: borrowed from archaeological terminology in Europe, this term refers to a construction of huge slabs or boulders forming sides and roofed by one or few single massive stones. Although not everywhere tested by excavation, dolmens generally served as funerary chambers.

***dom*:** Anglicized version of *dūm*, fruit from the *ʿilb* tree (*Ziziphus spina-christi, Ziziphus* spp.)

en écharpe: used in lithic analysis to describe blows that leave flake scars at oblique angles to the edge

en éperon: refers to a particular knapping technique that leaves characteristic flake scars from a distinctive style of platform preparation

façonnage: used in lithic technical analysis to describe the way of making a tool; developing a preform for a tool

Fartak: Raʿs Fartak, an imposing headland and peninsula along the South Arabian coastline between al-Mukallā and al-Ghayḍah

***futa*:** a sarong; a garment wrapped around a man's waist, rolled and held at the waist

***galabiya*:** Anglicized form of an Egyptian term describing a full-length man's robe, open with a slit at the neck

Ghayl Bin Yumain: Ghayl bin Yumayn or Ghaīl bin Yumaīn, the town and springs at the source of Wādī Sanā

GT: gravel terrace, one of the landform classes used by the RASA Project

hadith: prophetic tradition as a narrative relating of the deeds and utterances of the Prophet and his companions

Hadramawt: one of the governorates of Yemen, Anglicized from Ḥaḍamawt. See also Kingdom of Hadramawt.

Hadrami: describing a person or thing from Hadramawt; Anglicized from Ḥaḍramī

Hadramitic: an Ancient South Arabian language or script

HAE: height above ellipsoid, a measure of elevation above a mathematical model that approximates the shape of the earth. The most common ellipsoid model is World Geodetic System 1984 (WGS84), which is the native format for elevation readings from GPS. Height above ellipsoid differs from geoid models of the earth, such as Earth Gravitational Model 1996 (EGM96), which are often used to define mean sea level (MSL).

Halqoot: an Iron Age pastoral camp in Dhofar, Oman. Arabic orthography is not attempted here as the name is in the Shehri language and transliterated as "Halqoot" on the entrance sign to the village.

Halula: a Neolithic site in northern Syria

Hayd al-Ghalib: an archaeological site in the Shabwa region

HCT: high circular tombs; a South Arabian tomb originally constructed with vertical outer faces and a central burial chamber opening from the roof

Hismaic: a distinct variety of ancient script and language; see Thamudic

Hūd: one of the prophets appearing in Qurʿān. Hūd's grave is marked by a mosque and is an important site of annual pilgrimage in Hadramawt.

Ḥumūm: a large bedouin tribe occupying the southeastern mountains of Hadramawt, of which the Al-ʿAlīy are one branch

Ḥumūmī: of the Ḥumūm tribe

Hureidha: an archaeological site at al-Ḥuraīḍah in Wādī ʿAmd, Hadramawt. The spelling provided here is that used by Gertrude Caton-Thompson in her publication of excavations.

hyraceum: a desiccated urine and fecal pellet conglomerate accumulated over time by the multiple defecations of desert hyraxes; informally called a hyrax midden

hyrax: *Procavia capensis*, a desert herbivorous mammal dwelling in rock crevices, about 50–70 cm in length. Rock hyraxes feed within a limited radius of their dens.

Ietha Al-ʿmari: correctly transliterated as ʿrrḍah Al-ʿĀmirī; ethnographer and field representative of the General Organization for Antiquities and Museums, Hadramawt Province

ʿilb: a local name for *Ziziphus spina-christi*, a species of evergreen tree known as Christ's thorn

inselberg: an isolated hill or mountain rising abruptly from a plain

***istiqāʿ*:** intercession for rainfall, usually carried out in association with a high place in Arabian pre-Islamic and Islamic times

ITCZ: Inter-Tropical Convergence Zone, the circulation belt where northern and southern trade winds meet near the equator. The ITCZ migrates farther north and south annually, driven by solar insolation and wind strength.

Jābirī: one group of the bedouin tribes of Hadramawt

Jāhiliyah: an Islamic term for the Period of Ignorance, pre-Islamic paganism, and, more generally, pre-Islamic times

***jambiya*:** Anglicized form of *janbīyah*, a curving knife worn at the belt by Yemeni tribesmen

Jarf an-Nabīrah: also printed as "Jarf al-Nabīrah"; a site where rockart has been documented in Yemen

Jawf-Hadramawt: the continuous Tertiary-period drainage originating in northern Yemen and discharging through Wādī al-Masilah

Jebal al-Buhais: an archaeological site in United Arab Emirates

Jebel Faya: an archaeological name for a site in United Arab Emirates

Jeza Fm: the Jizāʾ Formation; Anglicized in geography to refer to one of the (mostly shale) formations of southern Ḥaḍramawt. In our geological maps and text, we use the conventional, Anglicized form.

Jibālī: a modern term for a dialect of a non-Arabic South Arabian language; more properly called Shehri

***jol*:** an Anglicized, widely printed version of *jūl*, meaning "plateau" in Yemen. The term is derived from a verb stem that connotes roving a circuit (as would bedouin in the high mountains).

Jubbah basin: an archaeological reference to site locations in Saudi Arabia

Kfar Hahoresh: a published site name in Israel

Kheshiya: an Anglicized name for the site SU151-001, derived from Shiʿb Khishiyah (Anglicized as Shiʿb Kheshiya), the small drainage in middle Wādī Sanā near the site

***khor*:** Anglicized orthography for *khūwr*, a brackish inlet at the mouth of a wadi

Khuzma or Khuzma as-Shumlya: a site name; the Anglicized version of Khuzmah as-Shumlyah, the inselberg at the confluence of Wādī Sanā and Wādī as-Shumlyah

Khuzmum: an Anglicized version of Khuzmūm, the RASA team's earlier reading of "Ḥzmm" on the eastern rock face of Khuzmah as-Shumlyah; used here as a site name, consistent with earlier publications

Kingdom of Hadramawt: one of the pre-Islamic kingdoms of Yemen. The name is Anglicized from Ḥaḍramawt.

Krif Magrad: Krīf Maqrāẓ, one of many vernal pools in the vicinity of Jibal Yuwān, surely named for the *Acacia nilotica* tree (*qaraẓ*) growing beside it. We use the Anglicized Krif Magrad as the site name for the settlement ruins.

***krif* or *krīf*:** a local term for a vernal pool. Many if not all have been artificially enhanced. In recent times, bulldozers deepen the pools and smooth the catchment; in premodern times, walls and diversions channeled surface flow into natural basins formed by faulting and uplifting. We use the Anglicized orthography when referring to the generic feature rather than a site name.

lacrimal: part of a skull

***laḥd*:** a niche for the corpse in the lateral wall of a grave or tomb

Land of ʿĀd: a pre-Islamic region and people of Southern Arabia, famous for ignoring the warnings of Prophet Hūd

Landsat: a long-running National Aeronautics and Space Administration (NASA) program of satellites bearing TM (thematic mapper) instruments.

Landsat-5: one of the TM versions of the Landsat program

Levallois: a technology of stone-tool knapping that produces one characteristic flake from a specially prepared core. Widely used in the Middle Paleolithic period (circa 50 kya), it has never been directly dated in Wādī Sanā.

***madhbaḥ*:** an Arabic term for a place of slaughter or a sacrificial altar. The plural is *madhābiḥ*.

Mahra: an Anglicized form of al-Mahrah, one of the largest and the easternmost provinces of Yemen

Makaynūn: an archaeological site in Wadi Hadramawt, a ruined city of the ancient Hadramawt Kingdom

mano: from the Spanish term for a hand stone

manuport: an item carried by humans to a location where it could not have arrived by natural means; a transported object that may not be otherwise modified by humans

Masila: an oil field sector named for the Masila Formation (geological)

Mawlā Maṭar: a hamlet with antiquities and refashioned structures high in the Southern Jol, northwest of al-Mukallā

Mehri: an Anglicized term for speakers of one of the remaining South Arabian languages. Most Mehri speakers live in Mahra Province and Dhofar Province (Oman).

Mukalla: an Anglicized orthography for al-Mukallā, coastal capital of Ḥaḍramawt Province

Munayder: the published name of the archaeological site Shiʿb Munaydar

Mundafan: the Anglicized, published name of a paleolake in southwestern Saudi Arabia, where archaeologists and geologists have documented and published Paleolithic and Neolithic campsites

Murāfiq: a guide; usually a tribal member whose company assures the safe passage of voyagers through tribal areas

***mūwqaḍ*:** a hearth, especially one built as a ring of small boulders filled with smooth gravel and used for grilling

Natufian: a culture group preceding the earliest Neolithic in the eastern Mediterranean

Nefud: a desert region of central-northern Saudi Arabia

Nexen: Canadian Nexen Petroleum Yemen (CPNY) and Nexen Petroleum were partners in developing the Masilah Block and Block 51 in the Yemen Petroleum concessions scheme.

onager: a wild equid

orthostat: a monolithic upright forming part of a structure, often in the doors or outer wall facings

OSL: optically stimulated luminescence, a technique used in obtaining dates (since burial) from buried quartz grains and other buried crystalline structures that trap electrons and release them as photons when exposed to laboratory stimulation

palatine: part of a skull

pareo: a cloth drape; a clothing wrap that ties over one shoulder

phytolith: a biogenic silica body forming in the interstices of plant cells and remnant in soils after the rest of the plant has decayed

pommel: the end grip of a weapon, dagger, or sword

PSI (paleo-stage indicator): in Wādī Sanā, an erosional scar as a notch in limestone cliff faces

***qanat*:** an Anglicized version of a Persian term for an underground water channel reached by intermittent shafts

Qārah Ḥabshiyah: a Himyaritic fortress atop an imposing promontory at the neck of Wādī Sanā, where the main stream channel is incised from the convergence of three tributaries

***qāt*:** *Catha edulis*, a medium tree planted for its leaves, which produce a mild stimulant when chewed

***qīr*:** a local term for a rock pool retaining water

QuickBird: a registered trademark for a satellite run by DigitalGlobe Inc

Qur'anic: an adjective, Anglicized from Qur'ān

Raidah al-Maʿārrah: a highland region of the Southern Jol with gently sloping, wide drainages with sufficient precipitation for spate-water farming

Raidāt: used colloquially to describe collective villages of the Raidah al-Maʿārrah

Ramlah as-Sabʿatayn: a Quaternary-formed sand dune field in the Jawf-Ḥaḍramawt basin between Ma'rib and Shabwah

Ras al-Hamra: Rʿās al-Ḥamrah, a prehistoric archaeological site in the Qūrum district of Muscat, Oman

RASA: the Roots of Agriculture in Southern Arabia Project, an archaeological and paleoecological field and analytical program from 1996 to 2005

Rāwik: a village in Wādī ʿIdim (when used as village name)

Rawk: an archaeological site at Rāwik, published with Anglicized orthography

reg: a geological term for a desert pavement formed of close-packed rock or pebbles. It forms where deflation removes finer surface particles.

rhizome: an underground stem of a plant that sends out shoots and roots at intervals (for example, bamboo and ginger)

Riqseh: the Anglicized, published name of an archaeological site in Jordan

rockart: images pecked, engraved, painted, or otherwise registered on natural rock faces

Rub' al Khali: Anglicized from the Arabic ar-Rubʿ al-Khālī, the Empty Quarter of Arabia, where the most arid sand desert lies

Rus: Rūs; Anglicized in geography to refer to one of the upper formations of southern Ḥaḍramawt. In our geological maps and text, we use the conventional, Anglicized form.

Sa'ada: Anglicized orthography for the Ṣaʿdah district of northwest Yemen

Sabaean: material, people. The adjective qualifies the culture and writing of the inhabitants of the Kingdom of Sabā, one of the ancient South Arabian kingdoms.

Sabr: published orthography for an archaeological site, Ṣabr, in al-Laḥij district, outside Aden

Safaitic: describes a distinct variety of South Semitic scripts; see also Thamudic, Hismaic

Sāh: a town in Wādī ʿIdim

Sana'a: Anglicized name of Ṣan'ā', the capital of modern Yemen

Say'un-Masilah: See entries for Say'ūn and Wādī Masilah (under Wadi Hadramawt).

saylan: an Arabic term for a torrential flood. In Arabia, these may occur with little warning, like flash flooding in the U.S. Southwest.

Say'ūn: a city in the Wādī Ḥaḍramawt with an important archaeological museum and collections.

scree: a geological term describing loose stones or gravel covering sloping ground

Shiʿb Kheshiya: See Kheshiya.

Shabwa: an Anglicized (and German orthography) name for Shabwah, the ruined capital of the ancient Kingdom of Ḥaḍramawt

***shahīd*:** literally, a witness; a term used to describe an upright stone set upon a grave

Shakeel: a local site name for an archaeological excavation site near Halqoot, Dhufar. No Arabic transliteration is attempted here because the name is in the Shehri language.

Sharia: as-Shīrāʿ, the revealed, canonical law of Islam

Sharma: Sharmah, a medieval archaeological site that was a port on the coast of Shi'bāt Diḥyah

Shiʿb Khishiyah: a small drainage in middle Wādī Sanā near the site

***shiʿb*:** a streamlet-sided tributary or side canyon of a drainage system

***shrūj*:** a local Arabic term for a low berm or diversion deflecting sheet-wash water toward a lower plot of agricultural land

Shurruj Bakeli: published orthography of an archaeological site, probably Shrūj al-Baklī

Sikdān: Wādī Sikdān, a minor tributary in Wādī ʿIdim

siliclastic: relating to or denoting clastic rocks consisting largely of silica or silicates. Clastic rocks have sharply fragmented—not water-worn—edges.

spherulite: small, round bodies formed in the digestive tract of animals. Their presence at archaeological sites can indicate animal dung and penning.

statue-menhirs: adopted from European terminology for upright monolithic installations (*menhirs*), perhaps erected to represent a human presence. The statue-menhirs of Arabia have pecked human features, certainly designed to anthropomorphize the upright stones.

tahir: a wild ungulate related to goats, native to Eastern Arabia

Tarīm: an important town in Wādī Ḥaḍramawt, famous as an Islamic intellectual and cultural center

teff: *Eragrostis teff*, an Ethiopian domesticated small-grained grass

Tell Brak: the published site name for Tall Brak, an archaeological site in northeastern Syria

Thalweg: a geological convention that describes an imaginary line connecting the lowest points of a drainage system along its length, tracing the stream channel

Thamūd: an oasis in the interior desert of northern Ḥaḍramawt Province

Thamudic: describes inscriptions and graffiti in ancient scripts and languages across the Arabian Peninsula

trilith: also known as āthafī, a distinctive form of prehistoric monument consisting of low platforms that support grouped upright stones (often triads), arrangements of four stones, and a parallel row of grilling hearths

tufa: a porous rock composed of calcium carbonate and formed by precipitation from water (for example, around mineral springs

tumulus: a deliberately formed pile of earth or stones, forming a heap or sine-shaped hill

Umm an-Nar: an Anglicized term for an archaeological period (third millennium BCE)

Umm er-Radhuma: Umm ar-Rudūmah or Umm ar-Rudmah; Anglicized in geography to refer to one of the limestone formations of southern Ḥaḍramawt. In our geological maps and text, we use the conventional, Anglicized form.

***ʻummah*:** the community, the people of Islam

ʿurf: legal practice according to (tribal) custom; an alternate legal system to Islamic law

UTM: Universal Transverse Mercator is a map grid system used to specify two-dimensional Cartesian coordinates (east and north) for locations on earth. The UTM system divides the world into 60 zones running north–south that are each 6 degrees of longitude wide.

Wadi Zalaqa: a published site name in the Negev, Israel; possibly also Wādī Zalaqah or Ẓalaqah

wadi: an Anglicized orthography of wādī, an Arabic term for an intermittently flowing or dry streambed, similar to the Spanish arroyo or the southwestern U.S. wash. Here we use *wadi* in geographic and geological context to refer to this portion of the landscape; where it is part of a place-name, Arabic orthography has been retained.

WGS84: World Geodetic System 1984, a terrestrial reference system and geodetic datum that describes the size and shape of the earth and is the standard global reference system for GPS

WS: wadi silts; one of the landform classifications used by the RASA Project

***wusūm*:** markings on rocks and animals that denote their association with a social group (for example, tribal *wusūm*)

Yemen: al-Yaman; the Anglicized name of this country

Zabid: Zabīd, an important center on the western coastal plain, known as a center for learning and government in medieval Yemen. Long-term archaeological excavations are published under the Anglicized name.

Zagros: a mountain range in the western region of Iran

zebu: the common name for *Bos indicus*, humped cattle

Index

UCLA COTSEN INSTITUTE OF ARCHAEOLOGY PRESS

MONUMENTA ARCHAEOLOGICA

Volume 43	*Landscape History of Hadramawt: The Roots of Agriculture in Southern Arabia (RASA) Project 1998–2008,* edited by Joy McCorriston and Michael J. Harrower
Volume 42	*Early Athens: Settlements and Cemeteries in the Submycenaean, Geometric, and Archaic Periods,* by Eirini Dimitriadou
Volume 41	*The History and Archaeology of Jaffa 2,* edited by Aaron A. Burke, Katherine Strange Burke, and Martin Peilstöcker
Volume 40	*Tangatatau Rockshelter: The Evolution of an Eastern Polynesian Socio-Ecosystem,* edited by Patrick Vinton Kirch
Volume 39	*The Desert Fayum Reinvestigated: The Early to Mid-Holocene Landscape Archaeology of the Fayum North Shore, Egypt,* edited by Simon J. Holdaway and Willeke Wendrich
Volume 38	*The Archaeology of Grotta Scaloria: Ritual in Neolithic Southeast Italy,* edited by Ernestine S. Elster, Eugenia Isetti, John Robb, and Antonella Traverso
Volume 37	*A Bronze Age Landscape in the Russian Steppes: The Samara Valley Project,* edited by David W. Anthony, Dorcas R. Brown, Aleksandr A. Khokhlov, Pavel F. Kuznetsov, and Oleg D. Mochalov
Volume 36	*Rural Archaeology in Early Urban Northern Mesopotamia: Excavations at Tell al-Raqa'i,* edited by Glenn M. Schwartz
Volume 35	*New Insights into the Iron Age Archaeology of Edom, Southern Jordan,* by Thomas E. Levy, Mohammad Najjar, and Erez Ben-Yosef
Volume 34	*The Excavation of the Prehistoric Burial Tumulus at Lofkënd, Albania,* by John K. Papadopoulos, Sarah P. Morris, Lorenc Bejko, and Lynne A. Schepartz
Volume 33	*Formative Lifeways in Central Tlaxcala: Volume 1: Excavations, Ceramics, and Chronology,* edited by Richard G. Lesure
Volume 32	*Integrating Çatalhöyük: The 2000-2008 Seasons (Çatal Research Project Volume 10),* edited by Ian Hodder
Volume 31	*Substantive Technologies at Çatalhöyük: Reports from the 2000-2008 Seasons (Çatal Research Project Volume 9),* edited by Ian Hodder
Volume 30	*Humans and Landscapes of Çatalhöyük* (Çatal Research Project Volume 8), edited by Ian Hodder
Volume 29	*Çatalhöyük Excavations: The 2000-2008 Seasons (Çatal Research Project Volume 7),* edited by Ian Hodder
Volume 28	*Light and Shadow: Isolation and Interaction in the Shala Valley of Northern Albania,* edited by Michael L. Galaty, Ols Lafe, Wayne E. Lee, and Zamir Tafilica

Volume 27	*Last House on the Hill: BACH Area Reports from Çatalhöyük, Turkey,* edited by Ruth Tringham and Mirjana Stevanović
Volume 26	*The History and Archaeology of Jaffa 1,* edited by Martin Peilstöcker and Aaron A. Burke
Volume 24	*The Early Iron Age Cemetery at Torone,* by John K. Papadopoulos
Volume 23	*The Plain of Phaistos: Cycles of Social Complexity in the Mesara Region of Crete,* by L. Vance Watrous, Despoina Hadzi-Vallianou, and Harriet Blitzer
Volume 22	*K'axob: Ritual, Work, and Family in an Ancient Maya Village,* edited by Patricia A. McAnany
Volume 21	*The Sydney Cyprus Survey Project: Social Approaches to Regional Archaeological Survey,* by Michael Given and A. Bernard Knapp
Volume 20	*Prehistoric Sitagroi: Excavations in Northeast Greece 1968–1970.* Volume 2. *Final Report,* edited by Ernestine S. Elster and Colin Renfrew
Volume 19	*Archaeology of Solvieux, An Upper Paleolithic Open Air Site in France,* by James Sackett
Volume 18	*Down by the Station: Los Angeles Chinatown, 1880–1933,* by Roberta S. Greenwood
Volume 17	*Rock Art of Easter Island: Symbols of Power, Prayers to the Gods,* by Georgia Lee
Volume 16	*Landscape Archaeology as Long-Term History: Northern Keos in the Cycladic Islands,* by J. E. Cherry, J. L. Davis, and E. Mantzourani
Volume 15	*Selevac: A Neolithic Village in Yugoslavia,* edited by Ruth Tringham and Dusan Krstic
Volume 14	*Achilleion: A Neolithic Settlement in Thessaly, Greece, 6400–5600 B.C.,* by Marija Gimbutas, Shan Winn, and Daniel Shimabuku
Volume 13	*Excavations at Sitagroi: A Prehistoric Village in Northeast Greece.* Volume 1. By Colin Renfrew, Marija Gimbutas, and Ernestine S. Elster
Volume 12	*Petroglyphs in the Guianas and Adjacent Areas of Brazil and Venezuela: An Inventory with a Comprehensive Bibliography of South American and Antillean Petroglyphs,* by C. N. Dubelaar
Volume 11	*Chinese Archaeological Abstracts 4,* edited by Albert E. Dien, Jeffrey K. Riegel, and Nancy T. Price
Volume 10	*Chinese Archaeological Abstracts 3,* edited by Albert E. Dien, Jeffrey K. Riegel, and Nancy T. Price
Volume 9	*Chinese Archaeological Abstracts 2,* edited by Albert E. Dien, Jeffrey K. Riegel, and Nancy T. Price
Volume 8	*Kolmakovskiy Redoubt: The Ethnoarchaeology of a Russian Fort in Alaska,* by Wendell Oswalt
Volume 7	*Prehistoric Trails of the Atacama: Archaeology of Northern Chile,* by Clement Meighan and D. L. True
Volume 6	*Chinese Archaeological Abstracts,* by Richard C. Rudolph
Volume 5	*The Stone and Plaster Sculpture: Excavations at Dura Europos,* by Susan B. Downey
Volume 4	*The Transition to Mycenaean,* by Sarah and Jeremy Rutter
Volume 3	*The Marine Thiasos in Greek Sculpture,* by Steven Lattimore
Volume 2	*The Archaeology of Amapa, Nayarit,* edited by Clement W. Meighan
Volume 1	*Neolithic Macedonia as Reflected by Excavations at Anza, Southeast Yugoslavia,* edited by Marija Gimbutas

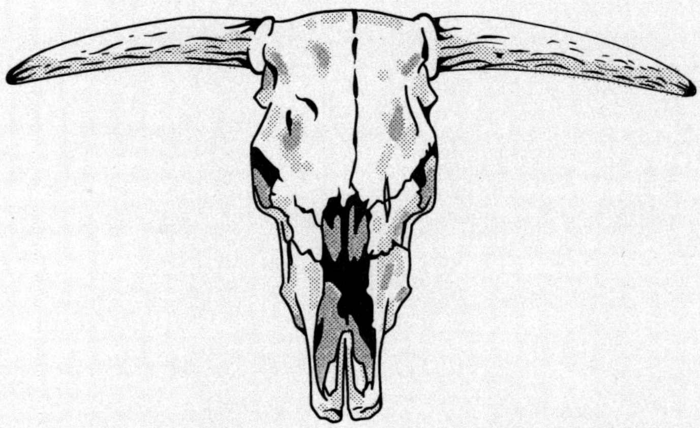